FOURTH EDITION   VOLUME 2

PHYSICS

JAMES S. WALKER

Western Washington University

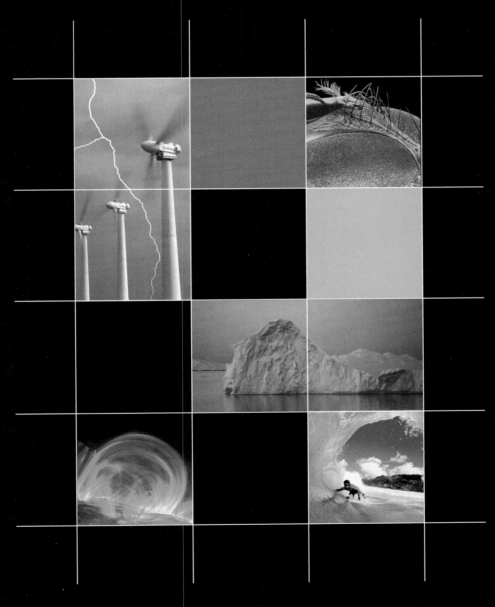

Addison-Wesley

San Francisco   Boston   New York   Cape Town   Hong Kong

London   Madrid   Mexico City   Montreal   Munich   Paris   Singapore   Sydney   Tokyo   Toronto

Publisher: Jim Smith
Director of Development: Michael Gillespie
Sr. Development Editor: Margot Otway
Editorial Manager: Laura Kenney
Sr. Project Editor: Katie Conley
Associate Editor: Grace Joo
Media Producer: David Huth
Director of Marketing: Christy Lawrence
Executive Marketing Manager: Scott Dustan
Executive Market Development Manager: Josh Frost
Managing Editor: Corinne Benson
Sr. Production Supervisor: Nancy Tabor
Production Management and Composition: Nesbitt Graphics, Inc.
Project Manager: Cindy Johnson
Illustrations: Rolin Graphics, Inc.
Cover and Text Design: Seventeenth Street Studios
Manufacturing Buyer: Jeff Sargent
Photo Research: Cypress Integrated Systems
Manager, Rights and Permissions: Zina Arabia
Image Permission Coordinator: Richard Rodrigues
Cover Printer: Phoenix Color Corporation
Text Printer and Binder: Quebecor World, Dubuque

Cover Images: Wind turbines with lightning: Mark Newman
(Photo Researchers, Inc.); scanning electron micrograph of head of
fly showing compound eye x96 (color enhanced): S. Lowry/
Univ Ulster (Getty Images); iceberg in the Errera Channel:
Seth Resnick (Getty Images); surfer in tube wave, North Shore,
Oahu, Hawaii, USA: Warren Bolster (Getty Images); solar coronal
loops: Science Source; light passing through triangular prism:
David Sutherland (Getty Images)

Photo Credits: See page C-1.

Library of Congress Cataloging-in-Publication Data
Walker, James S., 1950-
  Physics / James S. Walker. — 4th ed.
    p. cm.
  Includes index.
  ISBN 978-0-321-61111-6
  1. Physics—Textbooks. I. Title.

QC23.2.W35 2008
530—dc22

                        2008040978
ISBN: 978-0-321-61112-3 (student copy)

THIS BOOK IS DEDICATED TO MY PARENTS, IVAN AND JANET WALKER, AND TO MY WIFE, BETSY.

Addison-Wesley
is an imprint of

www.pearsonhighered.com

1 2 3 4 5 6 7 8 9 10—QWD—12 11 10 09 08

About the Author

JAMES S. WALKER

James Walker obtained his Ph.D. in theoretical physics from the University of Washington in 1978. He subsequently served as a post-doc at the University of Pennsylvania, the Massachusetts Institute of Technology, and the University of California at San Diego before joining the physics faculty at Washington State University in 1983. Professor Walker's research interests include statistical mechanics, critical phenomena, and chaos. His many publications on the application of renormalization-group theory to systems ranging from absorbed monolayers to binary-fluid mixtures have appeared in *Physical Review, Physical Review Letters, Physica*, and a host of other publications. He has also participated in observations on the summit of Mauna Kea, looking for evidence of extra-solar planets.

Jim Walker likes to work with students at all levels, from judging elementary school science fairs to writing research papers with graduate students, and has taught introductory physics for many years. His enjoyment of this course and his empathy for students have earned him a reputation as an innovative, enthusiastic, and effective teacher. Jim's educational publications include "Reappearing Phases" (*Scientific American*, May 1987) as well as articles in the *American Journal of Physics* and *The Physics Teacher*. In recognition of his contributions to the teaching of physics at Washington State University, Jim was named the Boeing Distinguished Professor of Science and Mathematics Education for 2001–2003. He currently teaches at Western Washington University.

When he is not writing, conducting research, teaching, or developing new classroom demonstrations and pedagogical materials, Jim enjoys amateur astronomy, eclipse chasing, bird and dragonfly watching, photography, juggling, unicycling, boogie boarding, and kayaking. Jim is also an avid jazz pianist and organist. He has served as ballpark organist for a number of Class A minor league baseball teams, including the Bellingham Mariners, an affiliate of the Seattle Mariners, and the Salem-Keizer Volcanoes, an affiliate of the San Francisco Giants. He can play "Take Me Out to the Ball Game" in his sleep.

About the Cover

The photographs on the cover of this book are a reminder of the wide spectrum of physics applications that are a part of our everyday lives.

Wind Turbines and Lightning Bolt: Wind turbines convert the mechanical energy of moving air into electrical energy to power our homes and cities. Electrical energy is also produced by nature, and occasionally unleashed in impressive bolts of lightning.

Scanning Electron Micrograph: Though electrons are usually thought of as "particles," they also have wave-like properties similar to light. The image of a fly's eye was taken with a beam of electrons.

Iceberg in the Errera Channel: A floating iceberg is a visual demonstration that ice has a lower density than liquid water.

Solar Coronal Loops: Magnetic storms often rage on the surface of the Sun. These glowing loops of ionized gas follow the curved lines of the magnetic field.

Surfer in the "Tube" on the North Shore of Oahu: The laws of physics determine the motion of the wave this surfer is riding.

As you study the material in this book, your understanding of physics will deepen, and your appreciation for the world around you will increase as you come to recognize the fundamental physical principles on which all of our lives are based.

Teaching introductory algebra-based physics can be a most challenging—and rewarding—experience. Students enter the course with a wide range of backgrounds, interests, and skills and we, the instructors, strive not only to convey the basic concepts and fundamental laws of physics but also to give students an appreciation of its relevance and appeal.

I wrote this book to help with that task. It incorporates a number of unique and innovative pedagogical features that evolved from years of teaching experience. The materials have been tested extensively in the classroom and in focus groups, and refined based on comments from students and teachers who used the earlier editions of the text. The enthusiastic response I received from users of the first three editions was both flattering and motivating. The fourth edition has been improved in response to this feedback.

Learning Tools in the Text

A key goal of this text is to help students make the connection between a conceptual understanding of physics and the various skills necessary to solve quantitative problems. One of the chief means to that end is the replacement of traditional "textbook" Examples with an integrated system of learning tools: fully worked Examples with Solutions in Two-Column Format, Active Examples, Conceptual Checkpoints, and Exercises. Each of these tools is specialized to meet the needs of students at a particular point in the development of a chapter.

These needs are not always the same. Sometimes students require a detailed explanation of how to tackle a particular problem; at other times, they must be allowed to take an active role and work out the details for themselves. Sometimes it is important for them to perform calculations and concentrate on numerical precision; at other times it is more fruitful for them to explore a key idea in a conceptual context. And sometimes, all that is required is practice using a new equation or definition.

This text attempts to emulate the teaching style of successful instructors by providing the right tool at the right place and the right time.

Perspective Across Chapters

It's easy for students to miss the forest for the trees—to overlook the unifying concepts that are central to physics and that will make the details easier to learn and retain. To address this difficulty, the fourth edition adds two features. At key **NEW** junctures in the text are six **Physics in Perspective** features, two-page spreads that take a highly visual look at core ideas whose significance students are now prepared to understand. For instance, after working through the energy chapters, do students *really* understand how conservation of energy relates to conservation of mechanical energy, and the role of work done by dissipative and nondissipative forces? And after working through the chapters on electricity and magnetism, do they have a clear view of how electric and magnetic forces relate to each other? These are two of the topics on which the Physics in Perspective pages focus. Each **NEW** chapter now ends with a **Big Picture box** that links ideas covered in the chapter to related material from earlier and later chapters in the text.

WORKED EXAMPLES WITH SOLUTIONS IN TWO-COLUMN FORMAT

Examples model the most complete and detailed method of solving a particular type of problem. The Examples in this text are presented in a format that focuses on the basic strategies and thought processes involved in problem solving. This focus on the intimate relationship between conceptual insights and problem-solving techniques encourages students to view the ability to solve problems as a logical outgrowth of conceptual understanding rather than a kind of parlor trick.

Each Example has the same basic structure:

- **Picture the Problem**   This first step discusses how the physical situation can be represented visually and what such a representation can tell us about how to analyze and solve the problem. At this step, always accompanied by a figure, we set up a coordinate system where appropriate, label important quantities, and indicate which values are known.

- **Strategy**   The Strategy addresses the commonly asked question, "How do I get started?" by providing a clear overview of the problem and helping students to identify the relevant physical principles. It then guides the student in using known relationships to map a step-by-step path to the solution.

- **Solution in Two-Column Format**   In the step-by-step Solution of the problem, each of the steps is presented with a prose statement in the left-hand column and the corresponding mathematical implementation in the right-hand column. Each step clearly translates the idea described in words into the appropriate equations.

- **Insight**   Each Example wraps up with an Insight—a comment regarding the solution just obtained. Some Insights deal with possible alternative solution techniques, others with new ideas suggested by the results.

- **Practice Problem**   Following the Insight is a Practice Problem, which gives the student a chance to practice the type of calculation just presented. The Practice Problems, always accompanied by their answers, provide students with a valuable check on their understanding of the material. Finally, each Example ends with a reference to some related end-of-chapter Problems to allow students to test their skills further.

ACTIVE EXAMPLES

Active Examples serve as a bridge between the fully worked Examples, in which every detail is fully discussed and every step is given, and the homework Problems, where no help is given at all. In an Active Example, students take an active role in solving the problem by thinking through the logic of the steps described on the left and checking their answers on the right. Students often find it useful to practice problem solving by covering one column of an Active Example with a sheet of paper and filling in the covered steps as they refer to the other column. Follow-up questions, called Your Turns, ask students to look at the problem in a slightly different way. Answers to Your Turns are provided at the end of the book.

CONCEPTUAL CHECKPOINTS

Conceptual Checkpoints help students sharpen their insight into key physical principles. A typical Conceptual Checkpoint presents a thought-provoking question that can be answered by logical reasoning based on physical concepts rather than by numerical calculations. The statement of the question is followed by a detailed discussion and analysis in the section titled Reasoning and Discussion, and the Answer is given at the end of the checkpoint for quick and easy reference.

EXERCISES

Exercises present brief calculations designed to illustrate the application of important new relationships, without the expenditure of time and space required by a fully worked Example. Exercises generally give students an opportunity to practice the use of a new equation, become familiar with the units of a new physical quantity, and get a feeling for typical magnitudes.

PROBLEM-SOLVING NOTES

Each chapter includes a number of Problem-Solving Notes in the margin. These practical hints are designed to highlight useful problem-solving methods while helping students avoid common pitfalls and misconceptions.

End-of-Chapter Learning Tools

The end-of-chapter material in this text also includes a number of innovations, along with refinements of more familiar elements.

- Each chapter concludes with a **Chapter Summary** presented in an easy-to-use outline style. Key concepts, equations, and important figures are organized by topic for convenient reference.

- A unique feature of this text is the **Problem-Solving Summary** at the end of the chapter. This summary addresses common sources of misconceptions in problem solving, and gives specific references to Examples and Active Examples illustrating the correct procedures.

- The homework for each chapter begins with a section of **Conceptual Questions**. Answers to the odd-numbered Questions can be found in the back of the book. Answers to even-numbered Conceptual Questions are available in the online Instructor Solutions Manual.

- **Conceptual Exercises (CE)** have been integrated into the homework section at the end of the chapter and consist of multiple-choice and ranking questions. These questions have been carefully selected and written for maximum effectiveness when used with classroom-response systems (clickers). Answers to the odd-numbered Exercises can be found in the back of the book. Answers to even-numbered Conceptual Exercises are available in the online Instructor Solutions Manual.

NEW
- **Predict/Explain problems** are new to this edition. These problems ask the student to predict what will happen in a given physical situation and then to choose an explanation for their prediction.

NEW
- Also new to this edition, **Passage Problems** are similar to those found on MCAT exams, with associated multiple-choice questions.

- **Interactive Problems** are based on the animations and simulations associated with the Interactive Figures and are found within MasteringPhysics.

- A popular feature within the homework section is the **Integrated Problem (IP).** These problems, labeled with the symbol **IP**, integrate a conceptual question with a numerical problem. Problems of this type, which stress the importance of reasoning from basic principles, show how conceptual insight and numerical calculation go hand in hand in physics.

- In addition, a section titled **General Problems** presents a variety of problems that use material from two or more sections within the chapter, or refer to material covered in earlier chapters.

- **Problems of special biological or medical relevance** are indicated with the symbol **BIO**.

Scope and Organization
TABLE OF CONTENTS

The presentation of physics in this text follows the standard practice for introductory courses, with only a few well-motivated refinements.

- First, note that Chapter 3 is devoted to **vectors and their application to physics.** My experience has been that students benefit greatly from a full discussion of vectors early in the course. Most students have seen vectors and trigonometric functions before, but rarely from the point of view of physics. Thus, including vectors in the text sends a message that this is important material, and it gives students an opportunity to brush up on their math skills.

- Note also that **additional time is given to some of the more fundamental aspects of physics,** such as Newton's laws and energy. Presenting such material in two chapters gives the student a better opportunity to assimilate and master these crucial topics. Sections considered optional are marked with an asterisk.

REAL-WORLD PHYSICS

Since physics applies to everything in nature, it is only reasonable to point out applications of physics that students may encounter in the real world. Each chapter presents a number of discussions focusing on "Real-World Physics." Those of general interest are designated by a globe icon in the margin. Applications that pertain more specifically to biology and medicine are indicated by a green frog icon in the margin.

REAL-WORLD PHYSICS

REAL-WORLD PHYSICS: BIO

The Illustration Program

DRAWINGS

Many physics concepts are best conveyed by graphic means. Figures do far more than illustrate a physics text—often, they bear the main burden of the exposition. Accordingly, great attention has been paid to the figures in this book, with the primary emphasis always on the clarity of the analysis. Color has been used consistently throughout the text to reinforce concepts and make the diagrams easier for students to understand. New to this edition, helpful **annotations in blue** are included on select figures to help guide students in "reading" graphs and other figures. This technique emulates what instructors do at the chalkboard when explaining figures.

NEW

PHOTOGRAPHS

One of the most fundamental ways in which we learn is by comparing and contrasting. Many **companion photos** are presented in groups of two or three that contrast opposing physical principles or illustrate a single concept in a variety of contexts. Grouping carefully chosen photographs in this way helps students to see the universality of physics. In this edition, we have added new **demonstration photos** that use high-speed time-lapse photography to dramatically illustrate topics, such as standing waves, static versus kinetic friction, and the motion of center of mass, in a way that reveals physical principles in the world around us.

NEW

Resources

The fourth edition is supplemented by an ancillary package developed to address the needs of both students and instructors.

FOR THE INSTRUCTOR

Instructor Solutions Manual by Kenneth L. Menningen (University of Wisconsin–Stevens Point) is available online at the Instructor Resource Center: www.pearsonhighered.com/educator
You will find detailed, worked solutions to every Problem and Conceptual Exercise in the text, all solved using the step-by-step problem-solving strategy of the in-chapter Examples (Picture the Problem, Strategy, two-column Solutions, and Insight). The solutions also contain answers to the even-numbered Conceptual Questions.

Instructor Resource Manual with Notes on ConcepTest Questions
Available at the Instructor Resource Center: www.pearsonhighered.com/educator, this online manual consists of two parts. The first part, prepared by Katherine Whatley and Judith Beck (both of University of North Carolina, Asheville), contains sample syllabi, lecture outlines, notes, demonstration suggestions, readings, and additional references and resources. The second part, prepared by Cornelius Bennhold and Gerald Feldman (both of George Washington University) contains an overview of the development and implementation of ConcepTests, as well as instructor notes for each ConcepTest found in the Instructor Resource Center and available on the Instructor Resource DVD.

Test Bank Available at the Instructor Resource Center: www.pearsonhighered.com/educator
Written by Delena Bell Gatch (Georgia Southern University), this online, cross-platform test bank contains approximately 3000 multiple-choice, short-answer, and true/false questions, many conceptual in nature. All are referenced to the corresponding text section and ranked by level of difficulty.

Instructor Resource DVD (ISBN 0-321-60193-9)
This cross-platform DVD provides virtually every electronic asset you'll need in and out of the classroom. The DVD is organized by chapter and includes all text illustrations and tables from *Physics,* Fourth Edition, in jpeg and PowerPoint formats. The IRDVD also contains the Interactive Figures, chapter-by-chapter lecture outlines in PowerPoint, ConcepTest "Clicker" Questions in PowerPoint, editable Word files of all numbered equations, the eleven "Physics You Can See" demonstration videos, and pdf files of the *Instructor Resource Manual with Notes on ConcepTest Questions.*

MasteringPhysics™ www.masteringphysics.com
This homework, tutorial, and assessment system is designed to assign, assess, and track each student's progress using a wide diversity of tutorials and extensively pretested problems. All the end-of-chapter problems from the text and the Interactive

NEW Figures are available in MasteringPhysics. MasteringPhysics provides instructors with a fast and effective way to assign uncompromising, wide-ranging online homework assignments of just the right difficulty and duration. The tutorials coach 90% of students to the correct answer with specific wrong-answer feedback. The powerful post-assignment diagnostics allow instructors to assess the progress of their class as a whole or to quickly identify individual students' areas of difficulty.

myeBook is available through MasteringPhysics either automatically when Mastering-Physics is packaged with new books, or available as a purchased upgrade online.

NEW Allowing students access to the text wherever they have access to the Internet, myeBook comprises the full text, including figures that can be enlarged for better viewing. Within myeBook, students are also able to pop up definitions and terms to help with vocabulary and the reading of the material. Students can also take notes in myeBook using the annotation feature at the top of each page.

ActivPhysics OnLine™ (accessed through the Self Study area within www.masteringphysics.com) provides a comprehensive library of more than 420 tried and tested *ActivPhysics* applets. In addition, it provides a suite of applet-based tutorials developed by education pioneers Alan Van Heuvelen and Paul D'Alessandris. The online exercises are designed to encourage students to confront misconceptions, reason qualitatively about physical processes, experiment quantitatively, and learn to think critically. They cover all topics from mechanics to electricity and magnetism and from optics to modern physics. The *ActivPhysics OnLine* companion workbooks help students work through complex concepts and understand them more clearly.

FOR THE STUDENT

Student Study Guide with Selected Solutions by David Reid (University of Chicago) Volume 1: ISBN 0-321-60200-5; Volume 2: ISBN 0-321-60199-8
The print study guide provides the following for each chapter:

Objectives; Warm-Up Questions from the Just-in-Time Teaching (JiTT) method by Gregor Novak and Andrew Gavrin (Indiana University–Purdue University, Indianapolis); Chapter Review with two-column Examples and integrated quizzes; Reference Tools & Resources (equation summaries, important tips, and tools); Puzzle Questions (also from Novak & Gavrin's JiTT method); Selected Solutions for several end-of-chapter questions and problems.

MasteringPhysics™ (www.masteringphysics.com)
This homework, tutorial, and assessment system is based on years of research into how students work physics problems and precisely where they need help. Studies show that students who use MasteringPhysics significantly increase their final scores compared to hand-written homework. MasteringPhysics achieves this improvement by providing students with instantaneous feedback specific to their wrong answers, simpler sub-problems upon request when they get stuck, and partial credit for their method(s) used. This individualized, 24/7 Socratic tutoring is recommended by nine out of ten students to their peers as the most effective and time-efficient way to study.

NEW

myeBook is available through MasteringPhysics either automatically when Mastering-Physics is packaged with new books, or available as a purchased upgrade online. Allowing students access to the text wherever they have access to the Internet, myeBook comprises the full text, including figures that can be enlarged for better viewing. Within myeBook, students are also able to pop up definitions and terms to help with vocabulary and the reading of the material. Students can also take notes in myeBook using the annotation feature at the top of each page.

NEW

ActivPhysics OnLine™ (accessed via www.masteringphysics.com) provides students with a suite of highly regarded applet-based self-study tutorials (see description on previous page). The following workbooks provide a range of tutorial problems designed to use the *ActivPhysics OnLine* simulations, helping students work through complex concepts and understand them more clearly:

- *ActivPhysics OnLine Workbook* Volume 1: Mechanics • Thermal Physics • Oscillations & Waves (ISBN 0-8053-9060-X)

- *ActivPhysics OnLine Workbook* Volume 2: Electricity & Magnetism • Optics • Modern Physics (ISBN 0-8053-9061-8)

Pearson Tutor Services (www.pearsontutorservices.com) Each student's subscription to MasteringPhysics also contains complimentary access to Pearson Tutor Services, powered by Smarthinking, Inc. By logging in with their Mastering-Physics ID and password, they will be connected to highly qualified e-structors™ who provide additional, interactive online tutoring on the major concepts of physics. Some restrictions apply; offer subject to change.

Acknowledgments

I would like to express sincere gratitude to my colleagues at Washington State University and Western Washington University, as well as to many others in the physics community, for their contributions to this project. In particular, I would like to thank Professor Ken Menningen of the University of Wisconsin–Stevens Point for his painstaking attention to detail in producing the Instructor Solutions Manual.

My thanks are due also to the many wonderful and talented people at Addison-Wesley who have been such a pleasure to work with during the development of the fourth edition, and especially to Katie Conley, Michael Gillespie, Margot Otway, and Jim Smith.

In addition, I am grateful for the dedicated efforts of Cindy Johnson, who choreographed a delightfully smooth production process.

Finally, I owe a great debt to all my students over the years. My interactions with them provided the motivation and inspiration that led to this book.

Reviewers

We are grateful to the following instructors for their thoughtful comments on the manuscript of this text.

REVIEWERS OF THE FOURTH EDITION

Raymond Benge
Tarrant County College–NE Campus

Matthew Bigelow
Saint Cloud University

Edward J. Brash
Christopher Newport University

Michaela Burkardt
New Mexico State University

Jennifer Chen
University of Illinois at Chicago

Eugenia Ciocan
Clemson University

Shahida Dar
University of Delaware

Joseph Dodoo
University of Maryland, Eastern Shore

Thomas Dooling
University of Northern Carolina at Pembroke

Hui Fang
Sam Houston State University

Carlos E. Figueroa
Cabrillo College

Lyle Ford
University of Wisconsin, Eau Claire

Darrin Johnson
University of Minnesota, Duluth

Paul Lee
California State University, Northridge

Sheng-Chiang (John) Lee
Mercer University

Nilanga Liyanage
University of Virginia

Michael Ottinger
Missouri Western State University

Melodi Rodrigue
University of Nevada

Claudiu Rusu
Richland College of DCCCD

Mark Sprague
East Carolina University

Richard Szwerc
Montgomery College

Lisa Will
San Diego City College

Guanghua Xu
University of Houston

Bill Yen
University of Georgia

REVIEWERS OF PREVIOUS EDITIONS

Daniel Akerib, *Case Western Reserve University*; Richard Akerib, *Queens College*; Alice M. Hawthorne Allen, *Virginia Tech*; Barbara S. Andereck, *Ohio Wesleyan University*; Eva Andrei, *Rutgers University*; Bradley C. Antanaitis, *Lafayette College*; Michael Arnett, *Kirkwood Community College*; Robert W. Arts, *Pikeville College*; David Balogh, *Fresno City College*; David T. Bannon, *Oregon State University*; Rama Bansil, *Boston University*; Anand Batra, *Howard University*; Paul Beale, *University of Colorado–Boulder*; Mike Berger, *Indiana University*; David Berman, *University of Iowa*; S. M. Bhagat, *University of Maryland*; James D. Borgardt, *Juniata College*; James P. Boyle, *Western Connecticut State University*; David Branning, *Trinity College*; Jeff Braun, *University of Evansville*; Matthew E. Briggs, *University of Wisconsin–Madison*; Jack Brockway, *State University of New York–Oswego*; Neal Cason, *University of Notre Dame*; Thomas B. Cobb, *Bowling Green State University*; Lattie Collins, *Eastern Tennessee State University*; James Cook, *Middle Tennessee State University*; Stephen Cotanch, *North Carolina State University*; David Craig, *LeMoyne College*; David Curott, *University of North Alabama*; William Dabby, *Edison Community College*; Robert Davie, *St. Petersburg Junior College*; Steven Davis, *University of Arkansas–Little Rock*; N. E. Davison, *University of Manitoba*; Duane Deardorff, *University of North Carolina at Chapel Hill*; Edward Derringh, *Wentworth Institute of Technology*; Martha Dickinson, *Maine Maritime Academy*; Anthony DiStefano, *University of Scranton*; David C. Doughty, Jr., *Christopher Newport University*; F. Eugene Dunnam, *University of Florida*; John J. Dykla, *Loyola University–Chicago*; Eldon Eckard, *Bainbridge College*; Donald Elliott, *Carroll College*; David Elmore, *Purdue University*; Robert Endorf, *University of Cincinnati*; Raymond Enzweiler, *Northern Kentucky University*; John Erdei, *University of Dayton*; David Faust, *Mt. Hood Community College*; Frank Ferrone, *Drexel University*; John Flaherty, *Yuba College*; Curt W. Foltz, *Clarion University*; Lewis Ford, *Texas A&M University*; Armin Fuchs, *Florida Atlantic University*; Joseph Gallant, *Kent State University, Trumbull*; Asim Gangopadhyaya, *Loyola University–Chicago*; Thor Garber, *Pensacola Junior College*; David Gerdes, *University of Michigan*; John D. Gieringer, *Alvernia College*; Karen Gipson, *Grand Valley State University*; Barry Gilbert, *Rhode Island College*; Fred Gittes, *Washington State University*; Michael Graf, *Boston College*; William Gregg, *Louisiana State University*; Rainer Grobe, *Illinois State University*; Steven Hagen, *University of Florida*; Mitchell Haeri, *Saddleback College*; Parameswar Hari, *California State University–Fresno*; Xiaochun He, *Georgia State University*;

Timothy G. Heil, *University of Georgia*; J. Erik Hendrickson, *University of Wisconsin–Eau Claire*; Scott Holmstrom, *University of Tulsa*; John Hopkins, *The Pennsylvania State University*; Manuel A. Huerta, *University of Miami*; Zafar Ismail, *Daemen College*; Adam Johnston, *Weber State University*; Gordon O. Johnson, *Washington State University*; Nadejda Kaltcheva, *University of Wisconsin–Oshkosh*; William Karstens, *Saint Michael's College*; Sanford Kern, *Colorado State University*; Dana Klinck, *Hillsborough Community College*; Ilkka Koskelo, *San Francisco State University*; Laird Kramer, *Florida International University*; R. Gary Layton, *Northern Arizona University*; Kevin M. Lee, *University of Nebraska–Lincoln*; Michael Lieber, *University of Arkansas*; Ian M. Lindevald, *Truman State University*; Mark Lindsay, *University of Louisville*; Jeff Loats, *Fort Lewis College*; Daniel Ludwigsen, *Kettering University*; Lorin Matthews, *Baylor University*; Hilliard Macomber, *University of Northern Iowa*; Trecia Markes, *University of Nebraska-Kearny*; William McNairy, *Duke University*; Kenneth L. Menningen, *University of Wisconsin–Stevens Point*; Joseph Mills, *University of Houston*; Anatoly Miroshnichenko, *University of Toledo*; Wouter Montfrooij, *University of Missouri*; Gary Morris, *Valparaiso University*; Paul Morris, *Abilene Christian University*; David Moyle, *Clemson University*; Ashok Muthukrishnan, *Texas A&M University*; K. W. Nicholson, *Central Alabama Community College*; Robert Oman, *University of South Florida*; Michael Ottinger, *Missouri Western State College*; Larry Owens, *College of the Sequoias*; A. Ray Penner, *Malaspina University*; Francis Pichanick, *University of Massachusetts, Amherst*; Robert Piserchio, *San Diego State University*; Anthony Pitucco, *Pima Community College*; William Pollard, *Valdosta State University*; Jerry Polson, *Southeastern Oklahoma State University*; Robert Pompi, *Binghamton University*; David Procopio, *Mohawk Valley Community College*; Earl Prohofsky, *Purdue University*; Jia Quan, *Pasadena City College*; David Raffaelle, *Glendale Community College*; Michele Rallis, *Ohio State University*; Michael Ram, *State University of New York–Buffalo*; Prabha Ramakrishnan, *North Carolina State University*; Rex Ramsier, *University of Akron*; John F. Reading, *Texas A&M University*; Lawrence B. Rees, *Brigham Young University*; M. Anthony Reynolds, *Embry Riddle University*; Dennis Rioux, *University of Wisconsin–Oshkosh*; John A. Rochowicz, Jr., *Alvernia College*; Bob Rogers, *San Francisco State University*; Gaylon Ross, *University of Central Arkansas*; Lawrence G. Rowan, *University of North Carolina at Chapel Hill*; Gerald Royce, *Mary Washington College*; Wolfgang Rueckner, *Harvard University*; Misa T. Saros, *Viterbo University*; C. Gregory Seab, *University of New Orleans*; Mats Selen, *University of Illinois*; Bartlett Sheinberg, *Houston Community College*; Peter Shull, *Oklahoma State University*; Christopher Sirola, *Tri-County Technical College*; Daniel Smith, *South Carolina State University*; Leigh M. Smith, *University of Cincinnati*; Soren Sorensen, *University of Tennessee–Knoxville*; Mark W. Sprague, *East Carolina University*; George Strobel, *University of Georgia*; Carey E. Stronach, *Virginia State University*; Irina Struganova, *Barry University*; Daniel Stump, *Michigan State University*; Leo Takahashi, *Penn State University–Beaver*; Harold Taylor, *Richard Stockton College*; Frederick Thomas, *Sinclair Community College*; Jack Tuszynski, *University of Alberta*; Lorin Vant Hull, *University of Houston*; John A. Underwood, *Austin Community College, Rio Grande*; Karl Vogler, *Northern Kentucky University*; Desmond Walsh, *Memorial University of Newfoundland*; Toby Ward, *College of Lake County*; Richard Webb, *Pacific Union College*; Lawrence Weinstein, *Old Dominion University*; Jeremiah Williams, *Illinois Wesleyan University*, Linda Winkler, *Moorhead State University*; Lowell Wood, *University of Houston*; Robert Wood, *University of Georgia*; Jeffrey L. Wragg, *College of Charleston*

STUDENT REVIEWERS

We wish to thank the following students at New Mexico State University and Chemetka Community College for providing helpful feedback during the development of the fourth edition of this text. Their comments offered us valuable insight into the student experience.

Teresa M. Abbott, Rachel Acuna, Sonia Arroyos, Joanna Beeson, Carl Bryce, Jennifer Currier, Juan Farias, Mark Ferra, Bonnie Galloway, Cameron Haider, Gina Hedberg, Kyle Kazsinas, Ty Keeney, Justin Kvenzi, Tannia Lau, Ann MaKarewicz, Jasmine Pando, Jenna Painter, Jonathan Romero, Aaron Ryther, Sarah Salaido, Ashley Slape, Christina Timmons, Christopher Torrez, Charmaine Vega, Elisa Wingerd

We would also like to thank the following students at Boston University, California State University–Chico, the University of Houston, Washington State University, and North Carolina State University for providing helpful feedback via review or focus group for the first three editions of this text:

Ali Ahmed, Joel Amato, Max Aquino, Margaret Baker, Tynisa Bennett, Joshua Carmichael, Sabrina Carrie, Suprna Chandra, Kara Coffee, Tyler Cumby, Rebecca Currell, Philip Dagostino, Andrew D. Fisher, Shadi Miri Ghomizadea, Colleen Hanlon, Jonas Hauptmann, Parker Havron, Jamie Helms, Robert Hubbard, Tamara Jones, Bryce Lewis, Michelle Lim, Candida Mejia, Roderick Minogue, Ryan Morrison, Hang Nguyen, Mary Nguyen, Julie Palakovich, Suraj Parekh, Scott Parsons, Peter Ploewski, Darren B. Robertson, Chris Simons, Tiffany Switzer, Steven Taylor, Monique Thomas, Khang Tran, Michael Vasquez, Jerod Williams, Nathan Witwer, Alexander Wood, Melissa Wright. ATTENDEE: Lynda Klein, *California State University–Chico*

Preface: To the Student

As a student preparing to take an algebra-based physics course, you are probably aware that physics applies to absolutely everything in the natural world, from raindrops and people to galaxies and atoms. Because physics is so wide-ranging and comprehensive, it can sometimes seem a bit overwhelming. This text, which reflects nearly two decades of classroom experience, is designed to help you deal with a large body of information and develop a working understanding of the basic concepts in physics. Now in its fourth edition, it incorporates many refinements that have come directly from interacting with students using the first three editions. As a result of these interactions, I am confident that as you develop a deeper understanding of physics, you will also enrich your experience of the world in which you live.

Now, I must admit that I like physics, and so I may be a bit biased in this respect. Still, the reason I teach and continue to study physics is that I enjoy the insight it gives into the physical world. I can't help but notice—and enjoy—aspects of physics all around me each and every day. As I always tell my students on the first day of class, I would like to share some of this enjoyment and delight in the natural world with you. It is for this reason that I undertook the task of writing this book.

To assist you in the process of studying physics, this text incorporates a number of learning aids, including Two-Column Examples, Active Examples, and Conceptual Checkpoints. These and other elements work together in a unified way to enhance your understanding of physics on both a conceptual and a quantitative level—they have been developed to give you the benefit of what we know about how students learn physics, and to incorporate strategies that have proven successful to students over the years. The pages that follow will introduce these elements to you, describe the purpose of each, and explain how they can help you.

As you progress through the text, you will encounter many interesting and intriguing applications of physics drawn from the world around you. Some of these, such as magnetically levitated trains or the satellite-based Global Positioning System that enables you to determine your position anywhere on Earth to within a few feet, are primarily technological in nature. Others focus on explaining familiar or not-so-familiar phenomena, such as why the Moon has no atmosphere, how sweating cools the body, or why flying saucer shaped clouds often hover over mountain peaks even when the sky is clear. Still others, such as countercurrent heat exchange in animals and humans or the use of sound waves to destroy kidney stones, are of particular relevance to students of biology and the other life sciences.

In many cases, you may find the applications to be a bit surprising. Did you know, for example, that you are shorter at the end of the day than when you first get up in the morning? (This is discussed in Chapter 5.) That an instrument called the ballistocardiograph can detect the presence of a person hiding in a truck, just by registering the minute recoil from the beating of the stowaway's heart? (This is discussed in Chapter 9.) That if you hum next to a spider's web at just the right pitch you can cause a resonance effect that sends the spider into a tizzy? (This is discussed in Chapter 13.) That powerful magnets can exploit the phenomenon of diamagnetism to levitate living creatures? (This is discussed in Chapter 22.)

Writing this textbook was a rewarding experience for me. I hope using it will prove equally rewarding to you, and that it will inspire an interest in and appreciation of physics that will last a lifetime.

James S. Walker
jamesswalker@physics.wwu.edu

Detailed Contents

19 Electric Charges, Forces, and Fields

Amber, a form of fossilized tree resin long used to make beautiful beads and other ornaments, has also made contributions to two different sciences. Pieces of amber have preserved prehistoric insects and pollen grains for modern students of evolution. And over 2500 years ago, amber provided Greek scientists with their first opportunity to study electric forces—the subject of this chapter.

We are all made up of electric charges. Every atom in every human body contains both positive and negative charges held together by an attractive force that is similar to gravity—only vastly stronger. Our atoms are bound together by electric forces to form molecules; these molecules, in turn, interact with one another to produce solid bones and liquid blood. In a very real sense, we are walking, talking manifestations of electricity.

In this chapter, we discuss the basic properties of electric charge. Among these are that electric charge comes in discrete units (quantization) and that the total amount of charge in the universe remains constant (conservation). In addition, we present the force law that describes the interactions between electric charges. Finally, we introduce the idea of an electric *field*, and show how it is related to the distribution of charge.

19–1 Electric Charge

The effects of electric charge have been known since at least 600 B.C. About that time, the Greeks noticed that amber—a solid, translucent material formed from the fossilized resin of extinct coniferous trees—has a peculiar property. If a piece of amber is rubbed with animal fur, it attracts small, lightweight objects. This phenomenon is illustrated in **Figure 19–1**.

For some time, it was thought that amber was unique in its ability to become "charged." Much later, it was discovered that other materials can behave in this way as well. For example, if glass is rubbed with a piece of silk, it too can attract small objects. In this respect, glass and amber seem to be the same. It turns out, however, that these two materials have different types of charge.

To see this, imagine suspending a small, charged rod of amber from a thread, as in **Figure 19–2**. If a second piece of charged amber is brought near the rod, as shown in Figure 19–2 (a), the rod rotates away, indicating a repulsive force between the two pieces of amber. Thus, "like" charges repel. On the other hand, if a piece of charged glass is brought near the amber rod, the amber rotates toward the glass, indicating an attractive force. This is illustrated in Figure 19–2 (b). Clearly, then, the *different* charges on the glass and amber attract one another. We refer to different charges as being the "opposite" of one another, as in the familiar expression "opposites attract."

We know today that the two types of charge found on amber and glass are, in fact, the only types that exist, and we still use the purely arbitrary names—**positive** (+) charge and **negative** (−) charge—proposed by Benjamin Franklin (1706–1790) in 1747. In accordance with Franklin's original suggestion, the charge of amber is negative, and the charge of glass is positive (the opposite of negative). Calling the different charges + and − is actually quite useful mathematically; for example, an object that contains an equal amount of positive and negative charge has zero net charge. Objects with zero net charge are said to be electrically **neutral**.

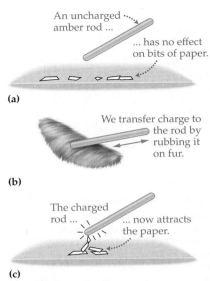

An uncharged amber rod ...

... has no effect on bits of paper.

(a)

We transfer charge to the rod by rubbing it on fur.

(b)

The charged rod ...

... now attracts the paper.

(c)

▲ **FIGURE 19–1 Charging an amber rod**
An uncharged amber rod **(a)** exerts no force on scraps of paper. When the rod is rubbed against a piece of fur **(b)**, it becomes charged and then attracts the paper **(c)**.

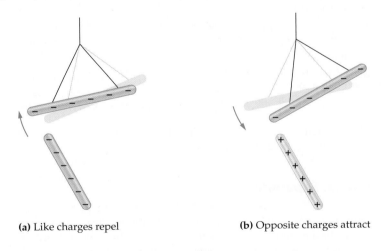

(a) Like charges repel

(b) Opposite charges attract

◀ **FIGURE 19–2 Likes repel, opposites attract**
A charged amber rod is suspended by a string. According to the convention introduced by Benjamin Franklin, the charge on the amber is designated as negative. **(a)** When another charged amber rod is brought near the suspended rod, it rotates away, indicating a repulsive force between like charges. **(b)** When a charged glass rod is brought close to the suspended amber rod, the amber rotates toward the glass, indicating an attractive force and the existence of a second type of charge, which we designate as positive.

A familiar example of an electrically neutral object is the atom. Atoms have a small, dense nucleus with a positive charge surrounded by a cloud of negatively charged electrons (from the Greek word for amber, *elektron*). A pictorial representation of an atom is shown in **Figure 19–3**.

All electrons have exactly the same electric charge. This charge is very small and is defined to have a magnitude, e, given by the following:

Magnitude of an Electron's Charge, e

$$e = 1.60 \times 10^{-19} \, \text{C}$$ 19–1

SI unit: coulomb, C

In this expression, C is a unit of charge referred to as the **coulomb**, named for the French physicist Charles-Augustin de Coulomb (1736–1806). (The precise definition

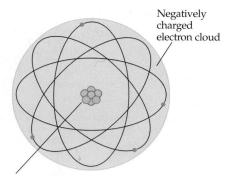

Negatively charged electron cloud

Positively charged nucleus

▲ **FIGURE 19–3 The structure of an atom**
A crude representation of an atom, showing the positively charged nucleus at its center and the negatively charged electrons orbiting about it. More accurately, the electrons should be thought of as forming a "cloud" of negative charge surrounding the nucleus.

of the coulomb is in terms of electric current, which we shall discuss in Chapter 21.) Clearly, the charge on an electron, which is negative, is $-e$. This is one of the defining, or *intrinsic*, properties of the electron. Another intrinsic property of the electron is its mass, m_e:

$$m_e = 9.11 \times 10^{-31} \text{ kg} \qquad 19\text{–}2$$

In contrast, the charge on a proton—one of the main constituents of nuclei—is *exactly* $+e$. Therefore, the net charge on atoms, which have equal numbers of electrons and protons, is precisely zero. The mass of the proton is

$$m_p = 1.673 \times 10^{-27} \text{ kg} \qquad 19\text{–}3$$

Note that this is about 2000 times larger than the mass of the electron. The other main constituent of the nucleus is the neutron, which, as its name implies, has zero charge. Its mass is slightly larger than that of the proton:

$$m_n = 1.675 \times 10^{-27} \text{ kg} \qquad 19\text{–}4$$

Since the magnitude of the charge per electron is 1.60×10^{-19} C/electron, it follows that the number of electrons in 1 C of charge is enormous:

$$\frac{1 \text{ C}}{1.60 \times 10^{-19} \text{ C/electron}} = 6.25 \times 10^{18} \text{ electrons}$$

As we shall see when we consider the force between charges, a coulomb is a significant amount of charge; even a powerful lightning bolt delivers only 20 to 30 C. A more common unit of charge is the microcoulomb, μC, where $1 \ \mu$C $= 10^{-6}$ C. Still, the amount of charge contained in everyday objects is very large, even in units of the coulomb, as we show in the following Exercise.

EXERCISE 19–1

Find the amount of positive electric charge in one mole of helium atoms. (Note that the nucleus of a helium atom consists of two protons and two neutrons.)

SOLUTION
Since each helium atom contains two positive charges of magnitude e, the total positive charge in a mole is

$$N_A(2e) = (6.02 \times 10^{23})(2)(1.60 \times 10^{-19} \text{ C}) = 1.93 \times 10^5 \text{ C}$$

Thus, a mere 4 g of helium contains almost 200,000 C of positive charge, and the same amount of negative charge, as well.

Charge Separation

How is it that rubbing a piece of amber with fur gives the amber a charge? Originally, it was thought that the friction of rubbing *created* the observed charge. We now know, however, that rubbing the fur across the amber simply results in a *transfer* of charge from the fur to the amber—with the total amount of charge remaining unchanged. This is indicated in **Figure 19–4**. Before charging, the fur and the amber are both neutral. During the rubbing process some electrons are transferred from the fur to the amber, giving the amber a net negative charge, and leaving the fur with a net positive charge. At no time during this process is charge ever created or destroyed. This, in fact, is an example of one of the fundamental conservation laws of physics:

Conservation of Electric Charge
The total electric charge of the universe is constant. No physical process can result in an increase or decrease in the total amount of electric charge in the universe.

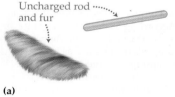

Uncharged rod and fur

(a)

Rubbing transfers charge.

(b)

The net charge is still zero ...

... but the rod and fur are now oppositely charged.

(c)

▲ **FIGURE 19–4 Charge transfer**
(a) Initially, an amber rod and a piece of fur are electrically neutral; that is, they each contain equal quantities of positive and negative charge. **(b)** As they are rubbed together, charge is transferred from one to the other. **(c)** In the end, the fur and the rod have charges of equal magnitude but opposite sign.

When charge is transferred from one object to another, it is generally due to the movement of electrons. In a typical solid, the nuclei of the atoms are fixed in position. The outer electrons of these atoms, however, are often weakly bound and fairly easily separated. As a piece of fur rubs across amber, for example, some of the electrons that were originally part of the fur are separated from their atoms and deposited onto the amber. The atom that loses an electron is now a **positive ion,** and the atom that receives an extra electron becomes a **negative ion.** This is charging by separation.

In general, when two materials are rubbed together, the magnitude *and* sign of the charge that each material acquires depend on how strongly it holds onto its electrons. For example, if silk is rubbed against glass, the glass acquires a positive charge, as was mentioned earlier in this section. It follows that electrons have moved from the glass to the silk, giving the silk a *negative* charge. If silk is rubbed against amber, however, the silk becomes *positively* charged, as electrons in this case pass from the silk to the amber.

These results can be understood by referring to Table 19–1, which presents the relative charging due to rubbing—also known as **triboelectric charging**—for a variety of materials. The more plus signs associated with a material, the more readily it gives up electrons and becomes positively charged. Similarly, the more minus signs for a material, the more readily it acquires electrons. For example, we know that amber becomes negatively charged when rubbed against fur, but a greater negative charge is obtained if rubber, PVC, or Teflon is rubbed with fur instead. In general, when two materials in Table 19–1 are rubbed together, the one higher in the list becomes positively charged, and the one lower in the list becomes negatively charged. The greater the separation on the list, the greater the magnitude of the charge.

Charge separation occurs not only when one object is rubbed against another, but also when objects collide. For example, colliding crystals of ice in a rain cloud can cause charge separation that may ultimately result in bolts of lightning to bring the charges together again. Similarly, particles in the rings of Saturn are constantly undergoing collisions and becoming charged as a result. In fact, when the *Voyager* spacecraft examined the rings of Saturn, it observed electrostatic discharges, similar to lightning bolts on Earth. In addition, ghostly radial "spokes" that extend across the rings of Saturn—which cannot be explained by gravitational forces alone—are also the result of electrostatic interactions.

TABLE 19–1 Triboelectric Charging

Material	Relative charging with rubbing
Rabbit fur	++++++
Glass	+++++
Human hair	++++
Nylon	+++
Silk	++
Paper	+
Cotton	−
Wood	−−
Amber	−−−
Rubber	−−−−
PVC	−−−−−
Teflon	−−−−−−

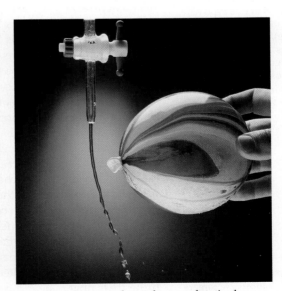

▲ The Van de Graaff generator (left) that these children are touching can produce very large charges of static electricity. Since they are clearly not frightened, why is their hair standing on end? On a smaller scale, if you rub a balloon against a cloth surface, the balloon acquires a negative electric charge. The balloon can then attract a stream of water (right), even though water molecules themselves are electrically neutral. This phenomenon occurs because the water molecules, though they have no net charge, are polar: one end of the molecule has a slight positive charge and the other a slight negative charge. Under the influence of the balloon's negative charge, the water molecules orient themselves so that their positive ends point toward the balloon. This alignment ensures that the electrical attraction between the balloon and the positive part of each molecule exceeds the repulsion between the balloon and the negative part of each molecule.

CONCEPTUAL CHECKPOINT 19-1 COMPARE THE MASS

Is the mass of an amber rod after charging with fur **(a)** greater than, **(b)** less than, or **(c)** the same as its mass before charging?

REASONING AND DISCUSSION

Since an amber rod becomes negatively charged, it has acquired electrons from the fur. Each electron has a small, but nonzero, mass. Therefore, the mass of the rod increases ever so slightly as it is charged.

ANSWER

(a) The mass of the amber rod is greater after charging.

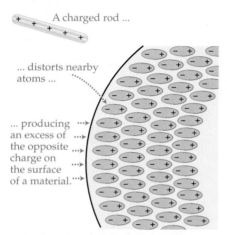

... distorts nearby atoms ...

... producing an excess of the opposite charge on the surface of a material.

▲ **FIGURE 19-5** Electrical polarization

When a charged rod is far from a neutral object, the atoms in the object are undistorted, as in Figure 19-3. As the rod is brought closer, however, the atoms distort, producing an excess of one type of charge on the surface of the object (in this case a negative charge). This induced charge is referred to as a polarization charge. Because the sign of the polarization charge is the opposite of the sign of the charge on the rod, there is an attractive force between the rod and the object.

REAL-WORLD PHYSICS: BIO

Bacterial infection from endoscopic surgery

Since electrons always have the charge $-e$, and protons always have the charge $+e$, it follows that all objects must have a net charge that is an integral multiple of e. This conclusion was confirmed early in the twentieth century by the American physicist Robert A. Millikan (1868–1953) in a classic series of experiments. He found that the charge on an object can be $\pm e$, $\pm 2e$, $\pm 3e$, and so on, but never $1.5e$ or $-9.3847e$, for example. We describe this restriction by saying that electric charge is **quantized.**

Polarization

We know that charges of opposite sign attract, but it is also possible for a charged rod to attract small objects that have zero net charge. The mechanism responsible for this attraction is called **polarization.**

To see how polarization works, consider **Figure 19-5**. Here we show a positively charged rod held close to an enlarged view of a neutral object. An atom near the surface of the neutral object will become elongated because the negative electrons in it are attracted to the rod while the positive protons are repelled. As a result, a net negative charge develops on the surface near the rod—the so-called polarization charge. The attractive force between the rod and this *induced* polarization charge leads to a net attraction between the rod and the entire neutral object.

Of course, the same conclusion is reached if we consider a negative rod held near a neutral object—except in this case the polarization charge is positive. Thus, the effect of polarization is to give rise to an attractive force regardless of the sign of the charged object. It is for this reason that both charged amber and charged glass attract neutral objects—even though their charges are opposite.

A potentially dangerous, and initially unsuspected, medical application of polarization occurs in endoscopic surgery. In these procedures, a tube carrying a small video camera is inserted into the body. The resulting video image is produced by electrons striking the inside surface of a computer monitor's screen, which is kept positively charged to attract the electrons. Minute airborne particles in the operating room—including dust, lint, and skin cells—are polarized by the positive charge on the screen, and are attracted to its exterior surface.

The problem comes when a surgeon touches the screen to point out an important feature to others in the medical staff. Even the slightest touch can transfer particles—many of which carry bacteria—from the screen to the surgeon's finger and from there to the patient. In fact, the surgeon's finger doesn't even have to touch the screen—as the finger approaches the screen, it too becomes polarized, and hence, it can attract particles from the screen, or directly from the air. Situations like these have resulted in infections, and surgeons are now cautioned not to bring their fingers near the video monitor.

19-2 Insulators and Conductors

Suppose you rub one end of an amber rod with fur, being careful not to touch the other end. The result is that the rubbed portion becomes charged, whereas the other end remains neutral. In particular, the negative charge transferred to the rubbed end stays put; it does not move about from one end of the rod to the other. Materials like amber, in which charges are not free to move, are referred to as **insulators.** Most insulators are nonmetallic substances, and most are also good thermal insulators.

In contrast, most metals are good **conductors** of electricity, in the sense that they allow charges to move about more or less freely. For example, suppose an uncharged metal sphere is placed on an insulating base. If a charged rod is brought into contact with the sphere, as in **Figure 19–6 (a)**, some charge will be transferred to the sphere at the point of contact. The charge does not stay put, however. Since the metal is a good conductor of electricity, the charges are free to move about the sphere, which they do because of their mutual repulsion. The result is a uniform distribution of charge over the surface of the sphere, as shown in **Figure 19–6 (b)**. Note that the insulating base prevents charge from flowing away from the sphere into the ground.

On a microscopic level, the difference between conductors and insulators is that the atoms in conductors allow one or more of their outermost electrons to become detached. These detached electrons, often referred to as "conduction electrons," can move freely throughout the conductor. In a sense, the conduction electrons behave almost like gas molecules moving about within a container. Insulators, in contrast, have very few, if any, free electrons; the electrons are bound to their atoms and cannot move from place to place within the material.

Some materials have properties that are intermediate between those of a good conductor and a good insulator. These materials, referred to as **semiconductors,** can be fine-tuned to display almost any desired degree of conductivity by controlling the concentration of the various components from which they are made. The great versatility of semiconductors is one reason they have found such wide areas of application in electronics and computers.

Exposure to light can sometimes determine whether a given material is an insulator or a conductor. An example of such a **photoconductive** material is selenium, which conducts electricity when light shines on it but is an insulator when in the dark. Because of this special property, selenium plays a key role in the production of photocopies. To see how, we first note that at the heart of every photocopier is a selenium-coated aluminum drum. Initially, the selenium is given a positive charge and kept in the dark—which causes it to retain its charge. When flash lamps illuminate a document to be copied, an image of the document falls on the drum. Where the document is light, the selenium is illuminated and becomes a conductor, and the positive charge flows away into the aluminum drum, leaving the selenium uncharged. Where the document is dark, the selenium is not illuminated, meaning that it is an insulator, and its charge remains in place. At this point, a negatively charged "toner" powder is wiped across the drum, where it sticks to those positively charged portions of the drum that were not illuminated. Next, the drum is brought into contact with paper, transferring the toner to it. Finally, the toner is fused into the paper fibers with heat, the drum is cleaned of excess toner, and the cycle repeats. Thus, a slight variation in electrical properties due to illumination is the basis of an entire technology.

The operation of a laser printer is basically the same as that of a photocopier, with the difference that in the laser printer the selenium-coated drum is illuminated with a computer-controlled laser beam. As the laser sweeps across the selenium, the computer turns the beam on and off to produce areas that will print light or dark, respectively.

19–3 Coulomb's Law

We have already discussed the fact that electric charges exert forces on one another. The precise law describing these forces was first determined by Coulomb in the late 1780s. His result is remarkably simple. Suppose, for example, that an idealized point charge q_1 is separated by a distance r from another point charge q_2. Both charges are at rest; that is, the system is **electrostatic.** According to Coulomb's law, the magnitude of the electrostatic force between these charges is proportional to the product of the magnitude of the charges, $|q_1||q_2|$, and inversely proportional to the square of the distance, r^2, between them:

▲ People who work with electricity must be careful to use gloves made of nonconducting materials. Rubber, an excellent insulator, is often used for this purpose.

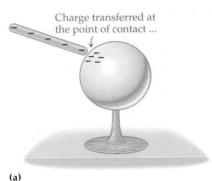

Charge transferred at the point of contact ...

(a)

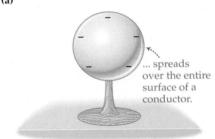

... spreads over the entire surface of a conductor.

(b)

▲ **FIGURE 19–6 Charging a conductor**
(a) When an uncharged metal sphere is touched by a charged rod, some charge is transferred at the point of contact. **(b)** Because like charges repel, and charges move freely on a conductor, the transferred charge quickly spreads out and covers the entire surface of the sphere.

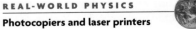

REAL-WORLD PHYSICS
Photocopiers and laser printers

> **Coulomb's Law for the Magnitude of the Electrostatic Force Between Point Charges**
>
> $$F = k\frac{|q_1||q_2|}{r^2}$$
>
> SI unit: newton, N

19–5

In this expression, the proportionality constant k has the value

$$k = 8.99 \times 10^9 \, \text{N} \cdot \text{m}^2/\text{C}^2$$

19–6

Note that the units of k are simply those required for the force F to have the units of newtons.

The direction of the force in Coulomb's law is along the line connecting the two charges. In addition, we know from the observations described in Section 19–1 that like charges repel and opposite charges attract. These properties are illustrated in **Figure 19–7**, where force vectors are shown for charges of various signs. Thus, when applying Coulomb's law, we first calculate the magnitude of the force using Equation 19–5, and then determine its direction with the "likes repel, opposites attract" rule.

▶ **FIGURE 19–7 Forces between point charges**

The forces exerted by two point charges on one another are always along the line connecting the charges. If the charges have the same sign, as in **(a)** and **(c)**, the forces are repulsive; that is, each charge experiences a force that points away from the other charge. Charges of opposite sign, as in **(b)**, experience attractive forces. Notice that in all cases the forces exerted on the two charges form an action–reaction pair. That is, $\vec{F}_{21} = -\vec{F}_{12}$.

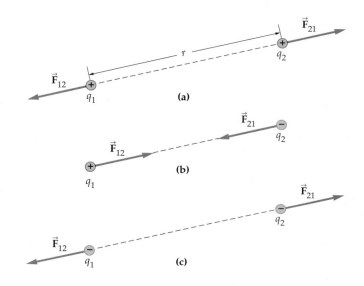

Finally, note how Newton's third law applies to each of the cases shown in Figure 19–7. For example, the force exerted on charge 1 by charge 2, $\vec{F}_{12}$, is always equal in magnitude and opposite in direction to the force exerted on charge 2 by charge 1, $\vec{F}_{21}$; that is, $\vec{F}_{21} = -\vec{F}_{12}$.

CONCEPTUAL CHECKPOINT 19–2   WHERE DO THEY COLLIDE?

An electron and a proton, initially separated by a distance d, are released from rest simultaneously. The two particles are free to move. When they collide, are they **(a)** at the midpoint of their initial separation, **(b)** closer to the initial position of the proton, or **(c)** closer to the initial position of the electron?

REASONING AND DISCUSSION

Because of Newton's third law, the forces exerted on the electron and proton are equal in magnitude and opposite in direction. For this reason, it might seem that the particles meet at the midpoint. The masses of the particles, however, are quite different. In fact, as mentioned in Section 19–1, the mass of the proton is about 2000 times greater than the mass of the electron; therefore, the proton's acceleration ($a = F/m$) is about 2000 times less than the electron's acceleration. As a result, the particles collide near the initial position of the proton. More specifically, they collide at the location of the center of mass of the system, which remains at rest throughout the process.

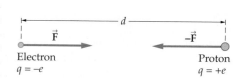

ANSWER

(b) The particles collide near the initial position of the proton.

It is interesting to note the similarities and differences between Coulomb's law, $F = k|q_1||q_2|/r^2$, and Newton's law of gravity, $F = Gm_1m_2/r^2$. In each case, the force decreases as the square of the distance between the two objects. In addition, both forces depend on a product of intrinsic quantities: in the case of the electric force the intrinsic quantity is the charge; in the case of gravity it is the mass.

Equally significant, however, are the differences. In particular, the force of gravity is always attractive, whereas the electric force can be attractive or repulsive. As a result, the net electric force between neutral objects, such as the Earth and the Moon, is essentially zero because attractive and repulsive forces cancel one another. Since gravity is always attractive, however, the net gravitational force between the Earth and the Moon is nonzero. Thus, in astronomy, gravity rules, and electric forces play hardly any role.

Just the opposite is true in atomic systems. To see this, let's compare the electric and gravitational forces between a proton and an electron in a hydrogen atom. Taking the distance between the two particles to be the radius of hydrogen, $r = 5.29 \times 10^{-11}$ m, we find that the gravitational force has a magnitude

$$F_g = G\frac{m_e m_p}{r^2}$$

$$= (6.67 \times 10^{-11}\,\text{N} \cdot \text{m}^2/\text{kg}^2)\frac{(9.11 \times 10^{-31}\,\text{kg})(1.673 \times 10^{-27}\,\text{kg})}{(5.29 \times 10^{-11}\,\text{m})^2}$$

$$= 3.63 \times 10^{-47}\,\text{N}$$

Similarly, the magnitude of the electric force between the electron and the proton is

$$F_e = k\frac{|q_1||q_2|}{r^2}$$

$$= (8.99 \times 10^9\,\text{N} \cdot \text{m}^2/\text{C}^2)\frac{|-1.60 \times 10^{-19}\,\text{C}||1.60 \times 10^{-19}\,\text{C}|}{(5.29 \times 10^{-11}\,\text{m})^2}$$

$$= 8.22 \times 10^{-8}\,\text{N}$$

Taking the ratio, we find that the electric force is greater than the gravitational force by a factor of

$$\frac{F_e}{F_g} = \frac{8.22 \times 10^{-8}\,\text{N}}{3.63 \times 10^{-47}\,\text{N}} = 2.26 \times 10^{39}$$

$$= 2{,}260{,}000{,}000{,}000{,}000{,}000{,}000{,}000{,}000{,}000{,}000{,}000{,}000$$

This huge factor explains why a small piece of charged amber can lift bits of paper off the ground, even though the entire mass of the Earth is pulling downward on the paper.

Clearly, then, the force of gravity plays essentially no role in atomic systems. The reason gravity dominates in astronomy is that, even though the force is incredibly weak, it always attracts, giving a larger net force the larger the astronomical body. The electric force, on the other hand, is very strong but cancels for neutral objects.

Next, we use the electric force to get an idea of the speed of an electron in a hydrogen atom and the frequency of its orbital motion.

PROBLEM-SOLVING NOTE

Distance Dependence of the Coulomb Force

The Coulomb force has an inverse-square dependence on distance. Be sure to divide the product of the charges, $k|q_1||q_2|$, by r^2 when calculating the force.

EXAMPLE 19–1   THE BOHR ORBIT

In an effort to better understand the behavior of atomic systems, the Danish physicist Niels Bohr (1885–1962) introduced a simple model for the hydrogen atom. In the Bohr model, as it is known today, the electron is imagined to move in a circular orbit about a stationary proton. The force responsible for the electron's circular motion is the electric force of attraction between the electron and the proton. **(a)** Given that the radius of the electron's orbit is 5.29×10^{-11} m, and its mass is $m_e = 9.11 \times 10^{-31}$ kg, find the electron's speed. **(b)** What is the frequency of the electron's orbital motion?

CONTINUED ON NEXT PAGE

CONTINUED FROM PREVIOUS PAGE

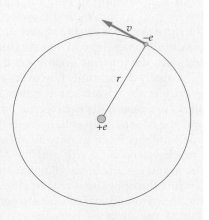

PICTURE THE PROBLEM
Our sketch shows the electron moving with a speed v in its orbit of radius r. Because the proton is so much more massive than the electron, it is essentially stationary at the center of the orbit. Note that the electron has a charge $-e$ and the proton has a charge $+e$.

STRATEGY

a. The idea behind this model is that a force is required to make the electron move in a circular path, and this force is provided by the electric force of attraction between the electron and the proton. Thus, as with any circular motion, we set the force acting on the electron equal to its mass times its centripetal acceleration. This allows us to solve for the centripetal acceleration, $a_{cp} = v^2/r$ (Equation 6–14), which in turn gives us the speed v.

b. The frequency of the electron's orbital motion is $f = 1/T$, where T is the period of the motion; that is, the time for one complete orbit. The time for an orbit, in turn, is the circumference divided by the speed, or $T = C/v = 2\pi r/v$. Taking the inverse immediately yields the frequency.

SOLUTION

Part (a)

1. Set the Coulomb force between the electron and proton equal to the centripetal force required for the electron's circular orbit:

$$k\frac{|q_1||q_2|}{r^2} = m_e a_{cp}$$

$$k\frac{e^2}{r^2} = m_e \frac{v^2}{r}$$

2. Solve for the speed of the electron, v:

$$v = e\sqrt{\frac{k}{m_e r}}$$

3. Substitute numerical values:

$$v = (1.60 \times 10^{-19}\,\text{C})\sqrt{\frac{8.99 \times 10^9\,\text{N}\cdot\text{m}^2/\text{C}^2}{(9.11 \times 10^{-31}\,\text{kg})(5.29 \times 10^{-11}\,\text{m})}}$$

$$= 2.19 \times 10^6\,\text{m/s}$$

Part (b)

4. Calculate the time for one orbit, T, which is the distance ($C = 2\pi r$) divided by the speed (v):

$$T = \frac{C}{v} = \frac{2\pi r}{v} = \frac{2\pi(5.29 \times 10^{-11}\,\text{m})}{2.19 \times 10^6\,\text{m/s}} = 1.52 \times 10^{-16}\,\text{s}$$

5. Take the inverse of T to find the frequency:

$$f = \frac{1}{T} = \frac{1}{1.52 \times 10^{-16}\,\text{s}} = 6.58 \times 10^{15}\,\text{Hz}$$

INSIGHT
If you could travel around the world at this speed, your trip would take only about 18 s, but your centripetal acceleration would be a more-than-lethal 75,000 times the acceleration of gravity. As it is, the centripetal acceleration of the electron in this "Bohr" orbit around the proton is about 10^{22} times greater than the acceleration of gravity on the surface of the Earth.

The frequency of the orbit is also incredibly large. We won't encounter frequencies this high again until we study light waves in Chapter 25.

PRACTICE PROBLEM
The second Bohr orbit has a radius that is four times the radius of the first orbit. What is the speed of an electron in this orbit?
[**Answer:** $v = 1.09 \times 10^6$ m/s]

Some related homework problems: Problem 19, Problem 28, Problem 37

Another indication of the strength of the electric force is given in the following Exercise.

EXERCISE 19–2

Find the electric force between two 1.00-C charges separated by 1.00 m.

SOLUTION
Substituting $q_1 = q_2 = 1.00$ C and $r = 1.00$ m in Coulomb's law, we find

$$F = k\frac{|q_1||q_2|}{r^2} = (8.99 \times 10^9\,\text{N}\cdot\text{m}^2/\text{C}^2)\frac{(1.00\,\text{C})(1.00\,\text{C})}{(1.00\,\text{m})^2} = 8.99 \times 10^9\,\text{N}$$

Exercise 19–2 shows that charges of one coulomb exert a force of about a million tons on one another when separated by a distance of a meter. If the charge in your body could be separated into a pile of positive charge on one side of the room and a pile of negative charge on the other side, the force needed to hold them apart would be roughly 10^{10} tons! Thus, everyday objects are never far from electrical neutrality, since disturbing neutrality requires such tremendous forces.

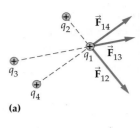

(a)

Superposition of Forces

The electric force, like all forces, is a vector quantity. Hence, when a charge experiences forces due to two or more other charges, the net force on it is simply the *vector* sum of the forces taken individually. For example, in **Figure 19–8**, the total force on charge 1, $\vec{F}_1$, is the vector sum of the forces due to charges 2, 3, and 4:

$$\vec{F}_1 = \vec{F}_{12} + \vec{F}_{13} + \vec{F}_{14}$$

This is referred to as the **superposition** of forces.

Notice that the total force acting on a given charge is the sum of interactions involving just *two* charges at a time, with the force between each pair of charges given by Coulomb's law. For example, the total force acting on charge 1 in Figure 19–8 is the sum of the forces between q_1 and q_2, q_1 and q_3, and q_1 and q_4. Therefore, superposition of forces can be thought of as the generalization of Coulomb's law to systems containing more than two charges. In our first numerical Example of superposition, we consider three charges in a line.

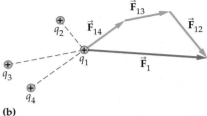

(b)

▲ **FIGURE 19–8 Superposition of forces**
(a) Forces are exerted on q_1 by the charges q_2, q_3, and q_4. These forces are $\vec{F}_{12}$, $\vec{F}_{13}$, and $\vec{F}_{14}$, respectively. **(b)** The net force acting on q_1, which we label $\vec{F}_1$, is the vector sum of $\vec{F}_{12}$, $\vec{F}_{13}$, and $\vec{F}_{14}$.

EXAMPLE 19–2 NET FORCE

A charge $q_1 = -5.4\ \mu\text{C}$ is at the origin, and a charge $q_2 = -2.2\ \mu\text{C}$ is on the x axis at $x = 1.00$ m. Find the net force acting on a charge $q_3 = +1.6\ \mu\text{C}$ located at $x = 0.75$ m.

PICTURE THE PROBLEM
The physical situation is shown in our sketch, with each charge at its appropriate location. Notice that the forces exerted on charge q_3 by the charges q_1 and q_2 are in opposite directions. We give the force on q_3 due to q_1 the label $\vec{F}_{31}$, and the force on q_3 due to q_2 the label $\vec{F}_{32}$.

STRATEGY
The net force on q_3 is the vector sum of the forces due to q_1 and q_2. In particular, note that $\vec{F}_{31}$ points in the negative x direction $(-\hat{\mathbf{x}})$, whereas $\vec{F}_{32}$ points in the positive x direction $(\hat{\mathbf{x}})$. The magnitude of $\vec{F}_{31}$ is $k|q_1||q_3|/r^2$, with $r = 0.75$ m. Similarly, the magnitude of $\vec{F}_{32}$ is $k|q_2||q_3|/r^2$, with $r = 0.25$ m.

SOLUTION

1. Find the force acting on q_3 due to q_1. Since this force is in the negative x direction, as indicated in the sketch, we give it a negative sign:

$$\vec{F}_{31} = -k\frac{|q_1||q_3|}{r^2}\hat{\mathbf{x}}$$
$$= -(8.99 \times 10^9\ \text{N}\cdot\text{m}^2/\text{C}^2)$$
$$\times \frac{(5.4 \times 10^{-6}\ \text{C})(1.6 \times 10^{-6}\ \text{C})}{(0.75\ \text{m})^2}\hat{\mathbf{x}}$$
$$= -0.14\ \text{N}\,\hat{\mathbf{x}}$$

2. Find the force acting on q_3 due to q_2. Since this force is in the positive x direction, as indicated in the sketch, we give it a positive sign:

$$\vec{F}_{32} = k\frac{|q_2||q_3|}{r^2}\hat{\mathbf{x}}$$
$$= (8.99 \times 10^9\ \text{N}\cdot\text{m}^2/\text{C}^2)$$
$$\times \frac{(2.2 \times 10^{-6}\ \text{C})(1.6 \times 10^{-6}\ \text{C})}{(0.25\ \text{m})^2}\hat{\mathbf{x}}$$
$$= 0.51\ \text{N}\,\hat{\mathbf{x}}$$

3. Superpose these forces to find the total force, $\vec{F}_3$, acting on q_3:

$$\vec{F}_3 = \vec{F}_{31} + \vec{F}_{32} = -0.14\ \text{N}\,\hat{\mathbf{x}} + 0.51\ \text{N}\,\hat{\mathbf{x}}$$
$$= 0.37\ \text{N}\,\hat{\mathbf{x}}$$

INSIGHT
The net force acting on q_3 has a magnitude of 0.37 N, and it points in the positive x direction. As usual, notice that we use only magnitudes for the charges in the numerator of Coulomb's law.

CONTINUED ON NEXT PAGE

CONTINUED FROM PREVIOUS PAGE

PRACTICE PROBLEM

Find the net force on q_3 if it is at the location $x = 0.25$ m.   [**Answer:** $\vec{F}_3 = -1.2$ N $\hat{x}$]

Some related homework problems: Problem 23, Problem 26, Problem 27

ACTIVE EXAMPLE 19–1 FIND THE LOCATION OF ZERO NET FORCE

In Example 19–2, the net force acting on the charge q_3 is to the right. To what value of x should q_3 be moved for the net force on it to be zero?

SOLUTION *(Test your understanding by performing the calculations indicated in each step.)*

1. Write the magnitude of the force due to q_1: $F_{31} = k|q_1||q_3|/x^2$

2. Write the magnitude of the force due to q_2: $F_{32} = k|q_2||q_3|/(1.00\text{ m} - x)^2$

3. Set these forces equal to one another, and cancel common terms: $|q_1|/x^2 = |q_2|/(1.00\text{ m} - x)^2$

4. Take the square root of both sides and solve for x: $x = 0.61$ m

INSIGHT

Therefore, if q_3 is placed between $x = 0.61$ m and $x = 1.00$ m, the net force acting on it is to the right, in agreement with Example 19–2. On the other hand, if q_3 is placed between $x = 0$ and $x = 0.61$ m, the net force acting on it is to the left. This agrees with the result in the Practice Problem of Example 19–2.

YOUR TURN

If the magnitude of each charge in this system is doubled, does the point of zero net force move to the right, move to the left, or remain in the same place? Explain.

*(Answers to **Your Turn** problems are given in the back of the book.)*

PROBLEM-SOLVING NOTE

Determining the Direction of the Electric Force

When determining the total force acting on a charge, begin by calculating the magnitude of each of the individual forces acting on it. Next, assign appropriate directions to the forces based on the principle that "opposites attract, likes repel" and perform a vector sum.

Next we consider systems in which the individual forces are not along the same line. In such cases, it is often useful to resolve the individual force vectors into components and then perform the required vector sum component by component. This technique is illustrated in the following Example and Conceptual Checkpoint.

EXAMPLE 19–3 SUPERPOSITION

Three charges, each equal to $+2.90\ \mu$C, are placed at three corners of a square 0.500 m on a side, as shown in the diagram. Find the magnitude and direction of the net force on charge 3.

PICTURE THE PROBLEM

The positions of the three charges are shown in the sketch. We also show the force produced by charge 1, $\vec{F}_{31}$, and the force produced by charge 2, $\vec{F}_{32}$. Note that $\vec{F}_{31}$ is 45.0° above the x axis and that $\vec{F}_{32}$ is in the positive x direction. Also, the distance from charge 2 to charge 3 is $r = 0.500$ m, and the distance from charge 1 to charge 3 is $\sqrt{2}\,r$.

STRATEGY

To find the net force, we first calculate the magnitudes of $\vec{F}_{31}$ and $\vec{F}_{32}$ and then their components. Summing these components yields the components of the net force, $\vec{F}_3$. Once we know the components of $\vec{F}_3$, we can calculate its magnitude and direction in the same way as for any other vector.

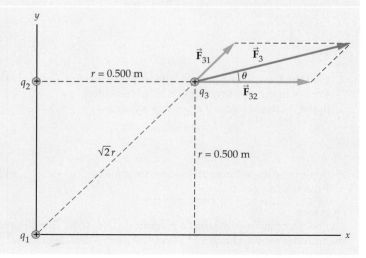

SOLUTION

1. Find the magnitude of $\vec{F}_{31}$:

$$F_{31} = k\frac{|q_1||q_3|}{(\sqrt{2}r)^2}$$

$$= (8.99 \times 10^9 \text{ N} \cdot \text{m}^2/\text{C}^2)\frac{(2.90 \times 10^{-6} \text{ C})^2}{2(0.500 \text{ m})^2}$$

$$= 0.151 \text{ N}$$

2. Find the magnitude of $\vec{F}_{32}$:

$$F_{32} = k\frac{|q_2||q_3|}{r^2}$$

$$= (8.99 \times 10^9 \text{ N} \cdot \text{m}^2/\text{C}^2)\frac{(2.90 \times 10^{-6} \text{ C})^2}{(0.500 \text{ m})^2}$$

$$= 0.302 \text{ N}$$

3. Calculate the components of $\vec{F}_{31}$ and $\vec{F}_{32}$:

$$F_{31,x} = F_{31} \cos 45.0° = (0.151 \text{ N})(0.707) = 0.107 \text{ N}$$
$$F_{31,y} = F_{31} \sin 45.0° = (0.151 \text{ N})(0.707) = 0.107 \text{ N}$$
$$F_{32,x} = F_{32} \cos 0° = (0.302 \text{ N})(1) = 0.302 \text{ N}$$
$$F_{32,y} = F_{32} \sin 0° = (0.302 \text{ N})(0) = 0$$

4. Find the components of $\vec{F}_3$:

$$F_{3,x} = F_{31,x} + F_{32,x} = 0.107 \text{ N} + 0.302 \text{ N} = 0.409 \text{ N}$$
$$F_{3,y} = F_{31,y} + F_{32,y} = 0.107 \text{ N} + 0 = 0.107 \text{ N}$$

5. Find the magnitude of $\vec{F}_3$:

$$F_3 = \sqrt{F_{3,x}^2 + F_{3,y}^2}$$

$$= \sqrt{(0.409 \text{ N})^2 + (0.107 \text{ N})^2} = 0.423 \text{ N}$$

6. Find the direction of $\vec{F}_3$:

$$\theta = \tan^{-1}\left(\frac{F_{3,y}}{F_{3,x}}\right) = \tan^{-1}\left(\frac{0.107 \text{ N}}{0.409 \text{ N}}\right) = 14.7°$$

INSIGHT
Thus, the net force on charge 3 has a magnitude of 0.423 N and points in a direction 14.7° above the x axis. Note that charge 1, which is $\sqrt{2}$ times farther away from charge 3 than is charge 2, produces only half as much force as charge 2.

PRACTICE PROBLEM
Find the magnitude and direction of the net force on charge 3 if its magnitude is doubled to 5.80 μC. Assume that charge 1 and charge 2 are unchanged.   [**Answer:** $F_3 = 2(0.423 \text{ N}) = 0.846 \text{ N}$, $\theta = 14.7°$. Note that the angle is unchanged.]

Some related homework problems: Problem 31, Problem 32

CONCEPTUAL CHECKPOINT 19-3 COMPARE THE FORCE

A charge $-q$ is to be placed at either point A or point B in the accompanying figure. Assume points A and B lie on a line that is midway between the two positive charges. Is the net force experienced at point A **(a)** greater than, **(b)** equal to, or **(c)** less than the net force experienced at point B?

REASONING AND DISCUSSION
Point A is closer to the two positive charges than is point B. As a result, the force exerted by each positive charge will be greater when the charge $-q$ is placed at A. The *net* force, however, is zero at point A, since the equal attractive forces due to the two positive charges cancel, as shown in the diagram.

At point B, on the other hand, the attractive forces combine to give a net downward force. Hence, the charge $-q$ will experience a greater net force at point B.

ANSWER
(c) The net force at point A is less than the net force at point B.

Net force is downward.

Net force equals zero.

Spherical Charge Distributions

Although Coulomb's law is stated in terms of point charges, it can be applied to any type of charge distribution by using the appropriate mathematics. For example, suppose a sphere has a charge Q distributed uniformly over its surface. If a point charge q is outside the sphere, a distance r from its center, the methods of calculus show that the magnitude of the force between the point charge and the sphere is simply

$$F = k\frac{|q||Q|}{r^2}$$

In situations like this, the spherical charge distribution behaves the same as if all its charge were concentrated in a point at its center. For point charges inside a charged spherical shell, the net force exerted by the shell is zero. In general, the electrical behavior of spherical *charge* distributions is analogous to the gravitational behavior of spherical *mass* distributions.

In the next Active Example, we consider a system in which a charge Q is distributed uniformly over the surface of a sphere. In such a case it is often convenient to specify the amount of *charge per area* on the sphere. This is referred to as the **surface charge density**, σ. If a sphere has an area A and a surface charge density σ, its total charge is

$$Q = \sigma A \qquad\qquad 19\text{–}7$$

Note that the SI unit of σ is C/m^2. If the radius of the sphere is R, then $A = 4\pi R^2$, and $Q = \sigma(4\pi R^2)$.

ACTIVE EXAMPLE 19–2   FIND THE FORCE EXERTED BY A SPHERE

An insulating sphere of radius $R = 0.10$ m has a uniform surface charge density equal to 5.9 μC/m^2. A point charge of magnitude 0.71 μC is 0.45 m from the center of the sphere. Find the magnitude of the force exerted by the sphere on the point charge.

SOLUTION *(Test your understanding by performing the calculations indicated in each step.)*

1. Find the area of the sphere: $\qquad\qquad\qquad\qquad A = 0.13$ m^2

2. Calculate the total charge on the sphere: $\qquad\qquad Q = 0.77\ \mu$C

3. Use Coulomb's law to calculate the magnitude $\qquad F = 0.024$ N
   of the force between the sphere and the point charge:

INSIGHT

As long as the point charge is outside the sphere, and the charge distribution remains spherically uniform, the sphere may be treated as a point charge.

YOUR TURN

Suppose the sphere in this problem is replaced by one with half the radius, but with the same surface charge density. Is the force exerted by this sphere greater than, less than, or the same as the force exerted by the original sphere? Explain.

*(Answers to **Your Turn** problems are given in the back of the book.)*

19–4 The Electric Field

You have probably encountered the notion of a "force field" in various science fiction novels and movies. A concrete example of a force field is provided by the force between electric charges. Consider, for example, a positive point charge q at the origin of a coordinate system, as in **Figure 19–9**. If a positive "test charge," q_0, is placed at point A, the force exerted on it by q is indicated by the vector $\vec{F}_A$. On the other hand, if the test charge is placed at point B, the force it experiences there is $\vec{F}_B$. At every point in space there is a corresponding force. In this sense, Figure 19–9 allows us to visualize the "force field" associated with the charge q.

Since the magnitude of the force at every point in Figure 19–9 is proportional to q_0 (due to Coulomb's law), it is convenient to divide by q_0 and define a *force per*

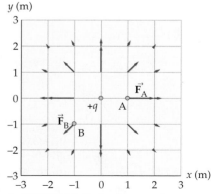

▲ **FIGURE 19–9** An electrostatic force field

The positive charge $+q$ at the origin of this coordinate system exerts a different force on a given charge at every point in space. Here we show the force vectors associated with q for a grid of points.

charge at every point in space that is independent of q_0. We refer to the force per charge as the electric field, $\vec{E}$. Its precise definition is as follows:

Definition of the Electric Field, $\vec{E}$
If a test charge q_0 experiences a force $\vec{F}$ at a given location, the electric field $\vec{E}$ at that location is

$$\vec{E} = \frac{\vec{F}}{q_0} \qquad\qquad 19\text{–}8$$

SI unit: N/C

It should be noted that this definition applies whether the force $\vec{F}$ is due to a single charge, as in Figure 19–9, or to a group of charges. In addition, it is assumed that the test charge is small enough that it does not disturb the position of any other charges in the system.

To summarize, *the electric field is the force per charge at a given location.* Therefore, if we know the electric field vector $\vec{E}$ at a given point, the force that a charge q experiences at that point is

$$\vec{F} = q\vec{E} \qquad\qquad 19\text{–}9$$

Notice that the direction of the force depends on the sign of the charge. In particular,

- A positive charge experiences a force *in the direction* of $\vec{E}$.
- A negative charge experiences a force *in the opposite direction* of $\vec{E}$.

Finally, the magnitude of the force is the product of the magnitudes of q and $\vec{E}$:

- The magnitude of the force acting on a charge q is $F = |q|E$.

As we continue in this chapter, we will determine the electric field for a variety of different charge distributions. In some cases, $\vec{E}$ will decrease with distance as $1/r^2$ (a point charge), in other cases as $1/r$ (a line of charge), and in others $\vec{E}$ will be a constant (a charged plane). Before we calculate the electric field itself, however, we first consider the force exerted on charges by a constant electric field.

PROBLEM-SOLVING NOTE

The Force Exerted by an Electric Field

The force exerted on a charge by an electric field can point in only one of two directions—parallel or antiparallel to the direction of the field.

EXAMPLE 19–4   FORCE FIELD

In a certain region of space, a uniform electric field has a magnitude of 4.60×10^4 N/C and points in the positive x direction. Find the magnitude and direction of the force this field exerts on a charge of **(a)** $+2.80\ \mu$C and **(b)** $-9.30\ \mu$C.

PICTURE THE PROBLEM
In our sketch we indicate the uniform electric field and the two charges mentioned in the problem. Note that the positive charge experiences a force in the positive x direction (the direction of $\vec{E}$), and the negative charge experiences a force in the negative x direction (opposite to $\vec{E}$).

STRATEGY
To find the magnitude of each force, we use $F = |q|E$. The direction has already been indicated in our sketch.

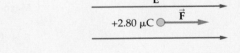
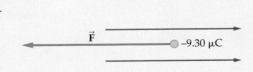

SOLUTION

Part (a)

1. Find the magnitude of the force on the $+2.80$-μC charge: $\qquad F = |q|E = (2.80 \times 10^{-6}\text{ C})(4.60 \times 10^4\text{ N/C}) = 0.129$ N

Part (b)

2. Find the magnitude of the force on the -9.30-μC charge: $\qquad F = |q|E = (9.30 \times 10^{-6}\text{ C})(4.60 \times 10^4\text{ N/C}) = 0.428$ N

INSIGHT
To summarize, the force on the $+2.80$-μC charge is of magnitude 0.129 N in the positive x direction; the force on the -9.30-μC charge is of magnitude 0.428 N in the negative x direction.

CONTINUED ON NEXT PAGE

CONTINUED FROM PREVIOUS PAGE

PRACTICE PROBLEM

If the $+2.80\text{-}\mu\text{C}$ charge experiences a force of 0.25 N, what is the magnitude of the electric field?   [**Answer:** $E = 8.9 \times 10^4\,\text{N/C}$]

Some related homework problems: Problem 44, Problem 47

REAL-WORLD PHYSICS

Electrodialysis for water purification

The fact that charges of opposite sign experience forces in opposite directions in an electric field is used to purify water in the process known as **electrodialysis.** This process depends on the fact that most minerals that dissolve in water dissociate into positive and negative ions. Probably the most common example is table salt (NaCl), which dissociates into positive sodium ions (Na^+) and negative chlorine ions (Cl^-). When brackish water is passed through a strong electric field in an electrodialysis machine, the mineral ions move in opposite directions and pass through two different types of semipermeable membrane—one that allows only positive ions to pass through, the other only negative ions. This process leaves water that is purified of dissolved minerals and suitable for drinking.

The Electric Field of a Point Charge

Perhaps the simplest example of an electric field is the field produced by an idealized point charge. To be specific, suppose a positive point charge q is at the origin in **Figure 19–10**. If a positive test charge q_0 is placed a distance r from the origin, the force it experiences is directed away from the origin and is of magnitude

$$F = k\frac{|q||q_0|}{r^2}$$

Applying our definition of the electric field in Equation 19–8, we find that the magnitude of the field is

$$E = \frac{F}{q_0} = \frac{\left(k\dfrac{|q||q_0|}{r^2}\right)}{q_0} = k\frac{|q|}{r^2}$$

Since a positive charge experiences a force that is radially outward, that too is the direction of $\vec{\mathbf{E}}$.

In general, then, we can say that the electric field a distance r from a point charge q has the following magnitude:

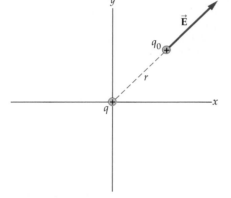

▲ **FIGURE 19–10 The electric field of a point charge**
The electric field $\vec{\mathbf{E}}$ due to a positive charge q at the origin is radially outward. Its magnitude is $E = k|q|/r^2$.

Magnitude of the Electric Field Due to a Point Charge

$$E = k\frac{|q|}{r^2}$$

19–10

If the charge q is positive, the field points radially outward from the charge; if it is negative, the field is radially inward. This is illustrated in **Figure 19–11**. Thus, to

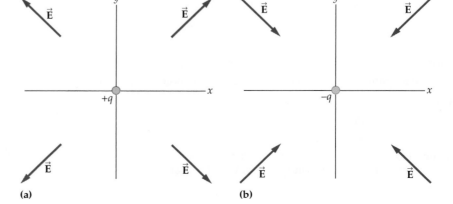

▶ **FIGURE 19–11 The direction of the electric field**

(a) The electric field due to a positive charge at the origin points radially outward. **(b)** If the charge at the origin is negative, the electric field is radially inward.

(a)   (b)

determine the electric field due to a point charge, we first use Equation 19–10 to find its magnitude, and then use the rule illustrated in Figure 19–11 to find its direction.

EXERCISE 19–3

Find the electric field produced by a 1.0-μC point charge at a distance of **(a)** 0.75 m and **(b)** 1.5 m.

SOLUTION

a. Applying Equation 19–10 with $q = 1.0 \ \mu$C and $r = 0.75$ m yields

$$E = k\frac{|q|}{r^2} = (8.99 \times 10^9 \ \text{N} \cdot \text{m}^2/\text{C}^2)\frac{(1.0 \times 10^{-6} \ \text{C})}{(0.75 \ \text{m})^2} = 1.6 \times 10^4 \ \text{N/C}$$

b. Noting that E depends on $1/r^2$, we see that doubling the distance from 0.75 m to 1.5 m results in a reduction in the electric field by a factor of 4:

$$E = \tfrac{1}{4}(1.6 \times 10^4 \ \text{N/C}) = 0.40 \times 10^4 \ \text{N/C}$$

Superposition of Fields

Many electrical systems consist of more than two charges. In such cases, the total electric field can be found by using superposition—just as when we find the total force due to a system of charges. In particular, the total electric field is found by calculating the vector sum—often using components—of the electric fields due to each charge separately.

For example, let's calculate the total electric field at point P in **Figure 19–12**. First we sketch the directions of the fields $\vec{E}_1$ and $\vec{E}_2$ due to the charges $q_1 = +q$ and $q_2 = +q$, respectively. In particular, if a positive test charge is at point P, the force due to q_1 is down and to the right, whereas the force due to q_2 is up and to the right. From the geometry of the figure we see that $\vec{E}_1$ is at an angle θ below the x axis, and—by symmetry—$\vec{E}_2$ is at the same angle θ above the axis. Since the two charges have the same magnitude, and the distances from P to the charges are the same, it follows that $\vec{E}_1$ and $\vec{E}_2$ have the same magnitude:

$$E_1 = E_2 = E = k\frac{|q|}{d^2}$$

To find the net electric field $\vec{E}_{\text{net}}$, we use components. First, consider the y direction. In this case, we have $E_{1,y} = -E \sin \theta$ and $E_{2,y} = +E \sin \theta$. Hence, the y component of the net electric field is zero:

$$E_{\text{net},y} = E_{1,y} + E_{2,y} = -E \sin \theta + E \sin \theta = 0$$

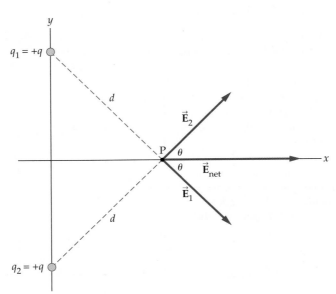

◀ **FIGURE 19–12 Superposition of the electric field**

The net electric field at the point P is the vector sum of the fields due to the charges q_1 and q_2. Note that $\vec{E}_1$ and $\vec{E}_2$ point away from the charges q_1 and q_2, respectively. This is as expected, since both of these charges are positive.

Referring again to Figure 19–12, it is apparent that this result could have been anticipated by symmetry considerations. Finally, we determine the x component of E_{net}:

$$E_{net,x} = E_{1,x} + E_{2,x} = E \cos \theta + E \cos \theta = 2E \cos \theta$$

Thus, the net electric field at P is in the positive x direction, as shown in Figure 19–12, and has a magnitude equal to $2E \cos \theta$.

CONCEPTUAL CHECKPOINT 19–4   THE SIGN OF THE CHARGES

Two charges, q_1 and q_2, have equal magnitudes q and are placed as shown in the figure to the right. The net electric field at point P is vertically upward. Do we conclude that **(a)** q_1 is positive, q_2 is negative; **(b)** q_1 is negative, q_2 is positive; or **(c)** q_1 and q_2 have the same sign?

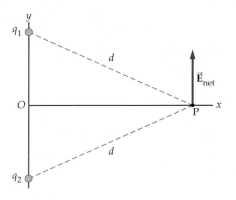

REASONING AND DISCUSSION
If the net electric field at P is vertically upward, the x components of $\vec{E}_1$ and $\vec{E}_2$ must cancel, and the y components must both be in the positive y direction. The only way for this to happen is to have q_1 negative and q_2 positive, as shown in the following diagram.

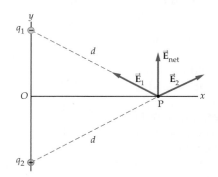

With this choice, a positive test charge at P is attracted to q_1 (so that $\vec{E}_1$ is up and to the left) and repelled from q_2 (so that $\vec{E}_2$ is up and to the right).

ANSWER
(b) q_1 is negative, q_2 is positive.

We conclude this section by considering the same physical system presented in Example 19–3, this time from the point of view of the electric field.

EXAMPLE 19–5   SUPERPOSITION IN THE FIELD

Two charges, each equal to $+2.90 \ \mu C$, are placed at two corners of a square 0.500 m on a side, as shown in the sketch. Find the magnitude and direction of the net electric field at a third corner of the square, the point labeled 3 in the sketch.

PICTURE THE PROBLEM
The positions of the two charges are shown in the sketch. We also show the electric field produced by each charge. The key difference between this sketch and the one in Example 19–3 is that in this case there is no charge at point 3; the electric field still exists there, even though it has no charge on which to exert a force.

STRATEGY
In analogy with Example 19–3, we first calculate the magnitudes of $\vec{E}_1$ and $\vec{E}_2$ and then their components. Summing these components yields the components of the net electric field, $\vec{E}_{net}$. Once we know the components of $\vec{E}_{net}$, we find its magnitude and direction in the same way as for $\vec{F}_{net}$ in Example 19–3.

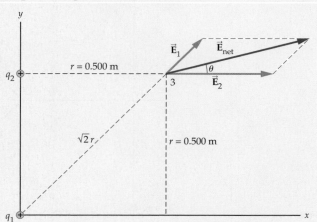

SOLUTION

1. Find the magnitude of $\vec{E}_1$:

$$E_1 = k\frac{|q_1|}{(\sqrt{2}r)^2}$$

$$= (8.99 \times 10^9 \text{ N} \cdot \text{m}^2/\text{C}^2)\frac{(2.90 \times 10^{-6} \text{ C})}{2(0.500 \text{ m})^2}$$

$$= 5.21 \times 10^4 \text{ N/C}$$

2. Find the magnitude of $\vec{E}_2$:

$$E_2 = k\frac{|q_2|}{r^2}$$

$$= (8.99 \times 10^9 \text{ N} \cdot \text{m}^2/\text{C}^2)\frac{(2.90 \times 10^{-6} \text{ C})}{(0.500 \text{ m})^2}$$

$$= 1.04 \times 10^5 \text{ N/C}$$

3. Calculate the components of $\vec{E}_1$ and $\vec{E}_2$:

$$E_{1,x} = E_1 \cos 45.0°$$
$$= (5.21 \times 10^4 \text{ N/C})(0.707) = 3.68 \times 10^4 \text{ N/C}$$
$$E_{1,y} = E_1 \sin 45.0°$$
$$= (5.21 \times 10^4 \text{ N/C})(0.707) = 3.68 \times 10^4 \text{ N/C}$$
$$E_{2,x} = E_2 \cos 0°$$
$$= (1.04 \times 10^5 \text{ N/C})(1) = 1.04 \times 10^5 \text{ N/C}$$
$$E_{2,y} = E_2 \sin 0° = (1.04 \times 10^5 \text{ N/C})(0) = 0$$

4. Find the components of $\vec{E}_{net}$:

$$E_{net,x} = E_{1,x} + E_{2,x}$$
$$= 3.68 \times 10^4 \text{ N/C} + 1.04 \times 10^5 \text{ N/C}$$
$$= 1.41 \times 10^5 \text{ N/C}$$
$$E_{net,y} = E_{1,y} + E_{2,y}$$
$$= 3.68 \times 10^4 \text{ N/C} + 0 = 3.68 \times 10^4 \text{ N/C}$$

5. Find the magnitude of $\vec{E}_{net}$:

$$E_{net} = \sqrt{E_{net,x}^2 + E_{net,y}^2}$$
$$= \sqrt{(1.41 \times 10^5 \text{ N/C})^2 + (3.68 \times 10^4 \text{ N/C})^2}$$
$$= 1.46 \times 10^5 \text{ N/C}$$

6. Find the direction of $\vec{E}_{net}$:

$$\theta = \tan^{-1}(E_{net,y}/E_{net,x})$$

$$= \tan^{-1}\left(\frac{3.68 \times 10^4 \text{ N/C}}{1.41 \times 10^5 \text{ N/C}}\right) = 14.6°$$

INSIGHT

Note that, as one would expect, the direction of the net electric field is the same as the direction of the net force in Example 19–3 (except for a small discrepancy in the last decimal place due to rounding off in the calculations). In addition, the magnitude of the force exerted by the electric field on a charge of 2.90 μC is $F = qE_{net} = (2.90 \; \mu\text{C})(1.46 \times 10^5 \text{ N/C}) = 0.423$ N, the same as was found in Example 19–3.

PRACTICE PROBLEM

Find the magnitude and direction of the net electric field at the bottom right corner of the square. [**Answer:** $E_{net} = 1.46 \times 10^5 \text{ N/C}$, $\theta = -14.6°$]

Some related homework problems: Problem 50, Problem 51

Many aquatic creatures are capable of producing electric fields. For example, African freshwater fishes in the family Mormyridae can generate weak electric fields from modified tail muscles and are able to detect variations in this field as they move through their environment. With this capability, these nocturnal feeders have an electrical guidance system that assists them in locating obstacles, enemies, and food. Much stronger electric fields are produced by electric eels and electric skates. In particular, the electric eel *Electrophorus electricus* generates

REAL-WORLD PHYSICS: BIO

Electric fish

electric fields great enough to kill small animals and to stun larger animals, including humans.

Sharks are well known for their sensitivity to weak electric fields in their surroundings. In fact, they possess specialized organs for this purpose, known as the ampullae of Lorenzini, which assist in the detection of prey. Recently, this sensitivity has been put to use as a method of repelling sharks in order to protect swimmers and divers. A device called the SharkPOD (Protective Oceanic Device) consists of two metal electrodes, one attached to a diver's air tank, the other to one of the diver's fins. These electrodes produce a strong electric field that completely surrounds the diver and causes sharks to turn away out to a distance of up to 7 m.

The SharkPOD was used in the 2000 Summer Olympic Games in Sydney to protect swimmers competing in the triathlon. The swimming part of the event was held in Sydney harbor, where great white sharks are a common sight. To protect the swimmers, divers wearing the SharkPOD swam along the course, a couple meters below the athletes. The race was completed without incident.

19–5 Electric Field Lines

When looking at plots like those in Figures 19–09 and 19–11, it is tempting to imagine a pictorial representation of the electric field. This thought is reinforced when one considers a photograph like **Figure 19–13**, which shows grass seeds suspended in oil. Because of polarization effects, the grass seeds tend to align in the direction of the electric field, much like the elongated atoms shown in Figure 19–5. In this case, the seeds are aligned radially, due to the electric field of the charged rod seen "end on" in the middle of the photograph. Clearly, a set of radial lines would seem to represent the electric field in this case.

In fact, an entirely consistent method of drawing electric field lines is obtained by using the following set of rules:

Rules for Drawing Electric Field Lines
Electric field lines:

1. Point in the *direction* of the electric field vector $\vec{E}$ at every point;
2. *Start* at positive (+) charges or at infinity;
3. *End* at negative (−) charges or at infinity;
4. Are more *dense* where $\vec{E}$ has a greater magnitude. In particular, the number of lines entering or leaving a charge is proportional to the magnitude of the charge.

We now show how these rules are applied.

For example, the electric field lines for two different point charges are presented in **Figure 19–14**. First, we know that the electric field points directly away from the charge (+q) in Figure 19–14 (a); hence from rule 1 the field lines are radial. In agreement with rule 2 the field lines start on a + charge, and in agreement

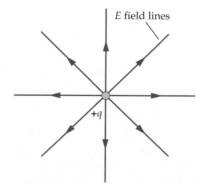

▲ **FIGURE 19–13 Grass seeds in an electric field**
Grass seeds aligning with electric field lines.

▶ **FIGURE 19–14 Electric field lines for a point charge**

(a) Near a positive charge the field lines point radially away from the charge. The lines start on the positive charge and end at infinity. **(b)** Near a negative charge the field lines point radially inward. They start at infinity and end on a negative charge and are more dense where the field is more intense. Notice that the number of lines drawn for part (b) is twice the number drawn for part (a), a reflection of the relative magnitudes of the charges.

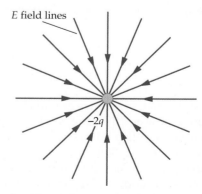

(a) *E* field lines point away from positive charges   **(b)** *E* field lines point toward negative charges

with rule 3 they end at infinity. Finally, as anticipated from rule 4, the field lines are closer together near the charge, where the field is more intense. Similar considerations apply to Figure 19–14 (b), where the charge is $-2q$. In this case, however, the direction of the field lines is reversed and the number of lines is doubled.

CONCEPTUAL CHECKPOINT 19–5   INTERSECT OR NOT?

Which of the following statements is correct: Electric field lines **(a)** can or **(b)** cannot intersect?

REASONING AND DISCUSSION
By definition, electric field lines are always tangent to the electric field. Since the electric force, and hence the electric field, can point in only one direction at any given location, it follows that field lines cannot intersect. If they did, the field at the intersection point would have two conflicting directions.

ANSWER
(b) Electric field lines cannot intersect.

Figure 19–15 shows examples of electric field lines for various combinations of charges. In systems like these, we draw a set of curved field lines that are tangent to the electric field vector, $\vec{E}$, at every point. This is illustrated for a variety of points in Figure 19–15 (a), and similar considerations apply to all such field diagrams. In addition, note that the magnitude of $\vec{E}$ is greater in those regions of Figure 19–15 where the field lines are more closely packed together. Clearly, then, we expect an intense electric field between the charges in Figure 19–15 (b) and a vanishing field between the charges in Figure 19–15 (c).

Of particular interest is the $+q$ and $-q$ charge combination in Figure 19–15 (a). In general, a system of equal and opposite charges separated by a nonzero distance is known as an **electric dipole.** The total charge of the dipole is zero, but because the charges are separated, the electric field does not vanish. Instead, the field lines form "loops" that are characteristic of dipoles.

Many molecules are polar—water is a common example—which means they have an excess of positive charge near one end and a corresponding excess of

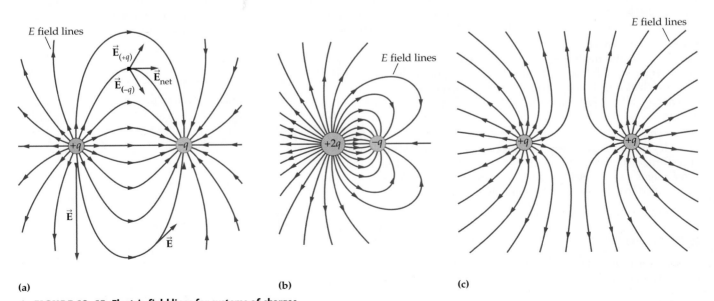

(a)   (b)   (c)

▲ **FIGURE 19–15 Electric field lines for systems of charges**
(a) The electric field lines for a dipole form closed loops that become more widely spaced with distance from the charges. Note that at each point in space, the electric field vector $\vec{E}$ is tangent to the field lines. **(b)** In a system with a net charge, some field lines extend to infinity. If the charges have opposite signs, some field lines start on one charge and terminate on the other charge. **(c)** All of the field lines in a system with charges of the same sign extend to infinity.

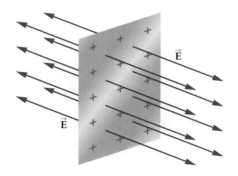

◀ **FIGURE 19–16 The electric field of a charged plate**
The electric field near a large charged plate is uniform
in direction and magnitude.

negative charge near the other end. As a result, they produce an electric dipole
field. Similarly, a typical bar magnet produces a *magnetic* dipole field, as we shall
see in Chapter 22.

Finally, the electric field representations in Figures 19–14 and 19–15 are two-
dimensional "slices" through the full field, which is three-dimensional. Therefore,
one should imagine a similar set of field lines in these figures coming out of the
page and going into the page.

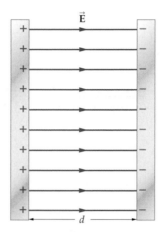

▲ **FIGURE 19–17 A parallel-plate
capacitor**
In the ideal case, the electric field is
uniform between the plates and zero
outside.

Parallel-Plate Capacitor

A particularly simple and important field picture results when charge is spread
uniformly over a large plate, as illustrated in **Figure 19–16**. At points that are not
near the edge of the plate, the electric field is uniform in both direction and magnitude. That is, the field points in a single direction—perpendicular to the plate—
and its magnitude is independent of the distance from the plate. This result can be
proved using Gauss's law, as we show in Section 19–7.

If two such conducting plates with opposite charge are placed parallel to one
another and separated by a distance d, as in **Figure 19–17**, the result is referred to as
a **parallel-plate capacitor.** The field for such a system is uniform between the
plates, and zero outside the plates. This is the ideal case, which is exactly true for
an infinite plate and a good approximation for real plates. The field lines are illustrated in Figure 19–17. Parallel-plate capacitors are discussed further in the
next chapter and will be of particular interest in Chapters 21 and 24, when we consider electric circuits.

EXAMPLE 19–6 DANGLING BY A THREAD

The electric field between the plates of a parallel-plate capacitor is horizontal, uniform, and has a magnitude E. A small object of
mass 0.0250 kg and charge $-3.10~\mu C$ is suspended by a thread between the plates, as shown in the sketch. The thread makes an
angle of 10.5° with the vertical. Find **(a)** the tension in the thread and **(b)** the magnitude of the electric field.

PICTURE THE PROBLEM
Our sketch shows the thread making an angle $\theta = 10.5°$ with the vertical.
The inset to the right shows the free-body diagram for the suspended object, as well as our choice of positive x and y directions. Note that we label
the charge of the object $-q$, where $q = 3.10~\mu C$, in order to clearly indicate
its sign.

STRATEGY
The relevant physical principle in this problem is that because the object is
at rest, the net force acting on it must vanish. Thus, setting the x and y
components of the net force equal to zero yields two conditions, which
can be used to solve for the two unknowns, T and E.

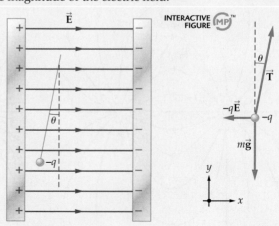

SOLUTION

1. Set the net force in the x direction equal to zero: $\qquad -qE + T\sin\theta = 0$

2. Set the net force in the y direction equal to zero: $\qquad T\cos\theta - mg = 0$

Part (a)

3. Because we know all the quantities in the y force
equation except for the tension, we use it to solve for T:
$$T = \frac{mg}{\cos\theta} = \frac{(0.0250~\text{kg})(9.81~\text{m/s}^2)}{\cos(10.5°)} = 0.249~\text{N}$$

Part (b)

4. Now use the x force equation to find the magnitude of the electric field, E:

$$E = \frac{T \sin \theta}{q} = \frac{(0.249 \text{ N}) \sin(10.5°)}{3.10 \times 10^{-6} \text{ C}} = 1.46 \times 10^4 \text{ N/C}$$

INSIGHT

As expected, the negatively charged object is attracted to the positively charged plate. This means that the electric force exerted on it is opposite in direction to the electric field.

PRACTICE PROBLEM

Suppose the electric field between the plates is 2.50×10^4 N/C, but the charge of the object and the angle of the thread with the vertical are the same as before. Find the tension in the thread and the mass of the object.   [**Answer:** $T = 0.425$ N, $m = 0.0426$ kg]

Some related homework problems: Problem 60, Problem 94

The fact that a charge experiences a force when it passes between two charged plates finds application in a wide variety of devices. For example, the image you see on many television screens is produced when a beam of electrons strikes the screen from behind and illuminates individual red, blue, or green *pixels*. Which pixels are illuminated and which remain dark is controlled by parallel charged plates that deflect the electron beam up or down and left or right. Thus, sending the appropriate electrical signals to the deflection plates makes the beam of electrons "paint" any desired picture on the screen.

Similar deflection plates are used in an ink-jet printer. In this case, the beam in question is not a beam of electrons but rather a beam of electrically charged ink droplets. The beam of droplets can be deflected as desired, so that individual letters can be constructed from a series of closely spaced dots on the page. A typical printer might produce as many as 600 dots per inch—that is, 600 dpi.

19–6 Shielding and Charging by Induction

In a perfect conductor there are enormous numbers of electrons completely free to move about within the conductor. This simple fact has some rather interesting consequences. Consider, for example, a solid metal sphere attached to an insulating base as in **Figure 19–18**. Suppose a positive charge Q is placed on the sphere. The question is: How does this charge distribute itself on the sphere when it is in equilibrium—that is, when all the charges are at rest? In particular, does the charge spread itself uniformly throughout the volume of the sphere, or does it concentrate on the surface?

The answer is that the charge concentrates on the surface, as shown in Figure 19–18 (a), but let's investigate why this should be the case. First, assume the opposite—that the charge is spread uniformly throughout the sphere's volume, as indicated in Figure 19–18 (b). If this were the case, a charge at location A would experience an outward force due to the spherical distribution of charge between it and the center of the sphere. Since charges are free to move, the charge at A would respond to this force by moving toward the surface. Clearly, then, a uniform distribution of charge within the sphere's volume is not in equilibrium. In fact, the argument that a charge at point A will move toward the surface can be applied to any charge within the sphere. Thus, the net result is that *all* the excess charge Q moves onto the surface of the sphere which, in turn, allows the individual charges to be spread as far from one another as possible.

The preceding result holds no matter what the shape of the conductor. In general,

Excess Charge on a Conductor

Excess charge placed on a conductor, whether positive or negative, moves to the exterior surface of the conductor.

Television screens and ink-jet printers

(a)

(b)

▲ **FIGURE 19–18 Charge distribution on a conducting sphere**

(a) A charge placed on a conducting sphere distributes itself uniformly on the surface of the sphere; none of the charge is within the volume of the sphere. **(b)** If the charge were distributed uniformly throughout the volume of a sphere, individual charges, like that at point A, would experience a force due to other charges in the volume. Since charges are free to move in a conductor, they will respond to these forces by moving as far from one another as possible—that is, to the surface of the conductor.

We specify the exterior surface in this statement because a conductor may contain one or more cavities. When an excess charge is applied to such a conductor, all the charge ends up on the exterior surface, and none on the interior surfaces.

Electrostatic Shielding

The ability of electrons to move freely within a conductor has another important consequence; namely, the electric field within a conductor vanishes.

Zero Field within a Conductor
When electric charges are at rest, the electric field within a conductor is zero; $E = 0$.

By *within* a conductor, we mean a location in the actual material of the conductor, as opposed to a location in a cavity within the material.

The best way to see the validity of this statement is to again consider the opposite. If there were a nonzero field within a conductor, electrons would move in response to the field. They would continue to move until finally the field was reduced to zero, at which point the system would be in equilibrium and no more charges would move. Thus, equilibrium and $E = 0$ within a conductor go hand in hand.

A straightforward extension of this idea explains the phenomenon of **shielding,** in which a conductor "shields" its interior from external electric fields. For example, in **Figure 19–19 (a)** we show an uncharged, conducting metal sphere placed in an electric field. Because the positive ions in the metal do not move, the field tends to move negative charges to the left and leave excess positive charges on the right; hence, it causes the sphere to have an **induced** negative charge on its left half and an induced positive charge on its right half. The total charge on the sphere, of course, is still zero. Since field lines end on (−) charges and begin on (+) charges, the external electric field ends on the left half of the sphere and starts up again on the right half. In between, within the conductor, the field is zero, as expected. Thus, the conductor has shielded its interior from the applied electric field.

Shielding occurs whether the conductor is solid, as in Figure 19–19, or hollow. In fact, even a thin sheet of metal foil formed into the shape of a box will shield its interior from external electric fields. This effect is put to use in numerous electrical devices, which often have a metal foil or wire mesh enclosure surrounding the sensitive electrical circuits. In this way, a given device can be isolated from the effects of other nearby devices that might otherwise interfere with its operation.

Notice also in Figure 19–19 that the field lines bend slightly near the surface of the sphere. In fact, on closer examination, as in **Figure 19–19 (b)**, we see that the field lines always contact the surface at right angles. This is true for any conductor:

Electric Fields at Conductor Surfaces
Electric field lines contact conductor surfaces at right angles.

If an electric field contacted a conducting surface at an angle other than 90°, the result would be a component of force parallel to the surface. This would result in a movement of electrons and, hence, would not correspond to equilibrium. Instead, electrons would move until the parallel component of the electric field was canceled.

▶ **FIGURE 19–19 Electric field near a conducting surface**

(a) When an uncharged conductor is placed in an electric field, the field induces opposite charges on opposite sides of the conductor. The net charge on the conductor is still zero, however. The induced charges produce a field within the conductor that *exactly* cancels the external field, leading to $E = 0$ inside the conductor. This is an example of electrical shielding. **(b)** Electric field lines meet the surface of a conductor at right angles.

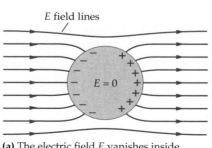

(a) The electric field E vanishes inside a conductor

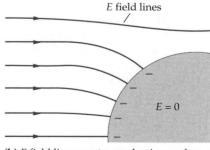

(b) E field lines meet a conducting surface at right angles

Further examples of field lines near conducting surfaces are shown in **Figure 19–20**. Notice that the field lines are more densely packed near a sharp point, indicating that the field is more intense in such regions. This effect illustrates the basic principle behind the operation of lightning rods. If you look closely, you will notice that all lightning rods have a sharply pointed tip. During an electrical storm, the electric field at the tip becomes so intense that electric charge is given off into the atmosphere. In this way, a lightning rod acts to discharge the area near a house—by giving off a steady stream of charge—thus preventing a strike by a bolt of lightning, which transfers charge in one sudden blast. Sharp points on the rigging of a ship at sea can also give off streams of charge during a storm, often producing glowing lights referred to as Saint Elmo's fire.

The same principle is used to clean the air we breathe, in devices known as *electrostatic precipitators*. In an electrostatic precipitator, smoke and other airborne particles in a smokestack are given a charge as they pass by sharply pointed electrodes—like lightning rods—within the stack. Once the particles are charged, they are removed from the air by charged plates that exert electrostatic forces on them. The resultant emission from the smokestack contains drastically reduced amounts of potentially harmful particulates.

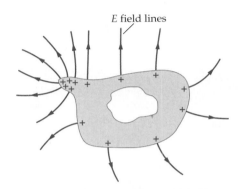

▲ **FIGURE 19–20 Intense electric field near a sharp point**

Electric charges and field lines are more densely packed near a sharp point. This means that the electric field is more intense in such regions as well. (Note that there are no electric charges on the interior surface surrounding the cavity.)

▲ In a dramatic science-museum demonstration of electrical shielding (left), the metal bars of a cage provide excellent protection from an artificially generated lightning bolt. A more practical safeguard is the lightning rod (right). Lightning rods always have sharp points, because that is where the electric field of a conductor is most intense. At the tip, the field can become so strong that charge leaks away into the atmosphere rather than building up to levels that will attract a lightning strike. If a strike does occur, it is conducted to the ground through the lightning rod, rather than through some part of the building itself.

One final note regarding shielding is that it works in one direction only: A conductor shields its interior from external fields, but it does not shield the external world from fields within it. This phenomenon is illustrated in **Figure 19–21** for the case of an uncharged conductor. First, the charge $+Q$ in the cavity induces a charge $-Q$ on the interior surface, in order for the field in the conductor to be zero. Since the conductor is uncharged, a charge $+Q$ will be induced on its exterior surface. As a result, the external world will experience a field due, ultimately, to the charge $+Q$ within the cavity of the conductor.

Charging by Induction

One way to charge an object is to touch it with a charged rod; but since electric forces can act at a distance, it is also possible to charge an object without making direct physical contact. This type of charging is referred to as **charging by induction.**

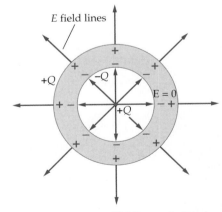

▲ **FIGURE 19–21 Shielding works in only one direction**

A conductor does not shield the external world from charges it encloses. Still, the electric field is zero within the conductor itself.

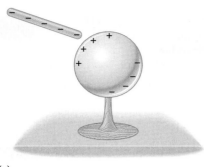

(a)

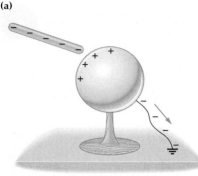

(b)

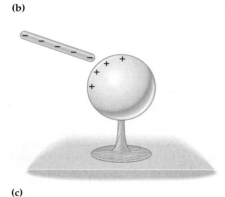

(c)

(d)

◀ **FIGURE 19–22 Charging by induction**
(a) A charged rod induces + and − charges on opposite sides of the conductor. **(b)** When the conductor is grounded, charges that are repelled by the rod enter the ground. There is now a net charge on the conductor. **(c)** Removing the grounding wire, with the charged rod still in place, traps the net charge on the conductor. **(d)** The charged rod can now be removed, and the conductor retains a charge that is opposite in sign to that on the charged rod.

To see how this type of charging works, consider an uncharged metal sphere on an insulating base. If a negatively charged rod is brought close to the sphere without touching it, as in **Figure 19–22 (a)**, electrons in the sphere are repelled. An induced positive charge is produced on the near side of the sphere, and an induced negative charge on the far side. At this point the sphere is still electrically neutral, however.

The key step in this charging process, shown in **Figure 19–22 (b)**, is to connect the sphere to the ground using a conducting wire. As one might expect, this is referred to as **grounding** the sphere, and is indicated by the symbol ⏚. (A table of electrical symbols can be found in Appendix D.) Since the ground is a fairly good conductor of electricity, and since the Earth can receive or give up practically unlimited numbers of electrons, the effect of grounding the sphere is that the electrons repelled by the charged rod enter the ground. Now the sphere has a net positive charge. With the rod kept in place, the grounding wire is removed, as in **Figure 19–22 (c)**, trapping the net positive charge on the sphere. The rod can now be pulled away, as in **Figure 19–22 (d)**.

Notice that the *induced* charge on the sphere is opposite in sign to the charge on the rod. In contrast, when the sphere is charged by *touch* as in Figure 19–6, it acquires a charge with the same sign as the charge on the rod.

19–7 Electric Flux and Gauss's Law

In this section, we introduce the idea of an electric flux and show that it can be used to calculate the electric field. The precise connection between electric flux and the charges that produce the electric field is provided by Gauss's law.

Electric Flux

Consider a uniform electric field $\vec{E}$, as in **Figure 19–23 (a)**, passing through an area A that is perpendicular to the field. Looking at the electric field lines with their arrows, we can easily imagine a "flow" of electric field through the area. Though there is no actual flow, of course, the analogy is a useful one. It is with this in mind that we define an **electric flux, Φ**, for this case as follows:

$$\Phi = EA$$

On the other hand, if the area A is parallel to the field lines, as in **Figure 19–23 (b)**, none of the $\vec{E}$ lines pierce the area, and hence there is no flux of electric field:

$$\Phi = 0$$

In an intermediate case, as shown in **Figure 19–23 (c)**, the $\vec{E}$ lines pierce the area A at an angle θ away from the perpendicular. As a result, the component of $\vec{E}$ perpendicular to the surface is $E \cos \theta$, and the component parallel to the surface is $E \sin \theta$. Since only the perpendicular component of $\vec{E}$ causes a flux (the parallel component does not pierce the area), the flux in the general case is the following:

Definition of Electric Flux, Φ

$$\Phi = EA \cos \theta \qquad \text{19–11}$$

SI unit: $\text{N} \cdot \text{m}^2/\text{C}$

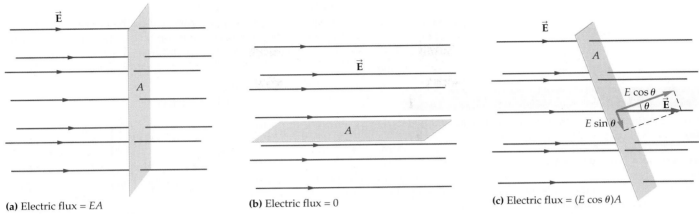

(a) Electric flux = EA

(b) Electric flux = 0

(c) Electric flux = $(E \cos \theta)A$

▲ **FIGURE 19–23 Electric flux**
(a) When an electric field $\vec{E}$ passes perpendicularly through the plane of an area A, the electric flux is $\Phi = EA$. **(b)** When the plane of an area is parallel to $\vec{E}$, so that no field lines "pierce" the area, the electric flux is zero, $\Phi = 0$. **(c)** When the normal to the plane of an area is tilted at an angle θ away from the electric field $\vec{E}$, only the perpendicular component of $\vec{E}$, $E \cos \theta$, contributes to the electric flux. Thus, the flux is $\Phi = (E \cos \theta)A$.

Finally, if the surface through which the flux is calculated is *closed*, the sign of the flux is as follows:

- The flux is *positive* for field lines that *leave* the enclosed volume of the surface.
- The flux is *negative* for field lines that *enter* the enclosed volume of the surface.

Gauss's Law

As a simple example of electric flux, consider a positive point charge q and a spherical surface of radius r centered on the charge, as in **Figure 19–24**. The electric field on the surface of the sphere has the constant magnitude

$$E = k\frac{q}{r^2}$$

Since the electric field is everywhere perpendicular to the spherical surface, it follows that the electric flux is simply E times the area $A = 4\pi r^2$ of the sphere:

$$\Phi = EA = \left(k\frac{q}{r^2}\right)(4\pi r^2) = 4\pi kq$$

We will often find it convenient to express k in terms of another constant, ε_0, as follows: $k = 1/(4\pi\varepsilon_0)$. This new constant, which we call the **permittivity of free space**, is

$$\varepsilon_0 = \frac{1}{4\pi k} = 8.85 \times 10^{-12}\,\text{C}^2/\text{N}\cdot\text{m}^2 \qquad 19\text{–}12$$

In terms of ε_0, the flux through the spherical surface reduces to

$$\Phi = 4\pi kq = \frac{q}{\varepsilon_0}$$

Thus, we find the very simple result that the electric flux through a sphere that encloses a charge q is the charge divided by the permittivity of free space, ε_0. This is a nice result, but what makes it truly remarkable is that it is equally true for *any* surface that encloses the charge q. For example, if one were to calculate the electric flux through the closed irregular surface also shown in Figure 19–24—which would be a difficult task—the result, nonetheless, would still be simply q/ε_0. This, in fact, is a special case of Gauss's law:

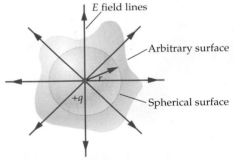

▲ **FIGURE 19–24 Electric flux for a point charge**
The electric flux through the spherical surface surrounding a positive point charge q is $\Phi = EA = (kq/r^2)(4\pi r^2) = q/\varepsilon_0$. The electric flux through an arbitrary surface is the same as for the sphere. The calculation of the flux, however, would be much more difficult for this surface.

> **Gauss's Law**
> If a charge q is enclosed by an arbitrary surface, the total electric flux through the surface, Φ, is
>
> $$\Phi = \frac{q}{\varepsilon_0} \qquad 19\text{–}13$$
>
> SI unit: $\text{N}\cdot\text{m}^2/\text{C}$

Note that we use q rather than $|q|$ in Equation 19–13. This is because the electric flux can be positive or negative, depending on the sign of the charge. In particular, if the charge q is positive, the field lines leave the enclosed volume and the flux is positive; if the charge is negative, the field lines enter the enclosed volume and the flux is negative.

CONCEPTUAL CHECKPOINT 19–6   SIGN OF THE ELECTRIC FLUX

Consider the surface S shown in the sketch. Is the electric flux through this surface **(a)** negative, **(b)** positive, or **(c)** zero?

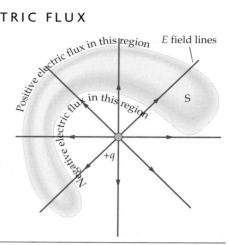

REASONING AND DISCUSSION
Because the surface S encloses no charge, the net electric flux through it must be zero, by Gauss's law. The fact that a charge $+q$ is nearby is irrelevant, because it is outside the volume enclosed by the surface.

We can explain why the flux vanishes in another way. Notice that the flux on portions of S near the charge is negative, since field lines enter the enclosed volume there. On the other hand, the flux is positive on the outer portions of S where field lines exit the volume. The combination of these positive and negative contributions is a net flux of zero.

ANSWER
(c) The electric flux through the surface S is zero.

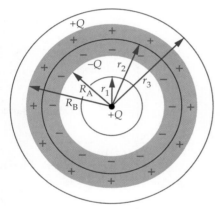

▲ **FIGURE 19–25 Gauss's law applied to a spherical shell**

A simple system with three different Gaussian surfaces.

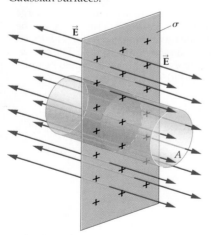

▲ **FIGURE 19–26 Gauss's law applied to a sheet of charge**

A charged sheet of infinite extent and the Gaussian surface used to calculate the electric field.

Although Gauss's law holds for an arbitrarily complex surface, its real utility is seen when the system in question has a simple symmetry. For example, consider a point charge $+Q$ in the center of a hollow, conducting spherical shell, as illustrated in **Figure 19–25**. The shell has an inside radius R_A and an outside radius R_B, and is uncharged. To calculate the field inside the cavity, where $r < R_A$, we consider an imaginary spherical surface—a so-called **Gaussian surface**—with radius r_1, and centered on the charge $+Q$, as indicated in Figure 19–25. The electric flux through this Gaussian surface is $\Phi = E(4\pi r_1^2) = Q/\varepsilon_0$. Therefore, the magnitude of the electric field, as expected, is

$$E = \frac{Q}{4\pi\varepsilon_0 r_1^2} = k\frac{Q}{r_1^2}$$

Note, in particular, that the charges on the spherical shell do not affect the electric flux through this Gaussian surface, since they are not contained within the surface.

Next, consider a Gaussian surface within the shell, with $R_A < r_2 < R_B$, as in Figure 19–25. Since the field within a conductor is zero, $E = 0$, it follows that the electric flux for this surface is zero: $\Phi = EA = 0$. This means that the net charge within the Gaussian surface is also zero; that is, the induced charge on the inner surface of the shell is $-Q$.

Finally, consider a spherical Gaussian surface that encloses the entire spherical shell, with a radius $r_3 > R_B$. In this case, also shown in Figure 19–25, the flux is $\Phi = E(4\pi r_3^2) = $ (enclosed charge)$/\varepsilon_0$. What is the enclosed charge? Well, we know that the spherical shell is uncharged—the induced charges of $+Q$ and $-Q$ on its outer and inner surfaces sum to zero—hence the total enclosed charge is simply $+Q$ from the charge at the center of the shell. Therefore, $\Phi = E(4\pi r_3^2) = Q/\varepsilon_0$, and

$$E = \frac{Q}{4\pi\varepsilon_0 r_3^2} = k\frac{Q}{r_3^2}$$

Note that the field outside the shell is the same as if the shell were not present, showing that the conducting shell does not shield the external world from charges within it, in agreement with our conclusions in the previous section.

Gaussian surfaces do not need to be spherical, however. Consider, for example, a thin sheet of charge that extends to infinity, as in **Figure 19–26**. We expect the field to be at right angles to the sheet because, by symmetry, there is no reason for it to tilt in one direction more than in any other direction. Hence, we choose our Gaussian surface to be a cylinder, as in Figure 19–26. With this choice, no field lines pierce the curved surface of the cylinder. The electric flux through this Gaussian surface, then, is due solely to the contributions of the two end caps, each of area A. Hence, the flux is $\Phi = E(2A)$. If the charge per area on the sheet is σ, the enclosed charge is σA, and hence Gauss's law states that

$$\Phi = E(2A) = \frac{(\sigma A)}{\varepsilon_0}$$

Canceling the area, we find

$$E = \frac{\sigma}{2\varepsilon_0}$$

Note that E does not depend in any way on the distance from the sheet, as was mentioned in Section 19–5.

We conclude this chapter with an additional example of Gauss's law in action.

PROBLEM-SOLVING NOTE

Applying Gauss's Law

Gauss's law is useful only when the electric field is constant on a given surface. It is only in such cases that the electric flux can be calculated with ease.

ACTIVE EXAMPLE 19–3   FIND THE ELECTRIC FIELD

Use the cylindrical Gaussian surface shown in the diagram to calculate the electric field between the metal plates of a parallel-plate capacitor. Each plate has a charge per area of magnitude σ.

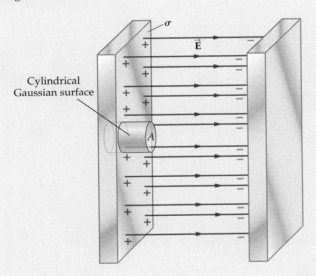

Cylindrical Gaussian surface

SOLUTION (Test your understanding by performing the calculations indicated in each step.)

1. Calculate the electric flux through the curved surface of the cylinder: 0

2. Calculate the electric flux through the end caps of the cylinder: $0 + EA$

3. Determine the charge enclosed by the cylinder: σA

4. Apply Gauss's law to find the field, E: $E = \sigma/\varepsilon_0$

INSIGHT
Note that the electric field is zero within the metal of the plates (because they are conductors). It is for this reason that the electric flux through the left end cap of the Gaussian surface is zero.

YOUR TURN
Suppose we extend the Gaussian surface so that its right end cap is within the metal of the right plate. The left end of the Gaussian surface remains in its original location. What is the electric flux through this new Gaussian surface? Explain.

*(Answers to **Your Turn** problems are given in the back of the book.)*

THE BIG PICTURE PUTTING PHYSICS IN CONTEXT

LOOKING BACK

The concept of a conserved quantity, like energy (Chapter 8) or momentum (Chapter 9), appears again in Section 19–1, where we show that electric charge is also a conserved quantity.

Coulomb's law for the electrostatic force between two charges, Equation 19–5, is virtually identical to Newton's law of universal gravitation between two masses (Chapter 12), but with electric charge replacing mass.

The electric force, like all forces, is a vector quantity. Therefore, the material on vector addition (Chapter 3) again finds use in Sections 19–3, 19–4, and 19–5.

LOOKING AHEAD

The electric force is conservative, and hence it has an associated electric potential energy. We will determine this potential energy in Chapter 20. We will also point out the close analogy between the electric potential energy and the gravitational potential energy of Chapter 12.

When electric charge flows from one location to another, it produces an electric current. We consider electric circuits with direct current (dc) in Chapter 21, and circuits with alternating current (ac) in Chapter 24.

Coulomb's law comes up again in atomic physics, where it plays a key role in the Bohr model of the hydrogen atom in Section 31–3.

CHAPTER SUMMARY

19–1 ELECTRIC CHARGE

Electric charge is one of the fundamental properties of matter. Electrons have a negative charge, $-e$, and protons have a positive charge, $+e$. An object with zero net charge, like a neutron, is said to be electrically neutral.

Magnitude of an Electron's Charge
The charge on an electron has the following magnitude:

$$e = 1.60 \times 10^{-19}\,\text{C} \qquad \text{19–1}$$

The SI unit of charge is the coulomb, C.

Charge Conservation
The total charge in the universe is constant.

Charge Quantization
Charge comes in quantized amounts that are always integer multiples of e.

19–2 INSULATORS AND CONDUCTORS

An insulator does not allow electrons within it to move from atom to atom. In conductors, each atom gives up one or more electrons that are then free to move throughout the material. Semiconductors have properties that are intermediate between those of insulators and conductors.

19–3 COULOMB'S LAW

Electric charges exert forces on one another along the line connecting them: Like charges repel, opposite charges attract.

Coulomb's Law
The magnitude of the force between two point charges, q_1 and q_2, separated by a distance r is

$$F = k\frac{|q_1||q_2|}{r^2} \qquad \text{19–5}$$

The constant k in this expression is

$$k = 8.99 \times 10^9\,\text{N}\cdot\text{m}^2/\text{C}^2 \qquad \text{19–6}$$

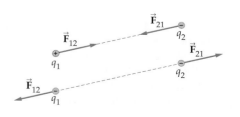

Superposition

The electric force on one charge due to two or more other charges is the vector sum of each individual force.

Spherical Charge Distributions

A spherical distribution of charge, when viewed from outside, behaves the same as an equivalent point charge at the center of the sphere.

19–4 THE ELECTRIC FIELD

The electric field is the force per charge at a given location in space.

Direction of $\vec{E}$

$\vec{E}$ points in the direction of the force experienced by a *positive* test charge.

Point Charge

The electric field a distance r from a point charge q has a magnitude given by

$$E = k\frac{|q|}{r^2} \qquad \text{19–10}$$

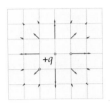

Superposition

The total electric field due to two or more charges is given by the vector sum of the fields due to each charge individually.

19–5 ELECTRIC FIELD LINES

The electric field can be visualized by drawing lines according to a given set of rules.

Rules for Drawing Electric Field Lines

Electric field lines (1) point in the direction of the electric field vector $\vec{E}$ at all points; (2) start at $+$ charges or infinity; (3) end at $-$ charges or infinity; and (4) are more dense the greater the magnitude of $\vec{E}$.

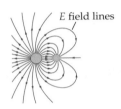

E field lines

Parallel-Plate Capacitor

A parallel-plate capacitor consists of two oppositely charged, conducting parallel plates separated by a finite distance. The field between the plates is uniform in direction (perpendicular to the plates) and magnitude.

19–6 SHIELDING AND CHARGING BY INDUCTION

Ideal conductors have a range of interesting behaviors that arise because they have an enormous number of electrons that are free to move.

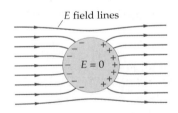

E field lines

Excess Charge

Any excess charge placed on a conductor moves to its exterior surface.

Zero Field in a Conductor

The electric field within a conductor in equilibrium is zero.

Shielding

A conductor shields a cavity within it from external electric fields.

Electric Fields at Conductor Surfaces

Electric field lines contact conductor surfaces at right angles.

Charging by Induction

A conductor can be charged without direct physical contact with another charged object. This is charging by induction.

Grounding

Connecting a conductor to the ground is referred to as grounding. The ground itself is a good conductor, and it can give up or receive an unlimited number of electrons.

19–7 ELECTRIC FLUX AND GAUSS'S LAW

Gauss's law relates the charge enclosed by a given surface to the electric flux through the surface.

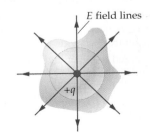

E field lines

Electric Flux

If an area A is tilted at an angle θ to an electric field $\vec{E}$, the electric flux through the area is

$$\Phi = EA\cos\theta \qquad \text{19–11}$$

Gauss's Law

Gauss's law states that if a charge q is enclosed by a surface, the electric flux through the surface is

$$\Phi = \frac{q}{\varepsilon_0} \qquad\qquad 19\text{--}13$$

The constant appearing in this equation is the permittivity of free space, ε_0:

$$\varepsilon_0 = \frac{1}{4\pi k} = 8.85 \times 10^{-12}\,\text{C}^2/\text{N} \cdot \text{m}^2 \qquad\qquad 19\text{--}12$$

Gauss's law is used to calculate the electric field in highly symmetric systems.

PROBLEM-SOLVING SUMMARY

Type of Problem	Relevant Physical Concepts	Related Examples
Find the electric force exerted by one or more point charges.	The magnitude of the electric force between point charges is $F = k\lvert q_1 \rvert \lvert q_2 \rvert / r^2$; the direction of the force is given by the expression "opposites attract, likes repel." When more than one charge exerts a force on a given charge, the net force is the vector sum of the individual forces.	Examples 19–1, 19–2, 19–3 Active Examples 19–1, 19–2
Find the electric force due to a spherical distribution of charge.	For points outside a spherical distribution of charge, the spherical distribution behaves the same as a point charge of the same amount at the center of the sphere.	Active Example 19–2
Calculate the force exerted by an electric field.	An electric field, $\vec{\mathbf{E}}$, produces a force, $\vec{\mathbf{F}} = q\vec{\mathbf{E}}$, on a point charge q. The direction of the force is in the direction of the field if the charge is positive and opposite to the field if the charge is negative.	Example 19–4
Find the electric field due to one or more point charges.	The electric field due to a point charge q has a magnitude given by $E = k\lvert q \rvert / r^2$. The direction of the field is radially outward if q is positive, and radially inward if q is negative. When a group of point charges is being considered, the total electric field is the vector sum of the individual fields.	Example 19–5
Calculate the electric field using Gauss's law.	The electric field can be found by setting the electric flux through a given surface equal to the charge enclosed by the surface divided by ε_0.	Active Example 19–3

CONCEPTUAL QUESTIONS

For instructor-assigned homework, go to www.masteringphysics.com

(Answers to odd-numbered Conceptual Questions can be found in the back of the book.)

1. When an object that was neutral becomes charged, does the total charge of the universe change? Explain.

2. The fact that the electron has a negative charge and the proton has a positive charge is due to a convention established by Benjamin Franklin. Would there have been any significant consequences if Franklin had chosen the opposite convention? Is there any advantage to naming charges plus and minus as opposed to, say, A and B?

3. Explain why a comb that has been rubbed through your hair attracts small bits of paper, even though the paper is uncharged.

4. Small bits of paper are attracted to an electrically charged comb, but as soon as they touch the comb they are strongly repelled. Explain this behavior.

5. A charged rod is brought near a suspended object, which is repelled by the rod. Can we conclude that the suspended object is charged? Explain.

6. A charged rod is brought near a suspended object, which is attracted to the rod. Can we conclude that the suspended object is charged? Explain.

7. Describe some of the similarities and differences between Coulomb's law and Newton's law of gravity.

8. A point charge $+Q$ is fixed at a height H above the ground. Directly below this charge is a small ball with a charge $-q$ and a mass m. When the ball is at a height h above the ground, the net force (gravitational plus electrical) acting on it is zero. Is this a stable equilibrium for the object? Explain.

9. Four identical point charges are placed at the corners of a square. A fifth point charge placed at the center of the square experiences zero net force. Is this a stable equilibrium for the fifth charge? Explain.

10. A proton moves in a region of constant electric field. Does it follow that the proton's velocity is parallel to the electric field? Does it follow that the proton's acceleration is parallel to the electric field? Explain.

11. Describe some of the differences between charging by induction and charging by contact.

12. A system consists of two charges of equal magnitude and opposite sign separated by a distance d. Since the total electric

charge of this system is zero, can we conclude that the electric field produced by the system is also zero? Does your answer depend on the separation d? Explain.

13. The force experienced by charge 1 at point A is different in direction and magnitude from the force experienced by charge 2 at point B. Can we conclude that the electric fields at points A and B are different? Explain.

14. Can an electric field exist in a vacuum? Explain.

15. Explain why electric field lines never cross.

16. Charge q_1 is inside a closed Gaussian surface; charge q_2 is just outside the surface. Does the electric flux through the surface depend on q_1? Does it depend on q_2? Explain.

17. In the previous question, does the electric field at a point on the Gaussian surface depend on q_1? Does it depend on q_2? Explain.

18. Gauss's law can tell us how much charge is contained within a Gaussian surface. Can it tell us where inside the surface it is located? Explain.

19. Explain why Gauss's law is not very useful in calculating the electric field of a charged disk.

PROBLEMS AND CONCEPTUAL EXERCISES

Note: Answers to odd-numbered Problems and Conceptual Exercises can be found in the back of the book. **IP** *denotes an integrated problem, with both conceptual and numerical parts;* **BIO** *identifies problems of biological or medical interest;* **CE** *indicates a conceptual exercise.* **Predict/Explain** *problems ask for two responses:* **(a)** *your prediction of a physical outcome, and* **(b)** *the best explanation among three provided. On all problems, red bullets (•, ••, •••) are used to indicate the level of difficulty.*

SECTION 19–1 ELECTRIC CHARGE

1. • **CE Predict/Explain** An electrically neutral object is given a positive charge. **(a)** In principle, does the object's mass increase, decrease, or stay the same as a result of being charged? **(b)** Choose the *best explanation* from among the following:
   I. To give the object a positive charge we must remove some of its electrons; this will reduce its mass.
   II. Since electric charges have mass, giving the object a positive charge will increase its mass.
   III. Charge is conserved, and therefore the mass of the object will remain the same.

2. • **CE Predict/Explain** An electrically neutral object is given a negative charge. **(a)** In principle, does the object's mass increase, decrease, or stay the same as a result of being charged? **(b)** Choose the *best explanation* from among the following:
   I. To give the object a negative charge we must give it more electrons, and this will increase its mass.
   II. A positive charge increases an object's mass; a negative charge decreases its mass.
   III. Charge is conserved, and therefore the mass of the object will remain the same.

3. • **CE (a)** Based on the materials listed in Table 19–1, is the charge of the rubber balloon shown on page 655 more likely to be positive or negative? Explain. **(b)** If the charge on the balloon is reversed, will the stream of water deflect toward or away from the balloon? Explain.

4. • **CE** This problem refers to the information given in Table 19–1. **(a)** If rabbit fur is rubbed against glass, what is the sign of the charge each acquires? Explain. **(b)** Repeat part (a) for the case of glass and rubber. **(c)** Comparing the situations described in parts (a) and (b), in which case is the magnitude of the triboelectric charge greater? Explain.

5. • Find the net charge of a system consisting of 4.9×10^7 electrons.

6. • Find the net charge of a system consisting of **(a)** 6.15×10^6 electrons and 7.44×10^6 protons or **(b)** 212 electrons and 165 protons.

7. • How much negative electric charge is contained in 2 moles of carbon?

8. • Find the total electric charge of 1.5 kg of **(a)** electrons and **(b)** protons.

9. • A container holds a gas consisting of 1.85 moles of oxygen molecules. One in a million of these molecules has lost a single electron. What is the net charge of the gas?

10. • **The Charge on Adhesive Tape** When adhesive tape is pulled from a dispenser, the detached tape acquires a positive charge and the remaining tape in the dispenser acquires a negative charge. If the tape pulled from the dispenser has $0.14 \ \mu C$ of charge per centimeter, what length of tape must be pulled to transfer 1.8×10^{13} electrons to the remaining tape?

11. •• **CE** Four pairs of conducting spheres, all with the same radius, are shown in **Figure 19–27**, along with the net charge placed on them initially. The spheres in each pair are now brought into contact, allowing charge to transfer between them. Rank the pairs of spheres in order of increasing magnitude of the charge transferred. Indicate ties where appropriate.

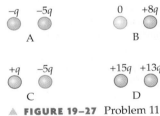

▲ **FIGURE 19–27** Problem 11

12. •• A system of 1525 particles, each of which is either an electron or a proton, has a net charge of -5.456×10^{-17} C. **(a)** How many electrons are in this system? **(b)** What is the mass of this system?

SECTION 19–3 COULOMB'S LAW

13. • **CE** A charge $+q$ and a charge $-q$ are placed at opposite corners of a square. Will a third point charge experience a greater force if it is placed at one of the empty corners of the square, or at the center of the square? Explain.

14. • **CE** Repeat the previous question, this time with charges $+q$ and $+q$ at opposite corners of a square.

15. • **CE** Consider the three electric charges, A, B, and C, shown in **Figure 19–28**. Rank the charges in order of increasing magnitude of the net force they experience. Indicate ties where appropriate.

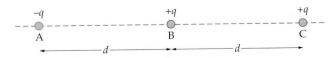

▲ **FIGURE 19–28** Problems 15 and 46

16. • **CE Predict/Explain** Suppose the charged sphere in Active Example 19–2 is made from a conductor, rather than an insulator. **(a)** Do you expect the magnitude of the force between the point charge and the conducting sphere to be greater than, less than, or equal to the force between the point charge and an insulating sphere? **(b)** Choose the *best explanation* from among the following:
   I. The conducting sphere will allow the charges to move, resulting in a greater force.
   II. The charge of the sphere is the same whether it is conducting or insulating, and therefore the force is the same.
   III. The charge on a conducting sphere will move as far away as possible from the point charge. This results in a reduced force.

17. • At what separation is the electrostatic force between a $+11.2$-μC point charge and a $+29.1$-μC point charge equal in magnitude to 1.57 N?

18. • The attractive electrostatic force between the point charges $+8.44 \times 10^{-6}$ C and Q has a magnitude of 0.975 N when the separation between the charges is 1.31 m. Find the sign and magnitude of the charge Q.

19. • If the speed of the electron in Example 19–1 were 7.3×10^5 m/s, what would be the corresponding orbital radius?

20. • **IP** Two point charges, the first with a charge of $+3.13 \times 10^{-6}$ C and the second with a charge of -4.47×10^{-6} C, are separated by 25.5 cm. **(a)** Find the magnitude of the electrostatic force experienced by the positive charge. **(b)** Is the magnitude of the force experienced by the negative charge greater than, less than, or the same as that experienced by the positive charge? Explain.

21. • When two identical ions are separated by a distance of 6.2×10^{-10} m, the electrostatic force each exerts on the other is 5.4×10^{-9} N. How many electrons are missing from each ion?

22. • A sphere of radius 4.22 cm and uniform surface charge density $+12.1 \, \mu$C/m^2 exerts an electrostatic force of magnitude 46.9×10^{-3} N on a point charge of $+1.95 \, \mu$C. Find the separation between the point charge and the center of the sphere.

23. • Given that $q = +12 \, \mu$C and $d = 16$ cm, find the direction and magnitude of the net electrostatic force exerted on the point charge q_1 in **Figure 19–29**.

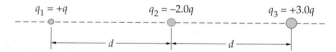

$q_1 = +q$ $q_2 = -2.0q$ $q_3 = +3.0q$

▲ **FIGURE 19–29** Problems 23, 26, 27, and 58

24. •• **CE** Five point charges, $q_1 = +q$, $q_2 = +2q$, $q_3 = -3q$, $q_4 = -4q$, and $q_5 = -5q$, are placed in the vicinity of an insulating spherical shell with a charge $+Q$, distributed uniformly over its surface, as indicated in **Figure 19–30**. Rank the point

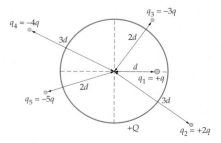

▲ **FIGURE 19–30** Problem 24

charges in order of increasing magnitude of the force exerted on them by the sphere. Indicate ties where appropriate.

25. •• **CE** Three charges, $q_1 = +q$, $q_2 = -q$, and $q_3 = +q$, are at the vertices of an equilateral triangle, as shown in **Figure 19–31**. **(a)** Rank the three charges in order of increasing magnitude of the electric force they experience. Indicate ties where appropriate. **(b)** Give the direction angle, θ, of the net electric force experienced by charge 1. Note that θ is measured counterclockwise from the positive x axis. **(c)** Repeat part (b) for charge 2. **(d)** Repeat part (b) for charge 3.

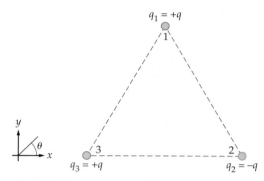

▲ **FIGURE 19–31** Problems 25 and 82

26. •• **IP** Given that $q = +12 \, \mu$C and $d = 19$ cm, **(a)** find the direction and magnitude of the net electrostatic force exerted on the point charge q_2 in Figure 19–29. **(b)** How would your answers to part (a) change if the distance d were tripled?

27. •• Suppose the charge q_2 in Figure 19–29 can be moved left or right along the line connecting the charges q_1 and q_3. Given that $q = +12 \, \mu$C, find the distance from q_1 where q_2 experiences a net electrostatic force of zero. (The charges q_1 and q_3 are separated by a fixed distance of 32 cm.)

28. •• Find the orbital radius for which the kinetic energy of the electron in Example 19–1 is 1.51 eV. (*Note:* 1 eV = 1 electron volt = 1.6×10^{-19} J.)

29. •• A point charge $q = -0.35$ nC is fixed at the origin. Where must a proton be placed in order for the electric force acting on it to be exactly opposite to its weight? (Let the y axis be vertical and the x axis be horizontal.)

30. •• A point charge $q = -0.35$ nC is fixed at the origin. Where must an electron be placed in order for the electric force acting on it to be exactly opposite to its weight? (Let the y axis be vertical and the x axis be horizontal.)

31. •• Find the direction and magnitude of the net electrostatic force exerted on the point charge q_2 in **Figure 19–32**. Let $q = +2.4 \, \mu$C and $d = 33$ cm.

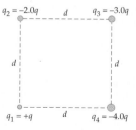

▲ **FIGURE 19–32** Problems 31 and 32

32. •• **IP (a)** Find the direction and magnitude of the net electrostatic force exerted on the point charge q_3 in Figure 19–32. Let

$q = +2.4 \, \mu C$ and $d = 27$ cm. **(b)** How would your answers to part (a) change if the distance d were doubled?

33. ••**IP** Two point charges lie on the x axis. A charge of $+9.9 \, \mu C$ is at the origin, and a charge of $-5.1 \, \mu C$ is at $x = 10.0$ cm. **(a)** At what position x would a third charge q_3 be in equilibrium? **(b)** Does your answer to part (a) depend on whether q_3 is positive or negative? Explain.

34. •• A system consists of two positive point charges, q_1 and $q_2 > q_1$. The total charge of the system is $+62.0 \, \mu C$, and each charge experiences an electrostatic force of magnitude 85.0 N when the separation between them is 0.270 m. Find q_1 and q_2.

35. •• **IP** The point charges in **Figure 19–33** have the following values: $q_1 = +2.1 \, \mu C$, $q_2 = +6.3 \, \mu C$, $q_3 = -0.89 \, \mu C$. **(a)** Given that the distance d in Figure 19–33 is 4.35 cm, find the direction and magnitude of the net electrostatic force exerted on the point charge q_1. **(b)** How would your answers to part (a) change if the distance d were doubled? Explain.

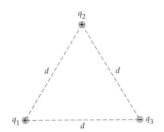

▲ **FIGURE 19–33** Problems 35, 36, 48, and 59

36. •• Referring to Problem 35, suppose that the magnitude of the net electrostatic force exerted on the point charge q_2 in Figure 19–33 is 0.65 N. **(a)** Find the distance d. **(b)** What is the direction of the net force exerted on q_2?

37. •• **IP** **(a)** If the nucleus in Example 19–1 had a charge of $+2e$ (as would be the case for a nucleus of helium), would the speed of the electron be greater than, less than, or the same as that found in the Example? Explain. (Assume the radius of the electron's orbit is the same.) **(b)** Find the speed of the electron for a nucleus of charge $+2e$.

38. •• Four point charges are located at the corners of a square with sides of length a. Two of the charges are $+q$, and two are $-q$. Find the magnitude and direction of the net electric force exerted on a charge $+Q$, located at the center of the square, for each of the following two arrangements of charge: **(a)** The charges alternate in sign $(+q, -q, +q, -q)$ as you go around the square; **(b)** the two positive charges are on the top corners, and the two negative charges are on the bottom corners.

39. •• **IP** Two identical point charges in free space are connected by a string 7.6 cm long. The tension in the string is 0.21 N. **(a)** Find the magnitude of the charge on each of the point charges. **(b)** Using the information given in the problem statement, is it possible to determine the sign of the charges? Explain. **(c)** Find the tension in the string if $+1.0 \, \mu C$ of charge is transferred from one point charge to the other. Compare with your result from part (a).

40. ••• Two spheres with uniform surface charge density, one with a radius of 7.2 cm and the other with a radius of 4.7 cm, are separated by a center-to-center distance of 33 cm. The spheres have a combined charge of $+55 \, \mu C$ and repel one another with a force of 0.75 N. What is the surface charge density on each sphere?

41. ••• Point charges, q_1 and q_2, are placed on the x axis, with q_1 at $x = 0$ and q_2 at $x = d$. A third point charge, $+Q$, is placed at $x = 3d/4$. If the net electrostatic force experienced by the charge $+Q$ is zero, how are q_1 and q_2 related?

SECTION 19–4   THE ELECTRIC FIELD

42. • **CE** Two electric charges are separated by a finite distance. Somewhere between the charges, on the line connecting them, the net electric field they produce is zero. **(a)** Do the charges have the same or opposite signs? Explain. **(b)** If the point of zero field is closer to charge 1, is the magnitude of charge 1 greater than or less than the magnitude of charge 2? Explain.

43. • What is the magnitude of the electric field produced by a charge of magnitude 7.50 μC at a distance of **(a)** 1.00 m and **(b)** 2.00 m?

44. • A $+5.0$-μC charge experiences a 0.44-N force in the positive y direction. If this charge is replaced with a -2.7-μC charge, what force will it experience?

45. • Two point charges lie on the x axis. A charge of $+6.2 \, \mu C$ is at the origin, and a charge of $-9.5 \, \mu C$ is at $x = 10.0$ cm. What is the net electric field at **(a)** $x = -4.0$ cm and at **(b)** $x = +4.0$ cm?

46. •• **CE** The electric field on the dashed line in Figure 19–28 vanishes at infinity, but also at two different points a finite distance from the charges. Identify the regions in which you can find $E = 0$ at a finite distance from the charges: region 1, to the left of point A; region 2, between points A and B; region 3, between points B and C; region 4, to the right of point C.

47. •• An object with a charge of $-3.6 \, \mu C$ and a mass of 0.012 kg experiences an upward electric force, due to a uniform electric field, equal in magnitude to its weight. **(a)** Find the direction and magnitude of the electric field. **(b)** If the electric charge on the object is doubled while its mass remains the same, find the direction and magnitude of its acceleration.

48. •• **IP** Figure 19–33 shows a system consisting of three charges, $q_1 = +5.00 \, \mu C$, $q_2 = +5.00 \, \mu C$, and $q_3 = -5.00 \, \mu C$, at the vertices of an equilateral triangle of side $d = 2.95$ cm. **(a)** Find the magnitude of the electric field at a point halfway between the charges q_1 and q_2. **(b)** Is the magnitude of the electric field halfway between the charges q_2 and q_3 greater than, less than, or the same as the electric field found in part (a)? Explain. **(c)** Find the magnitude of the electric field at the point specified in part (b).

49. •• Two point charges of equal magnitude are 7.5 cm apart. At the midpoint of the line connecting them, their combined electric field has a magnitude of 45 N/C. Find the magnitude of the charges.

50. •• **IP** A point charge $q = +4.7 \, \mu C$ is placed at each corner of an equilateral triangle with sides 0.21 m in length. **(a)** What is the magnitude of the electric field at the midpoint of any of the three sides of the triangle? **(b)** Is the magnitude of the electric field at the center of the triangle greater than, less than, or the same as the magnitude at the midpoint of a side? Explain.

51. ••• **IP** Four point charges, each of magnitude q, are located at the corners of a square with sides of length a. Two of the charges are $+q$, and two are $-q$. The charges are arranged in one of the following two ways: (1) The charges alternate in sign $(+q, -q, +q, -q)$ as you go around the square; (2) the top two corners of the square have positive charges $(+q, +q)$, and the bottom two corners have negative charges $(-q, -q)$. **(a)** In which case will the electric field at the center of the square have the greatest magnitude? Explain. **(b)** Calculate the electric field at the center of the square for each of these two cases.

52. ••• The electric field at the point $x = 5.00$ cm and $y = 0$ points in the positive x direction with a magnitude of 10.0 N/C. At the point $x = 10.0$ cm and $y = 0$ the electric field points in the positive x direction with a magnitude of 15.0 N/C. Assuming this

electric field is produced by a single point charge, find **(a)** its location and **(b)** the sign and magnitude of its charge.

SECTION 19–5 ELECTRIC FIELD LINES

53. • **IP** The electric field lines surrounding three charges are shown in **Figure 19–34**. The center charge is $q_2 = -10.0\ \mu C$. **(a)** What are the signs of q_1 and q_3? **(b)** Find q_1. **(c)** Find q_3.

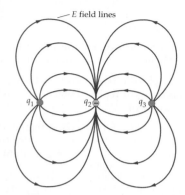

▲ **FIGURE 19–34** Problems 53 and 56

54. • Make a qualitative sketch of the electric field lines produced by two equal positive charges, $+q$, separated by a distance d.

55. • Make a qualitative sketch of the electric field lines produced by two charges, $+q$ and $-q$, separated by a distance d.

56. •• Referring to Figure 19–34, suppose q_2 is not known. Instead, it is given that $q_1 + q_2 = -2.5\ \mu C$. Find q_1, q_2, and q_3.

57. •• Make a qualitative sketch of the electric field lines produced by the four charges, $+q$, $-q$, $+q$, and $-q$, arranged clockwise on the four corners of a square with sides of length d.

58. •• Sketch the electric field lines for the system of charges shown in Figure 19–29.

59. •• Sketch the electric field lines for the system of charges described in Problem 35.

60. •• Suppose the magnitude of the electric field between the plates in Example 19–6 is changed, and a new object with a charge of $-2.05\ \mu C$ is attached to the string. If the tension in the string is 0.450 N, and the angle it makes with the vertical is 16°, what are **(a)** the mass of the object and **(b)** the magnitude of the electric field?

SECTION 19–7 ELECTRIC FLUX AND GAUSS'S LAW

61. • **CE Predict/Explain** Gaussian surface 1 has twice the area of Gaussian surface 2. Both surfaces enclose the same charge Q. **(a)** Is the electric flux through surface 1 greater than, less than, or the same as the electric flux through surface 2? **(b)** Choose the *best explanation* from among the following:

   I. Gaussian surface 2 is closer to the charge, since it has the smaller area. It follows that it has the greater electric flux.

   II. The two surfaces enclose the same charge, and hence they have the same electric flux.

   III. Electric flux is proportional to area. As a result, Gaussian surface 1 has the greater electric flux.

62. • **CE** Suppose the conducting shell in Figure 19–25—which has a point charge $+Q$ at its center—has a nonzero net charge. How much charge is on the inner and outer surface of the shell when the net charge of the shell is **(a)** $-2Q$, **(b)** $-Q$, and **(c)** $+Q$?

63. • **CE** Rank the Gaussian surfaces shown in **Figure 19–35** in order of increasing electric flux, starting with the most negative. Indicate ties where appropriate.

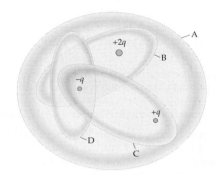

▲ **FIGURE 19–35** Problems 63 and 77

64. • A uniform electric field of magnitude 25,000 N/C makes an angle of 37° with a plane surface of area 0.0153 m². What is the electric flux through this surface?

65. • A surface encloses the charges $q_1 = 3.2\ \mu C$, $q_2 = 6.9\ \mu C$, and $q_3 = -4.1\ \mu C$. Find the electric flux through this surface.

66. • **IP** A uniform electric field of magnitude 6.00×10^3 N/C points upward. An empty, closed shoe box has a top and bottom that are 35.0 cm by 25.0 cm, vertical ends that are 25.0 cm by 20.0 cm, and vertical sides that are 20.0 cm by 35.0 cm. **(a)** Which side of the box has the greatest positive electric flux? Which side has the greatest negative electric flux? Which sides have zero electric flux? **(b)** Calculate the electric flux through each of the six sides of the box.

67. • **BIO Nerve Cells** Nerve cells are long, thin cylinders along which electrical disturbances (nerve impulses) travel. The cell membrane of a typical nerve cell consists of an inner and an outer wall separated by a distance of 0.10 μm. The electric field within the cell membrane is 7.0×10^5 N/C. Approximating the cell membrane as a parallel-plate capacitor, determine the magnitude of the charge density on the inner and outer cell walls.

68. •• The electric flux through each of the six sides of a rectangular box are as follows:

$$\Phi_1 = +150.0\ \text{N}\cdot\text{m}^2/\text{C}; \quad \Phi_2 = +250.0\ \text{N}\cdot\text{m}^2/\text{C};$$

$$\Phi_3 = -350.0\ \text{N}\cdot\text{m}^2/\text{C}; \quad \Phi_4 = +175.0\ \text{N}\cdot\text{m}^2/\text{C};$$

$$\Phi_5 = -100.0\ \text{N}\cdot\text{m}^2/\text{C}; \quad \Phi_6 = +450.0\ \text{N}\cdot\text{m}^2/\text{C}.$$

How much charge is in this box?

69. •• Consider a spherical Gaussian surface and three charges: $q_1 = 1.61\ \mu C$, $q_2 = -2.62\ \mu C$, and $q_3 = 3.91\ \mu C$. Find the electric flux through the Gaussian surface if it completely encloses **(a)** only charges q_1 and q_2, **(b)** only charges q_2 and q_3, and **(c)** all three charges. **(d)** Suppose a fourth charge, Q, is added to the situation described in part (c). Find the sign and magnitude of Q required to give zero electric flux through the surface.

70. ••• A thin wire of infinite extent has a charge per unit length of λ. Using the cylindrical Gaussian surface shown in **Figure 19–36**, show that the electric field produced by this wire at a radial distance r has a magnitude given by

$$E = \frac{\lambda}{2\pi\varepsilon_0 r}$$

Note that the direction of the electric field is always radially away from the wire.

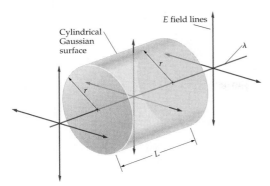

▲ FIGURE 19–36 Problems 70 and 87

GENERAL PROBLEMS

71. • **CE Predict/Explain** An electron and a proton are released from rest in space, far from any other objects. The particles move toward each other, due to their mutual electrical attraction. **(a)** When they meet, is the kinetic energy of the electron greater than, less than, or equal to the kinetic energy of the proton? **(b)** Choose the *best explanation* from among the following:
 I. The proton has the greater mass. Since kinetic energy is proportional to mass, it follows that the proton will have the greater kinetic energy.
 II. The two particles experience the same force, but the light electron moves farther than the massive proton. Therefore, the work done on the electron, and hence its kinetic energy, is greater.
 III. The same force acts on the two particles. Therefore, they will have the same kinetic energy and energy will be conserved.

72. • **CE Predict/Explain** In Conceptual Checkpoint 19–3, suppose the charge to be placed at either point A or point B is $+q$ rather than $-q$. **(a)** Is the magnitude of the net force experienced by the movable charge at point A greater than, less than, or equal to the magnitude of the net force at point B? **(b)** Choose the *best explanation* from among the following:
 I. Point B is farther from the two fixed charges. As a result, the net force at point B is less than at point A.
 II. The net force at point A cancels, just as it does in Conceptual Checkpoint 19–3. Therefore, the nonzero net force at point B is greater in magnitude than the zero net force at point A.
 III. The net force is greater in magnitude at point A because at that location the movable charge experiences a net repulsion from each of the fixed charges.

73. • **CE** An electron (charge $= -e$) orbits a helium nucleus (charge $= +2e$). Is the magnitude of the force exerted on the helium nucleus by the electron greater than, less than, or the same as the magnitude of the force exerted on the electron by the helium nucleus? Explain.

74. • **CE** In the operating room, technicians and doctors must take care not to create an electric spark, since the presence of the oxygen gas used during an operation increases the risk of a deadly fire. Should the operating-room personnel wear shoes that are conducting or nonconducting? Explain.

75. • **CE** Under normal conditions, the electric field at the surface of the Earth points downward, into the ground. What is the sign of the electric charge on the ground?

76. • **CE** Two identical spheres are made of conducting material. Initially, sphere 1 has a net charge of $+35Q$ and sphere 2 has a net charge of $-26Q$. If the spheres are now brought into contact, what is the final charge on sphere 1? Explain.

77. • **CE** A Gaussian surface for the charges shown in Figure 19–35 has an electric flux equal to $+3q/\varepsilon_0$. Which charges are contained within this Gaussian surface?

78. • A proton is released from rest in a uniform electric field of magnitude 1.08×10^5 N/C. Find the speed of the proton after it has traveled **(a)** 1.00 cm and **(b)** 10.0 cm.

79. • **BIO Ventricular Fibrillation** If a charge of 0.30 C passes through a person's chest in 1.0 s, the heart can go into ventricular fibrillation—a nonrhythmic "fluttering" of the ventricles that results in little or no blood being pumped to the body. If this rate of charge transfer persists for 4.5 s, how many electrons pass through the chest?

80. • A point charge at the origin of a coordinate system produces the electric field $\vec{E} = (36{,}000\ \text{N/C})\hat{x}$ on the x axis at the location $x = -0.75$ m. Determine the sign and magnitude of the charge.

81. •• **CE** Four lightweight, plastic spheres, labeled A, B, C, and D, are suspended from threads in various combinations, as illustrated in **Figure 19–37**. It is given that the net charge on sphere D is $+Q$, and that the other spheres have net charges of $+Q$, $-Q$, or 0. From the results of the four experiments shown in Figure 19–37, and the fact that the spheres have equal masses, determine the net charge of **(a)** sphere A, **(b)** sphere B, and **(c)** sphere C.

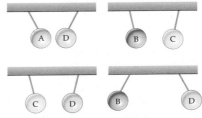

▲ FIGURE 19–37 Problem 81

82. •• Find **(a)** the direction and **(b)** the magnitude of the net electric field at the center of the equilateral triangle in Figure 19–31. Give your answers in terms of the angle θ, as defined in Figure 19–31, and E, the magnitude of the electric field produced by any *one* of the charges at the center of the triangle.

83. •• At the moment, the number of electrons in your body is essentially the same as the number of protons, giving you a net charge of zero. Suppose, however, that this balance of charges is off by 1% in both you and your friend, who is 1 meter away. Estimate the magnitude of the electrostatic force each of you experiences, and compare it with your weight.

84. •• A small object of mass 0.0150 kg and charge 3.1 μC hangs from the ceiling by a thread. A second small object, with a charge of 4.2 μC, is placed 1.2 m vertically below the first charge. Find **(a)** the electric field at the position of the upper charge due to the lower charge and **(b)** the tension in the thread.

85. •• **IP** Consider a system of three point charges on the x axis. Charge 1 is at $x = 0$, charge 2 is at $x = 0.20$ m, and charge 3 is at $x = 0.40$ m. In addition, the charges have the following values: $q_1 = -19\ \mu$C, $q_2 = q_3 = +19\ \mu$C. **(a)** The electric field vanishes at some point on the x axis between $x = 0.20$ m and $x = 0.40$ m. Is the point of zero field (i) at $x = 0.30$ m, (ii) to the left of $x = 0.30$ m, or (iii) to the right of $x = 0.30$ m? Explain. **(b)** Find the point where $E = 0$ between $x = 0.20$ m and $x = 0.40$ m.

86. •• **IP** Consider the system of three point charges described in the previous problem. **(a)** The electric field vanishes at two different points on the x axis. One point is between $x = 0.20$ m and $x = 0.40$ m. Is the second point located to the left of charge 1 or to the right of charge 3? Explain. **(b)** Find the value of x at the second point where $E = 0$.

87. •• The electric field at a radial distance of 47.7 cm from the thin charged wire shown in Figure 19–36 has a magnitude of 35,400 N/C. **(a)** Using the result given in Problem 70, what is the magnitude of the charge per length on this wire? **(b)** At what distance from the wire is the magnitude of the electric field equal to $\frac{1}{2}$(35,400 N/C)?

88. •• A system consisting entirely of electrons and protons has a net charge of 1.84×10^{-15} C and a net mass of 4.56×10^{-23} kg. How many **(a)** electrons and **(b)** protons are in this system?

89. •• **IP** Three charges are placed at the vertices of an equilateral triangle of side $a = 0.93$ m, as shown in **Figure 19–38**. Charges 1 and 3 are +7.3 μC; charge 2 is −7.3 μC. **(a)** Find the magnitude and direction of the net force acting on charge 3. **(b)** If charge 3 is moved to the origin, will the net force acting on it there be greater than, less than, or equal to the net force found in part (a)? Explain. **(c)** Find the net force on charge 3 when it is at the origin.

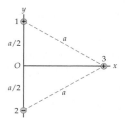

▲ **FIGURE 19–38** Problems 89 and 90

90. •• **IP** Consider the system of three charges described in the previous problem and shown in Figure 19–38. **(a)** Do you expect the net force acting on charge 1 to have a magnitude greater than, less than, or the same as the magnitude of the net force acting on charge 2? Explain. **(b)** Find the magnitude of the net force acting on charge 1. **(c)** Find the magnitude of the net force acting on charge 2.

91. •• **IP BIO Cell Membranes** The cell membrane in a nerve cell has a thickness of 0.12 μm. **(a)** Approximating the cell membrane as a parallel-plate capacitor with a surface charge density of 5.9×10^{-6} C/m^2, find the electric field within the membrane. **(b)** If the thickness of the membrane were doubled, would your answer to part (a) increase, decrease, or stay the same? Explain.

92. •• A square with sides of length L has a point charge at each of its four corners. Two corners that are diagonally opposite have charges equal to +2.25 μC; the other two diagonal corners have charges Q. Find the magnitude and sign of the charges Q such that each of the +2.25-μC charges experiences zero net force.

93. •• **IP** Suppose a charge $+Q$ is placed on the Earth, and another charge $+Q$ is placed on the Moon. **(a)** Find the value of Q needed to "balance" the gravitational attraction between the Earth and the Moon. **(b)** How would your answer to part (a) change if the distance between the Earth and the Moon were doubled? Explain.

94. •• Two small plastic balls hang from threads of negligible mass. Each ball has a mass of 0.14 g and a charge of magnitude q. The balls are attracted to each other, and the threads attached to the balls make an angle of 20.0° with the vertical, as shown in **Figure 19–39**. Find **(a)** the magnitude of the electric force acting on each ball, **(b)** the tension in each of the threads, and **(c)** the magnitude of the charge on the balls.

▲ **FIGURE 19–39** Problem 94

95. •• A small sphere with a charge of +2.44 μC is attached to a relaxed horizontal spring whose force constant is 89.2 N/m. The spring extends along the x axis, and the sphere rests on a frictionless surface with its center at the origin. A point charge $Q = -8.55$ μC is now moved slowly from infinity to a point $x = d > 0$ on the x axis. This causes the small sphere to move to the position $x = 0.124$ m. Find d.

96. •• Twelve identical point charges q are equally spaced around the circumference of a circle of radius R. The circle is centered at the origin. One of the twelve charges, which happens to be on the positive x axis, is now moved to the center of the circle. Find **(a)** the direction and **(b)** the magnitude of the net electric force exerted on this charge.

97. •• **BIO Nerve Impulses** When a nerve impulse propagates along a nerve cell, the electric field within the cell membrane changes from 7.0×10^5 N/C in one direction to 3.0×10^5 N/C in the other direction. Approximating the cell membrane as a parallel-plate capacitor, find the magnitude of the change in charge density on the walls of the cell membrane.

98. •• **IP The Electric Field of the Earth** The Earth produces an approximately uniform electric field at ground level. This electric field has a magnitude of 110 N/C and points radially inward, toward the center of the Earth. **(a)** Find the surface charge density (sign and magnitude) on the surface of the Earth. **(b)** Given that the radius of the Earth is 6.38×10^6 m, find the total electric charge on the Earth. **(c)** If the Moon had the same amount of electric charge distributed uniformly over its surface, would its electric field at the surface be greater than, less than, or equal to 110 N/C? Explain.

99. •• An object of mass $m = 3.1$ g and charge $Q = +48$ μC is attached to a string and placed in a uniform electric field that is inclined at an angle of 30.0° with the horizontal **(Figure 19–40)**. The object is in static equilibrium when the string is horizontal. Find **(a)** the magnitude of the electric field and **(b)** the tension in the string.

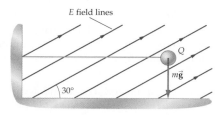

▲ **FIGURE 19–40** Problem 99

100. •• Four identical charges, $+Q$, occupy the corners of a square with sides of length a. A fifth charge, q, can be placed at any desired location. Find the location of the fifth charge, and the value of q, such that the net electric force acting on each of the original four charges, $+Q$, is zero.

101. •• **Figure 19–41** shows an electron entering a parallel-plate capacitor with a speed of 5.45×10^6 m/s. The electric field of the

2.25 cm

0.618 cm

$\vec{v}$

$\vec{E}$

▲ **FIGURE 19–41** Problem 101

capacitor has deflected the electron downward by a distance of 0.618 cm at the point where the electron exits the capacitor. Find **(a)** the magnitude of the electric field in the capacitor and **(b)** the speed of the electron when it exits the capacitor.

102. ••• Two identical conducting spheres are separated by a fixed center-to-center distance of 45 cm and have different charges. Initially, the spheres attract each other with a force of 0.095 N. The spheres are now connected by a thin conducting wire. After the wire is removed, the spheres are positively charged and repel one another with a force of 0.032 N. Find **(a)** the final and **(b)** the initial charges on the spheres.

PASSAGE PROBLEMS

Bumblebees and Static Cling

Have you ever pulled clothes from a dryer only to have them "cling" together? Have you ever walked across a carpet and had a "shocking" experience when you touched a doorknob? If so, you already know a lot about static electricity.

Ben Franklin showed that the same kind of spark we experience on a carpet, when scaled up in size, is responsible for bolts of lightning. His insight led to the invention of lightning rods to conduct electricity safely away from a building into the ground. Today, we employ static electricity in many technological applications, ranging from photocopiers to electrostatic precipitators that clean emissions from smokestacks. We even use electrostatic salting machines to give potato chips the salty taste we enjoy!

Living organisms also use static electricity—in fact, static electricity plays an important role in the pollination process. Imagine a bee busily flitting from flower to flower. As air rushes over its body and wings it acquires an electric charge—just as you do when your feet rub against a carpet. A bee might have only 93.0 pC of charge, but that's more than enough to attract grains of pollen from a distance, like a charged comb attracting bits of paper. The result is a bee covered with grains of pollen, as illustrated in the accompanying photo, unwittingly transporting pollen from one flower to another. So, the next time you experience annoying static cling in your clothes, just remember that the same force helps pollinate the plants that we all need for life on Earth.

▲ A white-tailed bumblebee with static cling. (Problems 103, 104, 105, and 106)

103. • How many electrons must be transferred away from a bee to produce a charge of +93.0 pC?

A. 1.72×10^{-9} **B.** 5.81×10^{8}

C. 1.02×10^{20} **D.** 1.49×10^{29}

104. • Suppose two bees, each with a charge of 93.0 pC, are separated by a distance of 1.20 cm. Treating the bees as point charges, what is the magnitude of the electrostatic force experienced by the bees? (In comparison, the weight of a 0.140-g bee is 1.37×10^{-3} N.)

A. 6.01×10^{-17} N   **B.** 6.48×10^{-9} N

C. 5.40×10^{-7} N   **D.** 5.81×10^{-3} N

105. • The force required to detach a grain of pollen from an avocado stigma is approximately 4.0×10^{-8} N. What is the maximum distance at which the electrostatic force between a bee and a grain of pollen is sufficient to detach the pollen? Treat the bee and pollen as point charges, and assume the pollen has a charge opposite in sign and equal in magnitude to the bee.

A. 4.7×10^{-7} m   **B.** 1.9 mm

C. 4.4 cm   **D.** 220 m

106. • The Earth produces an electric field of magnitude 110 N/C. What force does this electric field exert on a bee carrying a charge of 93.0 pC? (Again, for comparison, the weight of a bee is approximately 1.37×10^{-3} N.)

A. 1.76×10^{-17} N   **B.** 8.45×10^{-13} N

C. 1.02×10^{-8} N   **D.** 1.13×10^{-6} N

INTERACTIVE PROBLEMS

107. •• **IP** Referring to Example 19–5 Suppose $q_1 = +2.90 \ \mu\text{C}$ is no longer at the origin, but is now on the y axis between $y = 0$ and $y = 0.500$ m. The charge $q_2 = +2.90 \ \mu\text{C}$ is at $x = 0$ and $y = 0.500$ m, and point 3 is at $x = y = 0.500$ m. **(a)** Is the magnitude of the net electric field at point 3, which we call E_{net}, greater than, less than, or equal to its previous value? Explain. **(b)** Is the angle θ that E_{net} makes with the x axis greater than, less than, or equal to its previous value? Explain. Find the new values of **(c)** E_{net} and **(d)** θ if q_1 is at $y = 0.250$ m.

108. •• **IP** Referring to Example 19–5 In this system, the charge q_1 is at the origin, the charge q_2 is at $x = 0$ and $y = 0.500$ m, and point 3 is at $x = y = 0.500$ m. Suppose that $q_1 = +2.90 \ \mu\text{C}$, but that q_2 is increased to a value greater than $+2.90 \ \mu\text{C}$. As a result, do **(a)** E_{net} and **(b)** θ increase, decrease, or stay the same? Explain. If $E_{net} = 1.66 \times 10^{5}$ N/C, find **(c)** q_2 and **(d)** θ.

109. •• **IP** Referring to Example 19–6 The magnitude of the charge is changed until the angle the thread makes with the vertical is $\theta = 15.0°$. The electric field is 1.46×10^{4} N/C and the mass of the object is 0.0250 kg. **(a)** Is the new magnitude of q greater than or less than its previous value? Explain. **(b)** Find the new value of q.

110. •• **Referring to Example 19–6** Suppose the magnitude of the electric field is adjusted to give a tension of 0.253 N in the thread. This will also change the angle the thread makes with the vertical. **(a)** Find the new value of E. **(b)** Find the new angle between the thread and the vertical.

20 Electric Potential and Electric Potential Energy

Neon signs like the one shown here have been a familiar advertising tool since 1912, when the first commercial sign was sold to a Parisian barber. These signs do in fact contain neon, which gives the red color, as well as other noble gases such as argon, helium, krypton, and xenon to produce a wide range of brilliant colors. A typical neon sign operates with a difference in electric potential of about 10,000 volts. In contrast, muscle contractions in the human heart produce electric potentials that are only about a thousandth of a volt. The physics underlying electric potential—and its application to everything from neon signs to EKGs—is the subject of this chapter.

When paramedics try to revive a heart-attack victim, they apply a jolt of electrical energy to the person's heart with a device known as a defibrillator. Before they can use the defibrillator, however, they must wait a few seconds for it to be "charged up." What exactly is happening as the defibrillator charges? The answer is that electric charge is building up on a capacitor, and in the process storing electrical energy. When the defibrillator is activated, the energy stored in the capacitor is released in a sudden surge of power that is often capable of saving a person's life. In this chapter we develop the concept of electrical energy and discuss how both energy and charge can be stored in a capacitor.

20–1 Electric Potential Energy and the Electric Potential

Electric and gravitational forces have many similarities. One of the most important of these is that both forces are conservative. As a result, there must be an **electric potential energy**, U, associated with the electric force, just as there is a gravitational potential energy, $U = mgy$, due to the force of gravity. (Recall that conservative forces and potential energies are discussed in Chapter 8.)

To illustrate the concept of electric potential energy, consider a uniform electric field, $\vec{E}$, as shown in **Figure 20–1 (a)**. A positive test charge q_0 is placed in this field, where it experiences a downward electric force of magnitude $F = q_0E$. If the charge is moved upward through a distance d, the electric force and the displacement are in opposite directions; hence, the work done by the electric force is negative:

$$W = -q_0Ed$$

Using our definition of potential energy change given in Equation 8–1, $\Delta U = -W$, we find that the potential energy of the charge is changed by the amount

$$\Delta U = -W = q_0Ed \qquad \text{20–1}$$

Note that the electric potential energy increases, just as the gravitational potential energy of a ball increases when it is raised against the force of gravity to a higher altitude, as indicated in **Figure 20–1 (b)**.

On the other hand, if the charge q_0 is negative, the electric force acting on it will be upward. In this case, the electric force does *positive* work as the charge is raised through the distance d, and the change in potential energy is *negative*. Thus, the change in potential energy depends on the sign of a charge as well as on its magnitude.

The electric force depends on charge in the same way as does the change in electric potential energy. In the last chapter we found it convenient to define an electric field, $\vec{E}$, as the force per charge:

$$\vec{E} = \frac{\vec{F}}{q_0}$$

Similarly, it is useful to define a quantity that is equal to the change in electric potential energy per charge, $\Delta U/q_0$. This quantity, which is independent of the test charge q_0, is referred to as the **electric potential**, V:

Definition of Electric Potential, V

$$\Delta V = \frac{\Delta U}{q_0} = \frac{-W}{q_0} \qquad \text{20–2}$$

SI unit: joule/coulomb = volt, V

In common usage, the electric potential is often referred to simply as the "potential."

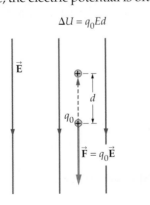

$\Delta U = q_0Ed$

$\vec{F} = q_0\vec{E}$

(a) Moving a charge in an electric field

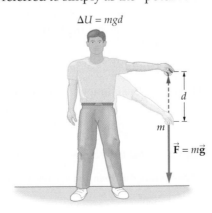

$\Delta U = mgd$

$\vec{F} = m\vec{g}$

(b) Moving a mass in a gravitational field

◀ **FIGURE 20–1 Change in electric potential energy**

(a) A positive test charge q_0 experiences a downward force due to the electric field $\vec{E}$. If the charge is moved upward a distance d, the work done by the electric field is $-q_0Ed$. At the same time, the electric potential energy of the system increases by q_0Ed. The situation is analogous to that of an object in a gravitational field. (b) If a ball is lifted against the force exerted by gravity, the gravitational potential energy of the system increases.

PROBLEM-SOLVING NOTE

Electric Potential and Its Unit of Measurement

Be careful to distinguish between the electric potential V and the corresponding unit of measurement, the volt V, since both are designated by the same symbol. For example, in the expression $\Delta V = 12$ V, the first V refers to the electric potential, the second V refers to the units of the volt. Similarly, when someone says he has a "12-volt battery," what he really means is that the battery produces a difference in electric potential, ΔV, of 12 volts.

Notice that our definition gives only the *change* in electric potential. As with gravitational potential energy, the electric potential can be set to zero at any desired location—only changes in electric potential are measurable. In addition, just as the potential energy is a scalar (that is, simply a number) so too is the potential V.

Potential energy is measured in joules, and charge is measured in coulombs; hence, the SI units of electric potential are joules per coulomb. This combination of units is referred to as the volt, in honor of Alessandro Volta (1745–1827), who invented a predecessor to the modern battery. To be specific, the volt (V) is defined as follows:

$$1\ \text{V} = 1\ \text{J/C} \qquad 20\text{–}3$$

Rearranging slightly, we see that energy can be expressed as charge times voltage: $1\ \text{J} = (1\ \text{C})(1\ \text{V})$.

A convenient and commonly used unit of energy in atomic systems is the **electron volt** (eV), defined as the product of the electron charge and a potential difference of 1 volt:

$$1\ \text{eV} = (1.60 \times 10^{-19}\ \text{C})(1\ \text{V}) = 1.60 \times 10^{-19}\ \text{J}$$

It follows that the electron volt is the energy change an electron experiences when it moves through a potential difference of 1 V. As we shall see in later chapters, typical atomic energies are in the eV range, whereas typical energies in nuclear systems are in the MeV (10^6 eV) range. Even the MeV is a minuscule energy, however, when compared to the joule.

EXERCISE 20–1

Find the change in electric potential energy, ΔU, as a charge of **(a)** 2.20×10^{-6} C or **(b)** -1.10×10^{-6} C moves from a point A to a point B, given that the change in electric potential between these points is $\Delta V = V_B - V_A = 24.0$ V.

SOLUTION

a. Solving $\Delta V = \Delta U/q_0$ for ΔU, we find

$$\Delta U = q_0\,\Delta V = (2.20 \times 10^{-6}\ \text{C})(24.0\ \text{V}) = 5.28 \times 10^{-5}\ \text{J}$$

b. Similarly, using $q_0 = -1.10 \times 10^{-6}$ C we obtain

$$\Delta U = q_0\,\Delta V = (-1.10 \times 10^{-6}\ \text{C})(24.0\ \text{V}) = -2.64 \times 10^{-5}\ \text{J}$$

Electric Field and the Rate of Change of Electric Potential

There is a connection between the electric field and the electric potential that is both straightforward and useful. To obtain this relation, let's apply the definition $\Delta V = -W/q_0$ to the case of a test charge that moves through a distance Δs in the direction of the electric field, as in **Figure 20–2**. The work done by the electric field in this case is simply the magnitude of the electric force, $F = q_0 E$, times the distance, Δs:

$$W = q_0 E \Delta s$$

Therefore, the change in electric potential is

$$\Delta V = \frac{-W}{q_0} = \frac{-(q_0 E \Delta s)}{q_0} = -E\Delta s$$

Solving for the electric field, we find

Connection Between the Electric Field and the Electric Potential

$$E = -\frac{\Delta V}{\Delta s} \qquad 20\text{–}4$$

SI unit: volts/meter, V/m

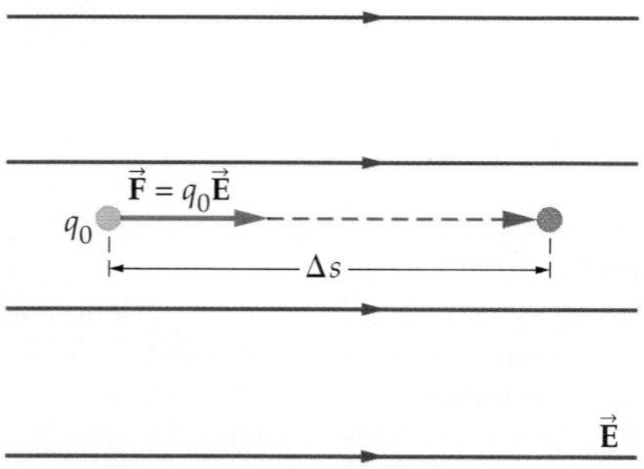

FIGURE 20–2 Electric field and electric potential

As a charge q_0 moves in the direction of the electric field, $\vec{E}$, the electric potential, V, decreases. In particular, if the charge moves a distance Δs, the electric potential decreases by the amount $\Delta V = -E\Delta s$.

▶ **FIGURE 20–3 The electric potential for a constant electric field**
As a general rule, the electric potential, V, decreases as one moves in the direction of the electric field. In the case shown here, the electric field is constant; as a result, the electric potential decreases uniformly with distance. We have arbitrarily set the potential equal to zero at the right-hand plate.

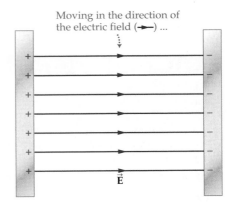
Moving in the direction of the electric field (→) ...

This relation shows that the electric field, which can be expressed in units of N/C, also has the units of volts per meter. That is,

$$1\,\text{N/C} = 1\,\text{V/m} \qquad \text{20–5}$$

To summarize, the electric field depends on *the rate of change of the electric potential with position.* In terms of our gravitational analogy, you can think of the potential V as the height of a hill and the electric field E as the slope of the hill.

In addition, it follows from Equation 20–4 that *the electric potential decreases as one moves in the direction of the electric field.* Specifically, notice that the change in potential, $\Delta V = -E\Delta s$, is negative when E and Δs are in the same direction—that is, when both E and Δs are positive or both are negative. For example, in cases where the electric field is *constant,* as between the plates of a parallel-plate capacitor (Section 19–5), the electric potential decreases *linearly* in the direction of the field. These observations are illustrated in **Figure 20–3**.

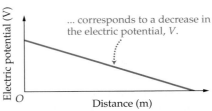
... corresponds to a decrease in the electric potential, V.

EXAMPLE 20–1   PLATES AT DIFFERENT POTENTIALS

A uniform electric field is established by connecting the plates of a parallel-plate capacitor to a 12-V battery. **(a)** If the plates are separated by 0.75 cm, what is the magnitude of the electric field in the capacitor? **(b)** A charge of $+6.24 \times 10^{-6}$ C moves from the positive plate to the negative plate. Find the change in electric potential energy for this charge. (In electrical systems we shall assume that gravity can be ignored, unless specifically instructed otherwise.)

PICTURE THE PROBLEM
Our sketch shows the parallel-plate capacitor connected to a 12-V battery. The battery guarantees that the potential difference between the plates is 12 V, with the positive plate at the higher potential. The separation of the plates is $d = 0.75$ cm $= 0.0075$ m, and the charge that moves from the positive to the negative plate is $q = +6.24 \times 10^{-6}$ C.

High V

Low V

STRATEGY

a. The electric field can be calculated using $E = -\Delta V/\Delta s$. Note that if one moves in the direction of the field from the positive plate to the negative plate ($\Delta s = 0.75$ cm), the electric potential decreases by 12 V; that is, $\Delta V = -12$ V.

b. The change in electric potential energy is $\Delta U = q\,\Delta V$.

SOLUTION

Part (a)

1. Substitute $\Delta s = 0.0075$ m and $\Delta V = -12$ V in $E = -\Delta V/\Delta s$:

$$E = -\frac{\Delta V}{\Delta s} = -\frac{(-12\,\text{V})}{0.0075\,\text{m}} = 1600\,\text{V/m}$$

Part (b)

2. Evaluate $\Delta U = q\Delta V$:

$$\Delta U = q\Delta V = (6.24 \times 10^{-6}\,\text{C})(-12\,\text{V}) = -7.5 \times 10^{-5}\,\text{J}$$

INSIGHT
Note that the electric potential energy of the system decreases as the positive charge moves in the direction of the electric field, just as the gravitational potential energy of a ball decreases when it falls. In the next section, we show how this decrease in electric potential energy shows up as an increase in the charge's kinetic energy, just as the kinetic energy of a falling ball increases.

PRACTICE PROBLEM
Find the separation of the plates that results in an electric field of 2.0×10^3 V/m.   [**Answer:** The separation should be 0.60 cm. Note that decreasing the separation *increases* the field.]

Some related homework problems: Problem 7, Problem 10

A similar problem is considered in the following Active Example.

ACTIVE EXAMPLE 20–1 FIND THE ELECTRIC FIELD AND POTENTIAL DIFFERENCE

The electric potential at point B in the parallel-plate capacitor shown here is less than the electric potential at point A by 4.50 V. The separation between points A and B is 0.120 cm, and the separation between the plates is 2.55 cm. Find **(a)** the electric field within the capacitor and **(b)** the potential difference between the plates.

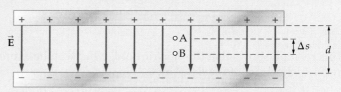

SOLUTION *(Test your understanding by performing the calculations indicated in each step.)*

Part (a)

1. Calculate the electric field using $E = -\Delta V/\Delta s$, with $\Delta V = V_B - V_A = -4.50$ V and $\Delta s = 0.120$ cm:

$E = 3750$ V/m

Part (b)

2. Calculate $\Delta V = V_{(-)} - V_{(+)}$ using $\Delta V = -E \, \Delta s$, with $E = 3750$ V/m and $\Delta s = 2.55$ cm:

$\Delta V = -95.6$ V

INSIGHT

Notice that in part (b) we use the same electric field found in part (a). This is valid because the field within an ideal parallel-plate capacitor is uniform; thus, the same value of E applies everywhere within the capacitor.

Finally, the negative value of ΔV in part (b) simply indicates that the electric potential of the negative plate is less than that of the positive plate by 95.6 V. Put another way, the electric potential increases as we move "against" the electric field, just as the gravitational potential energy increases when we move "against" gravity—that is, when we move uphill.

YOUR TURN

Is the electric potential energy of an electron at point A greater than or less than its electric potential energy at point B? Calculate the difference in electric potential energy as an electron moves from point A to point B.

(Answers to **Your Turn** *problems are given in the back of the book.)*

CONCEPTUAL CHECKPOINT 20–1 CONSTANT ELECTRIC POTENTIAL

In a certain region of space the electric potential V is known to be constant. Is the electric field in this region **(a)** positive, **(b)** zero, or **(c)** negative?

REASONING AND DISCUSSION

The electric field is related to the *rate of change* of the electric potential with position, not to the value of the potential. Since the rate of change of a constant potential is zero, so too is the electric field.

In particular, if one moves a distance Δs from one point to another in this region, the change in potential is zero; $\Delta V = 0$. Hence, the electric field vanishes; $E = -\Delta V/\Delta s = 0$.

ANSWER

(b) The electric field is zero.

Finally, in our calculations of the potential difference $\Delta V = -E\Delta s$ in the examples to this point, we have always taken Δs to be a displacement in the direction of the electric field. More generally, Δs can be a displacement in any direction, as long as we replace E with the component of $\vec{E}$ in the direction of Δs. Thus, for example, the change in potential as we move in the x direction is $\Delta V = -E_x \, \Delta x$, and the corresponding change as we move in the y direction, is $\Delta V = -E_y \, \Delta y$. We shall point out the utility of these expressions in Section 20–4.

20–2 Energy Conservation

When a ball is dropped in a gravitational field, its gravitational potential energy decreases as it falls. At the same time, its kinetic energy increases. If nonconservative forces such as air resistance can be ignored, we know that the decrease in gravitational potential energy is equal to the increase in kinetic energy—in other words, the total energy of the ball is conserved.

Because the electric force is also conservative, the same considerations apply to a charged object in an electric field. Therefore, ignoring other forces, we can say that the total energy of an electric charge must be conserved. As a result, the sum of its kinetic and electric potential energies must be the same at any two points, say A and B:

$$K_A + U_A = K_B + U_B$$

Expressing the kinetic energy as $\frac{1}{2}mv^2$, we can write energy conservation as

$$\tfrac{1}{2}mv_A{}^2 + U_A = \tfrac{1}{2}mv_B{}^2 + U_B$$

This expression applies to any conservative force. In the case of a uniform gravitational field the potential energy is $U = mgy$; for an ideal spring it is $U = \frac{1}{2}kx^2$. When dealing with an electrical system, we can use Equation 20–2 to express the electric potential energy in terms of the electric potential as follows:

$$U = qV$$

To be specific, suppose a particle of mass $m = 1.75 \times 10^{-5}$ kg and charge $q = 5.20 \times 10^{-5}$ C is released from rest at a point A. As the particle moves to another point, B, the electric potential decreases by 60.0 V; that is, $V_A - V_B = 60.0$ V. The particle's speed at point B can be found using energy conservation. To do so, we first solve for the kinetic energy at point B:

$$\tfrac{1}{2}mv_B{}^2 = \tfrac{1}{2}mv_A{}^2 + U_A - U_B$$

$$= \tfrac{1}{2}mv_A{}^2 + q(V_A - V_B)$$

Next we set $v_A = 0$, since the particle starts at rest, and solve for v_B:

$$\tfrac{1}{2}mv_B{}^2 = q(V_A - V_B)$$

$$v_B = \sqrt{\frac{2q(V_A - V_B)}{m}} = \sqrt{\frac{2(5.20 \times 10^{-5}\text{ C})(60.0\text{ V})}{1.75 \times 10^{-5}\text{ kg}}} = 18.9 \text{ m/s} \quad \textbf{20–6}$$

Thus, the decrease in electric potential energy appears as an increase in kinetic energy—and a corresponding increase in speed.

PROBLEM-SOLVING NOTE

Energy Conservation and Electric Potential Energy

Energy conservation applies to charges in an electric field. Therefore, to relate the speed or kinetic energy of a charge to its location simply requires setting the initial energy equal to the final energy. Note that the potential energy for a charge q is $U = qV$.

ACTIVE EXAMPLE 20–2   FIND THE SPEED WITH A RUNNING START

Suppose the particle described in the preceding discussion has an initial speed of 5.00 m/s at point A. What is its speed at point B?

SOLUTION *(Test your understanding by performing the calculations indicated in each step.)*

1. Write the equation for energy conservation: $\tfrac{1}{2}mv_A{}^2 + qV_A = \tfrac{1}{2}mv_B{}^2 + qV_B$

2. Solve for the kinetic energy at point B: $\tfrac{1}{2}mv_B{}^2 = \tfrac{1}{2}mv_A{}^2 + q(V_A - V_B)$

3. Solve for v_B: $v_B = \sqrt{v_A{}^2 + 2q(V_A - V_B)/m}$

4. Substitute numerical values: $v_B = 19.5$ m/s

INSIGHT
The final speed is not 5.00 m/s greater than 18.9 m/s; in fact, it is only 0.6 m/s greater. As usual, this is due to the fact that the kinetic energy depends on v^2 rather than on v.

YOUR TURN
What initial speed is required to obtain a final speed of 18.9 m/s + 5.00 m/s = 23.9 m/s?

*(Answers to **Your Turn** problems are given in the back of the book.)*

Notice that in the preceding discussions a positive charge moves to a region where the electric potential is less, and its speed increases. As one might expect, the situation is just the opposite for a negative charge. In particular, a negative charge will move to a region of higher electric potential with an increase in speed.

For example, if the charge in Equation 20–6 is changed in sign to -5.20×10^{-5} C, and the potential difference is changed in sign to $V_A - V_B = -60.0$ V, the final speed remains the same. In general:

> *Positive* charges accelerate in the direction of *decreasing* electric potential.
>
> *Negative* charges accelerate in the direction of *increasing* electric potential.

In both cases, however, the charge moves to a region of lower electric potential energy.

EXAMPLE 20–2 FROM PLATE TO PLATE

Suppose the charge in Example 20–1 is released from rest at the positive plate and that it reaches the negative plate with a speed of 3.4 m/s. What are **(a)** the mass of the charge and **(b)** its final kinetic energy?

PICTURE THE PROBLEM

The physical situation is the same as in Example 20–1. In this case, however, we know that the charge starts at rest at the positive plate, $v_A = 0$, and hits the negative plate with a speed of $v_B = 3.4$ m/s. Note that the electric potential of the positive plate is 12 V greater than that of the negative plate. Therefore $V_A - V_B = 12$ V.

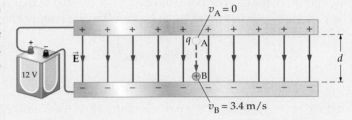

STRATEGY

a. The energy of the charge is conserved as it moves from one plate to the other. Setting the initial energy equal to the final energy gives us an equation in which there is only one unknown, the mass of the charge.

b. The final kinetic energy is simply $\frac{1}{2}mv_B{}^2$.

SOLUTION

Part (a)

1. Apply energy conservation to this system:
$$\tfrac{1}{2}mv_A{}^2 + qV_A = \tfrac{1}{2}mv_B{}^2 + qV_B$$

2. Solve for the mass, m:
$$m = \frac{2q(V_A - V_B)}{v_B{}^2 - v_A{}^2}$$

3. Substitute numerical values:
$$m = \frac{2(6.24 \times 10^{-6}\,\text{C})(12\,\text{V})}{(3.4\,\text{m/s})^2 - 0} = 1.3 \times 10^{-5}\,\text{kg}$$

Part (b)

4. Calculate the final kinetic energy:
$$K_B = \tfrac{1}{2}mv_B{}^2$$
$$= \tfrac{1}{2}(1.3 \times 10^{-5}\,\text{kg})(3.4\,\text{m/s})^2 = 7.5 \times 10^{-5}\,\text{J}$$

INSIGHT

We see that the final kinetic energy is precisely equal to the decrease in electric potential energy calculated in Example 20–1, as expected by energy conservation.

PRACTICE PROBLEM

If the mass of the charge had been 5.2×10^{-5} kg, what would be its **(a)** final speed and **(b)** final kinetic energy?
[Answer: (a) $v_B = 1.7$ m/s, **(b)** $K_B = 7.5 \times 10^{-5}$ J. Note that the final kinetic energy is the same, regardless of the mass, because energy, not speed, is conserved.]

Some related homework problems: Problem 18, Problem 19

CONCEPTUAL CHECKPOINT 20–2 FINAL SPEED

An electron, with a charge of -1.60×10^{-19} C, accelerates from rest through a potential difference V. A proton, with a charge of $+1.60 \times 10^{-19}$ C, accelerates from rest through a potential difference $-V$. Is the final speed of the electron **(a)** greater than, **(b)** less than, or **(c)** the same as the final speed of the proton?

REASONING AND DISCUSSION

The electron and proton have charges of equal magnitude, and therefore they have equal changes in electric potential energy. As a result, their final kinetic energies are equal. Since the electron has less mass than the proton, however, its speed must be greater.

ANSWER

(a) The electron is moving faster than the proton.

20–3 The Electric Potential of Point Charges

Consider a point charge, $+q$, that is fixed at the origin of a coordinate system, as in **Figure 20–4**. Suppose, in addition, that a positive test charge, $+q_0$, is held at rest at point A, a distance r_A from the origin. At this location the test charge experiences a repulsive force with a magnitude given by Coulomb's law, $F = k|q_0||q|/r_A^2$.

If the test charge is now released, the repulsive force between it and the charge $+q$ will cause it to accelerate away from the origin. When it reaches an arbitrary point B, its kinetic energy will have increased by the same amount that its electric potential energy has decreased. Thus, we conclude that the electric potential energy is greater at point A than at point B. In fact, the methods of integral calculus can be used to show that the difference in electric potential energy between points A and B is

$$U_A - U_B = \frac{kq_0q}{r_A} - \frac{kq_0q}{r_B}$$

Note that the electric potential energy for point charges depends inversely on their separation, the same as the distance dependence of the gravitational potential energy for point masses (Equation 12–9).

The corresponding change in electric potential is found by dividing the electric potential energy by the test charge, q_0:

$$V_A - V_B = \frac{1}{q_0}(U_A - U_B) = \frac{kq}{r_A} - \frac{kq}{r_B}$$

If the test charge is moved infinitely far away from the origin, so that $r_B \to \infty$, the term kq/r_B vanishes, and the difference in electric potential becomes

$$V_A - V_B = \frac{kq}{r_A}$$

Since the potential can be set to zero at any convenient location, we choose to set V_B equal to zero; in other words, *we choose the electric potential to be zero infinitely far from a given charge*. With this choice, the potential of a point charge is $V_A = kq/r_A$. Dropping the subscript, we see that the electric potential at an arbitrary distance r is given by the following:

Electric Potential for a Point Charge

$$V = \frac{kq}{r} \qquad\qquad 20\text{–}7$$

SI unit: volt, V

Recall that this expression for V actually represents a *change* in potential; in particular, V is the change in potential from a distance of infinity to a distance r. The corresponding difference in electric potential energy for the test charge q_0 is simply $U = q_0V$; that is,

Electric Potential Energy for Point Charges q and q_0 Separated by a Distance r

$$U = q_0V = \frac{kq_0q}{r} \qquad\qquad 20\text{–}8$$

SI unit: joule, J

Note that *the electric potential energy of two charges separated by an infinite distance is zero.*

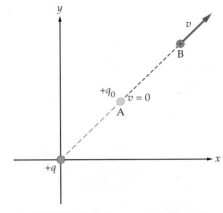

▲ **FIGURE 20–4 Energy conservation in an electrical system**

A test charge, $+q_0$, is released from rest at point A. When it reaches point B, its kinetic energy will have increased by the same amount that its electric potential energy has decreased.

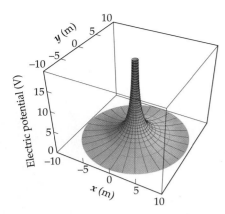

(a) Electric potential near a positive charge

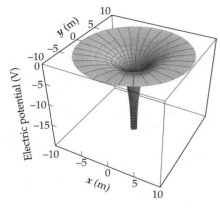

(b) Electric potential near a negative charge

▲ **FIGURE 20–5 The electric potential of a point charge**
Electric potential near **(a)** a positive and **(b)** a negative charge at the origin. In the case of the positive charge, the electric potential forms a "potential hill" near the charge. Near the negative charge we observe a "potential well."

One final point in regard to $V = kq/r$ and $U = kq_0q/r$ is that r is a *distance* and hence is always a positive quantity. For example, suppose a charge q is at the origin of a coordinate system. It follows that the potential at the point $x = 1$ m is the same as the potential at $x = -1$ m, since in both cases the distance is $r = 1$ m.

EXERCISE 20–2

Find the electric potential produced by a point charge of 6.80×10^{-7} C at a distance of 2.60 m.

SOLUTION
Substituting $q = 6.80 \times 10^{-7}$ C and $r = 2.60$ m in $V = kq/r$ yields

$$V = \frac{kq}{r} = \frac{(8.99 \times 10^9 \, \text{N} \cdot \text{m}^2/\text{C}^2)(6.80 \times 10^{-7} \, \text{C})}{2.60 \, \text{m}} = 2350 \, \text{V}$$

Thus, the potential at $r = 2.60$ m due to this point charge is 2350 V greater than the potential at infinity.

Note that V depends on the sign of the charge in question. This is shown in **Figure 20–5**, which represents the electric potential for a positive and a negative charge at the origin. The potential for the positive charge increases to positive infinity near the origin and decreases to zero far away, forming a "potential hill." On the other hand, the potential for the negative charge approaches negative infinity near the origin, forming a "potential well."

Thus, if the charge at the origin is positive, a positive test charge will move away from the origin, as if sliding "downhill" on the electric potential surface. Similarly, if the charge at the origin is negative, a positive test charge will move toward the origin, which again means that it slides downhill, this time into a potential well. Negative test charges, in contrast, always tend to slide "uphill" on electric potential curves, like bubbles rising in water.

Superposition of the Electric Potential

Like many physical quantities, the electric potential obeys a simple superposition principle. In particular:

> The total electric potential due to two or more charges is equal to the algebraic sum of the potentials due to each charge separately.

By *algebraic sum* we mean that the potential of a given charge may be positive or negative, and hence the *algebraic sign* of each potential must be taken into account when calculating the total potential. In particular, positive and negative contributions may cancel to give zero potential at a given location.

Finally, because the electric potential is a scalar, its superposition is as simple as adding numbers of various signs. This, in general, is easier than adding vectors, as is required for the superposition of electric fields.

EXAMPLE 20–3 TWO POINT CHARGES

A charge $q = 4.11 \times 10^{-9}$ C is placed at the origin, and a second charge equal to $-2q$ is placed on the x axis at the location $x = 1.00$ m. **(a)** Find the electric potential midway between the two charges. **(b)** The electric potential vanishes at some point between the charges; that is, for a value of x between 0 and 1.00 m. Find this value of x.

PICTURE THE PROBLEM
Our sketch shows the two charges placed on the x axis as described in the problem statement. Clearly, the electric potential is large and positive near the origin, and large and negative near $x = 1.00$ m. Thus, it follows that V must vanish at some point between $x = 0$ and $x = 1.00$ m.

A plot of V as a function of x is shown as an aid in visualizing the calculations given in this Example.

STRATEGY

a. As indicated in the sketch, an arbitrary point between $x = 0$ and $x = 1.00$ m is a distance x from the charge $+q$ and a distance 1.00 m $- x$ from the charge $-2q$. Thus, by superposition, the total electric potential at a point x is $V = kq/x + k(-2q)/(1.00$ m $- x)$.

b. Setting $V = 0$ allows us to solve for the unknown, x.

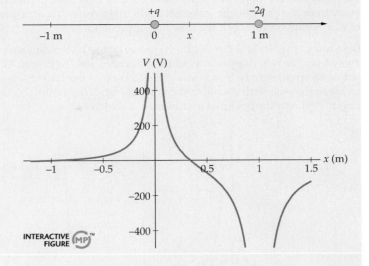

SOLUTION

Part (a)

1. Use superposition to write an expression for V at an arbitrary point x between $x = 0$ and $x = 1.00$ m:

$$V = \frac{kq}{x} + \frac{k(-2q)}{1.00 \text{ m} - x}$$

2. Substitute numerical values into the expression for V. Note that the midway point between the charges is $x = 0.500$ m:

$$V = \frac{(8.99 \times 10^9 \text{ N} \cdot \text{m}^2/\text{C}^2)(4.11 \times 10^{-9} \text{ C})}{0.500 \text{ m}}$$

$$+ \frac{(8.99 \times 10^9 \text{ N} \cdot \text{m}^2/\text{C}^2)(-2)(4.11 \times 10^{-9} \text{ C})}{1.00 \text{ m} - 0.500 \text{ m}}$$

$$= -73.9 \text{ N} \cdot \text{m/C} = -73.9 \text{ V}$$

Part (b)

3. Set the expression for V in Step 1 equal to zero, and simplify by canceling the common factor, kq:

$$V = \frac{kq}{x} + \frac{k(-2q)}{1.00 \text{ m} - x} = 0$$

$$\frac{1}{x} = \frac{2}{1.00 \text{ m} - x}$$

4. Solve for x:

$$1.00 \text{ m} - x = 2x$$

$$x = \tfrac{1}{3}(1.00 \text{ m}) = 0.333 \text{ m}$$

INSIGHT

Suppose a small positive test charge is released from rest at $x = 0.500$ m. In which direction will it move? From the point of view of the Coulomb force, we know it will move to the right—repelled by the positive charge at the origin and attracted to the negative charge at $x = 1.00$ m. We come to the same conclusion when considering the electric potential, since we know that a positive test charge will "slide downhill" on the potential curve.

PRACTICE PROBLEM

The electric potential in this system also vanishes at a point on the negative x axis. Find this point.   [**Answer:** $V = 0$ at $x = -1.00$ m. Note that the point $x = -1.00$ m is 1.00 m from the charge $+q$, and 2.00 m from the charge $-2q$. Hence, at $x = -1.00$ m we have $V = kq/(1.00$ m$) + k(-2q)/(2.00$ m$) = 0$.]

Some related homework problems: Problem 31, Problem 33

PROBLEM-SOLVING NOTE

Superposition for the Electric Potential

| CONCEPTUAL CHECKPOINT 20–3 | A PEAK OR A VALLEY? |

Two point charges, each equal to $+q$, are placed on the x axis at $x = -1$ m and $x = +1$ m. As one moves along the x axis, does the potential look like a peak or a valley near the origin?

Recall that to find the total electric potential due to a system of charges, you need only sum the potentials due to each charge separately—being careful to take into account the appropriate sign. Note that there are no components to deal with, as there are when we superpose electric fields.

REASONING AND DISCUSSION

We know that the potential is large and positive near each of the charges. As you move away from the charges, the potential tends to decrease. In particular, at very large positive or negative values of x the potential approaches zero.

Between $x = -1$ m and $x = +1$ m the potential has its lowest value when you are as far away from the two charges as possible. This occurs at the origin. Moving slightly to the left or the right simply brings you closer to one of the two charges, resulting in an increase in the potential; therefore, the potential has a minimum (bottom of a valley) at the origin. A plot of the potential for this case is shown below.

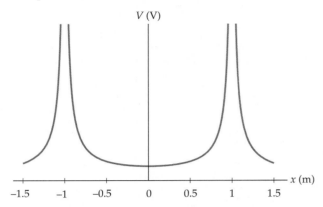

Finally, the electric field is associated with the slope of this curve, as mentioned in relation to Equation 20–4. Clearly, the field has a large magnitude near the charges and is zero at the origin.

ANSWER

Near the origin, the potential looks like a valley.

Superposition applies equally well to the electric potential energy. The following Example illustrates superposition for both the electric potential and the electric potential energy.

EXAMPLE 20–4 FLY AWAY

Two charges, $+q$ and $+2q$, are held in place on the x axis at the locations $x = -d$ and $x = +d$, respectively. A third charge, $+3q$, is released from rest on the y axis at $y = d$. **(a)** Find the electric potential due to the first two charges at the initial location of the third charge. **(b)** Find the initial electric potential energy of the third charge. **(c)** What is the kinetic energy of the third charge when it has moved infinitely far away from the other two charges?

PICTURE THE PROBLEM

As indicated in the sketch, the third charge, $+3q$, is separated from the other two charges by the initial distance, $\sqrt{2}d$. When the third charge is released, the repulsive forces due to $+q$ and $+2q$ will cause it to move away to an infinite distance.

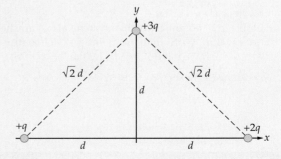

STRATEGY

a. The electric potential at the initial position of the third charge is the sum of the potentials due to $+q$ and $+2q$, each at a distance of $\sqrt{2}d$.

b. The initial electric potential energy of the third charge, U_i, is simply its charge, $+3q$, times the potential, V, found in part (a).

c. We can find the final kinetic energy using energy conservation; $U_i + K_i = U_f + K_f$. Because $K_i = 0$ (third charge starts at rest) and $U_f = 0$ (third charge infinitely far away), we find that $K_f = U_i$.

SOLUTION

Part (a)

1. Calculate the net electric potential at the initial position of the third charge:

$$V_i = \frac{k(+q)}{\sqrt{2}d} + \frac{k(+2q)}{\sqrt{2}d} = \frac{3kq}{\sqrt{2}d}$$

Part (b)

2. Multiply V by $(+3q)$ to find the initial electric potential energy, U_i, of the third charge:

$$U_i = (+3q)V_i = (+3q)\frac{3kq}{\sqrt{2d}} = \frac{9kq^2}{\sqrt{2d}}$$

Part (c)

3. Use energy conservation to find the final (infinite separation) kinetic energy:

$$U_i + K_i = U_f + K_f$$

$$\frac{9kq^2}{\sqrt{2d}} + 0 = 0 + K_f$$

$$K_f = \frac{9kq^2}{\sqrt{2d}}$$

INSIGHT

If the third charge had started closer to the other two charges, it would have been higher up on the "potential hill," and hence its kinetic energy at infinity would have been greater, as shown in the following Practice Problem.

PRACTICE PROBLEM

Suppose the third charge is released from rest just above the origin. What is its final kinetic energy in this case? **[Answer:** $K_f = U_i = 9kq^2/d$. As expected, this is greater than $9kq^2/\sqrt{2d}$.]

Some related homework problems: Problem 34, Problem 39

The electric potential energy for a pair of charges, q_1 and q_2, separated by a distance r is $U = kq_1q_2/r$. In systems that contain more than two charges, the total electric potential energy is the sum of terms like $U = kq_1q_2/r$ for each pair of charges in the system. This procedure is illustrated in the following Active Example.

ACTIVE EXAMPLE 20–3 FIND THE ELECTRIC POTENTIAL ENERGY

A system consists of the charges $-q$ at $(-d, 0)$, $+2q$ at $(d, 0)$, and $+3q$ at $(0, d)$. What is the total electric potential energy of the system?

SOLUTION *(Test your understanding by performing the calculations indicated in each step.)*

1. Write the electric potential energy between $-q$ and $+2q$: $-kq(2q)/2d$

2. Write the electric potential energy between $-q$ and $+3q$: $-kq(3q)/\sqrt{2d}$

3. Write the electric potential energy between $+2q$ and $+3q$: $k(2q)(3q)/\sqrt{2d}$

4. Sum the contributions from each pair of charges to find the total electric potential energy: $U = 3kq^2/\sqrt{2d} - kq^2/d$

YOUR TURN

What is the total electric potential energy of this system if the charge $+3q$ is moved from $(0, d)$ to $(0, -d)$?

*(Answers to **Your Turn** problems are given in the back of the book.)*

20–4 Equipotential Surfaces and the Electric Field

A contour map is a useful tool for serious hikers and backpackers. The first thing you notice when looking at a map such as the one shown in Figure 8–13 is a series of closed curves—the contours—each denoting a different altitude. When the contours are closely spaced, the altitude changes rapidly, indicating steep terrain. Conversely, widely spaced contours indicate a fairly flat surface.

A similar device can help visualize the electric potential due to one or more electric charges. Consider, for example, a single positive charge located at the origin. As we saw in the previous section, the electric potential due to this charge approaches zero far from the charge and rises to form an infinitely high "potential hill" near the charge. A three-dimensional representation of the potential is plotted

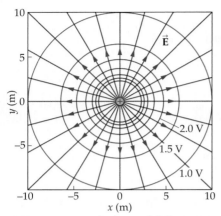

▲ **FIGURE 20–6 Equipotentials for a point charge**

Equipotential surfaces for a positive point charge located at the origin. Near the origin, where the equipotentials are closely spaced, the potential varies rapidly with distance and the electric field is large. This is a top view of Figure 20–5 (a).

in Figure 20–5. The same potential is shown as a contour map in **Figure 20–6**. In this case, the contours, rather than representing altitude, indicate the value of the potential. Since the value of the potential at any point on a given contour is equal to the value at any other point on the same contour, we refer to the contours as **equipotential surfaces,** or simply, **equipotentials.**

An equipotential plot also contains important information about the magnitude and direction of the electric field. For example, in Figure 20–6 we know that the electric field is more intense near the charge, where the equipotentials are closely spaced, than it is far from the charge, where the equipotentials are widely spaced. This simply illustrates that the electric field, $E = -\Delta V/\Delta s$, depends on the rate of change of the potential with position—the greater the change in potential, ΔV, over a given distance, Δs, the larger the magnitude of E.

The direction of the electric field is given by the minus sign in $E = -\Delta V/\Delta s$. For example, if the electric potential decreases over a distance Δs, it follows that ΔV is negative; $\Delta V < 0$. Thus, $-\Delta V$ is positive, and hence $E > 0$. To summarize:

The electric field points in the direction of decreasing electric potential.

This is also illustrated in Figure 20–6, where we see that the electric field points away from the charge $+q$, in the direction of decreasing electric potential.

Not only does the field point in the direction of decreasing potential, it is, in fact, *perpendicular* to the equipotential surfaces:

The electric field is always perpendicular to the equipotential surfaces.

To see why this is the case, we note that zero work is done when a charge is moved perpendicular to an electric field. That is, the work $W = Fd \cos \theta$ is zero when the angle θ is 90°. If zero work is done, it follows from $\Delta V = -W/q_0$ (Equation 20–2) that there is no change in potential. Therefore, the potential is constant (equipotential) in a direction perpendicular to the electric field.

CONCEPTUAL CHECKPOINT 20–4   EQUIPOTENTIAL SURFACES

Is it possible for equipotential surfaces to intersect?

REASONING AND DISCUSSION
To answer this question, it is useful to consider the analogous case of contours on a contour map (Figure 8–13). As we know, each contour corresponds to a different altitude. Because each point on the map has only a single value of altitude, it follows that it is impossible for contours to intersect.

Precisely the same reasoning applies to the electric potential, and hence equipotential surfaces never intersect.

ANSWER
No. Equipotential surfaces cannot intersect.

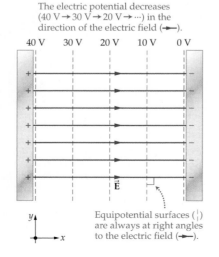

The electric potential decreases (40 V → 30 V → 20 V → ⋯) in the direction of the electric field (→).

Equipotential surfaces (┆) are always at right angles to the electric field (→).

The graphical relationship between equipotentials and electric fields can be illustrated with a few examples. We begin with the simple case of a uniform electric field, as between the plates of a parallel-plate capacitor. In **Figure 20–7**, we plot both the electric field lines, which point in the positive x direction, and the corresponding equipotential surfaces. As expected, the field lines are perpendicular to the equipotentials, and $\vec{\mathbf{E}}$ points in the direction of decreasing V.

To see this last result more clearly, note that $\vec{\mathbf{E}} = E\hat{\mathbf{x}}$ in Figure 20–7, from which it follows that $E_x = E > 0$ and $E_y = 0$. Referring to Section 20–1, we see

◀ **FIGURE 20–7 Equipotential surfaces for a uniform electric field**
The electric field is always perpendicular to equipotential surfaces, and points in the direction of decreasing electric potential. In this case the electric field is (i) uniform and (ii) horizontal. As a result, (i) the electric potential decreases at a uniform rate, and (ii) the equipotential surfaces are vertical.

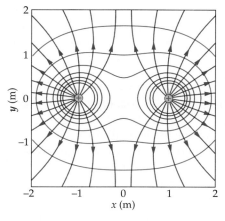
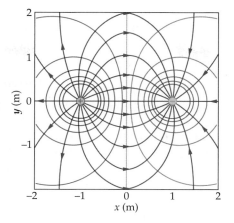

(a) Equipotentials for two positive charges   **(b)** Equipotentials for a dipole

◀ **FIGURE 20–8 Equipotential surfaces for two point charges**
(a) In the case of two equal positive charges, the electric field between them is weak because the field produced by one charge effectively cancels the field produced by the other. As a result, the electric potential is practically constant between the charges. **(b)** For equal charges of opposite sign (a dipole), the electric field is strong between the charges, and the potential changes rapidly.

that if we move in the positive x direction, the change in electric potential is negative, $\Delta V = -E_x \, \Delta x = -E \Delta x < 0$, as expected. On the other hand, if we move in the y direction, the change in potential is zero, $\Delta V = -E_y \, \Delta y = 0$. This is why the equipotentials, which are always perpendicular to $\vec{E}$, are parallel to the y axis in Figure 20–7.

Figure 20–8 (a) shows a similar plot for the case of two positive charges of equal magnitude. Notice that the electric field lines always cross the equipotentials at right angles. In addition, the electric field is more intense where the equipotential surfaces are closely spaced. In the region midway between the two charges, where the electric field is essentially zero, the potential is practically constant.

Finally, in **Figure 20–8 (b)** we show the equipotentials for two charges of opposite sign, one $+q$ (at $x = -1$ m) and the other $-q$ (at $x = +1$ m), forming an electric dipole. In this case, the amber-color equipotentials denote negative values of V. Teal-color equipotentials correspond to positive V, and the tan-color equipotential has the value $V = 0$. The electric field is nonzero between the charges, even though the potential is zero there—recall that $E = -\Delta V / \Delta s$ is related to the *rate of change of V*, not to its value. The relatively large number of equipotential surfaces between the charges shows that V is indeed *changing rapidly* in that region.

Ideal Conductors

A charge placed on an ideal conductor is free to move. As a result, a charge can be moved from one location on a conductor to another with no work being done. Since the work is zero, the change in potential is also zero; therefore, every point on or within an ideal conductor is at the same potential:

> Ideal conductors are equipotential surfaces; every point on or within such a conductor is at the same potential.

For example, when a charge Q is placed on a conductor, as in **Figure 20–9**, it distributes itself over the surface in such a way that the potential is the same everywhere. If the conductor has the shape of a sphere, the charge spreads uniformly over its surface, as in Figure 20–9 (a). If, however, the conductor has a sharp end and a blunt end, as in Figure 20–9 (b), the charge is more concentrated near the sharp end, which results in a large electric field. We pointed out this effect earlier, in Section 19–6.

To see why this should be the case, consider a conducting sphere of radius R with a charge Q distributed uniformly over its surface, as in **Figure 20–10 (a)**. The charge density on this sphere is $\sigma = Q/4\pi R^2$, and the potential at its surface is the same as for a point charge Q at the center of the sphere; that is, $V = kQ/R$. Writing this in terms of the surface charge density, we have

$$V = \frac{kQ}{R} = \frac{k\sigma(4\pi R^2)}{R} = 4\pi k\sigma R$$

(a) Uniform sphere, uniform charge distribution

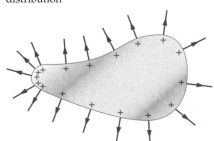

(b) Charge is concentrated at pointed end

▲ **FIGURE 20–9 Electric charges on the surface of ideal conductors**
(a) On a spherical conductor, the charge is distributed uniformly over the surface. **(b)** On a conductor of arbitrary shape, the charge is more concentrated, and the electric field is more intense, where the conductor is more sharply curved. Note that in all cases the electric field is perpendicular to the surface of the conductor.

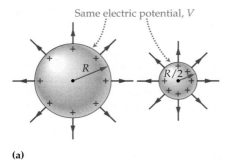

Same electric potential, V

(a)

High charge density, high electric field

Low charge density, low electric field

Equipotential surface

(b)

▲ **FIGURE 20–10 Charge concentration near points**

(a) If two spheres of different radii have the same electric potential at their surfaces, the sphere with the smaller radius of curvature has the greater charge density and the greater electric field. **(b)** An arbitrarily shaped conductor can be approximated by spheres with the same potential at the surface and varying radii of curvature. It follows that the more sharply curved end of a conductor has a greater charge density and a more intense field.

REAL-WORLD PHYSICS: BIO

Electrocardiograph

▶ **FIGURE 20–11 The electrocardiograph**

(a) A typical electrocardiograph tracing, with the major features labeled. The EKG records the electrical activity that accompanies the rhythmic contraction and relaxation of heart muscle tissue. The main pumping action of the heart is associated with the *QRS* complex: contraction of the ventricles begins just after the *R* peak. **(b)** The simplest arrangement of electrodes, or "leads," for an EKG. More precise information can be obtained by the use of additional leads—typically, a dozen in modern practice.

Now, consider a sphere of radius $R/2$. For this sphere to have the *same potential* as the large sphere, it must have twice the charge density, 2σ. In particular, letting σ go to 2σ and R go to $R/2$ in the expression $V = 4\pi k\sigma R$ yields the same result as for the large sphere:

$$V = 4\pi k(2\sigma)(R/2) = 4\pi k\sigma R$$

Clearly, then, the smaller the radius of a sphere with potential V, the greater its charge density.

The electric field is greater for the small sphere as well. At the surface of the large sphere in Figure 20–10 (a) the electric field has a magnitude given by

$$E = \frac{kQ}{R^2} = \frac{k\sigma(4\pi R^2)}{R^2} = 4\pi k\sigma$$

The small sphere in Figure 20–10 (a) has twice the charge density, hence it has twice the electric field at its surface.

The relevance of these results for a conductor of arbitrary shape can be seen in **Figure 20–10 (b)**. Here we show that even an arbitrarily shaped conductor has regions that approximate a spherical surface. In this case, the left end of the conductor is approximately a portion of a sphere of radius R, and the right end is approximately a sphere of radius $R/2$. Noting that every point on a conductor is at the *same potential*, we see that the situation in Figure 20–10 (b) is similar to that in Figure 20–10 (a). It follows that the sharper end of the conductor has the greater charge density and the greater electric field. If a conductor has a sharp point, as in a lightning rod, the corresponding radius of curvature is very small, which can result in an enormous electric field. We consider this possibility in more detail in the next section.

Finally, we note that because the surface of an ideal conductor is an equipotential, the electric field must meet the surface at right angles. This is also shown in Figures 20–9 and 20–10.

Electric Potential and the Human Body

The human body is a relatively good conductor of electricity, but it is not an ideal conductor. If it were, the entire body would be one large equipotential surface. Instead, muscle activity, the beating of the heart, and nerve impulses in the brain all lead to slight differences in electric potential from one point on the skin to another.

In the case of the heart, the powerful waves of electrical activity as the heart muscles contract result in potential differences that are typically in the range of 1 mV. These potential differences can be detected and displayed with an instrument known as an **electrocardiograph**, abbreviated ECG or EKG. (The K derives from the original Dutch name for the device.) A typical EKG signal is displayed in **Figure 20–11 (a)**.

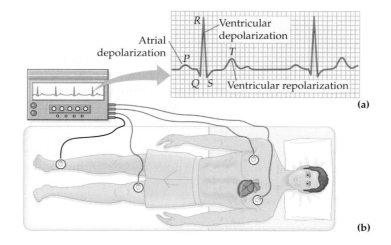

R Ventricular depolarization

Atrial depolarization

P

T

Q *S* Ventricular repolarization

(a)

(b)

To understand the various features of an EKG signal, we note that a heartbeat begins when the heart's natural "pacemaker," the sinoatrial (SA) node in the right atrium, triggers a wave of muscular contraction across both atria that pumps blood into the ventricles. This activity gives rise to the pulse known as the *P wave* in Figure 20–11 (a). Following this contraction, the atrioventricular (AV) node, located between the ventricles, initiates a more powerful wave of contraction in the ventricles, sending oxygen-poor blood to the lungs and oxygenated blood to the rest of the body. These events are reflected in the series of pulses known as the *QRS complex*. Other features in the EKG signal can be interpreted as well. For example, the *T wave* is associated with repolarization of the ventricles in preparation for the next contraction cycle. Even small irregularities in the magnitude, shape, sequence, or timing of these features can provide an experienced physician with essential clues for the diagnosis of many cardiac abnormalities and pathologies. For example, an inverted *T* wave often indicates inadequate blood supply to the heart muscle (ischemia), possibly caused by a heart attack (myocardial infarction).

To record these signals, electrodes are attached to the body in a variety of locations. In the simplest case, illustrated in **Figure 20–11 (b)**, three electrodes are connected in an arrangement known as *Einthoven's triangle*, after the Dutch pioneer of these techniques. Two of these three electrodes are attached to the shoulders, the third to the left groin. For convenience, the two shoulder electrodes are often connected to the wrists, and the groin electrode is connected to the left ankle—the signal at these locations is practically the same as for the locations in Figure 20–11 (b) because of the high conductivity of the human limbs. By convention, the electrode on the right leg is used as the ground. In current applications of the EKG, it is typical to use 12 electrodes to gain more detailed information about the heart's activity.

Electrical activity in the brain can be detected and displayed with an **electroencephalograph** or EEG. In a typical application of this technique, a regular array of 8 to 16 electrodes is placed around the head. The potential differences in this case—in the range of 0.1 mV—are much smaller than those produced by the heart, and much more complex to interpret. One of the key characteristics of an EEG signal, such as the one shown in **Figure 20–12**, is the frequency of the waves. For example, waves at 0.5 to 3.5 Hz, referred to as *D waves*, are common during sleep. A relaxed brain produces *a waves*, with frequencies in the range of 8 to 13 Hz, and an alert brain generates *b waves*, with frequencies greater than 13 Hz. Finally, *q waves*—with frequencies of 5 to 8 Hz—are common in newborns but indicate severe stress in adults.

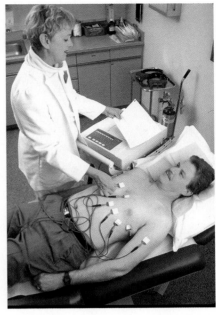

An electrocardiograph can be made with as few as three electrodes and a ground. However, more detailed information about the heart's electrical activity can be obtained by using additional "leads" at precise anatomical locations. Modern EKGs often use a 12-lead array.

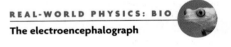

REAL-WORLD PHYSICS: BIO

The electroencephalograph

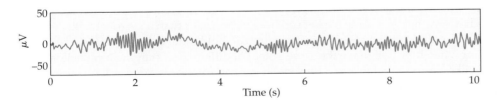

◄ **FIGURE 20–12** The electro-encephalograph
A typical EEG signal.

20-5 Capacitors and Dielectrics

A **capacitor** gets its name from the fact that it has a *capacity* to store both electric charge and energy. Capacitors are a common and important element in modern electronic devices. They can provide large bursts of energy to a circuit, or protect delicate circuitry from excess charge originating elsewhere.

In general, a capacitor is nothing more than two conductors, referred to as *plates*, separated by a finite distance. When the plates of a capacitor are connected to the terminals of a battery, they become charged—one plate acquires a charge $+Q$, the other an equal and opposite charge, $-Q$. The greater the charge Q for a given voltage V, the greater the **capacitance,** C, of the capacitor.

To be specific, suppose a certain battery produces a potential difference of V volts between its terminals. When this battery is connected to a capacitor, a charge of magnitude Q appears on each plate. If a different battery, with a voltage of $2V$, is connected to the same capacitor, the charge on the plates doubles in magnitude to $2Q$. Thus, the charge Q is proportional to the applied voltage V. We define the constant of proportionality to be the capacitance C:

$$Q = CV$$

Solving for the capacitance, we have

Definition of Capacitance, C

$$C = \frac{Q}{V}$$
20–9

SI unit: coulomb/volt = farad, F

In this expression, Q is the magnitude of the charge on either plate, and V is the magnitude of the voltage difference between the plates. By definition, then, the capacitance is always a positive quantity.

As we can see from the relation $C = Q/V$, the units of capacitance are coulombs per volt. In the SI system this combination of units is referred to as the farad (F), in honor of the English physicist Michael Faraday (1791–1867), a pioneering researcher into the properties of electricity and magnetism. In particular,

$$1\,\text{F} = 1\,\text{C/V}$$
20–10

Just as the coulomb is a rather large unit of charge, so too is the farad a rather large unit of capacitance. More typical values for capacitance are in the picofarad ($1\,\text{pF} = 10^{-12}\,\text{F}$) to microfarad ($1\,\mu\text{F} = 10^{-6}\,\text{F}$) range.

EXERCISE 20–3

A capacitor of 0.75 μF is charged to a voltage of 16 V. What is the magnitude of the charge on each plate of the capacitor?

SOLUTION
Using $Q = CV$, we find

$$Q = CV = (0.75 \times 10^{-6}\,\text{F})(16\,\text{V}) = 1.2 \times 10^{-5}\,\text{C}$$

A bucket of water provides a useful analogy when thinking about capacitors. In this analogy we make the following identifications: (i) The cross-sectional area of the bucket is the capacitance C; (ii) the amount of water in the bucket is the charge Q; and (iii) the depth of the water is the voltage difference V between the plates. Therefore, just as a wide bucket can hold more water than a narrow bucket—when filled to the same level—a large capacitor can hold more charge than a small capacitor when they both have the same potential difference. Charging a capacitor, then, is like pouring water into a bucket—if the capacitance is large, a large amount of charge can be placed on the plates with little potential difference between them.

Parallel-Plate Capacitor

A particularly simple capacitor is the parallel-plate capacitor, first introduced in Section 19–5. In this device, two parallel plates of area A are separated by a distance d, as indicated in **Figure 20–13**. We would like to determine the capacitance of such a capacitor. As we shall see, the capacitance is related in a simple way to the parameters A and d.

To begin, we note that when a charge $+Q$ is placed on one plate, and a charge $-Q$ is placed on the other, a uniform electric field is produced between the plates. The magnitude of this field is given by Gauss's law, and was determined in Active Example 19–3. There we found that $E = \sigma/\varepsilon_0$. Noting that the charge per area is $\sigma = Q/A$, we have

$$E = \frac{Q}{\varepsilon_0 A}$$
20–11

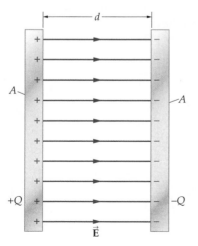

▲ **FIGURE 20–13 A parallel-plate capacitor**
A parallel-plate capacitor, with plates of area A, separation d, and charges of magnitude Q. The capacitance of such a capacitor is $C = \varepsilon_0 A/d$.

The corresponding potential difference between the plates is $\Delta V = -E\Delta s = -(Q/\varepsilon_0 A)d$. The magnitude of this potential difference is simply $V = (Q/\varepsilon_0 A)d$.

Now that we have determined the potential difference for a parallel-plate capacitor with a charge Q on its plates, we can apply $C = Q/V$ to find an expression for the capacitance:

$$C = \frac{Q}{V} = \frac{Q}{(Q/\varepsilon_0 A)d}$$

Canceling the charge Q and rearranging slightly give the final result:

Capacitance of a Parallel-Plate Capacitor

$$C = \frac{\varepsilon_0 A}{d} \qquad\qquad 20\text{–}12$$

As mentioned previously, the capacitance depends in a straightforward way on the area of the plates, A, and their separation, d. The capacitance does not depend separately on the amount of charge on the plates, Q, or the potential difference between the plates, V, but instead depends only on the *ratio* of these two quantities. We use this expression for capacitance in the next Example.

PROBLEM-SOLVING NOTE

Using Magnitudes with Capacitance

Note that the capacitance C is always positive. Therefore, when applying a relation like $C = Q/V$, we always use magnitudes for the charge and the electric potential.

EXAMPLE 20–5 ALL CHARGED UP

A parallel-plate capacitor is constructed with plates of area 0.0280 m^2 and separation 0.550 mm. **(a)** Find the magnitude of the charge on each plate of this capacitor when the potential difference between the plates is 20.1 V. **(b)** What is the magnitude of the electric field between the plates of the capacitor?

PICTURE THE PROBLEM

The capacitor is shown in our sketch. Note that the plates of the capacitor have an area $A = 0.0280$ m^2, a separation $d = 0.550$ mm, and a potential difference $V = 20.1$ V. The charge on each plate has a magnitude Q. The electric field is uniform, and points from the positive plate to the negative plate.

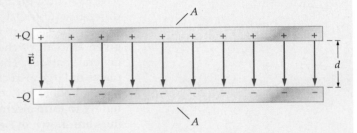

STRATEGY

a. The charge on the plates is given by $Q = CV$. We know the potential difference, V, but we must determine the capacitance, C. We can do this using the relation $C = \varepsilon_0 A/d$, along with the given information for A and d.

b. We can find the magnitude of the electric field with Equation 20–11; that is, $E = Q/\varepsilon_0 A$. As in part (a), the charge on the plates is $Q = CV$. In addition, using the symbolic expression $C = \varepsilon_0 A/d$ will allow us to simplify the final result for E.

SOLUTION

Part (a)

1. Calculate the capacitance of the capacitor:

$$C = \frac{\varepsilon_0 A}{d} = \frac{(8.85 \times 10^{-12}\ \text{C}^2/\text{N}\cdot\text{m}^2)(0.0280\ \text{m}^2)}{0.550 \times 10^{-3}\ \text{m}}$$
$$= 4.51 \times 10^{-10}\ \text{F}$$

2. Find the charge on the plates of the capacitor:

$$Q = CV = (4.51 \times 10^{-10}\ \text{F})(20.1\ \text{V}) = 9.06 \times 10^{-9}\ \text{C}$$

Part (b)

3. Calculate the magnitude of the electric field using $E = Q/\varepsilon_0 A$. Use symbols to simplify the final expression:

$$E = \frac{Q}{\varepsilon_0 A} = \frac{CV}{\varepsilon_0 A} = \frac{(\varepsilon_0 A/d)V}{\varepsilon_0 A} = \frac{V}{d}$$

4. Substitute numerical values:

$$E = \frac{V}{d} = \frac{20.1\ \text{V}}{0.550 \times 10^{-3}\ \text{m}} = 36{,}500\ \text{V/m}$$

CONTINUED ON NEXT PAGE

CONTINUED FROM PREVIOUS PAGE

INSIGHT

Notice the relatively small values for both the capacitance, C, and the magnitude of the charge on a plate, Q. The total amount of charge in the capacitor is zero, $+Q + (-Q) = 0$, but the fact that there is a charge separation—with one plate positive and the other negative—means that the capacitor stores energy, as we shall see in the next section.

If a capacitor is connected to a battery, the battery maintains a constant voltage between the plates of the capacitor. It follows from $Q = CV$ that any change in the capacitance—by changing the plate area, A, or separation, d, for example—results in a different amount of charge on the plates.

Finally, by simplifying the expression for E, and concerning ourselves only with magnitudes, we have obtained a result that is consistent with Equation 20–4, with $\Delta V = V$ and $\Delta s = d$.

PRACTICE PROBLEM

What separation d is necessary to give an increased charge of magnitude 2.00×10^{-8} C on each plate of this capacitor? Assume all other quantities in the system remain the same.   [**Answer:** $d = 0.249$ mm. Note that a *smaller* separation results in a *greater* amount of stored charge.]

Some related homework problems: Problem 50, Problem 51

We see from Equation 20–12 that the capacitance of a parallel-plate capacitor increases with the area A of its plates—basically, the greater the area, the more room there is in the capacitor to hold charge. Again, this is like pouring water into a bucket with a large cross-sectional area. On the other hand, the capacitance decreases with increasing separation d. The reason for this dependence is that with the electric field between the plates constant (as we saw in Section 19–5), the potential difference between the plates is proportional to their separation: $V = Ed$. Thus, the capacitance, which is inversely proportional to the potential difference ($C = Q/V$), is also inversely proportional to the plate separation—the greater the separation, the greater the potential difference required to store a given amount of charge. All capacitors, regardless of their design, share these general features. Thus, to produce a large capacitance, one would like to have plates of large area close together. This is often accomplished by inserting a thin piece of paper between two large sheets of metal foil. The foil is then rolled up tightly to form a compact, large-capacity capacitor. Examples of common capacitors are shown in **Figure 20–14**.

▶ **FIGURE 20–14 Capacitors**
Capacitors come in a variety of physical forms (left), with many different types of dielectric (insulating material) occupying the space between their plates. Capacitors are also rated for the maximum voltage that can be applied to them before the dielectric breaks down, allowing a spark to jump across the gap between the plates. A variable air capacitor (right), often used in early radios, has interleaved plates that can be rotated to vary their area of overlap. This changes the effective area of the capacitor, and hence its capacitance.

CONCEPTUAL CHECKPOINT 20–5 CHARGE ON THE PLATES

A parallel-plate capacitor is connected to a battery that maintains a constant potential difference V between the plates. If the plates are pulled away from each other, increasing their separation, does the magnitude of the charge on the plates **(a)** increase, **(b)** decrease, or **(c)** remain the same?

REASONING AND DISCUSSION
Since the capacitance of a parallel-plate capacitor is $C = \varepsilon_0 A/d$, increasing the separation, d, decreases the capacitance. With a smaller value of C, and a constant value for V, the charge $Q = CV$ will decrease. The same general behavior can be expected with any capacitor.

ANSWER
(b) The charge on the plates decreases.

Sometimes a capacitor is first connected to a battery to be charged and is then disconnected. In this case, the charge on the plates is "trapped"—it has no place to go—and hence Q must remain constant. If the capacitance is changed now, the result is a different potential difference, $V = Q/C$, between the plates.

Dielectrics

One way to increase the capacitance of a capacitor is to insert an insulating material, referred to as a **dielectric**, between its plates. With a dielectric in place, a capacitor can store more charge or energy in the same volume. Thus, dielectrics play an important role in miniaturizing electronic devices.

To see how this works, consider the parallel-plate capacitor shown in **Figure 20–15 (a)**. Initially the plates are separated by a vacuum and connected to a battery, giving the plates the charges $+Q$ and $-Q$. The battery is now removed, and the charge on the plates remains constant. The electric field between the plates is uniform and has a magnitude E_0. If the distance between the plates is d, the corresponding potential difference is $V_0 = E_0 d$, and the capacitance is

$$C_0 = \frac{Q}{V_0}$$

Now, insert a dielectric slab, as illustrated in **Figures 20–15 (b)** and **(c)**. If the molecules in the dielectric have a permanent dipole moment, they will align with the field, as shown in Figures 20–15 (b). Even without a permanent dipole moment, however, the molecules will become polarized by the field (see Section 19–1). This polarization leads to the same type of alignment, although the effect is weaker. The result of this alignment is a positive charge on the surface of the slab near the negative plate and a negative charge on the surface near the positive plate.

Recalling that electric field lines terminate on negative charges and start on positive charges, we can see from Figure 20–15 (c) that fewer field lines exist within

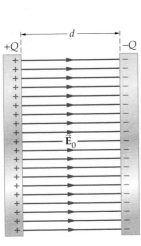

(a) Capacitor without dielectric

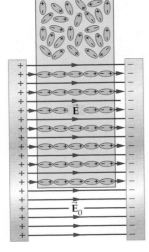

(b) Inserting a dielectric

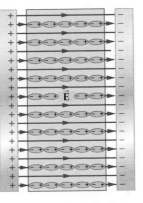

(c) A dielectric reduces the electric field within a capacitor

◄ **FIGURE 20–15 The effect of a dielectric on the electric field of a capacitor**

When a dielectric is placed in the electric field between the plates of a capacitor, the molecules of the dielectric tend to become oriented with their positive ends pointing toward the negatively charged plate and their negative ends pointing toward the positively charged plate. The result is a buildup of positive charge on one surface of the dielectric and of negative charge on the other. Since field lines start on positive charges and end on negative charges, we see that the number of field lines within the dielectric is reduced. Thus, within the dielectric the applied electric field $\vec{E}_0$ is partially canceled. Because the strength of the electric field is less, the voltage between the plates is less as well. Since V is smaller while Q remains the same, the capacitance, $C = Q/V$, is increased by the dielectric.

the dielectric. Consequently, there is a reduced field, E, in a dielectric, which we characterize with a dimensionless **dielectric constant, κ,** as follows:

$$E = \frac{E_0}{\kappa} \qquad \text{20–13}$$

TABLE 20–1 Dielectric Constants

Substance	Dielectric constant, κ
Water	80.4
Neoprene rubber	6.7
Pyrex glass	5.6
Mica	5.4
Paper	3.7
Mylar	3.1
Teflon	2.1
Air	1.00059
Vacuum	1

In the case of a vacuum, $\kappa = 1$, and $E = E_0$, as before. For an insulating material, however, the value of κ is greater than one. For example, paper has a dielectric constant of roughly 4, which means that the electric field within paper is about one-quarter what it would be in a vacuum. Typical values of κ are listed in Table 20–1.

Thus, a dielectric reduces the field between the plates of a capacitor by a factor of κ. This, in turn, decreases the *potential difference* between the plates by the same factor:

$$V = Ed = \left(\frac{E_0}{\kappa}\right)d = \frac{E_0 d}{\kappa} = \frac{V_0}{\kappa}$$

Finally, since the potential difference is smaller, the capacitance must be larger:

$$C = \frac{Q}{V} = \frac{Q}{(V_0/\kappa)} = \kappa\frac{Q}{V_0} = \kappa C_0 \qquad \text{20–14}$$

In effect, the dielectric partially shields one plate from the other, making it easier to build up a charge on the plates. If the space between the plates of a capacitor is filled with paper, for example, the capacitance will be about four times larger than if the space had been a vacuum.

The relation $C = \kappa C_0$ applies to any capacitor. For the special case of a parallel-plate capacitor filled with a dielectric, we have

PROBLEM-SOLVING NOTE

The Effects of a Dielectric

Dielectrics reduce the electric field in a capacitor, which results in a reduced potential difference between the plates. As a result, a dielectric always increases the capacitance.

Capacitance of a Parallel-Plate Capacitor Filled with a Dielectric

$$C = \frac{\kappa\varepsilon_0 A}{d} \qquad \text{20–15}$$

We apply this relation in the next Example.

EXAMPLE 20–6 EVEN MORE CHARGED UP

A parallel-plate capacitor is constructed with plates of area 0.0280 m² and separation 0.550 mm. The space between the plates is filled with a dielectric with dielectric constant κ. When the capacitor is connected to a 12.0-V battery, each of the plates has a charge of magnitude 3.62×10^{-8} C. What is the value of the dielectric constant, κ?

PICTURE THE PROBLEM
The sketch shows the capacitor with a dielectric material inserted between the plates. In other respects, the capacitor is the same as the one considered in Example 20–5.

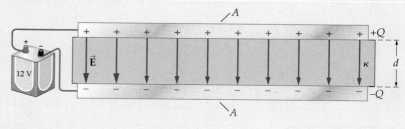

STRATEGY
Since we are given the potential difference V and the charge Q, we can find the capacitance using $C = Q/V$. Next, we relate the capacitance to the physical characteristics of the capacitor with $C = \kappa\varepsilon_0 A/d$. Using the given values for A and d, we solve for κ.

SOLUTION

1. Determine the value of the capacitance:

$$C = \frac{Q}{V} = \frac{(3.62 \times 10^{-8}\,\text{C})}{12.0\,\text{V}} = 3.02 \times 10^{-9}\,\text{F}$$

2. Solve $C = \kappa\varepsilon_0 A/d$ for the dielectric constant, κ:

$$C = \kappa\varepsilon_0 A/d$$
$$\kappa = Cd/\varepsilon_0 A$$
$$\kappa = \frac{Cd}{\varepsilon_0 A}$$

3. Substitute numerical values to find κ:

$$= \frac{(3.02 \times 10^{-9}\,\text{F})(0.550 \times 10^{-3}\,\text{m})}{(8.85 \times 10^{-12}\,\text{C}^2/\text{N}\cdot\text{m}^2)(0.0280\,\text{m}^2)} = 6.70$$

INSIGHT
Comparing our result with the dielectric constants given in Table 20–1, we see that the dielectric may be neoprene rubber.

PRACTICE PROBLEM
If a different dielectric with a smaller dielectric constant is inserted into the capacitor, does the charge on the plates increase, decrease, or remain the same? Find the charge on the plates for $\kappa = 3.5$. [**Answer:** The charge decreases to $Q = 1.89 \times 10^{-8}$ C.]

Some related homework problems: Problem 53, Problem 54

The fact that the capacitance of a capacitor depends on the separation of its plates finds a number of interesting applications. For example, if you have ever typed on a computer keyboard, you have probably been utilizing the phenomenon of capacitance without realizing it. Many computer keyboards are designed in such a way that each key is connected to the upper plate of a parallel-plate capacitor, as illustrated in **Figure 20–16**. When you depress a given key, the separation between the plates of that capacitor decreases, and the corresponding capacitance increases. The circuitry of the computer can detect this change in capacitance, thereby determining which key you have pressed.

Another, less well-known application of capacitance is the theremin, a musical instrument that you play without touching! Two antennas on the theremin are used to control the sound it makes; one antenna adjusts the volume, the other adjusts the pitch. When a person places a hand near one of the antennas, the effect is similar to that of a parallel-plate capacitor, with the hand playing the role of one plate and the antenna playing the role of the other plate. Changing the separation between hand and antenna changes the capacitance, which the theremin's circuitry then converts into a corresponding change of volume or pitch. Theremins have been used to provide "ethereal" music for a number of science fiction films, and some popular bands use theremins in their musical arrangements.

Dielectric Breakdown

If the electric field applied to a dielectric is large enough, it can literally tear the atoms apart, allowing the dielectric to conduct electricity. This condition is referred to as **dielectric breakdown.** The maximum field a dielectric can withstand before breakdown is called the **dielectric strength.** Typical values are given in Table 20–2.

For example, if the electric field in air exceeds about 3,000,000 V/m, dielectric breakdown will occur, leading to a spark on a small scale or a bolt of lightning on a larger scale. Next time you walk across a carpet and get a shock when reaching for the doorknob, think about the fact that you have just produced an electric field of roughly *3 million volts per meter!* The sharp tip of a lightning rod, which has a high electric field in its vicinity, helps initiate and guide lightning to the ground, or to dissipate charge harmlessly so that no lightning occurs at all. Saint Elmo's fire—the glow of light around the rigging of a ship in a storm—is another example of dielectric breakdown in air.

20–6 Electrical Energy Storage

As mentioned in the previous section, capacitors store more than just charge—they also store energy. To see how, consider a capacitor that has charges of magnitude Q on its plates, and a potential difference of V. Now, imagine transferring a small amount of charge, ΔQ, from one plate to the other, as in **Figure 20–17**. Since this charge must be moved across a potential difference of V, the change in electric potential energy is $\Delta U = (\Delta Q)V$. Thus, the potential energy of the capacitor increases by $(\Delta Q)V$ when the magnitude of the charge on its plates is increased from Q to $Q + \Delta Q$. As more charge is transferred from one plate to the other, more electric potential energy is stored in the capacitor.

To find the total electric energy stored in a capacitor, we must take into account the fact that the potential difference between the plates increases as the charge on

Computer keyboards

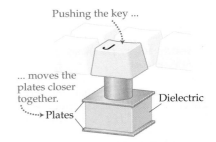

▲ **FIGURE 20–16 Capacitance and the computer keyboard**
The keys on many computer keyboards form part of a parallel-plate capacitor. Depressing the key changes the plate separation. The corresponding change in capacitance can be detected by the computer's circuitry.

The theremin—a musical instrument you play without touching

▲ A musician plays a theremin at an outdoor concert.

TABLE 20–2 Dielectric Strengths

Substance	Dielectric Strength (V/m)
Mica	100×10^6
Teflon	60×10^6
Paper	16×10^6
Pyrex glass	14×10^6
Neoprene rubber	12×10^6
Air	3.0×10^6

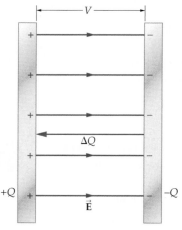

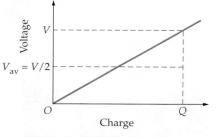

▲ **FIGURE 20–17 The energy required to charge a capacitor**

A capacitor has a charge of magnitude Q on its plates and a potential difference V between the plates. Transferring a small charge increment, $+\Delta Q$, from the negative plate to the positive plate increases the electric potential energy of the capacitor by the amount $\Delta U = (\Delta Q)V$.

REAL-WORLD PHYSICS

The electronic flash

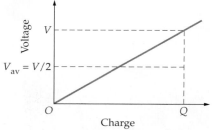

▲ **FIGURE 20–18 The voltage of a capacitor being charged**

The voltage V between the plates of a capacitor increases linearly with the charge Q on the plates, $V = Q/C$. Therefore, if a capacitor is charged to a final voltage of V, the average voltage during charging is $V_{av} = \frac{1}{2}V$.

the plates increases. In fact, recalling that the potential difference is given by $V = Q/C$, it is clear that V increases linearly with the charge, as illustrated in **Figure 20–18**. In particular, if the final potential difference is V, the average potential during charging is $\frac{1}{2}V$. Therefore, the total energy U stored in a capacitor with charge Q and potential difference V can be written as follows:

$$U = QV_{av} = \tfrac{1}{2}QV \qquad \text{20–16}$$

Equivalently, since $Q = CV$, the energy stored in a capacitor of capacitance C and voltage V is

$$U = \tfrac{1}{2}CV^2 \qquad \text{20–17}$$

Finally, using $V = Q/C$, we find that the energy stored in a capacitor of charge Q and capacitance C is

$$U = \frac{Q^2}{2C} \qquad \text{20–18}$$

All these expressions are equivalent; they simply give the energy in terms of different variables.

The energy stored in a capacitor can be put to a number of practical uses. Any time you take a flash photograph, for example, you are triggering the rapid release of energy from a capacitor. The flash unit typically contains a capacitor with a capacitance of 100 to 400 μF. When fully charged to a voltage of about 300 V, the capacitor contains roughly 15 J of energy. Activating the flash causes the stored energy, which took several seconds to accumulate, to be released in less than a millisecond. Because of the rapid release of energy, the power output of a flash unit is impressively large—about 10 to 20 kW. This is far in excess of the power provided by the battery that operates the unit. Similar considerations apply to the defibrillator used in the treatment of heart attack victims, as we show in the next Example.

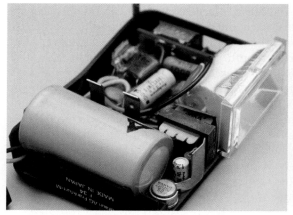

▲ An electronic flash unit like the one at left includes a capacitor (gray) that can store a large amount of charge. When the charge is released, the resulting flash can be as brief as a millisecond or less, allowing photographers to "freeze" motion, as in the photo at right. Even faster strobe units can be used to photograph explosions, shock waves, or speeding bullets.

EXAMPLE 20–7 THE DEFIBRILLATOR: DELIVERING A SHOCK TO THE SYSTEM

REAL WORLD PHYSICS: BIO When a person's heart undergoes ventricular fibrillation—a rapid, uncoordinated twitching of the heart muscles—it often takes a strong jolt of electrical energy to restore the heart's regular beating and save the person's life. The device that delivers this jolt of energy is known as a defibrillator, and it uses a capacitor to store the necessary energy. In a typical defibrillator, a 175-μF capacitor is charged until the potential difference between the plates is 2240 V. **(a)** What is the magnitude of the charge on each plate of the fully charged capacitor? **(b)** Find the energy stored in the charged-up defibrillator.

PICTURE THE PROBLEM
Our sketch shows a simplified representation of a capacitor. The values of
the capacitance and the potential difference are indicated.

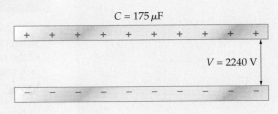

$C = 175\,\mu F$

$V = 2240\ V$

STRATEGY

a. We can find the charge stored on the capacitor plates using $Q = CV$.

b. The energy stored in the capacitor can be determined immediately
using $U = \frac{1}{2}CV^2$. In addition, now that we know the charge on each
plate of the capacitor, the energy can also be found with the relations
$U = \frac{1}{2}QV$ and $U = Q^2/2C$.

SOLUTION

Part (a)

1. Use $Q = CV$ to find the charge on the plates:

$$Q = CV = (175 \times 10^{-6}\ F)(2240\ V) = 0.392\ C$$

Part (b)

2. Find the stored energy using $U = \frac{1}{2}CV^2$:

$$U = \frac{1}{2}CV^2 = \frac{1}{2}(175 \times 10^{-6}\ F)(2240\ V)^2 = 439\ J$$

3. As a check, use $U = \frac{1}{2}QV$:

$$U = \frac{1}{2}QV = \frac{1}{2}(0.392\ C)(2240\ V) = 439\ J$$

4. Finally, use the relation $U = Q^2/2C$:

$$U = \frac{Q^2}{2C} = \frac{(0.392\ C)^2}{2(175 \times 10^{-6}\ F)} = 439\ J$$

INSIGHT
Of the 439 J stored in the defibrillator's capacitor, typically about 200 J will actually pass through the person's body in a pulse
lasting about 2 ms. The power delivered by the pulse is approximately $P = U/t = (200\ J)/(0.002\ s) = 100\ kW$. This is significantly larger than the power delivered by the battery, which can take up to 30 s to fully charge the capacitor.

PRACTICE PROBLEM
Suppose the defibrillator is "fired" when the voltage is only half its maximum value of 2240 V. How much energy is stored in
this case?   [**Answer:** $E = (439\ J)/4 = 110\ J$]

Some related homework problems: Problem 64, Problem 69

A defibrillator uses a capacitor to deliver a shock to a person's heart, restoring
it to normal function. Capacitors can have the opposite effect as well, and it is for
this reason that they can be quite dangerous, even in electrical devices that are
turned off and unplugged from the wall. For example, a television set contains a
number of capacitors, some of which store significant amounts of charge and energy. When a TV is unplugged, the capacitors retain their charge for long periods
of time. Therefore, if you reach into the back of an unplugged television set there
is a danger that you may come in contact with the terminals of a capacitor, which
would then discharge its stored energy through your body. The resulting shock
could be harmful or even fatal.

Finally, we have discussed many examples of energy stored in a capacitor, but
where exactly is the energy located? The answer is that the energy can be thought of
as stored in the electric field, E, between the plates. To be specific, consider the relation

$$\text{energy} = \frac{1}{2}QV$$

In the case of a parallel-plate capacitor of area A and separation d, we know that
$Q = \varepsilon_0 EA$ (Equation 20–11) and $V = Ed$. Thus, the energy stored in the capacitor
can be written as

$$U = \text{energy} = \frac{1}{2}(\varepsilon_0 EA)(Ed) = \frac{1}{2}\varepsilon_0 E^2(Ad)$$

We have grouped A and d together because the product Ad is simply the total volume between the plates. Therefore, the **energy density** (energy per volume) is
given by the following:

$$u_E = \text{electric energy density} = \frac{\text{electric energy}}{\text{volume}} = \frac{1}{2}\varepsilon_0 E^2 \qquad 20\text{–}19$$

This result, though derived for a capacitor, is valid for any electric field, whether
it occurs within a capacitor or anyplace else.

REAL-WORLD PHYSICS: BIO

Capacitor hazards

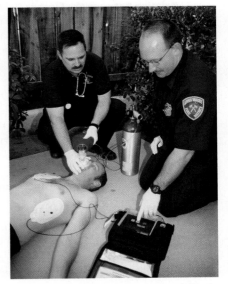

▲ A jolt of electric current from a defibrillator can restore normal heartbeat when
the heart muscle has begun to twitch
irregularly or has stopped beating
altogether. A capacitor is used to store
electricity, discharging it in a burst lasting
only a couple of milliseconds.

THE BIG PICTURE **PUTTING PHYSICS IN CONTEXT**

LOOKING BACK

In Section 20–1 we introduce the concepts of the electric potential and the electric potential energy. The development of these concepts parallels that of the gravitational potential energy for a uniform gravitational field (Chapter 8).

Energy conservation (Chapter 8) is used in Section 20–2, this time with the electric potential energy.

We also return to the idea of equipotential curves—curves along which the potential energy is constant—in direct analogy to the contour maps discussed at the end of Chapter 8.

The electric potential of a point charge is developed in Section 20–3. The results are almost identical to those obtained for the gravitational potential energy of a point mass in Chapter 12.

LOOKING AHEAD

Electrical energy is generalized to direct-current (dc) electric circuits in Chapter 21.

We will also see the important role that capacitors play in dc circuits in Chapter 21, and in alternating-current (ac) circuits in Chapter 24.

The concept of electrical energy is generalized yet again in Chapter 23, where we show how an electric motor can convert electrical energy to mechanical energy. We also show that the reverse process is possible, with a generator converting mechanical energy to electrical energy.

The electric potential energy for a point charge is applied to Bohr's model of the hydrogen atom in Chapter 31. With this energy we can determine the colors of light that hydrogen atoms emit.

CHAPTER SUMMARY

20–1 ELECTRIC POTENTIAL ENERGY AND THE ELECTRIC POTENTIAL

The electric force is conservative, just like the force of gravity. As a result, there is a potential energy U associated with the electric force.

Electric Potential Energy, U
The change in electric potential energy is defined by $\Delta U = -W$, where W is the work done by the electric field.

Electric Potential, V
The change in electric potential is defined to be $\Delta V = \Delta U / q_0$.

Relation Between the Electric Field and the Electric Potential
The electric field is related to the rate of change of the electric potential. In particular, if the electric potential changes by the amount ΔV with a displacement Δs, the electric field in the direction of the displacement is

$$E = -\frac{\Delta V}{\Delta s}$$
20–4

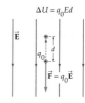

$\Delta U = q_0 E d$

20–2 ENERGY CONSERVATION

Another consequence of the fact that the electric force is conservative is that the total energy of an object is conserved—as long as nonconservative forces like friction can be ignored.

Energy Conservation
As usual, energy conservation can be expressed as follows:

$$\tfrac{1}{2}mv_A^2 + U_A = \tfrac{1}{2}mv_B^2 + U_B$$

In the case of the electric force, the potential energy is $U = q_0 V$.

Direction of Acceleration
Positive charges accelerate in the direction of decreasing electric potential; negative charges accelerate in the direction of increasing electric potential.

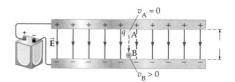

20–3 THE ELECTRIC POTENTIAL OF POINT CHARGES

If we define the electric potential of a point charge q to be zero at an infinite distance from the charge, the electric potential at a distance r is

$$V = \frac{kq}{r}$$
20–7

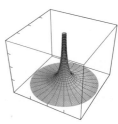

Electric potential of
a positive point charge

Electric Potential Energy

We define the electric potential energy of two charges, q_0 and q, to be zero when the separation between them is infinite. When the charges are separated by a distance r, the potential energy of the system is

$$U = \frac{kq_0q}{r} \qquad \text{20–8}$$

Superposition

The electric potential of two or more point charges is simply the algebraic sum of the potentials due to each charge separately.

The total electric potential energy of two or more point charges is the sum of the potential energies due to each pair of charges.

20–4 EQUIPOTENTIAL SURFACES AND THE ELECTRIC FIELD

Equipotential surfaces are defined as surfaces on which the electric potential is constant. Different equipotential surfaces correspond to different values of the potential.

Electric Field

The electric field is always perpendicular to the equipotential surfaces, and it points in the direction of decreasing electric potential.

Ideal Conductors

Ideal conductors are equipotential surfaces; every point on or within an ideal conductor is at the same potential. The electric field, therefore, is perpendicular to the surface of a conductor.

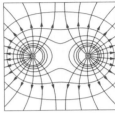

Equipotentials for two
positive charges

20–5 CAPACITORS AND DIELECTRICS

A capacitor is a device that stores electric charge.

Capacitance

Capacitance is defined as the amount of charge Q stored in a capacitor per volt of potential difference V between the plates of the capacitor. Thus,

$$C = \frac{Q}{V} \qquad \text{20–9}$$

Parallel-Plate Capacitor

The capacitance of a parallel-plate capacitor, with plates of area A and separation d, is

$$C = \frac{\varepsilon_0 A}{d} \qquad \text{20–12}$$

Dielectrics

A dielectric is an insulating material that increases the capacitance of a capacitor.

Dielectric Constant

A dielectric is characterized by the dimensionless dielectric constant, κ. In particular, the electric field in a dielectric is reduced by the factor κ, $E = E_0/\kappa$; the potential difference between capacitor plates is decreased by the factor κ, $V = V_0/\kappa$; and the capacitance is increased by the factor κ:

$$C = \kappa C_0 \qquad \text{20–14}$$

Dielectric Breakdown/Dielectric Strength

A large electric field can cause a dielectric material to conduct electricity. This condition is referred to as dielectric breakdown. The strength of electric field required for dielectric breakdown is called the dielectric strength of the material.

20–6 ELECTRICAL ENERGY STORAGE

A capacitor, in addition to storing charge, also stores electrical energy.

Energy Stored in a Capacitor

The electrical energy stored in a capacitor can be expressed as follows:

$$U = \tfrac{1}{2}QV = \tfrac{1}{2}CV^2 = Q^2/2C \qquad \text{20–16, 17, 18}$$

Electric Energy Density of an Electric Field

Electric energy can be thought of as stored in the electric field. The electrical energy per volume, referred to as the electric energy density, is given by the following relation:

$$u_E = \text{electric energy density} = \tfrac{1}{2}\varepsilon_0 E^2 \qquad\qquad 20\text{–}19$$

PROBLEM-SOLVING SUMMARY

Type of Problem	Relevant Physical Concepts	Related Examples
Find the electric field corresponding to a change in electric potential.	The electric field is related to the change of electric potential with distance. The precise relation is $E = -\Delta V/\Delta s$.	Example 20–1 Active Example 20–1
Find the kinetic energy or speed of a particle moving in an electric field.	Apply energy conservation, including the electric potential energy, $U = qV$.	Examples 20–2, 20–4 Active Example 20–2
Calculate the electric potential due to a system of point charges.	The electric potential of a single point charge q at a distance r is $V = kq/r$. For a system of point charges, the total electric potential is the algebraic sum of the potentials calculated for each charge separately.	Examples 20–3, 20–4 Active Example 20–3
Determine the charge on the plates of a capacitor, or the potential difference between the plates.	The charge Q and potential difference V are related to the capacitance C by the expression $C = Q/V$.	Examples 20–5, 20–6
Determine the amount of energy stored in a capacitor.	The energy stored in a capacitor is given by three equivalent expressions: $U = \tfrac{1}{2}QV$; $U = \tfrac{1}{2}CV^2$; $U = Q^2/2C$.	Example 20–7

CONCEPTUAL QUESTIONS

For instructor-assigned homework, go to www.masteringphysics.com

(Answers to odd-numbered Conceptual Questions can be found in the back of the book.)

1. In one region of space the electric potential has a positive constant value. In another region of space the potential has a negative constant value. What can be said about the electric field within each of these two regions of space?

2. Two like charges a distance r apart have a positive electric potential energy. Conversely, two unlike charges a distance r apart have a negative electric potential energy. Explain the physical significance of these observations.

3. If the electric field is zero in some region of space is the electric potential zero there as well? Explain.

4. Sketch the equipotential surface that goes through point 1 in **Figure 20–19**. Repeat for point 2 and for point 3.

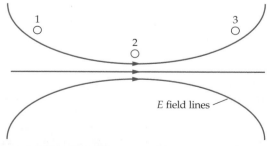

▲ **FIGURE 20–19** Conceptual Question 4 and Problems 43 and 44

5. How much work is required to move a charge from one location on an equipotential to another point on the same equipotential? Explain.

6. It is known that the electric potential is constant on a given two-dimensional surface. What can be said about the electric field on this surface?

7. Explain why equipotentials are always perpendicular to the electric field.

8. Two charges are at locations that have the same value of the electric potential. Is the electric potential energy the same for these charges? Explain.

9. A capacitor is connected to a battery and fully charged. What becomes of the charge on the capacitor when it is disconnected from the battery? What becomes of the charge when the two terminals of the capacitor are connected to one another?

10. It would be unwise to unplug a television set, take off the back, and reach inside. The reason for the danger is that if you happen to touch the terminals of a high-voltage capacitor you could receive a large electrical shock—even though the set is unplugged. Why?

11. On which of the following quantities does the capacitance of a capacitor depend: **(a)** the charge on the plates; **(b)** the separation of the plates; **(c)** the voltage difference between the plates; **(d)** the electric field between the plates; or **(e)** the area of the plates?

12. We say that a capacitor stores charge, yet the total charge in a capacitor is zero; that is, $Q + (-Q) = 0$. In what sense does a capacitor store charge if the net charge within it is zero?

13. The plates of a particular parallel-plate capacitor are uncharged. Is the capacitance of this capacitor zero? Explain.

PROBLEMS AND CONCEPTUAL EXERCISES

Note: Answers to odd-numbered Problems and Conceptual Exercises can be found in the back of the book. **IP** *denotes an integrated problem, with both conceptual and numerical parts;* **BIO** *identifies problems of biological or medical interest;* **CE** *indicates a conceptual exercise.* **Predict/Explain** *problems ask for two responses:* **(a)** *your prediction of a physical outcome, and* **(b)** *the best explanation among three provided. On all problems, red bullets (•, ••, •••) are used to indicate the level of difficulty.*

SECTION 20–1 ELECTRIC POTENTIAL ENERGY AND THE ELECTRIC POTENTIAL

1. • **CE** An electron is released from rest in a region of space with a nonzero electric field. As the electron moves, does it experience an increasing or decreasing electric potential? Explain.

2. • A uniform electric field of magnitude 4.1×10^5 N/C points in the positive x direction. Find the change in electric potential energy of a 4.5-μC charge as it moves from the origin to the points **(a)** (0, 6.0 m); **(b)** (6.0 m, 0); and **(c)** (6.0 m, 6.0 m).

3. • A uniform electric field of magnitude 6.8×10^5 N/C points in the positive x direction. Find the change in electric potential between the origin and the points **(a)** (0, 6.0 m); **(b)** (6.0 m, 0); and **(c)** (6.0 m, 6.0 m).

4. • **BIO Electric Potential Across a Cell Membrane** In a typical living cell, the electric potential inside the cell is 0.070 V lower than the electric potential outside the cell. The thickness of the cell membrane is 0.10 μm. What are the magnitude and direction of the electric field within the cell membrane?

5. • A computer monitor accelerates electrons and directs them to the screen in order to create an image. If the accelerating plates are 1.05 cm apart, and have a potential difference of 25,500 V, what is the magnitude of the uniform electric field between them?

6. • Find the change in electric potential energy for an electron that moves from one accelerating plate to the other in the computer monitor described in the previous problem.

7. • A parallel-plate capacitor has plates separated by 0.75 mm. If the electric field between the plates has a magnitude of **(a)** 1.2×10^5 V/m or **(b)** 2.4×10^4 N/C, what is the potential difference between the plates?

8. • When an ion accelerates through a potential difference of 2140 V, its electric potential energy decreases by 1.37×10^{-15} J. What is the charge on the ion?

9. • **The Electric Potential of the Earth** The Earth has a vertical electric field with a magnitude of approximately 100 V/m near its surface. What is the magnitude of the potential difference between a point on the ground and a point on the same level as the top of the Washington Monument (555 ft high)?

10. •• A uniform electric field with a magnitude of 6350 N/C points in the positive x direction. Find the change in electric potential energy when a +12.5-μC charge is moved 5.50 cm in **(a)** the positive x direction, **(b)** the negative x direction, and **(c)** the positive y direction.

11. ••**IP** A spark plug in a car has electrodes separated by a gap of 0.025 in. To create a spark and ignite the air-fuel mixture in the engine, an electric field of 3.0×10^6 V/m is required in the gap. **(a)** What potential difference must be applied to the spark plug to initiate a spark? **(b)** If the separation between electrodes is increased, does the required potential difference increase, decrease, or stay the same? Explain. **(c)** Find the potential difference for a separation of 0.050 in.

12. •• A uniform electric field with a magnitude of 1200 N/C points in the negative x direction, as shown in **Figure 20–20**.

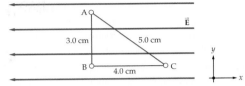

▲ **FIGURE 20–20** Problems 12 and 21

(a) What is the difference in electric potential, $\Delta V = V_B - V_A$, between points A and B? **(b)** What is the difference in electric potential, $\Delta V = V_B - V_C$, between points B and C? **(c)** What is the difference in electric potential, $\Delta V = V_C - V_A$, between points C and A? **(d)** From the information given in this problem, is it possible to determine the value of the electric potential at point A? If so, determine V_A; if not, explain why.

13. •• **A Charged Battery** A typical 12-V car battery can deliver 7.5×10^5 C of charge. If the energy supplied by the battery could be converted entirely to kinetic energy, what speed would it give to a 1400-kg car that is initially at rest?

14. •• **IP BIO The Sodium Pump** Living cells actively "pump" positive sodium ions (Na^+) from inside the cell to outside the cell. This process is referred to as pumping because work must be done on the ions to move them from the negatively charged inner surface of the membrane to the positively charged outer surface. Given that the electric potential is 0.070 V higher outside the cell than inside the cell, and that the cell membrane is 0.10 μm thick, **(a)** calculate the work that must be done (in joules) to move one sodium ion from inside the cell to outside. **(b)** If the thickness of the cell membrane is increased, does your answer to part (a) increase, decrease, or stay the same? Explain. (It is estimated that as much as 20% of the energy we consume in a resting state is used in operating this "sodium pump.")

15. •• **IP** The electric potential of a system as a function of position along the x axis is given in **Figure 20–21**. **(a)** In which of the regions, 1, 2, 3, or 4, do you expect E_x to be greatest? In which region does E_x have its greatest magnitude? Explain. **(b)** Calculate the value of E_x in each of the regions, 1, 2, 3, and 4.

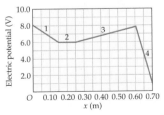

▲ **FIGURE 20–21** Problems 15 and 94

16. ••• Points A and B have electric potentials of 332 V and 149 V, respectively. When an electron released from rest at point A arrives at point C, its kinetic energy is K_A. When the electron is released from rest at point B, however, its kinetic energy when it reaches point C is $K_B = 2K_A$. What are **(a)** the electric potential at point C and **(b)** the kinetic energy K_A?

SECTION 20–2 ENERGY CONSERVATION

17. • **CE Predict/Explain** An electron is released from rest in a region of space with a nonzero electric field. **(a)** As the electron

moves, does the electric potential energy of the system increase, decrease, or stay the same? **(b)** Choose the *best explanation* from among the following:

  I. Because the electron has a negative charge its electric potential energy doesn't decrease, as one might expect, but increases instead.

  II. As the electron begins to move, its kinetic energy increases. The increase in kinetic energy is equal to the decrease in the electric potential energy of the system.

  III. The electron will move perpendicular to the electric field, and hence its electric potential energy will remain the same.

18. • Calculate the speed of **(a)** a proton and **(b)** an electron after each particle accelerates from rest through a potential difference of 275 V.

19. • The electrons in a TV picture tube are accelerated from rest through a potential difference of 25 kV. What is the speed of the electrons after they have been accelerated by this potential difference?

20. • Find the potential difference required to accelerate protons from rest to 10% of the speed of light. (At this point, relativistic effects start to become significant.)

21. •• **IP** A particle with a mass of 3.8 g and a charge of $+0.045 \ \mu C$ is released from rest at point A in Figure 20–20. **(a)** In which direction will this charge move? **(b)** What speed will it have after moving through a distance of 5.0 cm? The electric field has a magnitude of 1200 N/C. **(c)** Suppose the particle continues moving for another 5.0 cm. Will its increase in speed for the second 5.0 cm be greater than, less than, or equal to its increase in speed in the first 5.0 cm? Explain.

22. •• A proton has an initial speed of 4.0×10^5 m/s. **(a)** What potential difference is required to bring the proton to rest? **(b)** What potential difference is required to reduce the initial speed of the proton by a factor of 2? **(c)** What potential difference is required to reduce the initial kinetic energy of the proton by a factor of 2?

SECTION 20–3 THE ELECTRIC POTENTIAL OF POINT CHARGES

23. • In **Figure 20–22**, it is given that $q_1 = +Q$. **(a)** What value must q_2 have if the electric potential at point A is to be zero? **(b)** With the value for q_2 found in part (a), is the electric potential at point B positive, negative, or zero? Explain.

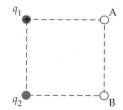

▲ **FIGURE 20–22** Problems 23, 24, and 25

24. • **CE** The charge q_1 in Figure 20–22 has the value $+Q$. **(a)** What value must q_2 have if the electric potential at point B is to be zero? **(b)** With the value for q_2 found in part (a), is the electric potential at point A positive, negative, or zero? Explain.

25. • **CE** It is given that the electric potential is zero at the center of the square in Figure 20–22. **(a)** If $q_1 = +Q$, what is the value of the charge q_2? **(b)** Is the electric potential at point A positive, negative, or zero? Explain. **(c)** Is the electric potential at point B positive, negative, or zero? Explain.

26. • The electric potential 1.1 m from a point charge q is 2.8×10^4 V. What is the value of q?

27. • A point charge of $-7.2 \ \mu C$ is at the origin. What is the electric potential at **(a)** (3.0 m, 0); **(b)** $(-3.0$ m, 0); and **(c)** (3.0 m, -3.0 m)?

28. • **The Bohr Atom** The hydrogen atom consists of one electron and one proton. In the Bohr model of the hydrogen atom, the electron orbits the proton in a circular orbit of radius 0.529×10^{-10} m. What is the electric potential due to the proton at the electron's orbit?

29. • How far must the point charges $q_1 = +7.22 \ \mu C$ and $q_2 = -26.1 \ \mu C$ be separated for the electric potential energy of the system to be -126 J?

30. •• **CE** Four different arrangements of point charges are shown in **Figure 20–23**. In each case the charges are the same distance from the origin. Rank the four arrangements in order of increasing electric potential at the origin, taking the potential at infinity to be zero. Indicate ties where appropriate.

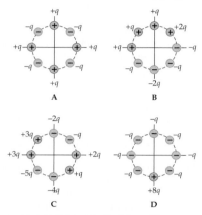

▲ **FIGURE 20–23** Problem 30

31. •• **IP** Point charges $+4.1 \ \mu C$ and $-2.2 \ \mu C$ are placed on the x axis at (11 m, 0) and $(-11$ m, 0), respectively. **(a)** Sketch the electric potential on the x axis for this system. **(b)** Your sketch should show one point on the x axis between the two charges where the potential vanishes. Is this point closer to the $+4.1$-μC charge or closer to the -2.2-μC charge? Explain. **(c)** Find the point referred to in part (b).

32. •• **IP (a)** In the previous problem, find the point to the left of the negative charge where the electric potential vanishes. **(b)** Is the electric field at the point found in part (a) positive, negative, or zero? Explain.

33. •• A dipole is formed by point charges $+3.6 \ \mu C$ and $-3.6 \ \mu C$ placed on the x axis at (0.25 m, 0) and $(-0.25$ m, 0), respectively. **(a)** Sketch the electric potential on the x axis for this system. **(b)** At what positions on the x axis does the potential have the value 7.5×10^5 V?

34. •• A charge of $3.05 \ \mu C$ is held fixed at the origin. A second charge of $3.05 \ \mu C$ is released from rest at the position (1.25 m, 0.570 m). **(a)** If the mass of the second charge is 2.16 g, what is its speed when it moves infinitely far from the origin? **(b)** At what distance from the origin does the second charge attain half the speed it will have at infinity?

35. •• **IP** A charge of $20.2 \ \mu C$ is held fixed at the origin. **(a)** If a -5.25-μC charge with a mass of 3.20 g is released from rest at the position (0.925 m, 1.17 m), what is its speed when it is halfway to the origin? **(b)** Suppose the -5.25-μC charge is released from rest at the point $x = \frac{1}{2}(0.925$ m$)$ and $y = \frac{1}{2}(1.17$ m$)$. When it is halfway to the origin, is its speed greater than, less than, or equal to the speed found in part (a)? Explain. **(c)** Find the speed of the charge for the situation described in part (b).

36. •• A charge of -2.205 μC is located at (3.055 m, 4.501 m), and a charge of 1.800 μC is located at $(-2.533$ m, 0). **(a)** Find the electric potential at the origin. **(b)** There is one point on the line connecting these two charges where the potential is zero. Find this point.

37. •• **IP Figure 20–24** shows three charges at the corners of a rectangle. **(a)** How much work must be done to move the $+2.7$-μC charge to infinity? **(b)** Suppose, instead, that we move the -6.1-μC charge to infinity. Is the work required in this case greater than, less than, or the same as when we moved the $+2.7$-μC charge to infinity? Explain. **(c)** Calculate the work needed to move the -6.1-μC charge to infinity.

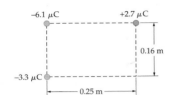

▲ **FIGURE 20–24** Problems 37 and 38

38. •• How much work must be done to move the three charges in Figure 20–24 infinitely far from one another?

39. •• **(a)** Find the electric potential at point P in **Figure 20–25**. **(b)** Suppose the three charges shown in Figure 20–25 are held in place. A fourth charge, with a charge of $+6.11$ μC and a mass of 4.71 g, is released from rest at point P. What is the speed of the fourth charge when it has moved infinitely far away from the other three charges?

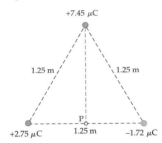

▲ **FIGURE 20–25** Problems 39 and 91

40. ••• A square of side a has a charge $+Q$ at each corner. What is the electric potential energy of this system of charges?

41. ••• A square of side a has charges $+Q$ and $-Q$ alternating from one corner to the next, as shown in **Figure 20–26**. Find the electric potential energy for this system of charges.

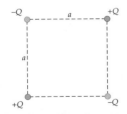

▲ **FIGURE 20–26** Problems 41 and 100

SECTION 20–4   EQUIPOTENTIAL SURFACES AND THE ELECTRIC FIELD

42. • **CE Predict/Explain** A positive charge is moved from one location on an equipotential to another point on the same equipotential. **(a)** Is the work done on the charge positive, negative, or zero? **(b)** Choose the *best explanation* from among the following:
  I. The electric field is perpendicular to an equipotential, therefore the work done in moving along an equipotential is zero.

  II. Because the charge is positive the work done on it is also positive.
  III. It takes negative work to keep the positive charge from accelerating as it moves along the equipotential.

43. • **CE Predict/Explain (a)** Is the electric potential at point 1 in **Figure 20–19** greater than, less than, or equal to the electric potential at point 3? **(b)** Choose the *best explanation* from among the following:
  I. The electric field lines point to the right, indicating that the electric potential is greater at point 3 than at point 1.
  II. The value of the electric potential is large where the electric field lines are close together, and small where they are widely spaced. Therefore, the electric potential is the same at points 1 and 3.
  III. The electric potential decreases as we move in the direction of the electric field, as shown in Figure 20–3. Therefore, the electric potential is greater at point 1 than at point 3.

44. • **CE Predict/Explain** Imagine sketching a large number of equipotential surfaces in Figure 20–19, with a constant difference in electric potential between adjacent surfaces. **(a)** Would the equipotentials at point 2 be more closely spaced, be less closely spaced, or have the same spacing as equipotentials at point 1? **(b)** Choose the *best explanation* from among the following:
  I. When electric field lines are close together, the corresponding equipotentials are far apart.
  II. Equipotential surfaces, by definition, always have equal spacing between them.
  III. The electric field is more intense at point 2 than at point 1, which means the equipotential surfaces are more closely spaced in that region.

45. • Two point charges are on the x axis. Charge 1 is $+q$ and is located at $x = -1.0$ m; charge 2 is $-2q$ and is located at $x = 1.0$ m. Make sketches of the equipotential surfaces for this system **(a)** out to a distance of about 2.0 m from the origin and **(b)** far from the origin. In each case, indicate the direction in which the potential increases.

46. • Two point charges are on the x axis. Charge 1 is $+q$ and is located at $x = -1.0$ m; charge 2 is $+2q$ and is located at $x = 1.0$ m. Make sketches of the equipotential surfaces for this system **(a)** out to a distance of about 2.0 m from the origin and **(b)** far from the origin. In each case, indicate the direction in which the potential increases.

47. •• **CE Figure 20–27** shows a series of equipotentials in a particular region of space, and five different paths along which an electron is moved. **(a)** Does the electric field in this region point to the right, to the left, up, or down? Explain. **(b)** For each path, indicate whether the work done on the electron by the electric field is positive, negative, or zero. **(c)** Rank the paths in order of increasing amount of work done on the electron by the electric field. Indicate ties where appropriate. **(d)** Is the electric field near path A greater than, less than, or equal to the electric field near path E? Explain.

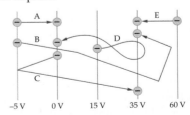

▲ **FIGURE 20–27** Problem 47

48. •• **IP** Consider a region in space where a uniform electric field $E = 6500$ N/C points in the negative x direction. **(a)** What is the orientation of the equipotential surfaces? Explain. **(b)** If you move in the positive x direction, does the electric potential increase or decrease? Explain. **(c)** What is the distance between the +14-V and the +16-V equipotentials?

49. •• A given system has the equipotential surfaces shown in **Figure 20–28**. **(a)** What are the magnitude and direction of the electric field? **(b)** What is the shortest distance one can move to undergo a change in potential of 5.00 V?

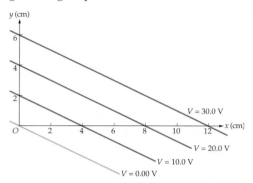

▲ **FIGURE 20–28** Problems 49 and 92

SECTION 20–5 CAPACITORS AND DIELECTRICS

50. • A 0.40-μF capacitor is connected to a 9.0-V battery. How much charge is on each plate of the capacitor?

51. • It is desired that 5.8 μC of charge be stored on each plate of a 3.2-μF capacitor. What potential difference is required between the plates?

52. • To operate a given flash lamp requires a charge of 32 μC. What capacitance is needed to store this much charge in a capacitor with a potential difference between its plates of 9.0 V?

53. •• A parallel-plate capacitor is made from two aluminum-foil sheets, each 6.3 cm wide and 5.4 m long. Between the sheets is a Teflon strip of the same width and length that is 0.035 mm thick. What is the capacitance of this capacitor? (The dielectric constant of Teflon is 2.1.)

54. •• A parallel-plate capacitor is constructed with circular plates of radius 0.056 m. The plates are separated by 0.25 mm, and the space between the plates is filled with a dielectric with dielectric constant κ. When the charge on the capacitor is 1.2 μC the potential difference between the plates is 750 V. Find the value of the dielectric constant, κ.

55. •• **IP** A parallel-plate capacitor has plates with an area of 0.012 m^2 and a separation of 0.88 mm. The space between the plates is filled with a dielectric whose dielectric constant is 2.0. **(a)** What is the potential difference between the plates when the charge on the capacitor plates is 4.7 μC? **(b)** Will your answer to part (a) increase, decrease, or stay the same if the dielectric constant is increased? Explain. **(c)** Calculate the potential difference for the case where the dielectric constant is 4.0.

56. •• **IP** Consider a parallel-plate capacitor constructed from two circular metal plates of radius R. The plates are separated by a distance of 1.5 mm. **(a)** What radius must the plates have if the capacitance of this capacitor is to be 1.0 μF? **(b)** If the separation between the plates is increased, should the radius of the plates be increased or decreased to maintain a capacitance of 1.0 μF? Explain. **(c)** Find the radius of the plates that gives a capacitance of 1.0 μF for a plate separation of 3.0 mm.

57. •• A parallel-plate capacitor has plates of area 3.45 $\times$ 10^{-4} m^2. What plate separation is required if the capacitance is to be 1630 pF? Assume that the space between the plates is filled with **(a)** air or **(b)** paper.

58. •• **IP** A parallel-plate capacitor filled with air has plates of area 0.0066 m^2 and a separation of 0.45 mm. **(a)** Find the magnitude of the charge on each plate when the capacitor is connected to a 12-V battery. **(b)** Will your answer to part (a) increase, decrease, or stay the same if the separation between the plates is increased? Explain. **(c)** Calculate the magnitude of the charge on the plates if the separation is 0.90 mm.

59. •• Suppose that after walking across a carpeted floor you reach for a doorknob and just before you touch it a spark jumps 0.50 cm from your finger to the knob. Find the minimum voltage needed between your finger and the doorknob to generate this spark.

60. •• **(a)** What plate area is required if an air-filled, parallel-plate capacitor with a plate separation of 2.6 mm is to have a capacitance of 22 pF? **(b)** What is the maximum voltage that can be applied to this capacitor without causing dielectric breakdown?

61. •• **Lightning** As a crude model for lightning, consider the ground to be one plate of a parallel-plate capacitor and a cloud at an altitude of 550 m to be the other plate. Assume the surface area of the cloud to be the same as the area of a square that is 0.50 km on a side. **(a)** What is the capacitance of this capacitor? **(b)** How much charge can the cloud hold before the dielectric strength of the air is exceeded and a spark (lightning) results?

62. ••• A parallel-plate capacitor is made from two aluminum-foil sheets, each 3.00 cm wide and 10.0 m long. Between the sheets is a mica strip of the same width and length that is 0.0225 mm thick. What is the maximum charge that can be stored in this capacitor? (The dielectric constant of mica is 5.4, and its dielectric strength is 1.00 $\times$ 10^8 V/m.)

SECTION 20–6 ELECTRICAL ENERGY STORAGE

63. • Calculate the work done by a 3.0-V battery as it charges a 7.8-μF capacitor in the flash unit of a camera.

64. • **BIO Defibrillator** An automatic external defibrillator (AED) delivers 125 J of energy at a voltage of 1050 V. What is the capacitance of this device?

65. •• **IP BIO Cell Membranes** The membrane of a living cell can be approximated by a parallel-plate capacitor with plates of area 4.75 $\times$ 10^{-9} m^2, a plate separation of 8.5 $\times$ 10^{-9} m, and a dielectric with a dielectric constant of 4.5. **(a)** What is the energy stored in such a cell membrane if the potential difference across it is 0.0725 V? **(b)** Would your answer to part (a) increase, decrease, or stay the same if the thickness of the cell membrane is increased? Explain.

66. •• A 0.22-μF capacitor is charged by a 1.5-V battery. After being charged, the capacitor is connected to a small electric motor. Assuming 100% efficiency, **(a)** to what height can the motor lift a 5.0-g mass? **(b)** What initial voltage must the capacitor have if it is to lift a 5.0-g mass through a height of 1.0 cm?

67. •• Find the electric energy density between the plates of a 225-μF parallel-plate capacitor. The potential difference between the plates is 345 V, and the plate separation is 0.223 mm.

68. •• What electric field strength would store 17.5 J of energy in every 1.00 mm^3 of space?

69. •• An electronic flash unit for a camera contains a capacitor with a capacitance of 890 μF. When the unit is fully charged and ready for operation, the potential difference between the capacitor plates is 330 V. **(a)** What is the magnitude of the charge on each plate of the fully charged capacitor? **(b)** Find the energy stored in the "charged-up" flash unit.

70. ••• A parallel-plate capacitor has plates with an area of 405 cm² and an air-filled gap between the plates that is 2.25 mm thick. The capacitor is charged by a battery to 575 V and then is disconnected from the battery. (a) How much energy is stored in the capacitor? (b) The separation between the plates is now increased to 4.50 mm. How much energy is stored in the capacitor now? (c) How much work is required to increase the separation of the plates from 2.25 mm to 4.50 mm? Explain your reasoning.

GENERAL PROBLEMS

71. • **CE** A proton is released from rest in a region of space with a nonzero electric field. As the proton moves, does it experience an increasing or decreasing electric potential? Explain.

72. • **CE Predict/Explain** A proton is released from rest in a region of space with a nonzero electric field. (a) As the proton moves, does the electric potential energy of the system increase, decrease, or stay the same? (b) Choose the *best explanation* from among the following:
   I. As the proton begins to move, its kinetic energy increases. The increase in kinetic energy is equal to the decrease in the electric potential energy of the system.
   II. Because the proton has a positive charge, its electric potential energy will always increase.
   III. The proton will move perpendicular to the electric field, and hence its electric potential energy will remain the same.

73. • **CE** In the Bohr model of the hydrogen atom, a proton and an electron are separated by a constant distance r. (a) Would the electric potential energy of the system increase, decrease, or stay the same if the electron is replaced with a proton? Explain. (b) Suppose, instead, that the proton is replaced with an electron. Would the electric potential energy of the system increase, decrease, or stay the same? Explain.

74. • **CE** The plates of a parallel-plate capacitor have constant charges of +Q and −Q. Do the following quantities increase, decrease, or remain the same as the separation of the plates is increased? (a) The electric field between the plates; (b) the potential difference between the plates; (c) the capacitance; (d) the energy stored in the capacitor.

75. • **CE** A parallel-plate capacitor is connected to a battery that maintains a constant potential difference V between the plates. If the plates of the capacitor are pulled farther apart, do the following quantities increase, decrease, or remain the same? (a) The electric field between the plates; (b) the charge on the plates; (c) the capacitance; (d) the energy stored in the capacitor.

76. • **CE** The plates of a parallel-plate capacitor have constant charges of +Q and −Q. Do the following quantities increase, decrease, or remain the same as a dielectric is inserted between the plates? (a) The electric field between the plates; (b) the potential difference between the plates; (c) the capacitance; (d) the energy stored in the capacitor.

77. • **CE** A parallel-plate capacitor is connected to a battery that maintains a constant potential difference V between the plates. If a dielectric is inserted between the plates of the capacitor, do the following quantities increase, decrease, or remain the same? (a) The electric field between the plates; (b) the charge on the plates; (c) the capacitance; (d) the energy stored in the capacitor.

78. • Find the difference in electric potential, $\Delta V = V_B - V_A$, between the points A and B for the following cases: (a) The electric field does 0.052 J of work as you move a +5.7-μC charge from A to B. (b) The electric field does −0.052 J of work as you move a −5.7-μC charge from A to B. (c) You perform 0.052 J of work as you slowly move a +5.7-μC charge from A to B.

79. • The separation between the plates of a parallel-plate capacitor is doubled and the area of the plates is halved. How is the capacitance affected?

80. • A parallel-plate capacitor is connected to a battery that maintains a constant potential difference between the plates. If the spacing between the plates is doubled, how is the magnitude of charge on the plates affected?

81. •• **CE** Two point charges are placed on the x axis. The charge +2q is at x = 1.5 m, and the charge −q is at x = −1.5 m. (a) There is a point on the x axis between the two charges where the electric potential is zero. Where is this point? (b) The electric potential also vanishes at a point in one of the following regions: region 1, x between 1.5 m and 5.0 m; region 2, x between −1.5 m and −3.0 m; region 3, x between −3.5 m and −5.0 m. Identify the appropriate region. (c) Find the value of x referred to in part (b).

82. •• A charge of 24.5 μC is located at (4.40 m, 6.22 m), and a charge of −11.2 μC is located at (−4.50 m, 6.75 m). What charge must be located at (2.23 m, −3.31 m) if the electric potential is to be zero at the origin?

83. •• **The Bohr Model** In the Bohr model of the hydrogen atom (see Problem 28) what is the smallest amount of work that must be done on the electron to move it from its circular orbit, with a radius of 0.529×10^{-10} m, to an infinite distance from the proton? This value is referred to as the ionization energy of hydrogen.

84. ••**IP** A +1.2-μC charge and a −1.2-μC charge are placed at (0.50 m, 0) and (−0.50 m, 0), respectively. (a) In **Figure 20–29**, at which of the points A, B, C, or D is the electric potential smallest in value? At which of these points does it have its greatest value? Explain. (b) Calculate the electric potential at points A, B, C, and D.

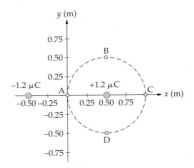

▲ **FIGURE 20–29** Problems 84, 85, and 103

85. •• Repeat Problem 84 for the case where both charges are +1.2 μC.

86. •• How much work is required to bring three protons, initially infinitely far apart, to a configuration where each proton is 1.5×10^{-15} m from the other two? (This is a typical separation for protons in a nucleus.)

87. •• A point charge Q = +87.1 μC is held fixed at the origin. A second point charge, with mass m = 0.0576 kg and charge q = −2.87 μC, is placed at the location (0.323 m, 0). (a) Find the electric potential energy of this system of charges. (b) If the second charge is released from rest, what is its speed when it reaches the point (0.121 m, 0)?

88. •• **Electron Escape Speed** An electron is at rest just above the surface of a sphere with a radius of 2.7 mm and a uniformly distributed positive charge of 1.8×10^{-15} C. Like a rocket blasting off from the Earth, the electron is given an initial speed v_e radially

outward from the sphere. If the electron coasts to infinity, where its kinetic energy drops to zero, what is the escape speed, v_e?

89. •• **Quark Model of the Neutron** According to the quark model of fundamental particles, neutrons—the neutral particles in an atom's nucleus—are composed of three quarks. Two of these quarks are "down" quarks, each with a charge of $-e/3$; the third quark is an "up" quark, with a charge of $+2e/3$. This gives the neutron a net charge of zero. What is the electric potential energy of these three quarks, assuming they are equidistant from one another, with a separation distance of 1.3×10^{-15} m? (Quarks are discussed in Chapter 32.)

90. •• A parallel-plate capacitor is charged to an electric potential of 325 V by moving 3.75×10^{16} electrons from one plate to the other. How much work is done in charging the capacitor?

91. •• **IP** The three charges shown in Figure 20–25 are held in place as a fourth charge, q, is brought from infinity to the point P. The charge q starts at rest at infinity and is also at rest when it is placed at the point P. **(a)** If q is a positive charge, is the work required to bring it to the point P positive, negative, or zero? Explain. **(b)** Find the value of q if the work needed to bring it to point P is -1.3×10^{-11} J.

92. •• **(a)** In Figure 20–28 we see that the electric potential increases by 10.0 V as one moves 4.00 cm in the positive x direction. Use this information to calculate the x component of the electric field. (Ignore the y direction for the moment.) **(b)** Apply the same reasoning as in part (a) to calculate the y component of the electric field. **(c)** Combine the results from parts (a) and (b) to find the magnitude and direction of the electric field for this system.

93. •• **IP BIO** **Electric Catfish** The electric catfish (*Malapterurus electricus*) is an aggressive fish, 1.0 m in length, found today in tropical Africa (and depicted in Egyptian hieroglyphics). The catfish is capable of generating jolts of electricity up to 350 V by producing a positively charged region of muscle near the head and a negatively charged region near the tail. **(a)** For the same amount of charge, can the catfish generate a higher voltage by separating the charge from one end of its body to the other, as it does, or from one side of the body to the other? Explain. **(b)** Estimate the charge generated at each end of a catfish as follows: Treat the catfish as a parallel-plate capacitor with plates of area 1.8×10^{-2} m^2, separation 1.0 m, and filled with a dielectric with a dielectric constant $\kappa = 95$.

94. •• As a $+6.2$-μC charge moves along the x axis from $x = 0$ to $x = 0.70$ m, the electric potential it experiences is shown in Figure 20–21. Find the approximate location(s) of the charge when its electric potential energy is **(a)** 2.6×10^{-5} J and **(b)** 4.3×10^{-5} J.

95. •• **IP** **Computer Keyboards** Many computer keyboards operate on the principle of capacitance. As shown in Figure 20–16, each key forms a small parallel-plate capacitor whose separation is reduced when the key is depressed. **(a)** Does depressing a key increase or decrease its capacitance? Explain. **(b)** Suppose the plates for each key have an area of 47.5 mm^2 and an initial separation of 0.550 mm. In addition, let the dielectric have a dielectric constant of 3.75. If the circuitry of the computer can detect a change in capacitance of 0.425 pF, what is the minimum distance a key must be depressed to be detected?

96. •• **IP** A point charge of mass 0.081 kg and charge $+6.77$ μC is suspended by a thread between the vertical parallel plates of a parallel-plate capacitor, as shown in **Figure 20–30**. **(a)** If the charge deflects to the right of vertical, as indicated in the figure, which of the two plates is at the higher electric potential? **(b)** If the angle of deflection is 22°, and the separation between the plates is 0.025 m, what is the potential difference between the plates?

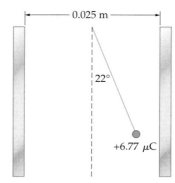

▲ **FIGURE 20–30** Problems 96 and 101

97. •• **BIO** **Cell Membranes and Dielectrics** Many cells in the body have a cell membrane whose inner and outer surfaces carry opposite charges, just like the plates of a parallel-plate capacitor. Suppose a typical cell membrane has a thickness of 8.1×10^{-9} m, and its inner and outer surfaces carry charge densities of -0.58×10^{-3} C/m^2 and $+0.58 \times 10^{-3}$ C/m^2, respectively. In addition, assume that the material in the cell membrane has a dielectric constant of 5.5. **(a)** Find the direction and magnitude of the electric field within the cell membrane. **(b)** Calculate the potential difference between the inner and outer walls of the membrane, and indicate which wall of the membrane has the higher potential.

98. •• Long, long ago, on a planet far, far away, a physics experiment was carried out. First, a 0.250-kg ball with zero net charge was dropped from rest at a height of 1.00 m. The ball landed 0.552 s later. Next, the ball was given a net charge of 7.75 μC and dropped in the same way from the same height. This time the ball fell for 0.680 s before landing. What is the electric potential at a height of 1.00 m above the ground on this planet, given that the electric potential at ground level is zero? (Air resistance can be ignored.)

99. •• **Rutherford's Planetary Model of the Atom** In 1911, Ernest Rutherford developed a planetary model of the atom, in which a small positively charged nucleus is orbited by electrons. The model was motivated by an experiment carried out by Rutherford and his graduate students, Geiger and Marsden. In this experiment, they fired alpha particles with an initial speed of 1.75×10^7 m/s at a thin sheet of gold. (Alpha particles are obtained from certain radioactive decays. They have a charge of $+2e$ and a mass of 6.64×10^{-27} kg.) How close can the alpha particles get to a gold nucleus (charge = $+79e$), assuming the nucleus remains stationary? (This calculation sets an upper limit on the size of the gold nucleus. See Chapter 31 for further details.)

100. ••• **IP** **(a)** One of the $-Q$ charges in Figure 20–26 is given an outward "kick" that sends it off with an initial speed v_0 while the other three charges are held at rest. If the moving charge has a mass m, what is its speed when it is infinitely far from the other charges? **(b)** Suppose the remaining $-Q$ charge, which also has a mass m, is now given the same initial speed, v_0. When it is infinitely far away from the two $+Q$ charges, is its speed greater than, less than, or the same as the speed found in part (a)? Explain.

101. ••• Figure 20–30 shows a charge $q = +6.77$ μC with a mass $m = 0.071$ kg suspended by a thread of length $L = 0.022$ m between the plates of a capacitor. **(a)** Plot the electric potential energy of the system as a function of the angle θ the thread makes with the vertical. (The electric field between the plates has a magnitude $E = 4.16 \times 10^4$ V/m.) **(b)** Repeat part (a) for

the case of the gravitational potential energy of the system. **(c)** Show that the total potential energy of the system (electric plus gravitational) is a minimum when the angle θ satisfies the equilibrium condition for the charge, $\tan \theta = qE/mg$. This relation implies that $\theta = 22°$.

102. ••• The electric potential a distance r from a point charge q is 2.70×10^4 V. One meter farther away from the charge the potential is 6140 V. Find the charge q and the initial distance r.

103. ••• Referring to Problem 84, calculate and plot the electric potential on the circle centered at (0.50 m, 0). Give your results in terms of the angle θ, defined as follows: θ is the angle measured counterclockwise from a vertex at the center of the circle, with $\theta = 0$ at point C.

104. ••• When the potential difference between the plates of a capacitor is increased by 3.25 V, the magnitude of the charge on each plate increases by 13.5 μC. What is the capacitance of this capacitor?

105. ••• The electric potential a distance r from a point charge q is 155 V, and the magnitude of the electric field is 2240 N/C. Find the values of q and r.

PASSAGE PROBLEMS

BIO The Electric Eel

Of the many unique and unusual animals that inhabit the rainforests of South America, including howler monkeys, freshwater dolphins, and deadly piranhas, one stands out because of its mastery of electricity. The electric eel (*Electrophorus electricus*), one of the few creatures on Earth able to generate, store, and discharge electricity, can deliver a powerful series of high-voltage discharges reaching 650 V. These jolts of electricity are so strong, in fact, that electric eels have been known to topple a horse crossing a stream 20 feet away, and to cause respiratory paralysis, cardiac arrhythmia, and even death in humans.

Though similar in appearance to an eel, the electric "eel" is actually more closely related to catfish. They are found primarily in the Amazon and Orinoco river basins, where they navigate the slow-moving, muddy water with low-voltage electric organ discharges (EOD), saving the high-voltage EODs for stunning prey and defending against predators. Obligate air breathers, electric eels obtain about 80% of their oxygen by gulping air at the water's surface. Even so, they are able to attain lengths of 2.5 m and a mass of 20 kg.

The organs that produce the eel's electricity take up most of its body, and consist of thousands of modified muscle cells—called electroplaques—stacked together like the cells in a battery. Each electroplaque is capable of generating a voltage of 0.15 V, and together they produce a positive charge near the head of the eel and a negative charge near its tail.

106. • Electric eels produce an electric field within their body. In which direction does the electric field point?

    A. toward the head   **B.** toward the tail

    C. upward   **D.** downward

107. • As a rough approximation, consider an electric eel to be a parallel-plate capacitor with plates of area 1.8×10^{-2} m^2 separated by 2.0 m and filled with a dielectric whose dielectric constant is $\kappa = 95$. What is the capacitance of the eel in this model?

    A. 8.0×10^{-14} F   **B.** 7.6×10^{-12} F

    C. 1.5×10^{-11} F   **D.** 9.3×10^{-8} F

108. • In terms of the parallel-plate model of the previous problem, how much charge does an electric eel generate at each end of its body when it produces a voltage of 650 V?

    A. 1.2×10^{-14} C   **B.** 5.2×10^{-11} C

    C. 4.9×10^{-9} C   **D.** 6.1×10^{-5} C

109. • How much energy is stored by an electric eel when it is charged up to 650 V. Use the same parallel-plate model discussed in the previous two problems.

    A. 1.8×10^{-17} J   **B.** 1.7×10^{-8} J

    C. 1.6×10^{-6} J   **D.** 2.0×10^{-2} J

INTERACTIVE PROBLEMS

110. •• **IP Referring to Example 20–3** Suppose the charge $-2q$ at $x = 1.00$ m is replaced with a charge $-3q$, where $q = 4.11 \times 10^{-9}$ C. The charge $+q$ is at the origin. **(a)** Is the electric potential positive, negative, or zero at the point $x = 0.333$ m? Explain. **(b)** Find the point between $x = 0$ and $x = 1.00$ m where the electric potential vanishes. **(c)** Is there a point in the region $x < 0$ where the electric potential passes through zero?

111. •• **Referring to Example 20–3** Suppose we can change the location of the charge $-2q$ on the x axis. The charge $+q$ (where $q = 4.11 \times 10^{-9}$C) is still at the origin. **(a)** Where should the charge $-2q$ be placed to ensure that the electric potential vanishes at $x = 0.500$ m? **(b)** With the location of $-2q$ found in part (a), where does the electric potential pass through zero in the region $x < 0$?

112. •• **IP Referring to Example 20–3** Suppose the charge $+q$ at the origin is replaced with a charge $+5q$, where $q = 4.11 \times 10^{-9}$ C. The charge $-2q$ is still at $x = 1.00$ m. **(a)** Is there a point in the region $x < 0$ where the electric potential passes through zero? **(b)** Find the location between $x = 0$ and $x = 1.00$ m where the electric potential passes through zero. **(c)** Find the location in the region $x > 1.00$ m where the electric potential passes through zero.

21 Electric Current and Direct-Current Circuits

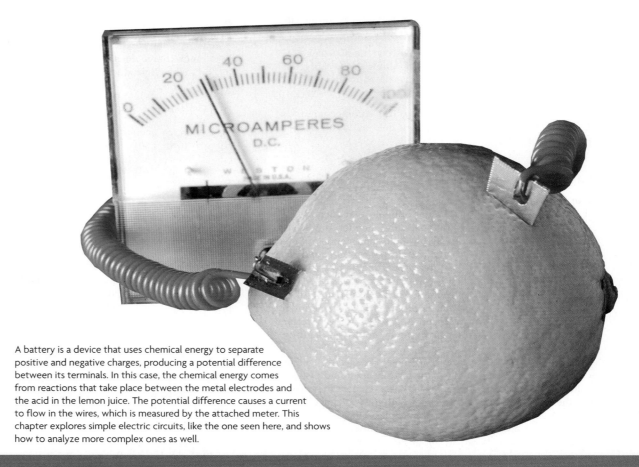

A battery is a device that uses chemical energy to separate positive and negative charges, producing a potential difference between its terminals. In this case, the chemical energy comes from reactions that take place between the metal electrodes and the acid in the lemon juice. The potential difference causes a current to flow in the wires, which is measured by the attached meter. This chapter explores simple electric circuits, like the one seen here, and shows how to analyze more complex ones as well.

As you read this paragraph, your heart is pumping blood through the arteries and veins in your body. In a way, your heart is acting like a battery in an electric circuit: A battery causes electric charge to flow through a closed circuit of wires; your heart causes blood to flow through your body. Just as the flow of blood is important to life, the flow of electric charge is of central importance to modern technology. In this chapter we consider some of the basic properties of moving electric charges, and we apply these results to simple electric circuits.

21–1 Electric Current

A flow of electric charge from one place to another is referred to as an **electric current.** Often, the charge is carried by electrons moving through a metal wire. Though the analogy should not be pushed too far, the electrons flowing through a wire are much like water molecules flowing through a garden hose or blood cells flowing through an artery.

To be specific, suppose a charge ΔQ flows past a given point in a wire in a time Δt. In such a case, we say that the electric current, I, in the wire is:

Definition of Electric Current, *I*

$$I = \frac{\Delta Q}{\Delta t}$$ 21–1

SI unit: coulomb per second, C/s = ampere, A

The unit of current, the ampere (A) or *amp* for short, is named for the French physicist André-Marie Ampère (1775–1836) and is defined simply as 1 coulomb per second:

$$1 \, A = 1 \, C/s$$

The following Example shows that the number of electrons involved in typical electric circuits, with currents of roughly an amp, is extremely large—not unlike the large number of water molecules flowing through a garden hose.

EXAMPLE 21–1   MEGA BLASTER

The disk drive in a portable CD player is connected to a battery that supplies it with a current of 0.22 A. How many electrons pass through the drive in 4.5 s?

PICTURE THE PROBLEM
Our sketch shows the CD drive with a current $I = 0.22$ A flowing through it. Also indicated is the time $\Delta t = 4.5$ s during which the current flows.

STRATEGY
Since we know both the current, I, and the length of time, Δt, we can use the definition of current, $I = \Delta Q/\Delta t$, to find the charge, ΔQ, that flows through the player. Once we know the charge, the number of electrons, N, is simply ΔQ divided by the magnitude of the electron's charge: $N = \Delta Q/e$.

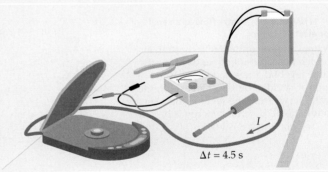

$\Delta t = 4.5$ s

SOLUTION

1. Calculate the charge, ΔQ, that flows through the drive:

$$\Delta Q = I \, \Delta t = (0.22 \, A)(4.5 \, s) = 0.99 \, C$$

2. Divide by the magnitude of the electron's charge, e, to find the number of electrons:

$$N = \frac{\Delta Q}{e} = \frac{0.99 \, C}{1.60 \times 10^{-19} \, C/electron}$$
$$= 6.2 \times 10^{18} \, electrons$$

INSIGHT
Thus, even a modest current flowing for a brief time corresponds to the transport of an extremely large number of electrons.

PRACTICE PROBLEM
How long must this current last if 7.5×10^{18} electrons are to flow through the disk drive?   [**Answer:** 5.5 s]

Some related homework problems: Problem 1, Problem 2

When charge flows through a closed path and returns to its starting point, we refer to the closed path as an *electric circuit.* In this chapter we consider **direct-current circuits,** also known as dc circuits, in which the current always flows in the same direction. Circuits with currents that periodically reverse their direction

▲ Electric currents are not confined to the wires in our houses and machines, but occur in nature as well. A lightning bolt is simply an enormous, brief current. It flows when the difference in electric potential between cloud and ground (or cloud and cloud) becomes so great that it exceeds the breakdown strength of air. An enormous quantity of charge then leaps across the gap in a fraction of a second. Some organisms, such as this electric torpedo ray, have internal organic "batteries" that can produce significant electric potentials. The resulting current is used to stun their prey.

are referred to as **alternating-current circuits.** These AC circuits are considered in detail in Chapter 24.

Batteries and Electromotive Force

Although electrons move rather freely in metal wires, they do not flow unless the wires are connected to a source of electrical energy. A close analogy is provided by water in a garden hose. Imagine that you and a friend each hold one end of a garden hose filled with water. If the two ends are held at the same level, as in **Figure 21–1 (a)**, the water does not flow. If, however, one end is raised above the other, as in **Figure 21–1 (b)**, water flows from the high end—where the gravitational potential energy is high—to the low end.

▶ **FIGURE 21–1 Water flow as an analogy for electric current**

Water can flow quite freely through a garden hose, but if both ends are at the same level **(a)**, there is no flow. If the ends are held at different levels **(b)**, the water flows from the region where the gravitational potential energy is high to the region where it is low.

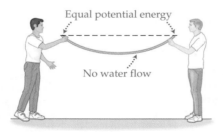

Equal potential energy

No water flow

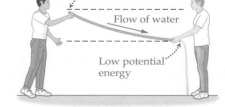

High potential energy

Flow of water

Low potential energy

(a) Equal potential energy → no flow

(b) Water flows from high potential energy to low

A **battery** performs a similar function in an electric circuit. To put it simply, a battery uses chemical reactions to produce a difference in electric potential between its two ends, or **terminals.** The symbol for a battery is ⊣⊢ . The terminal corresponding to a high electric potential is denoted by a +, and the terminal corresponding to a low electric potential is denoted by a −. When the battery is connected to a circuit, electrons move in a closed path from the negative terminal of the battery, through the circuit, and back to the positive terminal.

A simple example of an electrical system is shown in **Figure 21–2 (a)**, where we show a battery, a switch, and a lightbulb as they might be connected in a flashlight. In the schematic circuit shown in **Figure 21–2 (b)**, the switch is "open"—creating an **open circuit**—which means there is no closed path through which the electrons can flow. As a result, the light is off. When the switch is closed—which "closes" the circuit—charge flows around the circuit, causing the light to glow.

A mechanical analog to the flashlight circuit is shown in **Figure 21–3**. In this system, the person raising the water from a low to a high level is analogous to the battery, the paddle wheel is analogous to the lightbulb, and the water is analogous to the electric charge. Notice that the person does work in raising the water; later, as the water falls to its original level, it does work on the external world by turning the paddle wheel.

When a battery is disconnected from a circuit and carries no current, the difference in electric potential between its terminals is referred to as its *electromotive force*, or *emf* ($\mathcal{E}$). It follows that the units of emf are the same as those of electric potential, namely, volts. Clearly, then, the electromotive force is not really a force at all. Instead, the emf determines the amount of work a battery does to move a certain amount of charge around a circuit (like the person lifting water in Figure 21–3). To be specific, the magnitude of the work done by a battery of emf $\mathcal{E}$ as a charge ΔQ moves from one of its terminals to the other is given by Equation 20–2:

$$W = \Delta Q \mathcal{E}$$

We apply this relation to a flashlight circuit in the following Active Example.

(a) A simple flashlight

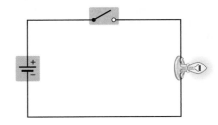

(b) Circuit diagram for flashlight

▲ **FIGURE 21–2** **The flashlight: A simple electric circuit**
(a) A simple flashlight, consisting of a battery, a switch, and a lightbulb. **(b)** When the switch is in the open position, the circuit is "broken," and no charge can flow. When the switch is closed, electrons flow through the circuit and the light glows.

ACTIVE EXAMPLE 21–1 OPERATING A FLASHLIGHT: FIND THE CHARGE AND THE WORK

A battery with an emf of 1.5 V delivers a current of 0.44 A to a flashlight bulb for 64 s (see Figure 21–2). Find **(a)** the charge that passes through the circuit and **(b)** the work done by the battery.

SOLUTION *(Test your understanding by performing the calculations indicated in each step.)*

Part (a)

1. Use the definition of current, $I = \Delta Q/\Delta t$, to find the charge that flows through the circuit: $\Delta Q = 28$ C

Part (b)

2. Once we know ΔQ, we can use $W = \Delta Q \mathcal{E}$ to find the work: $W = 42$ J

INSIGHT
Note that the more charge a battery moves through a circuit, the more work it does. Similarly, the greater the emf, the greater the work. We can see, then, that a car battery that operates at 12 volts and delivers several amps of current does much more work than a flashlight battery—as expected.

YOUR TURN
How long must the flashlight battery operate to do 150 J of work?

(Answers to **Your Turn** *problems are given in the back of the book.)*

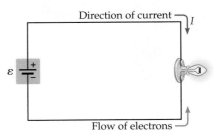

▲ **FIGURE 21–3** **A mechanical analog to the flashlight circuit**
The person lifting the water corresponds to the battery in Figure 21–2, and the paddle wheel corresponds to the lightbulb.

The emf of a battery is the potential difference it can produce between its terminals under ideal conditions. In real batteries, however, there is always some internal loss, leading to a potential difference that is less than the ideal value. In fact, the greater the current flowing through a battery, the greater the reduction in potential difference between its terminals, as we shall see in Section 21–4. Only when the current is zero can a real battery produce its full emf. Because most batteries have relatively small internal losses, we shall treat batteries as ideal—always producing a potential difference precisely equal to $\mathcal{E}$—unless specifically stated otherwise.

When we draw an electric circuit, it will be useful to draw an arrow indicating the flow of current. By convention, the direction of the current arrow is given in terms of a positive test charge, in much the same way that the direction of the electric field is determined:

> The direction of the current in an electric circuit is the direction in which a *positive* test charge would move.

Of course, in typical circuits the charges that flow are actually *negatively* charged electrons. As a result, the flow of electrons and the current arrow point in opposite directions, as indicated in **Figure 21–4**. Notice that a positive charge will flow from

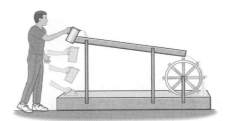

▲ **FIGURE 21–4** **Direction of current and electron flow**
In the flashlight circuit, electrons flow from the negative terminal of the battery to the positive terminal. The direction of the current, *I*, is just the opposite: from positive terminal to the negative terminal.

▲ **FIGURE 21–5 Path of an electron in a wire**
Typical path of an electron as it bounces off atoms in a metal wire. Because of the tortuous path the electron follows, its average velocity is rather small.

▲ A light-emitting diode (LED) is a relatively small, nonohmic device (top), but groups of LEDs can be used to form displays of practically any size (bottom). Because LEDs are extremely durable, and predicted to last 20 years or more, they are becoming the illumination of choice in high-reliability applications such as traffic lights, emergency exit signs, and brake lights. You'll probably see several on your way home today.

a region of high electric potential, near the positive terminal of the battery, to a region of low electric potential, near the negative terminal, as one would expect.

Finally, surprising as it may seem, electrons move rather slowly through a typical wire. They suffer numerous collisions with the atoms in the wire, and hence their path is rather tortuous and roundabout, as indicated in **Figure 21–5**. Like a car contending with a series of speed bumps, the electron's average speed, or **drift speed** as it is often called, is limited by the repeated collisions—in fact, their average speed is commonly about 10^{-4} m/s. Thus, if you switch on the headlights of a car, for example, an electron leaving the battery will take about an hour to reach the lightbulb, yet the lights seem to shine from the instant the switch is turned on. How is this possible?

The answer is that as an electron begins to move away from the battery, it exerts a force on its neighbors, causing them to move in the same general direction and, in turn, to exert a force on their neighbors, and so on. This process generates a propagating influence that travels through the wire at nearly the speed of light. The phenomenon is analogous to a bowling ball hitting one end of a line of balls; the effect of the colliding ball travels through the line at roughly the speed of sound, although the individual balls have very little displacement. Similarly, the electrons in a wire move with a rather small average velocity as they collide with and bounce off the atoms making up the wire, whereas the influence they have on one another races ahead and causes the light to shine.

21–2 Resistance and Ohm's Law

Electrons flow through metal wires with relative ease. In the ideal case, nothing about the wire would prevent their free motion. Real wires, however, under normal conditions, always affect the electrons to some extent, creating a **resistance** to their motion in much the same way that friction slows a box sliding across the floor.

In order to cause electrons to move against the resistance of a wire, it is necessary to apply a potential difference between its ends. For a wire with constant resistance, R, the potential difference, V, necessary to create a current, I, is given by Ohm's law:

Ohm's Law

$$V = IR \qquad\qquad 21\text{–}2$$

SI unit: volt, V

Ohm's law is named for the German physicist Georg Simon Ohm (1789–1854).

It should be noted at the outset that Ohm's law is not a law of nature but more on the order of a useful rule of thumb—like Hooke's law for springs or the ideal-gas laws that approximate the behavior of real gases. Materials that are well approximated by Ohm's law are said to be "ohmic" in their behavior; they show a simple linear relationship between the voltage applied to them and the current that results. In particular, if one plots current versus voltage for an ohmic material, the result is a straight line, with a constant slope equal to $1/R$. Nonohmic materials, on the other hand, have more complex relationships between voltage and current. A plot of current versus voltage for a nonohmic material is nonlinear; hence, the material does not have a constant resistance. (As an example, see Problem 9.) It is precisely these "nonlinearities," however, that can make such materials so useful in the construction of electronic devices, including the ubiquitous light-emitting diodes (LEDs).

Solving Ohm's law for the resistance, we find

$$R = \frac{V}{I}$$

From this expression it is clear that the units of resistance are volts per amp. In particular, we define 1 volt per amp to be 1 **ohm.** Letting the Greek letter omega, Ω, designate the ohm, we have

$$1\ \Omega = 1\ \text{V/A}$$

A device for measuring resistance is called an ohmmeter. We describe the operation of an ohmmeter in Section 21–8.

EXERCISE 21–1

A potential difference of 24 V is applied to a 150-Ω resistor. How much current flows through the resistor?

SOLUTION

Solving Ohm's law for the current, I, we find

$$I = \frac{V}{R} = \frac{24\ V}{150\ \Omega} = \frac{24\ V}{150\ V/A} = 0.16\ A$$

In an electric circuit a resistor is signified by a zigzag line: $\text{\small WWW}$. The straight lines in a circuit indicate ideal wires of zero resistance. To indicate the resistance of a real wire or device, we simply include a resistor of the appropriate value in the circuit.

Resistivity

Suppose you have a piece of wire of length L and cross-sectional area A. The resistance of this wire depends on the particular material from which it is constructed. If the wire is made of copper, for instance, its resistance will be less than if it is made from iron. The quantity that characterizes the resistance of a given material is its **resistivity**, ρ. For a wire of given dimensions, the greater the resistivity, the greater the resistance.

The resistance of a wire also depends on its length and area. To understand the dependence on L and A, consider again the analogy of water flowing through a hose. If the hose is very long, the resistance it presents to the water will be correspondingly large, whereas a wider hose—one with a greater cross-sectional area—will offer less resistance to the water. After all, water flows more easily through a short fire hose than through a long soda straw; hence, the resistance of a hose—and similarly a piece of wire—should be proportional to L and inversely proportional to A; that is, proportional to (L/A).

Combining these observations, we can write the resistance of a wire of length L, area A, and resistivity ρ in the following way:

Definition of Resistivity, ρ

$$R = \rho\left(\frac{L}{A}\right) \qquad \text{21–3}$$

Since the units of L are m and the units of A are m^2, it follows that the units of resistivity are $(\Omega \cdot m)$. Typical values for ρ are given in Table 21–1. Notice the enormous range in values of ρ, with the resistivity of an insulator like rubber about 10^{21} times greater than the resistivity of a good conductor like silver.

TABLE 21–1 Resistivities

Substance	Resistivity, $\rho\ (\Omega \cdot m)$
Insulators	
Quartz (fused)	7.5×10^{17}
Rubber	1 to 100×10^{13}
Glass	1 to $10,000 \times 10^{9}$
Semiconductors	
Silicon*	0.10 to 60
Germanium*	0.001 to 0.5
Conductors	
Lead	22×10^{-8}
Iron	9.71×10^{-8}
Tungsten	5.6×10^{-8}
Aluminum	2.65×10^{-8}
Gold	2.20×10^{-8}
Copper	1.68×10^{-8}
Silver	1.59×10^{-8}

*The resistivity of a semiconductor varies greatly with the type and amount of impurities it contains. This property makes them particularly useful in electronic applications.

CONCEPTUAL CHECKPOINT 21–1 COMPARE THE RESISTANCE

Wire 1 has a length L and a circular cross section of diameter D. Wire 2 is constructed from the same material as wire 1 and has the same shape, but its length is $2L$, and its diameter is $2D$. Is the resistance of wire 2 **(a)** the same as that of wire 1, **(b)** twice that of wire 1, or **(c)** half that of wire 1?

REASONING AND DISCUSSION

First, the resistance of wire 1 is

$$R_1 = \rho\left(\frac{L}{A}\right) = \rho\frac{L}{(\pi D^2/4)}$$

Note that we have used the fact that the area of a circle of diameter D is $\pi D^2/4$. For wire 2 we replace L with $2L$ and D with $2D$:

$$R_2 = \rho\frac{2L}{[\pi(2D)^2/4]} = \left(\tfrac{1}{2}\right)\rho\frac{L}{(\pi D^2/4)} = \tfrac{1}{2}R_1$$

CONTINUED ON NEXT PAGE

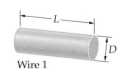

Wire 1

Wire 2

Thus, increasing the length by a factor of 2 increases the resistance by a factor of 2; on the other hand, increasing the diameter by a factor of 2 increases the area, and decreases the resistance, by a factor of 4. Overall, then, the resistance of wire 2 is half that of wire 1.

ANSWER
(c) The resistance of wire 2 is half that of wire 1; $R_2 = R_1/2$.

EXAMPLE 21–2 A CURRENT-CARRYING WIRE

A current of 1.82 A flows through a copper wire 1.75 m long and 1.10 mm in diameter. Find the potential difference between the ends of the wire. (The value of ρ for copper may be found in Table 21–1.)

PICTURE THE PROBLEM
The wire carries a current $I = 1.82$ A, and its total length L is 1.75 m. We assume that the wire has a circular cross section, with a diameter $D = 1.10$ mm.

STRATEGY
We know from Ohm's law that the potential difference associated with a current I and a resistance R is $V = IR$. We are given the current in the wire, but not the resistance. The resistance is easily determined, however, using $R = \rho(L/A)$ with $A = \pi D^2/4$. Thus, we first calculate R and then substitute the result into $V = IR$ to obtain the potential difference.

SOLUTION

1. Calculate the resistance of the wire:

$$R = \rho\left(\frac{L}{A}\right) = \rho\left(\frac{L}{\pi D^2/4}\right)$$

$$= (1.72 \times 10^{-8} \ \Omega \cdot m)\left[\frac{1.75 \ m}{\pi(0.00110 \ m)^2/4}\right] = 0.0317 \ \Omega$$

2. Multiply R by the current, I, to find the potential difference: $V = IR = (1.82 \ A)(0.0317 \ \Omega) = 0.0577 \ V$

INSIGHT
Copper is an excellent conductor; therefore, both the resistance and the potential difference are quite small.

PRACTICE PROBLEM
What diameter of copper wire is needed for there to be a potential difference of 0.100 V? Assume that all other quantities remain the same.   [**Answer:** 0.835 mm]

Some related homework problems: Problem 17, Problem 18

Temperature Dependence and Superconductivity

We know from everyday experience that a wire carrying an electric current can become warm—even quite hot, as in the case of a burner on a stove or the filament in an incandescent lightbulb. This follows from our earlier discussion of the fact that electrons collide with the atoms in a wire as they flow through an electric circuit. These collisions cause the atoms to jiggle with greater kinetic energy about their equilibrium positions. As a result, the temperature of the wire increases (see Section 17–2, and Equation 17–21 in particular). For example, the wire filament in an incandescent lightbulb can reach temperatures of roughly 2800 °C (in comparison, the surface of the Sun has a temperature of about 5500 °C), and the heating coil on a stove has a temperature of about 750 °C.

As a wire is heated, its resistivity tends to increase. This is because atoms that are jiggling more rapidly are more likely to collide with electrons and slow their progress through the wire. In fact, many metals show an approximately linear increase of ρ over a wide range of temperature. Once the dependence of ρ on T is known for a given material, the change in resistivity can be used as a means of measuring temperature.

REAL-WORLD PHYSICS

The bolometer

The first practical application of this principle was in a device known as the **bolometer.** Invented in 1880, the bolometer is an extremely sensitive thermometer that uses the temperature variation in the resistivity of platinum, nickel, or bismuth as a means of detecting temperature changes as small as 0.0001 C°. Soon after its invention, a bolometer was used to detect infrared radiation from the stars.

Some materials, like semiconductors, actually show a drop in resistivity as temperature is increased. This is because the resistivity of a semiconductor is strongly dependent on the number of electrons that are free to move about and conduct a current. As the temperature is increased in a semiconductor, more electrons are able to break free from their atoms, leading to an increased current and a reduced resistivity. Electronic devices incorporating such temperature-dependent semiconductors are known as **thermistors.** The digital fever thermometer so common in today's hospitals uses a thermistor to provide accurate measurements of a patient's temperature.

Since resistivity typically increases with temperature, it follows that if a wire is cooled below room temperature, its resistivity will *decrease*. Quite surprising, however, was a discovery made in the laboratory of Heike Kamerlingh-Onnes in 1911. Measuring the resistance of a sample of mercury at temperatures just a few degrees above absolute zero, researchers found that at about 4.2 K the resistance of the mercury suddenly dropped to zero—not just to a very small value, but to *zero.* At this temperature, we say that the mercury becomes **superconducting,** a hitherto unknown phase of matter. Since that time many different superconducting materials have been discovered, with various different **critical temperatures,** T_c, at which superconductivity begins. Today we know that superconductivity is a result of quantum effects (Chapter 30).

When a material becomes superconducting, a current can flow through it with absolutely no resistance. In fact, if a current is initiated in a superconducting ring of material, it will flow undiminished for as long as the ring is kept cool enough. In some cases, circulating currents have been maintained for years, with absolutely no sign of diminishing.

In 1986 a new class of superconductors was discovered that has zero resistance at temperatures significantly greater than those of any previously known superconducting materials. At the moment, the highest temperature at which superconductivity has been observed is about 125 K. Since the discovery of these "high-T_c" superconductors, hopes have been raised that it may one day be possible to produce room-temperature superconductors. The practical benefits of such a breakthrough, including power transmission with no losses, improved MRI scanners, and magnetically levitated trains, could be immense.

REAL-WORLD PHYSICS

Thermistors and fever thermometers

REAL-WORLD PHYSICS

Superconductors and high-temperature superconductivity

▲ When cooled below their critical temperature, superconductors not only lose their resistance to current flow but also exhibit new magnetic properties, such as repelling an external magnetic field. Here, a superconductor (bottom) levitates a small permanent magnet.

21–3 Energy and Power in Electric Circuits

When a charge ΔQ moves across a potential difference V, its electrical potential energy, U, changes by the amount

$$\Delta U = (\Delta Q)V$$

Recalling that power is the rate at which energy changes, $P = \Delta U / \Delta t$, we can write the electrical power as follows:

$$P = \frac{\Delta U}{\Delta t} = \frac{(\Delta Q)V}{\Delta t}$$

Since the electric current is given by $I = \Delta Q / \Delta t$, we have:

Electrical Power

$$P = IV \qquad\qquad 21\text{–}4$$

SI unit: watt, W

Thus, a current of 1 amp flowing across a potential difference of 1 V produces a power of 1 W.

▲ The heating element of an electric space heater is nothing more than a length of resistive wire coiled up for compactness. As electric current flows through the wire, the power it dissipates ($P = I^2R$) is converted to heat and light. The coils near the center are the hottest, and hence they glow with a higher-frequency, yellowish light.

EXERCISE 21–2

A handheld electric fan operates on a 3.00-V battery. If the power generated by the fan is 2.24 W, what is the current supplied by the battery?

SOLUTION

Solving $P = IV$ for the current, we obtain

$$I = \frac{P}{V} = \frac{2.24 \text{ W}}{3.00 \text{ V}} = 0.747 \text{ A}$$

The expression $P = IV$ applies to any electrical system. In the special case of a resistor, the electrical power is dissipated in the form of heat. Applying Ohm's law ($V = IR$) to this case, we can write the power dissipated in a resistor as

$$P = IV = I(IR) = I^2R \qquad \text{21–5}$$

Similarly, using Ohm's law to solve for the current, $I = V/R$, we have

$$P = IV = \left(\frac{V}{R}\right)V = \frac{V^2}{R} \qquad \text{21–6}$$

These relations also apply to incandescent lightbulbs, which are basically resistors that become hot enough to glow.

CONCEPTUAL CHECKPOINT 21–2   COMPARE LIGHTBULBS

A battery that produces a potential difference V is connected to a 5-W lightbulb. Later, the 5-W lightbulb is replaced with a 10-W lightbulb. **(a)** In which case does the battery supply more current? **(b)** Which lightbulb has the greater resistance?

REASONING AND DISCUSSION

a. To compare the currents, we need consider only the relation $P = IV$. Solving for the current yields $I = P/V$. When the voltage V is the same, it follows that the greater the power, the greater the current. In this case, then, the current in the 10-W bulb is twice the current in the 5-W bulb.

b. We now consider the relation $P = V^2/R$, which gives resistance in terms of voltage and power. In fact, $R = V^2/P$. Again, with V the same, it follows that the smaller the power, the greater the resistance. Thus, the resistance of the 5-W bulb is twice that of the 10-W bulb.

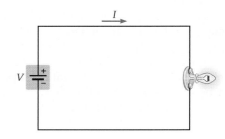

ANSWER

(a) When the battery is connected to the 10-W bulb, it delivers twice as much current as when it is connected to the 5-W bulb. **(b)** The 5-W bulb has twice as much resistance as the 10-W bulb.

On a microscopic level, the power dissipated by a resistor is the result of incessant collisions between electrons moving through the circuit and atoms making up the resistor. Specifically, the electric potential difference produced by the battery causes electrons to accelerate until they bounce off an atom of the resistor. At this point the electrons transfer energy to the atoms, causing them to jiggle more rapidly. The increased kinetic energy of the atoms is reflected in an increased temperature of the resistor (see Section 17–2). After each collision, the potential difference accelerates the electrons again and the process repeats—like a car bouncing through a series of speed bumps—resulting in a continuous transfer of energy from the electrons to the atoms.

EXAMPLE 21–3   HEATED RESISTANCE

A battery with an emf of 12 V is connected to a 545-Ω resistor. How much energy is dissipated in the resistor in 65 s?

PICTURE THE PROBLEM

The circuit, consisting of a battery and a resistor, is shown in our sketch. We show the current flowing from the positive terminal of the 12-V battery, through the 545-Ω resistor, and into the negative terminal of the battery.

STRATEGY
We know that a current flowing through a resistor dissipates power (energy per time), which means that the energy it dissipates in a given time is simply the power multiplied by the time: $\Delta U = P \, \Delta t$. The time is given ($\Delta t = 65$ s), and the power can be found using $P = IV$, $P = I^2R$, or $P = V^2/R$. The last expression is most convenient in this case, because the problem statement gives us the voltage and resistance.

To summarize, we first calculate the power, then multiply by the time.

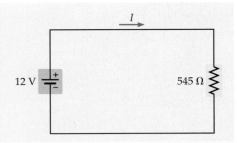

SOLUTION

1. Calculate the power dissipated in the resistor:

$$P = V^2/R = (12 \text{ V})^2/(545 \ \Omega) = 0.26 \text{ W}$$

2. Multiply the power by the time to find the energy dissipated:

$$\Delta U = P \, \Delta t = (0.26 \text{ W})(65 \text{ s}) = 17 \text{ J}$$

INSIGHT
The current in this circuit is $I = V/R = 0.022$ A. Using this result, we find that the power is $P = I^2R = IV = 0.26$ W, as expected.

PRACTICE PROBLEM
How much energy is dissipated in the resistor if the voltage is doubled to 24 V?   [**Answer:** 4(17 J) = 68 J]

Some related homework problems: Problem 29, Problem 32

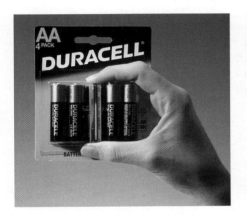

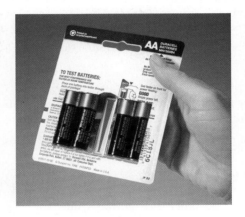

◀ The battery testers now often built into battery packages (left) employ a tapered graphite strip. The narrow end (at bottom in the right-hand photo) has the highest resistance, and thus produces the most heat when a current flows through the strip. The heat is used to produce the display on the front that indicates the strength of the battery—if the current is sufficient to warm even the top of the strip, where the resistance is lowest, the battery is fresh.

A commonly encountered application of resistance heating is found in the "battery check" meters often included with packs of batteries. To operate one of these meters, you simply press the contacts on either end of the meter against the corresponding terminals of the battery to be checked. This allows a current to flow through the main working element of the meter—a tapered strip of graphite.

The reason the strip is tapered is to provide a variation in resistance. According to the relation given in Equation 21–3, $R = \rho(L/A)$, the smaller the cross-sectional area A of the strip, the larger the resistance R. It follows that the narrow end has a higher resistance than the wide end. Because the same current I flows through all parts of the strip, the power dissipated is expressed most conveniently in the form $P = I^2R$. It follows that at the narrow end of the strip, where R is largest, the heating due to the current will be the greatest. Pressing the meter against the terminals of the battery, then, results in an overall warming of the graphite strip, with the narrow end warmer than the wide end.

The final element in the meter is a thin layer of liquid crystal (similar to the material used in LCD displays) that responds to small increases in temperature. In particular, this liquid crystal is black and opaque at room temperature but transparent when heated slightly. The liquid crystal is placed in front of a colored background, which can be seen in those regions where the graphite strip is warm enough to make the liquid crystal transparent. If the battery is weak, only the narrow portion of the strip becomes warm enough, and the meter shows only a small stripe of color. A strong battery, on the other hand, heats the entire strip enough to make the liquid crystal transparent, resulting in a colored stripe the full length of the meter.

REAL-WORLD PHYSICS

"Battery check" meters

Energy Usage

When you get a bill from the local electric company, you will find the number of kilowatt-hours of electricity that you have used. Notice that a kilowatt-hour (kWh) has the units of energy:

$$1 \text{ kilowatt-hour} = (1000 \text{ W})(3600 \text{ s}) = (1000 \text{ J/s})(3600 \text{ s})$$
$$= 3.6 \times 10^6 \text{ J}$$

Thus, the electric company is charging for the amount of energy you use—as one would expect—and not for the rate at which you use it. The following Example considers the energy and monetary cost for a typical everyday situation.

EXAMPLE 21–4 YOUR GOOSE IS COOKED

A holiday goose is cooked in the kitchen oven for 4.00 h. Assume that the stove draws a current of 20.0 A, operates at a voltage of 220.0 V, and uses electrical energy that costs $0.068 per kWh. How much does it cost to cook your goose?

PICTURE THE PROBLEM
We show a schematic representation of the stove cooking the goose in our sketch. The current in the circuit is 20.0 A, and the voltage difference across the heating coils is 220 V.

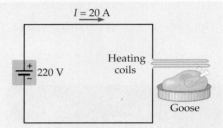

STRATEGY
The cost is simply the energy usage (in kWh) times the cost per kilowatt-hour ($0.068). To find the energy used, we note that energy is power multiplied by time. The time is given, and the power associated with a current I and a voltage V is $P = IV$.

Thus, we find the power, multiply by the time, and then multiply by $0.068 to find the cost.

SOLUTION

1. Calculate the power delivered to the stove: $P = IV = (20.0 \text{ A})(220.0 \text{ V}) = 4.40 \text{ kW}$

2. Multiply power by time to determine the total energy supplied to the stove during the 4.00 h of cooking: $\Delta U = P \, \Delta t = (4.40 \text{ kW})(4.00 \text{ h}) = 17.6 \text{ kWh}$

3. Multiply by the cost per kilowatt-hour to find the total cost of cooking: $\text{cost} = (17.6 \text{ kWh})(\$0.068/\text{kWh}) = \$1.20$

INSIGHT
Thus, your goose can be cooked for just over a dollar.

PRACTICE PROBLEM
If the voltage and current are reduced by a factor of 2 each, how long must the goose be cooked to use the same amount of energy? [**Answer:** $4(4.00 \text{ h}) = 16.0 \text{ h}$. *Note*: You should be able to answer a question like this by referring to your previous solution, without repeating the calculation in detail.]

Some related homework problems: Problem 30, Problem 31

21–4 Resistors in Series and Parallel

Electric circuits often contain a number of resistors connected in various ways. In this section we consider simple circuits containing only resistors and batteries. For each type of circuit considered, we calculate the **equivalent resistance** produced by a group of individual resistors.

Resistors in Series

When resistors are connected one after the other, end to end, we say that they are in *series*. **Figure 21–6 (a)** shows three resistors, R_1, R_2, and R_3, connected in series. The three resistors acting together have the same effect—that is, they draw the same current—as a single resistor, referred to as the equivalent resistor, R_{eq}. This equivalence is illustrated in **Figure 21–6 (b)**. We now calculate the value of the equivalent resistance.

The first thing to notice about the circuit in Figure 21–6 (a) is that the same current I must flow through each of the resistors—there is no other place for the current to go. As a result, the potential differences across the three resistors are

$V_1 = IR_1$, $V_2 = IR_2$, and $V_3 = IR_3$, respectively. Since the total potential difference from point A to point B must be the emf of the battery, $\mathcal{E}$, it follows that

$$\mathcal{E} = V_1 + V_2 + V_3$$

Writing each of the potentials in terms of the current and resistance, we find

$$\mathcal{E} = IR_1 + IR_2 + IR_3 = I(R_1 + R_2 + R_3)$$

Now, let's compare this expression with the result we obtain for the equivalent circuit in Figure 21–6 (b). In this circuit, the potential difference across the battery is $V = IR_{eq}$. Since this potential must be the same as the emf of the battery, we have

$$\mathcal{E} = IR_{eq}$$

Comparing this expression with $\mathcal{E} = I(R_1 + R_2 + R_3)$, we see that the equivalent resistance is simply the sum of the individual resistances:

$$R_{eq} = R_1 + R_2 + R_3$$

In general, for any number of resistors in series, the equivalent resistance is

Equivalent Resistance for Resistors in Series

$$R_{eq} = R_1 + R_2 + R_3 + \cdots = \sum R \qquad \text{21–7}$$

SI unit: ohm, Ω

Note that the equivalent resistance is greater than the greatest resistance of any of the individual resistors. Connecting the resistors in series is like making a single resistor increasingly longer; as its length increases so does its resistance.

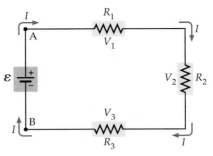

(a) Three resistors in series

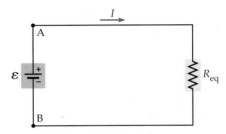

(b) Equivalent resistance has the same current

▲ **FIGURE 21–6   Resistors in series**
(a) Three resistors, R_1, R_2, and R_3, connected in series. Note that the same current I flows through each resistor.
(b) The equivalent resistance, $R_{eq} = R_1 + R_2 + R_3$, has the same current I flowing through it as the current I in the original circuit.

EXAMPLE 21–5   THREE RESISTORS IN SERIES

A circuit consists of three resistors connected in series to a 24.0-V battery. The current in the circuit is 0.0320 A. Given that $R_1 = 250.0\ \Omega$ and $R_2 = 150.0\ \Omega$, find **(a)** the value of R_3 and **(b)** the potential difference across each resistor.

PICTURE THE PROBLEM
The circuit is shown in our sketch. Note that the 24.0-V battery delivers the same current, $I = 0.0320$ A, to each of the three resistors. This is the key characteristic of a series circuit.

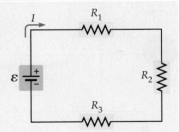

STRATEGY

a. First, we can obtain the equivalent resistance of the circuit using Ohm's law (as in Equation 21–2); $R_{eq} = \mathcal{E}/I$. Since the resistors are in series, we also know that $R_{eq} = R_1 + R_2 + R_3$. We can solve this relation for the only unknown, R_3.

b. We can then calculate the potential difference across each resistor using Ohm's law, $V = IR$.

SOLUTION

Part (a)

1. Use Ohm's law to find the equivalent resistance of the circuit:

$$R_{eq} = \frac{\mathcal{E}}{I} = \frac{24.0\ \text{V}}{0.0320\ \text{A}} = 7.50 \times 10^2\ \Omega$$

2. Set R_{eq} equal to the sum of the individual resistances, and solve for R_3:

$$R_{eq} = R_1 + R_2 + R_3$$
$$R_3 = R_{eq} - R_1 - R_2$$
$$= 7.50 \times 10^2\ \Omega - 250.0\ \Omega - 150.0\ \Omega = 3.50 \times 10^2\ \Omega$$

Part (b)

3. Use Ohm's law to determine the potential difference across R_1:

$$V_1 = IR_1 = (0.0320\ \text{A})(250.0\ \Omega) = 8.00\ \text{V}$$

4. Find the potential difference across R_2:

$$V_2 = IR_2 = (0.0320\ \text{A})(150.0\ \Omega) = 4.80\ \text{V}$$

5. Find the potential difference across R_3:

$$V_3 = IR_3 = (0.0320\ \text{A})(3.50 \times 10^2\ \Omega) = 11.2\ \text{V}$$

CONTINUED ON NEXT PAGE

CONTINUED FROM PREVIOUS PAGE

INSIGHT
Note that the greater the resistance, the greater the potential difference. In addition, the sum of the individual potential differences is 8.00 V + 4.80 V + 11.2 V = 24.0 V, as expected.

PRACTICE PROBLEM
Find the power dissipated in each resistor.   [**Answer:** $P_1 = 0.256$ W, $P_2 = 0.154$ W, $P_3 = 0.358$ W]

Some related homework problems: Problem 43, Problem 44

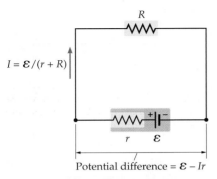

▲ **FIGURE 21–7 The internal resistance of a battery**
Real batteries always dissipate some energy in the form of heat. These losses can be modeled by a small "internal" resistance, r, within the battery. As a result, the potential difference between the terminals of a real battery is less than its ideal emf, $\mathcal{E}$. For example, if a battery produces a current I, the potential difference between its terminals is $\mathcal{E} - Ir$. In the case shown here, a battery is connected in series with the resistor R. Instead of producing the current $I = \mathcal{E}/R$, as in the ideal case, it produces the current $I = \mathcal{E}/(r + R)$.

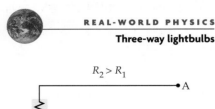

REAL-WORLD PHYSICS
Three-way lightbulbs

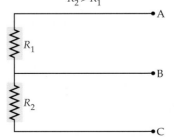

▲ **FIGURE 21–8 A three-way bulb**
The circuit diagram for a three-way lightbulb. For the brightest light, terminals A and B are connected to the household electrical line, so the current passes through the low-resistance filament R_1. For intermediate brightness, terminals B and C are used, so the current passes through the higher-resistance filament R_2. For the lowest light output, terminals A and C are used, so the current passes through both R_1 and R_2 in series.

An everyday example of resistors in series is the **internal resistance,** r, of a battery. As was mentioned in Section 21–1, real batteries have internal losses that cause the potential difference between their terminals to be less than $\mathcal{E}$ and to depend on the current in the battery. The simplest way to model a real battery is to imagine it to consist of an ideal battery of emf $\mathcal{E}$ in series with an internal resistance r, as shown in **Figure 21–7**. If this battery is then connected to an external resistance, R, the equivalent resistance of the circuit is $r + R$. As a result, the current flowing through the circuit is $I = \mathcal{E}/(r + R)$, and the potential difference between the terminals of the battery is $\mathcal{E} - Ir$. Thus, we see that the potential difference produced by the battery is less than $\mathcal{E}$ by an amount that is proportional to the current I. Only in the limit of zero current, or zero internal resistance, will the battery produce its full emf. (See Problems 51, 54, 116, and 121 for examples of batteries with internal resistance.)

Another application of resistors in series is the three-way lightbulb circuit shown in **Figure 21–8**. In this circuit, the two resistors represent two different filaments within a single bulb that are connected to a constant potential difference V. At the "high" setting, the lower-resistance filament, R_1, is connected to the electrical outlet via terminals A and B, and the brightest light is obtained ($P = V^2/R$). At the "middle" setting, the higher-resistance filament, R_2, is connected to the outlet via terminals B and C, resulting in a dimmer light. Finally, at the "low" setting, both filaments are connected in series via terminals A and C. This setting gives the greatest equivalent resistance, and thus the lowest light output.

An alternative method of producing a three-way lightbulb is to connect the resistors in parallel. This will be discussed in the next subsection.

Resistors in Parallel

Resistors are in *parallel* when they are connected across the same potential difference, as in **Figure 21–9 (a)**. In a case like this, the current has parallel paths through which it can flow. As a result, the total current in the circuit, I, is equal to the sum of the currents through each of the three resistors:

$$I = I_1 + I_2 + I_3$$

Since the potential difference is the same for each of the resistors, it follows that the currents flowing through them are as follows:

$$I_1 = \frac{\mathcal{E}}{R_1}, \quad I_2 = \frac{\mathcal{E}}{R_2}, \quad I_3 = \frac{\mathcal{E}}{R_3}$$

Summing these three currents, we find

$$I = \frac{\mathcal{E}}{R_1} + \frac{\mathcal{E}}{R_2} + \frac{\mathcal{E}}{R_3} = \mathcal{E}\left(\frac{1}{R_1} + \frac{1}{R_2} + \frac{1}{R_3}\right) \qquad 21\text{–}8$$

Now, in the equivalent circuit shown in **Figure 21–9 (b)**, Ohm's law gives $\mathcal{E} = IR_{eq}$, or

$$I = \mathcal{E}\left(\frac{1}{R_{eq}}\right) \qquad 21\text{–}9$$

Comparing Equations 21–8 and 21–9, we find that the equivalent resistance for three resistors in parallel is

$$\frac{1}{R_{eq}} = \frac{1}{R_1} + \frac{1}{R_2} + \frac{1}{R_3}$$

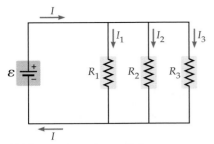

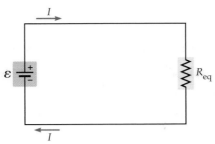

(a) Three resistors in parallel

(b) Equivalent resistance has the same current

▶ **FIGURE 21–9 Resistors in parallel**
(a) Three resistors, R_1, R_2, and R_3, connected in parallel. Note that each resistor is connected across the same potential difference $\mathcal{E}$. **(b)** The equivalent resistance, $1/R_{eq} = 1/R_1 + 1/R_2 + 1/R_3$, has the same current flowing through it as the total current I in the original circuit.

In general, for any number of resistors in parallel, we have:

Equivalent Resistance for Resistors in Parallel

$$\frac{1}{R_{eq}} = \frac{1}{R_1} + \frac{1}{R_2} + \frac{1}{R_3} + \cdots = \sum \frac{1}{R}$$

21–10

SI unit: ohm, Ω

As a simple example, consider a circuit with two identical resistors R connected in parallel. The equivalent resistance in this case is given by

$$\frac{1}{R_{eq}} = \frac{1}{R} + \frac{1}{R} = \frac{2}{R}$$

Solving for R_{eq}, we find $R_{eq} = \frac{1}{2}R$. If we connect three such resistors in parallel, the corresponding result is

$$\frac{1}{R_{eq}} = \frac{1}{R} + \frac{1}{R} + \frac{1}{R} = \frac{3}{R}$$

In this case, $R_{eq} = \frac{1}{3}R$. Thus, the more resistors we connect in parallel, the smaller the equivalent resistance. Each time we add a new resistor in parallel with the others, we give the battery a new path through which current can flow—analogous to opening an additional lane of traffic on a busy highway. Stated another way, giving the current multiple paths through which it can flow is equivalent to using a wire with a greater cross-sectional area. From either point of view, the fact that more current flows with the same potential difference means that the equivalent resistance has been reduced.

Finally, if any one of the resistors in a parallel connection is equal to zero, the equivalent resistance is also zero. This situation is referred to as a **short circuit**, and is illustrated in **Figure 21–10**. In this case, all of the current flows through the path of zero resistance.

PROBLEM-SOLVING NOTE

The Equivalent Resistance of Resistors in Parallel

After summing the inverse of resistors in parallel, remember to take one more inverse at the end of your calculation to find the equivalent resistance.

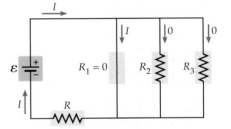

▲ **FIGURE 21–10 A short circuit**
If one of the resistors in parallel with others is equal to zero, all the current flows through that portion of the circuit, giving rise to a short circuit. In this case, resistors R_2 and R_3 are "shorted out," and the current in the circuit is $I = \mathcal{E}/R$.

EXAMPLE 21–6 THREE RESISTORS IN PARALLEL

Consider a circuit with three resistors, $R_1 = 250.0 \ \Omega$, $R_2 = 150.0 \ \Omega$, and $R_3 = 350.0 \ \Omega$, connected in parallel with a 24.0-V battery. Find **(a)** the total current supplied by the battery and **(b)** the current through each resistor.

PICTURE THE PROBLEM
The accompanying sketch indicates the parallel connection of the resistors with the battery. Notice that each of the resistors experiences precisely the same potential difference; namely, the 24.0 V produced by the battery. This is the feature that characterizes parallel connections.

STRATEGY

a. We can find the total current from $I = \mathcal{E}/R_{eq}$, where $1/R_{eq} = 1/R_1 + 1/R_2 + 1/R_3$.

b. For each resistor, the current is given by Ohm's law, $I = \mathcal{E}/R$.

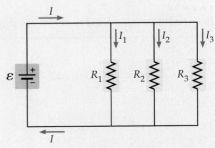

CONTINUED ON NEXT PAGE

CONTINUED FROM PREVIOUS PAGE

SOLUTION

Part (a)

1. Find the equivalent resistance of the circuit:

$$\frac{1}{R_{eq}} = \frac{1}{R_1} + \frac{1}{R_2} + \frac{1}{R_3}$$

$$= \frac{1}{250.0\ \Omega} + \frac{1}{150.0\ \Omega} + \frac{1}{350.0\ \Omega} = 0.01352\ \Omega^{-1}$$

$$R_{eq} = (0.01352\ \Omega^{-1})^{-1} = 73.96\ \Omega$$

2. Use Ohm's law to find the total current:

$$I = \frac{\mathcal{E}}{R_{eq}} = \frac{24.0\ V}{73.96\ \Omega} = 0.325\ A$$

Part (b)

3. Calculate I_1 using $I_1 = \mathcal{E}/R_1$ with $\mathcal{E} = 24.0$ V:

$$I_1 = \frac{\mathcal{E}}{R_1} = \frac{24.0\ V}{250.0\ \Omega} = 0.0960\ A$$

4. Repeat the preceding calculation for resistors 2 and 3:

$$I_2 = \frac{\mathcal{E}}{R_2} = \frac{24.0\ V}{150.0\ \Omega} = 0.160\ A$$

$$I_3 = \frac{\mathcal{E}}{R_3} = \frac{24.0\ V}{350.0\ \Omega} = 0.0686\ A$$

INSIGHT

As expected, the smallest resistor, R_2, carries the greatest current. The three currents combined yield the total current, as they must; that is, $I_1 + I_2 + I_3 = 0.0960\ A + 0.160\ A + 0.0686\ A = 0.325\ A = I$.

PRACTICE PROBLEM

Find the power dissipated in each resistor.   [**Answer:** $P_1 = 2.30$ W, $P_2 = 3.84$ W, $P_3 = 1.65$ W]

Some related homework problems: Problem 45, Problem 46

In comparing Examples 21–5 and 21–6 note the differences in the power dissipated in each circuit. First, the total power dissipated in the parallel circuit is much greater than that dissipated in the series circuit. This is due to the fact that the equivalent resistance of the parallel circuit is smaller than the equivalent resistance of the series circuit, and the power delivered by a voltage V to a resistance R is inversely proportional to the resistance ($P = V^2/R$). In addition, note that the smallest resistor, R_2, has the smallest power in the series circuit but the largest power in the parallel circuit. These issues are explored further in the following Conceptual Checkpoint.

CONCEPTUAL CHECKPOINT 21–3   SERIES VERSUS PARALLEL

Two identical lightbulbs are connected to a battery, either in series or in parallel. Are the bulbs in series **(a)** brighter than, **(b)** dimmer than, or **(c)** the same brightness as the bulbs in parallel?

REASONING AND DISCUSSION

Both sets of lightbulbs are connected to the same potential difference, V; hence, the power delivered to the bulbs is V^2/R_{eq}, where R_{eq} is twice the resistance of a bulb in the series circuit and half the resistance of a bulb in the parallel circuit. As a result, more power is converted to light in the parallel circuit.

ANSWER

(b) The bulbs connected in series are dimmer than the bulbs connected in parallel.

Finally, note that a three-way lightbulb can also be produced by simply wiring two filaments in parallel. For example, one filament might have a power of 50 W and the second filament a power of 100 W. One setting of the switch sends current through the 50-W filament, the next setting sends current through the 100-W filament, and the third setting connects the two filaments in parallel. With the third connection, each filament produces the same power as before—since each is connected to the same potential difference—giving a total power of 50 W + 100 W = 150 W.

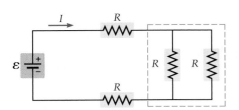 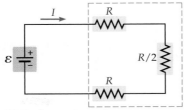 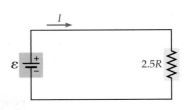

(a) Replace parallel resistors   (b) Replace series resistors   (c) Final equivalent resistance

▲ **FIGURE 21–11 Analyzing a complex circuit of resistors**
(a) The two vertical resistors are in parallel with one another; hence, they can be replaced with their equivalent resistance, $R/2$. **(b)** Now the circuit consists of three resistors in series. The equivalent resistance of these three resistors is 2.5 R. **(c)** The original circuit reduced to a single equivalent resistance.

Combination Circuits

The rules we have developed for series and parallel resistors can be applied to more complex circuits as well. For example, consider the circuit shown in **Figure 21–11 (a)**, where four resistors, each equal to R, are connected in a way that combines series and parallel features. To analyze this circuit, we first note that the two vertically oriented resistors are connected in parallel with one another. Therefore, the equivalent resistance of this unit is given by $1/R_{eq} = 1/R + 1/R$, or $R_{eq} = R/2$. Replacing these two resistors with $R/2$ yields the circuit shown in **Figure 21–11 (b)**, which consists of three resistors in series. As a result, the equivalent resistance of the entire circuit is $R + R/2 + R = 2.5R$, as indicated in **Figure 21–11 (c)**. Similar methods can be applied to a wide variety of circuits.

PROBLEM-SOLVING NOTE

Analyzing a Complex Circuit

When considering an electric circuit with resistors in series and parallel, work from the smallest units of the circuit outward to ever larger units.

EXAMPLE 21–7   COMBINATION SPECIAL

In the circuit shown in the diagram, the emf of the battery is 12.0 V, and each resistor has a resistance of 200.0 Ω. Find **(a)** the current supplied by the battery to this circuit and **(b)** the current through the lower two resistors.

PICTURE THE PROBLEM
The circuit for this problem has three resistors connected to a battery. Note that the lower two resistors are in series with one another, and in parallel with the upper resistor. The battery has an emf of 12.0 V.

STRATEGY

a. The current supplied by the battery, I, is given by Ohm's law, $I = \mathcal{E}/R_{eq}$, where R_{eq} is the equivalent resistance of the three resistors. To find R_{eq}, we first note that the lower two resistors are in series, giving a net resistance of $2R$. Next, the upper resistor, R, is in parallel with $2R$. Calculating this equivalent resistance yields the desired R_{eq}.

b. Because the voltage across the lower two resistors is $\mathcal{E}$, the current through them is $I_{lower} = \mathcal{E}/R_{eq,lower} = \mathcal{E}/2R$.

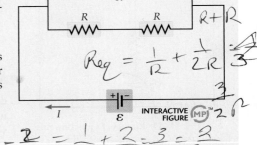

SOLUTION

Part (a)

1. Calculate the equivalent resistance of the lower two resistors:

$$R_{eq,lower} = R + R = 2R$$

2. Calculate the equivalent resistance of R in parallel with $2R$:

$$\frac{1}{R_{eq}} = \frac{1}{R} + \frac{1}{2R} = \frac{3}{2R}$$

$$R_{eq} = \tfrac{2}{3}R = \tfrac{2}{3}(200.0\ \Omega) = 133.3\ \Omega$$

3. Find the current supplied by the battery, I:

$$I = \frac{\mathcal{E}}{R_{eq}} = \frac{12.0\ \text{V}}{133.3\ \Omega} = 0.0900\ \text{A}$$

Part (b)

4. Use $\mathcal{E}$ and $R_{eq,lower}$ to find the current in the lower two resistors:

$$I_{lower} = \frac{\mathcal{E}}{R_{eq,lower}} = \frac{12.0\ \text{V}}{2(200.0\ \Omega)} = 0.0300\ \text{A}$$

INSIGHT
Note that the total resistance of the three 200.0-Ω resistors is less than 200.0 Ω—in fact, it is only 133.3 Ω. We also see that 0.0300 A flows through the lower two resistors, and therefore twice that much—0.0600 A—flows through the upper resistor.

CONTINUED ON NEXT PAGE

CONTINUED FROM PREVIOUS PAGE

PRACTICE PROBLEM

Suppose the upper resistor is changed from R to $2R$, and the lower two resistors remain the same. **(a)** Will the current supplied by the battery increase, decrease, or stay the same? **(b)** Find the new current.   [**Answer:** (a) The current will decrease because there is greater resistance to its flow; (b) 0.0600 A.]

Some related homework problems: Problem 48, Problem 49, Problem 51

▶ The electric circuit in these photos starts with two identical lightbulbs (1 and 2) in series with a battery, as we see on the left. The bulbs are equally bright. Now, *before you examine the photo to the right*, consider the effect of adding a third identical bulb (3) to the circuit by placing it in the empty socket. What happens to the brightness of bulbs 1 and 2? As you can see, adding bulb 3 creates a new path for the current and increases the total current in the circuit by a factor of 4/3 (check this yourself). The current passing through bulb 1 is equally split between bulbs 2 and 3, however, and the new current in bulb 2 is now only $\frac{1}{2}(4/3) = 2/3$ of its original value. Thus, bulb 1 brightens and bulb 2 becomes dimmer.

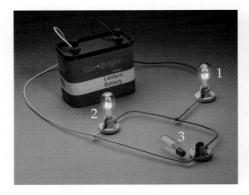

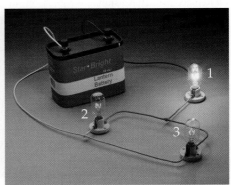

21–5 Kirchhoff's Rules

To find the currents and voltages in a general electric circuit, we use two rules first introduced by the German physicist Gustav Kirchhoff (1824–1887). The *Kirchhoff rules* are simply ways of expressing charge conservation (the junction rule) and energy conservation (the loop rule) in a closed circuit. Since these conservation laws are always obeyed in nature, the Kirchhoff rules are completely general.

The Junction Rule

The junction rule follows from the observation that the current entering any point in a circuit must equal the current leaving that point. If this were not the case, charge would either build up or disappear from a circuit.

As an example, consider the circuit shown in **Figure 21–12**. At point A, three wires join to form **a junction.** (In general, a *junction* is any point in a circuit where three or more wires meet.) The current carried by each of the three wires is indicated in the figure. Notice that the current entering the junction is I_1; the current leaving the junction is $I_2 + I_3$. Setting the incoming and outgoing currents equal, we have $I_1 = I_2 + I_3$, or equivalently

$$I_1 - I_2 - I_3 = 0$$

This is Kirchhoff's junction rule applied to the junction at point A.

In general, if we associate a $+$ sign with currents entering a junction and a $-$ sign with currents leaving a junction, Kirchhoff's junction rule can be stated as follows:

> The algebraic sum of all currents meeting at any junction in a circuit must equal zero.

In the example just discussed, I_1 enters the junction $(+)$, I_2 and I_3 leave the junction $(-)$; hence, the algebraic sum of currents at the junction is $I_1 - I_2 - I_3$. Setting this sum equal to zero recovers our previous result.

In some cases we may not know the direction of all the currents meeting at a junction in advance. When this happens, we simply choose a direction for the unknown currents, apply the junction rule, and continue as usual. If the value we obtain for a given current is negative, it simply means that the direction we chose was wrong; the current actually flows in the opposite direction.

For example, suppose we know both the direction and magnitude of the currents I_1 and I_2 in **Figure 21–13**. To find the third current, we apply the junction

The current I_1 entering junction A ...

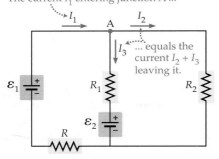

... equals the current $I_2 + I_3$ leaving it.

▲ **FIGURE 21–12 Kirchhoff's junction rule**
Kirchhoff's junction rule states that the sum of the currents entering a junction must equal the sum of the currents leaving the junction. In this case, for the junction labeled A, $I_1 = I_2 + I_3$, or $I_1 - I_2 - I_3 = 0$.

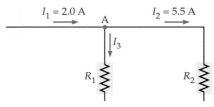

▲ **FIGURE 21–13 A specific application of Kirchhoff's junction rule**
Applying Kirchhoff's junction rule to the junction A, $I_1 - I_2 - I_3 = 0$, yields the result $I_3 = -3.5$ A. The minus sign indicates that I_3 flows opposite to the direction shown; that is, I_3 is actually upward.

rule—but first we must choose a direction for I_3. If we choose I_3 to point downward, as shown in the figure, the junction rule gives

$$I_1 - I_2 - I_3 = 0$$

Solving for I_3, we have

$$I_3 = I_1 - I_2 = 2.0\ \text{A} - 5.5\ \text{A} = -3.5\ \text{A}$$

Since I_3 is negative, we conclude that the actual direction of this current is upward; that is, the 2.0-A current and the 3.5-A current enter the junction and combine to yield the 5.5-A current that leaves the junction.

The Loop Rule

Imagine taking a day hike on a mountain path. First, you gain altitude to reach a scenic viewpoint; later you descend below your starting point into a valley; finally, you gain altitude again and return to the trailhead. During the hike you sometimes increase your gravitational potential energy, and sometimes you decrease it, but the net change at the end of the hike is zero—after all, you return to the same altitude from which you started. Kirchhoff's loop rule is an application of the same idea to an electric circuit.

For example, consider the simple circuit shown in **Figure 21–14**. The electric potential increases by the amount $\mathcal{E}$ in going from point A to point B, since we move from the low-potential ($-$) terminal of the battery to the high-potential ($+$) terminal. This is like gaining altitude in the hiking analogy. Next, there is no potential change as we go from point B to point C, since these points are connected by an ideal wire. As we move from point C to point D, however, the potential does change—recall that a potential difference is required to force a current through a resistor. We label the potential difference across the resistor ΔV_{CD}. Finally, there is no change in potential between points D and A, since they too are connected by an ideal wire.

We can now apply the idea that the net change in electric potential (the analog to gravitational potential energy in the hike) must be zero around any closed loop. In this case, we have

$$\mathcal{E} + \Delta V_{CD} = 0$$

Thus, we find that $\Delta V_{CD} = -\mathcal{E}$; that is, the electric potential *decreases* as one moves across the resistor *in the direction of the current*. To indicate this drop in potential, we label the side where the current enters the resistor with a $+$ (indicating high potential) and the side where the current leaves the resistor with a $-$ (indicating low potential). Finally, we can use Ohm's law to set the magnitude of the potential drop equal to IR and find the current in the circuit:

$$|\Delta V_{CD}| = \mathcal{E} = IR$$
$$I = \frac{\mathcal{E}}{R}$$

This, of course, is the expected result.

In general, Kirchhoff's loop rule can be stated as follows:

> The algebraic sum of all potential differences around any closed loop in a circuit is zero.

We now consider a variety of applications in which both the junction rule and the loop rule are used to find the various currents and potentials in a circuit.

Applications

We begin by considering the relatively simple circuit shown in **Figure 21–15**. The currents and voltages in this circuit can be found by considering various parallel and series combinations of the resistors, as we did in the previous section. Thus, Kirchhoff's rules are not strictly needed in this case. Still, applying the rules to this circuit illustrates many of the techniques that can be used when studying more complex circuits.

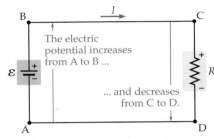

▲ **FIGURE 21–14   Kirchhoff's loop rule**
Kirchhoff's loop rule states that as one moves around a closed loop in a circuit, the algebraic sum of all potential differences must be zero. The electric potential increases as one moves from the $-$ to the $+$ plate of a battery; it decreases as one moves through a resistor in the direction of the current.

PROBLEM-SOLVING NOTE

Applying Kirchhoff's Rules

When applying Kirchhoff's rules, be sure to use the appropriate sign for currents and potential differences.

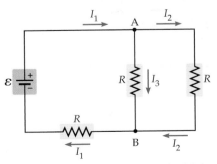

▲ **FIGURE 21–15   Analyzing a simple circuit**
A simple circuit that can be studied using either equivalent resistance or Kirchhoff's rules.

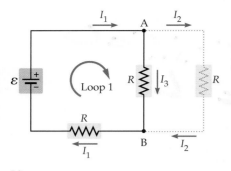

(a)

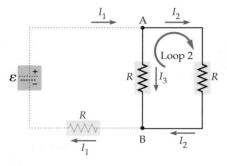

(b)

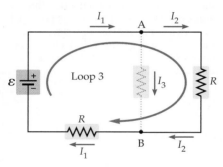

(c)

▲ **FIGURE 21–16 Using loops to analyze a circuit**

Three loops associated with the circuit in Figure 21–15.

Let's suppose that all the resistors have the value $R = 100.0 \ \Omega$, and that the emf of the battery is $\mathcal{E} = 15.0$ V. The equivalent resistance of the resistors can be obtained by noting that the vertical resistors are connected in parallel with one another and in series with the horizontal resistor. The vertical resistors combine to give a resistance of $R/2$, which, when added to the horizontal resistor, gives an equivalent resistance of $R_{eq} = 3R/2 = 150.0 \ \Omega$. The current in the circuit, then, is $I = \mathcal{E}/R_{eq} = 15.0$ V$/150.0 \ \Omega = 0.100$ A.

Now we approach the same problem from the point of view of Kirchhoff's rules. First, we apply the junction rule to point A:

$$I_1 - I_2 - I_3 = 0 \qquad \text{(junction A)} \qquad 21\text{–}11$$

Note that current I_1 splits at point A into currents I_2 and I_3, which combine again at point B to give I_1 flowing through the horizontal resistor. We can apply the junction rule to point B, which gives $-I_1 + I_2 + I_3 = 0$, but since this differs from Equation 21–11 by only a minus sign, no new information is gained.

Next, we apply the loop rule. Since there are three unknowns, I_1, I_2, and I_3, we need three independent equations for a full solution. One has already been given by the junction rule; thus, we expect that two loop equations will be required to complete the solution. To begin, we consider loop 1, which is shown in **Figure 21–16 (a)**. We choose to move around this loop in the clockwise direction. (If we were to choose the counterclockwise direction instead, the same information would be obtained.) For loop 1, then, we have an increase in potential as we move across the battery, a drop in potential across the vertical resistor of I_3R, and another drop in potential across the horizontal resistor, this time of magnitude I_1R. Applying the loop rule, we find the following:

$$\mathcal{E} - I_3R - I_1R = 0 \qquad \text{(loop 1)} \qquad 21\text{–}12$$

Similarly, we can apply the loop rule to loop 2, shown in **Figure 21–16 (b)**. In this case we cross the right-hand vertical resistor in the direction of the current, implying a drop in potential, and we cross the left-hand vertical resistor against the current, implying an increase in potential. Therefore, the loop rule gives

$$I_3R - I_2R = 0 \qquad \text{(loop 2)} \qquad 21\text{–}13$$

There is a third possible loop, shown in **Figure 21–16 (c)**, but the information it gives is not different from that already obtained. In fact, *any two of the three loops* complete our solution.

Note that R cancels in Equation 21–13; hence, we see that $I_3 - I_2 = 0$, or $I_3 = I_2$. Substituting this result into the junction rule (Equation 21–11), we obtain

$$I_1 - I_2 - I_3 = I_1 - I_2 - I_2$$
$$= I_1 - 2I_2 = 0$$

Solving this equation for I_2 gives us $I_2 = I_1/2 = I_3$. Finally, using the first loop equation (Equation 21–12), we find

$$\mathcal{E} - (I_1/2)R - I_1R = \mathcal{E} - \tfrac{3}{2}I_1R = 0$$

Note that the only unknown in this equation is current I_1. Solving for this current, we find

$$I_1 = \frac{\mathcal{E}}{\frac{3}{2}R} = \frac{15.0 \text{ V}}{\frac{3}{2}(100.0 \ \Omega)} = 0.100 \text{ A}$$

As expected, our result using Kirchhoff's rules agrees with the result obtained previously. Finally, the other two currents in the circuit are $I_2 = I_3 = I_1/2 = 0.0500$ A.

EXERCISE 21–3

Write the loop equation for loop 3 in Figure 21–16 (c).

SOLUTION

Proceeding in a clockwise direction, as indicated in the figure, we find

$$\mathcal{E} - I_2R - I_1R = 0$$

Since I_2 and I_3 are equal (loop 2), it follows that loop 1 ($\mathcal{E} - I_3R - I_1R = 0$) and loop 3 ($\mathcal{E} - I_2R - I_1R = 0$) give the same information. If we proceed in a counterclockwise direction around loop 3, we find

$$-\mathcal{E} + I_2R + I_1R = 0$$

Notice that this result is the same as the clockwise result except for an overall minus sign, and, therefore, it contains no new information. In general, it does not matter in which direction we choose to go around a loop.

Clearly, the Kirchhoff approach is more involved than the equivalent-resistance method. However, it is not possible to analyze all circuits in terms of equivalent resistances. In such cases, Kirchhoff's rules are the only option, as illustrated in the next Active Example.

ACTIVE EXAMPLE 21–2   TWO LOOPS, TWO BATTERIES: FIND THE CURRENTS

Find the currents in the circuit shown.

SOLUTION *(Test your understanding by performing the calculations indicated in each step.)*

1. Apply the junction rule to point A: $I_1 - I_2 - I_3 = 0$

2. Apply the loop rule to loop 1 (let $R = 100.0\ \Omega$): $15\ \text{V} - I_3R - I_1R = 0$

3. Apply the loop rule to loop 2 (let $R = 100.0\ \Omega$): $-9.0\ \text{V} - I_2R + I_3R = 0$

4. Solve for I_1, I_2, and I_3: $I_1 = 0.070\ \text{A}$, $I_2 = -0.010\ \text{A}$, $I_3 = 0.080\ \text{A}$

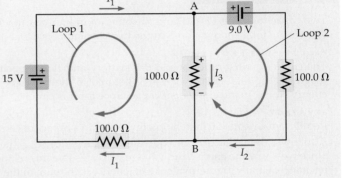

INSIGHT
Note that I_2 is negative. This means that its direction is opposite to that shown in the circuit diagram.

YOUR TURN
Suppose the polarity of the 9.0-V battery is reversed. What are the currents in this case?

*(Answers to **Your Turn** problems are given in the back of the book.)*

21–6 Circuits Containing Capacitors

To this point we have considered only resistors and batteries in electric circuits. Capacitors, which can also play an important role, are represented by a set of parallel lines (reminiscent of a parallel-plate capacitor): ⊣⊢. We now investigate simple circuits involving batteries and capacitors, leaving for the next section circuits that combine all three circuit elements.

Capacitors in Parallel

The simplest way to combine capacitors, as we shall see, is by connecting them in parallel. For example, **Figure 21–17 (a)** shows three capacitors connected in parallel with a battery of emf ε. As a result, each capacitor has the same potential difference, ε, between its plates. The magnitudes of the charges on each capacitor are as follows:

$$Q_1 = C_1\varepsilon, \quad Q_2 = C_2\varepsilon, \quad Q_3 = C_3\varepsilon$$

As a result, the total charge on the three capacitors is

$$Q = Q_1 + Q_2 + Q_3 = \varepsilon C_1 + \varepsilon C_2 + \varepsilon C_3 = (C_1 + C_2 + C_3)\varepsilon$$

If an equivalent capacitor is used to replace the three in parallel, as in **Figure 21–17 (b)**, the charge on its plates must be the same as the total charge on the individual capacitors:

$$Q = C_{eq}\varepsilon$$

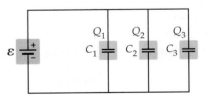

(a) Three capacitors in parallel

(b) Equivalent capacitance with same total charge

▲ **FIGURE 21–17 Capacitors in parallel**
(a) Three capacitors, C_1, C_2, and C_3, connected in parallel. Note that each capacitor is connected across the same potential difference, $\mathcal{E}$. **(b)** The equivalent capacitance, $C_{eq} = C_1 + C_2 + C_3$, has the same charge on its plates as the total charge on the three original capacitors.

Comparing $Q = C_{eq}\mathcal{E}$ with $Q = (C_1 + C_2 + C_3)\mathcal{E}$, we see that the equivalent capacitance is simply

$$C_{eq} = C_1 + C_2 + C_3$$

In general, the equivalent capacitance of capacitors connected in parallel is the sum of the individual capacitances:

Equivalent Capacitance for Capacitors in Parallel

$$C_{eq} = C_1 + C_2 + C_3 + \cdots = \sum C \qquad \text{21–14}$$

SI unit: farad, F

Thus, connecting capacitors in parallel produces an equivalent capacitance greater than the greatest individual capacitance. It is as if the plates of the individual capacitors are connected together to give one large set of plates, with a correspondingly large capacitance.

EXAMPLE 21–8   ENERGY IN PARALLEL

Two capacitors, one 12.0 μF and the other of unknown capacitance C, are connected in parallel across a battery with an emf of 9.00 V. The total energy stored in the two capacitors is 0.0115 J. What is the value of the capacitance C?

PICTURE THE PROBLEM
The circuit, consisting of one 9.00-V battery and two capacitors, is illustrated in the diagram. The total energy of 0.0115 J stored in the two capacitors is the same as the energy stored in the equivalent capacitance for this circuit.

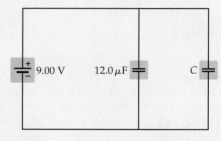

STRATEGY
Recall from Chapter 20 that the energy stored in a capacitor can be written as $U = \frac{1}{2}CV^2$. It follows, then, that for an equivalent capacitance, C_{eq}, the energy is $U = \frac{1}{2}C_{eq}V^2$. Since we know the energy and voltage, we can solve this relation for the equivalent capacitance. Finally, the equivalent capacitance is the sum of the individual capacitances, $C_{eq} = 12.0\ \mu$F$ + C$. We use this relation to solve for C.

SOLUTION

1. Solve $U = \frac{1}{2}C_{eq}V^2$ for the equivalent capacitance:

$$U = \frac{1}{2}C_{eq}V^2$$
$$C_{eq} = \frac{2U}{V^2}$$

2. Substitute numerical values to find C_{eq}:

$$C_{eq} = \frac{2U}{V^2} = \frac{2(0.0115\ \text{J})}{(9.00\ \text{V})^2} = 284\ \mu\text{F}$$

3. Solve for C in terms of the equivalent capacitance:

$$C_{eq} = 12.0\ \mu\text{F} + C$$

$$C = C_{eq} - 12.0\ \mu\text{F} = 284\ \mu\text{F} - 12.0\ \mu\text{F} = 272\ \mu\text{F}$$

INSIGHT
The energy stored in the 12.0-μF capacitor is $U = \frac{1}{2}CV^2 = 0.000486$ J. In comparison, the 272-μF capacitor stores an energy equal to 0.0110 J. Thus, the larger capacitor stores the greater amount of energy. Though this may seem only natural, one needs to be careful. When we examine capacitors in *series* later in this section, we shall find exactly the opposite result.

PRACTICE PROBLEM
What is the total charge stored on the two capacitors?   [**Answer:** $Q = C_{eq}\mathcal{E} = 2.56 \times 10^{-3}$ C]

Some related homework problems: Problem 72, Problem 73

REAL-WORLD PHYSICS
"Touch-sensitive" lamps

Although you probably haven't realized it, when you turn on a "touch sensitive" lamp, you are part of a circuit with capacitors in parallel. In fact, you are one of the capacitors! When you touch such a lamp, a small amount of charge moves onto your body—your body is like the plate of a capacitor. Because you have effectively increased the plate area—as always happens when capacitors are connected in parallel—the capacitance of the circuit increases. The electronic circuitry in the lamp senses this increase in capacitance and triggers the switch to turn the light on or off.

Capacitors in Series

You have probably noticed from Equation 21–14 that capacitors connected in *parallel* combine in the same way as resistors connected in *series*. Similarly, capacitors connected in *series* obey the same rules as resistors connected in *parallel*, as we now show.

Consider three capacitors—initially uncharged—connected in series with a battery, as in **Figure 21–18 (a)**. The battery causes the left plate of C_1 to acquire a positive charge, $+Q$. This charge, in turn, attracts a negative charge $-Q$ onto the right plate of the capacitor. Because the capacitors start out uncharged, there is zero net charge between C_1 and C_2. As a result, the negative charge $-Q$ on the right plate of C_1 leaves a corresponding positive charge $+Q$ on the upper plate of C_2. The charge $+Q$ on the upper plate of C_2 attracts a negative charge $-Q$ onto its lower plate, leaving a corresponding positive charge $+Q$ on the right plate of C_3. Finally, the positive charge on the right plate of C_3 attracts a negative charge $-Q$ onto its left plate. The result is that all three capacitors have charge of the same magnitude on their plates.

With the same charge Q on all the capacitors, the potential difference for each is as follows:

$$V_1 = \frac{Q}{C_1}, \quad V_2 = \frac{Q}{C_2}, \quad V_3 = \frac{Q}{C_3}$$

Since the total potential difference across the three capacitors must equal the emf of the battery, we have

$$\mathcal{E} = V_1 + V_2 + V_3 = \frac{Q}{C_1} + \frac{Q}{C_2} + \frac{Q}{C_3} = Q\left(\frac{1}{C_1} + \frac{1}{C_2} + \frac{1}{C_3}\right) \qquad 21\text{–}15$$

An equivalent capacitor connected to the same battery, as in **Figure 21–18 (b)**, will satisfy the relation $Q = C_{eq}\mathcal{E}$, or

$$\mathcal{E} = Q\left(\frac{1}{C_{eq}}\right) \qquad 21\text{–}16$$

A comparison of Equations 21–15 and 21–16 yields the result

$$\frac{1}{C_{eq}} = \frac{1}{C_1} + \frac{1}{C_2} + \frac{1}{C_3}$$

Thus, in general, we have the following rule for combining capacitors in series:

Equivalent Capacitance for Capacitors in Series

$$\frac{1}{C_{eq}} = \frac{1}{C_1} + \frac{1}{C_2} + \frac{1}{C_3} + \cdots = \sum \frac{1}{C} \qquad 21\text{–}17$$

SI unit: farad, F

It follows, then, that the equivalent capacitance of a group of capacitors connected in series is less than the smallest individual capacitance. In this case, it is as if the plate separations of the individual capacitors add to give a larger effective separation, and a correspondingly smaller capacitance.

More complex circuits, with some capacitors in series and others in parallel, can be handled in the same way as was done earlier with resistors. This is illustrated in the following Active Example.

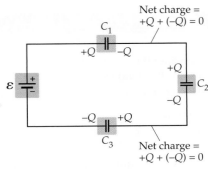

Net charge =
$+Q + (-Q) = 0$

Net charge =
$+Q + (-Q) = 0$

(a) Three capacitors in series

(b) Equivalent capacitance with same total charge

▲ **FIGURE 21–18   Capacitors in series**
(a) Three capacitors, C_1, C_2, and C_3, connected in series. Note that each capacitor has the same magnitude charge on its plates. **(b)** The equivalent capacitance, $1/C_{eq} = 1/C_1 + 1/C_2 + 1/C_3$, has the same charge as the original capacitors.

PROBLEM-SOLVING NOTE

Finding the Equivalent Capacitance of a Circuit

When calculating the equivalent capacitance of capacitors in series, be sure to take one final inverse at the end of your calculation to find C_{eq}. Also, when considering circuits with capacitors in both series and parallel, start with the smallest units of the circuit and work your way out to the larger units.

ACTIVE EXAMPLE 21–3 FIND THE EQUIVALENT CAPACITANCE AND THE STORED ENERGY

Consider the electric circuit shown here, consisting of a 12.0-V battery and three capacitors connected partly in series and partly in parallel. Find **(a)** the equivalent capacitance of this circuit and **(b)** the total energy stored in the capacitors.

SOLUTION (Test your understanding by performing the calculations indicated in each step.)

Part (a)

1. Find the equivalent capacitance of a 10.0-μF capacitor in series with a 5.00-μF capacitor: 3.33 μF

2. Find the equivalent capacitance of a 3.33-μF capacitor in parallel with a 20.0-μF capacitor: $C_{eq} = 23.3\ \mu$F

Part (b)

3. Calculate the stored energy using $U = \frac{1}{2}C_{eq}V^2$: $U = 1.68 \times 10^{-3}\,$J

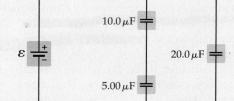

INSIGHT

Notice that the 10.0-μF capacitor and the 5.00-μF capacitor are connected in series. As you might expect, one of these capacitors stores twice as much energy as the other. Which is it? Check the Your Turn question for the answer.

YOUR TURN

Is the energy stored in the 10.0-μF capacitor greater than or less than the energy stored in the 5.0-μF capacitor? Explain. Check your answer by calculating the energy stored in each of the capacitors.

(Answers to **Your Turn** problems are given in the back of the book.)

21–7 RC Circuits

When the switch is closed on a circuit containing only batteries and capacitors, the charge on the capacitor plates appears almost instantaneously—essentially at the speed of light. This is not the case, however, in circuits that also contain resistors. In these situations, the resistors limit the rate at which charge can flow, and an appreciable amount of time may be required before the capacitors acquire a significant charge. A useful analogy is the amount of time needed to fill a bucket with water. If you use a fire hose, which has little resistance to the flow of water, the bucket fills almost instantly. If you use a garden hose, which presents a much greater resistance to the water, filling the bucket may take a minute or more.

The simplest example of such a circuit, a so-called **RC circuit,** is shown in **Figure 21–19.** Initially (before $t = 0$) the switch is open, and there is no current in the resistor or charge on the capacitor. At $t = 0$ the switch is closed and current begins to flow. If the resistor was not present, the capacitor would immediately take on the charge $Q = C\mathcal{E}$. The effect of the resistor, however, is to slow the charging process—in fact, the larger the resistance, the longer it takes for the capacitor to charge. One way to think of this is to note that as long as a current flows in the circuit, as in Figure 21–19 (b), there is a potential drop across the resistor; hence, the potential difference between the plates of the capacitor is less than the emf of the battery. With less voltage across the capacitor there will be less charge on its plates compared with the charge that would result if the plates were connected directly to the battery.

The methods of calculus can be used to show that the charge on the capacitor in Figure 21–19 varies with time as follows:

$$q(t) = C\mathcal{E}(1 - e^{-t/\tau}) \qquad\qquad 21\text{–}18$$

In this expression, e is Euler's number ($e = 2.718\ldots$) or, more precisely, the base of natural logarithms (see Appendix A). The quantity τ is referred to as the **time constant** of the circuit. The time constant is related to the resistance and capacitance of a circuit by the following simple relation: $\tau = RC$. As we shall see, τ can be thought of as a characteristic time for the behavior of an RC circuit.

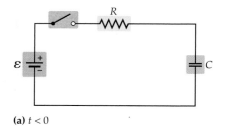

(a) $t < 0$

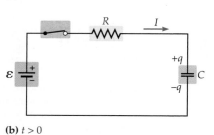

(b) $t > 0$

▲ **FIGURE 21–19 A typical RC circuit**
(a) Before the switch is closed ($t < 0$) there is no current in the circuit and no charge on the capacitor. **(b)** After the switch is closed ($t > 0$), current flows and the charge on the capacitor builds up over a finite time. As $t \rightarrow \infty$ the charge on the capacitor approaches $Q = C\mathcal{E}$.

For example, at time $t = 0$ the exponential term is $e^{-0/\tau} = e^0 = 1$; therefore, the charge on the capacitor is zero at $t = 0$, as expected:

$$q(0) = C\mathcal{E}(1 - 1) = 0$$

In the opposite limit, $t \to \infty$, the exponential vanishes: $e^{-\infty/\tau} = 0$. Thus the charge in this limit is $C\mathcal{E}$:

$$q(t \to \infty) = C\mathcal{E}(1 - 0) = C\mathcal{E}$$

This is just the charge Q the capacitor would have had from $t = 0$ on if there had been no resistor in the circuit. Finally, at time $t = \tau$ the charge on the capacitor is $q = C\mathcal{E}(1 - e^{-1}) = C\mathcal{E}(1 - 0.368) = 0.632C\mathcal{E}$, which is 63.2% of its final charge. The charge on the capacitor as a function of time is plotted in **Figure 21–20**.

Before we continue, let's check to see that the quantity $\tau = RC$ is in fact a time. Suppose, for example, that the resistor and capacitor in an RC circuit have the values $R = 120\ \Omega$ and $C = 3.5\ \mu F$, respectively. Multiplying R and C we find

$$\tau = RC = (120\ \text{ohm})(3.5 \times 10^{-6}\ \text{farad})$$

$$= \left(\frac{120\ \text{volt}}{\text{coulomb/second}}\right)\left(\frac{3.5 \times 10^{-6}\ \text{coulomb}}{\text{volt}}\right) = 4.2 \times 10^{-4}\ \text{second}$$

The tick marks on the horizontal axis in Figure 21–20 indicate the times $\tau, 2\tau, 3\tau$, and 4τ. Notice that the capacitor is almost completely charged by the time $t = 4\tau$.

Figure 21–20 also shows that the charge on the capacitor increases rapidly initially, indicating a large current in the circuit. Eventually, the charging slows down, because the greater the charge on the capacitor, the harder it is to transfer additional charge against the electrical repulsive force. Later, the charge barely changes with time, which means that the current is essentially zero. In fact, the mathematical expression for the current—again derived from calculus—is the following:

$$I(t) = \left(\frac{\mathcal{E}}{R}\right)e^{-t/\tau} \qquad \text{21–19}$$

This expression is plotted in **Figure 21–21**, where we see that significant variation in the current occurs over times ranging from $t = 0$ to $t \sim 4\tau$. At time $t = 0$ the current is $I(0) = \mathcal{E}/R$, which is the value it would have if the capacitor were replaced by an ideal wire. As $t \to \infty$, the current approaches zero, as expected: $I(t \to \infty) \to 0$. In this limit, the capacitor is essentially fully charged, so that no more charge can flow onto its plates. Thus, in this limit, the capacitor behaves like an open switch.

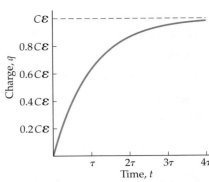

▲ **FIGURE 21–20 Charge versus time for the *RC* circuit in Figure 21–19**

The horizontal axis shows time in units of the characteristic time, $\tau = RC$. The vertical axis shows the magnitude of the charge on the capacitor in units of $C\mathcal{E}$.

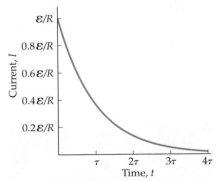

▲ **FIGURE 21–21 Current versus time for the *RC* circuit in Figure 21–19**

Initially the current is $\mathcal{E}/R$, the same as if the capacitor were not present. The current approaches zero after a period equal to several time constants, $\tau = RC$.

EXAMPLE 21–9   CHARGING A CAPACITOR

A circuit consists of a 126-Ω resistor, a 275-Ω resistor, a 182-μF capacitor, a switch, and a 3.00-V battery all connected in series. Initially the capacitor is uncharged and the switch is open. At time $t = 0$ the switch is closed. **(a)** What charge will the capacitor have a long time after the switch is closed? **(b)** At what time will the charge on the capacitor be 80.0% of the value found in part (a)?

PICTURE THE PROBLEM
The circuit described in the problem statement is shown with the switch in the open position. Once the switch is closed at $t = 0$, current will flow in the circuit and charge will begin to accumulate on the capacitor plates.

STRATEGY

a. A long time after the switch is closed, the current stops and the capacitor is fully charged. At this point, the voltage across the capacitor is equal to the emf of the battery. Therefore, the charge on the capacitor is $Q = C\mathcal{E}$.

b. To find the time when the charge will be 80.0% of the full charge, $Q = C\mathcal{E}$, we can set $q(t) = C\mathcal{E}(1 - e^{-t/\tau}) = 0.800C\mathcal{E}$ and solve for the desired time, t.

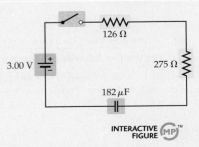

INTERACTIVE FIGURE

CONTINUED ON NEXT PAGE

CONTINUED FROM PREVIOUS PAGE
SOLUTION

Part (a)

1. Evaluate $Q = C\mathcal{E}$ for this circuit:

$$Q = C\mathcal{E} = (182\ \mu\text{F})(3.00\ \text{V}) = 546\ \mu\text{C}$$

Part (b)

2. Set $q(t) = 0.800C\mathcal{E}$ in $q(t) = C\mathcal{E}(1 - e^{-t/\tau})$ and cancel $C\mathcal{E}$:

$$q(t) = 0.800C\mathcal{E} = C\mathcal{E}(1 - e^{-t/\tau})$$
$$0.800 = 1 - e^{-t/\tau}$$

3. Solve for t in terms of the time constant τ:

$$e^{-t/\tau} = 1 - 0.800 = 0.200$$
$$t = -\tau \ln(0.200)$$

4. Calculate τ and use the result to find the time t:

$$\tau = RC = (126\ \Omega + 275\ \Omega)(182\ \mu\text{F}) = 73.0\ \text{ms}$$
$$t = -(73.0\ \text{ms}) \ln(0.200)$$
$$= -(73.0\ \text{ms})(-1.61) = 118\ \text{ms}$$

INSIGHT
Note that the time required for the charge on a capacitor to reach 80.0% of its final value is 1.61 time constants. This result is independent of the values of R and C in an RC circuit.

PRACTICE PROBLEM
What is the current in this circuit at the time found in part (b)? [**Answer:** $I(t) = (\mathcal{E}/R)e^{-t/\tau} = [(3.00\ \text{V})/(126\ \Omega + 275\ \Omega)](0.200) = (7.48\ \text{mA})(0.200) = 1.50\ \text{mA}]$

Some related homework problems: Problem 79, Problem 82

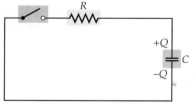

(a) $t < 0$ **(b)** $t > 0$

▲ **FIGURE 21–22 Discharging a capacitor**
(a) A charged capacitor is connected to a resistor. Initially the circuit is open, and no current can flow. **(b)** When the switch is closed, current flows from the $+$ plate of the capacitor to the $-$ plate. The charge remaining on the capacitor approaches zero after several time units, RC.

Similar behavior occurs when a charged capacitor is allowed to discharge, as in **Figure 21–22**. In this case, the initial charge on the capacitor is Q. If the switch is closed at $t = 0$, the charge for later times is

$$q(t) = Qe^{-t/\tau} \qquad\qquad 21\text{–}20$$

Like charging, the discharging of a capacitor occurs with a characteristic time $\tau = RC$.

To summarize, circuits with resistors and capacitors have the following general characteristics:

▲ A modern-day circuit board incorporates numerous resistors (cylinders with colored bands) and capacitors (yellow cylinders and metal container).

PROBLEM-SOLVING NOTE

The Limiting Behavior of Capacitors

Capacitors in dc circuits act like short circuits at $t = 0$ and open circuits as $t \to \infty$.

- Charging and discharging occur over a finite, characteristic time given by the time constant, $\tau = RC$.
- At $t = 0$ current flows freely through a capacitor being charged; it behaves like a short circuit.
- As $t \to \infty$ the current flowing into a capacitor approaches zero. In this limit, a capacitor behaves like an open switch.

We explore these features further in the following Conceptual Checkpoint.

CONCEPTUAL CHECKPOINT 21–4   CURRENT IN AN *RC* CIRCUIT

What current flows through the battery in this circuit **(a)** immediately after the switch is closed and **(b)** a long time after the switch is closed?

REASONING AND DISCUSSION

a. Immediately after the switch is closed, the capacitor acts like a short circuit; that is, as if the battery were connected to two resistors R in parallel. The equivalent resistance in this case is $R/2$; therefore, the current is $I = \mathcal{E}/(R/2) = 2\mathcal{E}/R$.

b. After current has been flowing in the circuit for a long time, the capacitor acts like an open switch. Now current can flow only through the one resistor, R; hence, the current is $I = \mathcal{E}/R$, half of its initial value.

ANSWER

(a) The current is $2\mathcal{E}/R$; **(b)** the current is $\mathcal{E}/R$.

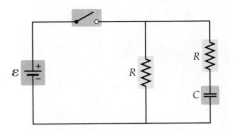

The fact that *RC* circuits have a characteristic time makes them useful in a variety of different applications. On a rather mundane level, *RC* circuits are used to determine the time delay on windshield wipers. When you adjust the delay knob in your car, you change a resistance or a capacitance, which in turn changes the time constant of the circuit. This results in a greater or a smaller delay. The blinking rate of turn signals is also determined by the time constant of an *RC* circuit.

A more critical application of *RC* circuits is the heart pacemaker. In the simplest case, these devices use an *RC* circuit to deliver precisely timed pulses directly to the heart. The more sophisticated pacemakers available today can even "sense" when a patient's heart rate falls below a predetermined value. The pacemaker then begins sending appropriate pulses to the heart to increase its rate. Many pacemakers can even be reprogrammed after they are surgically implanted to respond to changes in a patient's condition.

Normally, the heart's rate of beating is determined by its own natural pacemaker, the sinoatrial or SA node, located in the upper right chamber of the heart. If the SA node is not functioning properly, it may cause the heart to beat slowly or irregularly. To correct the problem, a pacemaker is implanted just under the collarbone, and an electrode is introduced intravenously via the cephalic vein. The distal end of the electrode is positioned, with the aid of fluoroscopic guidance, in the right ventricular apex. From that point on, the operation of the pacemaker follows the basic principles of electric circuits, as described in this chapter.

REAL-WORLD PHYSICS

Delay circuits in windshield wipers and turn signals

REAL-WORLD PHYSICS: BIO

Pacemakers

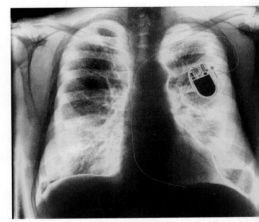

▲ An X-ray showing a pacemaker installed in a person's chest. The timing of the electrical pulses that keep the heart beating regularly is determined by an *RC* circuit powered by a small, long-lived battery.

*21–8 Ammeters and Voltmeters

Devices for measuring currents and voltages in a circuit are referred to as **ammeters** and **voltmeters,** respectively. In each case, the ideal situation is for the meter to measure the desired quantity without altering the characteristics of the circuit being studied. This is accomplished in different ways for these two types of meters, as we shall see.

First, the ammeter is designed to measure the flow of current through a particular portion of a circuit. For example, we may want to know the current flowing between points A and B in the circuit shown in **Figure 21–23 (a)**. To measure this current, we insert the ammeter into the circuit in such a way that all the current flowing from A to B must also flow through the meter. This is done by connecting the meter "in series" with the other circuit elements between A and B, as indicated in **Figure 21–23 (b)**.

If the ammeter has a finite resistance—which must be the case for real meters—the presence of the meter in the circuit will alter the current it is intended to measure. Thus, an *ideal* ammeter would be one with zero resistance. In practice, if the resistance of the ammeter is much less than the other resistances in the circuit, its reading will be reasonably accurate.

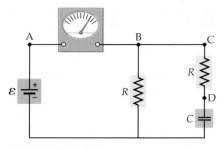

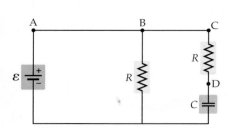

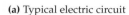

(a) Typical electric circuit

(b) Measuring the current between A and B

▲ **FIGURE 21–23 Measuring the current in a circuit**
To measure the current flowing between points A and B in **(a)**, an ammeter is inserted into the circuit, as shown in **(b)**. An ideal ammeter would have zero resistance.

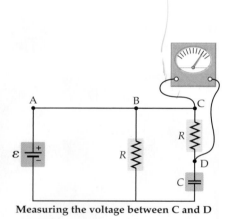

▲ A typical digital multimeter, which can measure resistance (teal settings), current (yellow settings), or voltage (red settings). This meter is measuring the voltage of a "9 volt" battery.

Measuring the voltage between C and D

Second, a voltmeter measures the potential drop between any two points in a circuit. Referring again to the circuit in Figure 21–23 (a), we may wish to know the difference in potential between points C and D. To measure this voltage, we connect the voltmeter "in parallel" to the circuit at the appropriate points, as in **Figure 21–24**.

A real voltmeter always allows some current to flow through it, which means that the current flowing through the circuit is less than before the meter was connected. As a result, the measured voltage is altered from its ideal value. An *ideal* voltmeter, then, would be one in which the resistance is infinite, so that the current it draws from the circuit is negligible. In practical situations it is sufficient that the resistance of the meter be much greater than the resistances in the circuit.

Sometimes the functions of an ammeter, voltmeter, and ohmmeter are combined in a single device called a **multimeter.** Adjusting the settings on a multimeter allows a variety of circuit properties to be measured.

◀ **FIGURE 21–24 Measuring the voltage in a circuit**
The voltage difference between points C and D can be measured by connecting a voltmeter in parallel to the original circuit. An ideal voltmeter would have infinite resistance.

THE BIG PICTURE	PUTTING PHYSICS IN CONTEXT
LOOKING BACK	**LOOKING AHEAD**
The concept of electric potential energy (Chapter 20) is used in Section 21–3, where we talk about the energy associated with an electric circuit.	A dc circuit with a current flowing through it will play an important role in our discussion of magnetism in Chapter 22. We will also consider the magnetic force exerted on a current-carrying wire in Chapter 22.
We also discuss the power of an electric circuit in Section 21–3. For this we refer back to mechanics, where power was originally introduced in Chapter 7.	In Chapter 24 we extend our discussion of electric circuits from those in which the current flows in only one direction (dc) to circuits in which the current alternates in direction (ac, or alternating current). We will again use resistors and capacitors in the ac circuits.
Capacitors, first introduced in Chapter 20, are used in dc circuits in Section 21–6.	
	A simple dc circuit appears in Chapter 30, where we discuss the photoelectric effect and its importance in the development of quantum mechanics.

CHAPTER SUMMARY

21–1 ELECTRIC CURRENT

Electric current is the flow of electric charge.

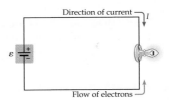

Definition
If a charge ΔQ passes a given point in the time Δt, the corresponding electric current is

$$I = \frac{\Delta Q}{\Delta t} \qquad\qquad 21\text{–}1$$

Ampere
The unit of current is the ampere, or amp for short. By definition, 1 amp is one coulomb per second; $1\ \text{A} = 1\ \text{C/s}$.

Battery
A battery is a device that uses chemical reactions to produce a potential difference between its two terminals.

Electromotive Force
The electromotive force, or emf, $\mathcal{E}$, is the potential difference between the terminals of a battery under ideal conditions.

Work Done by a Battery
As a battery moves a charge ΔQ around a circuit, it does the work $W = (\Delta Q)\mathcal{E}$.

Direction of Current
By definition, the direction of the current I in a circuit is the direction in which *positive* charges would move. The actual charge carriers, however, are generally electrons; hence, they move in the opposite direction to I.

21–2 RESISTANCE AND OHM'S LAW

When electrons move through a wire, they encounter resistance to their motion. In order to move electrons against this resistance, it is necessary to apply a potential difference between the ends of the wire.

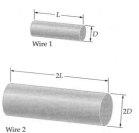

Ohm's Law
To produce a current I through a wire with resistance R the following potential difference, V, is required:

$$V = IR \qquad\qquad 21\text{–}2$$

Resistivity
The resistivity ρ of a material determines how much resistance it gives to the flow of electric current.

Resistance of a Wire
The resistance of a wire of length L, cross-sectional area A, and resistivity ρ is

$$R = \rho\!\left(\frac{L}{A}\right) \qquad\qquad 21\text{–}3$$

Temperature Dependence
The resistivity of most metals increases approximately linearly with temperature.

Superconductivity
Below a certain critical temperature, T_c, certain materials lose all electrical resistance. A current flowing in a superconductor can continue undiminished as long as its temperature is maintained below T_c.

21–3 ENERGY AND POWER IN ELECTRIC CIRCUITS

In general, energy is required to cause an electric current to flow through a circuit. The rate at which the energy must be supplied is the power.

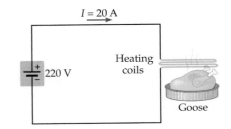

Electrical Power
If a current I flows across a potential difference V, the corresponding electrical power is

$$P = IV \qquad\qquad 21\text{–}4$$

Power Dissipation in a Resistor

If a potential difference V produces a current I in a resistor R, the electrical power converted to heat is

$$P = I^2 R = V^2/R \qquad \text{21–5, 21–6}$$

Energy Usage and the Kilowatt-Hour

The energy equivalent of one kilowatt-hour (kWh) is

$$1 \text{ kWh} = 3.6 \times 10^6 \text{ J}$$

21–4 RESISTORS IN SERIES AND PARALLEL

Resistors connected end to end—so that the same current flows through each one—are said to be in series. Resistors connected across the same potential difference—allowing parallel paths for the current to flow—are said to be connected in parallel.

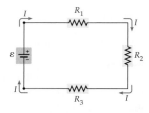

Series

The equivalent resistance, R_{eq}, of resistors connected in series is equal to the sum of the individual resistances:

$$R_{eq} = R_1 + R_2 + R_3 + \cdots = \sum R \qquad \text{21–7}$$

Parallel

The equivalent resistance, R_{eq}, of resistors connected in parallel is given by the following:

$$\frac{1}{R_{eq}} = \frac{1}{R_1} + \frac{1}{R_2} + \frac{1}{R_3} + \cdots = \sum \frac{1}{R} \qquad \text{21–10}$$

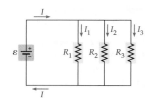

21–5 KIRCHHOFF'S RULES

Kirchhoff's rules are statements of charge conservation and energy conservation as applied to closed electric circuits.

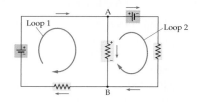

Junction Rule (Charge Conservation)

The algebraic sum of all currents meeting at a junction must equal zero. Currents entering the junction are taken to be positive; currents leaving are taken to be negative.

Loop Rule (Energy Conservation)

The algebraic sum of all potential differences around a closed loop is zero. The potential increases in going from the $-$ to the $+$ terminal of a battery and decreases when crossing a resistor in the direction of the current.

21–6 CIRCUITS CONTAINING CAPACITORS

Capacitors connected end to end—so that the same charge is on each one—are said to be in series. Capacitors connected across the same potential difference are said to be connected in parallel.

Parallel

The equivalent capacitance, C_{eq}, of capacitors connected in parallel is equal to the sum of the individual capacitances:

$$C_{eq} = C_1 + C_2 + C_3 + \cdots = \sum C \qquad \text{21–14}$$

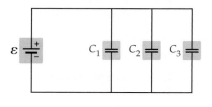

Series

The equivalent capacitance, C_{eq}, of capacitors connected in series is given by

$$\frac{1}{C_{eq}} = \frac{1}{C_1} + \frac{1}{C_2} + \frac{1}{C_3} + \cdots = \sum \frac{1}{C} \qquad \text{21–17}$$

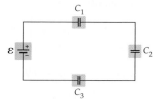

21–7 RC CIRCUITS

In circuits containing both resistors and capacitors, there is a characteristic time, $\tau = RC$, during which significant changes occur. This time is referred to as the time constant. The simplest such circuit, known as an RC circuit, consists of one resistor and one capacitor connected in series.

Charging a Capacitor

The charge on a capacitor in an RC circuit varies with time as follows:

$$q(t) = C\mathcal{E}(1 - e^{-t/\tau})$$ 21–18

The corresponding current is given by

$$I(t) = \left(\frac{\mathcal{E}}{R}\right)e^{-t/\tau}$$ 21–19

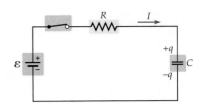

Discharging a Capacitor

If a capacitor in an RC circuit starts with a charge Q at time $t = 0$, its charge at all later times is

$$q(t) = Qe^{-t/\tau}$$ 21–20

Behavior near $t = 0$

Just after the switch is closed in an RC circuit, capacitors behave like ideal wires—that is, they offer no resistance to the flow of current.

Behavior as $t \rightarrow \infty$

Long after the switch is closed in an RC circuit, capacitors behave like open circuits.

*21–8 AMMETERS AND VOLTMETERS

Ammeters and voltmeters are devices for measuring currents and voltages, respectively, in electric circuits.

Ammeter

An ammeter is connected in series with the section of the circuit in which the current is to be measured. In the ideal case, an ammeter's resistance is zero.

Voltmeter

A voltmeter is connected in parallel with the portion of the circuit to be measured. In the ideal case, a voltmeter's resistance is infinite.

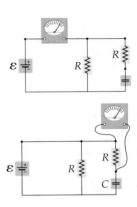

PROBLEM-SOLVING SUMMARY

Type of Problem	Relevant Physical Concepts	Related Examples
Find the work done by a battery.	The work done by a battery is the charge that passes through the battery times the emf of the battery: $W = \Delta Q \mathcal{E}$.	Active Example 21–1
Relate resistance to resistivity.	The resistance of a wire is its resistivity, ρ, times its length, divided by its cross-sectional area: $R = \rho(L/A)$.	Example 21–2
Relate the power in an electric circuit to the current, voltage, and resistance.	The basic definition of electrical power is current times voltage: $P = IV$. Using Ohm's law when appropriate, the power can also be expressed as $P = I^2R$ and $P = V^2/R$.	Examples 21–3, 21–4
Determine the equivalent resistance of resistors in series and parallel.	Resistors in series simply add: $R_{eq} = R_1 + R_2 + \cdots$; resistors in parallel add in terms of inverses: $1/R_{eq} = 1/R_1 + 1/R_2 + \cdots$.	Examples 21–5, 21–6, 21–7
Find the current in a circuit containing resistors that are not simply in series or parallel.	Apply Kirchhoff's junction rule (the algebraic sum of currents at a junction must be zero) and loop rule (the algebraic sum of potential difference around a loop is zero).	Active Example 21–2
Determine the equivalent capacitance of capacitors in series and parallel.	Capacitors in parallel simply add: $C_{eq} = C_1 + C_2 + \cdots$; capacitors in series add in terms of inverses: $1/C_{eq} = 1/C_1 + 1/C_2 + \cdots$.	Example 21–8 Active Example 21–3
Find the charge and the current in an RC circuit as a function of time.	The charge and current in an RC circuit during charging vary exponentially with time as follows: $q(t) = C\mathcal{E}(1 - e^{-t/\tau})$; $I(t) = (\mathcal{E}/R)e^{-t/\tau}$. The characteristic time is $\tau = RC$.	Example 21–9

CONCEPTUAL QUESTIONS

For instructor-assigned homework, go to www.masteringphysics.com

(Answers to odd-numbered Conceptual Questions can be found in the back of the book.)

1. What is the direction of the electric current produced by an electron that falls toward the ground?

2. Your body is composed of electric charges. Does it follow, then, that you produce an electric current when you walk?

3. Suppose you charge a comb by rubbing it through your hair. Do you produce a current when you walk across the room carrying the comb?

4. Suppose you charge a comb by rubbing it through the fur on your dog's back. Do you produce a current when you walk across the room carrying the comb?

5. An electron moving through a wire has an average drift speed that is very small. Does this mean that its instantaneous velocity is also very small?

6. Are car headlights connected in series or parallel? Give an everyday observation that supports your answer.

7. Give an example of how four resistors of resistance R can be combined to produce an equivalent resistance of R.

8. Is it possible to connect a group of resistors of value R in such a way that the equivalent resistance is less than R? If so, give a specific example.

9. What physical quantity do resistors connected in series have in common?

10. What physical quantity do resistors connected in parallel have in common?

11. Explain how electrical devices can begin operating almost immediately after you throw a switch, even though individual electrons in the wire may take hours to reach the device.

12. Explain the difference between resistivity and resistance.

13. Explain why birds can roost on high-voltage wire without being electrocuted.

14. List two electrical applications that would benefit from room-temperature superconductors. List two applications for which room-temperature superconductivity would not be beneficial.

15. On what basic conservation laws are Kirchhoff's rules based?

16. What physical quantity do capacitors connected in series have in common?

17. What physical quantity do capacitors connected in parallel have in common?

18. Consider the circuit shown in **Figure 21–25**, in which a light of resistance R and a capacitor of capacitance C are connected in series. The capacitor has a large capacitance, and is initially uncharged. The battery provides enough power to light the bulb when connected to the battery directly. Describe the behavior of the light after the switch is closed.

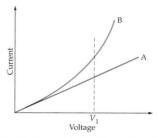

▲ **FIGURE 21–25** Conceptual Question 18

PROBLEMS AND CONCEPTUAL EXERCISES

Note: Answers to odd-numbered Problems and Conceptual Exercises can be found in the back of the book. **IP** *denotes an integrated problem, with both conceptual and numerical parts;* **BIO** *identifies problems of biological or medical interest;* **CE** *indicates a conceptual exercise.* **Predict/Explain** *problems ask for two responses:* **(a)** *your prediction of a physical outcome, and* **(b)** *the best explanation among three provided. On all problems, red bullets (•, ••, •••) are used to indicate the level of difficulty.*

SECTION 21–1 ELECTRIC CURRENT

1. • How many coulombs of charge are in one ampere-hour?

2. • A flashlight bulb carries a current of 0.18 A for 78 s. How much charge flows through the bulb in this time? How many electrons?

3. • The picture tube in a particular television draws a current of 15 A. How many electrons strike the viewing screen every second?

4. • **IP** A car battery does 260 J of work on the charge passing through it as it starts an engine. **(a)** If the emf of the battery is 12 V, how much charge passes through the battery during the start? **(b)** If the emf is doubled to 24 V, does the amount of charge passing through the battery increase or decrease? By what factor?

5. • Highly sensitive ammeters can measure currents as small as 10.0 fA. How many electrons per second flow through a wire with a 10.0-fA current?

6. •• A television set connected to a 120-V outlet consumes 78 W of power. **(a)** How much current flows through the television? **(b)** How long does it take for 10 million electrons to pass through the TV?

7. ••**BIO** **Pacemaker Batteries** Pacemakers designed for long-term use commonly employ a lithium–iodine battery capable of

supplying 0.42 A·h of charge. **(a)** How many coulombs of charge can such a battery supply? **(b)** If the average current produced by the pacemaker is 5.6 μA, what is the expected lifetime of the device?

SECTION 21–2 RESISTANCE AND OHM'S LAW

8. • **CE** A conducting wire is quadrupled in length and tripled in diameter. **(a)** Does its resistance increase, decrease, or stay the same? Explain. **(b)** By what factor does its resistance change?

9. • **CE** Figure 21–26 shows a plot of current versus voltage for two different materials, A and B. Which of these materials satisfies Ohm's law? Explain.

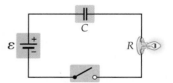

▲ **FIGURE 21–26** Problems 9 and 10

10. • **CE Predict/Explain** Current-versus-voltage plots for two materials, A and B, are shown in Figure 21–26. **(a)** Is the resistance of material A greater than, less than, or equal to the resistance of material B at the voltage V_1? **(b)** Choose the *best explanation* from among the following:
    I. Curve B is higher in value than curve A.
    II. A larger slope means a larger value of I/V, and hence a smaller value of R.
    III. Curve B has the larger slope at the voltage V_1 and hence the larger resistance.

11. • **CE** Two cylindrical wires are made of the same material and have the same length. If wire B is to have nine times the resistance of wire A, what must be the ratio of their radii, r_B/r_A?

12. • A silver wire is 5.9 m long and 0.49 mm in diameter. What is its resistance?

13. • When a potential difference of 18 V is applied to a given wire, it conducts 0.35 A of current. What is the resistance of the wire?

14. • The tungsten filament of a lightbulb has a resistance of 0.07 Ω. If the filament is 27 cm long, what is its diameter?

15. • What is the resistance of 6.0 mi of copper wire with a diameter of 0.55 mm?

16. •• **CE** The four conducting cylinders shown in **Figure 21–27** are all made of the same material, though they differ in length and/or diameter. They are connected to four different batteries, which supply the necessary voltages to give the circuits the same current, I. Rank the four voltages, V_1, V_2, V_3, and V_4, in order of increasing value. Indicate ties where appropriate.

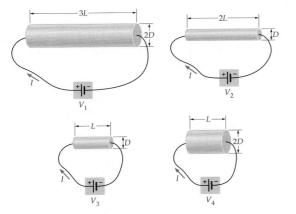

▲ **FIGURE 21–27** Problem 16

17. •• **IP** A bird lands on a bare copper wire carrying a current of 32 A. The wire is 8 gauge, which means that its cross-sectional area is 0.13 cm². **(a)** Find the difference in potential between the bird's feet, assuming they are separated by a distance of 6.0 cm. **(b)** Will your answer to part (a) increase or decrease if the separation between the bird's feet increases? Explain.

18. •• A current of 0.96 A flows through a copper wire 0.44 mm in diameter when it is connected to a potential difference of 15 V. How long is the wire?

19. •• **IP BIO Current Through a Cell Membrane** A typical cell membrane is 8.0 nm thick and has an electrical resistivity of 1.3×10^7 Ω · m. **(a)** If the potential difference between the inner and outer surfaces of a cell membrane is 75 mV, how much current flows through a square area of membrane 1.0 μm on a side? **(b)** Suppose the thickness of the membrane is doubled, but the resistivity and potential difference remain the same. Does the current increase or decrease? By what factor?

20. •• When a potential difference of 12 V is applied to a wire 6.9 m long and 0.33 mm in diameter, the result is an electric current of 2.1 A. What is the resistivity of the wire?

21. •• **IP (a)** What is the resistance per meter of an aluminum wire with a cross-sectional area of 2.4×10^{-7} m². **(b)** Would your answer to part (a) increase, decrease, or stay the same if the diameter of the wire were increased? Explain. **(c)** Repeat part (a) for a wire with a cross-sectional area of 3.6×10^{-7} m².

22. •• **BIO Resistance and Current in the Human Finger** The interior of the human body has an electrical resistivity of 0.15 Ω · m. **(a)** Estimate the resistance for current flowing the length of your index finger. (For this calculation, ignore the much higher resistivity of your skin.) **(b)** Your muscles will contract when they carry a current greater than 15 mA. What voltage is required to produce this current through your finger?

23. ••• Consider a rectangular block of metal of height A, width B, and length C, as shown in **Figure 21–28**. If a potential difference V is maintained between the two $A \times B$ faces of the block, a current I_{AB} is observed to flow. Find the current that flows if the same potential difference V is applied between the two $B \times C$ faces of the block. Give your answer in terms of I_{AB}.

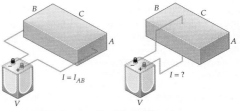

▲ **FIGURE 21–28** Problem 23

SECTION 21–3 ENERGY AND POWER IN ELECTRIC CIRCUITS

24. • **CE** Light A has four times the power rating of light B when operated at the same voltage. **(a)** Is the resistance of light A greater than, less than, or equal to the resistance of light B? Explain. **(b)** What is the ratio of the resistance of light A to the resistance of light B?

25. • **CE** Two lightbulbs operate on the same potential difference. Bulb A has four times the power output of bulb B. **(a)** Which bulb has the greater current passing through it? Explain. **(b)** What is the ratio of the current in bulb A to the current in bulb B?

26. • **CE** Two lightbulbs operate on the same current. Bulb A has four times the power output of bulb B. **(a)** Is the potential difference across bulb A greater than or less than the potential difference across bulb B? Explain. **(b)** What is the ratio of the potential difference across bulb A to that across bulb B?

27. • A 75-V generator supplies 3.8 kW of power. How much current does the generator produce?

28. • A portable CD player operates with a current of 22 mA at a potential difference of 4.1 V. What is the power usage of the player?

29. • Find the power dissipated in a 25-Ω electric heater connected to a 120-V outlet.

30. • The current in a 120-V reading lamp is 2.6 A. If the cost of electrical energy is $0.075 per kilowatt-hour, how much does it cost to operate the light for an hour?

31. • It costs 2.6 cents to charge a car battery at a voltage of 12 V and a current of 15 A for 120 minutes. What is the cost of electrical energy per kilowatt-hour at this location?

32. •• **IP** A 75-W lightbulb operates on a potential difference of 95 V. Find **(a)** the current in the bulb and **(b)** the resistance of the bulb. **(c)** If this bulb is replaced with one whose resistance is half the value found in part (b), is its power rating greater than or less than 75 W? By what factor?

33. •• **Rating Car Batteries** Car batteries are rated by the following two numbers: **(1)** cranking amps = current the battery can produce for 30.0 seconds while maintaining a terminal voltage of at least 7.2 V and **(2)** reserve capacity = number of minutes the battery can produce a 25-A current while maintaining a terminal voltage of at least 10.5 V. One particular battery is advertised as having 905 cranking amps and a 155-minute reserve capacity. Which of these two ratings represents the greater amount of energy delivered by the battery?

SECTION 21–4 RESISTORS IN SERIES AND PARALLEL

34. • **CE Predict/Explain** A dozen identical lightbulbs are connected to a given emf. **(a)** Will the lights be brighter if they are connected in series or in parallel? **(b)** Choose the *best explanation* from among the following:
I. When connected in parallel each bulb experiences the maximum emf and dissipates the maximum power.
II. Resistors in series have a larger equivalent resistance and dissipate more power.
III. Resistors in parallel have a smaller equivalent resistance and dissipate less power.

35. • **CE Predict/Explain** A *fuse* is a device to protect a circuit from the effects of a large current. The fuse is a small strip of metal that burns through when the current in it exceeds a certain value, thus producing an open circuit. **(a)** Should a fuse be connected in series or in parallel with the circuit it is intended to protect? **(b)** Choose the *best explanation* from among the following:
I. Either connection is acceptable; the main thing is to have a fuse in the circuit.
II. The fuse should be connected in parallel, otherwise it will interrupt the current in the circuit.
III. With the fuse connected in series, the current in the circuit drops to zero as soon as the fuse burns through.

36. • **CE** A circuit consists of three resistors, $R_1 < R_2 < R_3$, connected in series to a battery. Rank these resistors in order of increasing **(a)** current through them and **(b)** potential difference across them. Indicate ties where appropriate.

37. • **CE Predict/Explain** Two resistors are connected in parallel. **(a)** If a third resistor is now connected in parallel with the original two, does the equivalent resistance of the circuit increase, decrease, or remain the same? **(b)** Choose the *best explanation* from among the following:
I. Adding a resistor generally tends to increase the resistance, but putting it in parallel tends to decrease the resistance; therefore the effects offset and the resistance stays the same.
II. Adding more resistance to the circuit will increase the equivalent resistance.
III. The third resistor gives yet another path for current to flow in the circuit, which means that the equivalent resistance is less.

38. • Find the equivalent resistance between points A and B for the group of resistors shown in **Figure 21–29**.

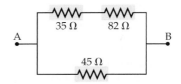

▲ **FIGURE 21–29** Problems 38 and 115

39. • What is the minimum number of 65-Ω resistors that must be connected in parallel to produce an equivalent resistance of 11 Ω or less?

40. •• Four lightbulbs (A, B, C, D) are connected together in a circuit of unknown arrangement. When each bulb is removed one at a time and replaced, the following behavior is observed:

	A	B	C	D
A removed	*	on	on	on
B removed	on	*	on	off
C removed	off	off	*	off
D removed	on	off	on	*

Draw a circuit diagram for these bulbs.

41. •• Your toaster has a power cord with a resistance of 0.020 Ω connected in series with a 9.6-Ω nichrome heating element. If the potential difference between the terminals of the toaster is 120 V, how much power is dissipated in **(a)** the power cord and **(b)** the heating element?

42. •• A hobbyist building a radio needs a 150-Ω resistor in her circuit, but has only a 220-Ω, a 79-Ω, and a 92-Ω resistor available. How can she connect these resistors to produce the desired resistance?

43. •• A circuit consists of a 12.0-V battery connected to three resistors (42 Ω, 17 Ω, and 110 Ω) in series. Find **(a)** the current that flows through the battery and **(b)** the potential difference across each resistor.

44. •• **IP** Three resistors, 11 Ω, 53 Ω, and R, are connected in series with a 24.0-V battery. The total current flowing through the battery is 0.16 A. **(a)** Find the value of resistance R. **(b)** Find the potential difference across each resistor. **(c)** If the voltage of the battery had been greater than 24.0 V, would your answer to part (a) have been larger or smaller? Explain.

45. •• A circuit consists of a battery connected to three resistors (65 Ω, 25 Ω, and 170 Ω) in parallel. The total current through the resistors is 1.8 A. Find **(a)** the emf of the battery and **(b)** the current through each resistor.

46. •• **IP** Three resistors, 22 Ω, 67 Ω, and R, are connected in parallel with a 12.0-V battery. The total current flowing through the battery is 0.88 A. **(a)** Find the value of resistance R. **(b)** Find the current through each resistor. **(c)** If the total current in the battery had been greater than 0.88 A, would your answer to part (a) have been larger or smaller? Explain.

47. •• An 89-Ω resistor has a current of 0.72 A and is connected in series with a 130-Ω resistor. What is the emf of the battery to which the resistors are connected?

48. •• The equivalent resistance between points A and B of the resistors shown in **Figure 21–30** is 26 Ω. Find the value of resistance R.

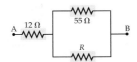

▲ **FIGURE 21–30** Problems 48, 52, and 98

49. •• Find the equivalent resistance between points A and B shown in **Figure 21–31**.

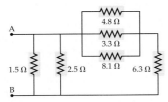

▲ **FIGURE 21–31** Problems 49 and 53

50. •• How many 65-W lightbulbs can be connected in parallel across a potential difference of 85 V before the total current in the circuit exceeds 2.1 A?

51. •• The circuit in **Figure 21–32** includes a battery with a finite internal resistance, $r = 0.50 \, \Omega$. **(a)** Find the current flowing through the 7.1-Ω and the 3.2-Ω resistors. **(b)** How much current flows through the battery? **(c)** What is the potential difference between the terminals of the battery?

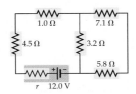

▲ **FIGURE 21–32** Problems 51 and 54

52. •• **IP** A 12-V battery is connected to terminals A and B in Figure 21–30. **(a)** Given that $R = 85 \, \Omega$, find the current in each resistor. **(b)** Suppose the value of R is increased. For each resistor in turn, state whether the current flowing through it increases or decreases. Explain.

53. •• **IP** The terminals A and B in Figure 21–31 are connected to a 9.0-V battery. **(a)** Find the current flowing through each resistor. **(b)** Is the potential difference across the 6.3-Ω resistor greater than, less than, or the same as the potential difference across the 1.5-Ω resistor? Explain.

54. •• **IP** Suppose the battery in Figure 21–32 has an internal resistance $r = 0.25 \, \Omega$. **(a)** How much current flows through the battery? **(b)** What is the potential difference between the terminals of the battery? **(c)** If the 3.2-Ω resistor is increased in value, will the current in the battery increase or decrease? Explain.

55. ••• **IP** The current flowing through the 8.45-Ω resistor in **Figure 21–33** is 1.52 A. **(a)** What is the voltage of the battery? **(b)** If the

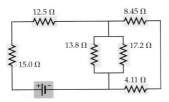

▲ **FIGURE 21–33** Problems 55 and 56

17.2-Ω resistor is increased in value, will the current provided by the battery increase, decrease, or stay the same? Explain.

56. ••• The current in the 13.8-Ω resistor in Figure 21–33 is 0.795 A. Find the current in the other resistors in the circuit.

57. ••• **IP** Four identical resistors are connected to a battery as shown in **Figure 21–34**. When the switch is open, the current through the battery is I_0. **(a)** When the switch is closed, will the current through the battery increase, decrease, or stay the same? Explain. **(b)** Calculate the current that flows through the battery when the switch is closed. Give your answer in terms of I_0.

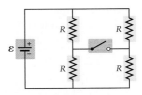

▲ **FIGURE 21–34** Problem 57

SECTION 21–5 KIRCHHOFF'S RULES

58. • Find the magnitude and direction (clockwise or counterclockwise) of the current in **Figure 21–35**.

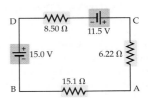

▲ **FIGURE 21–35** Problems 58, 59, and 60

59. • **IP** Suppose the polarity of the 11.5-V battery in Figure 21–35 is reversed. **(a)** Do you expect this to increase or decrease the amount of current flowing in the circuit? Explain. **(b)** Calculate the magnitude and direction (clockwise or counterclockwise) of the current in this case.

60. •• **IP** It is given that point A in Figure 21–35 is grounded ($V = 0$). **(a)** Is the potential at point B greater than or less than zero? Explain. **(b)** Is the potential at point C greater than or less than zero? Explain. **(c)** Calculate the potential at point D.

61. •• Consider the circuit shown in **Figure 21–36**. Find the current through each resistor using **(a)** the rules for series and parallel resistors and **(b)** Kirchhoff's rules.

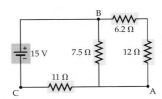

▲ **FIGURE 21–36** Problems 61 and 62

62. •• Suppose point A is grounded ($V = 0$) in Figure 21–36. Find the potential at points B and C.

63. •• **IP** **(a)** Find the current in each resistor in **Figure 21–37**. **(b)** Is the potential at point A greater than, less than, or equal to the potential at point B? Explain. **(c)** Determine the potential difference between the points A and B.

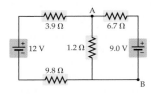

▲ **FIGURE 21–37** Problem 63

64. ••• Two batteries and three resistors are connected as shown in **Figure 21–38**. How much current flows through each battery when the switch is **(a)** closed and **(b)** open?

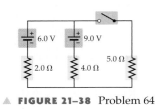

▲ **FIGURE 21–38** Problem 64

SECTION 21–6 CIRCUITS CONTAINING CAPACITORS

65. • **CE** Two capacitors, $C_1 = C$ and $C_2 = 2C$, are connected to a battery. **(a)** Which capacitor stores more energy when they are connected to the battery in series? Explain. **(b)** Which capacitor stores more energy when they are connected in parallel? Explain.

66. • **CE Predict/Explain** Two capacitors are connected in series. **(a)** If a third capacitor is now connected in series with the original two, does the equivalent capacitance increase, decrease, or remain the same? **(b)** Choose the *best explanation* from among the following:
    I. Adding a capacitor generally tends to increase the capacitance, but putting it in series tends to decrease the capacitance; therefore, the net result is no change.
    II. Adding a capacitor in series will increase the total amount of charge stored, and hence increase the equivalent capacitance.
    III. Adding a capacitor in series decreases the equivalent capacitance since each capacitor now has less voltage across it, and hence stores less charge.

67. • **CE Predict/Explain** Two capacitors are connected in parallel. **(a)** If a third capacitor is now connected in parallel with the original two, does the equivalent capacitance increase, decrease, or remain the same? **(b)** Choose the *best explanation* from among the following:
    I. Adding a capacitor tends to increase the capacitance, but putting it in parallel tends to decrease the capacitance; therefore, the net result is no change.
    II. Adding a capacitor in parallel will increase the total amount of charge stored, and hence increase the equivalent capacitance.
    III. Adding a capacitor in parallel decreases the equivalent capacitance since each capacitor now has less voltage across it, and hence stores less charge.

68. • Find the equivalent capacitance between points A and B for the group of capacitors shown in **Figure 21–39**.

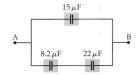

▲ **FIGURE 21–39** Problems 68 and 72

69. • A 12-V battery is connected to three capacitors in series. The capacitors have the following capacitances: $4.5\ \mu F$, $12\ \mu F$, and $32\ \mu F$. Find the voltage across the 32-μF capacitor.

70. •• **CE** You conduct a series of experiments in which you connect the capacitors C_1 and $C_2 > C_1$ to a battery in various ways. The experiments are as follows: **A**, C_1 alone connected to the battery; **B**, C_2 alone connected to the battery; **C**, C_1 and C_2 connected to the battery in series; **D**, C_1 and C_2 connected to the battery in parallel. Rank these four experiments in order of increasing equivalent capacitance. Indicate ties where appropriate.

71. •• **CE** Three different circuits, each containing a switch and two capacitors, are shown in **Figure 21–40**. Initially, the plates of the capacitors are charged as shown. The switches are then closed, allowing charge to move freely between the capacitors. Rank the circuits in order of increasing final charge on the left plate of **(a)** the upper capacitor and **(b)** the lower capacitor. Indicate ties where appropriate.

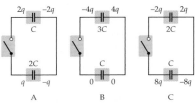

▲ **FIGURE 21–40** Problem 71

72. •• Terminals A and B in Figure 21–39 are connected to a 9.0-V battery. Find the energy stored in each capacitor.

73. •• **IP** Two capacitors, one $7.5\ \mu F$ and the other $15\ \mu F$, are connected in parallel across a 15-V battery. **(a)** Find the equivalent capacitance of the two capacitors. **(b)** Which capacitor stores more charge? Explain. **(c)** Find the charge stored on each capacitor.

74. •• **IP** Two capacitors, one $7.5\ \mu F$ and the other $15\ \mu F$, are connected in series across a 15-V battery. **(a)** Find the equivalent capacitance of the two capacitors. **(b)** Which capacitor stores more charge? Explain. **(c)** Find the charge stored on each capacitor.

75. •• The equivalent capacitance of the capacitors shown in **Figure 21–41** is $9.22\ \mu F$. Find the value of capacitance C.

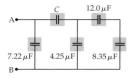

▲ **FIGURE 21–41** Problems 75 and 118

76. ••• Two capacitors, C_1 and C_2, are connected in series and charged by a battery. Show that the energy stored in C_1 plus the energy stored in C_2 is equal to the energy stored in the equivalent capacitor, C_{eq}, when it is connected to the same battery.

77. ••• With the switch in position A, the 11.2-μF capacitor in **Figure 21–42** is fully charged by the 12.0-V battery, and the

9.50-μF capacitor is uncharged. The switch is now moved to position B. As a result, charge flows between the capacitors until they have the same voltage across their plates. Find this voltage.

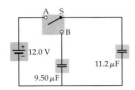

▲ **FIGURE 21–42** Problem 77

SECTION 21–7 RC CIRCUITS

78. • The switch on an *RC* circuit is closed at $t = 0$. Given that $\mathcal{E} = 9.0$ V, $R = 150\ \Omega$, and $C = 23\ \mu$F, how much charge is on the capacitor at time $t = 4.2$ ms?

79. • The capacitor in an *RC* circuit ($R = 120\ \Omega$, $C = 45\ \mu$F) is initially uncharged. Find **(a)** the charge on the capacitor and **(b)** the current in the circuit one time constant ($\tau = RC$) after the circuit is connected to a 9.0-V battery.

80. •• **CE** Three *RC* circuits have the emf, resistance, and capacitance given in the accompanying table. Initially, the switch on the circuit is open and the capacitor is uncharged. Rank these circuits in order of increasing **(a)** initial current (immediately after the switch is closed) and **(b)** time for the capacitor to acquire half its final charge. Indicate ties where appropriate.

	$\mathcal{E}$ (V)	$R\ (\Omega)$	$C\ (\mu F)$
Circuit A	12	4	3
Circuit B	9	3	1
Circuit C	9	9	2

81. •• Consider an *RC* circuit with $\mathcal{E} = 12.0$ V, $R = 175\ \Omega$, and $C = 55.7\ \mu$F. Find **(a)** the time constant for the circuit, **(b)** the maximum charge on the capacitor, and **(c)** the initial current in the circuit.

82. •• The resistor in an *RC* circuit has a resistance of 145 Ω. **(a)** What capacitance must be used in this circuit if the time constant is to be 3.5 ms? **(b)** Using the capacitance determined in part (a), calculate the current in the circuit 7.0 ms after the switch is closed. Assume that the capacitor is uncharged initially and that the emf of the battery is 9.0 V.

83. •• A flash unit for a camera has a capacitance of 1500 μF. What resistance is needed in this *RC* circuit if the flash is to charge to 90% of its full charge in 21 s?

84. •• **Figure 21–43** shows a simplified circuit for a photographic flash unit. This circuit consists of a 9.0-V battery, a 50.0-kΩ resistor, a 140-μF capacitor, a flashbulb, and two switches. Initially, the capacitor is uncharged and the two switches are open. To charge the unit, switch S_1 is closed; to fire the flash, switch S_2

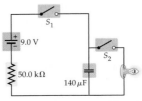

▲ **FIGURE 21–43** Problem 84

(which is connected to the camera's shutter) is closed. How long does it take to charge the capacitor to 5.0 V?

85. •• **IP** Consider the *RC* circuit shown in **Figure 21–44**. Find **(a)** the time constant and **(b)** the initial current for this circuit. **(c)** It is desired to increase the time constant of this circuit by adjusting the value of the 6.5-Ω resistor. Should the resistance of this resistor be increased or decreased to have the desired effect? Explain.

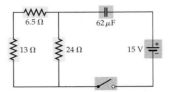

▲ **FIGURE 21–44** Problems 85 and 119

86. ••• The capacitor in an *RC* circuit is initially uncharged. In terms of R and C, determine **(a)** the time required for the charge on the capacitor to rise to 50% of its final value and **(b)** the time required for the initial current to drop to 10% of its initial value.

GENERAL PROBLEMS

87. • **CE** A given car battery is rated as 250 amp-hours. Is this rating a measure of energy, power, charge, voltage, or current? Explain.

88. • **CE Predict/Explain** The resistivity of tungsten increases with temperature. **(a)** When a light containing a tungsten filament heats up, does its power consumption increase, decrease, or stay the same? **(b)** Choose the *best explanation* from among the following:
  I. The voltage is unchanged, and therefore an increase in resistance implies a reduced power, as we can see from $P = V^2/R$.
  II. Increasing the resistance increases the power, as is clear from $P = I^2R$.
  III. The power consumption is independent of resistance, as we can see from $P = IV$.

89. • **CE** A cylindrical wire is to be doubled in length, but it is desired that its resistance remain the same. **(a)** Must its radius be increased or decreased? Explain. **(b)** By what factor must the radius be changed?

90. • **CE Predict/Explain** An electric space heater has a power rating of 500 W when connected to a given voltage V. **(a)** If two of these heaters are connected in series to the same voltage, is the power consumed by the two heaters greater than, less than, or equal to 1000 W? **(b)** Choose the *best explanation* from among the following:
  I. Each heater consumes 500 W; therefore two of them will consume $500\ W + 500\ W = 1000\ W$.
  II. The voltage is the same, but the resistance is doubled by connecting the heaters in series. Therefore, the power consumed ($P = V^2/R$) is less than 1000 W.
  III. Connecting two heaters in series doubles the resistance. Since power depends on the resistance squared, it follows that the power consumed is greater than 1000 W.

91. • **CE** Two resistors, $R_1 = R$ and $R_2 = 2R$, are connected to a battery. **(a)** Which resistor dissipates more power when they are connected to the battery in series? Explain. **(b)** Which resistor dissipates more power when they are connected in parallel? Explain.

92. • **CE** Consider the circuit shown in **Figure 21–45**, in which three lights, each with a resistance R, are connected in series. The circuit also contains an open switch. **(a)** When the switch is closed, does the intensity of light 2 increase, decrease, or stay the same? Explain. **(b)** Do the intensities of lights 1 and 3 increase, decrease, or stay the same when the switch is closed? Explain.

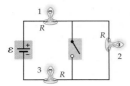

▲ **FIGURE 21–45** Problems 92, 93, and 94

93. • **CE Predict/Explain (a)** Referring to Problem 92 and the circuit in Figure 21–45, does the current supplied by the battery increase, decrease, or remain the same when the switch is closed? **(b)** Choose the *best explanation* from among the following:

  I. The current decreases because only two resistors can draw current from the battery when the switch is closed.

  II. Closing the switch makes no difference to the current since the second resistor is still connected to the battery as before.

  III. Closing the switch shorts out the second resistor, decreases the total resistance of the circuit, and increases the current.

94. • **CE Predict/Explain (a)** Referring to Problem 92 and the circuit in Figure 21–45, does the total power dissipated in the circuit increase, decrease, or remain the same when the switch is closed? **(b)** Choose the *best explanation* from among the following:

  I. Closing the switch shorts out one of the resistors, which means that the power dissipated decreases.

  II. The equivalent resistance of the circuit is reduced by closing the switch, but the voltage remains the same. Therefore, from $P = V^2/R$ we see that the power dissipated increases.

  III. The power dissipated remains the same because power, $P = IV$, is independent of resistance.

95. • **CE** Consider the circuit shown in **Figure 21–46**, in which three lights, each with a resistance R, are connected in parallel. The circuit also contains an open switch. **(a)** When the switch is closed, does the intensity of light 3 increase, decrease, or stay the same? Explain. **(b)** Do the intensities of lights 1 and 2 increase, decrease, or stay the same when the switch is closed? Explain.

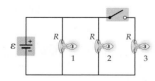

▲ **FIGURE 21–46** Problems 95, 96, and 97

96. • **CE Predict/Explain (a)** When the switch is closed in the circuit shown in Figure 21–46, does the current supplied by the battery increase, decrease, or stay the same? **(b)** Choose the *best explanation* from among the following:

  I. The current increases because three resistors are drawing current from the battery when the switch is closed, rather than just two.

  II. Closing the switch makes no difference to the current because the voltage is the same as before.

  III. Closing the switch decreases the current because an additional resistor is added to the circuit.

97. • **CE Predict/Explain (a)** When the switch is closed in the circuit shown in Figure 21–46, does the total power dissipated in the circuit increase, decrease, or stay the same? **(b)** Choose the *best explanation* from among the following:

  I. Closing the switch adds one more resistor to the circuit. This makes it harder for the battery to supply current, which decreases the power dissipated.

  II. The equivalent resistance of the circuit is reduced by closing the switch, but the voltage remains the same. Therefore, from $P = V^2/R$ we see that the power dissipated increases.

  III. The power dissipated remains the same because power, $P = IV$, is independent of resistance.

98. • Suppose that points A and B in Figure 21–30 are connected to a 12-V battery. Find the power dissipated in each of the resistors assuming that $R = 65\ \Omega$.

99. • You are given resistors of 413 Ω, 521 Ω, and 146 Ω. Describe how these resistors must be connected to produce an equivalent resistance of 255 Ω.

100. • You are given capacitors of 18 μF, 7.2 μF, and 9.0 μF. Describe how these capacitors must be connected to produce an equivalent capacitance of 22 μF.

101. • Suppose your car carries a charge of 85 μC. What current does it produce as it travels from Dallas to Fort Worth (35 mi) in 0.75 h?

102. •• **CE** The circuit shown in **Figure 21–47** shows a resistor and two capacitors connected in series with a battery of voltage V. The circuit also has an ammeter and a switch. Initially, the switch is open and both capacitors are uncharged. The following questions refer to a time long after the switch is closed and current has ceased to flow. **(a)** In terms of V, what is the voltage across the capacitor C_1? **(b)** In terms of CV, what is the charge on the right plate of C_2? **(c)** What is the net charge that flowed through the ammeter during charging? Give your answer in terms of CV.

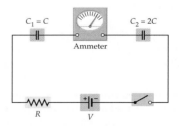

▲ **FIGURE 21–47** Problem 102

103. •• **CE** The three circuits shown in **Figure 21–48** have identical batteries, resistors, and capacitors. Initially, the switches are open and the capacitors are uncharged. Rank the circuits in order of increasing **(a)** final charge on the capacitor and **(b)** time for the current to drop to 90% of its initial value. Indicate ties where appropriate.

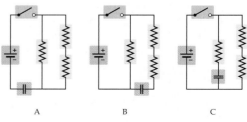

▲ **FIGURE 21–48** Problem 103

104. •• It is desired to construct a 5.0-Ω resistor from a 1.2-m length of tungsten wire. What diameter is needed for this wire?

105. •• **Electrical Safety Codes** For safety reasons, electrical codes have been established that limit the amount of current a wire of a given size can carry. For example, an 18-gauge (cross-sectional area $= 1.17$ mm^2), rubber-insulated extension cord with copper wires can carry a maximum current of 5.0 A. Find the voltage drop in a 12-ft, 18-gauge extension cord carrying a current of 5.0 A. (*Note:* In an extension cord, the current must flow through two lengths—down and back.)

106. •• **A Three-Way Lightbulb** A three-way lightbulb has two filaments with resistances R_1 and R_2 connected in series. The resistors are connected to three terminals, as indicated in **Figure 21–49**, and the light switch determines which two of the three terminals are connected to a potential difference of 120 V at any given time. When terminals A and B are connected to 120 V the bulb uses 75.0 W of power. When terminals A and C are connected to 120 V the bulb uses 50.0 W of power. **(a)** What is the resistance R_1? **(b)** What is the resistance R_2? **(c)** How much power does the bulb use when 120 V is connected to terminals B and C?

▲ **FIGURE 21–49** Problem 106

107. •• A portable CD player uses a current of 7.5 mA at a potential difference of 3.5 V. **(a)** How much energy does the player use in 35 s? **(b)** Suppose the player has a mass of 0.65 kg. For what length of time could the player operate on the energy required to lift it through a height of 1.0 m?

108. •• An electrical heating coil is immersed in 4.6 kg of water at 22 °C. The coil, which has a resistance of 250 Ω, warms the water to 32 °C in 15 min. What is the potential difference at which the coil operates?

109. •• **IP** Consider the circuit shown in **Figure 21–50**. **(a)** Is the current flowing through the battery immediately after the switch is closed greater than, less than, or the same as the current flowing through the battery long after the switch is closed? Explain. **(b)** Find the current flowing through the battery immediately after the switch is closed. **(c)** Find the current in the battery long after the switch is closed.

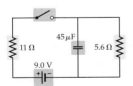

▲ **FIGURE 21–50** Problems 109 and 116

110. •• A silver wire and a copper wire have the same volume and the same resistance. Find the ratio of their radii, $r_{\text{silver}}/r_{\text{copper}}$.

111. •• Two resistors are connected in series to a battery with an emf of 12 V. The voltage across the first resistor is 2.7 V and the current through the second resistor is 0.15 A. Find the resistance of the two resistors.

112. •• **BIO Pacemaker Pulses** A pacemaker sends a pulse to a patient's heart every time the capacitor in the pacemaker charges to a voltage of 0.25 V. It is desired that the patient receive 75 pulses per minute. Given that the capacitance of the pacemaker is 110 μF and that the battery has a voltage of 9.0 V, what value should the resistance have?

113. •• A long, thin wire has a resistance R. The wire is now cut into three segments of equal length, which are connected in parallel. In terms of R, what is the equivalent resistance of the three wire segments?

114. •• Three resistors $(R, \frac{1}{2}R, 2R)$ are connected to a battery. **(a)** If the resistors are connected in series, which one has the greatest rate of energy dissipation? **(b)** Repeat part (a), this time assuming that the resistors are connected in parallel.

115. •• **IP** Suppose we connect a 12.0-V battery to terminals A and B in Figure 21–29. **(a)** Is the current in the 45-Ω resistor greater than, less than, or the same as the current in the 35-Ω resistor? Explain. **(b)** Calculate the current flowing through each of the three resistors in this circuit.

116. •• **IP** Suppose the battery in Figure 21–50 has an internal resistance of 0.73 Ω. **(a)** What is the potential difference across the terminals of the battery when the switch is open? **(b)** When the switch is closed, does the potential difference of the battery increase or decrease? Explain. **(c)** Find the potential difference across the battery after the switch has been closed a long time.

117. •• **National Electric Code** In the United States, the National Electric Code sets standards for maximum safe currents in insulated copper wires of various diameters. The accompanying table gives a portion of the code. Notice that wire diameters are identified by the *gauge* of the wire, and that 1 mil $= 10^{-3}$ in. Find the maximum power dissipated per length in **(a)** an 8-gauge wire and **(b)** a 10-gauge wire.

Gauge	Diameter (mils)	Safe current (A)
8	129	35
10	102	25

118. ••• **IP** A 15.0-V battery is connected to terminals A and B in Figure 21–41. **(a)** Given that $C = 15.0$ μF, find the charge on each of the capacitors. **(b)** Find the total energy stored in this system. **(c)** If the 7.22-μF capacitor is increased in value, will the total energy stored in the circuit increase or decrease? Explain.

119. ••• **IP** The switch in the RC circuit shown in Figure 21–44 is closed at $t = 0$. **(a)** How much power is dissipated in each resistor just after $t = 0$ and in the limit $t \rightarrow \infty$? **(b)** What is the charge on the capacitor at the time $t = 0.35$ ms? **(c)** How much energy is stored in the capacitor in the limit $t \rightarrow \infty$? **(d)** If the voltage of the battery is doubled, by what factor does your answer to part (c) change? Explain.

120. ••• Two resistors, R_1 and R_2, are connected in parallel and connected to a battery. Show that the power dissipated in R_1 plus the power dissipated in R_2 is equal to the power dissipated in the equivalent resistor, R_{eq}, when it is connected to the same battery.

121. ••• A battery has an emf $\mathcal{E}$ and an internal resistance r. When the battery is connected to a 25-Ω resistor, the current through the battery is 0.65 A. When the battery is connected to a 55-Ω resistor, the current is 0.45 A. Find the battery's emf and internal resistance.

122. ••• When two resistors, R_1 and R_2, are connected in series across a 6.0-V battery, the potential difference across R_1 is 4.0 V. When R_1 and R_2 are connected in parallel to the same battery, the current through R_2 is 0.45 A. Find the values of R_1 and R_2.

123. ••• The circuit shown in **Figure 21–51** is known as a Wheatstone bridge. Find the value of the resistor R such that the current through the 85.0-Ω resistor is zero.

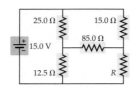

▲ **FIGURE 21–51** Problem 123

PASSAGE PROBLEMS

BIO Footwear Safety

The American National Standards Institute (ANSI) specifies safety standards for a number of potential workplace hazards. For example, ANSI requires that footwear provide protection against the effects of compression from a static weight, impact from a dropped object, puncture from a sharp tool, and cuts from saws. In addition, to protect against the potentially lethal effects of an electrical shock, ANSI provides standards for the electrical resistance that a person and footwear must offer to the flow of electric current.

Specifically, regulation ANSI Z41-1999 states that the resistance of a person and his or her footwear must be tested with the circuit shown in **Figure 21–52**. In this circuit, the voltage supplied by the battery is $\mathcal{E} = 50.0$ V and the resistance in the circuit is $R = 1.00$ MΩ. Initially the circuit is open and no current flows. When a person touches the metal sphere attached to the battery, however, the circuit is closed and a small current flows through the person, the shoes, and back to the battery. The amount of current flowing through the person can be determined by using a voltmeter to measure the voltage drop V across the resistor R. To be safe, the current should not exceed 150 μA.

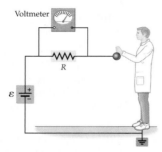

▲ **FIGURE 21–52** Problems 124, 125, 126, and 127

Notice that the experimental setup in Figure 21–52 is a dc circuit with two resistors in series—the resistance R and the resistance of the person and footwear, R_{pf}. It follows that the current in the circuit is $I = \mathcal{E}/(R + R_{pf})$. We also know that the current is $I = V/R$, where V is the reading of the voltmeter. These relations can be combined to relate the voltage V to the resistance R_{pf}, with the result shown in **Figure 21–53**. According to ANSI regulations, Type II footwear must give a resistance R_{pf} in the range of 0.1×10^7 Ω to 100×10^7 Ω.

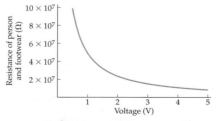

▲ **FIGURE 21–53** Problems 124, 125, 126, and 127

124. • Suppose the voltmeter measures a potential difference of 3.70 V across the resistor. What is the current that flows through the person's body?

   A. 3.70×10^{-6} A   **B.** 5.00×10^{-5} A

   C. 0.0740 A   **D.** 3.70 A

125. • What is the resistance of the person and footwear when the voltmeter reads 3.70 V?

   A. 1.25×10^7 Ω   **B.** 1.35×10^7 Ω

   C. 4.63×10^7 Ω   **D.** 1.71×10^8 Ω

126. • The resistance of a given person and footwear is 4.00×10^7 Ω. What is the reading on the voltmeter when this person is tested?

   A. 0.976 V   **B.** 1.22 V

   C. 1.25 V   **D.** 50.0 V

127. • Suppose that during one test a person's shoes become wet when water spills onto the floor. When this happens, do you expect the reading on the voltmeter to increase, decrease, or stay the same?

INTERACTIVE PROBLEMS

128. •• **Referring to Example 21–7** Suppose the three resistors in this circuit have the values $R_1 = 100.0$ Ω, $R_2 = 200.0$ Ω, and $R_3 = 300.0$ Ω, and that the emf of the battery is 12.0 V. (The resistor numbers are given in the Interactive Figure.) **(a)** Find the potential difference across each resistor. **(b)** Find the current that flows through each resistor.

129. •• **Referring to Example 21–7** Suppose $R_1 = R_2 = 225$ Ω and $R_3 = R$. The emf of the battery is 12.0 V. (The resistor numbers are given in the Interactive Figure.) **(a)** Find the value of R such that the current supplied by the battery is 0.0750 A. **(b)** Find the value of R that gives a potential difference of 2.65 V across resistor 2.

130. •• **IP Referring to Example 21–9** Suppose the resistance of the 126-Ω resistor is reduced by a factor of 2. The other resistor is 275 Ω, the capacitor is 182 μF, and the battery has an emf of 3.00 V. **(a)** Does the final value of the charge on the capacitor increase, decrease, or stay the same? Explain. **(b)** Does the time for the capacitor to charge to 80.0% of its final value increase, decrease, or stay the same? Explain. **(c)** Find the time referred to in part (b).

131. •• **IP Referring to Example 21–9** Suppose the capacitance of the 182-μF capacitor is reduced by a factor of 2. The two resistors are 126 Ω and 275 Ω, and the battery has an emf of 3.00 V. **(a)** Find the final value of the charge on the capacitor. **(b)** Does the time for the capacitor to charge to 80.0% of its final value increase, decrease, or stay the same? Explain. **(c)** Find the time referred to in part (b).

22 Magnetism

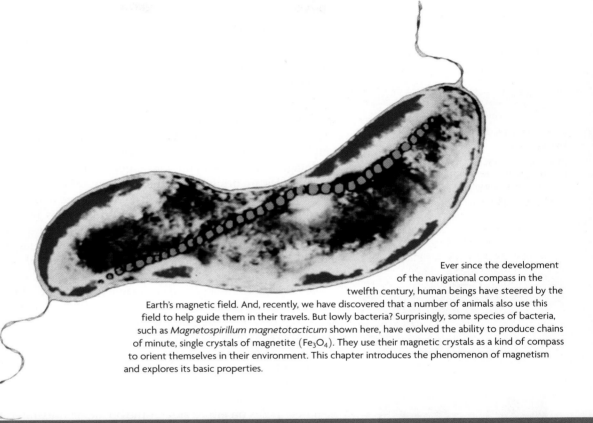

Ever since the development of the navigational compass in the twelfth century, human beings have steered by the Earth's magnetic field. And, recently, we have discovered that a number of animals also use this field to help guide them in their travels. But lowly bacteria? Surprisingly, some species of bacteria, such as *Magnetospirillum magnetotacticum* shown here, have evolved the ability to produce chains of minute, single crystals of magnetite (Fe_3O_4). They use their magnetic crystals as a kind of compass to orient themselves in their environment. This chapter introduces the phenomenon of magnetism and explores its basic properties.

The effects of magnetism have been known since antiquity. For example, a piece of lodestone, a naturally occurring iron oxide mineral, can behave just like a modern-day bar magnet. What has been discovered only much more recently, however, is the intimate connection that exists between electricity and magnetism. The first evidence of such a connection was obtained quite by accident, in 1820, as the Danish scientist Hans Christian Oersted (1777–1851) performed a demonstration during one of his popular public lectures. His observation, and the conclusions that followed from it, are the primary focus of this chapter.

22–1 The Magnetic Field

We begin our study of magnetism with a few general observations regarding magnets and the fields they produce. These observations apply over a wide range of scales, from the behavior of small, handheld bar magnets to the global effects associated with the magnetic field of the Earth. As we shall see later in this chapter, the ultimate source of any magnetic field is nothing more than the motion of electric charge.

(a) Opposite poles attract   (b) Like poles repel

▲ **FIGURE 22–1 The force between two bar magnets**

Permanent Magnets

Our first direct experience with magnets is often a playful exploration of the way small permanent magnets, called bar magnets, affect one another. As we know, a bar magnet can attract another magnet or repel it, depending on which ends of the magnets are brought together. One end of a magnet is referred to as its *north pole;* the other end is its *south pole.* The poles are defined as follows: Imagine suspending a bar magnet from a string so that it is free to rotate in a horizontal plane—much like the needle of a compass. The end of the magnet that points toward the north geographic pole of the Earth we refer to as the "north-seeking pole," or simply the north pole. The opposite end of the magnet is the "south-seeking pole," or the south pole.

The rule for whether two bar magnets attract or repel each other is learned at an early age: opposites attract; likes repel. Thus, if two magnets are brought together so that opposite poles approach each other, as in **Figure 22–1 (a)**, the force they experience is attractive. Like poles brought close together, as in **Figure 22–1 (b)**, experience a repulsive force.

An interesting aspect of magnets is that they *always* have two poles. You might think that if you break a bar magnet in two, each of the halves will have just one pole. Instead, breaking a magnet in half results in the appearance of two new poles on either side of the break, as illustrated in **Figure 22–2**. This behavior is fundamentally different from that found in electricity, where the two types of charge can exist separately. Though physicists continue to look for the elusive "magnetic monopole," and speculate as to its possible properties, none has been found.

Magnetic Field Lines

Just as an electric charge creates an electric "field" in its vicinity, so too does a magnet create a magnetic field. As in the electric case, the magnetic field represents the effect a magnet has on its surroundings. For example, in Figure 19–15 we saw a visual indication of the electric field $\vec{E}$ of a point charge using grass seeds suspended in oil. Similarly, the magnetic field, which we represent with the symbol $\vec{B}$, can be visualized using small iron filings sprinkled onto a smooth surface. In **Figure 22–3 (a)**, for example, a sheet of glass or plastic is placed on top of a bar magnet. When iron filings are dropped onto the sheet they align with the magnetic field in their vicinity, giving a good representation of the overall field produced by the magnet. Similar effects are seen in **Figure 22–3 (b)**, with a magnet that has been bent to bring its poles close together. Because of its shape, this type of magnet is referred to as a horseshoe magnet.

In both cases shown in Figure 22–3, notice that the filings are bunched together near the poles of the magnets. This is where the magnetic field is most intense. We illustrate this by drawing field lines that are densely packed near the poles, as

Breaking a bar magnet in two ...

... produces two new poles.

▲ **FIGURE 22–2 Magnets always have two poles**

When a bar magnet is broken in half, two new poles appear. Each half has both a north pole and a south pole, just like any other bar magnet.

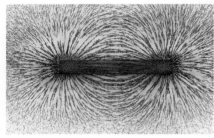

(a)

(b)

▲ **FIGURE 22–3 Magnetic field lines**

The field of a bar magnet **(a)** or horseshoe magnet **(b)** can be visualized using iron filings on a sheet of glass or paper. The filings orient themselves with the field lines, creating a "snapshot" of the magnetic field.

in **Figure 22–4**. As one moves away from a magnet in any direction the field weakens, as indicated by the increasingly wider separations between field lines. Thus, we indicate the magnitude of the vector $\vec{B}$ by the spacing of the field lines.

We can also assign a direction to the magnetic field. In particular, the direction of $\vec{B}$ is defined as follows:

> The direction of the magnetic field, $\vec{B}$, at a given location is the direction in which the north pole of a compass points when placed at that location.

We apply this definition to the bar magnet in Figure 22–4. Imagine, for example, placing a compass near its south pole. Because opposites attract, the north pole of the compass needle—the end with the arrowhead—will point toward the south pole of the magnet. Hence, according to our definition, the direction of the magnetic field at that location is toward the bar magnet's south pole. Similarly, one can see that the magnetic field must point away from the north pole of the bar magnet. In general,

> Magnetic field lines exit from the north pole of a magnet and enter at the south pole.

The field lines continue even within the body of a magnet. In fact, magnetic field lines always form closed loops; they never start or stop anywhere, in contrast to electric field lines. Again, this is related to the fact that there are no magnetic monopoles, where a magnetic field line could start or stop, whereas electric field lines start on positive charges and stop on negative charges.

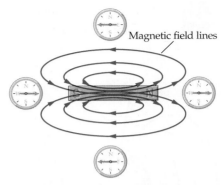

▲ **FIGURE 22–4 Magnetic field lines for a bar magnet**
The field lines are closely spaced near the poles, where the magnetic field $\vec{B}$ is most intense. In addition, the lines form closed loops that leave at the north pole of the magnet and enter at the south pole.

REAL-WORLD PHYSICS
Refrigerator magnets

CONCEPTUAL CHECKPOINT 22–1   MAGNETIC FIELD LINES

Can magnetic field lines cross one another?

REASONING AND DISCUSSION
Recall that the direction in which a compass points at any given location is the direction of the magnetic field at that point. Since a compass can point in only one direction, there must be only one direction for the field $\vec{B}$. If field lines were to cross, however, there would be two directions for $\vec{B}$ at the crossing point, and this is not allowed.

ANSWER
No. Magnetic field lines never cross.

An interesting example of magnetic field lines is provided by the humble refrigerator magnet. The flexible variety of these popular magnets has the unusual property that one side sticks quite strongly to the refrigerator, whereas the other side (the printed side) does not stick at all. Clearly, the magnetic field produced by such a magnet is not like that of a simple bar magnet, which generates a symmetrical field. Instead, these refrigerator magnets are composed of multiple magnetic stripes of opposite polarity, as indicated in **Figure 22–5**. The net effect is a magnetic field similar to the field that would be produced by a large number of tiny horseshoe magnets placed side by side, pointing in the same direction. In this way, the field is intense on the side containing the poles of the tiny magnets, as in Figure 22–3 (b), and very weak on the other side.

Geomagnetism

The Earth, like many planets, produces its own magnetic field. In many respects, the Earth's magnetic field is like that of a giant bar magnet, as illustrated in **Figure 22–6**, with a pole near each geographic pole of the Earth. The magnetic poles are not perfectly aligned with the rotational axis of the Earth, however, but are inclined at an angle that varies slowly with time. Presently, the magnetic poles are tilted away from the rotational axis by an angle of about 11.5°. The current location of the north magnetic pole is just west of Ellef Ringnes Island, one of the Queen Elizabeth Islands of extreme northern Canada.

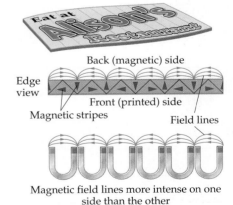

▲ **FIGURE 22–5 Refrigerator magnet**
A flexible refrigerator magnet is made from a large number of narrow magnetic stripes with magnetic fields in different directions. The net effect is a field similar to that of a series of parallel horseshoe magnets placed side by side.

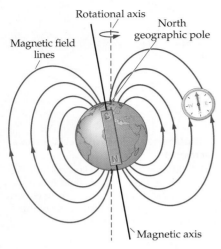

▲ **FIGURE 22–6 Magnetic field of the Earth**
The Earth's magnetic field is similar to that of a giant bar magnet tilted slightly from the rotational axis. Notice that the north geographic pole is actually near the south pole of the Earth's magnetic field.

Since the north pole of a compass needle points toward the north magnetic pole of the Earth, and since opposites attract, it follows that

> the *north* geographic pole of the Earth is actually near the *south* pole of the Earth's magnetic field.

This is indicated in Figure 22–6. The figure also shows that the field lines are essentially horizontal (parallel to the Earth's surface) near the equator but enter or leave the Earth vertically near the poles. Thus, for example, if you were to stand near the north geographic pole, your compass would try to point straight down.

Although the Earth's magnetic field is similar in many respects to that of a huge bar magnet, it is far more complex in both its shape and behavior—in fact, even the mechanism by which the field is produced is still not completely understood. It seems likely, however, that flowing currents of molten material in the Earth's core are the primary cause of the field, as expressed in the **dynamo theory** of magnetism.

One of the reasons for lingering uncertainty about the Earth's magnetic field is that its behavior over time is rather complicated. For example, we have already mentioned that the magnetic poles drift about slowly with time. This is just the beginning of the field's interesting behavior, however. We have learned in the last century that the Earth's field has actually *reversed* direction many times over the ages. In fact, the last reversal occurred about 780,000 years ago—from 980,000 years ago to 780,000 years ago your compass would have pointed opposite to its direction today.

These ancient field reversals have left a permanent record in the rocks of the ocean floors, among other places. By analyzing the evidence of these reversals, geologists have found strong support for the theories of continental drift and plate tectonics and have developed a whole new branch of geology referred to as **paleomagnetism.** We shall return to these topics again briefly near the end of this chapter.

22–2 The Magnetic Force on Moving Charges

We have discussed briefly the familiar forces that act between one magnet and another. In this section we consider the force a magnetic field exerts on a moving electric charge. As we shall see, both the magnitude and the direction of this force have rather unusual characteristics. We begin with the magnitude.

Magnitude of the Magnetic Force

Consider a magnetic field $\vec{B}$ that points from left to right in the plane of the page, as indicated in **Figure 22–7**. A particle of charge q moves through this region with a velocity $\vec{v}$. Note that the angle between $\vec{v}$ and $\vec{B}$ has a magnitude denoted by the symbol θ. Experiment shows that the magnitude of the force $\vec{F}$ experienced by this particle is given by the following:

Magnitude of the Magnetic Force, *F*

$$F = |q|vB \sin \theta \qquad \qquad 22\text{–}1$$

SI unit: newton, N

Thus, the magnetic force depends on several factors. Two of these are the same as for the electric force:

> (i) the charge of the particle, q;
> (ii) the magnitude of the field, in this case the magnetic field, B;

However, the magnetic force also depends on two factors that do not affect the strength of the electric force:

> (iii) the speed of the particle, v;
> (iv) the magnitude of the angle between the velocity vector and the magnetic field vector, θ.

It follows that the behavior of particles in magnetic fields is significantly different from their behavior in electric fields.

▲ **FIGURE 22–7 The magnetic force on a moving charged particle**
A particle of charge q moves through a region of magnetic field $\vec{B}$ with a velocity $\vec{v}$. The magnitude of the force experienced by the charge is $F = |q|vB \sin \theta$. Note that the force is a maximum when the velocity is perpendicular to the field and is zero when the velocity is parallel to the field.

In particular, a particle must have a charge and must be moving if the magnetic field is to exert a force on it. Even then the force will vanish if the particle moves in the direction of the field (that is, $\theta = 0$), or in the direction opposite to the field ($\theta = 180°$). The maximum force is experienced when the particle moves at right angles to the field, so that $\theta = 90°$, and $\sin \theta = 1$.

Finally, the expression in Equation 22–1 gives only the magnitude of the force, and hence depends on the magnitude of the charge, $|q|$. As we shall see later in this section, the *sign* of q is important in determining the *direction* of the magnetic force. In addition, the angle θ in Equation 22–1 is the magnitude of the angle between $\vec{v}$ and $\vec{B}$, and always has a value in the range $0 \le \theta \le 180°$.

In practice, the magnitude—or strength—of a magnetic field is *defined* by the relation given in Equation 22–1. Thus, B is given by the following:

Definition of the Magnitude of the Magnetic Field, B

$$B = \frac{F}{|q|v \sin \theta}$$ 22–2

SI unit: 1 tesla $= 1\,\mathrm{T} = 1\,\mathrm{N/(A \cdot m)}$

The units of B are those of force divided by the product of charge and speed; that is, $\mathrm{N/[C \cdot (m/s)]}$. Rearranging slightly, we can write these units as $\mathrm{N/[(C/s) \cdot m]}$. Finally, noting that $1\,\mathrm{A} = 1\,\mathrm{C/s}$, we find that $1\,\mathrm{N/[C \cdot (m/s)]} = 1\,\mathrm{N/(A \cdot m)}$. The latter combination of units is named the **tesla,** in recognition of the pioneering electrical and magnetic studies of the Croatian-born American engineer Nikola Tesla (1856–1943). In particular,

$$1\text{ tesla} = 1\,\mathrm{T} = 1\,\mathrm{N/(A \cdot m)}$$

We now give an example of the application of Equation 22–1.

EXAMPLE 22-1   A TALE OF TWO CHARGES

Particle 1, with a charge $q_1 = 3.60\,\mu\mathrm{C}$ and a speed $v_1 = 862\,\mathrm{m/s}$, travels at right angles to a uniform magnetic field. The magnetic force it experiences is $4.25 \times 10^{-3}\,\mathrm{N}$. Particle 2, with a charge $q_2 = 53.0\,\mu\mathrm{C}$ and a speed $v_2 = 1.30 \times 10^3\,\mathrm{m/s}$, moves at an angle of $55.0°$ relative to the same magnetic field. Find **(a)** the strength of the magnetic field and **(b)** the magnitude of the magnetic force exerted on particle 2.

PICTURE THE PROBLEM
The two charged particles are shown in our sketch, along with the magnetic field lines. Notice that $q_1 = 3.60\,\mu\mathrm{C}$ moves at right angles to $\vec{B}$; the charge $q_2 = 53.0\,\mu\mathrm{C}$ moves at an angle of $55.0°$ with respect to $\vec{B}$. The magnetic force also depends on the speeds of the particles, $v_1 = 862\,\mathrm{m/s}$ and $v_2 = 1.30 \times 10^3\,\mathrm{m/s}$.

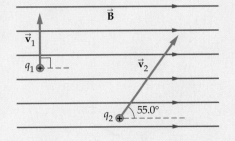

STRATEGY

a. We can find the strength of the magnetic field using the information given for particle 1. In particular, the particle moves at right angles to the magnetic field ($\theta_1 = 90°$), and, therefore, the magnetic force it experiences is $F_1 = |q_1|v_1 B \sin \theta_1 = q_1 v_1 B$. Given that we know F_1, q_1, and v_1, we can solve for B.

b. Once we have determined B in part (a), it is straightforward to calculate the magnetic force on particle 2 using $F_2 = |q_2|v_2 B \sin \theta_2$.

SOLUTION
Part (a)

1. Set $\theta_1 = 90°$ in $F_1 = |q_1|v_1 B \sin \theta_1$, then solve for B. Use q_1 and v_1 for the magnitude of the charge and speed, respectively:

$$F_1 = |q_1|v_1 B \sin 90° = q_1 v_1 B$$
$$B = \frac{F_1}{q_1 v_1}$$

2. Substitute numerical values:

$$B = \frac{F_1}{q_1 v_1} = \frac{4.25 \times 10^{-3}\,\mathrm{N}}{(3.60 \times 10^{-6}\,\mathrm{C})(862\,\mathrm{m/s})} = 1.37\,\mathrm{T}$$

CONTINUED ON NEXT PAGE

CONTINUED FROM PREVIOUS PAGE

Part (b)

3. Use $B = 1.37$ T in $F_2 = |q_2|v_2B \sin \theta_2$ to find the magnetic force exerted on particle 2:

$$F_2 = |q_2|v_2B \sin \theta_2$$
$$= (53.0 \times 10^{-6}\,\text{C})(1.30 \times 10^3\,\text{m/s})(1.37\,\text{T}) \sin 55.0°$$
$$= 0.0773\,\text{N}$$

INSIGHT

Notice that the charge and speed of a particle are not enough to determine the magnetic force acting on it—its direction of motion relative to the magnetic field is needed as well.

PRACTICE PROBLEM

At what angle relative to the magnetic field must particle 2 move if the magnetic force it experiences is to be 0.0500 N? [**Answer:** $\theta = 32.0°$]

Some related homework problems: Problem 8, Problem 11

TABLE 22–1 Typical Magnetic Fields

Physical system	Magnetic field (G)
Magnetar (a magnetic neutron star formed in a supernova explosion)	10^{15}
Strongest man-made magnetic field	6×10^5
High-field MRI	15,000
Low-field MRI	2000
Sunspots	1000
Bar magnet	100
Earth	0.50

The tesla is a fairly large unit of magnetic strength, especially when compared with the magnetic field at the surface of the Earth, which is roughly 5.0×10^{-5} T. Thus, another commonly used unit of magnetism is the gauss (G), defined as follows:

$$1\ \text{gauss} = 1\ \text{G} = 10^{-4}\ \text{T}$$

In terms of the gauss, the Earth's magnetic field on the surface of the Earth is approximately 0.5 G. It should be noted, however, that the gauss is not an SI unit of magnetic field. Even so, it finds wide usage because of its convenient magnitude. The magnitudes of some typical magnetic fields are given in Table 22–1.

Magnetic Force Right-Hand Rule (RHR)

We now consider the *direction* of the magnetic force, which is rather interesting and unexpected. Instead of pointing in the direction of the magnetic field, $\vec{B}$, or the velocity, $\vec{v}$, as one might expect, the following behavior is observed:

The magnetic force $\vec{F}$ points in a direction that is perpendicular to both $\vec{B}$ and $\vec{v}$.

As an example, consider the vectors $\vec{B}$ and $\vec{v}$ in **Figure 22–8 (a)**, which lie in the indicated plane. The force on a positive charge, $\vec{F}$, as shown in the figure, is perpendicular to this plane and hence to both $\vec{B}$ and $\vec{v}$.

Notice that $\vec{F}$ could equally well point downward in Figure 22–8 (a) and still be perpendicular to the plane. The way we determine the precise direction of $\vec{F}$ is with a right-hand rule—similar to the right-hand rules used in calculating torques and angular momentum in Chapter 11. To be specific, the direction of $\vec{F}$ is found using the *magnetic force* right-hand rule:

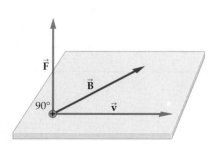

(a) $\vec{F}$ is perpendicular to both $\vec{v}$ and $\vec{B}$

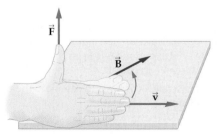

(b) Curl fingers from $\vec{v}$ to $\vec{B}$; thumb points in direction of $\vec{F}$

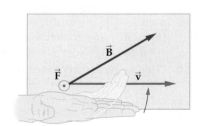

(c) Top view, looking down on $\vec{F}$

▲ **FIGURE 22–8 The magnetic force right-hand rule**
(a) The magnetic force, $\vec{F}$, is perpendicular to both the velocity, $\vec{v}$, and the magnetic field, $\vec{B}$. (The force vectors shown in this figure are for the case of a positive charge. The force on a negative charge would be in the opposite direction.) **(b)** As the fingers of the right hand are curled from $\vec{v}$ to $\vec{B}$, the thumb points in the direction of $\vec{F}$. **(c)** An overhead view, looking down on the plane defined by the vectors $\vec{v}$ and $\vec{B}$. In this two-dimensional representation, the force vector comes out of the page and is indicated by a circle with a dot inside. If the charge was negative, the force would point into the page, and the symbol indicating $\vec{F}$ would be a circle with an X inside.

Magnetic Force Right-Hand Rule (RHR)

To find the direction of the magnetic force on a positive charge, start by pointing the fingers of your right hand in the direction of the velocity, $\vec{v}$. Now curl your fingers toward the direction of $\vec{B}$, as illustrated in **Figure 22–8 (b)**. Your thumb points in the direction of $\vec{F}$. If the charge is negative, the force points opposite to the direction of your thumb.

Applying this rule to Figure 22–8 (a), we see that $\vec{F}$ does indeed point upward, as indicated.

A mathematical way to write the magnetic force that includes both its magnitude and direction is in terms of the vector **cross product**. Details on the cross product are given in Appendix A, but basically we can write the magnetic force $\vec{F}$ as follows:

$$\vec{F} = q\vec{v} \times \vec{B}$$

In this expression, the term $\vec{v} \times \vec{B}$ is referred to as the cross product of $\vec{v}$ and $\vec{B}$. From the definition of the cross product, the magnitude of the force is $F = |q|vB \sin \theta$ where θ is the magnitude of the angle between $\vec{v}$ and $\vec{B}$, precisely as in Equation 22–1. In addition, the direction of $\vec{F}$ is given by the magnetic force RHR. Thus, Equation 22–1 plus the magnetic force RHR are all we need to calculate magnetic forces—the cross product simply provides a more compact way of saying the same thing.

Since three-dimensional plots like the one in Figure 22–8 (b) are somewhat difficult to sketch, we often use a shorthand for indicating vectors that point into or out of the page. (See Appendix A for a discussion and examples.) Imagining a vector as an arrow, with a pointed tip at one end and crossed feathers at the other end, we say that if a vector points out of the page—toward us—we see its tip. This situation is indicated by drawing a circle with a dot inside it on the page, as in **Figure 22–8 (c)**. If, on the other hand, the vector points into the page, we see the crossed feathers of its base; hence, we draw a circle with an X inside it to symbolize this situation.

As an example, **Figure 22–9** indicates a uniform magnetic field $\vec{B}$ that points into the page. A particle with a positive charge q moves to the right. Using the magnetic force RHR—extending our fingers to the right and then curling them into the page—we see that the force exerted on this particle is upward, as indicated. If the charge is negative, the direction of $\vec{F}$ is reversed, as is also illustrated in Figure 22–9.

PROBLEM-SOLVING NOTE

The Magnetic Force for Positive and Negative Charges

Note that the magnetic force right-hand rule is stated for the case of a positively charged particle. If the particle has a negative charge, the direction of the force is reversed.

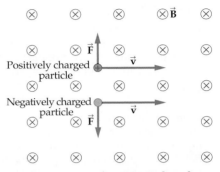

▲ **FIGURE 22–9 The magnetic force for positive and negative charges**

The direction of the magnetic force depends on the sign of the charge. Specifically, the force exerted on a negatively charged particle is opposite in direction to the force exerted on a positively charged particle.

▲ A cathode-ray tube (CRT) produces images by deflecting a beam of electrons with electric fields. For more details, see the discussion and figure given in the Passage Problem for Chapter 29. When a bar magnet is moved close to the screen of a CRT, as shown in this photo, the magnetic field it produces exerts a force on the moving electrons in the beam. This force changes the direction of motion of the electrons, resulting in a distorted picture. The distortion will vanish if the magnetic field is reduced by moving the bar magnet farther away from the screen.

CONCEPTUAL CHECKPOINT 22–2   CHARGE OF A PARTICLE

Three particles travel through a region of space where the magnetic field is out of the page, as shown below in the sketch to the left. For each of the three particles, state whether the particle's charge is positive, negative, or zero.

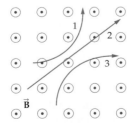   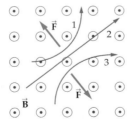

REASONING AND DISCUSSION

In the second sketch, we indicate the general direction of the force required to cause the observed motion. The force indicated for particle 3 is in the direction given by the magnetic force RHR; hence, particle 3 must have a positive charge. The force acting on particle 1 is in the opposite direction; hence, that particle must be negatively charged. Finally, particle 2 is undeflected; hence, its charge, and the force acting on it, must be zero.

ANSWER

Particle 1, negative; particle 2, zero; particle 3, positive.

We conclude this section with an Example illustrating the magnetic force RHR.

EXAMPLE 22–2 ELECTRIC AND MAGNETIC FIELDS

A particle with a charge of 7.70 μC and a speed of 435 m/s is acted on by both an electric and a magnetic field. The particle moves along the x axis in the positive direction, the magnetic field has a strength of 3.20 T and points in the positive y direction, and the electric field points in the positive z direction with a magnitude of 8.10×10^3 N/C. Find the magnitude and direction of the net force acting on the particle.

PICTURE THE PROBLEM
The physical situation is shown in our sketch. In particular, we show a three-dimensional co-ordinate system with each of the three relevant vectors, $\vec{v}$, $\vec{B}$, and $\vec{E}$, indicated. Note that the charge is positive, and therefore $|q| = q$.

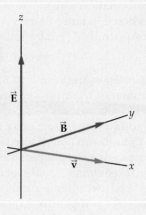

STRATEGY
First, the force due to the electric field is in the direction of $\vec{E}$ (positive z direction), and it has a magnitude qE. Second, the direction of the magnetic force is given by the RHR. In this case curling the fingers of a right hand from $\vec{v}$ to $\vec{B}$ results in the thumb (the direction of $\vec{F}$) pointing in the positive z direction. Thus, in this system the forces due to the electric and magnetic fields are in the same direction. Finally, because the angle between $\vec{v}$ and $\vec{B}$ is 90°, the magnitude of the magnetic force is qvB.

SOLUTION

1. Calculate the magnitude of the electric force exerted on the particle:

$$F_E = qE$$
$$= (7.70 \times 10^{-6}\,\text{C})(8.10 \times 10^3\,\text{N/C}) = 6.24 \times 10^{-2}\,\text{N}$$

2. Calculate the magnitude of the magnetic force exerted on the particle:

$$F_B = qvB$$
$$= (7.70 \times 10^{-6}\,\text{C})(435\,\text{m/s})(3.20\,\text{T}) = 1.07 \times 10^{-2}\,\text{N}$$

3. Since both forces are in the positive z direction, simply add them to obtain the net force:

$$F_{net} = F_E + F_B = (6.24 \times 10^{-2}\,\text{N}) + (1.07 \times 10^{-2}\,\text{N})$$
$$= 7.31 \times 10^{-2}\,\text{N}$$

INSIGHT
Note that the net force on the particle will be in the positive z direction as long as the velocity vector, $\vec{v}$, points in *any* direction in the x-y plane. The magnitude of the net force, however, will depend on the direction of $\vec{v}$.

In addition, the coordinate system shown in the Picture the Problem is a *right-handed* coordinate system. This means, for example, that the axes are set up so that $\hat{x} \times \hat{y} = \hat{z}$, as you can verify with the right-hand rule. Further details are given in Appendix A. We will always use right-handed coordinate systems in this text.

PRACTICE PROBLEM
Find the net force on the particle if its velocity is reversed and it moves in the negative x direction.
[**Answer:** $F_{net} = F_E + F_B = (6.24 \times 10^{-2}\,\text{N}) - (1.07 \times 10^{-2}\,\text{N}) = 5.17 \times 10^{-2}\,\text{N}$ in the positive z direction.]

Some related homework problems: Problem 14, Problem 16

22–3 The Motion of Charged Particles in a Magnetic Field

As we have seen, the magnetic force has characteristics that set it apart from the force exerted by electric or gravitational fields. In particular, the magnetic force depends not only on the speed of a particle but on its direction of motion as well. We now explore some of the consequences that follow from these characteristics and relate them to the type of motion that occurs in magnetic fields.

Electric Versus Magnetic Forces

We begin by investigating the motion of a charged particle as it is projected into a region with either an electric or a magnetic field. For example, in **Figure 22–10 (a)** we consider a uniform electric field pointing downward. A positively charged particle moving horizontally into this region experiences a constant downward

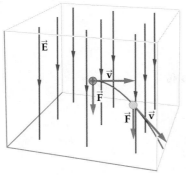
(a) Motion in an electric field

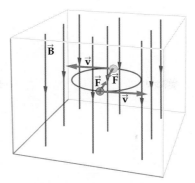
(b) Motion in a magnetic field

◀ **FIGURE 22–10 Differences between motion in electric and magnetic fields**
(a) A positively charged particle moving into a region with an electric field experiences a downward force that causes it to accelerate. **(b)** A positively charged particle entering a magnetic field experiences a horizontal force at right angles to its direction of motion. In this case, the speed of the particle remains constant.

force—much like a mass in a uniform gravitational field. As a result, the particle begins to accelerate downward and follow a parabolic path.

If the same particle, moving to the right, encounters a magnetic field instead, as in **Figure 22–10 (b)**, the resulting motion is quite different. We again assume that the field is uniform and pointing downward. In this case, however, the magnetic force RHR shows that $\vec{F}$ points into the page. Our particle now begins to follow a horizontal path into the page. As we shall see, this path is circular.

Perhaps even more significant than these differences in motion is the fact that an electric field can do work on a charged particle, whereas a constant magnetic field cannot. In Figure 22–10 (a), for example, as soon as the particle begins to move downward, a component of its velocity is in the direction of the electric force. This means that the electric force does work on the particle, and its speed increases—again, just like a mass falling in a gravitational field. If the particle moves through a magnetic field, however, no work is done on it because the magnetic force is *always* at right angles to the direction of motion. Thus, the speed of the particle in the magnetic field remains constant.

CONCEPTUAL CHECKPOINT 22–3   DIRECTION OF THE MAGNETIC FIELD

In a device called a **velocity selector**, charged particles move through a region of space with both an electric and a magnetic field. If the speed of the particle has a particular value, the net force acting on it is zero. Assume that a positively charged particle moves in the positive x direction, as shown in the sketch, and the electric field is in the positive y direction. Should the magnetic field be in **(a)** the positive z direction, **(b)** the negative y direction, or **(c)** the negative z direction in order to give zero net force?

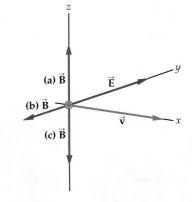

REASONING AND DISCUSSION
The force exerted by the electric field is in the positive y direction; hence, the magnetic force must be in the negative y direction if it is to cancel the electric force. If we simply try the three possible directions for $\vec{B}$ one at a time, applying the magnetic force RHR in each case, we find that only a magnetic field along the positive z axis gives rise to a force in the negative y direction, as desired.

ANSWER
(a) $\vec{B}$ should point in the positive z direction.

To follow up on Conceptual Checkpoint 22–3, let's determine the speed for which the net force is zero. First, note that the electric force has a magnitude qE. Second, the magnitude of the magnetic force is qvB. With $\vec{E}$ in the positive y direction and $\vec{B}$ in the positive z direction, these forces are in opposite directions. Setting the magnitudes of the forces equal yields $qE = qvB$. Canceling the charge q and solving for v we find

$$v = \frac{E}{B}$$ (velocity selector)

A particle with this speed, regardless of its charge, passes through the velocity selector with zero net force and hence no deflection.

The physical principle illustrated in the velocity selector can be used to measure the speed of blood with a device known as an *electromagnetic flowmeter*.

REAL-WORLD PHYSICS: BIO
The electromagnetic flowmeter

▶ **FIGURE 22–11** **The electromagnetic flowmeter**

As blood flows through a magnetic field, the charged ions it contains are deflected. The deflection of charged particles results in an electric field opposing the deflection. If the speed of the blood is v, the electric field generated by ions moving through the magnetic field satisfies the relation $v = E/B$.

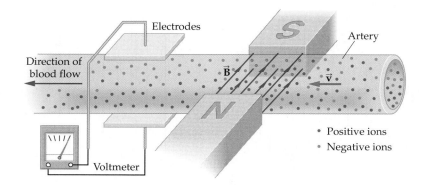

- Positive ions
- Negative ions

Electrodes

Artery

Direction of blood flow

$\vec{B}$

$\vec{v}$

Voltmeter

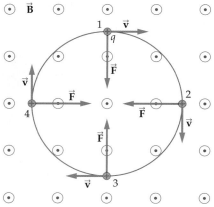

▲ **FIGURE 22–12** **Circular motion in a magnetic field**

A charged particle moves in a direction perpendicular to the magnetic field. At each point on the particle's path the magnetic force is at right angles to the velocity and hence toward the center of a circle.

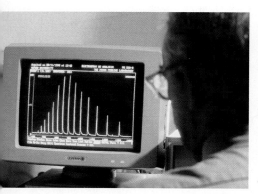

▲ In addition to separating and "weighing" isotopes, mass spectrometers can be used to study the chemical composition and structure of large, biologically important molecules. Here, a researcher studies the spectrum of myoglobin, an oxygen-carrying protein found in muscle tissue. Each peak represents a fragment of the molecule with its own characteristic combination of mass and charge.

Suppose an artery passes between the poles of a magnet, as shown in **Figure 22–11**. Charged ions in the blood will be deflected at right angles to the artery by the magnetic field. The resulting charge separation produces an electric field that opposes the magnetic deflection. When the electric field is strong enough, the deflection ceases and the blood flows normally through the artery. If the electric field is measured, and the magnetic field is known, the speed of the blood flow is simply $v = E/B$, as in a standard velocity selector.

Constant-Velocity, Straight-Line Motion

Recall that the magnetic force is zero if a particle's velocity $\vec{v}$ is parallel (or antiparallel) to the magnetic field $\vec{B}$. In such a case the particle's acceleration is zero; therefore, its velocity remains constant. Thus, the simplest type of motion in a magnetic field is a constant-velocity, straight-line drift along the magnetic field lines.

Circular Motion

The next simplest case is a particle with a velocity that is perpendicular to the magnetic field. Consider, for example, the situation shown in **Figure 22–12**. Here a particle of mass m, charge $+q$, and speed v moves in a region with a constant magnetic field $\vec{B}$ pointing out of the page. Since $\vec{v}$ is at right angles to $\vec{B}$, the magnitude of the magnetic force is $F = |q|vB \sin 90° = |q|vB$.

Now we consider the particle's motion. At point 1 the particle is moving to the right; hence, the magnetic force is downward, causing the particle to accelerate in the downward direction. When the particle reaches point 2, it is moving downward, and now the magnetic force is to the left. This causes the particle to accelerate to the left. At point 3, the force exerted on it is upward, and so on, as the particle continues moving.

Thus, at every point on the particle's path the magnetic force is at right angles to the velocity, pointing toward a common center—but this is just the condition required for circular motion. As we saw in Section 6–5, in circular motion the acceleration is toward the center of the circle; for this reason, a centripetal force is required to cause the motion. In this case the centripetal force is supplied by the magnetic force—in the same way that a string exerts a centripetal force on a ball being whirled about in a circle.

Recall that the centripetal acceleration of a particle moving with a speed v in a circle of radius r is

$$a_{\mathrm{cp}} = \frac{v^2}{r}$$

Therefore, setting ma_{cp} equal to the magnitude of the magnetic force, $|q|vB$, yields the following condition:

$$m\frac{v^2}{r} = |q|vB$$

Canceling one power of the speed, v, we find that the radius of the circular orbit is

$$r = \frac{mv}{|q|B} \qquad \text{22–3}$$

We can see, therefore, that the faster and more massive the particle, the larger the circle. Conversely, the stronger the magnetic field and the greater the charge, the smaller the circle.

EXERCISE 22–1

An electron moving perpendicular to a magnetic field of 4.60×10^{-3} T follows a circular path of radius 2.80 mm. What is the electron's speed?

SOLUTION
Solving Equation 22–3 for v, we find

$$v = \frac{r|q|B}{m} = \frac{(2.80 \times 10^{-3}\,\text{m})(1.60 \times 10^{-19}\,\text{C})(4.60 \times 10^{-3}\,\text{T})}{9.11 \times 10^{-31}\,\text{kg}} = 2.26 \times 10^6\,\text{m/s}$$

Thus, the speed of this electron is about 1% of the speed of light.

One of the applications of circular motion in a magnetic field is in a device known as a **mass spectrometer.** A mass spectrometer can be used to separate isotopes (atoms of the same element with different masses) and to measure atomic masses. It finds many uses in medicine (anesthesiologists use it to measure respiratory gases), biology (to determine reaction mechanisms in photosynthesis), geology (to date fossils), space science (to determine the atmospheric composition of Mars), and a variety of other fields.

The basic principles of a mass spectrometer are illustrated in **Figure 22–13.** Here we see a beam of ions of mass m and charge $+q$ entering a region of constant magnetic field with a speed v. The field causes the ions to move along a circular path, with a radius that depends on the mass and charge of the ion, as described by Equation 22–3. Thus, different isotopes follow different paths and hence can be separated and identified. A specific example is considered next.

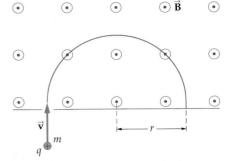

▲ **FIGURE 22–13 The operating principle of a mass spectrometer**

In a mass spectrometer, a beam of charged particles enters a region with a magnetic field perpendicular to the velocity. The particles then follow a circular orbit of radius $r = mv/|q|B$. Particles of different mass will follow different paths.

REAL-WORLD PHYSICS

The mass spectrometer

EXAMPLE 22–3   URANIUM SEPARATION

Two isotopes of uranium, ^{235}U ($m = 3.90 \times 10^{-25}$ kg) and ^{238}U ($m = 3.95 \times 10^{-25}$ kg), are sent into a mass spectrometer with a speed of 1.05×10^5 m/s, as indicated in the accompanying sketch. Given that each isotope is singly ionized, and that the strength of the magnetic field is 0.750 T, what is the distance d between the two isotopes after they complete half a circular orbit?

PICTURE THE PROBLEM
The relevant features of the mass spectrometer are indicated in our sketch. Note that both isotopes enter at the same location with the same speed. Because they have different masses, however, they follow circular paths of different radii. This difference results in the separation d after half an orbit.

STRATEGY
We begin by calculating the radius of each isotope's circular path using $r = mv/|q|B$. The masses, speeds, and magnetic field are given. We can infer the charge from the fact that the isotopes are "singly ionized," which means that a single electron has been removed from each atom. Since the atoms were electrically neutral before the electron was removed, it follows that the charge of the isotopes is $q = e = 1.60 \times 10^{-19}$ C.

Finally, once we know the radii, the separation between the isotopes is given by $d = 2r_{238} - 2r_{235}$, as indicated in the sketch.

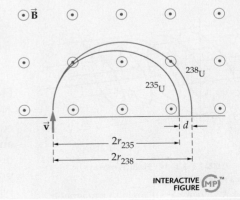

INTERACTIVE FIGURE MPJ™

SOLUTION

1. Determine the radius of the circular path of ^{235}U:

$$r_{235} = \frac{mv}{|q|B}$$

$$= \frac{(3.90 \times 10^{-25}\,\text{kg})(1.05 \times 10^5\,\text{m/s})}{(1.60 \times 10^{-19}\,\text{C})(0.750\,\text{T})} = 34.1\,\text{cm}$$

CONTINUED ON NEXT PAGE

CONTINUED FROM PREVIOUS PAGE

2. Determine the radius of the circular path of ^{238}U:

$$r_{238} = \frac{mv}{|q|B}$$

$$= \frac{(3.95 \times 10^{-25}\,\text{kg})(1.05 \times 10^5\,\text{m/s})}{(1.60 \times 10^{-19}\,\text{C})(0.750\,\text{T})} = 34.6\,\text{cm}$$

3. Calculate the separation between the isotopes:

$$d = 2r_{238} - 2r_{235} = 2(34.6\,\text{cm} - 34.1\,\text{cm}) = 1\,\text{cm}$$

INSIGHT

Notice that although the difference in masses is very small, the mass spectrometer converts this small difference into an easily measurable distance of separation.

PRACTICE PROBLEM

Does the separation d increase or decrease if the magnetic field is increased? Check your answer by calculating the separation for $B = 1.00$ T, everything else remaining the same.   [**Answer:** A stronger field will decrease the radii and hence decrease the separation d. In particular, if B changes by a factor x, the separation changes by the factor $1/x$. In this case, the factor is $x = 4/3$. That is, $B = (4/3)(0.750\,\text{T}) = 1.00$ T; therefore, $d = (3/4)(1\,\text{cm}) = 3/4\,\text{cm}$. Note that it is not necessary to repeat the entire calculation in detail.]

Some related homework problems: Problem 19, Problem 27

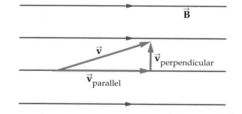

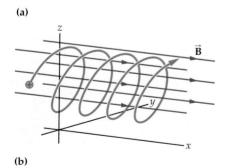

(a)

(b)

▲ **FIGURE 22–14 Helical motion in a magnetic field**

(a) A velocity at an angle to a magnetic field $\vec{B}$ can be resolved into parallel and perpendicular components. The parallel component gives a constant-velocity drift in the direction of the field. The perpendicular component gives circular motion perpendicular to the field. **(b)** Helical motion is a combination of linear motion and circular motion.

ACTIVE EXAMPLE 22–1   FIND THE TIME FOR ONE ORBIT

Calculate the time T required for a particle of mass m and charge q to complete a circular orbit in a magnetic field.

SOLUTION *(Test your understanding by performing the calculations indicated in each step.)*

1. Write the speed, v, of a particle in terms of the time T to complete an orbit of radius r:

$$v = 2\pi r/T$$

2. Substitute this expression for v into the condition for a circular orbit, $r = \dfrac{mv}{|q|B}$:

$$r = \frac{m(2\pi r/T)}{|q|B}$$

3. Cancel r and solve for T:

$$T = \frac{2\pi m}{|q|B}$$

INSIGHT

Notice that the time required for an orbit is independent of its radius.

YOUR TURN

The circumference of an orbit depends on its radius, r. Why is it, then, that a particle completes an orbit in a time, T, that is independent of r?

*(Answers to **Your Turn** problems are given in the back of the book.)*

Helical Motion

The final type of motion we consider is a combination of the two motions discussed already. Suppose, for example, that a particle has an initial velocity at an angle to the magnetic field, as in **Figure 22–14 (a)**. In this case there is a component of velocity parallel to $\vec{B}$ and a component perpendicular to $\vec{B}$. The parallel component of the velocity remains constant with time (zero force in this direction), whereas the perpendicular component results in a circular motion, as just discussed. Combining the two motions, we can see that the particle follows a helical path, as shown in **Figure 22–14 (b)**.

If a magnetic field is curved, as in the case of a bar magnet or the magnetic field of the Earth, the helical motion of charged particles will be curved as well. Specifically, the axis of the helical motion will follow the direction of $\vec{B}$. For example, electrons and protons emitted by the "solar wind" frequently encounter the Earth's magnetic field and begin to move in helical paths following the field lines.

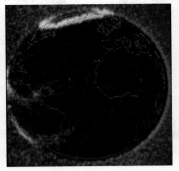

▲ (Left) An enormous eruption of matter from the surface of the Sun, large enough to encompass the entire Earth many times over. The loop structure is created by the Sun's own complex magnetic field. Charged particles from such eruptions often reach the Earth and become trapped in its magnetic field, producing auroras (see text). (Center) A glowing aurora borealis surrounds the Earth's north geographic pole in this photograph from space. (Right) An auroral display seen from Earth. The characteristic red and green colors are produced by ionized nitrogen and oxygen atoms, respectively.

Near the poles, where the field lines are concentrated as they approach the Earth's surface, the circulating electrons begin to collide with atoms and molecules in the atmosphere. These collisions can excite and ionize the atmospheric atoms and molecules, resulting in the emission of light known as the **aurora borealis** (northern lights) in the northern hemisphere and the **aurora australis** (southern lights) in the southern hemisphere. The most common color of the auroras is a pale green, the color of light given off by excited oxygen atoms.

22–4 The Magnetic Force Exerted on a Current-Carrying Wire

A charged particle experiences a force when it moves across magnetic field lines. This is true whether it travels in a vacuum or in a current-carrying wire. Thus, a wire with a current will experience a force that is simply the resultant of all the forces experienced by the individual moving charges responsible for the current.

Specifically, consider a straight wire segment of length L with a current I flowing from left to right, as in **Figure 22–15 (a)**. Also present in this region of space is a magnetic field $\vec{B}$ at an angle θ to the length of the wire. If the conducting charges move through the wire with an average drift speed v, the time required for them to move from one end of the wire segment to the other is $\Delta t = L/v$. The amount of charge that flows through the wire in this time is $q = I \Delta t = IL/v$. Therefore, the force exerted on the wire is

$$F = qvB \sin \theta = \left(\frac{IL}{v}\right)vB \sin \theta$$

Canceling v, we find that the force on a wire segment of length L with a current I at an angle θ to the magnetic field $\vec{B}$ is

Magnetic Force on a Current-Carrying Wire

$F = ILB \sin \theta$ 22–4

SI unit: newton, N

As with single charges, the maximum force occurs when the current is perpendicular to the magnetic field ($\theta = 90°$) and is zero if the current is in the same direction as $\vec{B}$ ($\theta = 0$).

The direction of the magnetic force on a wire is given by the same right-hand rule used earlier for single charges. Thus, to find the direction of the force in **Figure 22–15 (b)**, start by pointing the fingers of your right hand in the direction of the current I. This assumes that positive charges are flowing in the direction of I, consistent with our convention from Chapter 21. Now, curl your fingers toward the direction of $\vec{B}$. Your thumb, which points out of the page, indicates the direction of $\vec{F}$.

The aurora borealis and aurora australis

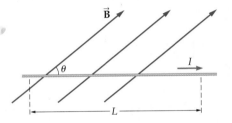

(a)

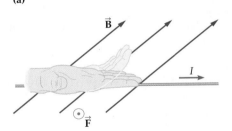

(b)

▲ **FIGURE 22–15 The magnetic force on a current-carrying wire**

A current-carrying wire in a magnetic field experiences a force, unless the current is parallel or antiparallel to the field. **(a)** For a wire segment of length L the magnitude of the force is $F = ILB \sin \theta$. **(b)** The direction of the force is given by the magnetic force RHR; the only difference is that you start by pointing the fingers of your right hand in the direction of the current I. In this case the force points out of the page.

PROBLEM-SOLVING NOTE

The Magnetic Force on a Current-Carrying Wire

Note that a current-carrying wire experiences no force when it is in the same direction as the magnetic field. The maximum force occurs when the wire is perpendicular to the magnetic field.

Of course, the current is actually caused by negatively charged electrons flowing in the opposite direction to the current. The magnetic force on these negatively charged particles moving in the opposite direction is the same as the force on positively charged particles moving in the direction of I. Thus, in all cases pertaining to current-carrying wires, we can simply think of the current as the direction in which positively charged particles move.

CONCEPTUAL CHECKPOINT 22–4   MAGNETIC POLES

When the switch is closed in the circuit shown in the sketch on the left, the wire between the poles of the horseshoe magnet deflects downward. Is the left end of the magnet **(a)** a north magnetic pole or **(b)** a south magnetic pole?

REASONING AND DISCUSSION
Once the switch is closed, the current in the wire is into the page, as shown in the sketch below.

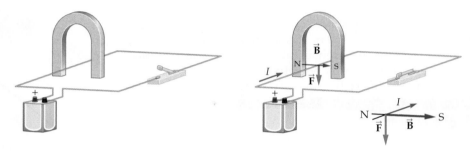

Applying the magnetic force RHR, we see that the magnetic field must point from left to right in order for the force to be downward. Since magnetic field lines leave from north poles and enter at south poles, it follows that the left end of the magnet must be a north magnetic pole.

ANSWER
(a) The left end of the magnet is a north magnetic pole.

The magnetic force exerted on a current-carrying wire can be quite substantial. In the following Example, we consider the current necessary to levitate a copper rod.

EXAMPLE 22–4   MAGNETIC LEVITY

A copper rod 0.150 m long and with a mass of 0.0500 kg is suspended from two thin, flexible wires, as shown in the sketch. At right angles to the rod is a uniform magnetic field of 0.550 T pointing into the page. Find **(a)** the direction and **(b)** magnitude of the electric current needed to levitate the copper rod.

PICTURE THE PROBLEM
Our sketch shows the physical situation with relevant quantities labeled. The direction of the current has been indicated as well. Note that if the current is in this direction, the RHR gives an upward magnetic force, as required for the rod to be levitated.

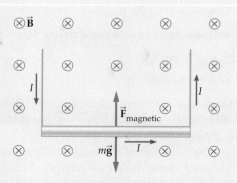

STRATEGY
To find the magnitude of the current, we set the magnetic force equal in magnitude to the force of gravity. For the magnetic force we have $F = ILB \sin \theta$. In this case the field is at right angles to the current; hence, $\theta = 90°$, and the force simplifies to $F = ILB$. Thus, setting ILB equal to mg determines the current.

SOLUTION

Part (a)

1. Determine the direction of I:

The current, I, points to the right. To verify, point the fingers of your right hand to the right, curl into the page, and your thumb will point upward, as desired.

Part (b)

2. Set the magnitude of the magnetic force equal to the magnitude of the force of gravity:

$$ILB = mg$$

3. Solve for the current, I:

$$I = \frac{mg}{LB} = \frac{(0.0500 \text{ kg})(9.81 \text{ m/s}^2)}{(0.150 \text{ m})(0.550 \text{ T})} = 5.95 \text{ A}$$

INSIGHT
Magnetic forces, like electric forces, can easily exceed the force of gravity. In fact, when we consider atomic systems in Chapter 31 we shall see that gravity plays no role in the behavior of an atom—only electric and magnetic forces are important on the atomic level.

PRACTICE PROBLEM
Suppose the rod is doubled in length, which also doubles its mass. Does the current needed to levitate the rod increase, decrease, or stay the same?   [**Answer:** Since the levitation current is given by $I = mg/LB$, it is clear that doubling both m and L has no effect on I; the current needed to levitate remains the same.]

Some related homework problems: Problem 30, Problem 33

22–5 Loops of Current and Magnetic Torque

The fact that a current-carrying wire experiences a force when placed in a magnetic field is one of the fundamental discoveries that makes modern applications of electric power possible. In most of these applications, including electric motors and generators, the wire is shaped into a current-carrying loop. We will examine some of these applications further in Chapter 23; in this section we lay the groundwork by considering what happens when a simple current loop is placed in a magnetic field.

Rectangular Current Loops

Consider a rectangular loop of height h and width w carrying a current I, as shown in **Figure 22–16**. The loop is placed in a region of space with a uniform magnetic field $\vec{B}$ that is parallel to the plane of the loop. From Figure 22–16, it is clear that the horizontal segments of the loop experience zero force, since they are parallel to the field. The vertical segments, on the other hand, are perpendicular to the field; hence, they experience forces of magnitude $F = IhB$. One of these forces is into the page (left side); the other is out of the page (right side).

Perhaps the best way to visualize the torque caused by these forces is to use a top view and look directly down on the loop, as in **Figure 22–17 (a)**. Here we can see more clearly that the forces on the vertical segments are equal in magnitude and opposite in direction. If we imagine an axis of rotation through the center of the loop, at the point O, it is clear that the forces exert a torque that tends to rotate the loop clockwise. The magnitude of this torque for each vertical segment is the force ($F = IhB$) times the moment arm ($w/2$). Noting that both vertical segments exert a torque in the same direction, we see that the total torque is simply the sum of the torque produced by each segment:

$$\tau = (IhB)\left(\frac{w}{2}\right) + (IhB)\left(\frac{w}{2}\right) = IB(hw)$$

Finally, observing that the area of the rectangular loop is $A = hw$, we can express the torque as follows:

$$\tau = IAB$$

As the loop begins to rotate, the situation will be like that shown in **Figure 22–17 (b)**. Here we see that the forces still have the same magnitude, IhB, but now the moment arms are $(w/2)\sin\theta$ rather than $w/2$. Thus, for a general angle, the torque must include the factor $\sin\theta$:

Torque Exerted on a Rectangular Loop of Area A

$$\tau = IAB \sin\theta$$

$$22\text{–}5$$

SI unit: N·m

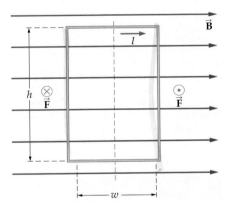

▲ **FIGURE 22–16   Magnetic forces on a current loop**

A rectangular current loop in a magnetic field. Only the vertical segments of the loop experience forces, and they tend to rotate the loop about a vertical axis.

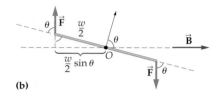

(a)

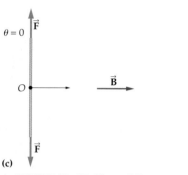

(b)

(c)

▲ **FIGURE 22–17 Magnetic torque on a current loop**
A current loop placed in a magnetic field produces a torque. **(a)** The torque is greatest when the plane of the loop is parallel to the magnetic field; that is, when the normal to the loop is perpendicular to the magnetic field. **(b)** As the loop rotates, the torque decreases by a factor of sin θ. **(c)** The torque vanishes when the plane of the loop is perpendicular to the magnetic field.

Note that the angle θ is the angle between the plane of the loop and the magnetic force exerted on each side of the loop. Equivalently, θ is the angle between the normal and the magnetic field.

In the case shown in Figure 22–17 (a), the angle θ is 90°, and the normal to the plane of the loop is perpendicular to the magnetic field. In this orientation the torque attains its maximum value, $\tau = IBA$. When θ is zero, as in **Figure 22–17 (c)**, the torque vanishes because the moment arm of the magnetic forces is zero. Thus, there is no torque when the magnetic field is perpendicular to the plane of a loop.

General Loops

We have shown that the torque exerted on a rectangular loop of area A is $\tau = IBA \sin \theta$. A more detailed derivation shows that the same relation applies to *any* planar loop, no matter what its shape. For example, a circular loop of radius r and area πr^2 experiences a torque given by the expression $\tau = IB\pi r^2 \sin \theta$.

In many applications it is desirable to produce as large a torque as possible. A simple way to increase the torque is to wrap a long wire around a loop N times, creating a coil of N "turns." Each of the N turns produces the same torque as a single loop; hence, the total torque is increased by a factor of N. In general, then, the torque produced by a loop with N turns is

Torque Exerted on a General Loop of Area A and N Turns

$$\tau = NIAB \sin \theta \qquad\qquad 22\text{–}6$$

SI unit: N · m

Notice that the torque depends on a number of factors in the system. First, it depends on the strength of the magnetic field, B, and on its orientation, θ, with respect to the normal of the loop. In addition, the torque depends on the current in the loop, I, the area of the loop, A, and the number of turns in the loop, N. The product of these "loop factors," NIA, is referred to as the **magnetic moment** of the loop. The magnetic moment, which has units of A · m², is proportional to the amount of torque a given loop can exert.

EXAMPLE 22–5 TORQUE ON A COIL

A rectangular coil with 200 turns is 5.0 cm high and 4.0 cm wide. When the coil is placed in a magnetic field of 0.35 T, its maximum torque is 0.22 N · m. What is the current in the coil?

PICTURE THE PROBLEM
The coil, along with its dimensions, is shown in the sketch. The sketch also reflects the fact that the maximum torque is produced when the magnetic field is in the plane of the coil.

STRATEGY
The torque is given by the expression $\tau = NIAB \sin \theta$. Clearly, the maximum torque occurs when $\sin \theta = 1$; that is, $\tau_{\text{max}} = NIAB$. Solving this relation for I yields the current in the coil.

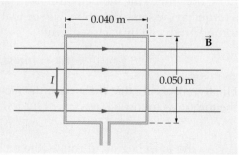

SOLUTION

1. Write an expression for the maximum torque:

$$\tau_{\text{max}} = NIAB$$

2. Solve for the current, I:

$$I = \frac{\tau_{\text{max}}}{NAB}$$

3. Substitute numerical values:

$$I = \frac{0.22 \text{ N} \cdot \text{m}}{(200)(0.050 \text{ m})(0.040 \text{ m})(0.35 \text{ T})} = 1.6 \text{ A}$$

INSIGHT

Note that this calculation gives the magnitude of the current and not its direction. The direction of the current will determine whether the torque on the coil is clockwise or counterclockwise.

PRACTICE PROBLEM

If the shape of this coil were changed to circular, keeping a constant perimeter, would the maximum torque increase, decrease, or stay the same? [**Answer:** In general, a circle has the greatest area for a given perimeter. Therefore, the maximum torque would increase if the coil were made circular, because its area would be larger.]

Some related homework problems: Problem 39, Problem 40

Applications of Torque

The torque exerted by a magnetic field finds a number of useful applications. For example, if a needle is attached to a coil, as in **Figure 22–18**, it can be used as part of a meter. When the coil is connected to an electric circuit, it experiences a torque, and a corresponding deflection of the needle, that is proportional to the current. The result is that the current in a circuit can be indicated by the reading on the meter. A simple device of this type is referred to as a **galvanometer.**

Notice that the galvanometer can be a very sensitive instrument for two different reasons. First, the coil can have a large number of turns. Since the torque exerted by a coil is proportional to the number of turns, it follows that a coil with many turns produces a significant torque even when the current is small. Second, the needle amplifies the motion of the coil. Specifically, if the coil turns through an angle θ, and the needle has a length L, the tip of the needle moves through a distance $L\theta$. Thus, both the number of turns and the length of the needle increase the sensitivity of the meter.

Of even greater practical importance is the fact that magnetic torque can be used to power a motor. For example, an electric current passing through the coils of a motor causes a torque that rotates the axle of the motor. As the coils rotate, a device known as the commutator reverses the direction of the current as the orientation of the coil reverses, which ensures that the torque is always in the same direction and that the coils continue to turn in the same direction. Electric motors, which will be discussed in greater detail in Chapter 23, are used in everything from CD players to electric razors to electric cars.

22–6 Electric Currents, Magnetic Fields, and Ampère's Law

In the introduction to this chapter, we mentioned that a previously unexpected connection between electricity and magnetism was discovered accidentally by Hans Christian Oersted in 1820. Specifically, Oersted was giving a public lecture on various aspects of science when, at one point, he closed a switch and allowed a current to flow through a wire. What he noticed was that a nearby compass needle deflected from its usual orientation when the switch was closed—Oersted had just discovered that electric currents can create magnetic fields. In this section we focus on the connection between electric currents and magnetic fields. In so doing, our attention shifts from the effects of magnetic fields—the subject of the previous sections—to their production.

A Long, Straight Wire

We start with the simplest possible case—a straight, infinitely long wire that carries a current I. To visualize the magnetic field such a wire produces, we shake iron filing onto a sheet of paper that is pierced by the wire, as indicated in **Figure 22–19 (a)**. The result is that the filings form into circular patterns centered on the wire—evidently, the magnetic field "circulates" around the wire.

We can gain additional information about the field by placing a group of small compasses about the wire, as in **Figure 22–19 (b)**. In addition to confirming the

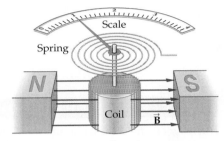

▲ **FIGURE 22–18 The galvanometer**

The basic elements of a galvanometer are a coil, a magnetic field, a spring, and a needle attached to the coil. As current passes through the coil, a torque acts on it, causing it to rotate. The spring ensures that the angle of rotation is proportional to the current in the coil.

REAL-WORLD PHYSICS

The galvanometer

(a)

(b)

▲ **FIGURE 22–19 The magnetic field of a current-carrying wire**

(a) An electric current flowing through a wire produces a magnetic field. In the case of a long, straight wire, the field circulates around the wire. **(b)** Compass needles point along the circumference of a circle centered on the wire.

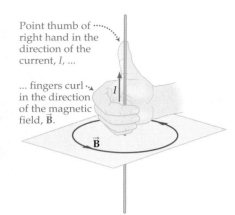

Point thumb of right hand in the direction of the current, I, ...

... fingers curl in the direction of the magnetic field, $\vec{B}$.

I

$\vec{B}$

◀ **FIGURE 22–20** The magnetic-field right-hand rule
The magnetic field right-hand rule determines the direction of the magnetic field produced by a current-carrying wire. With the thumb of the right hand pointing in the direction of the current, the fingers curl in the direction of the field.

circular shape of the field lines, the compass needles show the field's direction. To understand this direction, we must again utilize a right-hand rule—this time, we refer to the rule as the *magnetic field* right-hand rule:

> **Magnetic Field Right-Hand Rule**
> To find the direction of the magnetic field due to a current-carrying wire, point the thumb of your right hand along the wire in the direction of the current I. Your fingers are now curling around the wire in the direction of the magnetic field.

This rule is illustrated in **Figure 22–20**, where we see that it predicts the same direction as that indicated by the compass needles in Figure 22–19 (b).

CONCEPTUAL CHECKPOINT 22–5 DIRECTION OF THE CURRENT

The magnetic field shown in the sketch is due to the horizontal, current-carrying wire. Does the current in the wire flow to the left or to the right?

REASONING AND DISCUSSION
If you point the thumb of your right hand along the wire to the left, your fingers curl into the page above the wire and out of the page below the wire, as shown in the figure. Thus, the current flows to the left.

ANSWER
The current in the wire flows to the left.

$\otimes$ $\otimes$ $\otimes$ $\otimes$ $\otimes$ $\otimes$ $\otimes$

$\vec{B}$

$\otimes$ $\otimes$ $\otimes$ $\otimes$ $\otimes$ $\otimes$ $\otimes$

$\otimes$ $\otimes$ $\otimes$ $\otimes$ $\otimes$ $\otimes$ $\otimes$

$\odot$ $\odot$ $\odot$ $\odot$ $\odot$ $\odot$ $\odot$

$\odot$ $\odot$ $\odot$ $\odot$ $\odot$ $\odot$ $\odot$

$\vec{B}$

$\odot$ $\odot$ $\odot$ $\odot$ $\odot$ $\odot$ $\odot$

Experiment shows that the field produced by a current-carrying wire doubles if the current I is doubled. In addition, the field decreases by a factor of 2 if the distance from the wire, r, is doubled. Hence, we conclude that the magnetic field B must be proportional to I/r that is,

$$B = (\text{constant})\frac{I}{r}$$

The precise expression for B will now be derived using a law of nature known as Ampère's law.

Ampère's Law

Ampère's law relates the magnetic field along a closed path to the electric current enclosed by the path. Specifically, consider the current-carrying wires shown in **Figure 22–21**. These wires are enclosed by the closed path P, which can be divided into many small, straight-line segments of length ΔL. On each of these segments, the magnetic field $\vec{B}$ can be resolved into a component parallel to the segment, $B_{\parallel}$, and a component perpendicular to the segment, $B_{\perp}$. Of particular interest is the product $B_{\parallel}\Delta L$. According to Ampère's law, the sum of $B_{\parallel}\Delta L$ over all segments of a closed path is equal to a constant times the current enclosed by the path. This relationship can be written symbolically as follows:

> **Ampère's Law**
> $$\sum B_{\parallel}\Delta L = \mu_0 I_{\text{enclosed}}$$

22–7

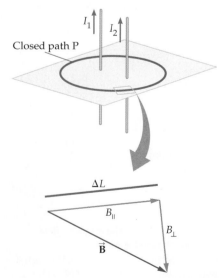

I_1 I_2

Closed path P

ΔL

$B_{\parallel}$

$B_{\perp}$

$\vec{B}$

▲ **FIGURE 22–21** Illustrating Ampère's law
A closed path P encloses the currents I_1 and I_2. According to Ampère's law, the sum of $B_{\parallel}\Delta L$ around the path P is equal to $\mu_0 I_{\text{enclosed}}$. In this case, $I_{\text{enclosed}} = I_1 + I_2$.

In this expression, μ_0 is a constant called the **permeability of free space**. Its value is

$$\mu_0 = 4\pi \times 10^{-7}\,\text{T}\cdot\text{m/A}$$

22–8

We should emphasize that Ampère's law is a law of nature—it is valid for all magnetic fields and currents that are constant in time.

Let's apply Ampère's law to the case of a long, straight wire carrying a current I. We already know that the field circulates around the wire, as illustrated in **Figure 22–22**. It is reasonable, then, to choose a circular path of radius r (and circumference $2\pi r$) to enclose the wire. Since the magnetic field is parallel to the circular path at every point, and all points on the path are the same distance from the wire, it follows that $B_\parallel = B = $ constant. Therefore, the sum of $B_\parallel \Delta L$ around the closed path gives

$$\sum B_\parallel \Delta L = B \sum \Delta L = B(2\pi r)$$

According to Ampère's law, this sum must equal $\mu_0 I_{enclosed} = \mu_0 I$. Therefore, we have

$$B(2\pi r) = \mu_0 I$$

Solving for B, we obtain

Magnetic Field for a Long, Straight Wire

$$B = \frac{\mu_0 I}{2\pi r} \qquad \text{22–9}$$

SI unit: tesla, T

As expected, the field is equal to a constant times I/r.

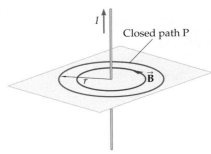

▲ **FIGURE 22–22 Applying Ampère's law**
To apply Ampère's law to a long, straight wire, we consider a circular path centered on the wire. Since the magnetic field is everywhere parallel to this path, and has constant magnitude B at all points on it, the sum of $B_\parallel \Delta L$ over the path is $B(2\pi r)$. Setting this equal to $\mu_0 I$ yields the magnetic field of the wire: $B = \mu_0 I/2\pi r$.

EXERCISE 22–2

Find the magnitude of the magnetic field 1 m from a long, straight wire carrying a current of 1 A.

SOLUTION

Straightforward substitution in Equation 22–9 yields

$$B = \frac{\mu_0 I}{2\pi r} = \frac{(4\pi \times 10^{-7} \text{ T} \cdot \text{m/A})(1 \text{ A})}{2\pi(1 \text{ m})} = 2 \times 10^{-7} \text{ T}$$

This is a weak field, less than one hundredth the strength of the Earth's magnetic field.

EXAMPLE 22–6 AN ATTRACTIVE WIRE

A 52-μC charged particle moves parallel to a long wire with a speed of 720 m/s. The separation between the particle and the wire is 13 cm, and the magnitude of the force exerted on the particle is 1.4×10^{-7} N. Find **(a)** the magnitude of the magnetic field at the location of the particle and **(b)** the current in the wire.

PICTURE THE PROBLEM
The physical situation is illustrated in our sketch. Note that the charged particle moves parallel to the current-carrying wire; hence, its velocity is at right angles to the magnetic field. This means that it will experience the maximum magnetic force. Finally, by pointing the thumb of your right hand in the direction of I, you can verify that $\vec{B}$ points out of the page above the wire and into the page below it.

STRATEGY

a. The maximum magnetic force on the charged particle is $F = qvB$. This relation can be solved for B.

b. The magnetic field is produced by the current in the wire. Therefore, $B = \mu_0 I/2\pi r$. Using B from part (a), we can solve for the current I.

SOLUTION

Part (a)

1. Use $F = qvB$ to solve for the magnetic field:

$$F = qvB$$
$$B = \frac{F}{qv} = \frac{1.4 \times 10^{-7} \text{ N}}{(5.2 \times 10^{-5} \text{ C})(720 \text{ m/s})} = 3.7 \times 10^{-6} \text{ T}$$

CONTINUED ON NEXT PAGE

CONTINUED FROM PREVIOUS PAGE

Part (b)

2. Use the relation for the magnetic field of a wire to solve for the current:

$$B = \frac{\mu_0 I}{2\pi r}$$

$$I = \frac{2\pi r B}{\mu_0}$$

3. Substitute the value for B found in part (a) and evaluate I:

$$I = \frac{2\pi r B}{\mu_0} = \frac{2\pi (0.13 \text{ m})(3.7 \times 10^{-6} \text{ T})}{4\pi \times 10^{-7} \text{ T} \cdot \text{m/A}} = 2.4 \text{ A}$$

INSIGHT

Using the magnetic force RHR, we can see that the direction of the force exerted on the charged particle is toward the wire, as illustrated in our sketch. If the particle had been negatively charged, the force would have been away from the wire.

PRACTICE PROBLEM

Suppose a particle with a charge of 52 μC is 13 cm above the wire and moving with a speed of 720 m/s to the left. Find the magnitude and direction of the force acting on this particle.   [**Answer:** $F = 1.4 \times 10^{-7}$ N, away from wire]

Some related homework problems: Problem 46, Problem 52

ACTIVE EXAMPLE 22–2   FIND THE MAGNETIC FIELD

Two wires separated by a distance of 22 cm carry currents in the same direction. The current in one wire is 1.5 A, and the current in the other wire is 4.5 A. Find the magnitude of the magnetic field halfway between the wires.

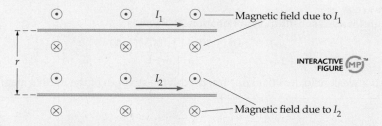

SOLUTION   *(Test your understanding by performing the calculations indicated in each step.)*

1. Find the magnitude and direction of the magnetic field produced by wire 1: $B_1 = 2.7 \times 10^{-6}$ T, into page

2. Find the magnitude and direction of the magnetic field produced by wire 2: $B_2 = 8.2 \times 10^{-6}$ T, out of page

3. Calculate the magnitude of the net field: $B = B_2 - B_1 = 5.5 \times 10^{-6}$ T

INSIGHT

Since the field produced by I_2 has the greater magnitude, the net field is out of the page. If the currents were equal, the net field midway between the wires would be zero.

YOUR TURN

Find the magnitude and direction of the magnetic field 11 cm *below* wire 2.

*(Answers to **Your Turn** problems are given in the back of the book.)*

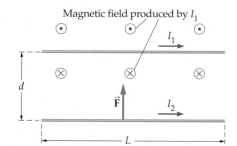

▲ **FIGURE 22–23 The magnetic force between current-carrying wires**

A current in wire 1 produces a magnetic field, $B_1 = \mu_0 I_1 / 2\pi d$, at the location of wire 2. The result is a force exerted on a length L of wire 2 of magnitude $F = \mu_0 I_1 I_2 L / 2\pi d$.

Forces Between Current-Carrying Wires

We know that a current-carrying wire in a magnetic field experiences a force. We also know that a current-carrying wire produces a magnetic field. It follows, then, that one current-carrying wire will exert a force on another.

To work this relationship out in detail, consider the two wires with parallel currents and separation d shown in **Figure 22–23**. The magnetic field produced by wire 1 circulates around it, coming out of the page above the wire and entering the page below the wire. Thus, wire 2 experiences a magnetic field pointing into the page with a magnitude given by Equation 22–9: $B = \mu_0 I_1/2\pi d$. The force experienced by wire 2, therefore, has a magnitude given by Equation 22–4 with $\theta = 90°$:

$$F = I_2 LB = I_2 L\left(\frac{\mu_0 I_1}{2\pi d}\right) = \frac{\mu_0 I_1 I_2}{2\pi d}L \qquad 22\text{--}10$$

The direction of the force acting on wire 2, as given by the magnetic force RHR, is upward, that is, toward wire 1. A similar calculation starting with the field produced by wire 2 gives a force of the same magnitude acting downward on wire 1. This is to be expected, because the forces acting on wires 1 and 2 form an action–reaction pair, as indicated in **Figure 22–24 (a)**. Hence, wires with parallel currents attract one another.

If the currents in wires 1 and 2 are in opposite directions, as in **Figure 22–24 (b)**, the situation is similar to that discussed in the preceding paragraph, except that the direction of the forces is reversed. Thus, wires with opposite currents repel one another.

22–7 Current Loops and Solenoids

We now consider the magnetic fields produced when a current-carrying wire has a circular or a helical geometry—as opposed to the straight wires considered in the previous section. As we shall see, there are many practical applications for such geometries.

Current Loop

We begin by considering the magnetic field produced by a current-carrying wire that is formed into the shape of a circular loop. In **Figure 22–25 (a)** we show a wire loop connected to a battery producing a current in the direction indicated. Using the magnetic field RHR, as shown in the figure, we see that $\vec{B}$ points from left to right as it passes through the loop. Notice also that the field lines are bunched together within the loop, indicating an intense field there, but are more widely spaced outside the loop.

The most interesting aspect of the field produced by the loop is its close resemblance to the field of a bar magnet. This similarity is illustrated in **Figure 22–25 (b)**, where we see that one side of the loop behaves like a north magnetic pole and the other side like a south magnetic pole. Thus, if two loops with identical currents are placed near each other, as in **Figure 22–26 (a)**, the force between them will be similar to the force between two bar magnets pointing in the same direction—that is, they will attract each other. If the loops have oppositely directed currents, as in **Figure 22–26 (b)**, the force between them is repulsive. Note the similarity of the results for straight wires, Figure 22–24, and for loops, Figure 22–26.

On a more fundamental level, the connection between a current loop and a permanent magnet is much more than accidental. In fact, the atomic origin of magnetic fields is due to circulating currents produced by electrons. We discuss this connection in greater detail in the next section.

The magnitude of the magnetic field produced by a circular loop of N turns, radius R, and current I varies from point to point; however, it can be shown that at the center of the loop, the field is given by the following simple expression:

$$B = \frac{N\mu_0 I}{2R} \qquad \text{(center of circular loop of radius } R) \qquad 22\text{--}11$$

▶ **FIGURE 22–25 The magnetic field of a current loop**
(a) The magnetic field produced by a current loop is relatively intense within the loop and falls off rapidly outside the loop. **(b)** A permanent magnet produces a field that is very similar to the field of a current loop.

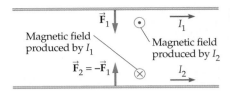
(a) Currents in same direction

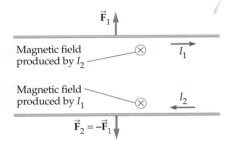
(b) Currents in opposite directions

▲ **FIGURE 22–24 The direction of the magnetic force between current-carrying wires**
The forces between current-carrying wires depend on the relative direction of their currents. **(a)** If the currents are in the same direction, the force is attractive. **(b)** Wires with oppositely directed currents experience repulsive forces.

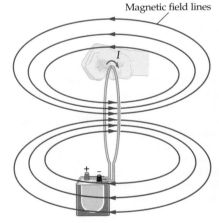
(a) Magnetic field of a current loop

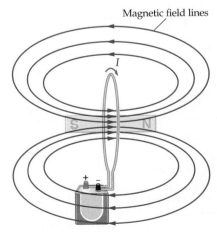
(b) Magnetic field of bar magnet is similar

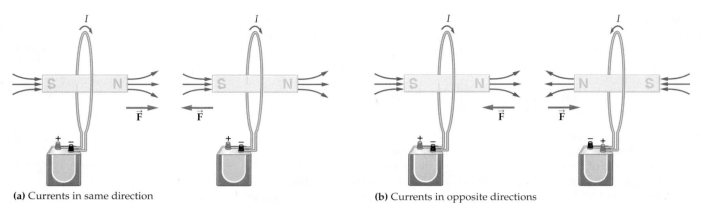

(a) Currents in same direction   **(b)** Currents in opposite directions

▲ **FIGURE 22–26   Magnetic forces between current loops**

To decide whether current loops will experience an attractive or repulsive force, it is useful to think in terms of the corresponding permanent magnets. **(a)** Current loops with currents in the same direction are like two bar magnets lined up in the same direction; they attract each other. **(b)** Current loops with opposite currents act like bar magnets with opposite orientations; they repel each other.

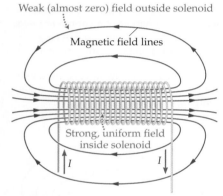

▲ **FIGURE 22–27   The solenoid**

A solenoid is formed from a long wire that is wound into a succession of current loops. The magnetic field inside the solenoid is relatively strong and uniform. Outside the solenoid the field is weak. In the ideal case, we consider the field outside the solenoid to be zero and that inside to be uniform and parallel to the solenoid's axis.

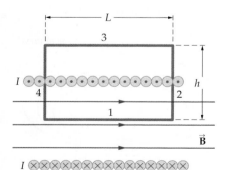

▲ **FIGURE 22–28   Ampère's law and the magnetic field in a solenoid**

To calculate the magnetic field inside a solenoid, we apply Ampère's law to the rectangular path shown here. The only side of the rectangle that has a nonzero, parallel component of $\vec{\mathbf{B}}$ is side 1.

Notice that the field is proportional to the current in the loop and inversely proportional to its radius.

Solenoid

A **solenoid** is an electrical device in which a long wire has been wound into a succession of closely spaced loops with the geometry of a helix. Also referred to as an **electromagnet,** a solenoid carrying a current produces an intense, nearly uniform, magnetic field inside the loops, as indicated in **Figure 22–27**. Notice that each loop of the solenoid carries a current in the same direction; therefore, the magnetic force between loops is attractive and serves to hold the loops tightly together.

The magnetic field lines in Figure 22–27 are tightly packed inside the solenoid but are widely spaced outside. In the ideal case of a very long, tightly packed solenoid, the magnetic field outside is practically zero—especially when compared with the intense field inside the solenoid. We can use this idealization, in combination with Ampère's law, to calculate the magnitude of the field inside the solenoid.

To do so, consider the rectangular path of width L and height h shown in **Figure 22–28**. Notice that the parallel component of the field on side 1 is simply $\vec{\mathbf{B}}$. On sides 2 and 4 the parallel component is zero, since $\vec{\mathbf{B}}$ is perpendicular to those sides. Finally, on side 3 (which is outside the solenoid) the magnetic field is zero. Using these results, we obtain the sum of $B_\parallel \Delta L$ over the rectangular loops:

$$\sum B_\parallel \Delta L = \sum_{\text{side 1}} B_\parallel \Delta L + \sum_{\text{side 2}} B_\parallel \Delta L + \sum_{\text{side 3}} B_\parallel \Delta L + \sum_{\text{side 4}} B_\parallel \Delta L$$

$$= BL + 0 + 0 + 0 = BL$$

Next, the current enclosed by the rectangular circuit is NI, where N is the number of loops in the length L. Therefore, Ampère's law gives

$$BL = \mu_0 NI$$

Solving for B and letting the number of loops per length be $n = N/L$, we find

Magnetic Field of a Solenoid

$$B = \mu_0\left(\frac{N}{L}\right)I = \mu_0 nI \qquad \qquad 22\text{–}12$$

SI unit: tesla, T

Note that this result is independent of the cross-sectional area of the solenoid.

When used as an electromagnet, a solenoid has many useful properties. First and foremost, it produces a strong magnetic field that can be turned on or off at the flip of a switch—unlike the field of a permanent magnet. In addition, the magnetic field can be further intensified by filling the core of the solenoid with an iron bar. In such a case, the magnetic field of the solenoid magnetizes the iron bar, which then adds its field to the overall field of the system. These properties and others make solenoids useful devices in a variety of electric circuits. Further examples are given in the next chapter.

PROBLEM-SOLVING NOTE

The Magnetic Field Inside a Solenoid

In an ideal solenoid, the magnetic field inside the solenoid points along the axis and is uniform. Outside the solenoid the field is zero. Therefore, when calculating the field produced by a solenoid, it is not necessary to specify a particular point inside the solenoid; all inside points have the same field.

CONCEPTUAL CHECKPOINT 22–6   MAGNETIC FIELD IN A SOLENOID

If you want to increase the strength of the magnetic field inside a solenoid, is it better to **(a)** double the number of loops, keeping the length the same, or **(b)** double the length, keeping the number of loops the same?

REASONING AND DISCUSSION
Referring to the expression $B = \mu_0(N/L)I$, we see that doubling the number of loops $(N \rightarrow 2N)$ while keeping the length the same $(L \rightarrow L)$ results in a doubled magnetic field $(B \rightarrow 2B)$. On the other hand, doubling the length $(L \rightarrow 2L)$ while keeping the number of loops the same $(N \rightarrow N)$ reduces the magnetic field by a factor of two $(B \rightarrow B/2)$. Hence, to increase the field one should pack more loops into the same length.

ANSWER
(a) Double the number of loops with the same length.

EXAMPLE 22–7   THROUGH THE CORE OF A SOLENOID

A solenoid is 20.0 cm long, has 200 loops, and carries a current of 3.25 A. Find the magnitude of the force exerted on a 15.0-μC charged particle moving at 1050 m/s through the interior of the solenoid, at an angle of 11.5° relative to the solenoid's axis.

PICTURE THE PROBLEM
Our sketch shows the solenoid, along with the uniform magnetic field it produces parallel to its axis. Inside the solenoid, a positively charged particle moves at an angle of $\theta = 11.5°$ relative to the magnetic field.

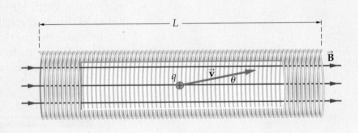

STRATEGY
The force exerted on the charged particle is magnetic; hence, we start by calculating the magnetic field produced by the solenoid, $B = \mu_0(N/L)I$. Next, we note that the magnetic field in a solenoid is parallel to its axis. It follows that the magnitude of the force exerted on the charge is given by $F = |q|vB \sin \theta$, with $\theta = 11.5°$.

SOLUTION

1. Calculate the magnetic field inside the solenoid:

$$B = \mu_0 \left(\frac{N}{L}\right)I$$

$$= (4\pi \times 10^{-7}\,\text{T}\cdot\text{m/A})\left(\frac{200}{0.200\,\text{m}}\right)(3.25\,\text{A}) = 4.08 \times 10^{-3}\,\text{T}$$

2. Use B to find the force exerted on the charged particle:

$$F = |q|vB \sin \theta$$

$$= (15.0 \times 10^{-6}\,\text{C})(1050\,\text{m/s})(4.08 \times 10^{-3}\,\text{T}) \sin 11.5°$$

$$= 1.28 \times 10^{-5}\,\text{N}$$

INSIGHT
Note that the magnetic field strength inside this modest solenoid is approximately 100 times greater than the magnetic field at the surface of the Earth.

PRACTICE PROBLEM
What current would be required to double the force acting on the particle to 2.56×10^{-5} N?   [**Answer:** $I = 2(3.25\,\text{A}) = 6.50\,\text{A}$]

Some related homework problems: Problem 57, Problem 58

REAL-WORLD PHYSICS: BIO
MRI instruments

REAL-WORLD PHYSICS: BIO
Magnetic reed switches in pacemakers

Magnetic resonance imaging (MRI) instruments utilize solenoids large enough to accommodate a person within their coils. Not only are these solenoids large in size, they are also capable of producing extremely powerful magnetic fields on the order of 1 or 2 tesla (see Table 22–1). So powerful are these fields, in fact, that metallic objects such as mop buckets and stretchers have been known to be pulled from across the room into the bore of the magnet. A metal oxygen bottle in the same room can be turned into a dangerous, high-speed projectile.

In some cases, artificial pacemakers have been affected. Many types of pacemakers have what are known as *magnetic reed switches*. These switches allow a physician to change the operating mode of a pacemaker, without surgery, by simply placing a magnet at the appropriate location on a patient's chest. If a person with one of these pacemakers comes anywhere near an operating MRI instrument, the results can be serious.

22–8 Magnetism in Matter

Some materials have strong magnetic fields; others do not. To understand these differences, we must consider the behavior of matter on the atomic level.

To begin, recall that circulating electric currents can produce magnetic fields much like those produced by bar magnets. This is significant because atoms have electrons orbiting their nucleus, which means they have circulating electric currents. The fields produced by these orbiting electrons can be sizable—on the order of 10 T or so at the nucleus of the atom. In most atoms, though, the fields produced by the various individual electrons tend to cancel one another, resulting in a very small or zero net magnetic field.

Another type of circulating electric current present in atoms produces fields that occasionally do not cancel. These currents are associated with the *spin* that all electrons have. A simple model of the electron—which should *not* be taken literally—is of a spinning sphere of charge. The circulating charge associated with the spinning motion gives rise to a magnetic field. Electrons tend to "pair up" in atoms in such a way that their spins are opposite to one another, again resulting in zero net field. However, in some atoms—such as iron, nickel, and cobalt—the net field due to the spinning electrons is nonzero, and strong magnetic effects can occur as the magnetic field of one atom tends to align with the magnetic field of another. A full understanding of these types of effects requires the methods of quantum mechanics, which will be discussed in Chapters 30 and 31.

Ferromagnetism

If the tendency of magnetic atoms to self-align is strong enough, as it is in materials such as iron and nickel, the result can be an intense magnetic field—as in a bar magnet. Materials with this type of behavior are called **ferromagnets.** Counteracting the tendency of atoms to align, however, is the disorder caused by increasing temperature. In fact, all ferromagnets lose their magnetic field if the temperature is high enough to cause their atoms to orient in random directions. For example, the magnetic field of a bar magnet made of iron vanishes if its temperature exceeds 770 °C.

This type of temperature behavior shows that the simple picture of a large bar magnet within the Earth cannot be correct. As we know, temperature increases with depth below the Earth's surface. In fact, at depths of about 15 miles the temperature is already above 770 °C, so the magnetism of an iron magnet would be lost due to thermal effects. Of course, at even greater depths an iron magnet would melt to form a liquid. In fact, it appears that the magnetic field of the Earth is caused by circulating currents of molten iron, nickel, and other metals. These circulating currents create a magnetic field in much the same way as the circulating current in a solenoid.

Temperature also plays a key role in the magnetization that is observed in rocks on the ocean floor. As molten rock is extruded from mid-ocean ridges, it has no net magnetization because of its high temperature. When the rock cools, however, it becomes magnetized in the direction of the Earth's magnetic field. In effect, the Earth's magnetic field becomes "frozen" in the solidified rock. As the seafloor spreads, and more material is formed along the ridge, a continuous record of the Earth's magnetic field is formed. In particular, if the Earth's field reverses at some point in time, the field in the solidified rocks will record that fact. **Figure 22–29** shows a record of the magnetization of rock that has been formed by seafloor spreading, showing the regions of oppositely magnetized rocks that indicate the field reversals.

Another important feature of ferromagnets is that their magnetism is characterized by magnetic **domains** within the material, as illustrated in **Figure 22–30 (a)**. Each domain has a strong magnetic field, but different domains are oriented differently, so that the net effect may be small. The typical size of these domains is on the order of 10^{-4} cm to 10^{-1} cm. When an external field is applied to such a material, the magnetic domains that are pointing in the direction of the applied field often grow in size at the expense of domains with different orientations, as indicated in **Figure 22–30 (b)**. The result is that the applied external field produces a net magnetization in the material.

Many living organisms are known to incorporate small ferromagnetic crystals, consisting of *magnetite*, in their bodies. For example, some species of bacteria use magnetite crystals to help orient themselves with respect to the Earth's magnetic field. Magnetite has also been found in the brains of bees and pigeons, where it is suspected to play a role in navigation. It is even found in human brains, though its possible function there is unclear.

Whatever the role of magnetite in people, observations show that a magnetic field can affect the way the human brain operates. In recent experiments involving people viewing optical illusions, it has been found that if the parietal lobe on one side of the brain is exposed to a highly focused magnetic field of about 1 T, a temporary interruption in much of the neural activity in that hemisphere results. Fortunately, fields of 1 T are generally not encountered in everyday life.

Paramagnetism and Diamagnetism

Not all magnetic materials are ferromagnetic, however. In some cases a ferromagnet has zero magnetic field simply because it is at too high a temperature. In other cases, the tendency for the self-alignment of individual atoms in a given material is too weak to produce a net magnetic field, even at low temperatures. In either case, a strong external magnetic field applied to the material can cause alignment of the atoms and result in a magnetic field. Magnetic effects of this type are referred to as **paramagnetism.**

Finally, all materials display a further magnetic effect referred to as **diamagnetism.** In the diamagnetic effect, an applied magnetic field on a material produces an oppositely directed field in response. The resulting repulsive force is usually too weak to be noticed, except in superconductors. The basic mechanism responsible for diamagnetism will be discussed in greater detail in the next chapter.

If a magnetic field is strong enough, however, even the relatively weak repulsion of diamagnetism can lead to significant effects. For example, researchers at the Nijmegen High Field Magnet Laboratory have used a field of 16 T to levitate a strawberry, a cricket, and even a frog. The diamagnetic repulsion of the water in these organisms is great enough, in a field that strong, to counteract the gravitational force of the Earth. The researchers reported that the living frog showed no visible signs of discomfort during its levitation, and that it hopped away normally after the experiment.

A similar diamagnetic effect can be used to levitate a small magnet between a person's fingertips. In the case shown in the photograph, a powerful magnet about 8 ft above the person's hand applies a magnetic field strong enough to counteract the weight of the small magnet. By itself, this system of two magnets is not stable—if the small magnet is displaced upward slightly, the increased attractive force of the upper magnet raises it farther; if it is displaced downward, the

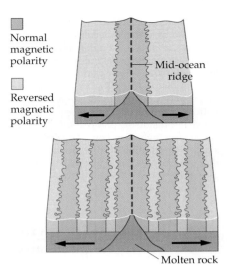

▲ **FIGURE 22–29 Mid-ocean ridge**
Molten rock extruded at a mid-ocean ridge magnetizes in the direction of the Earth's magnetic field when it cools to temperatures below about 770 °C. After cooling, the direction of the Earth's field remains "frozen" in the rocks. As these rocks move away from the ridge, due to seafloor spreading, newly extruded material near the ridge undergoes the same process. As a result, the magnetization of the rocks on either side of a mid-ocean ridge produces a geological record of the polarity of the Earth's magnetic field over time, as well as convincing confirmation of seafloor spreading.

REAL-WORLD PHYSICS: BIO
Magnetite in living organisms

REAL-WORLD PHYSICS: BIO
Magnetism and the brain

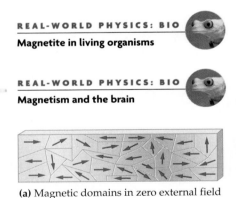

(a) Magnetic domains in zero external field

$\vec{B}$

(b) Domains in direction of external magnetic field grow in size

▲ **FIGURE 22–30 Magnetic domains**
(a) A ferromagnetic material tends to form into numerous domains with magnetization pointing in different directions. **(b)** When an external field is applied, the domains pointing in the direction of the field often grow in size at the expense of domains pointing in other directions.

▶ (Left) The repulsive diamagnetic forces produced by water molecules in the body of a frog are strong enough to levitate it in an intense magnetic field of 16 T. (Animals levitated in this way appear to suffer no harm or discomfort.) (Right) This small magnet is suspended in the strong magnetic field produced by a larger magnet above it (outside the photo). Ordinarily, such an arrangement would be highly unstable. The addition of small, repulsive diamagnetic forces due to a person's fingers, however, is sufficient to convert it into a stable equilibrium.

REAL-WORLD PHYSICS
Magnetic levitation

force of gravity pulls it down even farther. With the diamagnetic effect of the fingers, however, the small magnet is stabilized. If it is displaced upward now, the repulsive diamagnetic effect of the finger pushes it back down; if it is displaced downward, the diamagnetic repulsion of the thumb pushes it back up. The result is a stable, magnetic levitation—with no smoke or mirrors.

THE BIG PICTURE   PUTTING PHYSICS IN CONTEXT

LOOKING BACK

Newton's third law (Chapter 5) is applied to the force between bar magnets in Section 22–1. It is important to note that Newton's laws apply to all types of forces, including electric and magnetic forces.

The direction of the force exerted on a moving charge by a magnetic field is given by the right-hand rule (Chapter 11), as we see in Section 22–2. The magnetic field also exerts a force on current-carrying wires. This can result in a torque (Chapter 11) on a current loop, as shown in Section 22–5.

The force exerted by a magnetic field on a moving charge is at right angles to the velocity. This can lead to circular motion (Chapter 6). Several examples of circular motion are discussed in Section 22–3.

LOOKING AHEAD

The magnetic field is central to the concept of Faraday's law of induction, which is presented in Chapter 23. It also plays a key role in alternating-current (ac) circuits, as we shall see in Chapter 24.

In Chapter 23 we introduce the idea of an inductor. In particular, we will show that an inductor can store energy in the form of a magnetic field, just as a capacitor (Chapter 20) can store energy in an electric field.

In Chapter 23 we also show that a changing magnetic field is central to the operation of electric motors and generators, as well as transformers.

We study light and the electromagnetic spectrum in Chapter 25. As we shall see, light is a wave formed of oscillating electric and magnetic fields. Both electric and magnetic fields must be present for light to exist.

CHAPTER SUMMARY

22–1 THE MAGNETIC FIELD

The magnetic field gives an indication of the effect a magnet will have in a given region.

Magnetic Poles
A magnet is characterized by two poles, referred to as the north pole and the south pole. These poles cannot be separated—all magnets have both poles.

Magnetic Field Lines
Magnetic fields can be represented with lines in much the same way as electric fields can be portrayed. In particular, the more closely spaced the lines, the more intense the field. Magnetic field lines, which point away from north poles and toward south poles, always form closed loops.

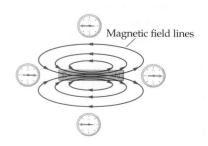

Magnetic field lines

Geomagnetism

The Earth produces its own magnetic field, which is inclined at an angle of about 11.5° with its rotational axis. The geographic north pole of the Earth is actually the south magnetic pole of the Earth's magnetic field.

22–2 THE MAGNETIC FORCE ON MOVING CHARGES

In order for a magnetic field to exert a force on a particle, the particle must have charge and must be moving.

Magnitude of the Magnetic Force

The magnitude of the magnetic force is

$$F = |q|vB \sin \theta \qquad \text{22–1}$$

where q is the charge of the particle, v is its speed, B is the magnitude of the magnetic field, and θ is the angle between the velocity vector $\vec{v}$ and the magnetic field vector $\vec{B}$.

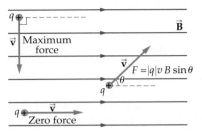

Magnetic Force Right-Hand Rule (RHR)

The magnetic force $\vec{F}$ points in a direction that is perpendicular to both $\vec{B}$ and $\vec{v}$. For a positive charge, point the fingers of your right hand in the direction of $\vec{v}$ and curl them toward the direction of $\vec{B}$. Your thumb points in the direction of the force $\vec{F}$. The force on a negative charge is in the opposite direction to that on a positive charge.

22–3 THE MOTION OF CHARGED PARTICLES IN A MAGNETIC FIELD

The motion of a charged particle in a magnetic field is quite different from that in an electric field.

Electric Versus Magnetic Forces

A charged particle in an electric field accelerates in the direction of the field; in a magnetic field the acceleration is perpendicular to the field and to the velocity. The electric field does work on a particle and changes its speed; a magnetic field does no work on a particle, and its speed remains constant.

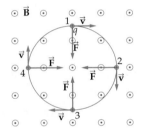

Constant-Velocity Motion

If a charged particle moves parallel or antiparallel to a magnetic field, it experiences no force; hence, its velocity remains constant.

Circular Motion

If a charged particle moves perpendicular to a magnetic field, it will orbit with constant speed in a circle of radius $r = mv/|q|B$.

Helical Motion

When a particle's velocity has components both parallel and perpendicular to a magnetic field, it will follow a helical path.

22–4 THE MAGNETIC FORCE EXERTED ON A CURRENT-CARRYING WIRE

An electric current in a wire is caused by the movement of electric charges. Since moving electric charges experience magnetic forces, it follows that a current-carrying wire will as well.

Force on a Current-Carrying Wire

A wire of length L carrying a current I at an angle θ to a magnetic field B experiences a force given by

$$F = ILB \sin \theta \qquad \text{22–4}$$

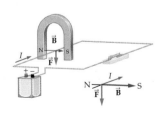

22–5 LOOPS OF CURRENT AND MAGNETIC TORQUE

A current loop placed in a magnetic field experiences a torque that depends on the relative orientation of the plane of the loop and the magnetic field.

Torque on a General Loop

The magnetic torque exerted on a current loop is given by

$$\tau = NIAB \sin \theta \qquad \text{22–6}$$

where N is the number of turns around the loop, I is the current, A is the area of the loop, B is the strength of the magnetic field, and θ is the angle between the plane of the loop and the magnetic force.

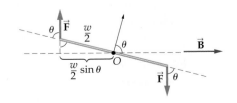

22–6 ELECTRIC CURRENTS, MAGNETIC FIELDS, AND AMPÈRE'S LAW

The key observation that serves to unify electricity and magnetism is that electric currents cause magnetic fields.

Magnetic Field Right-Hand Rule

The direction of the magnetic field produced by a current is found by pointing the thumb of the right hand in the direction of the current. The fingers of the right hand curl in the direction of the field.

Ampère's Law

Ampère's law can be expressed as follows:

$$\sum B_{\parallel} \Delta L = \mu_0 I_{\text{enclosed}} \qquad 22\text{–}7$$

where $B_{\parallel}$ is the component of the magnetic field parallel to a segment of a closed path of length ΔL, I is the current enclosed by the path, and $\mu_0 = 4\pi \times 10^{-7}$ T·m/A is a constant called the permeability of free space.

Magnetic Field of a Long, Straight Wire

A long, straight wire carrying a current I produces a magnetic field of magnitude B given by

$$B = \frac{\mu_0 I}{2\pi r} \qquad 22\text{–}9$$

In this expression, r is the radial distance from the wire.

Forces Between Current-Carrying Wires

Two wires, carrying the currents I_1 and I_2, exert forces on each other. If the wires are separated by a distance d, the force exerted on a length L is

$$F = \frac{\mu_0 I_1 I_2}{2\pi d} L \qquad 22\text{–}10$$

Wires that carry current in the same direction attract one another; wires with opposite-directed currents repel one another.

22–7 CURRENT LOOPS AND SOLENOIDS

A single loop of current produces a magnetic field much like that of a permanent magnet. A succession of loops grouped together in a coil forms a solenoid.

Current Loop

The magnetic field at the center of a current loop of N turns, radius R, and current I is

$$B = \frac{N\mu_0 I}{2R} \qquad 22\text{–}11$$

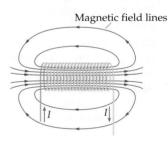

Magnetic field lines

Magnetic Field of a Solenoid

The magnetic field inside a solenoid is nearly uniform and aligned along the solenoid's axis. If the solenoid has N loops in a length L and carries a current I, its magnetic field is

$$B = \mu_0 \left(\frac{N}{L}\right) I = \mu_0 n I \qquad 22\text{–}12$$

In this expression, n is the number of loops per length; $n = N/L$. The magnetic field outside a solenoid is small, and in the ideal case can be considered to be zero.

22–8 MAGNETISM IN MATTER

The ultimate origin of the magnetic fields we observe around us is circulating electric currents on the atomic level.

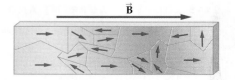

Paramagnetism

A paramagnetic material has no magnetic field unless an external magnetic field is applied to it. In this case, it develops a magnetization in the direction of the external field.

Ferromagnetism

A ferromagnetic material produces a magnetic field even in the absence of an external magnetic field. Permanent magnets are constructed of ferromagnetic materials.

Diamagnetism
Diamagnetism is the effect of the production by a material of a magnetic field
in the opposite direction to an external magnetic field that is applied to it. All
materials show at least a small diamagnetic effect.

PROBLEM-SOLVING SUMMARY

Type of Problem	Relevant Physical Concepts	Related Examples
Find the magnetic force exerted on a moving charge.	The magnitude of the magnetic force depends on the charge, the speed, the magnetic field, and the angle between the magnetic field and the velocity. The direction of the magnetic force is given by the right-hand rule.	Examples 22–1, 22–2
Determine the radius of the path followed by a charged particle in a magnetic field.	A charged particle moving at right angles to a uniform magnetic field moves on a circular path with a radius given by $r = mv/|q|B$.	Example 22–3
Calculate the magnetic force on a current-carrying wire.	If a wire segment of length L carries a current I at an angle θ to a magnetic field of strength B, the force it experiences is $F = ILB \sin \theta$.	Example 22–4
Find the torque exerted on a current loop.	If a current loop with current I, cross-sectional area A, and N turns is in a magnetic field B, the maximum torque is $\tau = NIAB$. The maximum torque occurs when the plane of the loop is parallel to the magnetic field ($\theta = 90°$). When the plane of the loop is at an angle θ to the magnetic force, the torque is $\tau = NIAB \sin \theta$.	Example 22–5
Determine the magnetic field produced by a long, straight wire carrying a current I.	The magnetic field produced by a long, straight wire with a current I circulates around the wire in a direction given by the right-hand rule. The magnitude of the magnetic field a radial distance r from the wire is $B = \mu_0 I/2\pi r$. This result follows from Ampère's law.	Example 22–6
Find the magnetic field inside a solenoid.	An ideal solenoid with n turns per length and current I produces a uniform magnetic field of magnitude $B = \mu_0 nI$. The field is parallel to the axis of the solenoid. In the ideal case, the magnetic field outside the solenoid is zero.	Example 22–7

CONCEPTUAL QUESTIONS

For instructor-assigned homework, go to www.masteringphysics.com

(Answers to odd-numbered Conceptual Questions can be found in the back of the book.)

1. Two charged particles move at right angles to a magnetic field and deflect in opposite directions. Can one conclude that the particles have opposite charges?

2. An electron moves with constant velocity through a region of space that is free of magnetic fields. Can one conclude that the electric field is zero in this region? Explain.

3. An electron moves with constant velocity through a region of space that is free of electric fields. Can one conclude that the magnetic field is zero in this region? Explain.

4. Describe how the motion of a charged particle can be used to distinguish between an electric and a magnetic field.

5. Explain how a charged particle moving in a circle of small radius can take the same amount of time to complete an orbit as an identical particle orbiting in a circle of large radius.

6. A current-carrying wire is placed in a region with a uniform magnetic field. The wire experiences zero magnetic force. Explain.

PROBLEMS AND CONCEPTUAL EXERCISES

Note: Answers to odd-numbered Problems and Conceptual Exercises can be found in the back of the book. **IP** *denotes an integrated problem, with both conceptual and numerical parts;* **BIO** *identifies problems of biological or medical interest;* **CE** *indicates a conceptual exercise.* **Predict/Explain** *problems ask for two responses:* **(a)** *your prediction of a physical outcome, and* **(b)** *the best explanation among three provided. On all problems, red bullets (•, ••, •••) are used to indicate the level of difficulty.*

SECTION 22–2 THE MAGNETIC FORCE ON MOVING CHARGES

1. • **CE Predict/Explain** Proton 1 moves with a speed v from the east coast to the west coast in the continental United States; proton 2 moves with the same speed from the southern United States toward Canada. **(a)** Is the magnitude of the magnetic force experienced by proton 2 greater than, less than, or equal to the force experienced by proton 1? **(b)** Choose the *best explanation* from among the following:
   I. The protons experience the same force because the magnetic field is the same and their speeds are the same.
   II. Proton 1 experiences the greater force because it moves at right angles to the magnetic field.
   III. Proton 2 experiences the greater force because it moves in the same direction as the magnetic field.

2. • **CE** An electron moves west to east in the continental United States. Does the magnetic force experienced by the electron point in a direction that is generally north, south, east, west, upward, or downward? Explain.

3. • **CE** An electron moving in the positive x direction, at right angles to a magnetic field, experiences a magnetic force in the positive y direction. What is the direction of the magnetic field?

4. • **CE** Suppose particles A, B, and C in **Figure 22–31** have identical masses and charges of the same magnitude. Rank the particles in order of increasing speed. Indicate ties where appropriate.

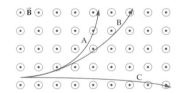

▲ **FIGURE 22–31** Problems 4, 5, and 6

5. • **CE** Referring to Figure 22–31, what is the sign of the charge for each of the three particles? Explain.

6. • **CE** Suppose the three particles in Figure 22–31 have the same mass and speed. Rank the particles in order of increasing magnitude of their charge. Indicate ties where appropriate.

7. • What is the acceleration of a proton moving with a speed of 6.5 m/s at right angles to a magnetic field of 1.6 T?

8. • An electron moves at right angles to a magnetic field of 0.18 T. What is its speed if the force exerted on it is 8.9×10^{-15} N?

9. • A negatively charged ion moves due north with a speed of 1.5×10^6 m/s at the Earth's equator. What is the magnetic force exerted on this ion?

10. • A proton high above the equator approaches the Earth moving straight downward with a speed of 355 m/s. Find the acceleration of the proton, given that the magnetic field at its altitude is 4.05×10^{-5} T.

11. •• A 0.32-μC particle moves with a speed of 16 m/s through a region where the magnetic field has a strength of 0.95 T. At what angle to the field is the particle moving if the force exerted on it is **(a)** 4.8×10^{-6} N, **(b)** 3.0×10^{-6} N, or **(c)** 1.0×10^{-7} N?

12. •• A particle with a charge of 14 μC experiences a force of 2.2×10^{-4} N when it moves at right angles to a magnetic field with a speed of 27 m/s. What force does this particle experience when it moves with a speed of 6.3 m/s at an angle of 25° relative to the magnetic field?

13. •• An ion experiences a magnetic force of 6.2×10^{-16} N when moving in the positive x direction but no magnetic force when moving in the positive y direction. What is the magnitude of the magnetic force exerted on the ion when it moves in the x-y plane along the line $x = y$? Assume that the ion's speed is the same in all cases.

14. •• An electron moving with a speed of 4.2×10^5 m/s in the positive x direction experiences zero magnetic force. When it moves in the positive y direction, it experiences a force of 2.0×10^{-13} N that points in the negative z direction. What are the direction and magnitude of the magnetic field?

15. •• **IP** Two charged particles with different speeds move one at a time through a region of uniform magnetic field. The particles move in the same direction and experience equal magnetic forces. **(a)** If particle 1 has four times the charge of particle 2, which particle has the greater speed? Explain. **(b)** Find the ratio of the speeds, v_1/v_2.

16. •• A 6.60-μC particle moves through a region of space where an electric field of magnitude 1250 N/C points in the positive x direction, and a magnetic field of magnitude 1.02 T points in the positive z direction. If the net force acting on the particle is 6.23×10^{-3} N in the positive x direction, find the magnitude and direction of the particle's velocity. Assume the particle's velocity is in the x-y plane.

17. ••• When at rest, a proton experiences a net electromagnetic force of magnitude 8.0×10^{-13} N pointing in the positive x direction. When the proton moves with a speed of 1.5×10^6 m/s in the positive y direction, the net electromagnetic force on it decreases in magnitude to 7.5×10^{-13} N, still pointing in the positive x direction. Find the magnitude and direction of **(a)** the electric field and **(b)** the magnetic field.

SECTION 22–3 THE MOTION OF CHARGED PARTICLES IN A MAGNETIC FIELD

18. • **CE** A velocity selector is to be constructed using a magnetic field in the positive y direction. If positively charged particles move through the selector in the positive z direction, **(a)** what must be the direction of the electric field? **(b)** Repeat part (a) for the case of negatively charged particles.

19. • Find the radius of an electron's orbit when it moves perpendicular to a magnetic field of 0.66 T with a speed of 6.27×10^5 m/s.

20. • Find the radius of a proton's orbit when it moves perpendicular to a magnetic field of 0.66 T with a speed of 6.27×10^5 m/s.

21. • Charged particles pass through a velocity selector with electric and magnetic fields at right angles to each other, as shown in **Figure 22–32**. If the electric field has a magnitude of 450 N/C and the magnetic field has a magnitude of 0.18 T, what speed must the particles have to pass through the selector undeflected?

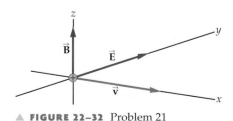

▲ **FIGURE 22–32** Problem 21

22. • The velocity selector in **Figure 22–33** is designed to allow charged particles with a speed of 4.5×10^3 m/s to pass through

undeflected. Find the direction and magnitude of the required electric field, given that the magnetic field has a magnitude of 0.96 T.

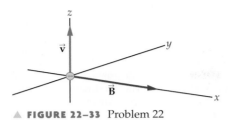

▲ **FIGURE 22–33** Problem 22

23. •• **IP BIO** The artery in Figure 22–11 has an inside diameter of 2.75 mm and passes through a region where the magnetic field is 0.065 T. **(a)** If the voltage difference between the electrodes is 195 μV, what is the speed of the blood? **(b)** Which electrode is at the higher potential? Does your answer depend on the sign of the ions in the blood? Explain.

24. •• An electron accelerated from rest through a voltage of 550 V enters a region of constant magnetic field. If the electron follows a circular path with a radius of 17 cm, what is the magnitude of the magnetic field?

25. •• A 12.5-μC particle with a mass of 2.80×10^{-5} kg moves perpendicular to a 1.01-T magnetic field in a circular path of radius 21.8 m. **(a)** How fast is the particle moving? **(b)** How long will it take the particle to complete one orbit?

26. •• **IP** When a charged particle enters a region of uniform magnetic field, it follows a circular path, as indicated in **Figure 22–34**. **(a)** Is this particle positively or negatively charged? Explain. **(b)** Suppose that the magnetic field has a magnitude of 0.180 T, the particle's speed is 6.0×10^6 m/s, and the radius of its path is 52.0 cm. Find the mass of the particle, given that its charge has a magnitude of 1.60×10^{-19} C. Give your result in atomic mass units, u, where $1 \text{ u} = 1.67 \times 10^{-27}$ kg.

▲ **FIGURE 22–34** Problem 26

27. •• A proton with a kinetic energy of 4.9×10^{-16} J moves perpendicular to a magnetic field of 0.26 T. What is the radius of its circular path?

28. •• **IP** An alpha particle (the nucleus of a helium atom) consists of two protons and two neutrons, and has a mass of 6.64×10^{-27} kg. A horizontal beam of alpha particles is injected with a speed of 1.3×10^5 m/s into a region with a vertical magnetic field of magnitude 0.155 T. **(a)** How long does it take for an alpha particle to move halfway through a complete circle? **(b)** If the speed of the alpha particle is doubled, does the time found in part (a) increase, decrease, or stay the same? Explain. **(c)** Repeat part (a) for alpha particles with a speed of 2.6×10^5 m/s.

29. ••• An electron and a proton move in circular orbits in a plane perpendicular to a uniform magnetic field $\vec{\textbf{B}}$. Find the ratio of the radii of their circular orbits when the electron and the proton have **(a)** the same momentum and **(b)** the same kinetic energy.

SECTION 22–4 THE MAGNETIC FORCE EXERTED ON A CURRENT-CARRYING WIRE

30. • What is the magnetic force exerted on a 2.15-m length of wire carrying a current of 0.899 A perpendicular to a magnetic field of 0.720 T?

31. • A wire with a current of 2.8 A is at an angle of 36.0° relative to a magnetic field of 0.88 T. Find the force exerted on a 2.25-m length of the wire.

32. • The magnetic force exerted on a 1.2-m segment of straight wire is 1.6 N. The wire carries a current of 3.0 A in a region with a constant magnetic field of 0.50 T. What is the angle between the wire and the magnetic field?

33. •• A 0.45-m copper rod with a mass of 0.17 kg carries a current of 11 A in the positive x direction. What are the magnitude and direction of the minimum magnetic field needed to levitate the rod?

34. •• The long, thin wire shown in **Figure 22–35** is in a region of constant magnetic field $\vec{\textbf{B}}$. The wire carries a current of 6.2 A and is oriented at an angle of 7.5° to the direction of the magnetic field. **(a)** If the magnetic force exerted on this wire per meter is 0.033 N, what is the magnitude of the magnetic field? **(b)** At what angle will the force exerted on the wire per meter be equal to 0.015 N?

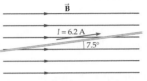

▲ **FIGURE 22–35** Problem 34

35. •• A wire with a length of 3.6 m and a mass of 0.75 kg is in a region of space with a magnetic field of 0.84 T. What is the minimum current needed to levitate the wire?

36. •• A high-voltage power line carries a current of 110 A at a location where the Earth's magnetic field has a magnitude of 0.59 G and points to the north, 72° below the horizontal. Find the direction and magnitude of the magnetic force exerted on a 250-m length of wire if the current in the wire flows **(a)** horizontally toward the east or **(b)** horizontally toward the south.

37. ••• A metal bar of mass m and length L is suspended from two conducting wires, as shown in **Figure 22–36**. A uniform magnetic field of magnitude B points vertically downward. Find the angle θ the suspending wires make with the vertical when the bar carries a current I.

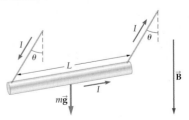

▲ **FIGURE 22–36** Problem 37

SECTION 22–5 LOOPS OF CURRENT AND MAGNETIC TORQUE

38. • **CE** For each of the three situations shown in **Figure 22–37**, indicate whether there will be a tendency for the square current loop to rotate clockwise, counterclockwise, or not at all, when viewed from above the loop along the indicated axis.

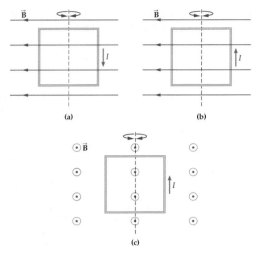

(a)   (b)

(c)

▲ **FIGURE 22–37** Problems 38, 42, and 68

39. • A rectangular loop of 260 turns is 33 cm wide and 16 cm high. What is the current in this loop if the maximum torque in a field of 0.48 T is 23 N·m?

40. • A single circular loop of radius 0.23 m carries a current of 2.6 A in a magnetic field of 0.95 T. What is the maximum torque exerted on this loop?

41. •• In the previous problem, find the angle the plane of the loop must make with the field if the torque is to be half its maximum value.

42. •• Consider a current loop in a region of uniform magnetic field, as shown in Figure 22–37 (a) (Problem 38). Find the magnitude of the torque exerted on the loop about the vertical axis of rotation, using the data given in Problem 68.

43. •• **IP** Two current loops, one square the other circular, have one turn made from wires of the same length. **(a)** If these loops carry the same current and are placed in magnetic fields of equal magnitude, is the maximum torque of the square loop greater than, less than, or the same as the maximum torque of the circular loop? Explain. **(b)** Calculate the ratio of the maximum torques, $\tau_{\text{square}}/\tau_{\text{circle}}$.

44. ••• **IP** Each of the 10 turns of wire in a vertical, rectangular loop carries a current of 0.22 A. The loop has a height of 8.0 cm and a width of 15 cm. A horizontal magnetic field of magnitude 0.050 T is oriented at an angle of $\theta = 65°$ relative to the normal to the plane of the loop, as indicated in **Figure 22–38**. Find **(a)** the magnetic force on each side of the loop, **(b)** the net magnetic force on the loop, and **(c)** the magnetic torque on the loop. **(d)** If the loop can rotate about a vertical axis with only a small amount of friction, will it end up with an orientation given by $\theta = 0, \theta = 90°$, or $\theta = 180°$? Explain.

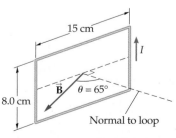

▲ **FIGURE 22–38** Problem 44

SECTION 22–6 ELECTRIC CURRENTS, MAGNETIC FIELDS, AND AMPÈRE'S LAW

45. • Find the magnetic field 6.25 cm from a long, straight wire that carries a current of 7.81 A.

46. • A long, straight wire carries a current of 7.2 A. How far from this wire is the magnetic field it produces equal to the Earth's magnetic field, which is approximately 5.0×10^{-5} T?

47. • You travel to the north magnetic pole of the Earth, where the magnetic field points vertically downward. There, you draw a circle on the ground. Applying Ampère's law to this circle, show that zero current passes through its area.

48. • Two power lines, each 270 m in length, run parallel to each other with a separation of 25 cm. If the lines carry parallel currents of 110 A, what are the magnitude and direction of the magnetic force each exerts on the other?

49. • **BIO Pacemaker Switches** Some pacemakers employ magnetic reed switches to enable doctors to change their mode of operation without surgery. A typical reed switch can be switched from one position to another with a magnetic field of 5.0×10^{-4} T. What current must a wire carry if it is to produce a 5.0×10^{-4} T field at a distance of 0.50 m?

50. •• **IP** Consider the long, straight, current-carrying wires shown in **Figure 22–39**. One wire carries a current of 6.2 A in the positive y direction; the other wire carries a current of 4.5 A in the positive x direction. **(a)** At which of the two points, A or B, do you expect the magnitude of the net magnetic field to be greater? Explain. **(b)** Calculate the magnitude of the net magnetic field at points A and B.

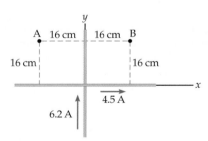

▲ **FIGURE 22–39** Problems 50 and 51

51. •• **IP** Repeat Problem 50 for the case where the 6.2-A current is reversed in direction.

52. •• In Oersted's experiment, suppose that the compass was 0.25 m from the current-carrying wire. If a magnetic field of half the Earth's magnetic field of 5.0×10^{-5} T was required to give a noticeable deflection of the compass needle, what current must the wire have carried?

53. •• **IP** Two long, straight wires are separated by a distance of 9.25 cm. One wire carries a current of 2.75 A, the other carries a current of 4.33 A. **(a)** Find the force per meter exerted on the 2.75-A wire. **(b)** Is the force per meter exerted on the 4.33-A wire greater than, less than, or the same as the force per meter exerted on the 2.75-A wire? Explain.

54. ••• Two long, straight wires are oriented perpendicular to the page, as shown in **Figure 22–40**. The current in one wire is $I_1 = 3.0$ A, pointing into the page, and the current in the other wire is $I_2 = 4.0$ A, pointing out of the page. Find the magnitude and direction of the net magnetic field at point P.

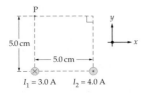

▲ **FIGURE 22–40** Problems 54 and 106

SECTION 22–7   CURRENT LOOPS AND SOLENOIDS

55. • **CE** A loop of wire is connected to the terminals of a battery, as indicated in **Figure 22–41**. If the loop is to attract the bar magnet, which of the terminals, A or B, should be the positive terminal of the battery? Explain.

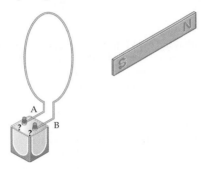

▲ **FIGURE 22–41** Problem 55

56. • **CE Predict/Explain** The number of turns in a solenoid is doubled, and at the same time its length is doubled. Does the magnetic field within the solenoid increase, decrease, or stay the same? **(b)** Choose the *best explanation* from among the following:
   I. Doubling the number of turns in a solenoid doubles its magnetic field, and hence the field increases.
   II. Making a solenoid longer decreases its magnetic field, and therefore the field decreases.
   III. The magnetic field remains the same because the number of turns per length is unchanged.

57. • It is desired that a solenoid 38 cm long and with 430 turns produce a magnetic field within it equal to the Earth's magnetic field (5.0×10^{-5} T). What current is required?

58. • A solenoid that is 62 cm long produces a magnetic field of 1.3 T within its core when it carries a current of 8.4 A. How many turns of wire are contained in this solenoid?

59. • The maximum current in a superconducting solenoid can be as large as 3.75 kA. If the number of turns per meter in such a solenoid is 3650, what is the magnitude of the magnetic field it produces?

60. •• To construct a solenoid, you wrap insulated wire uniformly around a plastic tube 12 cm in diameter and 55 cm in length. You would like a 2.0-A current to produce a 2.5-kG magnetic field inside your solenoid. What is the total length of wire you will need to meet these specifications?

GENERAL PROBLEMS

61. • **CE** At a point near the equator, the Earth's magnetic field is horizontal and points to the north. If an electron is moving vertically upward at this point, does the magnetic force acting on it point north, south, east, west, upward, or downward? Explain.

62. • **CE** A proton is to orbit the Earth at the equator using the Earth's magnetic field to supply part of the necessary centripetal force. Should the proton move eastward or westward? Explain.

63. • **CE** The accompanying photograph shows an electron beam whose initial direction of motion is horizontal, from right to left. A magnetic field deflects the beam downward. What is the direction of the magnetic field?

A horizontal electron beam is deflected downward by a magnetic field. (Problem 63)

64. • **CE** The three wires shown in **Figure 22–42** are long and straight, and they each carry a current of the same magnitude, I. The currents in wires 1 and 3 are out of the page; the current in wire 2 is into the page. What is the direction of the magnetic force experienced by wire 3?

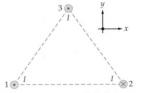

▲ **FIGURE 22–42** Problems 64 and 65

65. • **CE** Each of the current-carrying wires in Figure 22–42 is long and straight, and carries the current I either into or out of the page, as shown. What is the direction of the net magnetic field produced by these three wires at the center of the triangle?

66. • **CE** The four wires shown in **Figure 22–43** are long and straight, and they each carry a current of the same magnitude, I. The currents in wires 1, 2, and 3 are out of the page; the current in wire 4 is into the page. What is the direction of the magnetic force experienced by wire 2?

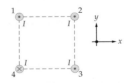

▲ **FIGURE 22–43** Problems 66 and 67

67. • **CE** Each of the current-carrying wires in Figure 22–43 is long and straight, and carries the current I either into or out of the page, as shown. What is the direction of the net magnetic field produced by these four wires at the center of the square?

68. • Consider a current loop immersed in a magnetic field, as in Figure 22–37 (a) (Problem 38). It is given that $B = 0.34$ T and $I = 9.5$ A. In addition, the loop is a square 0.46 m on a side. Find the magnitude of the magnetic force exerted on each side of the loop.

69. • A stationary proton ($q = 1.60 \times 10^{-19}$ C) is located between the poles of a horseshoe magnet, where the magnetic field is 0.35 T. What is the magnitude of the magnetic force acting on the proton?

70. • **BIO Brain Function and Magnetic Fields** Experiments have shown that thought processes in the brain can be affected if the parietal lobe is exposed to a magnetic field with a strength of 1.0 T. How much current must a long, straight wire carry if it is to produce a 1.0-T magnetic field at a distance of 0.50 m? (For comparison, a typical lightning bolt carries a current of about 20,000 A, which would melt most wires.)

71. • A mixture of two isotopes is injected into a mass spectrometer. One isotope follows a curved path of radius $R_1 = 48.9$ cm; the other follows a curved path of radius $R_2 = 51.7$ cm. Find the mass ratio, m_1/m_2, assuming that the two isotopes have the same charge and speed.

72. • High above the surface of the Earth, charged particles (such as electrons and protons) can become trapped in the Earth's magnetic field in regions known as Van Allen belts. A typical electron in a Van Allen belt has an energy of 45 keV and travels in a roughly circular orbit with an average radius of 220 m. What is the magnitude of the Earth's magnetic field where such an electron orbits?

73. • **Credit-Card Magnetic Strips** Experiments carried out on the television show *Mythbusters* determined that a magnetic field of 1000 gauss is needed to corrupt the information on a credit card's magnetic strip. (They also busted the myth that a credit card can be demagnetized by an electric eel or an eelskin wallet.) Suppose a long, straight wire carries a current of 3.5 A. How close can a credit card be held to this wire without damaging its magnetic strip?

74. • **Superconducting Solenoid** Cryomagnetics, Inc., advertises a high-field, superconducting solenoid that produces a magnetic field of 17 T with a current of 105 A. What is the number of turns per meter in this solenoid?

75. •• **CE** A positively charged particle moves through a region with a uniform electric field pointing toward the top of the page and a uniform magnetic field pointing into the page. The particle can have one of the four velocities shown in **Figure 22–44**. **(a)** Rank the four possibilities in order of increasing magnitude of the net force (F_1, F_2, F_3, and F_4) the particle experiences. Indicate ties where appropriate. **(b)** Which of the four velocities could potentially result in zero net force?

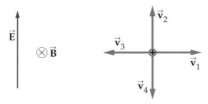

▲ **FIGURE 22–44** Problems 75 and 76

76. •• **CE** Suppose the fields in Figure 22–44 are interchanged, with the magnetic field pointing toward the top of the page and the electric field pointing into the page. **(a)** Rank the four possibilities in order of increasing magnitude of the net force (F_1, F_2, F_3, and F_4) the particle experiences. Indicate ties where appropriate. **(b)** Which of the four velocities could potentially result in zero net force?

77. •• **CE** A proton follows the path shown in **Figure 22–45** as it moves through three regions with different uniform magnetic

fields, B_1, B_2, and B_3. In each region the proton completes a half-circle, and the magnetic field is perpendicular to the page. **(a)** Rank the three fields in order of increasing magnitude. Indicate ties where appropriate. **(b)** Give the direction (into or out of the page) for each of the fields.

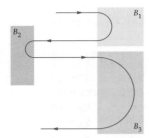

▲ **FIGURE 22–45** Problems 77, 78, and 79

78. •• **CE Predict/Explain** Suppose the initial speed of the proton in Figure 22–45 is increased. **(a)** Does the radius of each half-circular path segment increase, decrease, or stay the same? **(b)** Choose the *best explanation* from among the following:
  I. The radius of a circular orbit in a magnetic field is proportional to the speed of the proton; therefore, the radius of the half-circular path will increase.
  II. A greater speed means the proton will experience more force from the magnetic field, resulting in a decrease of the radius.
  III. The increase in speed offsets the increase in magnetic force, resulting in no change of the radius.

79. •• **CE Predict/Explain** Suppose the initial speed of the proton in Figure 22–45 is decreased. **(a)** Does the time spent in each of the field regions increase, decrease, or stay the same? **(b)** Choose the *best explanation* from among the following:
  I. The proton moves more slowly, and therefore the time spent moving through each of magnetic regions will increase.
  II. With a smaller speed the proton is forced out of each magnetic region more quickly, which results in a decrease in the time.
  III. The time for an orbit in a magnetic field is independent of speed. Therefore, the time the proton spends in each of magnetic regions is the same no matter what its speed.

80. •• **BIO Magnetic Resonance Imaging** An MRI (magnetic resonance imaging) solenoid produces a magnetic field of 1.5 T. The solenoid is 2.5 m long, 1.0 m in diameter, and wound with insulated wires 2.2 mm in diameter. Find the current that flows in the solenoid. (Your answer should be rather large. A typical MRI solenoid uses niobium–titanium wire kept at liquid helium temperatures, where it is superconducting.)

81. •• **IP** A long, straight wire carries a current of 14 A. Next to the wire is a square loop with sides 1.0 m in length, as shown in **Figure 22–46**. The loop carries a current of 2.5 A in the direction indicated. **(a)** What is the direction of the net force exerted on

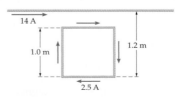

▲ **FIGURE 22–46** Problems 81 and 82

the loop? Explain. **(b)** Calculate the magnitude of the net force acting on the loop.

82. •• Suppose the 14-A current in the straight wire in Figure 22–46 is reversed in direction, but the current in the loop is unchanged. **(a)** Calculate the magnitude and direction of the net force acting on the loop. **(b)** If the loop is extended in the horizontal direction, so that it is 1.0 m high and 2.0 m wide, does the net force exerted on the loop increase or decrease? By what factor? Explain. **(c)** If, instead, the loop is extended in the vertical direction, so it is 2.0 m high and 1.0 m wide, does the net force exerted on the loop increase or decrease? Explain.

83. •• A charged particle moves through a region of space containing both electric and magnetic fields. The velocity of the particle is $\vec{V} = (4.4 \times 10^3 \text{m/s})\hat{x} + (2.7 \times 10^3 \text{ m/s})\hat{y}$ and the magnetic field is $\vec{B} = (0.73 \text{ T})\hat{z}$. Find the electric field vector $\vec{E}$ necessary to yield zero net force on the particle. (*Note:* You may wish to use cross products in this problem. They are discussed in Appendix A.)

84. •• **IP Medical X-rays** An electron in a medical X-ray machine is accelerated from rest through a voltage of 10.0 kV. **(a)** Find the maximum force a magnetic field of 0.957 T can exert on this electron. **(b)** If the voltage of the X-ray machine is increased, does the maximum force found in part (a) increase, decrease, or stay the same? Explain. **(c)** Repeat part (a) for an electron accelerated through a potential of 25.0 kV.

85. •• A particle with a charge of 34 μC moves with a speed of 73 m/s in the positive x direction. The magnetic field in this region of space has a component of 0.40 T in the positive y direction, and a component of 0.85 T in the positive z direction. What are the magnitude and direction of the magnetic force on the particle?

86. •• **IP** A beam of protons with various speeds is directed in the positive x direction. The beam enters a region with a uniform magnetic field of magnitude 0.52 T pointing in the negative z direction, as indicated in **Figure 22–47**. It is desired to use a uniform electric field (in addition to the magnetic field) to select from this beam only those protons with a speed of 1.42×10^5 m/s—that is, only these protons should be undeflected by the two fields. **(a)** Determine the magnitude and direction of the electric field that yields the desired result. **(b)** Suppose the electric field is to be produced by a parallel-plate capacitor with a plate separation of 2.5 cm. What potential difference is required between the plates? **(c)** Which plate in Figure 22–47 (top or bottom) should be positively charged? Explain.

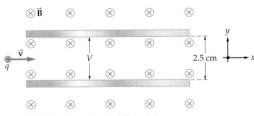

▲ **FIGURE 22–47** Problem 86

87. •• **IP** A charged particle moves in a horizontal plane with a speed of 8.70×10^6 m/s. When this particle encounters a uniform magnetic field in the vertical direction it begins to move on a circular path of radius 15.9 cm. **(a)** If the magnitude of the magnetic field is 1.21 T, what is the charge-to-mass ratio (q/m) of this particle? **(b)** If the radius of the circular path were greater than 15.9 cm, would the corresponding charge-to-mass ratio be greater than, less than, or the same as that found in part (a)? Explain. (Assume that the magnetic field remains the same.)

88. •• Two parallel wires, each carrying a current of 2.2 A in the same direction, are shown in **Figure 22–48**. Find the direction and magnitude of the net magnetic field at points A, B, and C.

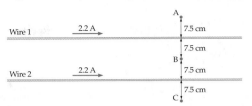

▲ **FIGURE 22–48** Problems 88 and 89

89. •• Repeat Problem 88 for the case where the current in wire 1 is reversed in direction.

90. •• **Lightning Bolts** A powerful bolt of lightning can carry a current of 225 kA. **(a)** Treating a lightning bolt as a long, thin wire, calculate the magnitude of the magnetic field produced by such a bolt of lightning at a distance of 35 m. **(b)** If two such bolts strike simultaneously at a distance of 35 m from each other, what is the magnetic force per meter exerted by one bolt on the other?

91. •• **IP** Consider the two current-carrying wires shown in **Figure 22–49**. The current in wire 1 is 3.7 A; the current in wire 2 is adjusted to make the net magnetic field at point A equal to zero. **(a)** Is the magnitude of the current in wire 2 greater than, less than, or the same as that in wire 1? Explain. **(b)** Find the magnitude and direction of the current in wire 2.

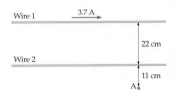

▲ **FIGURE 22–49** Problems 91 and 92

92. •• **IP** Consider the physical system shown in Figure 22–49, which consists of two current-carrying wires each with a length of 71 cm. **(a)** If the net magnetic field at the point A is out of the page, is the force between the wires attractive or repulsive? Explain. **(b)** Calculate the magnitude of the force exerted by each wire on the other wire, given that the magnetic field at point A is out of the page with a magnitude of 2.1×10^{-6} T.

93. •• **Magnetars** The astronomical object 4U014+61 has the distinction of creating the most powerful magnetic field ever observed. This object is referred to as a "magnetar" (a subclass of pulsars), and its magnetic field is 1.3×10^{15} times greater than the Earth's magnetic field. **(a)** Suppose a 2.5-m straight wire carrying a current of 1.1 A is placed in this magnetic field at an angle of 65° to the field lines. What force does this wire experience? **(b)** A field this strong can significantly change the behavior of an atom. To see this, consider an electron moving with a speed of 2.2×10^6 m/s. Compare the maximum magnetic force exerted on the electron to the electric force a proton exerts on an electron in a hydrogen atom. The radius of the hydrogen atom is 5.29×10^{-11} m.

94. •• Consider a system consisting of two concentric solenoids, as illustrated in **Figure 22–50**. The current in the outer solenoid is $I_1 = 1.25$ A, and the current in the inner solenoid is $I_2 = 2.17$ A. Given that the number of turns per centimeter is 105 for the outer solenoid and 125 for the inner solenoid, find the magnitude and direction of magnetic field **(a)** between the solenoids and **(b)** inside the inner solenoid.

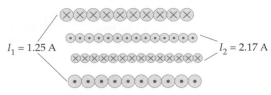

$I_1 = 1.25$ A $I_2 = 2.17$ A

▲ **FIGURE 22–50** Problem 94

95. •• **IP** A long, straight wire on the x axis carries a current of 3.12 A in the positive x direction. The magnetic field produced by the wire combines with a uniform magnetic field of 1.45×10^{-6} T that points in the positive z direction. **(a)** Is the net magnetic field of this system equal to zero at a point on the positive y axis or at a point on the negative y axis? Explain. **(b)** Find the distance from the wire to the point where the field vanishes.

96. •• Find the angle between the plane of a loop and the magnetic field for which the magnetic torque acting on the loop is equal to x times its maximum value, where $0 \leq x \leq 1$.

97. •• Solenoids produce magnetic fields that are relatively intense for the amount of current they carry. To make a direct comparison, consider a solenoid with 55 turns per centimeter, a radius of 1.05 cm, and a current of 0.622 A. **(a)** Find the magnetic field at the center of the solenoid. **(b)** What current must a long, straight wire carry to have the same magnetic field as that found in part (a)? Let the distance from the wire be the same as the radius of the solenoid, 1.05 cm.

98. •• The current in a solenoid with 22 turns per centimeter is 0.50 A. The solenoid has a radius of 1.5 cm. A long, straight wire runs along the axis of the solenoid, carrying a current of 13 A. Find the magnitude of the net magnetic field a radial distance of 0.75 cm from the straight wire.

99. •• **IP BIO Transcranial Magnetic Stimulation** A recently developed method to study brain function is to produce a rapidly changing magnetic field within the brain. When this technique, known as transcranial magnetic stimulation (TMS), is applied to the prefrontal cortex, for example, it can reduce a person's ability to conjugate verbs, though other thought processes are unaffected. The rapidly varying magnetic field is produced with a circular coil of 21 turns and a radius of 6.0 cm placed directly on the head. The current in this loop increases at the rate of 1.2×10^7 A/s (by discharging a capacitor). **(a)** At what rate does the magnetic field at the center of the coil increase? **(b)** Suppose a second coil with half the area of the first coil is used instead. Would your answer to part (a) increase, decrease, or stay the same? By what factor?

100. ••• An electron with a velocity given by $\vec{v} = (1.5 \times 10^5$ m/s$)\hat{x} + (0.67 \times 10^4$ m/s$)\hat{y}$ moves through a region of space with a magnetic field $\vec{B} = (0.25$ T$)\hat{x} - (0.11$ T$)\hat{z}$ and an electric field $\vec{E} = (220$ N/C$)\hat{x}$. Using cross products, find the magnitude of the net force acting on the electron. (Cross products are discussed in Appendix A.)

101. ••• A thin ring of radius R and charge per length λ rotates with an angular speed ω about an axis perpendicular to its plane and passing through its center. Find the magnitude of the magnetic field at the center of the ring.

102. ••• A solenoid is made from a 25-m length of wire of resistivity $2.3 \times 10^{-8} \Omega \cdot$ m. The wire, whose radius is 2.1 mm, is wrapped uniformly onto a plastic tube 4.5 cm in diameter and 1.65 m long. Find the emf to which the ends of the wire must

be connected to produce a magnetic field of 0.015 T within the solenoid.

103. ••• **IP** A single current-carrying circular loop of radius R is placed next to a long, straight wire, as shown in **Figure 22–51**. The current in the wire points to the right and is of magnitude I. **(a)** In which direction must current flow in the loop to produce zero magnetic field at its center? Explain. **(b)** Calculate the magnitude of the current in part (a).

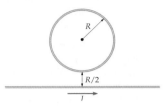

▲ **FIGURE 22–51** Problem 103

104. ••• **Magnetic Fields in the Bohr Model** In the Bohr model of the hydrogen atom, the electron moves in a circular orbit of radius 5.29×10^{-11} m about the nucleus. Given that the charge on the electron is -1.60×10^{-19} C, and that its speed is 2.2×10^6 m/s, find the magnitude of the magnetic field the electron produces at the nucleus of the atom.

105. ••• A single-turn square loop carries a current of 18 A. The loop is 15 cm on a side and has a mass of 0.035 kg. Initially the loop lies flat on a horizontal tabletop. When a horizontal magnetic field is turned on, it is found that only one side of the loop experiences an upward force. Find the minimum magnetic field, $B_{\min}$, necessary to start tipping the loop up from the table.

106. ••• Consider the physical system shown in Figure 22–40. **(a)** Find the net magnetic field (direction and magnitude) at an arbitrary point on the bottom side of the square, a distance $0 < x < 5.0$ cm to the right of wire 1. **(b)** Find the magnitude of the net magnetic field at an arbitrary point on the left side of the square, a distance $0 < y < 5.0$ cm above wire 1.

PASSAGE PROBLEMS

BIO Magnetoencephalography

To read and understand this sentence your brain must process visual input from your eyes and translate it into words and thoughts. As you do so, minute electric currents flow through the neurons in your visual cortex. These currents, like any electric current, produce magnetic fields. In fact, even your innermost thoughts and dreams produce magnetic fields that can be detected outside your head.

Magnetoencephalography (MEG) is the study of magnetic fields produced by electrical activity in the brain. Though completely noninvasive, MEG can provide detailed information on spontaneous brain function—like alpha waves and pathological epileptic spikes—as well as brain activity that is evoked by visual, auditory, and tactile stimuli.

The magnetic fields produced by brain activity are incredibly weak—roughly 100 million times smaller than the Earth's magnetic field. Even so, sensitive detectors called SQUIDS (superconducting quantum interference devices), which were invented by physicists as a research tool, can detect fields as small as 1.0×10^{-15} T. Coupled with sophisticated electronics and software, and operating at liquid helium temperatures (-269 °C), SQUIDS can localize the source of brain activity to within millimeters. When the information from MEG is overlaid with the anatomical data from an MRI scan, the result is a richly detailed "map" of the electrical activity within the brain.

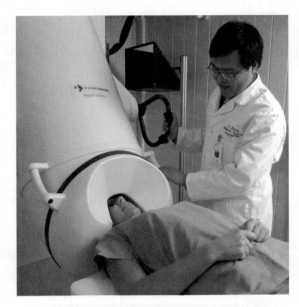

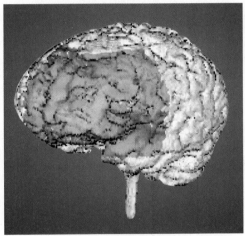

A magnetoencephalograph (MEG) is made by measuring the magnetic fields produced by the brain. It is a completely noninvasive process. The result is a map of the electrical activity within the brain. (Problems 107, 108, 109, and 110)

107. • Approximating a neuron by a straight wire, what electric current is needed to produce a magnetic field of 1.0×10^{-15} T at a distance of 5.0 cm?

A. 4.0×10^{-22} A   **B.** 7.9×10^{-11} A

C. 2.5×10^{-10} A   **D.** 1.0×10^{-7} A

108. • Suppose a neuron in the brain carries a current of 5.0×10^{-8} A. Treating the neuron as a straight wire, what is the magnetic field it produces at a distance of 7.5 cm?

A. 1.3×10^{-13} T   **B.** 4.2×10^{-12} T

C. 1.1×10^{-10} T   **D.** 3.3×10^{-7} T

109. • A given neuron in the brain carries a current of 3.1×10^{-8} A. If the SQUID detects a magnetic field of 2.8×10^{-14} T, how far away is the neuron? Treat the neuron as a straight wire.

A. 22 cm   **B.** 70 cm

C. 140 cm   **D.** 176 cm

110. •• A SQUID detects a magnetic field of 1.8×10^{-14} T at a distance of 13 cm. How many electrons flow through the neuron per second? Treat the neuron as a straight wire.

A. 1.2×10^{10} **B.** 2.3×10^{10}

C. 7.3×10^{10} **D.** 9.2×10^{10}

INTERACTIVE PROBLEMS

111. •• **IP Referring to Example 22–3** Suppose the speed of the isotopes is doubled. **(a)** Does the separation distance, d, increase, decrease, or stay the same? Explain. **(b)** Find the separation distance for this case.

112. •• **IP Referring to Example 22–3** Suppose we change the initial speed of ^{238}U, leaving everything else the same. **(a)** If we want the separation distance to be zero, should the initial speed of ^{238}U be increased or decreased? Explain. **(b)** Find the required initial speed.

113. •• **Referring to Active Example 22–2** The current I_1 is adjusted until the magnetic field halfway between the wires has a magnitude of 2.5×10^{-6} T and points *into* the page. Everything else in the system remains the same as in Active Example 22–2. Find the magnitude and direction of I_1.

114. •• **Referring to Active Example 22–2** The current I_2 is adjusted until the magnetic field 5.5 cm below wire 2 has a magnitude of 2.5×10^{-6} T and points *out* of page. Everything else in the system remains the same as in Active Example 22–2. Find the magnitude and direction of I_2.

23 Magnetic Flux and Faraday's Law of Induction

If you mention an electric guitar, everyone knows what you mean, whereas a reference to a "magnetic guitar" would probably evoke some very puzzled looks. Yet it could be argued that the second term is just as appropriate as the first. If the steel strings of the guitar didn't respond to the magnetic field of the pickups, there would be nothing to be "picked up"—no signal would go to the amp, and no sound would emerge from the speaker. This chapter focuses on the intimate connection between electricity and magnetism—more specifically, on the phenomenon of electromagnetic induction, which makes possible not only electric guitars but also an enormous number of other devices that we use every day.

In the last chapter we saw a number of connections between electricity and magnetism. All the cases we considered, however, involved magnetic fields that were constant in time; that is, static fields. We now explore some of the interesting new phenomena that are produced when magnetic fields change with time.

Of particular interest is that changing magnetic fields can create electric fields, which, in turn, can cause stationary charges to move. When applied to an electric circuit, a changing magnetic field can produce an electric current, just like a battery. Effects of this nature have led to

a host of important applications in everyday life, from electric motors and generators to microphones, speakers, and computer disk drives.

Finally, a recurrent theme throughout this chapter is energy—its conservation and the various forms in which it can appear. As we shall see, mechanical energy can be converted to electrical energy—with the help of a magnetic field—and the voltage and current in one electric circuit can be converted, or "transformed," into different values in another circuit. Just as in earlier chapters, we again find that energy is one of the most fundamental and versatile of all physical concepts.

23–1 Induced Electromotive Force

When Oersted observed that an electric current produces a magnetic field, it was pure serendipity. In contrast, Michael Faraday (1791–1867), an English chemist and physicist who was aware of Oersted's results, purposefully set out to determine whether a similar effect operates in the reverse direction, that is, whether a magnetic field can produce an electric field. His careful experimentation showed that such a connection does indeed exist.

In **Figure 23–1** we show a simplified version of the type of experiment performed by Faraday. Two electric circuits are involved. The first, called the **primary circuit,** consists of a battery, a switch, a resistor to control the current, and a coil of several turns around an iron bar. When the switch is closed on the primary circuit, a current flows through the coil, producing a magnetic field that is particularly intense within the iron bar.

The **secondary circuit** also has a coil wrapped around the iron bar, and this coil is connected to an ammeter to detect any current in the circuit. Note, however, that there is no battery in the secondary circuit, and no direct physical contact between the two circuits. What does link the circuits, instead, is the magnetic field in the iron bar—it ensures that the field experienced by the secondary coil is approximately the same as that produced by the primary coil.

Now, let's look at the experimental results. When the switch is closed on the primary circuit, the magnetic field in the iron bar rises from zero to some finite amount, and the ammeter in the secondary coil deflects to one side briefly, and then returns to zero. As long as the current in the primary circuit is maintained at a constant value, the ammeter in the secondary circuit gives zero reading. If the switch on the primary circuit is now opened, so that the magnetic field decreases again to zero, the ammeter in the secondary circuit deflects briefly in the opposite direction, and then returns to zero. We can summarize these observations as follows:

- The current in the secondary circuit is zero as long as the current in the primary circuit is constant—which means, in turn, that the magnetic field in the iron bar is constant. It does not matter whether the constant value of the magnetic field is zero or nonzero.
- When the magnetic field passing through the secondary coil increases, a current is observed to flow in one direction in the secondary circuit; when the magnetic field decreases, a current is observed in the opposite direction.

Recall that the current in the secondary circuit appears without any direct contact between the primary and secondary circuits. For this reason we refer to the secondary current as an **induced current.** Since the induced current behaves the same as a current produced by a battery with an electromotive force (emf), we say that the changing magnetic field creates an **induced emf** in the secondary circuit. This leads to our final experimental observation:

- The magnitudes of the induced current and induced emf are found to be proportional to the rate of change of the magnetic field—the more rapidly the magnetic field changes, the greater the induced emf.

In Section 23–3 we return to these observations and state them in the form of a mathematical law.

Finally, in Faraday's experiment the changing magnetic field is caused by a changing current in the primary circuit. Any means of altering the magnetic field is just as effective, however. For example, in **Figure 23–2** we show a common classroom demonstration of induced emf. In this case, there is no primary circuit; instead, the magnetic field is changed by simply moving a permanent magnet toward or away from a coil connected to an ammeter. When the magnet is moved toward the coil, the meter deflects in one direction; when it is pulled away from the coil, the meter deflects in the opposite direction. There is no induced emf, however, when the magnet is held still.

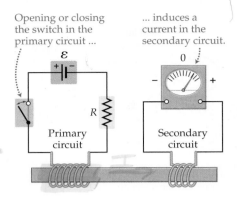

Opening or closing the switch in the primary circuit ...

... induces a current in the secondary circuit.

▲ **FIGURE 23–1 Magnetic induction**

Basic setup of Faraday's experiment on magnetic induction. When the position of the switch on the primary circuit is changed from open to closed or from closed to open, an electromotive force (emf) is induced in the secondary circuit. The induced emf causes a current in the secondary circuit, and the current is detected by the ammeter. If the current in the primary circuit does not change, no matter how large it may be, there is no induced current in the secondary circuit.

▲ When a magnet moves toward or away from a coil, the magnetic field within the coil changes. This, in turn, induces an electric current in the coil, which is detected by the meter.

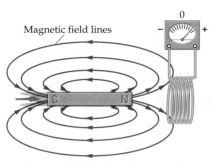 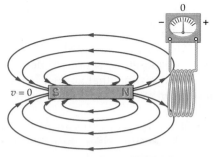 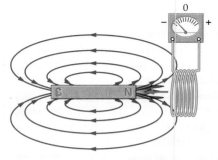

(a) Moving magnet toward coil induces current in one direction

(b) No motion, no induced current

(c) Moving magnet away from coil induces current in opposite direction

▲ **FIGURE 23–2 Induced current produced by a moving magnet**

A coil experiences an induced current when the magnetic field passing through it varies. **(a)** When the magnet moves toward the coil, the current is in one direction. **(b)** No current is induced while the magnet is held still. **(c)** When the magnet is pulled away from the coil, the current is in the other direction.

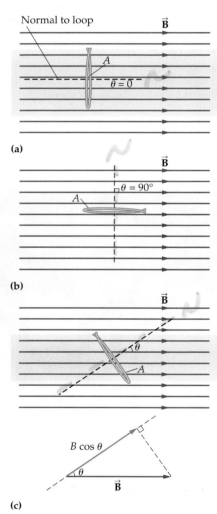

▲ **FIGURE 23–3 The magnetic flux through a loop**

The magnetic flux through a loop of area A is $\Phi = BA \cos \theta$, where θ is the angle between the normal to the loop and the magnetic field. **(a)** The loop is perpendicular to the field; hence, $\theta = 0$, and $\Phi = BA$. **(b)** The loop is parallel to the field; therefore, $\theta = 90°$ and $\Phi = 0$. **(c)** For a general angle θ, the component of the field that is perpendicular to the loop is $B \cos \theta$; hence, the flux is $\Phi = BA \cos \theta$.

23–2 Magnetic Flux

In the previous section, we saw that an emf can be induced in a coil of wire by simply changing the strength of the magnetic field that passes through the coil. This isn't the only way to induce an emf, however. More generally, we can change the direction of the magnetic field rather than its magnitude. We can even change the orientation of the coil in a constant magnetic field, or change the cross-sectional area of the coil. All of these situations can be described in terms of the change of a single physical quantity, the magnetic flux.

Basically, **magnetic flux** is a measure of the number of magnetic field lines that cross a given area, in complete analogy with the electric flux discussed in Section 19–7. Suppose, for example, that a magnetic field $\vec{B}$ crosses a surface area A at right angles, as in **Figure 23–3 (a)**. The magnetic flux, Φ, in this case is simply the magnitude of the magnetic field times the area:

$$\Phi = BA$$

If, on the other hand, the magnetic field is parallel to the surface, as in **Figure 23–3 (b)**, we see that *no field lines cross the surface*. In a case like this the magnetic flux is zero:

$$\Phi = 0$$

In general, only the component of $\vec{B}$ that is *perpendicular* to a surface contributes to the magnetic flux. The magnetic field in **Figure 23–3 (c)**, for example, crosses the surface at an angle θ relative to the normal, and hence its perpendicular component is $B \cos \theta$. The magnetic flux, then, is simply $B \cos \theta$ times the area A:

Definition of Magnetic Flux, Φ

$$\Phi = (B \cos \theta)A = BA \cos \theta \qquad \text{23–1}$$

SI unit: $1 \text{ T} \cdot \text{m}^2 = 1 \text{ weber} = 1 \text{ Wb}$

Note that $\Phi = BA \cos \theta$ gives $\Phi = BA$ when $\vec{B}$ is perpendicular to the surface ($\theta = 0$), and $\Phi = 0$ when $\vec{B}$ is parallel to the surface ($\theta = 90°$), as expected. Finally, the unit of magnetic flux is the **weber (Wb),** named after the German physicist Wilhelm Weber (1804–1891). It is defined as follows:

$$1 \text{ Wb} = 1 \text{ T} \cdot \text{m}^2 \qquad \text{23–2}$$

As we have seen, magnetic flux depends on the magnitude of the magnetic field, B, its orientation with respect to a surface, θ, and the area of the surface, A. A change in any of these variables results in a change in the flux. For example, in the case of the permanent magnet moved toward or away from the coil in the previous section, it is the change in *magnitude* of the field, B, that results in a change in flux. In the following Example, we consider the effect of changing the *orientation* of a wire loop in a region of constant magnetic field.

EXAMPLE 23–1 A SYSTEM IN FLUX

Consider a circular loop with a 2.50-cm radius in a constant magnetic field of 0.625 T. Find the magnetic flux through this loop when its normal makes an angle of **(a)** 0°, **(b)** 30.0°, **(c)** 60.0°, and **(d)** 90.0° with the direction of the magnetic field $\vec{\mathbf{B}}$.

PICTURE THE PROBLEM

Our sketch shows four top views of the system, looking directly down on the top edge of the circular loop. The orientation of the loop and its normal is shown in each of the four panels. Note that the specified angles between the normal and the magnetic field direction are $\theta = 0°$, 30.0°, 60.0°, and 90.0°.

STRATEGY

To find the magnetic flux, we use the following relation $\Phi = BA\cos\theta$. In this expression, B is the magnitude of the magnetic field and $A = \pi r^2$ is the area of a circular loop of radius r. The values of B, r, and θ are given in the problem statement.

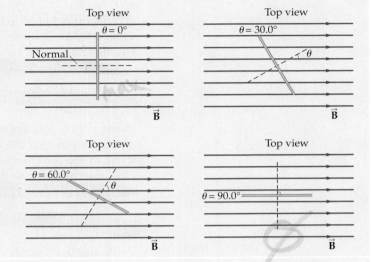

SOLUTION

Part (a)

1. Substitute $\theta = 0°$ in $\Phi = BA\cos\theta$:

$$\Phi = BA\cos\theta$$
$$= (0.625\text{ T})\pi(0.0250\text{ m})^2\cos 0° = 1.23 \times 10^{-3}\text{ T}\cdot\text{m}^2$$

Part (b)

2. Substitute $\theta = 30.0°$ in $\Phi = BA\cos\theta$:

$$\Phi = BA\cos\theta$$
$$= (0.625\text{ T})\pi(0.0250\text{ m})^2\cos 30.0° = 1.06 \times 10^{-3}\text{ T}\cdot\text{m}^2$$

Part (c)

3. Substitute $\theta = 60.0°$ in $\Phi = BA\cos\theta$:

$$\Phi = BA\cos\theta$$
$$= (0.625\text{ T})\pi(0.0250\text{ m})^2\cos 60.0° = 6.14 \times 10^{-4}\text{ T}\cdot\text{m}^2$$

Part (d)

4. Substitute $\theta = 90.0°$ in $\Phi = BA\cos\theta$:

$$\Phi = BA\cos\theta$$
$$= (0.625\text{ T})\pi(0.0250\text{ m})^2\cos 90.0° = 0$$

INSIGHT

Thus, even if a magnetic field is uniform in space and constant in time, the magnetic flux through a given area will change if the orientation of the area changes. This is particularly relevant to the case of a coil that rotates in a field and constantly changes its orientation. As we shall see in the next section, a changing magnetic flux can create an electric current, and Section 23–6 shows how this applies to generators and motors.

PRACTICE PROBLEM

At what angle is the flux in this system equal to $1.00 \times 10^{-4}\text{ T}\cdot\text{m}^2$? **[Answer:** $\theta = 85.3°$**]**

Some related homework problems: Problem 1, Problem 3

CONCEPTUAL CHECKPOINT 23–1 MAGNETIC FLUX

The three loops of wire shown in the sketch are all in a region of space with a uniform, constant magnetic field. Loop 1 swings back and forth as the bob on a pendulum; loop 2 rotates about a vertical axis; and loop 3 oscillates vertically on the end of a spring. Which loop or loops have a magnetic flux that changes with time?

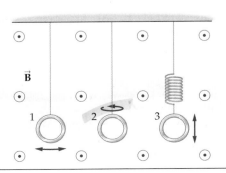

REASONING AND DISCUSSION

Loop 1 moves back and forth, and loop 3 moves up and down, but since the magnetic field is uniform, the flux doesn't depend on the loop's position. Loop 2, on the other hand, changes its orientation relative to the field as it rotates; hence, its flux does change with time.

ANSWER

Only loop 2 has a changing magnetic flux.

23–3 Faraday's Law of Induction

Now that the magnetic flux is defined, it is possible to be more precise about the experimental observations described in Section 23–1. In particular, Faraday found that the secondary coil in Figure 23–1 experiences an induced emf only when the *magnetic flux* through it changes with time. Furthermore, the induced emf for a given loop is found to be proportional to the *rate* at which the flux changes with time, $\Delta\Phi/\Delta t$. If there are N loops in a coil, each with the same magnetic flux, Faraday found that the induced emf is given by the following relation:

Faraday's Law of Induction

$$\mathcal{E} = -N\frac{\Delta\Phi}{\Delta t} = -N\frac{\Phi_{final} - \Phi_{initial}}{t_{final} - t_{initial}}$$ 23–3

In this expression, known as **Faraday's law of induction,** the minus sign in front of N indicates that the induced emf opposes the change in magnetic flux, as we show in the next section. When we are concerned only with magnitudes, as is often the case, we use the form of Faraday's law given next:

Magnitude of Induced emf

$$|\mathcal{E}| = N\left|\frac{\Delta\Phi}{\Delta t}\right| = N\left|\frac{\Phi_{final} - \Phi_{initial}}{t_{final} - t_{initial}}\right|$$ 23–4

Notice that Faraday's law gives the *emf* that is induced in a circuit or a loop of wire. The current that is induced as a result of the emf depends on the characteristics of the circuit itself—for example, how much resistance it contains.

EXERCISE 23–1

The induced emf in a single loop of wire has a magnitude of 1.48 V when the magnetic flux is changed from $0.850\ T\cdot m^2$ to $0.110\ T\cdot m^2$. How much time is required for this change in flux?

SOLUTION

Solving Equation 23–4 for the time, we obtain

$$|\Delta t| = N\frac{|\Delta\Phi|}{|\mathcal{E}|} = (1)\frac{|0.110\ T\cdot m^2 - 0.850\ T\cdot m^2|}{1.48\ V} = 0.500\ s$$

In the next Example we continue to deal with magnitudes, so we use Equation 23–4. We also consider a system in which the magnetic field varies in both magnitude and direction over the area of a coil. We can still calculate the magnetic flux in such a case if we simply replace the perpendicular component of the magnetic field $(B\cos\theta)$ with its average value, B_{av}. With this substitution, the magnetic flux, $\Phi = BA\cos\theta$, becomes $\Phi = B_{av}A$.

EXAMPLE 23–2 BAR MAGNET INDUCTION

A bar magnet is moved rapidly toward a 40-turn circular coil of wire. As the magnet moves, the average value of $B\cos\theta$ over the area of the coil increases from 0.0125 T to 0.450 T in 0.250 s. If the radius of the coil is 3.05 cm, and the resistance of its wire is 3.55 Ω, find the magnitude of **(a)** the induced emf and **(b)** the induced current.

PICTURE THE PROBLEM
The motion of the bar magnet relative to the coil is shown in our sketch. As the magnet approaches the coil, the average value of the perpendicular component of the magnetic field increases. As a result, the magnetic flux through the coil increases with time.

STRATEGY
To find the induced emf and the induced current we must first calculate the rate of change of the magnetic flux. To find the flux, we use $\Phi = B_{av}A$, where $A = \pi r^2$, and B_{av} is the average value of $B\cos\theta$. Once we have found the induced emf using Faraday's law, $|\mathcal{E}| = N|\Delta\Phi/\Delta t|$, we obtain the induced current using Ohm's law, $I = V/R$, with $V = |\mathcal{E}|$.

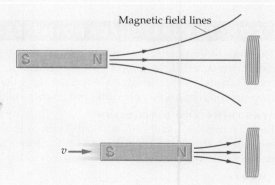
Magnetic field lines

SOLUTION

Part (a)

1. Calculate the initial and final values of the magnetic flux through the coil:

$$\Phi_i = B_i A$$
$$= (0.0125 \text{ T})\pi(0.0305 \text{ m})^2 = 3.65 \times 10^{-5} \text{ T} \cdot \text{m}^2$$
$$\Phi_f = B_f A$$
$$= (0.450 \text{ T})\pi(0.0305 \text{ m})^2 = 1.32 \times 10^{-3} \text{ T} \cdot \text{m}^2$$

2. Use Faraday's law to find the magnitude of the induced emf:

$$|\mathcal{E}| = N\left|\frac{\Phi_{final} - \Phi_{initial}}{t_{final} - t_{initial}}\right|$$
$$= (40)\left|\frac{1.32 \times 10^{-3} \text{ T} \cdot \text{m}^2 - 3.65 \times 10^{-5} \text{ T} \cdot \text{m}^2}{0.250 \text{ s}}\right|$$
$$= 0.205 \text{ V}$$

Part (b)

3. Use Ohm's law to calculate the induced current:

$$I = \frac{V}{R} = \frac{0.205 \text{ V}}{3.55 \text{ }\Omega} = 0.0577 \text{ A}$$

INSIGHT
If the magnet is now pulled back to its original position in the same amount of time, the induced emf and current will have the same magnitudes; their directions will be reversed, however.

PRACTICE PROBLEM
How many turns of wire would be required in this coil to give an induced current of at least 0.100 A? [**Answer:** $N = 70$]

Some related homework problems: Problem 9, Problem 18, Problem 19

An everyday application of Faraday's law of induction can be found in a type of microphone known as a *dynamic microphone*. These devices employ a stationary permanent magnet and a wire coil attached to a movable diaphragm, as illustrated in **Figure 23–4**. When a sound wave strikes the microphone, it causes the diaphragm to oscillate, moving the coil alternately closer to and farther from the magnet. This movement changes the magnetic flux through the coil, which in turn produces an induced emf. Connecting the coil to an amplifier increases the magnitude of the induced emf enough that it can power a set of large speakers. The same principle is employed in the *seismograph*, except in this case the oscillations that produce the induced emf are generated by earthquakes and the vibrations they send through the ground.

REAL-WORLD PHYSICS

Dynamic microphones and seismographs

The American pop-jazz guitarist Les Paul (1915–) applied the same basic physics to musical instruments when he made the first solid-body *electric guitar* in 1941. The pickup in an electric guitar is simply a small permanent magnet with a coil wrapped around it, as shown in **Figure 23–5**. This magnet produces a field that is strong enough to produce a magnetization in the steel guitar string, which is the moving part in the system. When the string is plucked, the oscillating string changes the magnetic flux in the coil, inducing an emf that can be amplified.

REAL-WORLD PHYSICS

Electric guitar pickups

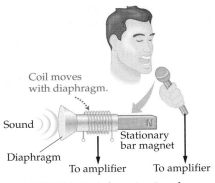

Coil moves with diaphragm.

Sound

Diaphragm

Stationary bar magnet

To amplifier   To amplifier

▲ **FIGURE 23–4 A dynamic microphone**

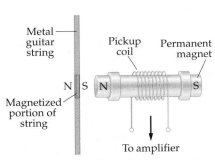

Metal guitar string

Pickup coil   Permanent magnet

Magnetized portion of string

To amplifier

▲ **FIGURE 23–5 The pickup on an electric guitar**

▶ Many devices take advantage of Faraday's law to translate physical oscillations into electrical impulses. In all such devices, the oscillations are used to change the relative position of a coil and a magnet, thus causing the electric current in the coil to vary. For example, the vibrations produced by an earthquake or underground explosion produce an oscillating current in a seismograph (left) that can be amplified and used to drive a plotting pen. The magnetic pickups located under the strings of an electric guitar (right) function in much the same way. In this case, plucking a string (which is magnetized by the pickup) causes it to vibrate and generate an oscillating electrical signal in a pickup. This signal is amplified and fed to speakers, which produce most of the sound we hear.

Magnetic disk drives and credit card readers

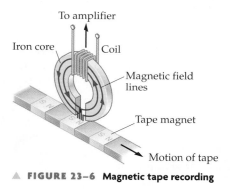

▲ **FIGURE 23-6 Magnetic tape recording**

T coils and induction loops

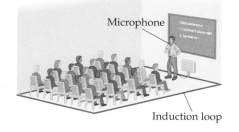

A typical electric guitar will have two or three sets of pickups, each positioned to amplify a different harmonic of the vibrating strings.

Faraday's law of induction also applies to the operation of computer disk drives, credit card readers at grocery stores, audio tape players, and ATM machines. In these devices, information is recorded on a magnetic material by using an electromagnet to produce regions of opposite magnetic polarity, as indicated in **Figure 23-6**. When the information is to be "read out," the magnetic material is moved past the gap in an iron core that is wrapped with many turns of a wire coil. As magnetic regions of different polarity pass between the poles of the magnet, its field is alternately increased or decreased—resulting in a changing magnetic flux through the coil. The corresponding induced emf is sent to electronic circuitry to be translated into information a computer can process.

Finally, induced emf can be of assistance to the hearing impaired. Many hearing aids have a "T switch" that activates a coil of wire in the device known as the telecoil, or **T coil** for short. This coil is designed to pick up the varying magnetic field produced by the speaker in a telephone and to convert the corresponding induced emf into sound for the user of the hearing aid. It should be emphasized that the varying magnetic field of the telephone speaker is simply a natural byproduct of the way it operates and is not produced specifically for the benefit of persons with hearing loss. Nonetheless, a person using the T coil on a hearing aid can usually hear more clearly than if the hearing aid merely amplifies the sound that comes from the speaker.

This principle is now being used on a larger scale in churches, auditoriums, and other public places. A single loop of wire, known as an **induction loop**, is placed on the floor of a room around its perimeter, as shown in **Figure 23-7**. When a person gives a talk in the room, the signal from the microphone is sent not only to speakers but to the induction loop as well. The loop produces a varying magnetic field throughout the room that hearing aids with the T switch can convert directly to sound. Ironically, it is often the hearing impaired in such a case who are able to hear the speech most clearly.

◀ **FIGURE 23-7 An induction loop for the hearing impaired**
A single loop of wire placed around the perimeter of a room can be used as an induction loop that creates a varying magnetic field. This field can be picked up by T coils in hearing aids.

23–4 Lenz's Law

In this section we discuss **Lenz's law,** a physical way of expressing the meaning of the minus sign in Faraday's law of induction. The basic idea of Lenz's law, first stated by the Estonian physicist Heinrich Lenz (1804–1865), is the following:

Lenz's Law

An induced current always flows in a direction that *opposes* the change that caused it.

This general physical principle is closely related to energy conservation, as we shall see at several points in our discussion. The remainder of this section, how-ever, is devoted to specific applications of Lenz's law.

As a first example, consider a bar magnet that is moved toward a conducting ring, as in **Figure 23–8 (a)**. If the north pole of the magnet approaches the ring, a cur-rent is induced that tends to oppose the motion of the magnet. To be specific, the current in the ring creates a magnetic field that has a north pole (that is, diverging field lines) facing the north pole of the magnet, as indicated in the figure. There is now a repulsive force acting on the magnet, opposing its motion.

On the other hand, if the magnet is pulled away from the ring, as in **Figure 23–8 (b)**, the induced current creates a south pole facing the north pole of the mag-net. The resulting attractive force again opposes the magnet's motion. We can ob-tain the same results using Faraday's law directly, but Lenz's law gives a simpler, more physical way to approach the problem.

◀ **FIGURE 23–8 Applying Lenz's law to a magnet moving toward and away from a current loop**
(a) If the north pole of a magnet is moved toward a conducting loop, the induced current produces a north pole pointing toward the magnet's north pole. This creates a repulsive force opposing the change that caused the current. **(b)** If the north pole of a magnet is pulled away from a conducting loop, the induced cur-rent produces a south magnetic pole near the magnet's north pole. The result is an attractive force opposing the motion of the magnet.

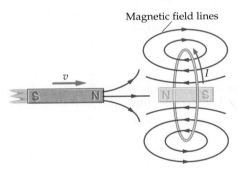
Magnetic field lines

(a) Moving magnet toward coil induces a field that repels the magnet

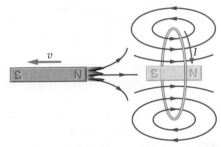

(b) Moving magnet away from coil induces a field that attracts the magnet

CONCEPTUAL CHECKPOINT 23–2   FALLING MAGNETS

The magnets shown in the sketch are dropped from rest through the middle of conduct-ing rings. Notice that the ring on the right has a small break in it, whereas the ring on the left forms a closed loop. As the magnets drop toward the rings, does the magnet on the left have an acceleration that is **(a)** more than, **(b)** less than, or **(c)** the same as that of the magnet on the right?

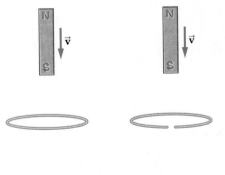

REASONING AND DISCUSSION
As the magnet on the left approaches the ring, it induces a circulating current. According to Lenz's law, this current produces a magnetic field that exerts a repulsive force on the magnet—to oppose its motion. In contrast, the ring on the right has a break, so it cannot have a circulating current. As a result, it exerts no force on its magnet. Therefore, the magnet on the right falls with the acceleration of gravity; the magnet on the left falls with a smaller acceleration.

ANSWER
(b) The magnet on the left has the smaller acceleration.

In the examples given so far in this section, the "change" involved the motion of a bar magnet, but Lenz's law applies no matter how the magnetic flux is changed. Suppose, for example, that a magnetic field is decreased with time, as

▶ **FIGURE 23–9 Lenz's law applied to a decreasing magnetic field**

As the magnetic field is decreased, the induced current produces a magnetic field that passes through the ring in the same direction as $\vec{B}$.

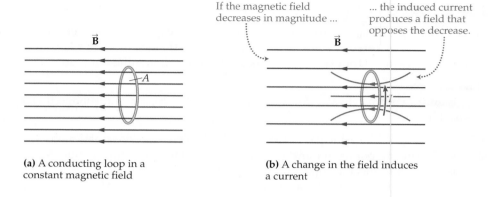

If the magnetic field decreases in magnitude ... ... the induced current produces a field that opposes the decrease.

(a) A conducting loop in a constant magnetic field

(b) A change in the field induces a current

illustrated in **Figure 23–9**. In this case the change is a decrease of magnetic flux through the ring. The induced current flows in a direction that tends to oppose this change; that is, the induced current must produce a field within the ring that is in the same direction as the decreasing B field, as we see in Figure 23–9 (b).

Motional emf: Qualitative

Next, we consider the physical situation shown in **Figure 23–10**. Here we see a metal rod that is free to move in the vertical direction, and a constant magnetic field that points out of the page. The rod is in frictionless contact with two vertical wires; hence, a current can flow in a loop through the rod, the wires, and the lightbulb. Since the motion of the rod produces an emf in this system, it is referred to as **motional emf.** The question to be answered at this point is this: What is the direction of the current when the rod is released from rest and allowed to fall?

In this system the magnetic field is constant, but the magnetic flux through the current loop, $\Phi = BA$, decreases anyway. The reason is that as the rod falls, the area A enclosed by the loop decreases. To oppose this decrease in flux, Lenz's law implies that the induced current must flow in a direction that strengthens the magnetic field $\vec{B}$ within the loop, thus increasing the flux. This direction is counterclockwise, as indicated in **Figure 23–11**.

Notice that the current in the rod flows from right to left. This direction is also in agreement with Lenz's law, since it opposes the change that caused it. To be specific, the change for the rod is that it begins to fall downward. In order to oppose this change, the current through the rod must produce an upward magnetic force. As we see in Figure 23–11, using the right-hand rule, a current from right to left does just that.

When the rod is first released, there is no current and its downward acceleration is the acceleration of gravity—the rod speeds up as its gravitational potential energy is converted to kinetic energy. As the rod falls, the induced current begins to flow, however, and the rod is acted on by the upward magnetic force. This causes its acceleration to decrease. Ultimately, the magnetic force cancels the force of gravity, and the rod's acceleration vanishes. From this point on it falls with constant speed and hence constant kinetic energy. Now you may wonder what happens to the rod's gravitational potential energy. In fact, all the rod's gravitational potential energy is now being converted to heat and light in the lightbulb—none of it goes to increasing the kinetic energy of the rod.

Imagine for a moment the consequences that would follow if Lenz's law were not obeyed. If the current in the rod were from left to right, for example, the magnetic force would be downward, causing the rod to increase its downward acceleration. As the rod sped up, the induced current would become even larger and the downward force would increase as well. Thus, the rod would simply move faster and faster, without limit. Clearly, energy would not be conserved in such a case. The connections between mechanical and electrical energy are explored in greater detail in the next section.

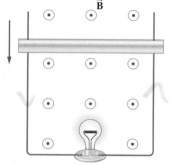

▲ **FIGURE 23–10 Motional emf**

Motional emf is created in this system as the rod falls. The result is an induced current, which causes the light to shine.

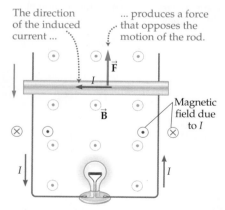

The direction of the induced current ... ... produces a force that opposes the motion of the rod.

Magnetic field due to I

▲ **FIGURE 23–11 Determining the direction of an induced current**

The direction of the current induced by the rod's downward motion is counterclockwise, because this direction produces a magnetic field within the loop that points out of the page—in the same direction as the original field. Notice that a current flowing in this direction through the rod interacts with the original magnetic field to give an upward force, opposing the downward motion of the rod.

CONCEPTUAL CHECKPOINT 23–3   THE DIRECTION OF INDUCED CURRENT

Consider a system in which a metal ring is falling out of a region with a magnetic field and into a field-free region, as shown in our sketch to the right. According to Lenz's law, is the induced current in the ring **(a)** clockwise or **(b)** counterclockwise?

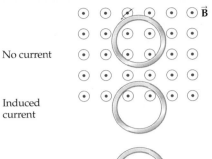

No current

Induced current

No current

INTERACTIVE FIGURE

REASONING AND DISCUSSION

The induced current must be in a direction that opposes the change in the system. In this case, the change is that fewer magnetic field lines are piercing the area of the loop and pointing out of the page. The induced current can oppose this change by generating more field lines out of the page *within the loop*. As shown to the left in the following figure, the induced current must be counterclockwise to accomplish this.

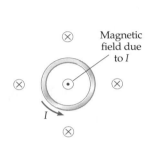

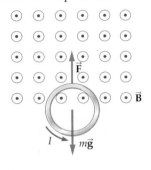

Finally, to the right in the preceding figure, note that the induced current generates an upward magnetic force at the top of the ring, but no magnetic force on the bottom, where the magnetic field is zero. Hence, the motion of the ring is retarded as it drops out of the field.

ANSWER

(b) The induced current is counterclockwise.

The retarding effect on a ring leaving a magnetic field, as just described in the Conceptual Checkpoint, allows us to understand the behavior of **eddy currents.** Suppose, for example, that a sheet of metal falls from a region with a magnetic field to a region with no field, as in **Figure 23–12**. In the portion of the sheet that is just leaving the field, a circulating current—an eddy current—is induced in the metal, just like the circulating current set up in the ring in Conceptual Checkpoint 23–3. As in the case of the ring, the current will retard the motion of the sheet of metal, somewhat like a frictional force.

The frictionlike effect of eddy currents is the basis for the phenomenon known as *magnetic braking.* An important advantage of this type of braking is that no direct physical contact is needed, thus eliminating frictional wear. In addition, the force of magnetic braking is stronger the greater the speed of the metal with respect to the magnetic field. In contrast, kinetic friction is independent of the relative speed of the surfaces. Magnetic braking is used in everything from fishing reels to exercise bicycles to roller coasters.

Perhaps the most dramatic manifestation of magnetic braking is the slowing down of rotating magnetic stars. White dwarf stars, with intense magnetic fields, produce eddy currents in the clouds of ionized material that surround them. The result is a rapid slowing of their rotational speed, almost as if they were rotating in a viscous liquid. In some stars, magnetic braking has increased the rotational period from as short as a few hours to as long as 200 years.

On a more down-to-earth scale, magnetic braking is the operating principle behind the common analog *speedometer.* In these devices, a cable is connected at one end to a gear in the car's transmission and at the other end to a small permanent magnet. When the car is under way, the magnet rotates within a metal cup. The resulting "magnetic friction" between the rotating magnet and the metal cup— which is proportional to the speed of rotation—causes the cup to rotate in the same direction. A hairspring attached to the cup stops the rotation when the force of the spring equals the force of magnetic friction. Thus, the cup—and the needle attached to it—rotate through an angle that is proportional to the speed of the car.

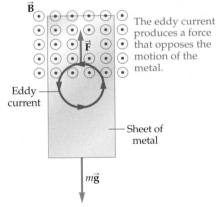

The eddy current produces a force that opposes the motion of the metal.

Eddy current

Sheet of metal

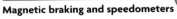

▲ **FIGURE 23–12 Eddy currents**
A circulating current is induced in a sheet of metal near where it leaves a region with a magnetic field. This "eddy current" is similar to the current induced in the ring described in Conceptual Checkpoint 23–3. In both cases, the induced current exerts a retarding force on the moving object, opposing its motion.

REAL-WORLD PHYSICS
Magnetic braking and speedometers

Finally, eddy currents are also used in the kitchen. In an *induction stove,* a metal coil is placed just beneath the cooking surface. When an alternating current is sent through the coil, it sets up an alternating magnetic field that, in turn, induces eddy currents in nearby metal objects. The finite resistance of a metal pan, for instance, will result in heating as the eddy current dissipates power according to the relation $P = I^2R$. Nonmetallic objects, like the surface of the stove and glassware, remain cool to the touch because they are poor electrical conductors and thus have negligible eddy currents.

23–5 Mechanical Work and Electrical Energy

To see how mechanical work can be converted to electrical energy, we turn to a detailed calculation of the forces and currents involved in a simple physical system. This analysis will form the basis for our understanding of electrical generators in the next section.

Motional emf: Quantitative

We again consider motional emf, though this time in a somewhat simpler system where gravity plays no role. As shown in **Figure 23–13**, our setup consists of a rod that slides horizontally without friction on a U-shaped wire connected to a light-bulb of resistance R. A constant magnetic field points out of the page, and the rod is pushed by an external agent—your hand, for instance—so that it moves to the right with a constant speed v.

▶ **FIGURE 23–13 Force and induced current**
A conducting rod slides without friction on horizontal wires in a region where the magnetic field is uniform and pointing out of the page. The motion of the rod to the right induces a clockwise current and a corresponding magnetic force to the left. An external force of magnitude $F = B^2v\ell^2/R$ is required to offset the magnetic force and to keep the rod moving with a constant speed v.

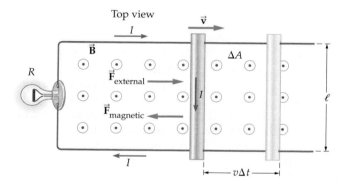

Let's begin by calculating the rate at which the magnetic flux changes with time. Since $\vec{\mathbf{B}}$ is perpendicular to the area of the loop, it follows that $\Phi = BA$. Now the area increases with time because of the motion of the rod. In fact, the rod moves a horizontal distance $v\Delta t$ in the time Δt; therefore, the area of the loop increases by the amount $\Delta A = (v\Delta t)\ell$, as indicated in Figure 23–13. As a result, the increase in magnetic flux is

$$\Delta\Phi = B\,\Delta A = Bv\ell\,\Delta t$$

With this result in hand, we can use Faraday's law to calculate the magnitude of the induced emf. We find that for this single-turn loop

$$|\mathcal{E}| = N\left|\frac{\Delta\Phi}{\Delta t}\right| = (1)\frac{Bv\ell\,\Delta t}{\Delta t} = Bv\ell \qquad\qquad 23\text{–}5$$

As one might expect, the induced emf is proportional to the strength of the magnetic field and to the speed of the rod.

Now let's calculate the electric field caused by the motion of the rod. We know that the emf from one end of the rod to the other is $\mathcal{E} = Bv\ell$, where $\mathcal{E}$ is an electric potential difference measured in volts. Recall that the difference in electric potential caused by an electric field E over a length ℓ is $V = E\ell$. Applying this to the emf in the rod, we have $E\ell = Bv\ell$, or

$$E = Bv \qquad\qquad 23\text{–}6$$

Thus, a changing magnetic flux does indeed create an electric field, as Faraday suspected. In particular, moving through a magnetic field B at a speed v causes an electric field of magnitude $E = Bv$ in a direction that is at right angles to both the velocity and the magnetic field.

EXERCISE 23–2

As the rod in Figure 23–13 moves through a 0.445-T magnetic field, it experiences an induced electric field of 0.668 V/m. How fast is the rod moving?

SOLUTION
Solving Equation 23–6 for v, we find

$$v = \frac{E}{B} = \frac{0.668 \text{ V/m}}{0.445 \text{ T}} = 1.50 \text{ m/s}$$

If we assume that the only resistance in the system is the lightbulb, R, it follows that the current in the circuit has the following magnitude:

$$I = \frac{|\mathcal{E}|}{R} = \frac{Bv\ell}{R} \qquad\qquad 23\text{–}7$$

Using Lenz's law, we can determine that the direction of the current is clockwise, as indicated in Figure 23–13. Next, we show how both the magnitude and direction of the current play a key role in the energy aspects of this system.

Mechanical Work/Electrical Energy

We begin with a calculation of the force and power required to keep the rod moving with a constant speed v. First, note that the rod carries a current of magnitude $I = Bv\ell/R$ at right angles to a magnetic field of strength B. Since the length of the rod is ℓ, it follows that the magnetic force exerted on it has a magnitude

$$F = I\ell B = \left(\frac{Bv\ell}{R}\right)(\ell)B = \frac{B^2 v\ell^2}{R} \qquad\qquad 23\text{–}8$$

The direction of the current in the rod is downward on the page in Figure 23–13; hence, the magnetic-force RHR shows that $\vec{F}_{\text{magnetic}}$ is directed to the left. As a result, an external force, $\vec{F}_{\text{external}}$, must act to the right with equal magnitude to maintain the rod's constant speed. The mechanical power delivered by the external force is simply the force times the speed:

$$P_{\text{mechanical}} = Fv = \left(\frac{B^2 v\ell^2}{R}\right)v = \frac{B^2 v^2 \ell^2}{R} \qquad\qquad 23\text{–}9$$

Now let's compare this mechanical power with the electrical power converted to light and heat in the lightbulb. Recalling that the power dissipated in a resistor with a resistance R is $P = I^2 R$, we have

$$P_{\text{electrical}} = I^2 R = \left(\frac{Bv\ell}{R}\right)^2 R = \frac{B^2 v^2 \ell^2}{R} \qquad\qquad 23\text{–}10$$

Note that this is *exactly* the same as the mechanical power. Thus, with the aid of a magnetic field, we can convert *mechanical* power directly to *electrical* power. This simple example of motional emf illustrates the basic principle behind the generation of virtually all the world's electrical energy.

EXAMPLE 23–3 LIGHT POWER

The lightbulb in the circuit shown here has a resistance of 12 Ω and consumes 5.0 W of power; the rod is 1.25 m long and moves to the left with a constant speed of 3.1 m/s. **(a)** What is the strength of the magnetic field? **(b)** What external force is required to maintain the rod's constant speed?

CONTINUED ON NEXT PAGE

CONTINUED FROM PREVIOUS PAGE

PICTURE THE PROBLEM

Our sketch shows the rod being pushed to the left by the force $\vec{F}_{external}$. The motion of the rod, in turn, generates an electric current whose direction, according to Lenz's law, produces a magnetic force, $\vec{F}_{magnetic}$, that opposes the motion of the rod. We know that the rod moves with constant speed, and therefore the applied force $\vec{F}_{external}$ exactly cancels the magnetic force $\vec{F}_{magnetic}$.

STRATEGY

a. The power consumed by the lightbulb is given by the expression $P = B^2 v^2 \ell^2 / R$. This relation can be solved to give the strength of the magnetic field, B.

b. The magnitude of the external force is given by $F = B^2 v \ell^2 / R$, where B has the value obtained in part (a). Alternatively, we can solve for the force using the power relation, $P = Fv$.

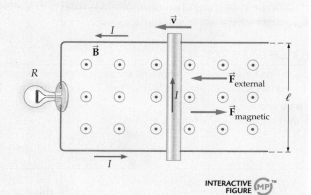

INTERACTIVE FIGURE MP™

SOLUTION

Part (a)

1. Solve the power expression for the magnetic field, B:

$$P = \frac{B^2 v^2 \ell^2}{R}$$

$$B = \frac{\sqrt{PR}}{v\ell}$$

2. Substitute numerical values:

$$B = \frac{\sqrt{PR}}{v\ell} = \frac{\sqrt{(5.0\ \text{W})(12\ \Omega)}}{(3.1\ \text{m/s})(1.25\ \text{m})} = 2.0\ \text{T}$$

Part (b)

3. Calculate the magnitude of the external force using $F = B^2 v \ell^2 / R$:

$$F = \frac{B^2 v \ell^2}{R}$$

$$= \frac{(2.0\ \text{T})^2 (3.1\ \text{m/s})(1.25\ \text{m})^2}{12\ \Omega} = 1.6\ \text{N}$$

4. Find the magnitude of the external force using $P = Fv$:

$$F = \frac{P}{v} = \frac{5.0\ \text{W}}{3.1\ \text{m/s}} = 1.6\ \text{N}$$

INSIGHT

Note that a constant external *force* is required to move the rod with constant *speed* and that the magnitude of the external force is proportional to the speed. This is similar to the behavior of a "drag" force like air resistance. Also, note that the *net* force acting on the rod is zero—as expected for an object that moves with constant speed.

PRACTICE PROBLEM

How fast will the rod move if the external force exerted on it is 1.5 N? **[Answer: $v = 2.9$ m/s]**

Some related homework problems: Problem 39, Problem 40, Problem 41

ACTIVE EXAMPLE 23–1	FIND THE WORK CONVERTED TO LIGHT ENERGY

Referring to Example 23–3, find **(a)** the work done by the external force in 0.58 s and **(b)** the energy consumed by the lightbulb in the same time.

SOLUTION *(Test your understanding by performing the calculations indicated in each step.)*

Part (a)

1. Find the distance covered by the rod in 0.58 s: 1.8 m

2. Find the work done using $W = Fd$: $(1.6\ \text{N})(1.8\ \text{m}) = 2.9\ \text{J}$

Part (b)

3. Calculate the energy used in the lightbulb by multiplying the power by the time: $(5.0\ \text{W})(0.58\ \text{s}) = 2.9\ \text{J}$

Motional emfs are generated whenever a magnet moves relative to a conductor or when a conductor moves relative to a magnetic field. For example, *magnetic antitheft devices* in libraries and stores detect the motional emf produced by a small magnetic strip placed in a book or other item. If this strip is not demagnetized, it will trigger an alarm as it moves past a sensitive set of conductors designed to pick up the induced emf.

REAL-WORLD PHYSICS
Magnetic antitheft devices

In numerous biomedical applications of the same principle, it is the conductor that moves while the magnetic field remains stationary. For example, researchers have tracked the head and body movements of several flying insects, including blowflies, hover flies, and honeybees. They attach lightweight, flexible wires to a small metal coil on the insect's head, and another on its thorax, and then allow the insect to fly in a stationary magnetic field. As the coils move through the field, they experience induced emfs that can be analyzed by computer to determine the corresponding orientation of the head and thorax.

REAL-WORLD PHYSICS: BIO
Tracking the movement of insects

The same technique is used by researchers investigating the movements of the human eye. You may not be aware of it, but your eyes are jiggling back and forth rather rapidly at this very moment. This constant motion, referred to as **saccadic motion,** prevents an image from appearing at the same position on the retina for a period of time. Without saccadic motion, our brains would tend to ignore stationary images that we are focused on in favor of moving images of little interest. To study the dynamics of saccadic motion, a small coil of wire is attached to a modified contact lens. As the eye moves the coil through a stationary magnetic field, the induced motional emf gives the eye's orientation and rotational speed.

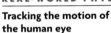

REAL-WORLD PHYSICS: BIO
Tracking the motion of
the human eye

23–6 Generators and Motors

We have just seen that a changing magnetic flux can serve as a means of converting mechanical work to electrical energy. In this section we discuss this concept in greater detail as we explore the workings of an electric generator. We also point out that the energy conversion can run in the other direction as well, with electrical energy being converted to mechanical work in an electric motor.

Electric Generators

An electric **generator** is a device designed to efficiently convert mechanical energy to electrical energy. The mechanical energy can be supplied by any of a number of sources, such as falling water in a hydroelectric dam, expanding steam in a coal-fired power plant, or the output of a small gasoline motor that powers a portable generator. In all cases the basic operating principle is the same—mechanical energy is used to move a conductor through a magnetic field to produce motional emf.

REAL-WORLD PHYSICS
Electric generators

The linear motion of a metal rod through a magnetic field, as in Figure 23–13, results in motional emf, but to continue producing an electric current in this way the rod must move through ever greater distances. A practical way to employ the same effect is to use a coil of wire that can be *rotated* in a magnetic field. By rotating the coil, which in turn changes the magnetic flux, we can continue the process indefinitely.

▲ Hydroelectric power provides an excellent example of energy transformations. First, the gravitational potential energy of water held behind a dam is converted to kinetic energy as the water descends through pipes in the generating station (left). Next, the kinetic energy of the falling water is used to spin turbines (center). Each turbine rotates the coil of a generator in a strong magnetic field, thus converting mechanical energy into an electric current in the coil (right). Eventually the electrical energy is delivered to homes, factories, and cities where it is converted to light (in incandescent bulbs and fluorescent tubes), heat (in devices such as toasters and broilers), and work (in various machines).

PROBLEM-SOLVING NOTE

Maximum and Minimum Values of Induced emf

Note that the induced emf of a rotating coil has a maximum value of $NBA\omega$ and a minimum value of $-NBA\omega$. This follows from Equation 23–11 and the fact that the maximum and minimum values of $\sin \omega t$ are $+1$ and -1, respectively.

Imagine, then, a coil of area A located in the magnetic field between the poles of a magnet, as illustrated in **Figure 23–14**. Metal rings are attached to either end of the wire that makes up the coil, and carbon brushes are in contact with the rings to allow the induced emf to be communicated to the outside world. As the coil rotates with the angular speed ω, the emf produced in it is given by Faraday's law, $\mathcal{E} = -N(\Delta\Phi/\Delta t)$. Referring to the similar case of uniform circular motion in Chapter 13, and Equation 13–6 in particular, it can be shown that the explicit expression for the emf of a rotating coil is as follows:

$$\mathcal{E} = NBA\omega \sin \omega t \qquad 23\text{–}11$$

This result is plotted in **Figure 23–15**. Notice that the induced emf in the coil alternates in sign, which means that the current in the coil alternates in direction. For this reason, this type of generator is referred to as an **alternating-current generator** or, simply, an ac generator.

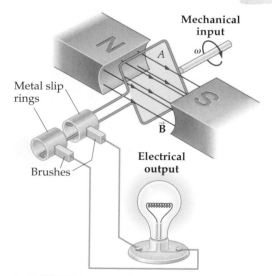

▲ **FIGURE 23–14** **An electric generator**
The basic operating elements of an electric generator are shown in a schematic representation. As the coil is rotated by an external source of mechanical work, it produces an emf that can be used to power an electric circuit.

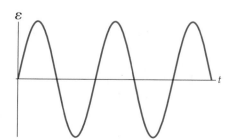

▲ **FIGURE 23–15** **Induced emf of a rotating coil**
Induced emf produced by an electric generator like the one shown in Figure 23–14. The emf alternates in sign, so the current changes sign as well. We say that a generator of this type produces an "alternating current."

EXAMPLE 23–4 GENERATOR NEXT

The coil in an electric generator has 100 turns and an area of 2.5×10^{-3} m^2. It is desired that the maximum emf of the coil be 120 V when it rotates at the rate of 60.0 cycles per second. **(a)** What is the angular speed of the generator? **(b)** Find the strength of the magnetic field B that is required for this generator to operate at the desired voltage.

PICTURE THE PROBLEM

In our sketch, a square conducting coil of area A and angular speed ω rotates between the poles of a magnet, where the magnetic field has a strength B. As the coil rotates, the magnetic flux through it changes continuously. The result is an induced emf that varies sinusoidally with time.

STRATEGY

a. We need to find the angular speed, ω, since it is not given directly in the problem statement. What we are given, however, is that the frequency of the coil's rotation is f = 60.0 cycles per second = 60.0 Hz. The corresponding angular speed is $\omega = 2\pi f$. (See Chapter 13.)

b. The induced emf of the generator is given by $\mathcal{E} = NBA\omega \sin \omega t$. Therefore, the maximum emf occurs when $\sin \omega t$ has its maximum value of 1. It follows that $\mathcal{E}_{max} = NBA\omega$. This relation can be solved for the magnetic field, B.

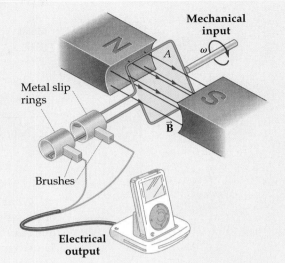

Mechanical input

Metal slip rings

$\vec{B}$

Brushes

Electrical output

SOLUTION

Part (a)

1. Calculate the angular speed of the coil:

$$\omega = 2\pi f = 2\pi(60.0 \text{ Hz}) = 377 \text{ rad/s}$$

Part (b)

2. Solve for the magnetic field in terms of the maximum emf:

$$\mathcal{E}_{max} = NBA\omega \quad \text{or} \quad B = \frac{\mathcal{E}_{max}}{NA\omega}$$

3. Substitute numerical values into the expression for B:

$$B = \frac{\mathcal{E}_{max}}{NA\omega} = \frac{120 \text{ V}}{(100)(2.5 \times 10^{-3} \text{ m}^2)(377 \text{ rad/s})} = 1.3 \text{ T}$$

INSIGHT

Note that the emf of a generator coil depends on its area A but not on its shape. In addition, each turn of the coil generates the same induced emf; therefore, the total emf of the coil is directly proportional to the number of turns. Finally, the more rapidly the coil rotates, the more rapidly the magnetic flux through it changes. As a result, the induced emf is also proportional to the angular speed of the generator.

PRACTICE PROBLEM

What is the maximum emf of this generator if its rotation frequency is reduced to 50.0 Hz?   **[Answer: 100 V]**

Some related homework problems: Problem 42, Problem 43

Electric Motors

The basic physical principle behind the operation of electric motors has already been discussed. In particular, we saw in Section 22–5 that a current-carrying loop in a magnetic field experiences a torque that tends to make it rotate. If such a loop is mounted on an axle, as in **Figure 23–16**, it is possible for the magnetic torque to be applied to the external world.

REAL-WORLD PHYSICS

Electric motors

To see just how this works in practice, notice that the torque exerted on the loop at the moment shown in Figure 23–16 causes it to rotate clockwise *toward* the vertical position. Once it reaches this orientation, and continues past it due to its angular momentum, the alternating current from the electrical input reverses direction. This reverses the torque on the loop and causes the loop to rotate *away* from the vertical—which means it is still rotating in the clockwise sense. The next time the loop becomes vertical, the current again reverses, causing the loop to continue rotating clockwise. The result is an axle continually turning in the same direction.

▶ **FIGURE 23–16 A simple electric motor**
An electric current causes the coil of a motor to rotate and deliver mechanical work to the outside world.

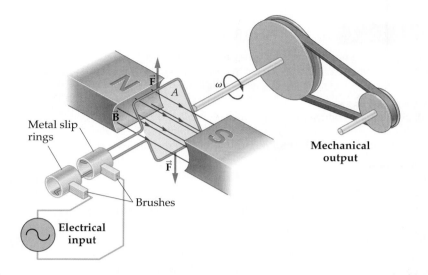

Metal slip rings

Brushes

Electrical input

Mechanical output

What we have just described is precisely the reverse of a generator—instead of doing work to turn a coil and produce an electric current, as in a generator, a motor uses an electric current to produce rotation in a coil, which then does work. Thus, a motor, in some sense, is a generator run in reverse. In fact, it is possible to have one electric motor turn the axle of another identical motor and thereby produce electricity from the "motor"-turned generator.

Similarly, if a car is powered by an electric motor, its motor can double as a generator when it helps to slow the car during braking. This results in a more efficient mode of transportation, because some of the car's kinetic energy that is normally converted to heat during braking can instead be used to recharge the batteries and extend the range of the car. Some of the newer "hybrid" cars, such as the Honda Insight and the Toyota Prius, make use of this "energy recovery" technology.

REAL-WORLD PHYSICS
Energy recovery technology in cars

23–7 Inductance

We began this chapter with a description of Faraday's induction experiment, in which a changing current in one coil induces a current in another coil. This type of interaction between coils is referred to as **mutual inductance.** It is also possible, however, for a single, isolated coil to have a similar effect—that is, a coil with a changing current can induce an emf, and hence a current, in itself. This type of process is known as **self-inductance.**

To be more precise, consider a coil of wire in a region free of magnetic fields, as in **Figure 23–17 (a).** Initially the current in the coil is zero, so the magnetic flux through the coil is zero as well. The current in the coil is now steadily increased from zero, as indicated in **Figure 23–17 (b).** As a result, the magnetic flux through the coil increases with time, which, according to Faraday's law, induces an emf in the coil.

(a) No current, no field in coil

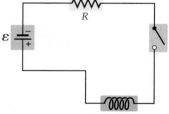

(b) Increasing current produces increasing magnetic field in coil

▲ **FIGURE 23–17 A changing current in an inductor**
(a) A coil of wire with no current and no magnetic flux. (b) The current is now increasing with time, which produces a magnetic flux that also increases with time.

The direction of the induced emf is given by Lenz's law, which says that the induced current opposes the change that caused it. In this case, the change is the increasing current in the coil; hence, the induced emf is in a direction opposite to the current. This situation is illustrated in **Figure 23–18,** where we schematically show the effect of the induced emf (often referred to as a "back" emf) in the circuit. Note that in order to increase the current in the coil it is necessary to *force* the current against an opposing emf.

This is a rather interesting result. It says that even if the wires in a coil are ideal, with no resistance whatsoever, work must still be done to increase the current. Conversely, if a current already exists in a coil and it decreases with time, Lenz's law ensures that the self-induced current will oppose this change as well, in an attempt to keep the current constant. Hence, a coil tends to resist changes in its current, regardless of whether the change is an increase or a decrease.

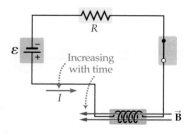

To make a quantitative connection between a changing current and the associated self-induced emf, we start with Faraday's law in magnitude form:

$$|\mathcal{E}| = N\left|\frac{\Delta\Phi}{\Delta t}\right|$$

Now, since the magnetic field is proportional to the current in a coil, it follows that the magnetic flux Φ is also proportional to I. As a result, the time rate of change of the magnetic flux, $\Delta\Phi/\Delta t$, must be proportional to the time rate of change of the current, $\Delta I/\Delta t$. Therefore, the induced emf can be written as

Definition of Inductance, L

$$|\mathcal{E}| = N\left|\frac{\Delta\Phi}{\Delta t}\right| = L\left|\frac{\Delta I}{\Delta t}\right|$$
23–12

SI unit: $1\ \text{V} \cdot \text{s/A} = 1\ \text{henry} = 1\ \text{H}$

This expression defines the constant of proportionality, L, which is referred to as the **inductance** of a coil. Rearranging Equation 23–12, we see that

$$L = N\left|\frac{\Delta\Phi}{\Delta I}\right|$$

The SI unit of inductance is the **henry (H)**, named to honor the work of Joseph Henry (1797–1878), an American physicist. As can be seen from Equation 23–12, the henry is one volt-second per amp:

$$1\ \text{H} = 1\ \text{V} \cdot \text{s/A}$$
23–13

Common inductances are generally in the millihenry (mH) range.

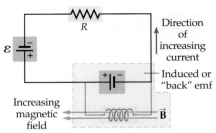

▲ **FIGURE 23–18 The back emf of an inductor**
The effect of an increasing current in a coil is an induced emf that opposes the increase. This is indicated schematically by replacing the coil with the opposing, or "back," emf.

EXERCISE 23–3

The coil in an electromagnet has an inductance of 2.9 mH and carries a constant direct current of 5.6 A. A switch is suddenly opened, allowing the current to drop to zero over a small interval of time, Δt. If the magnitude of the average induced emf during this time is 7.3 V, what is Δt?

SOLUTION
Solving Equation 23–12 for Δt, and using only magnitudes, we find

$$\Delta t = \frac{L|\Delta I|}{|\mathcal{E}|} = \frac{(2.9\times 10^{-3}\ \text{H})(5.6\ \text{A})}{7.3\ \text{V}} = 2.2\ \text{ms}$$

Of particular interest is the inductance of an ideal solenoid with N turns and length ℓ. For example, suppose the initial current in a solenoid is zero and the final current is I. It follows that the initial magnetic flux is zero. The final flux is simply $\Phi_f = BA\cos\theta$, where, according to Equation 22–12, the magnetic field in the solenoid is $B = \mu_0(N/\ell)I$. Finally, we know that the field in a solenoid is perpendicular to its cross-sectional area, A; thus $\theta = 0$ and $\Phi_f = BA$. Combining these results, we obtain

$$L = \frac{N(BA - 0)}{(I - 0)} = \frac{N[\mu_0(N/\ell)I]A}{I} = \mu_0\left(\frac{N^2}{\ell}\right)A$$

Denoting the number of turns per length as $n = N/\ell$, we can express the inductance of a solenoid in the following forms:

Inductance of a Solenoid

$$L = \mu_0\left(\frac{N^2}{\ell}\right)A = \mu_0 n^2 A\ell$$
23–14

Notice that doubling the number of turns per length quadruples the inductance. In addition, the inductance is proportional to the volume of a solenoid, $V = A\ell$.

In the following Examples we consider problems involving solenoids and their inductance.

EXAMPLE 23–5 SOLENOID SELF-INDUCTION

A 500-turn solenoid is 8.0 cm long. When the current in this solenoid is increased from 0 to 2.5 A in 0.35 s, the magnitude of the induced emf is 0.012 V. Find **(a)** the inductance and **(b)** the cross-sectional area of the solenoid.

PICTURE THE PROBLEM
The solenoid described in the problem statement is shown in our sketch. Note that it is 8.0 cm in length and contains 500 turns.

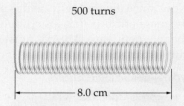

STRATEGY
a. We can use the definition of inductance, $|\mathcal{E}| = L|\Delta I/\Delta t|$, to solve for the inductance.
b. Once we know L from part (a), we can find the cross-sectional area from $L = \mu_0 n^2 A\ell$, where $n = N/\ell$.

SOLUTION

Part (a)

1. Using magnitudes, solve $|\mathcal{E}| = L|\Delta I/\Delta t|$ for the inductance:

$$L = \frac{|\mathcal{E}|}{|\Delta I/\Delta t|} = \frac{0.012 \text{ V}}{(2.5 \text{ A}/0.35 \text{ s})} = 1.7 \text{ mH}$$

Part (b)

2. Solve $L = \mu_0 n^2 A\ell$ for the cross-sectional area of the solenoid:

$$A = \frac{L}{\mu_0 n^2 \ell} = \frac{L\ell}{\mu_0 N^2}$$

3. Substitute numerical values:

$$A = \frac{(1.7 \times 10^{-3} \text{ H})(0.080 \text{ m})}{(4\pi \times 10^{-7} \text{ T} \cdot \text{m/A})(500)^2} = 4.3 \times 10^{-4} \text{ m}^2$$

INSIGHT
This cross-sectional area corresponds to a radius of about 1.2 cm. Therefore, it would take about 37 m of wire to construct this inductor.

PRACTICE PROBLEM
Find the inductance and induced emf for a solenoid with a cross-sectional area of 1.3×10^{-3} m^2. The solenoid, which is 9.0 cm long, is made from a piece of wire that is 44 m in length. Assume that the rate of change of current remains the same. **[Answer:** $L = 2.2$ mH, $|\mathcal{E}| = 0.016$ V**]**

Some related homework problems: Problem 49, Problem 50

ACTIVE EXAMPLE 23–2 FIND THE NUMBER OF TURNS

The inductance of a solenoid 5.00 cm long is 0.157 mH. If its cross-sectional area is 1.00×10^{-4} m^2, how many turns does the solenoid have?

SOLUTION *(Test your understanding by performing the calculations indicated in each step.)*

1. Solve $L = \mu_0(N^2/\ell)A$ for the number of turns, N: $N = (L\ell/\mu_0 A)^{1/2}$

2. Substitute numerical values: $N = 2.50 \times 10^2$

YOUR TURN
Which of the following changes would affect the value of the inductance more: **(a)** doubling the number of turns or **(b)** tripling the cross-sectional area? Verify your answer by giving the factor by which the inductance changes for each of these cases.

*(Answers to **Your Turn** problems are given in the back of the book.)*

A coil with a finite inductance—an **inductor** for short—resists changes in its electric current. Similarly, a particle of finite mass has the property of inertia, by which it resists changes in its velocity. These analogies, between inductance and mass on the one hand and current and velocity on the other, are useful ones, as we shall see in greater detail when we consider the behavior of ac circuits in the next chapter.

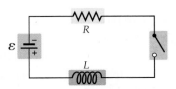

23–8 *RL* Circuits

The circuit shown in **Figure 23–19** is referred to as an **RL circuit;** it consists of a resistor, R, and an inductor, L, in series with a battery of emf $\mathcal{E}$. Note that the inductor is represented by a symbol reminiscent of a coil with several loops, as shown below:

We assume the inductor to be ideal, which means that the wires forming its loops have no resistance—the only resistance in the circuit is the resistor R. It follows, then, that after the switch has been closed for a long time, the current in the circuit is simply $I = \mathcal{E}/R$.

You may have noticed that we said the current would be $\mathcal{E}/R$ after the switch had been closed "for a long time." Why did we make this qualification? The point is that because of the inductor, and its tendency to resist changes in the current, it is not possible for the current to rise immediately to its final value. The inductor slows the process, causing the current to rise over a finite period of time, just as the charge on a capacitor builds up over the characteristic time $\tau = RC$ in an RC circuit.

The corresponding characteristic time for an RL circuit is given by

$$\tau = \frac{L}{R}$$ 23–15

As expected, the larger the inductance, the longer the time required for the current to build up. In addition, the current approaches its final value of $\mathcal{E}/R$ with an exponential time dependence, much like the case of an RC circuit. With the aid of calculus, it can be shown that the current as a function of time is given by

$$I = \frac{\mathcal{E}}{R}(1 - e^{-t/\tau}) = \frac{\mathcal{E}}{R}(1 - e^{-tR/L})$$ 23–16

Note the similarity between this expression, which is plotted in **Figure 23–20**, and the expression given in Equation 21–18 for an RC circuit.

EXERCISE 23–4

Find the inductance in the RL circuit of Figure 23–19, given that the resistance is 35 Ω and the characteristic time is 7.5 ms.

SOLUTION
Solving $\tau = L/R$ for the inductance, we get

$$L = \tau R = (7.5 \times 10^{-3}\,\text{s})(35\,\Omega) = 0.26\,\text{H}$$

In the next Example we consider a circuit with one inductor and two resistors. By replacing the resistors with their equivalent resistance, however, we can still treat the circuit as a simple RL circuit.

▲ **FIGURE 23–19 An *RL* circuit**

The characteristic time for an RL circuit is $\tau = L/R$. After the switch has been closed for a time much greater than τ, the current in the circuit is simply $I = \mathcal{E}/R$. That is, when the current is steady, the inductor behaves like a zero-resistance wire.

PROBLEM-SOLVING NOTE

Opening or Closing a Switch on an *RL* Circuit

When the switch is first closed in an RL circuit, the inductor has a back emf equal to that of the battery. Long after the switch is closed current flows freely through the inductor, with no resistance at all.

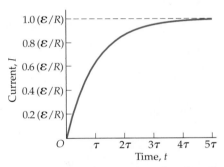

▲ **FIGURE 23–20 Current as a function of time in an *RL* circuit**

Notice that the current approaches $\mathcal{E}/R$ after several characteristic time intervals, $\tau = L/R$, have elapsed.

EXAMPLE 23–6 *RL* IN PARALLEL

The circuit shown in the sketch on the next page consists of a 12-V battery, a 56-mH inductor, and two 150-Ω resistors in parallel. **(a)** Find the characteristic time for this circuit. What is the current in this circuit **(b)** one characteristic time interval after the switch is closed and **(c)** a long time after the switch is closed?

PICTURE THE PROBLEM
The circuit in question is shown in diagram (a). Noting that two resistors R connected in parallel have an equivalent resistance equal to $R/2$, we can replace the original circuit with the equivalent RL circuit shown in diagram (b).

CONTINUED ON NEXT PAGE

CONTINUED FROM PREVIOUS PAGE

STRATEGY

a. We can find the characteristic time using $\tau = L/R_{eq}$, where $R_{eq} = R/2 = 75\ \Omega$.

b. To find the current after one characteristic time interval, we substitute $t = \tau$ in Equation 23–16.

c. A long time after the switch has been closed $(t \rightarrow \infty)$, the exponential term in Equation 23–16 is essentially zero; hence, the current in this case is simply $I = \mathcal{E}/R_{eq}$.

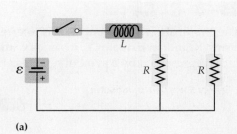

(a)

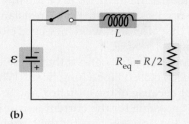

(b)

SOLUTION

Part (a)

1. Calculate the characteristic time, $\tau = L/R_{eq}$, using 75 Ω for the resistance:

$$\tau = \frac{L}{R_{eq}} = \frac{56 \times 10^{-3}\ \text{H}}{75\ \Omega} = 7.5 \times 10^{-4}\ \text{s}$$

Part (b)

2. Substitute $t = \tau$ in Equation 23–16:

$$I = \frac{\mathcal{E}}{R_{eq}}(1 - e^{-t/\tau}) = \left(\frac{12\ \text{V}}{75\ \Omega}\right)(1 - e^{-1}) = 0.10\ \text{A}$$

Part (c)

3. Substitute $t \rightarrow \infty$ in Equation 23–16:

$$I = \frac{\mathcal{E}}{R_{eq}}(1 - e^{-t/\tau}) = \frac{\mathcal{E}}{R_{eq}}(1 - e^{-\infty}) = \frac{\mathcal{E}}{R_{eq}} = \frac{12\ \text{V}}{75\ \Omega} = 0.16\ \text{A}$$

INSIGHT

Notice that the current rises to almost two-thirds of its final value after just one characteristic time interval. It takes only a few more time intervals for the current to essentially "saturate" at its final value, $\mathcal{E}/R_{eq}$.

PRACTICE PROBLEM

If L is doubled to $2(56\ \text{mH}) = 112\ \text{mH}$, will the current after one characteristic time interval be greater than, less than, or the same as that found in part (b)?   [**Answer:** The current is the same when $t = \tau$ because $\mathcal{E}$ and R are unchanged. Note that the value of τ is doubled, however, so it actually takes twice as long for the current to rise to 0.10 A.]

Some related homework problems: Problem 53, Problem 55

23–9 Energy Stored in a Magnetic Field

Work is required to establish a current in an inductor. For example, consider a circuit that consists of just two elements—a battery and an inductor. Even though there are no resistors in the circuit, the battery must still do electrical work to force charge to flow through the inductor, in opposition to its self-induced back emf. What happens to the energy expended to increase the current? It is not lost, nor is it dissipated, as there are no losses in this circuit. Instead, it is stored in the magnetic field, just as the energy in a capacitor is stored in its electric field.

To see how this energy storage works, imagine increasing the current in an inductor L from $I_i = 0$ to $I_f = I$ in the time $\Delta t = T$. The magnitude of the emf required to accomplish this increase in current is

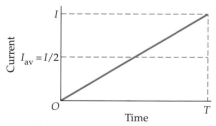

▲ **FIGURE 23–21 Current in an RL circuit with R = 0**

In a circuit with only a battery and an inductor, the current rises linearly with time, because the characteristic time $(\tau = L/R)$ with $R = 0$ is infinite. If the current after a time T is I, the average current during this time is $I_{av} = I/2$.

$$\mathcal{E} = L\frac{\Delta I}{\Delta t} = L\frac{I - 0}{T} = \frac{LI}{T} \qquad \text{23–17}$$

During this time the average current is $I/2$, as indicated in **Figure 23–21**. Thus, the average power supplied by the emf, $P_{av} = I_{av}V$, is

$$P_{av} = \tfrac{1}{2}I\mathcal{E} = \tfrac{1}{2}LI^2/T \qquad \text{23–18}$$

The total *energy* delivered by the emf is simply the average power times the total time; hence, the energy U required to increase the current in the inductor from 0 to I is

$$U = P_{av}T$$

Using the result given in Equation 23–18, we find

Energy Stored in an Inductor

$$U = \tfrac{1}{2}LI^2 \qquad\qquad\qquad 23\text{–}19$$

SI unit: joule, J

Note the similarity of Equation 23–19 to $U = \tfrac{1}{2}CV^2$ for the energy stored in a capacitor.

EXERCISE 23–5

An inductor carrying a current of 2.5 A stores 0.10 J of energy. What is the inductance of this inductor?

SOLUTION

Solving $U = \tfrac{1}{2}LI^2$ for the inductance, we find

$$L = \frac{2U}{I^2} = \frac{2(0.10 \text{ J})}{(2.5 \text{ A})^2} = 0.032 \text{ H}$$

As mentioned previously, the energy of an inductor is stored in its magnetic field. To be specific, consider a solenoid of length ℓ with n turns per unit length. The inductance of this solenoid, as determined in Section 23–7, is

$$L = \mu_0 n^2 A\ell$$

Therefore, the energy stored in the solenoid's magnetic field when it carries a current I is

$$U = \tfrac{1}{2}LI^2 = \tfrac{1}{2}(\mu_0 n^2 A\ell)I^2$$

Recalling that the magnetic field inside a solenoid has a magnitude given by $B = \mu_0 nI$, we can write the energy as follows:

$$U = \frac{1}{2\mu_0}B^2 A\ell$$

Finally, we note that the volume where the magnetic field exists—that is, the volume inside the solenoid—is area times length, $A\ell$. Therefore, the magnetic energy per volume is

$$u_B = \frac{\text{magnetic energy}}{\text{volume}} = \frac{B^2}{2\mu_0} \qquad\qquad 23\text{–}20$$

Although derived for the case of an ideal solenoid, this expression applies to any magnetic field no matter how it is generated.

Finally, recall that in the case of an electric field we found that the energy density was $u_E = \varepsilon_0 E^2/2$. Note, in both cases, that the energy density depends on the magnitude of the field squared. We shall return to these expressions in the next chapter and again when we study electromagnetic waves, such as radio waves and light, in Chapter 25.

PROBLEM-SOLVING NOTE

Energy Stored in an Inductor

Remember that an inductor carrying a current always stores energy in its magnetic field. Thus, energy is required to establish a current in an inductor, and energy is released when the current in an inductor decreases.

EXAMPLE 23–7   AN UNKNOWN RESISTANCE

After the switch in the circuit shown on the next page has been closed a long time, the energy stored in the inductor is 3.11×10^{-3} J. What is the value of the resistance R?

PICTURE THE PROBLEM

Our sketch shows the circuit for this system, with a 36.0-V battery, a 92.5-Ω resistor, and a 75.0-mH inductor all connected in series with an unknown resistor, R. Initially, the switch is open, but after it has been closed a long time the current settles down to a constant value, as does the amount of energy stored in the inductor.

CONTINUED ON NEXT PAGE

CONTINUED FROM PREVIOUS PAGE

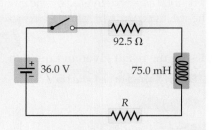

STRATEGY
First, the energy stored in the inductor, $U = \frac{1}{2}LI^2$, gives us the current in the circuit.

Second, we note that after the switch has been closed a long time the current has a constant value. Thus, there is no opposing emf produced by the inductor. As a result, the final current is simply the emf divided by the equivalent resistance of two resistors in series; $I = \mathcal{E}/(92.5\ \Omega + R)$. This relation can be used to solve for R.

SOLUTION

1. Use the energy stored in the inductor to solve for the current in the circuit:

$$U = \frac{1}{2}LI^2$$

$$I = \sqrt{\frac{2U}{L}}$$

2. Substitute numerical values:

$$I = \sqrt{\frac{2U}{L}} = \sqrt{\frac{2(3.11 \times 10^{-3}\ \text{J})}{75.0 \times 10^{-3}\ \text{H}}} = 0.288\ \text{A}$$

3. Use Ohm's law to solve for the resistance R:

$$I = \frac{\mathcal{E}}{92.5\ \Omega + R}$$

$$R = \frac{\mathcal{E}}{I} - 92.5\ \Omega$$

4. Substitute numerical values:

$$R = \frac{36.0\ \text{V}}{0.288\ \text{A}} - 92.5\ \Omega = 32.5\ \Omega$$

INSIGHT
Note that as long as the switch is closed, energy is dissipated in the resistors at a constant rate. On the other hand, *no* energy is dissipated in the inductor—it simply stores a constant amount of energy, like a mass moving with a constant speed.

PRACTICE PROBLEM
If R is increased, does the energy stored in the inductor increase, decrease, or stay the same? Calculate the stored energy with $R = 106\ \Omega$ as a check on your answer.   [**Answer:** With a larger R, the current is less; hence U is less. With $R = 106\ \Omega$ we find $U = 1.23 \times 10^{-3}\ \text{J}$.]

Some related homework problems: Problem 59, Problem 62

23–10 Transformers

In electrical applications it is often useful to be able to change the voltage from one value to another. For example, high-voltage power lines may operate at voltages as high as 750,000 V, but before the electrical power can be used in the home it must be "stepped down" to 120 V. Similarly, the 120 V from a wall socket may be stepped down again to 9 V or 12 V to power a portable CD player, or stepped up to give the 15,000 V needed in a television tube. The electrical device that performs these voltage conversions is called a **transformer.**

The basic idea of a transformer has already been described in connection with Faraday's induction experiment in Section 23–1. In fact, the device shown in Figure 23–1 can be thought of as a crude transformer. Basically, the magnetic flux created by a primary coil induces a voltage in a secondary coil. If the number of turns is different in the two coils, however, the voltages will be different as well.

To see how this works, consider the system shown in **Figure 23–22**. Here an ac generator produces an alternating current in the primary circuit of voltage V_p. The primary circuit then loops around an iron core with N_p turns. The iron core intensifies and concentrates the magnetic flux and ensures, at least to a good approximation, that the secondary coil experiences the same magnetic flux as the primary coil. The secondary coil has N_s turns around its iron core and is connected to a secondary circuit that may operate a CD player, a lightbulb, or some other device. We would now like to determine the voltage, V_s, in the secondary circuit.

▲ **FIGURE 23–22 The basic elements of a transformer**
An alternating current in the primary circuit creates an alternating magnetic flux and, hence, an alternating induced emf in the secondary circuit. The ratio of the emfs in the two circuits, V_s/V_p, is equal to the ratio of the number of turns in each circuit, N_s/N_p.

To find this voltage, we apply Faraday's law of induction to each of the coils. For the primary coil we have

$$\mathcal{E}_p = -N_p \frac{\Delta \Phi_p}{\Delta t}$$

Similarly, the emf of the secondary coil is

$$\mathcal{E}_s = -N_s \frac{\Delta \Phi_s}{\Delta t}$$

Now, recall that the magnetic fluxes are equal, $\Delta \Phi_p = \Delta \Phi_s$, because of the iron core. As a result, we can divide these two equations (which cancels the flux) to find

$$\frac{\mathcal{E}_p}{\mathcal{E}_s} = \frac{N_p}{N_s}$$

Assuming that the resistance in the coils is negligible, so that the emf in each is essentially the same as its voltage, we obtain the following result:

Transformer Equation

$$\frac{V_p}{V_s} = \frac{N_p}{N_s}$$ 23–21

Equation 23–21 is known as the **transformer equation.**
    It follows that the voltage in the secondary circuit is simply

$$V_s = V_p \left(\frac{N_s}{N_p} \right)$$

Therefore, if the number of turns in the secondary coil is less than the number of turns in the primary coil, the voltage is stepped down, and $V_s < V_p$. Conversely, if the number of turns in the secondary coil is higher, the voltage is stepped up, and $V_s > V_p$.
    Now, before you begin to think that transformers give us something for nothing, we note that there is more to the story. Because energy must always be conserved, the average power in the primary circuit must be the same as the average power in the secondary circuit. Since power can be written as $P = IV$, it follows that

$$I_p V_p = I_s V_s$$

Therefore, the ratio of the current in the secondary coil to that in the primary coil is

Transformer Equation (Current and Voltage)

$$\frac{I_s}{I_p} = \frac{V_p}{V_s} = \frac{N_p}{N_s}$$ 23–22

Thus, if the voltage is stepped up, the current is stepped down.
    Suppose, for example, that the number of turns in the secondary coil is twice the number of turns in the primary coil. As a result, the transformer doubles the voltage in the secondary circuit, $V_s = 2 V_p$, and at the same time halves the current, $I_s = I_p/2$. In general, there is always a tradeoff between voltage and current in a transformer:

> If a transformer increases the voltage by a given factor, it decreases
> the current by the same factor. Similarly, if it decreases the voltage,
> it increases the current.

    This behavior is similar to that of a lever, in which there is a tradeoff between the force that can be exerted and the distance through which it is exerted. In the case of a lever, as in the case of a transformer, the tradeoff is due to energy conservation: The work done on one end of the lever, $F_1 d_1$, must be equal to the work done on the other end of the lever, $F_2 d_2$.

PROBLEM-SOLVING NOTE

**Energy Conservation
in Transformers**

Energy conservation requires that the product of I and V be the same for each coil of a transformer. Thus, as the voltage of the secondary goes up with the number of turns, the current decreases by the same factor.

ACTIVE EXAMPLE 23–3   FIND THE NUMBER OF TURNS ON THE SECONDARY

A common summertime sound in many backyards is the *zap* heard when an unfortunate insect flies into a high-voltage "bug zapper." Typically, such devices operate with voltages of about 4000 V obtained from a transformer plugged into a standard 120-V outlet. How many turns are on the secondary coil of the transformer if the primary coil has 27 turns?

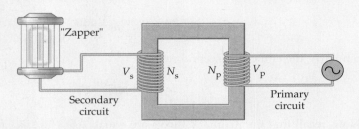

SOLUTION *(Test your understanding by performing the calculations indicated in each step.)*

1. Solve the transformer equation for the number of turns in the secondary coil: $N_s = N_p(V_s/V_p)$

2. Substitute numerical values: $N_s = 900$

INSIGHT

Of course, any transformer with a turns ratio of $N_s/N_p = 900/27$ will produce the same secondary voltage.

YOUR TURN

Suppose the primary voltage is doubled and the number of turns in the primary circuit is quadrupled. By what factor must the number of turns in the secondary circuit be changed if the secondary voltage is to remain the same?

*(Answers to **Your Turn** problems are given in the back of the book.)*

Notice that the operation of a transformer depends on a *changing* magnetic flux to create an induced emf in the secondary coil. If the current is constant in direction and magnitude, there is simply no induced emf. This is an important advantage that ac circuits have over dc circuits and is one reason that most electrical power systems today operate with alternating currents.

Finally, transformers play an important role in the *transmission* of electrical energy from the power plants that produce it to the communities and businesses where it is used. As was pointed out in Equation 21–3, the resistance of a wire is directly proportional to its length. Therefore, when electrical energy is transmitted over a large distance, the finite resistivity of the wires that carry the current becomes significant. If the wire carries a current I and has a resistance R, the power dissipated as waste heat is $P = I^2R$. One way to reduce this energy loss in a given wire is to reduce the current, I. A transformer that steps up the voltage of a power plant by a factor of 15 (from 10,000 V to 150,000 V, for example) at the same time reduces the current by a factor of 15 and reduces the energy dissipation by a factor of $15^2 = 225$. When the electrical energy reaches the location where it is to be used, step-down transformers can convert the voltage to levels (such as 120 V or 240 V) typically used in the home or workplace.

REAL-WORLD PHYSICS

High-voltage electric power transmission

▲ The transmission of electric power over long distances would not be feasible without transformers. A step-up transformer near the power plant (left) boosts the plant's output voltage from 12,000 V to the 240,000 V carried by high-voltage lines that crisscross the country (center). A series of step-down transformers near the destination reduce the voltage, first to 2400 V at local substations for distribution to neighborhoods, and then to the 240 V supplied to most houses. The familiar gray cylinders commonly seen on utility poles (right) are the transformers responsible for this last voltage reduction.

THE BIG PICTURE PUTTING PHYSICS IN CONTEXT

LOOKING BACK	LOOKING AHEAD

The concept of electric flux, so useful in Gauss's law in Section 19–7, is extended to magnetic flux in this chapter, where it plays a key role in Faraday's law.

Work, energy, and power (Chapters 7 and 8) are discussed in relation to electrical energy in Section 23–5. These concepts are also important in understanding the operation of electric motors and generators in Section 23–6.

In Chapter 20 we introduced the concept of a capacitor (C), and showed how a capacitor stores energy in its electric field. Here we generalize this concept to inductors (L), which store energy in their magnetic fields. We also used our experience with RC circuits (Chapter 21) to gain a better understanding of RL circuits.

Inductors are important elements in any electric circuit, but especially in alternating current (ac) circuits. This is discussed in detail in Section 24–4.

A particular type of circuit containing an inductor—the RLC circuit—plays an important role in Chapter 24. For example, we show in Section 24–5 how an RLC circuit behaves at high and low frequencies. Then, in Section 24–6, we see that an RLC circuit can show resonance, in analogy with a mass on a spring. In this analogy, the inductor plays the role of the mass.

The detailed connections between electric and magnetic fields developed in Chapters 22 and 23 are central to understanding electromagnetic waves, as described in Chapter 25. It would be hard to overstate the importance of electromagnetic waves, which cover a broad spectrum that includes radio waves, microwaves, infrared rays, visible light, ultraviolet light, X-rays, and gamma rays.

CHAPTER SUMMARY

23–1 INDUCED ELECTROMOTIVE FORCE

A changing magnetic field can induce a current in a circuit. The driving force behind the current is referred to as the induced electromotive force.

Basic Features of Magnetic Induction
An induced current occurs when there is a *change* in the magnetic field. The magnitude of the induced current depends on the *rate* of change of the magnetic field.

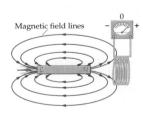

23–2 MAGNETIC FLUX

Magnetic flux is a measure of the number of magnetic field lines that cross a given area.

Magnetic Flux, Defined
If a magnetic field of strength B crosses an area A at an angle θ relative to the area's normal, the magnetic flux is

$$\Phi = (B \cos \theta)A = BA \cos \theta \qquad \text{23–1}$$

Units of Magnetic Flux
Magnetic flux is measured in webers (Wb) defined as follows:

$$1 \text{ Wb} = 1 \text{ T} \cdot \text{m}^2 \qquad \text{23–2}$$

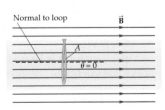

23–3 FARADAY'S LAW OF INDUCTION

Faraday's law of induction gives precise mathematical form to the basic features of magnetic induction discussed in Section 23–1. Faraday's law is a law of nature, not an approximate rule of thumb.

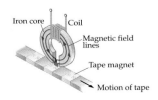

Faraday's Law

If the magnetic flux in a coil of N turns changes by the amount $\Delta\Phi$ in the time Δt, the induced emf is

$$\mathcal{E} = -N\frac{\Delta\Phi}{\Delta t} = -N\frac{\Phi_{\text{final}} - \Phi_{\text{initial}}}{t_{\text{final}} - t_{\text{initial}}} \qquad \text{23–3}$$

Magnitude of the Induced emf

In many cases we are concerned only with the magnitude of the magnetic flux and the induced current. In such cases we use the following form of Faraday's law:

$$|\mathcal{E}| = N\left|\frac{\Delta\Phi}{\Delta t}\right| = N\left|\frac{\Phi_{\text{final}} - \Phi_{\text{initial}}}{t_{\text{final}} - t_{\text{initial}}}\right| \qquad \text{23–4}$$

23–4 LENZ'S LAW

Lenz's law states that an induced current flows in the direction that opposes the change that caused the current.

Eddy Currents

Eddy currents are circulating electric currents formed in a conducting material that experiences a changing magnetic flux.

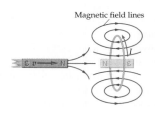

Magnetic field lines

23–5 MECHANICAL WORK AND ELECTRICAL ENERGY

With the help of a magnetic field, mechanical work can be converted into electrical energy in the form of currents flowing through a circuit.

Motional emf

If a conductor of length ℓ is moved at right angles to a magnetic field of strength B with a speed v, the magnitude of the induced emf in the conductor is

$$|\mathcal{E}| = Bv\ell \qquad \text{23–5}$$

The corresponding electric field is

$$E = Bv \qquad \text{23–6}$$

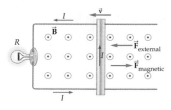

Mechanical Work/Electrical Energy

Assuming no frictional losses, the work done in moving a conductor through a magnetic field is equal to the electrical energy that can be delivered to a circuit.

23–6 GENERATORS AND MOTORS

Generators and motors are practical devices for converting between electrical and mechanical energy.

Electric Generators

Electric generators use mechanical work to produce electrical energy. The emf produced by a generator is

$$\mathcal{E} = NBA\omega \sin \omega t \qquad \text{23–11}$$

where N is the number of turns in the coil, B is the magnetic field, A is the area of the coil, and ω is the angular speed of the coil.

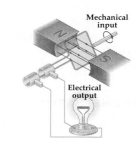

Mechanical input

Electrical output

Electric Motors

An electric motor is basically an electric generator operated in reverse; it uses electrical energy to produce mechanical work.

23–7 INDUCTANCE

Self-inductance, or inductance for short, is the effect produced when a coil with a changing current induces an emf in itself.

Inductance

Inductance, L, is defined as follows:

$$|\mathcal{E}| = L\left|\frac{\Delta I}{\Delta t}\right| \qquad \text{23–12}$$

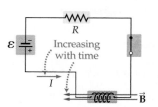

Increasing with time

The SI unit of inductance is the henry (H): $1\,\text{H} = 1\,\text{V}\cdot\text{s/A}$.

Inductance of a Solenoid

The inductance of a solenoid of length ℓ with N turns of wire and a cross-sectional area A is

$$L = \mu_0 n^2 A\ell = \mu_0\left(\frac{N^2}{\ell}\right)A \qquad 23\text{–}14$$

Note that the number of turns per length is $n = N/\ell$.

23–8 *RL* CIRCUITS

A simple *RL* circuit consists of a battery, a switch, a resistor R, and an inductor L connected in series.

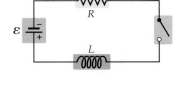

Time Constant

The characteristic time over which the current changes in an *RL* circuit is

$$\tau = \frac{L}{R} \qquad 23\text{–}15$$

Increasing Current Versus Time

When the switch is closed in an *RL* circuit at time $t = 0$, the current increases with time as follows:

$$I = \frac{\mathcal{E}}{R}(1 - e^{-t/\tau}) = \frac{\mathcal{E}}{R}(1 - e^{-tR/L}) \qquad 23\text{–}16$$

In this expression ε is the emf of the battery.

23–9 ENERGY STORED IN A MAGNETIC FIELD

Energy can be stored in a magnetic field, just as it is in an electric field.

Energy Stored in an Inductor

An inductor of inductance L carrying a current I stores the energy

$$U = \tfrac{1}{2}LI^2 \qquad 23\text{–}19$$

Energy Density

Whenever a magnetic field of strength B exists in a given region, there is a corresponding magnetic energy density given by

$$u_B = \frac{\text{magnetic energy}}{\text{volume}} = \frac{B^2}{2\mu_0} \qquad 23\text{–}20$$

23–10 TRANSFORMERS

A transformer is an electrical device used to convert the voltage and current in one circuit to a different voltage and current in another circuit using magnetic induction.

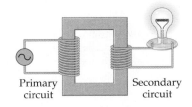

Primary circuit   Secondary circuit

Transformer Equation

The equation relating the voltage V, current I, and number of turns N in the primary (p) and secondary (s) circuits of a transformer is

$$\frac{I_s}{I_p} = \frac{V_p}{V_s} = \frac{N_p}{N_s} \qquad 23\text{–}22$$

PROBLEM-SOLVING SUMMARY

Type of Problem	Relevant Physical Concepts	Related Examples
Calculate the magnetic flux.	The magnetic flux of a field B passing through an area A at an angle θ to the normal is $\Phi = BA \cos \theta$.	Example 23–1
Find the induced emf.	Faraday's law states that the induced emf is proportional to the rate of change of magnetic flux with time, $\mathcal{E} = -N\dfrac{\Delta\Phi}{\Delta t}$. In the special case of a rotating coil, the induced emf is $\mathcal{E} = NBA\omega \sin \omega t$.	Examples 23–2, 23–4
Calculate the inductance of a solenoid.	A solenoid with n turns per length, area A, and length ℓ has an inductance given by $L = \mu_0 n^2 A\ell$.	Example 23–5 Active Example 23–2

Find the current as a function of time in an RL circuit.	The time constant of an RL circuit is $\tau = L/R$. When the current increases from zero, its time dependence is $$I = \frac{\mathcal{E}}{R}(1 - e^{-t/\tau}).$$	Example 23–6
Relate the energy stored in an inductor to its inductance and current.	The energy stored in an inductor L carrying a current I is $U = \frac{1}{2}LI^2$.	Example 23–7
Relate voltage and current to the number of turns in the primary and secondary coils of a transformer.	The voltage in the secondary coil is proportional to the number of turns in the secondary coil and inversely proportional to the number of turns in the primary coil.	Active Example 23–3
	The current and voltage in the secondary coil are inversely proportional to each other.	

CONCEPTUAL QUESTIONS

For instructor-assigned homework, go to www.masteringphysics.com

(Answers to odd-numbered Conceptual Questions can be found in the back of the book.)

1. Explain the difference between a magnetic field and a magnetic flux.

2. A metal ring with a break in its perimeter is dropped from a field-free region of space into a region with a magnetic field. What effect does the magnetic field have on the ring?

3. In a common classroom demonstration, a magnet is dropped down a long, vertical copper tube. The magnet moves very slowly as it moves through the tube, taking several seconds to reach the bottom. Explain this behavior.

4. Many equal-arm balances have a small metal plate attached to one of the two arms. The plate passes between the poles of a magnet mounted in the base of the balance. Explain the purpose of this arrangement.

5. **Figure 23–23** shows a vertical iron rod with a wire coil of many turns wrapped around its base. A metal ring slides over the rod and rests on the wire coil. Initially the switch connecting the coil to a battery is open, but when it is closed, the ring flies into the air. Explain why this happens.

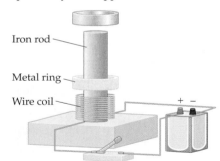

Iron rod

Metal ring

Wire coil

▲ **FIGURE 23–23** Conceptual Questions 5 and 6

6. Referring to Conceptual Question 5, suppose the metal ring has a break in its circumference. Describe what happens when the switch is closed in this case.

7. A metal rod of resistance R can slide without friction on two zero-resistance rails, as shown in **Figure 23–24**. The rod and the rails are immersed in a region of constant magnetic field pointing out of the page. Describe the motion of the rod when the switch is closed. Your discussion should include the effects of motional emf.

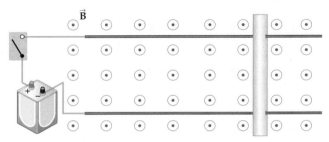

▲ **FIGURE 23–24** Conceptual Question 7

8. A penny is placed on edge in the powerful magnetic field of an MRI solenoid. If the penny is tipped over, it takes several seconds for it to land on one of its faces. Explain.

9. Recently, NASA tested a power generation system that involves connecting a small satellite to the space shuttle with a conducting wire several miles long. Explain how such a system can generate electrical power.

10. Explain what happens when the angular speed of the coil in an electric generator is increased.

11. The inductor in an RL circuit determines how long it takes for the current to reach a given value, but it has no effect on the final value of the current. Explain.

12. When the switch in a circuit containing an inductor is opened, it is common for a spark to jump across the contacts of the switch. Why?

PROBLEMS AND CONCEPTUAL EXERCISES

Note: Answers to odd-numbered Problems and Conceptual Exercises can be found in the back of the book. **IP** *denotes an integrated problem, with both conceptual and numerical parts;* **BIO** *identifies problems of biological or medical interest;* **CE** *indicates a conceptual exercise.* **Predict/Explain** *problems ask for two responses:* **(a)** *your prediction of a physical outcome, and* **(b)** *the best explanation among three provided. On all problems, red bullets (•, ••, •••) are used to indicate the level of difficulty.*

SECTION 23–2 MAGNETIC FLUX

1. • A 0.055-T magnetic field passes through a circular ring of radius 3.1 cm at an angle of 16° with the normal. Find the magnitude of the magnetic flux through the ring.

2. • A uniform magnetic field of 0.0250 T points vertically upward. Find the magnitude of the magnetic flux through each of the five sides of the open-topped rectangular box shown in **Figure 23–25**, given that the dimensions of the box are $L = 32.5$ cm, $W = 12.0$ cm, and $H = 10.0$ cm.

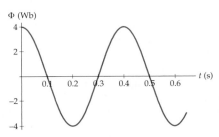

▲ **FIGURE 23–25** Problem 2

3. • A magnetic field is oriented at an angle of 47° to the normal of a rectangular area 5.1 cm by 6.8 cm. If the magnetic flux through this surface has a magnitude of 4.8×10^{-5} T·m², what is the strength of the magnetic field?

4. • Find the magnitude of the magnetic flux through the floor of a house that measures 22 m by 18 m. Assume that the Earth's magnetic field at the location of the house has a horizontal component of 2.6×10^{-5} T pointing north, and a downward vertical component of 4.2×10^{-5} T.

5. • **MRI Solenoid** The magnetic field produced by an MRI solenoid 2.5 m long and 1.2 m in diameter is 1.7 T. Find the magnitude of the magnetic flux through the core of this solenoid.

6. •• At a certain location, the Earth's magnetic field has a magnitude of 5.9×10^{-5} T and points in a direction that is 72° below the horizontal. Find the magnitude of the magnetic flux through the top of a desk at this location that measures 130 cm by 82 cm.

7. •• **IP** A solenoid with 385 turns per meter and a diameter of 17.0 cm has a magnetic flux through its core of magnitude 1.28×10^{-4} T·m². **(a)** Find the current in this solenoid. **(b)** How would your answer to part (a) change if the diameter of the solenoid were doubled? Explain.

8. ••• A single-turn square loop of side L is centered on the axis of a long solenoid. In addition, the plane of the square loop is perpendicular to the axis of the solenoid. The solenoid has 1250 turns per meter and a diameter of 6.00 cm, and carries a current of 2.50 A. Find the magnetic flux through the loop when **(a)** $L = 3.00$ cm, **(b)** $L = 6.00$ cm, and **(c)** $L = 12.0$ cm.

SECTION 23-3   FARADAY'S LAW OF INDUCTION

9. • A 0.45-T magnetic field is perpendicular to a circular loop of wire with 53 turns and a radius of 15 cm. If the magnetic field is reduced to zero in 0.12 s, what is the magnitude of the induced emf?

10. • Figure 23–26 shows the magnetic flux through a coil as a function of time. At what times shown in this plot do **(a)** the magnetic flux and **(b)** the induced emf have the greatest magnitude?

Φ (Wb)

▲ **FIGURE 23–26** Problems 10 and 15

11. • Figure 23–27 shows the magnetic flux through a single-loop coil as a function of time. What is the induced emf in the coil at **(a)** $t = 0.050$ s, **(b)** $t = 0.15$ s, and **(c)** $t = 0.50$ s?

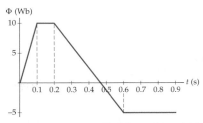

▲ **FIGURE 23–27** Problems 11 and 14

12. •• **CE** A wire loop is placed in a magnetic field that is perpendicular to its plane. The field varies with time as shown in **Figure 23–28.** Rank the six regions of time in order of increasing magnitude of the induced emf. Indicate ties where appropriate.

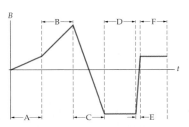

▲ **FIGURE 23–28** Problem 12

13. •• **CE Figure 23–29** shows four different situations in which a metal ring moves to the right with constant speed through a region with a varying magnetic field. The intensity of the color indicates the intensity of the field, and in each case the field either increases or decreases at a uniform rate from the left edge of the colored region to the right edge. The direction of the field in each region is indicated. For each of the four cases, state whether the induced emf is clockwise, counterclockwise, or zero.

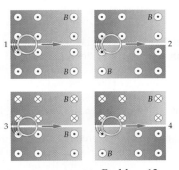

▲ **FIGURE 23–29** Problem 13

14. •• **IP** The magnetic flux through a single-loop coil is given by Figure 23–27. **(a)** Is the magnetic flux at $t = 0.25$ s greater than, less than, or the same as the magnetic flux at $t = 0.55$ s? Explain. **(b)** Is the induced emf at $t = 0.25$ s greater than, less than, or the same as the induced emf at $t = 0.55$ s? Explain. **(c)** Calculate the induced emf at the times $t = 0.25$ s and $t = 0.55$ s.

15. •• **IP** Consider a single-loop coil whose magnetic flux is given by Figure 23–26. **(a)** Is the magnitude of the induced emf in this coil greater near $t = 0.4$ s or near $t = 0.5$ s? Explain. **(b)** At what times in this plot do you expect the induced emf in the coil to have a maximum magnitude? Explain. **(c)** Estimate the induced emf in the coil at times near $t = 0.3$ s, $t = 0.4$ s, and $t = 0.5$ s.

16. •• A single conducting loop of wire has an area of 7.2×10^{-2} m² and a resistance of 110 Ω. Perpendicular to the plane of the loop is a magnetic field of strength 0.48 T. At what rate (in T/s) must this field change if the induced current in the loop is to be 0.32 A?

17. •• The area of a 120-turn coil oriented with its plane perpendicular to a 0.20-T magnetic field is 0.050 m². Find the average induced emf in this coil if the magnetic field reverses its direction in 0.34 s.

18. •• An emf is induced in a conducting loop of wire 1.22 m long as its shape is changed from square to circular. Find the average magnitude of the induced emf if the change in shape occurs in 4.25 s and the local 0.125-T magnetic field is perpendicular to the plane of the loop.

19. •• A magnetic field increases from 0 to 0.25 T in 1.8 s. How many turns of wire are needed in a circular coil 12 cm in diameter to produce an induced emf of 6.0 V?

SECTION 23–4 LENZ'S LAW

20. • **CE Predict/Explain** A metal ring is dropped into a localized region of constant magnetic field, as indicated in **Figure 23–30**. The magnetic field is zero above and below the region where it is finite. **(a)** For each of the three indicated locations (1, 2, and 3), is the induced current clockwise, counterclockwise, or zero? **(b)** Choose the *best explanation* from among the following:
   I. Clockwise at 1 to oppose the field; zero at 2 because the field is uniform; counterclockwise at 3 to try to maintain the field.
   II. Counterclockwise at 1 to oppose the field; zero at 2 because the field is uniform; clockwise at 3 to try to maintain the field.
   III. Clockwise at 1 to oppose the field; clockwise at 2 to maintain the field; clockwise at 3 to oppose the field.

▲ **FIGURE 23–30** Problems 20 and 21

21. • **CE Predict/Explain** A metal ring is dropped into a localized region of constant magnetic field, as indicated in Figure 23–30. The magnetic field is zero above and below the region where it is finite. **(a)** For each of the three indicated locations (1, 2, and 3), is the magnetic force exerted on the ring upward, downward, or zero? **(b)** Choose the *best explanation* from among the following:
   I. Upward at 1 to oppose entering the field; zero at 2 because the field is uniform; downward at 3 to help leaving the field.
   II. Upward at 1 to oppose entering the field; upward at 2 where the field is strongest; upward at 3 to oppose leaving the field.
   III. Upward at 1 to oppose entering the field; zero at 2 because the field is uniform; upward at 3 to oppose leaving the field.

22. • **CE Predict/Explain Figure 23–31** shows two metal disks of the same size and material oscillating in and out of a region with a magnetic field. One disk is solid; the other has a series of slots.

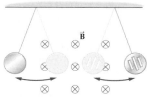

▲ **FIGURE 23–31** Problems 22, 23, and 24

(a) Is the retarding effect of eddy currents on the solid disk greater than, less than, or equal to the retarding effect on the slotted disk? **(b)** Choose the *best explanation* from among the following:
   I. The solid disk experiences a greater retarding force because eddy currents in it flow freely and are not interrupted by the slots.
   II. The slotted disk experiences the greater retarding force because the slots allow more magnetic field to penetrate the disk.
   III. The disks are the same size and made of the same material; therefore, they experience the same retarding force.

23. • **CE** Consider the solid disk in Figure 23–31. When this disk has swung to the right as far as it can go, is the induced current in it a maximum or a minimum? Explain.

24. • **CE Predict/Explain** **(a)** As the solid metal disk in Figure 23–31 swings to the right, from the region with no field into the region with a finite magnetic field, is the induced current in the disk clockwise, counterclockwise, or zero? **(b)** Choose the *best explanation* from among the following:
   I. The induced current is clockwise, since this produces a field within the disk in the same direction as the magnetic field that produced the current.
   II. The induced current is counterclockwise to generate a field within the disk that points out of the page.
   III. The induced current is zero because the disk enters a region where the magnetic field is uniform.

25. • A bar magnet with its north pole pointing downward is falling toward the center of a horizontal conducting ring. As viewed from above, is the direction of the induced current in the ring clockwise or counterclockwise? Explain.

26. • **A Wire Loop and a Magnet** A loop of wire is dropped and allowed to fall between the poles of a horseshoe magnet, as shown in **Figure 23–32**. State whether the induced current in the loop is clockwise or counterclockwise when **(a)** the loop is above the magnet and **(b)** the loop is below the magnet.

▲ **FIGURE 23–32** Problems 26, 27, and 28

27. •• Suppose we change the situation shown in Figure 23–32 as follows: Instead of allowing the loop to fall on its own, we attach a string to it and lower it with constant speed along the path indicated by the dashed line. Is the tension in the string greater than, less than, or equal to the weight of the loop? Give specific answers for times when **(a)** the loop is above the magnet and **(b)** the loop is below the magnet. Explain in each case.

28. •• Rather than letting the loop fall downward in Figure 23–32, suppose we attach a string to it and raise it upward with constant speed along the path indicated by the dashed line. Is the tension in the string greater than, less than, or equal to the weight of the loop? Give specific answers for times when **(a)** the loop is below the magnet and **(b)** the loop is above the magnet. Explain in each case.

29. •• **Figure 23–33** shows a current-carrying wire and a circuit containing a resistor R. **(a)** If the current in the wire is constant, is

the induced current in the circuit clockwise, counterclockwise, or zero? Explain. **(b)** If the current in the wire increases, is the induced current in the circuit clockwise, counterclockwise, or zero? Explain.

▲ **FIGURE 23–33** Problems 29 and 30

30. •• Consider the physical system shown in Figure 23–33. If the current in the wire changes direction, is the induced current in the circuit clockwise, counterclockwise, or zero? Explain.

31. •• A long, straight, current-carrying wire passes through the center of a circular coil. The wire is perpendicular to the plane of the coil. **(a)** If the current in the wire is constant, is the induced emf in the coil zero or nonzero? Explain. **(b)** If the current in the wire increases, is the induced emf in the coil zero or nonzero? Explain. **(c)** Does your answer to part (b) change if the wire no longer passes through the center of the coil but is still perpendicular to its plane? Explain.

32. •• **Figure 23–34** shows a circuit containing a resistor and an uncharged capacitor. Pointing into the plane of the circuit is a uniform magnetic field $\vec{B}$. If the magnetic field reverses direction in a short period of time, which plate of the capacitor (top or bottom) becomes positively charged? Explain.

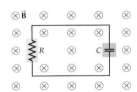

▲ **FIGURE 23–34** Problems 32 and 33

33. •• Referring to Problem 32, which plate of the capacitor (top or bottom) becomes positively charged if the magnetic field increases in magnitude with time? Explain.

34. •• A long, straight wire carries a current I, as indicated in **Figure 23–35**. Three small metal rings are placed near the current-carrying wire (A and C) or directly on top of it (B). If the current in the wire is increasing with time, indicate whether the induced emf in each of the rings is clockwise, counterclockwise, or zero. Explain your answer for each ring.

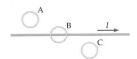

▲ **FIGURE 23–35** Problem 34

SECTION 23–5   MECHANICAL WORK AND ELECTRICAL ENERGY

35. • **CE** A conducting rod slides on two wires in a region with a magnetic field. The two wires are not connected. Is a force required to keep the rod moving with constant speed? Explain.

36. • A metal rod 0.76 m long moves with a speed of 2.0 m/s perpendicular to a magnetic field. If the induced emf between the ends of the rod is 0.45 V, what is the strength of the magnetic field?

37. •• **Airplane emf** A Boeing KC-135A airplane has a wingspan of 39.9 m and flies at constant altitude in a northerly direction with a speed of 850 km/h. If the vertical component of the Earth's magnetic field is 5.0×10^{-6} T, and its horizontal component is 1.4×10^{-6} T, what is the induced emf between the wing tips?

38. •• **IP** Figure 23–36 shows a zero-resistance rod sliding to the right on two zero-resistance rails separated by the distance $L = 0.450$ m. The rails are connected by a 12.5-Ω resistor, and the entire system is in a uniform magnetic field with a magnitude of 0.750 T. **(a)** Find the speed at which the bar must be moved to produce a current of 0.155 A in the resistor. **(b)** Would your answer to part (a) change if the bar was moving to the left instead of to the right? Explain.

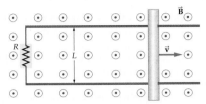

▲ **FIGURE 23–36** Problems 38 and 39

39. •• Referring to part (a) of Problem 38, **(a)** find the force that must be exerted on the rod to maintain a constant current of 0.155 A in the resistor. **(b)** What is the rate of energy dissipation in the resistor? **(c)** What is the mechanical power delivered to the rod?

40. •• **(a)** Find the current that flows in the circuit shown in Example 23–3. **(b)** What speed must the rod have if the current in the circuit is to be 1.0 A?

41. •• Suppose the mechanical power delivered to the rod in Example 23–3 is 8.9 W. Find **(a)** the current in the circuit and **(b)** the speed of the rod.

SECTION 23–6   GENERATORS AND MOTORS

42. • The maximum induced emf in a generator rotating at 210 rpm is 45 V. How fast must the rotor of the generator rotate if it is to generate a maximum induced emf of 55 V?

43. • A rectangular coil 25 cm by 35 cm has 120 turns. This coil produces a maximum emf of 65 V when it rotates with an angular speed of 190 rad/s in a magnetic field of strength B. Find the value of B.

44. • A 1.6-m wire is wound into a coil with a radius of 3.2 cm. If this coil is rotated at 85 rpm in a 0.075-T magnetic field, what is its maximum emf?

45. •• **IP** A circular coil with a diameter of 22.0 cm and 155 turns rotates about a vertical axis with an angular speed of 1250 rpm. The only magnetic field in this system is that of the Earth. At the location of the coil, the horizontal component of the magnetic field is 3.80×10^{-5} T, and the vertical component is 2.85×10^{-5} T. **(a)** Which component of the magnetic field is important when calculating the induced emf in this coil? Explain. **(b)** Find the maximum emf induced in the coil.

46. •• A generator is designed to produce a maximum emf of 170 V while rotating with an angular speed of 3600 rpm. Each coil of the generator has an area of 0.016 m². If the magnetic field used in the generator has a magnitude of 0.050 T, how many turns of wire are needed?

SECTION 23–7   INDUCTANCE

47. • Find the induced emf when the current in a 45.0-mH inductor increases from 0 to 515 mA in 16.5 ms.

48. • How many turns should a solenoid of cross-sectional area 0.035 m² and length 0.22 m have if its inductance is to be 45 mH?

49. •• The inductance of a solenoid with 450 turns and a length of 24 cm is 7.3 mH. **(a)** What is the cross-sectional area of the solenoid? **(b)** What is the induced emf in the solenoid if its current drops from 3.2 A to 0 in 55 ms?

50. •• Determine the inductance of a solenoid with 640 turns in a length of 25 cm. The circular cross section of the solenoid has a radius of 4.3 cm.

51. •• A solenoid with a cross-sectional area of $1.81 \times 10^{-3} \text{ m}^2$ is 0.750 m long and has 455 turns per meter. Find the induced emf in this solenoid if the current in it is increased from 0 to 2.00 A in 45.5 ms.

52. ••• **IP** A solenoid has N turns of area A distributed uniformly along its length, ℓ. When the current in this solenoid increases at the rate of 2.0 A/s, an induced emf of 75 mV is observed. **(a)** What is the inductance of this solenoid? **(b)** Suppose the spacing between coils is doubled. The result is a solenoid that is twice as long but with the same area and number of turns. Will the induced emf in this new solenoid be greater than, less than, or equal to 75 mV when the current changes at the rate of 2.0 A/s? Explain. **(c)** Calculate the induced emf for part (b).

SECTION 23–8   RL CIRCUITS

53. • How long does it take for the current in an RL circuit with $R = 130 \text{ } \Omega$ and $L = 68 \text{ mH}$ to reach half its final value?

54. •• **CE** The four electric circuits shown in **Figure 23–37** have identical batteries, resistors, and inductors. Rank the circuits in order of increasing current supplied by the battery long after the switch is closed. Indicate ties where appropriate.

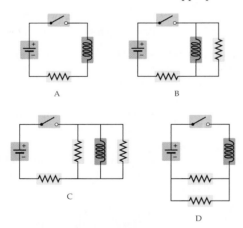

A   B   C   D

▲ **FIGURE 23–37** Problem 54

55. •• The circuit shown in **Figure 23–38** consists of a 6.0-V battery, a 37-mH inductor, and four 55-Ω resistors. **(a)** Find the characteristic time for this circuit. What is the current supplied by this battery **(b)** two characteristic time intervals after closing the switch and **(c)** a long time after the switch is closed?

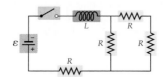

▲ **FIGURE 23–38** Problems 55 and 64

56. •• The current in an RL circuit increases to 95% of its final value 2.24 s after the switch is closed. **(a)** What is the time constant for this circuit? **(b)** If the inductance in the circuit is 0.275 H, what is the resistance?

57. ••• Consider the RL circuit shown in **Figure 23–39**. When the switch is closed, the current in the circuit is observed to increase from 0 to 0.32 A in 0.15 s. **(a)** What is the inductance L? **(b)** How long after the switch is closed does the current have the value 0.50 A? **(c)** What is the maximum current that flows in this circuit?

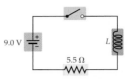

9.0 V   5.5 Ω L

▲ **FIGURE 23–39** Problems 57 and 59

SECTION 23–9   ENERGY STORED IN A MAGNETIC FIELD

58. • **CE** The number of turns per meter in a solenoid of fixed length is doubled. At the same time, the current in the solenoid is halved. Does the energy stored in the inductor increase, decrease, or stay the same? Explain.

59. • Consider the circuit shown in Figure 23–39. Assuming the inductor in this circuit has the value $L = 6.1 \text{ mH}$, how much energy is stored in the inductor after the switch has been closed a long time?

60. • A solenoid is 1.5 m long and has 470 turns per meter. What is the cross-sectional area of this solenoid if it stores 0.31 J of energy when it carries a current of 12 A?

61. •• **Alcator Fusion Experiment** In the Alcator fusion experiment at MIT, a magnetic field of 50.0 T is produced. **(a)** What is the magnetic energy density in this field? **(b)** Find the magnitude of the electric field that would have the same energy density found in part (a).

62. •• **IP** After the switch in **Figure 23–40** has been closed for a long time, the energy stored in the inductor is 0.110 J. **(a)** What is the value of the resistance R? **(b)** If it is desired that more energy be stored in the inductor, should the resistance R be greater than or less than the value found in part (a)? Explain.

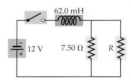

62.0 mH   12 V   7.50 Ω R

▲ **FIGURE 23–40** Problems 62 and 63

63. •• **IP** Suppose the resistor in Figure 23–40 has the value $R = 14 \text{ } \Omega$ and that the switch is closed at time $t = 0$. **(a)** How much energy is stored in the inductor at the time $t = \tau$? **(b)** How much energy is stored in the inductor at the time $t = 2\tau$? **(c)** If the value of R is increased, does the characteristic time, τ, increase or decrease? Explain.

64. •• **IP** Consider the circuit shown in Figure 23–38, which contains a 6.0-V battery, a 37-mH inductor, and four 55-Ω resistors. **(a)** Is more energy stored in the inductor just after the switch is closed or long after the switch is closed? Explain. **(b)** Calculate the energy stored in the inductor one characteristic time interval after the switch is closed. **(c)** Calculate the energy stored in the inductor long after the switch is closed.

65. ••• You would like to store 9.9 J of energy in the magnetic field of a solenoid. The solenoid has 580 circular turns of diameter 7.2 cm distributed uniformly along its 28-cm length. **(a)** How much current is needed? **(b)** What is the magnitude of the magnetic field inside the solenoid? **(c)** What is the energy density (energy/volume) inside the solenoid?

SECTION 23–10   TRANSFORMERS

66. • **CE** Transformer 1 has a primary voltage V_p and a secondary voltage V_s. Transformer 2 has twice the number of turns on both its primary and secondary coils compared with transformer 1. If the primary voltage on transformer 2 is $2V_p$, what is its secondary voltage? Explain.

67. • **CE** Transformer 1 has a primary current I_p and a secondary current I_s. Transformer 2 has twice as many turns on its primary coil as transformer 1, and both transformers have the same number of turns on the secondary coil. If the primary current on transformer 2 is $3I_p$, what is its secondary current? Explain.

68. • The electric motor in a toy train requires a voltage of 3.0 V. Find the ratio of turns on the primary coil to turns on the secondary coil in a transformer that will step the 110-V household voltage down to 3.0 V.

69. • **IP** A disk drive plugged into a 120-V outlet operates on a voltage of 9.0 V. The transformer that powers the disk drive has 125 turns on its primary coil. **(a)** Should the number of turns on the secondary coil be greater than or less than 125? Explain. **(b)** Find the number of turns on the secondary coil.

70. • A transformer with a turns ratio (secondary/primary) of 1:18 is used to step down the voltage from a 120-V wall socket to be used in a battery recharging unit. What is the voltage supplied to the recharger?

71. • A neon sign that requires a voltage of 11,000 V is plugged into a 120-V wall outlet. What turns ratio (secondary/primary) must a transformer have to power the sign?

72. •• A step-down transformer produces a voltage of 6.0 V across the secondary coil when the voltage across the primary coil is 120 V. What voltage appears across the primary coil of this transformer if 120 V is applied to the secondary coil?

73. •• A step-up transformer has 25 turns on the primary coil and 750 turns on the secondary coil. If this transformer is to produce an output of 4800 V with a 12-mA current, what input current and voltage are needed?

GENERAL PROBLEMS

74. • **CE Predict/Explain** An airplane flies level to the ground toward the north pole. **(a)** Is the induced emf from wing tip to wing tip when the plane is at the equator greater than, less than, or equal to the wing-tip-to-wing-tip emf when it is at the latitude of New York? **(b)** Choose the *best explanation* from among the following:
   I. The induced emf is the same because the strength of the Earth's magnetic field is the same at the equator and at New York.
   II. The induced emf is greater at New York because the vertical component of the Earth's magnetic field is greater there than at the equator.
   III. The induced emf is less at New York because at the equator the plane is flying parallel to the magnetic field lines.

75. • **CE** You hold a circular loop of wire at the north magnetic pole of the Earth. Consider the magnetic flux through this loop due to the Earth's magnetic field. Is the flux when the normal to the loop points horizontally greater than, less than, or equal to the flux when the normal points vertically downward? Explain.

76. • **CE** You hold a circular loop of wire at the equator. Consider the magnetic flux through this loop due to the Earth's magnetic field. Is the flux when the normal to the loop points north greater than, less than, or equal to the flux when the normal points vertically upward? Explain.

77. • **CE** The inductor shown in **Figure 23–41** is connected to an electric circuit with a changing current. At the moment in question, the inductor has an induced emf with the indicated direction. Is the current in the circuit at this time increasing and to the right, increasing and to the left, decreasing and to the right, or decreasing and to the left?

▲ **FIGURE 23–41**   Problem 77

78. • **Interstellar Magnetic Field** The *Voyager I* spacecraft moves through interstellar space with a speed of 8.0×10^3 m/s. The magnetic field in this region of space has a magnitude of 2.0×10^{-10} T. Assuming that the 5.0-m-long antenna on the spacecraft is at right angles to the magnetic field, find the induced emf between its ends.

79. • **BIO Blowfly Flight** The coils used to measure the movements of a blowfly, as described in Section 23–5, have a diameter of 2.0 mm. In addition, the fly is immersed in a magnetic field of magnitude 0.15 mT. Find the maximum magnetic flux experienced by one of these coils.

80. • **Electrognathography** Computerized jaw tracking, or electrognathography (EGN), is an important tool for diagnosing and treating temporomandibular disorders (TMDs) that affect a person's ability to bite effectively. The first step in applying EGN is to attach a small permanent magnet to the patient's gum below the lower incisors. Then, as the jaw undergoes a biting motion, the resulting change in magnetic flux is picked up by wire coils placed on either side of the mouth, as shown in **Figure 23–42**. Suppose this person's jaw moves to her right and that the north pole of the permanent magnet also points to her right. From her point of view, is the induced current in the coil to **(a)** her right and **(b)** her left clockwise or counterclockwise? Explain.

▲ **FIGURE 23–42**   Problem 80

81. •• A rectangular loop of wire 24 cm by 72 cm is bent into an L shape, as shown in **Figure 23–43**. The magnetic field in the vicinity of the loop has a magnitude of 0.035 T and points in a direction 25° below the y axis. The magnetic field has no x component. Find the magnitude of the magnetic flux through the loop.

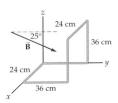

▲ **FIGURE 23–43**   Problem 81

82. •• **IP** A circular loop with a radius of 3.7 cm lies in the x-y plane. The magnetic field in this region of space is uniform and given by $\vec{\mathbf{B}} = (0.43 \text{ T})\hat{\mathbf{x}} + (-0.11 \text{ T})\hat{\mathbf{y}} + (0.52 \text{ T})\hat{\mathbf{z}}$. **(a)** What is

the magnitude of the magnetic flux through this loop? **(b)** Suppose we now increase the x component of $\vec{\mathbf{B}}$, leaving the other components unchanged. Does the magnitude of the magnetic flux increase, decrease, or stay the same? Explain. **(c)** Suppose, instead, that we increase the z component of $\vec{\mathbf{B}}$, leaving the other components unchanged. Does the magnitude of the magnetic flux increase, decrease, or stay the same? Explain.

83. •• Consider a rectangular loop of wire 5.8 cm by 8.2 cm in a uniform magnetic field of magnitude 1.3 T. The loop is rotated from a position of zero magnetic flux to a position of maximum flux in 21 ms. What is the average induced emf in the loop?

84. •• **IP** A car with a vertical radio antenna 85 cm long drives due east at 25 m/s. The Earth's magnetic field at this location has a magnitude of 5.9×10^{-5} T and points northward, 72° below the horizontal. **(a)** Is the top or the bottom of the antenna at the higher potential? Explain. **(b)** Find the induced emf between the ends of the antenna.

85. •• The rectangular coils in a 325-turn generator are 11 cm by 17 cm. What is the maximum emf produced by this generator when it rotates with an angular speed of 525 rpm in a magnetic field of 0.45 T?

86. •• A cubical box 22 cm on a side is placed in a uniform 0.35-T magnetic field. Find the net magnetic flux through the box.

87. •• **BIO Transcranial Magnetic Stimulation** Transcranial magnetic stimulation (TMS) is a noninvasive method for studying brain function, and possibly for treatment as well. In this technique, a conducting loop is held near a person's head, as shown in **Figure 23–44**. When the current in the loop is changed rapidly, the magnetic field it creates can change at the rate of 3.00×10^4 T/s. This rapidly changing magnetic field induces an electric current in a restricted region of the brain that can cause a finger to twitch, bright spots to appear in the visual field (magnetophosphenes), or a feeling of complete happiness to overwhelm a person. If the magnetic field changes at the previously mentioned rate over an area of 1.13×10^{-2} m², what is the induced emf?

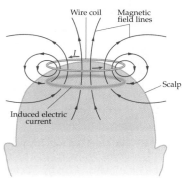

▲ **FIGURE 23–44** Transcranial magnetic stimulation (Problem 87)

88. •• A magnetic field with the time dependence shown in **Figure 23–45** is at right angles to a 155-turn circular coil with a diameter of 3.75 cm. What is the induced emf in the coil at **(a)** $t = 2.50$ ms, **(b)** $t = 7.50$ ms, **(c)** $t = 15.0$ ms, and **(d)** $t = 25.0$ ms?

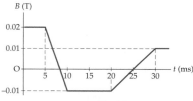

▲ **FIGURE 23–45** Problem 88

89. •• You would like to construct a 50.0-mH inductor by wrapping insulated copper wire (diameter = 0.0332 cm) onto a tube with a circular cross section of radius 2.67 cm. What length of wire is required if it is wrapped onto the tube in a single, close-packed layer?

90. •• The time constant of an RL circuit with $L = 25$ mH is twice the time constant of an RC circuit with $C = 45\ \mu$F. Both circuits have the same resistance R. Find **(a)** the value of R and **(b)** the time constant of the RL circuit.

91. •• A 6.0-V battery is connected in series with a 29-mH inductor, a 110-Ω resistor, and an open switch. **(a)** How long after the switch is closed will the current in the circuit be equal to 12 mA? **(b)** How much energy is stored in the inductor when the current reaches its maximum value?

92. •• A 9.0-V battery is connected in series with a 31-mH inductor, a 180-Ω resistor, and an open switch. **(a)** What is the current in the circuit 0.120 ms after the switch is closed? **(b)** How much energy is stored in the inductor at this time?

93. •• **BIO Blowfly Maneuvers** Suppose the fly described in Problem 79 turns through an angle of 90° in 37 ms. If the magnetic flux through one of the coils on the insect goes from a maximum to zero during this maneuver, and the coil has 85 turns of wire, find the magnitude of the induced emf.

94. ••• **IP** A conducting rod of mass m is in contact with two vertical conducting rails separated by a distance L, as shown in **Figure 23–46**. The entire system is immersed in a magnetic field of magnitude B pointing out of the page. Assuming the rod slides without friction, **(a)** describe the motion of the rod after it is released from rest. **(b)** What is the direction of the induced current (clockwise or counterclockwise) in the circuit? **(c)** Find the speed of the rod after it has fallen for a long time.

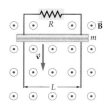

▲ **FIGURE 23–46** Problem 94

95. ••• **IP** A single-turn rectangular loop of width W and length L moves parallel to its length with a speed v. The loop moves from a region with a magnetic field $\vec{\mathbf{B}}$ perpendicular to the plane of the loop to a region where the magnetic field is zero, as shown in **Figure 23–47**. Find the rate of change in the magnetic flux through the loop **(a)** before it enters the region of zero field, **(b)** just after it enters the region of zero field, and **(c)** once it is fully within the region of zero field. **(d)** For each of the cases considered in parts (a), (b), and (c), state whether the induced current in the loop is clockwise, counterclockwise, or zero. Explain in each case.

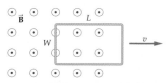

▲ **FIGURE 23–47** Problem 95

96. ••• **IP** The switch in the circuit shown in **Figure 23–48** is open initially. **(a)** Find the current in the circuit a long time after the switch is closed. **(b)** Describe the behavior of the lightbulb from the time the switch is closed until the current reaches the value

found in part (a). **(c)** Now, suppose the switch is opened after having been closed for a long time. If the inductor is large, it is observed that the light flashes brightly and then burns out. Explain this behavior. **(d)** Find the voltage across the lightbulb just before and just after the switch is opened.

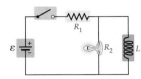

▲ **FIGURE 23–48** Problem 96

97. ••• **Energy Density in *E* and *B* Fields** An electric field *E* and a magnetic field *B* have the same energy density. **(a)** Express the ratio E/B in terms of the fundamental constants ε_0 and μ_0. **(b)** Evaluate E/B numerically, and compare your result with the speed of light.

PASSAGE PROBLEMS

Loop Detectors on Roadways

"Smart" traffic lights are controlled by loops of wire embedded in the road (**Figure 23–49**). These "loop detectors" sense the change in magnetic field as a large metal object—such as a car or a truck—moves over the loop. Once the object is detected, electric circuits in the controller check for cross traffic, and then turn the light from red to green.

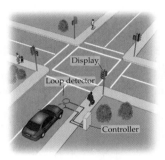

▲ **FIGURE 23–49** Problems 98, 99, 100, and 101

A typical loop detector consists of three or four loops of 14-gauge wire buried 3 in. below the pavement. You can see the marks on the road where the pavement has been cut to allow for installation of the wires. There may be more than one loop detector at a given intersection; this allows the system to recognize that an object is moving as it activates first one detector and then another over a short period of time. If the system determines that a car has entered the intersection while the light is red, it can activate one camera to take a picture of the car from the front—to see the driver's face—and then a second camera to take a picture of the car and its license plate from behind. This red-light camera system was used to good effect during an exciting chase scene through the streets of London in the movie *National Treasure: Book of Secrets*.

Motorcycles are small enough that they often fail to activate the detectors, leaving the cyclist waiting and waiting for a green light. Some companies have begun selling powerful neodymium magnets to mount on the bottom of a motorcycle to ensure that they are "seen" by the detectors.

98. • Suppose the downward vertical component of the magnetic field increases as a car drives over a loop detector. As viewed from above, is the induced current in the loop clockwise, counterclockwise, or zero?

99. • A car drives onto a loop detector and increases the downward component of the magnetic field within the loop from 1.2×10^{-5} T to 2.6×10^{-5} T in 0.38 s. What is the induced emf in the detector if it is circular, has a radius of 0.67 m, and consists of four loops of wire?

   A. 0.66×10^{-4} V   **B.** 1.5×10^{-4} V

   C. 2.1×10^{-4} V   **D.** 6.2×10^{-4} V

100. •• A truck drives onto a loop detector and increases the downward component of the magnetic field within the loop from 1.2×10^{-5} T to the larger value *B* in 0.38 s. The detector is circular, has a radius of 0.67 m, and consists of three loops of wire. What is *B*, given that the induced emf is 8.1×10^{-4} V?

   A. 3.6×10^{-5} T   **B.** 7.3×10^{-5} T

   C. 8.5×10^{-5} T   **D.** 24×10^{-5} T

101. •• Suppose a motorcycle increases the downward component of the magnetic field within a loop only from 1.2×10^{-5} T to 1.9×10^{-5} T. The detector is square, is 0.75 m on a side, and has four loops of wire. Over what period of time must the magnetic field increase if it is to induce an emf of 1.4×10^{-4} V?

   A. 0.028 s   **B.** 0.11 s

   C. 0.35 s   **D.** 0.60 s

INTERACTIVE PROBLEMS

102. •• **Referring to Conceptual Checkpoint 23–3** Suppose the ring is initially to the left of the field region, where there is no field, and is moving to the right. When the ring is partway into the field region, **(a)** is the induced current in the ring clockwise, counterclockwise, or zero, and **(b)** is the magnetic force exerted on the ring to the right, to the left, or zero? Explain.

103. •• **Referring to Conceptual Checkpoint 23–3** Suppose the ring is completely inside the field region initially and is moving to the right. **(a)** Is the induced current in the ring clockwise, counterclockwise, or zero, and **(b)** is the magnetic force on the ring to the right, to the left, or zero? Explain. The ring now begins to emerge from the field region, still moving to the right. **(c)** Is the induced current in the ring clockwise, counterclockwise, or zero, and **(d)** is the magnetic force on the ring to the right, to the left, or zero? Explain.

104. •• **Referring to Example 23–3** **(a)** What external force is required to give the rod a speed of 3.49 m/s, everything else remaining the same? **(b)** What is the current in the circuit in this case?

105. •• **IP Referring to Example 23–3** Suppose the direction of the magnetic field is reversed. Everything else in the system remains the same. **(a)** Is the magnetic force exerted on the rod to the right, to the left, or zero? Explain. **(b)** Is the direction of the induced current clockwise, counterclockwise, or zero? Explain. **(c)** Suppose we now adjust the strength of the magnetic field until the speed of the rod is 2.49 m/s, keeping the force equal to 1.60 N. What is the new magnitude of the magnetic field?

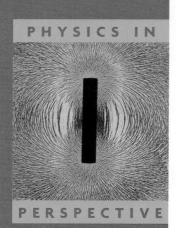

Electricity and Magnetism

Electricity and magnetism at first seemed quite unrelated. These pages follow the detective story by which physicists discovered that they are actually facets of a single force. This discovery led to much of our current technology, including telecommunications and electronics.

❶ Gravity, electricity, magnetism: Three distinct forces?

In daily life, gravity, electrostatic force, and magnetism seem distinct: socks out of the dryer cling to each other but not to refrigerators; magnets don't stick to socks; gravity pays no attention to charge or magnetism. We seem to have three forces with distinct characteristics.

Gravitational force	Electrostatic force	Magnetic force
• Acts between masses	• Acts between electric charges	• Acts between magnetic poles
• Always attractive	• Unlike charges attract; like charges repel	• Unlike poles attract; like poles repel
• Mass comes in only one "kind"	• Positive and negative charges can be isolated	• N and S poles cannot be isolated; occur as dipoles

Gravitational field	Electrostatic field	Magnetic field
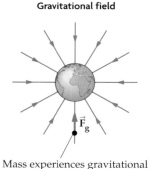	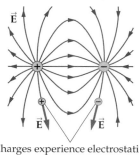	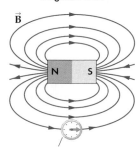
Mass experiences gravitational force tangent to field line	Charges experience electrostatic forces tangent to field lines	Magnetic dipole experiences a torque that aligns it with the field

But are these forces really as distinct as they seem?

❷ First clue: Moving charges generate and respond to magnetic fields

A motionless charge does not respond to magnetism. However, *moving* charges both generate a magnetic field and experience a force when moving at an angle to an external magnetic field.

Moving charges generate magnetic fields. Magnetic field lines form circles around a moving charge.

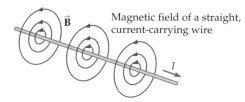

Magnetic field of a straight, current-carrying wire

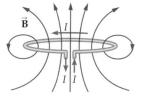

Magnetic field of a current-carrying loop—analogous to field of bar magnet

Moving charges feel a force from a magnetic field. This force causes charged particles to circle or spiral around magnetic field lines. For example, Earth's magnetic field guides charged particles from the solar wind toward Earth's magnetic poles, where they produce the auroras.

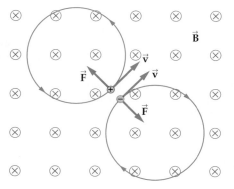

Paths of charged particles in a uniform magnetic field

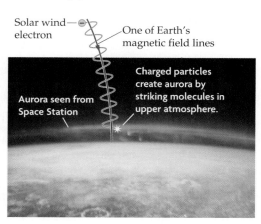

Solar wind electron

One of Earth's magnetic field lines

Charged particles create aurora by striking molecules in upper atmosphere.

Aurora seen from Space Station

❸ Second clue: A changing magnetic field generates an electric field

Holding a magnet next to a wire loop does not generate a current. However, if you do something that changes the magnetic flux Φ through the loop, the changing flux will induce an electric field within the wire, which will drive an induced current. The induced electric field is the emf for the induced current.

Notice that the field lines of the induced electric field do not start or end on charged particles—the electric field is generated by the magnetic field, not by charged objects.

Moving the magnet away from the loop changes the magnetic flux Φ through the loop.

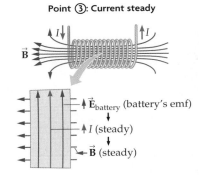

The changing magnetic flux induces an electric field in the wire, which drives an induced current.

Back emf in an *RL* circuit The following graph and diagrams explore how an induced electric field produces the back emf by which a solenoid (inductor) retards changes in current through a circuit.

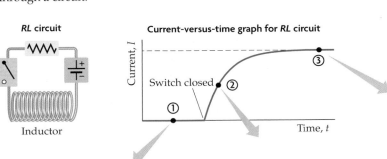

RL circuit

Inductor

Current-versus-time graph for *RL* circuit

Current, I

Switch closed

① ② ③

Time, t

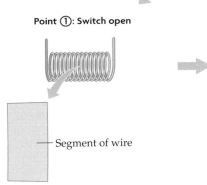

Point ①: Switch open

Segment of wire

No current, so no magnetic field in solenoid.

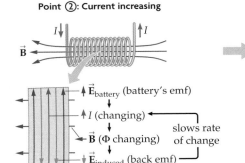

Point ②: Current increasing

$\vec{B}$

$\vec{E}_{battery}$ (battery's emf)

I (changing)

$\vec{B}$ (Φ changing)

$\vec{E}_{induced}$ (back emf)

slows rate of change

The current I reflects both the emf from the battery and the solenoid's induced back emf, which slows the current's rate of change.

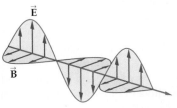

Point ③: Current steady

$\vec{B}$

$\vec{E}_{battery}$ (battery's emf)

I (steady)

$\vec{B}$ (steady)

Because the current is no longer changing, the magnetic field is steady, so there is no induced emf.

---

❹ Final clue and conclusions

Final clue: As we'll see in Chapter 25, a changing *electric* field can generate a *magnetic* field. In fact, electromagnetic waves (including light, radio waves, and microwaves) consist of oscillating, mutually generating electric and magnetic fields.

Conclusions: Physicists realized that electric and magnetic phenomena are facets of a single *electromagnetic force,* which is one of the four fundamental forces. (The others are gravity and two nuclear forces called the weak and strong forces.)

A central goal of physics is to "unify" the fundamental forces—that is, to show that they are really all facets of a single force. So far, electromagnetism has been unified with the weak force and possibly with the strong force. Unification of these forces with gravity is the elusive goal of string theory.

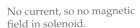

$\vec{E}$

$\vec{B}$

Electromagnetic waves (including the microwaves put out by a cell phone) consist of mutually generating electric and magnetic fields.

24 Alternating-Current Circuits

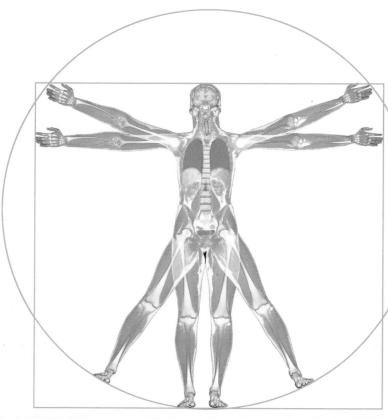

Magnetic resonance imaging (MRI), which produced the modern version of Leonardo da Vinci's *Vitruvian Man* shown here, would not be possible without alternating-current (ac) electric circuits. In fact, the key component of all MRI machines is an ac circuit that oscillates with a frequency in the range of 15 to 80 MHz. At these frequencies, the circuit resonates with hydrogen atoms within a person's body. By detecting how much each portion of the body responds to the resonance effect, a computer can construct detailed images of the body's internal structure. In this chapter, we explore many aspects of ac circuits, including resonance, and point out the close connections between inductors, capacitors, and resistors in electric circuits, and masses, springs, and friction in mechanical systems.

We conclude our study of electric circuits with a consideration of the type of circuit most common in everyday usage—a circuit with an alternating current. In an alternating-current (ac) circuit, the polarity of the voltage and the direction of the current undergo periodic reversals. Typically, the time dependence of the voltage and current is sinusoidal, with a frequency of 60 cycles per second being the standard in the United States.

As we shall see, ac circuits require that we generalize the notion of resistance to include the time-dependent effects associated with capacitors and inductors. In particular, the voltage and current in an ac circuit may not vary with time in exactly the same way—they are often "out of step" with each other. These effects and others are found in ac circuits but not in direct-current circuits.

Finally, we also consider the fact that some electric circuits have natural frequencies of oscillation, much like a pendulum or a mass on a spring. One of the results of having a natural frequency is that resonance effects are to be expected. Indeed, electric circuits do show resonance—in fact, this is how you are able to tune your radio or television to the desired station. Clearly, then, ac circuits have a rich variety of behaviors that make them not only very useful but also most interesting from a physics point of view.

24–1 Alternating Voltages and Currents

When you plug a lamp into a wall socket, the voltage and current supplied to the lightbulb vary sinusoidally with a steady frequency, completing 60 cycles each second. Because the current periodically reverses direction, we say that the wall socket provides an **alternating current.** A simplified alternating-current (ac) circuit diagram for the lamp is shown in **Figure 24–1**. In this circuit, we indicate the bulb by its equivalent resistance, R, and the wall socket by an **ac generator,** represented as a circle enclosing one cycle of a sine wave.

The voltage delivered by an ac generator, which is plotted in **Figure 24–2 (a)**, can be represented mathematically as follows:

$$V = V_{max} \sin \omega t \qquad\qquad 24\text{–}1$$

In this expression, V_{max} is the largest value attained by the voltage during a cycle, and the angular frequency is $\omega = 2\pi f$, where $f = 60$ Hz. (Note the similarity between the sinusoidal time dependence in an ac circuit and the time dependence given in Section 13–2 for simple harmonic motion.) From Ohm's law, $I = V/R$, it follows that the current in the lightbulb is

$$I = \frac{V}{R} = \left(\frac{V_{max}}{R}\right) \sin \omega t = I_{max} \sin \omega t \qquad\qquad 24\text{–}2$$

This result is plotted in **Figure 24–2 (b)**.

Notice that the voltage and current plots have the same time variation. In particular, the voltage reaches its maximum value at precisely the same time as the current. We express this relationship between the voltage and current in a resistor by saying that they are *in phase* with one another. As we shall see later in this chapter, other circuit elements, like capacitors and inductors, have different phase relationships between the current and voltage. For these elements, the current and voltage reach maximum values at different times.

Phasors

A convenient way to represent an alternating voltage and the corresponding current is with counterclockwise rotating vectors referred to as **phasors.** Phasors allow us to take advantage of the connection between uniform circular motion and linear, sinusoidal motion—just as we did when we studied oscillations in Chapter 13. In that case, we related the motion of a peg on a rotating turntable to the oscillating motion of a mass on a spring by projecting the position of the peg onto a screen. We use a similar projection with phasors.

For example, **Figure 24–3** shows a vector of magnitude V_{max} rotating about the origin with an angular speed ω. This rotating vector is the voltage phasor. If the voltage phasor makes an angle $\theta = \omega t$ with the x axis, it follows that its y component is $V_{max} \sin \omega t$; that is, if we project the voltage phasor onto the y axis, the projection gives the instantaneous value of the voltage, $V = V_{max} \sin \omega t$, in agreement with Equation 24–1. In general,

> the instantaneous value of a quantity represented by a phasor is the projection of the phasor onto the y axis.

Figure 24–3 also shows the current phasor, represented by a rotating vector of magnitude $I_{max} = V_{max}/R$. The current phasor points in the same direction as the voltage phasor; hence, the instantaneous current is $I = I_{max} \sin \omega t$, as in Equation 24–2.

The fact that the voltage and current phasors always point in the same direction for a resistor is an equivalent way of saying that the voltage and current are in phase:

> The voltage phasor for a resistor points in the same direction as the current phasor.

In circuits that also contain capacitors or inductors, the current and voltage phasors will usually point in different directions.

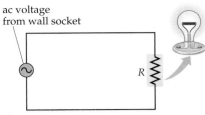

▲ **FIGURE 24–1   An ac generator connected to a lamp**

Simplified alternating-current circuit diagram for a lamp plugged into a wall socket. The lightbulb is replaced in the circuit with its equivalent resistance, R.

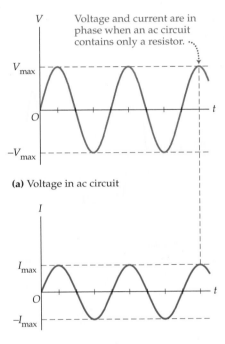

(a) Voltage in ac circuit

(b) Current in ac circuit with resistance only

▲ **FIGURE 24–2   Voltage and current for an ac resistor circuit**

(a) An ac voltage described by $V = V_{max} \sin \omega t$. **(b)** The alternating current, $I = I_{max} \sin \omega t = (V_{max}/R) \sin \omega t$, corresponding to the ac voltage in (a). Note that the voltage of the generator and the current in the resistor are in phase with each other; that is, their maxima and minima occur at precisely the same times.

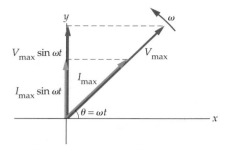

▲ **FIGURE 24–3 Phasor diagram for an ac resistor circuit**

Since the current and voltage are in phase in a resistor, the corresponding phasors point in the same direction at all times. Both phasors rotate about the origin with an angular speed ω, and the vertical component of each is the instantaneous value of that quantity.

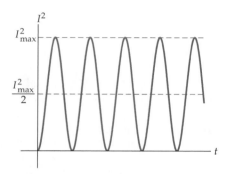

▲ **FIGURE 24–4 The square of a sinusoidally varying current**

Note that I^2 varies symmetrically about the value $\frac{1}{2}I^2_{max}$. The average of I^2 over time, then, is $\frac{1}{2}I^2_{max}$.

Root Mean Square (rms) Values

Notice that both the voltage and the current in Figure 24–2 have average values that are zero. Thus, V_{av} and I_{av} give very little information about the actual behavior of V and I. A more useful type of average, or mean, is the **root mean square,** or **rms** for short.

To see the significance of a *root mean square*, consider the current as a function of time, as given in Equation 24–2. First, we *square* the current to obtain

$$I^2 = I^2_{max} \sin^2 \omega t$$

Clearly, I^2 is always positive; hence, its average value will not vanish. Next, we calculate the *mean* value of I^2. This can be done by inspecting **Figure 24–4**, where we plot I^2 as a function of time. As we see, I^2 varies *symmetrically* between 0 and I^2_{max}; that is, it spends equal amounts of time above and below the value $\frac{1}{2}I^2_{max}$. Hence, the mean value of I^2 is half of I^2_{max}:

$$(I^2)_{av} = \frac{1}{2}I^2_{max}$$

Finally, we take the square *root* of this average so that our final result is a current rather than a current squared. This calculation yields the rms value of the current:

$$I_{rms} = \sqrt{(I^2)_{av}} = \frac{1}{\sqrt{2}}I_{max} \qquad 24\text{–}3$$

In general, any quantity x that varies with time as $x = x_{max} \sin \omega t$ or $x = x_{max} \cos \omega t$ obeys the same relationships among its average, maximum, and rms values:

RMS Value of a Quantity with Sinusoidal Time Dependence

$$(x^2)_{av} = \frac{1}{2}x^2_{max}$$

$$x_{rms} = \frac{1}{\sqrt{2}}x_{max} \qquad 24\text{–}4$$

As an example, the rms value of the voltage in an ac circuit is

$$V_{rms} = \frac{1}{\sqrt{2}}V_{max} \qquad 24\text{–}5$$

This result is applied to standard household voltages in the following Exercise.

EXERCISE 24–1

Typical household circuits operate with an rms voltage of 120 V. What is the maximum, or peak, value of the voltage in these circuits?

SOLUTION

Solving Equation 24–5 for the maximum voltage, V_{max}, we find

$$V_{max} = \sqrt{2}V_{rms} = \sqrt{2}(120 \text{ V}) = 170 \text{ V}$$

Whenever we refer to an ac generator in this chapter, we shall assume that the time variation is sinusoidal, so that the relations given in Equation 24–4 hold. If a different form of time variation is to be considered, it will be specified explicitly. Because the rms and maximum values of a sinusoidally varying quantity are proportional, it follows that *any* relation between rms values, like $I_{rms} = V_{rms}/R$, is equally valid as a relation between maximum values, $I_{max} = V_{max}/R$.

We now show how rms values are related to the average power consumed by a circuit. Referring again to the lamp circuit shown in Figure 24–1, we see that the instantaneous power dissipated in the resistor is $P = I^2R$. Using the time dependence $I = I_{max} \sin \omega t$, we find

$$P = I^2R = I^2_{max}R \sin^2 \omega t$$

Note that P is always positive, as one might expect; after all, a current always dissipates energy as it passes through a resistor, regardless of its direction. To find the average power dissipated in the resistor, we note again that the average of $\sin^2 \omega t$ is $\frac{1}{2}$, thus

$$P_{av} = I^2_{max}R(\sin^2 \omega t)_{av} = \tfrac{1}{2}I^2_{max}R$$

In terms of the rms current, $I_{rms} = I_{max}/\sqrt{2}$, it follows that

$$P_{av} = I^2_{rms}R$$

Therefore, we arrive at the following conclusion:

> $P = I^2R$ gives the instantaneous power consumption in both ac and dc circuits. To find the *average* power in an ac circuit, we simply re-place I with I_{rms}.

Similar conclusions apply to many other dc and instantaneous formulas as well. For example, the power can also be written as $P = V^2/R$. Using the time-dependent voltage of an ac circuit, we have

$$P = \frac{V^2}{R} = \left(\frac{V^2_{max}}{R}\right) \sin^2 \omega t$$

Clearly, the average power is

$$P_{av} = \left(\frac{V^2_{max}}{R}\right)\left(\tfrac{1}{2}\right) = \frac{V^2_{rms}}{R} \qquad\qquad 24\text{–}6$$

As before, we convert the dc power, $P = V^2/R$, to an average ac power by using the rms value of the voltage.

The close similarity between dc expressions and the corresponding ac expressions with rms values is one of the advantages of working with rms values. Another is that electrical meters, such as ammeters and voltmeters, generally give readings of rms values, rather than peak values, when used in ac circuits. In the remainder of this chapter, we shall make extensive use of rms quantities.

PROBLEM-SOLVING NOTE

Maximum Versus RMS Values

When reading a problem statement, be sure to determine whether a given voltage or current is a maximum value or an rms value. If an rms current is given, for example, it follows that the maximum current is $I_{max} = \sqrt{2}I_{rms}$.

EXAMPLE 24–1 A RESISTOR CIRCUIT

An ac generator with a maximum voltage of 24.0 V and a frequency of 60.0 Hz is connected to a resistor with a resistance $R = 265\ \Omega$. Find **(a)** the rms voltage and **(b)** the rms current in the circuit. In addition, determine **(c)** the average and **(d)** the maximum power dissipated in the resistor.

PICTURE THE PROBLEM

The circuit in this case consists of a 60.0-Hz ac generator connected directly to a 265-Ω resistor. The maximum voltage of the generator is 24.0 V.

STRATEGY

a. The rms voltage is simply $V_{rms} = V_{max}/\sqrt{2}$.

b. Ohm's law gives the rms current, $I_{rms} = V_{rms}/R$.

c. The average power can be found using $P_{av} = I^2_{rms}R$ or $P_{av} = V^2_{rms}/R$.

d. The maximum power, $P_{max} = V^2_{max}/R$, is twice the average power, from Equation 24–6.

SOLUTION

Part (a)

1. Use $V_{rms} = V_{max}/\sqrt{2}$ to find the rms voltage:

$$V_{rms} = \frac{V_{max}}{\sqrt{2}} = \frac{24.0\ V}{\sqrt{2}} = 17.0\ V$$

Part (b)

2. Divide the rms voltage by the resistance, R, to find the rms current:

$$I_{rms} = \frac{V_{rms}}{R} = \frac{17.0\ V}{265\ \Omega} = 0.0642\ A$$

CONTINUED ON NEXT PAGE

CONTINUED FROM PREVIOUS PAGE

Part (c)

3. Use $I_{rms}^2 R$ to find the average power:

$$P_{av} = I_{rms}^2 R = (0.0642\ A)^2(265\ \Omega) = 1.09\ W$$

4. Use V_{rms}^2/R to find the average power:

$$P_{av} = \frac{V_{rms}^2}{R} = \frac{(17.0\ V)^2}{265\ \Omega} = 1.09\ W$$

Part (d)

5. Use $P_{max} = V_{max}^2/R$ to find the maximum power:

$$P_{max} = \frac{V_{max}^2}{R} = 2\left(\frac{V_{rms}^2}{R}\right) = 2(1.09\ W) = 2.18\ W$$

INSIGHT
Note that the instantaneous power in the resistor, $P = (V_{max}^2/R)\sin^2 \omega t$, oscillates symmetrically between 0 and twice the average power. This is another example of the type of averaging illustrated in Figure 24–4.

PRACTICE PROBLEM
Suppose we would like the average power dissipated in the resistor to be 5.00 W. Should the resistance be increased or decreased, assuming the same ac generator? Find the required value of R.   [**Answer:** The resistance should be decreased. The required resistance is $R = 57.8\ \Omega$.]

Some related homework problems: Problem 3, Problem 4

CONCEPTUAL CHECKPOINT 24–1   COMPARE AVERAGE POWER

If the frequency of the ac generator in Example 24–1 is increased, does the average power dissipated in the resistor **(a)** increase, **(b)** decrease, or **(c)** stay the same?

REASONING AND DISCUSSION
None of the results in Example 24–1 depend on the frequency of the generator. For example, the relation $V_{rms} = V_{max}/\sqrt{2}$ depends only on the fact that the voltage varies sinusoidally with time and not at all on the frequency of the oscillations. The same frequency independence applies to the rms current and the average power.

These results are due to the fact that resistance is independent of frequency. In contrast, we shall see later in this chapter that the behavior of capacitors and inductors does indeed depend on frequency.

ANSWER
(c) The average power remains the same.

Safety Features in Household Electric Circuits

In today's technological world, with electrical devices as common as a horse and buggy in an earlier era, we sometimes forget that household electric circuits pose potential dangers to homes and their occupants. For example, if several electrical devices are plugged into a single outlet, the current in the wires connected to that outlet may become quite large. The corresponding power dissipation in the wires ($P = I^2 R$) can turn them red hot and lead to a fire.

To protect against this type of danger, household circuits use fuses and circuit breakers. In the case of a fuse, the current in a circuit must flow through a thin metal strip enclosed within the fuse. If the current exceeds a predetermined amount (typically 15 A) the metal strip becomes so hot that it melts and breaks the circuit. Thus, when a fuse "burns out," it is an indication that too many devices are operating on that circuit.

Circuit breakers provide similar protection with a switch that incorporates a bimetallic strip (Figure 16–5). When the bimetallic strip is cool, it closes the switch, allowing current to flow. When the strip is heated by a large current, however, it bends enough to open the switch and stop the current. Unlike the fuse, which cannot be used after it burns out, the circuit breaker can be reset when the bimetallic strip cools and returns to its original shape.

Household circuits also pose a threat to the occupants of a home, and at much lower current levels than the 15 A it takes to trigger a fuse or a circuit breaker. For example, it takes a current of only about 0.001 A to give a person a mild tingling

REAL-WORLD PHYSICS: BIO

Electric shock hazard

sensation. Currents in the range of 0.01 A to 0.02 A produce muscle spasms that make it difficult to let go of the wire delivering the current, or may even result in respiratory arrest. When currents reach 0.1 A to 0.2 A, the heartbeat is interrupted by an uncoordinated twitching referred to as ventricular fibrillation. As a result, currents in this range can prove to be fatal in a matter of seconds.

Several strategies are employed to reduce the danger of electrical shock. The first line of defense is the *polarized plug*, in which one prong of the plug is wider than the other prong. The corresponding wall socket will accept the plug in only one orientation, with the wide prong in the wide receptacle. The narrow receptacle of the outlet is wired to the high-potential side of the circuit; the wide receptacle is connected to the low-potential side, which is essentially at "ground" potential. A polarized plug provides protection by ensuring that the case of an electrical appliance, which manufacturers design to be connected to the wide prong, is at low potential. Furthermore, when an electrical device with a polarized plug is turned off, the high potential extends only from the wall outlet to the switch, leaving the rest of the device at zero potential.

The next line of defense against accidental shock is the three-prong grounded plug. In this plug, the rounded third prong is connected directly to ground when plugged into a three-prong receptacle. In addition, the third prong is wired to the case of an electrical appliance. If something goes wrong within the appliance, and a high-potential wire comes into contact with the case, the resulting current flows through the third prong, rather than through the body of a person who happens to touch the case.

An even greater level of protection is provided by a device known as a *ground fault circuit interrupter* (GFCI). The basic operating principle of an interrupter is illustrated in **Figure 24–5**. Note that the wires carrying an ac current to the protected appliance pass through a small iron ring. When the appliance operates normally, the two wires carry equal currents in opposite directions—in one wire the current goes to the appliance, in the other the same current returns from the appliance. Each of these wires produces a magnetic field (Equation 22–9), but because their currents are in opposite directions the magnetic fields are in opposite directions as well. As a result, the magnetic fields of the two wires cancel. If a malfunction occurs in the appliance—say, a wire frays and contacts the case of the appliance—current that would ordinarily return through the power cord may pass through the user's body instead and into the ground. In such a situation, the wire carrying current to the appliance now produces a net magnetic field within the iron ring that varies with the frequency of the ac generator. The changing magnetic field in the ring induces a current in the sensing coil wrapped around the ring, and the induced current triggers a circuit breaker in the interrupter. This cuts the flow of current to the appliance within a millisecond, protecting the user. In newer homes, interrupters are built directly into the wall sockets. The same protection can be obtained, however, by plugging a ground fault interrupter into an unprotected wall socket and then plugging an appliance into the interrupter.

REAL-WORLD PHYSICS
Polarized plugs and grounded plugs

REAL-WORLD PHYSICS
Ground fault circuit interrupter

◀ **FIGURE 24–5 The ground fault circuit interrupter**
A short circuit in an appliance will ordinarily blow the fuse or trip the circuit breaker in the electrical line supplying power to the circuit. However, although fuses and circuit breakers help prevent electrical fires, these devices are not very good at preventing electric shock. This is because they are activated by the heat produced when the current becomes abnormally high—for example, when a "hot" wire touches a conductor that is in contact with the ground. If you happen to be that conductor, the fuse or breaker may act too slowly to prevent serious injury or even death. A ground fault circuit interrupter, by contrast, can cut off the current in a shorted circuit in less than a millisecond, before any harm can be done.

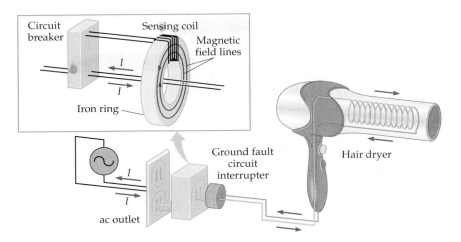

▲ Electricity is everywhere in the modern house, and electricity can be dangerous. Many common safety devices help us to minimize the risks associated with this very convenient form of energy. Whereas older homes typically have fuse boxes, most new houses are protected by circuit breaker panels (left). Both function in a similar way—if too much current is being drawn (perhaps as the result of a short circuit), the heat produced "blows" the fuse (by melting a metal strip) or "trips" the circuit breaker (by bending a bimetallic strip). Either way, the circuit is interrupted and current is cut off before the wires can become hot enough to start a fire.

The danger of electric shock can be reduced by the use of polarized plugs (center, top) or three-prong, grounded plugs (center, bottom) on appliances. Each of these provides a low-resistance path to ground that can be wired to the case of an appliance. In the event that a "hot" wire touches the case, most of the current will flow through the grounded wire rather than through the user. Even more protection is afforded by a ground fault circuit interrupter, or GFCI (right). This device, which is much faster and more sensitive than an ordinary circuit breaker, utilizes magnetic induction to interrupt the current in a circuit that has developed a short.

24–2 Capacitors in ac Circuits

In an ac circuit containing only a generator and a capacitor, the relationship between current and voltage is different in many important respects from the behavior seen with a resistor. In this section we explore these differences and show that the average power consumed by a capacitor is zero.

Capacitive Reactance

Consider a simple circuit consisting of an ac generator and a capacitor, as in **Figure 24–6**. The generator supplies an rms voltage, V_{rms}, to the capacitor. We would like to answer the following question: How is the rms current in this circuit related to the capacitance of the capacitor, C, and the frequency of the generator, ω?

To answer this question requires the methods of calculus, but the final result is quite straightforward. In fact, the rms current is simply

$$I_{rms} = \omega C V_{rms} \qquad \text{24–7}$$

In analogy with the expression $I_{rms} = V_{rms}/R$ for a resistor, we will find it convenient to rewrite this result in the following form:

$$I_{rms} = \frac{V_{rms}}{X_C} \qquad \text{24–8}$$

In this expression, X_C is referred to as the **capacitive reactance;** it plays the same role for a capacitor that resistance does for a resistor. In particular, to find the rms current in a resistor we divide V_{rms} by R; to find the rms current that flows into one side of a capacitor and out the other side, we divide V_{rms} by X_C. Comparing Equations 24–7 and 24–8, we see that the capacitive reactance can be written as follows:

Capacitive Reactance, X_C

$$X_C = \frac{1}{\omega C} \qquad \text{24–9}$$

SI unit: ohm, Ω

▲ **FIGURE 24–6 An ac generator connected to a capacitor**

The rms current in an ac capacitor circuit is $I_{rms} = V_{rms}/X_C$, where the capacitive reactance is $X_C = 1/\omega C$. Therefore, the rms current in this circuit is proportional to the frequency of the generator.

It is straightforward to show that the unit of $X_C = 1/\omega C$ is the ohm, the same unit as for resistance.

EXERCISE 24–2

Find the capacitive reactance of a 22-μF capacitor in a 60.0-Hz circuit.

SOLUTION

Using Equation 24–9, and recalling that $\omega = 2\pi f$, we find

$$X_C = \frac{1}{\omega C} = \frac{1}{2\pi(60.0 \text{ s}^{-1})(22 \times 10^{-6} \text{ F})} = 120 \text{ }\Omega$$

Unlike resistance, the capacitive reactance of a capacitor depends on the frequency of the ac generator. For example, at low frequencies the capacitive reactance becomes very large, and hence the rms current, $I_{rms} = V_{rms}/X_C$, is very small. This is to be expected. In the limit of low frequency, the current in the circuit becomes constant; that is, the ac generator becomes a dc battery. In this case we know that the capacitor becomes fully charged, and the current ceases to flow.

In the limit of large frequency, the capacitive reactance is small and the current is large. The reason is that at high frequency the current changes direction so rapidly that there is never enough time to fully charge the capacitor. As a result, the charge on the capacitor is never very large, and therefore it offers essentially no resistance to the flow of charge.

The behavior of a capacitor as a function of frequency is considered in the next Exercise.

EXERCISE 24–3

Suppose the capacitance in Figure 24–6 is 4.5 μF and the rms voltage of the generator is 120 V. Find the rms current in the circuit when the frequency of the generator is **(a)** 60.0 Hz and **(b)** 6.00 Hz.

SOLUTION

Applying Equations 24–8 and 24–9, we find for part (a)

$$I_{rms} = \frac{V_{rms}}{X_C} = \omega C V_{rms} = (2\pi)(60.0 \text{ s}^{-1})(4.5 \times 10^{-6} \text{ F})(120 \text{ V}) = 0.20 \text{ A}$$

Similarly, for part (b)

$$I_{rms} = \frac{V_{rms}}{X_C} = \omega C V_{rms} = (2\pi)(6.00 \text{ s}^{-1})(4.5 \times 10^{-6} \text{ F})(120 \text{ V}) = 0.020 \text{ A}$$

When the frequency is reduced, the reactance becomes larger and the current decreases, as expected.

Notice that the rms current in a capacitor is proportional to its capacitance, C, for all frequencies. A capacitor with a large capacitance can store and release large amounts of charge, which results in a large current. Finally, recall that rms expressions like $I_{rms} = V_{rms}/X_C$ (Equation 24–8) are equally valid in terms of maximum quantities; that is, $I_{max} = V_{max}/X_C$.

Phasor Diagrams: Capacitor Circuits

Figure 24–7 (a) shows the time dependence of the voltage and current in an ac capacitor circuit. Notice that the voltage and current are not in phase. For example, the current has its maximum value at the time $\omega t = \pi/2 = 90°$, whereas the voltage does not reach its maximum value until a later time, when $\omega t = \pi = 180°$. In general,

the voltage across a capacitor *lags* the current by 90°.

This 90° difference between current and voltage can probably best be seen in a phasor diagram. For example, in **Figure 24–8** we show both the current and the

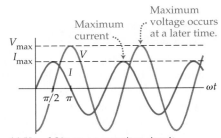

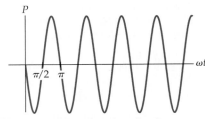

▲ **FIGURE 24–7 Time variation of voltage, current, and power for a capacitor in an ac circuit**

(a) Note that the time dependences are different for the voltage and the current. In particular, the voltage reaches its maximum value $\pi/2$ rad, or 90°, *after* the current. Thus we say that the voltage *lags* the current by 90°. **(b)** The power consumed by a capacitor in an ac circuit, $P = IV$, has an average value of zero.

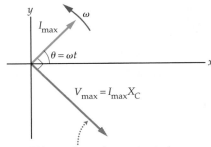

Voltage phasor lags current phasor by 90° in an ac capacitor circuit.

▲ **FIGURE 24–8 Current and voltage phasors for a capacitor**

Both the current and voltage phasors rotate counterclockwise about the origin. The voltage phasor lags the current phasor by 90°. This means that the voltage points in a direction that is 90° clockwise from the direction of the current.

When drawing a voltage phasor for a capacitor, always draw it at right angles to the current phasor. In addition, be sure the capacitor phasor is 90° *clockwise* from the current phasor.

voltage phasors for a capacitor. The current phasor, with a magnitude I_{max}, is shown at the angle $\theta = \omega t$; it follows that the instantaneous current is $I = I_{max} \sin \omega t$. (As a matter of consistency, all phasor diagrams in this chapter will show the current phasor at the angle $\theta = \omega t$.) The voltage phasor, with a magnitude $V_{max} = I_{max}X_C$, is at right angles to the current phasor, pointing in the direction $\theta = \omega t - 90°$. The instantaneous value of the voltage is $V = V_{max} \sin(\omega t - 90°)$. Because the phasors rotate counterclockwise, we see that the voltage phasor lags the current phasor.

EXAMPLE 24–2 INSTANTANEOUS VOLTAGE

Consider a capacitor circuit, as in Figure 24–6, where the capacitance is $C = 50.0 \ \mu F$, the maximum current is 2.10 A, and the frequency of the generator is 60.0 Hz. At a given time, the current in the capacitor is 0.500 A and increasing. What is the voltage across the capacitor at this time?

PICTURE THE PROBLEM
The appropriate phasor diagram for this system is shown in our sketch. To see how this diagram was drawn, we first recall that the current has a positive value and is increasing. The fact that it is positive means it must point in a direction θ between 0 and 180°. The additional fact that it is increasing means θ must be between 0 and 90°; if it were between 90° and 180°, it would be decreasing with time. Finally, the voltage phasor lags the current phasor by 90° in a capacitor.

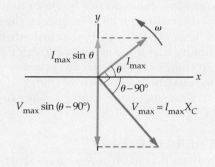

STRATEGY
There are several pieces that we must put together to complete this solution. First, we know the maximum current, I_{max}, and the instantaneous current, $I = I_{max} \sin \omega t = I_{max} \sin \theta$. We can solve this relation for the angle θ.

Second, the maximum voltage is given by $V_{max} = I_{max}X_C = I_{max}/\omega C$.

Finally, the instantaneous voltage is $V = V_{max} \sin(\theta - 90°)$, as indicated in the sketch.

SOLUTION

1. Use $I = I_{max} \sin \theta$ to find the angle θ of the current phasor above the x axis:

$$\theta = \sin^{-1}\left(\frac{I}{I_{max}}\right) = \sin^{-1}\left(\frac{0.500 \text{ A}}{2.10 \text{ A}}\right) = 13.8°$$

2. Find the maximum voltage in the circuit using $V_{max} = I_{max}/\omega C$:

$$V_{max} = \frac{I_{max}}{\omega C} = \frac{2.10 \text{ A}}{2\pi(60.0 \text{ s}^{-1})(50.0 \times 10^{-6} \text{ F})} = 111 \text{ V}$$

3. Calculate the instantaneous voltage across the capacitor with $V = V_{max} \sin(\theta - 90°)$:

$$V = V_{max} \sin(\theta - 90°)$$
$$= (111 \text{ V}) \sin(13.8° - 90°) = -108 \text{ V}$$

INSIGHT
We can see from the phasor diagram that the current is positive and increasing in magnitude, whereas the voltage is negative and decreasing in magnitude. It follows that both the current and the voltage are becoming more positive with time.

PRACTICE PROBLEM
What is the voltage across the capacitor when the current in the circuit is 2.10 A? Is the voltage increasing or decreasing at this time? [**Answer:** $V = 0$, increasing]

Some related homework problems: Problem 12, Problem 13

To gain a qualitative understanding of the phase relation between the voltage and current in a capacitor, note that at the time when $\omega t = \pi/2$ the voltage across the capacitor in Figure 24–7 (a) is zero. Since the capacitor offers no resistance to the flow of current at this time, the current in the circuit is now a maximum. As the current continues to flow, charge builds up on the capacitor, and its voltage increases. This causes the current to decrease. At the time when $\omega t = \pi$ the capacitor voltage reaches a maximum and the current vanishes. As the current begins to flow in the opposite direction, charge flows out of the capacitor and its voltage decreases. When the voltage goes to zero, the current is once again a maximum, though this time in the opposite direction. It follows, then, that the variations of current and voltage are 90° out of phase—that is, when one is a maximum or a minimum, the other is zero—just like position and velocity in simple harmonic motion.

Power

As a final observation on the behavior of a capacitor in an ac circuit, we consider the power it consumes. Recall that the instantaneous power for any circuit can be written as $P = IV$. In **Figure 24–7 (b)** we plot the power $P = IV$ corresponding to the current and voltage shown in Figure 24–7 (a). The result is a power that changes sign with time.

In particular, note that the power is negative when the current and voltage have opposite signs, as between $\omega t = 0$ and $\omega t = \pi/2$, but is positive when they have the same sign, as between $\omega t = \pi/2$ and $\omega t = \pi$. This means that between $\omega t = 0$ and $\omega t = \pi/2$ the capacitor draws energy from the generator, but between $\omega t = \pi/2$ and $\omega t = \pi$ it delivers energy back to the generator. As a result, the average power as a function of time is zero, as can be seen by the symmetry about zero power in Figure 24–7 (b). Thus, *a capacitor in an ac circuit consumes zero net energy.*

24–3 RC Circuits

We now consider a resistor and a capacitor in series in the same ac circuit. This leads to a useful generalization of resistance known as the impedance.

Impedance

The circuit shown in **Figure 24–9** consists of an ac generator, a resistor, R, and a capacitor, C, connected in series. It is assumed that the values of R and C are known, as well as the maximum voltage, V_{max}, and angular frequency, ω, of the generator. In terms of these quantities we would like to determine the maximum current in the circuit and the maximum voltages across both the resistor and the capacitor.

To begin, we note that the magnitudes of the voltages are readily determined. For instance, the magnitude of the maximum voltage across the resistor is $V_{max,R} = I_{max}R$, and for the capacitor it is $V_{max,C} = I_{max}X_C = I_{max}/\omega C$. The total voltage in this circuit is *not* the sum of these two voltages, however, because they are not in phase—they do not attain their maximum values at the same time. To take these phase differences into account we turn to a phasor diagram.

Figure 24–10 shows the phasor diagram for a simple RC circuit. To construct this diagram, we start by drawing the current phasor with a length I_{max} at the angle $\theta = \omega t$, as indicated in the figure. Next, we draw the voltage phasor associated with the resistor. This has a magnitude of $I_{max}R$ and points in the same direction as the current phasor. Finally, we draw the voltage phasor for the capacitor. This phasor has a magnitude of $I_{max}X_C$ and points in a direction that is rotated 90° clockwise from the current phasor.

Now, to obtain the total voltage in the circuit in terms of the individual voltages, we perform a vector sum of the resistor-voltage phasor and the capacitor-voltage phasor, as indicated in Figure 24–10. The total voltage, then, is the hypotenuse of the right triangle formed by these two phasors. Its magnitude, using the Pythagorean theorem, is simply

$$V_{max} = \sqrt{V_{max,R}^2 + V_{max,C}^2} \qquad 24\text{–}10$$

Substituting the preceding expressions for the voltages across R and C, we find

$$V_{max} = \sqrt{(I_{max}R)^2 + (I_{max}X_C)^2} = I_{max}\sqrt{R^2 + X_C^2}$$

The last quantity in the preceding expression is given a special name; it is called the **impedance**, Z:

Impedance in an *RC* Circuit

$$Z = \sqrt{R^2 + X_C^2} = \sqrt{R^2 + \left(\frac{1}{\omega C}\right)^2} \qquad 24\text{–}11$$

SI unit: ohm, Ω

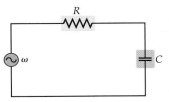

▲ **FIGURE 24–9 An alternating-current *RC* circuit**

An ac *RC* circuit consists of a generator, a resistor, and a capacitor connected in series.

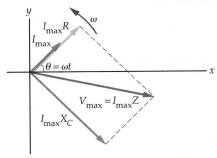

▲ **FIGURE 24–10 Phasor diagram for an *RC* circuit**

The maximum current in an *RC* circuit is $I_{max} = V_{max}/Z$, where V_{max} is the maximum voltage of the ac generator and $Z = \sqrt{R^2 + X_C^2}$ is the impedance of the circuit. The maximum voltage across the resistor is $V_{max,R} = I_{max}R$ and the maximum voltage across the capacitor is $V_{max,C} = I_{max}X_C$.

Clearly, the impedance has units of ohms, the same as those of resistance and reactance.

EXERCISE 24–4

A given RC circuit has $R = 135\ \Omega$, $C = 28.0\ \mu F$, and $f = 60.0$ Hz. What is the impedance of the circuit?

SOLUTION

Applying Equation 24–11 with $\omega = 2\pi f$, we obtain

$$Z = \sqrt{R^2 + X_C^2} = \sqrt{R^2 + \left(\frac{1}{\omega C}\right)^2}$$

$$= \sqrt{(135\ \Omega)^2 + \left(\frac{1}{2\pi(60.0\ \text{s}^{-1})(28.0 \times 10^{-6}\ \text{F})}\right)^2} = 165\ \Omega$$

We are now in a position to calculate the maximum current in an RC circuit. First, given R, C, and $\omega = 2\pi f$, we can determine the value of Z using Equation 24–11. Next, we solve for the maximum current using the relation

$$V_{\text{max}} = I_{\text{max}}\sqrt{R^2 + X_C^2} = I_{\text{max}}Z$$

which yields

$$I_{\text{max}} = \frac{V_{\text{max}}}{Z}$$

In a circuit containing only a resistor, the maximum current is $I_{\text{max}} = V_{\text{max}}/R$; thus, we see that the impedance Z is indeed a generalization of resistance that can be applied to more complex circuits.

To check some limits of Z, note that in the capacitor circuit discussed in the previous section (Figure 24–6), the resistance is zero. Hence, in that case the impedance is

$$Z = \sqrt{0 + X_C^2} = X_C$$

The maximum voltage across the capacitor, then, is $V_{\text{max}} = I_{\text{max}}/Z = V_{\text{max}}/X_C$, as expected. Similarly, in a resistor circuit with no capacitor, as in Figure 24–1, the capacitive reactance is zero; hence,

$$Z = \sqrt{R^2 + 0} = R$$

Thus, Z includes X_C and R as special cases.

EXAMPLE 24–3 FIND THE FREQUENCY

An ac generator with an rms voltage of 110 V is connected in series with a 35-Ω resistor and a 11-μF capacitor. The rms current in the circuit is 1.2 A. What are **(a)** the impedance and **(b)** the capacitive reactance of this circuit? **(c)** What is the frequency, f, of the generator?

PICTURE THE PROBLEM

The appropriate circuit is shown in our sketch. We are given the rms voltage of the generator, $V_{\text{rms}} = 110$ V; the resistance, $R = 35\ \Omega$; and the capacitance, $C = 11\ \mu$F. The only remaining variable that affects the current is the frequency of the generator, f.

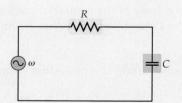

STRATEGY

a. The impedance, Z, can be solved directly from $V_{\text{rms}} = I_{\text{rms}}Z$.

b. Once the impedance is known, we can use $Z = \sqrt{R^2 + X_C^2}$ to find the capacitive reactance, X_C.

c. Now that the reactance is known, we can use the relation $X_C = 1/\omega C = 1/2\pi f C$ to solve for the frequency.

SOLUTION

Part (a)

1. Use $V_{\text{rms}} = I_{\text{rms}}Z$ to find the impedance, Z:
$$Z = \frac{V_{\text{rms}}}{I_{\text{rms}}} = \frac{110\ \text{V}}{1.2\ \text{A}} = 92\ \Omega$$

Part (b)

2. Solve for X_C in terms of Z and R:

$$Z = \sqrt{R^2 + X_C^2}$$

$$X_C = \sqrt{Z^2 - R^2} = \sqrt{(92\ \Omega)^2 - (35\ \Omega)^2} = 85\ \Omega$$

Part (c)

3. Use the value of X_C from part (b) to find the frequency, f:

$$X_C = \frac{1}{\omega C} = \frac{1}{2\pi f C}$$

$$f = \frac{1}{2\pi X_C C} = \frac{1}{2\pi(85\ \Omega)(11 \times 10^{-6}\ \text{F})} = 170\ \text{Hz}$$

INSIGHT

Note that the rms voltage across the resistor is $V_{\text{rms},R} = I_{\text{rms}} R = 42$ V, and the rms voltage across the capacitor is $V_{\text{rms},C} = I_{\text{rms}} X_C = 102$ V. As expected, these voltages *do not* add up to the generator voltage of 110 V—in fact, they add up to a considerably larger voltage. The point is, however, that the sum of 42 V and 102 V is physically meaningless because the voltages are 90° out of phase, and hence do not occur at the same time. If the voltages are combined as in Equation 24–10, which takes into account the 90° phase difference, we find $V_{\text{rms}} = \sqrt{(V_{\text{rms},R}^2 + V_{\text{rms},C}^2)} = \sqrt{(42\ \text{V})^2 + (102\ \text{V})^2} = 110$ V, as expected.

PRACTICE PROBLEM

(a) If the frequency in this circuit is increased, do you expect the rms current to increase, decrease, or stay the same? **(b)** What is the rms current if the frequency is increased to 250 Hz?   [**Answer: (a)** The rms current increases with frequency because the capacitive reactance decreases as the frequency is increased. **(b)** At 250 Hz the current is 1.6 A.]

Some related homework problems: Problem 20, Problem 21

ACTIVE EXAMPLE 24–1   FIND THE RESISTANCE

An ac generator with an rms voltage of 110 V and a frequency of 60.0 Hz is connected in series with a 270-μF capacitor and a resistor of resistance R. What value must R have if the rms current in this circuit is to be 1.7 A?

SOLUTION   *(Test your understanding by performing the calculations indicated in each step.)*

1. Solve for the resistance: $R = \sqrt{Z^2 - X_C^2}$

2. Find the impedance of the circuit: $Z = 65\ \Omega$

3. Find the capacitive reactance: $X_C = 9.8\ \Omega$

4. Substitute numerical values: $R = 64\ \Omega$

INSIGHT

At this frequency the capacitor has relatively little effect in the circuit, as can be seen by comparing X_C, R, and Z.

YOUR TURN

At what frequency is the capacitive reactance equal to 64 Ω? What is the current in the circuit at this frequency, assuming all other variables remain the same?

(Answers to **Your Turn** *problems are given in the back of the book.)*

We can also gain considerable insight about an ac circuit with qualitative reasoning—that is, without going through detailed calculations like those just given. This is illustrated in the following Conceptual Checkpoint.

CONCEPTUAL CHECKPOINT 24–2   COMPARE BRIGHTNESS: CAPACITOR CIRCUIT

Shown in the sketch are two circuits with identical ac generators and lightbulbs. Circuit 2 differs from circuit 1 by the addition of a capacitor in series with the light. Does the lightbulb in circuit 2 shine **(a)** more brightly, **(b)** less brightly, or **(c)** with the same intensity as that in circuit 1?

CONTINUED ON NEXT PAGE

CONTINUED FROM PREVIOUS PAGE

REASONING AND DISCUSSION

Since circuit 2 has both a resistance (in the lightbulb) and a capacitive reactance, its impedance, $Z = \sqrt{R^2 + X_C^2}$, is greater than that of circuit 1, which has only the resistance R. Therefore, the current in circuit 2, $I_{rms} = V_{rms}/Z$, is less than the current in circuit 1. As a result, the average power dissipated in the lightbulb, $P_{av} = I_{rms}^2 R$, is less in circuit 2, so its bulb shines less brightly.

Note that the bulb dims, even though no power is consumed by the capacitor.

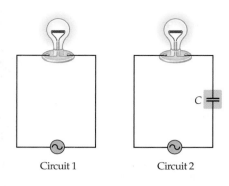

Circuit 1   Circuit 2

ANSWER

(b) The light bulb in circuit 2 shines less brightly.

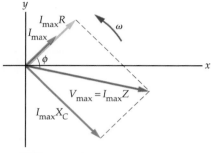

(a) Phase angle ϕ

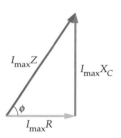

(b) $\cos \phi = R/Z$

▲ **FIGURE 24–11 Phase angle for an RC circuit**

(a) The phase angle, ϕ, is the angle between the voltage phasor, $V_{max} = I_{max}Z$, and the current phasor, I_{max}. Because I_{max} and $I_{max}R$ are in the same direction, we can say that ϕ is the angle between $I_{max}Z$ and $I_{max}R$. **(b)** From this triangle we can see that $\cos \phi = R/Z$, a result that is valid for any circuit.

PROBLEM-SOLVING NOTE

The Angle in the Power Factor

The angle in the power factor, $\cos \phi$, is the angle between the voltage phasor and the current phasor. Be careful not to identify ϕ with the angle between the voltage phasor and the x or y axis.

Phase Angle and Power Factor

We have seen how to calculate the current in an RC circuit and how to find the voltages across each element. Next we consider the phase relation between the total voltage in the circuit and the current. As we shall see, there is a direct connection between this phase relation and the power consumed by a circuit.

To do this, consider the phasor diagram shown in **Figure 24–11 (a)**. The **phase angle, ϕ**, between the voltage and the current can be read off the diagram as indicated in **Figure 24–11 (b)**. Clearly, the cosine of ϕ is given by the following:

$$\cos \phi = \frac{I_{max}R}{I_{max}Z} = \frac{R}{Z} \qquad 24\text{--}12$$

As we shall see, both the magnitude of the voltage and its phase angle relative to the current play important roles in the behavior of a circuit.

Consider, for example, the power consumed by the circuit shown in Figure 24–9. One way to obtain this result is to recall that no power is consumed by the capacitor at all, as was shown in the previous section. Thus the total power of the circuit is simply the power dissipated by the resistor. This power can be written as

$$P_{av} = I_{rms}^2 R$$

An equivalent expression for the power is obtained by replacing one power of I_{rms} with the expression $I_{rms} = V_{rms}/Z$. This replacement yields

$$P_{av} = I_{rms}^2 R = I_{rms}I_{rms}R = I_{rms}\left(\frac{V_{rms}}{Z}\right)R$$

Finally, recalling that $R/Z = \cos \phi$, we can write the power as follows:

$$P_{av} = I_{rms}V_{rms} \cos \phi \qquad 24\text{--}13$$

Thus, a knowledge of the current and voltage in a circuit, along with the value of $\cos \phi$, gives the power. The multiplicative factor $\cos \phi$ is referred to as the **power factor**.

Writing the power in terms of ϕ allows one to get a feel for the power in a circuit simply by inspecting the phasor diagram. For example, in the case of a circuit with only a capacitor, the phasor diagram (Figure 24–8) shows that the angle between the current and voltage has a magnitude of 90°, so the power factor is zero; $\cos \phi = \cos 90° = 0$. Thus no power is consumed in this circuit, as expected. In contrast, in a purely resistive circuit the phasor diagram shows that $\phi = 0$, as in Figure 24–3, giving a power factor of 1. In this case, the power is simply $P_{av} = I_{rms}V_{rms}$. Therefore, the angle between the current and the voltage in a phasor diagram gives an indication of the power being used by the circuit. We explore this feature in greater detail in the next Example and in **Figure 24–12**.

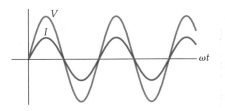
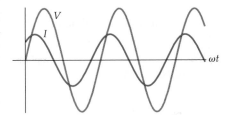
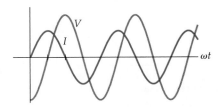

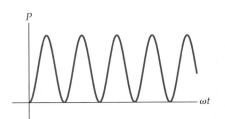
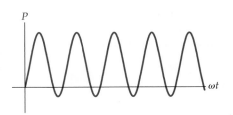
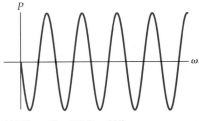

(a) *V* and *I* in phase ($\phi = 0$)   **(b)** *V* lags *I* by 45° ($\phi = 45°$)   **(c)** *V* lags *I* by 90° ($\phi = 90°$)

▲ **FIGURE 24–12 Voltage, current, and power for various phase angles, ϕ**
(a) $\phi = 0$. In this case the power is always positive. **(b)** $\phi = 45°$. The power oscillates between positive and negative values, with an average that is positive. **(c)** $\phi = 90°$. The power oscillates symmetrically about its average value of zero.

EXAMPLE 24–4 POWER AND RESISTANCE

A certain *RC* circuit has an ac generator with an rms voltage of 240 V. The rms current in the circuit is 2.5 A, and it leads the voltage by 56°. Find **(a)** the value of the resistance, *R*, and **(b)** the average power consumed by the circuit.

PICTURE THE PROBLEM
The phasor diagram appropriate to this circuit is shown in the sketch. Note that the voltage lags the current by the angle $\phi = 56°$.

STRATEGY

a. To find the resistance in the circuit, recall that $\cos \phi = R/Z$; thus, $R = Z \cos \phi$. We can find the impedance, *Z*, from $V_{rms} = I_{rms}Z$, since we know both V_{rms} and I_{rms}.

b. The average power consumed by the circuit is simply $P_{av} = I_{rms}V_{rms} \cos \phi$.

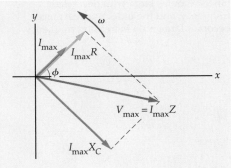

SOLUTION

Part (a)

1. Use $V_{rms} = I_{rms}Z$ to find the impedance of the circuit:

$$Z = \frac{V_{rms}}{I_{rms}} = \frac{240 \text{ V}}{2.5 \text{ A}} = 96 \ \Omega$$

2. Now use $\cos \phi = R/Z$ to find the resistance:

$$R = Z \cos \phi = (96 \ \Omega) \cos 56° = 54 \ \Omega$$

Part (b)

3. Calculate the average power with the expression $P_{av} = I_{rms}V_{rms} \cos \phi$:

$$P_{av} = I_{rms}V_{rms} \cos \phi = (2.5 \text{ A})(240 \text{ V}) \cos 56° = 340 \text{ W}$$

INSIGHT
Of course, the average power consumed by this circuit is simply the average power dissipated in the resistor: $P_{av} = I_{rms}^2 R = (2.5 \text{ A})^2 (54 \ \Omega) = 340$ W. Note that care must be taken if the average power dissipated in the resistor is calculated using $P_{av} = V_{rms}^2/R$. The potential pitfall is that one might use 240 V for the rms voltage; but this is the rms voltage of the generator, *not* the rms voltage across the resistor. This problem didn't arise with $P_{av} = I_{rms}^2 R$ because the same current flows through both the generator and the resistor.

PRACTICE PROBLEM
What is the capacitive reactance in this circuit?   [**Answer:** $X_C = 79 \ \Omega$]

Some related homework problems: Problem 22, Problem 23, Problem 24

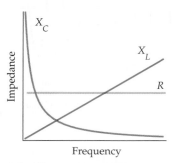

▲ **FIGURE 24–13** **An ac generator connected to an inductor**

The rms current in an ac inductor circuit is $I_{rms} = V_{rms}/X_L$, where the inductive reactance is $X_L = \omega L$.

▲ **FIGURE 24–14** **Frequency variation of the inductive reactance, X_L, the capacitive reactance, X_C, and the resistance, R**

The resistance is independent of frequency. In contrast, the capacitive reactance becomes large with decreasing frequency, and the inductive reactance becomes large with increasing frequency.

24–4 Inductors in ac Circuits

We turn now to the case of ac circuits containing inductors. As we shall see, the behavior of inductors is, in many respects, just the opposite of that of capacitors.

Inductive Reactance

The voltage across a capacitor is given by $V = IX_C$, where $X_C = 1/\omega C$ is the capacitive reactance. Similarly, the voltage across an inductor connected to an ac generator, as in **Figure 24–13**, can be written as $V = IX_L$, which defines the **inductive reactance,** X_L. The precise expression for X_L, in terms of frequency and inductance, can be derived using the methods of calculus. The result of such a calculation is the following:

Inductive Reactance, X_L

$$X_L = \omega L \qquad\qquad 24\text{–}14$$

SI unit: ohm, Ω

A plot of X_L versus frequency is given in **Figure 24–14**, where it is compared with X_C and R. The rms current in an ac inductor circuit is $I_{rms} = V_{rms}/X_L = V_{rms}/\omega L$.

Note that X_L increases with frequency, in contrast with the behavior of X_C. This is easily understood when one recalls that the voltage across an inductor has a magnitude given by $\mathcal{E} = L\,\Delta I/\Delta t$. Thus the higher the frequency, the more rapidly the current changes with time, and hence the greater the voltage across the inductor.

EXERCISE 24–5

A 21-mH inductor is connected to an ac generator with an rms voltage of 24 V and a frequency of 60.0 Hz. What is the rms current in the inductor?

SOLUTION

The rms current is $I_{rms} = V_{rms}/X_L = V_{rms}/\omega L$. Substituting numerical values, we find

$$I_{rms} = \frac{V_{rms}}{\omega L} = \frac{24\text{ V}}{2\pi(60.0\text{ s}^{-1})(21 \times 10^{-3}\text{ H})} = 3.0\text{ A}$$

Phasor Diagrams: Inductor Circuits

Now that we can calculate the magnitude of the voltage across an inductor, we turn to the question of the phase of the inductor's voltage relative to the current in the circuit. With both phase and magnitude determined, we can construct the appropriate phasor diagram for an inductor circuit.

Suppose the current in an inductor circuit is as shown in **Figure 24–15 (a)**. At time zero the current is zero, but increasing at its maximum rate. Since the voltage across an inductor depends on the rate of change of current, it follows that the inductor's voltage is a maximum at $t = 0$. When the current reaches a maximum value, at the time $\omega t = \pi/2$, its rate of change becomes zero; hence, the voltage across the inductor falls to zero at that point, as is also indicated in Figure 24–15 (a).

▶ **FIGURE 24–15** **Voltage, current, and power in an ac inductor circuit**

(a) Voltage V and current I in an inductor. Note that V reaches a maximum $\pi/2$ rad = 90° *before* the current. Thus, the voltage *leads* the current by 90°.
(b) The power consumed by an inductor. Note that the average power is zero.

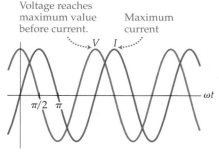

(a) V and I in an ac inductor circuit

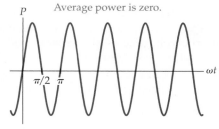

(b) Power in an ac inductor circuit

Thus we see that the current and the inductor's voltage are a quarter of a cycle (90°) out of phase. More specifically, since the voltage reaches its maximum *before* the current reaches its maximum, we say that the voltage *leads* the current:

> The voltage across an inductor *leads* the current by 90°.

Note that this, again, is just the opposite of the behavior in a capacitor.

The phase relation between current and voltage in an inductor is shown with a phasor diagram in **Figure 24–16**. Here we see the current, I_{max}, and the corresponding inductor voltage, $V_{max} = I_{max}X_L$. Note that the voltage is rotated counterclockwise (that is, ahead) of the current by 90°.

Because of the 90° angle between the current and voltage, the power factor for an inductor is zero; cos 90° = 0. Thus, an ideal inductor—like an ideal capacitor—consumes zero average power. That is,

$$P_{av} = I_{rms}V_{rms}\cos\phi = I_{rms}V_{rms}\cos 90° = 0$$

The instantaneous power in an inductor alternates in sign, as shown in **Figure 24–15 (b)**. Thus energy enters the inductor at one time, only to be given up at a later time, for a net gain on average of zero energy.

RL Circuits

Next, we consider an ac circuit containing both a resistor and an inductor, as shown in **Figure 24–17**. The corresponding phasor diagram is drawn in **Figure 24–18**, where we see the resistor voltage in phase with the current, and the inductor voltage 90° ahead. The total voltage, of course, is given by the vector sum of these two phasors. Therefore, the magnitude of the total voltage is

$$V_{max} = \sqrt{(I_{max}R)^2 + (I_{max}X_L)^2} = I_{max}\sqrt{R^2 + X_L^2} = I_{max}Z$$

This expression defines the impedance, just as with an *RC* circuit. In this case, the impedance is

Impedance in an *RL* Circuit

$$Z = \sqrt{R^2 + X_L^2} = \sqrt{R^2 + (\omega L)^2} \qquad \text{24–15}$$

SI unit: ohm, Ω

Note that the impedance for an *RL* circuit has the same form as for an *RC* circuit, except that X_L replaces X_C. Similarly, the power factor for an *RL* circuit can be written as follows:

$$\cos\phi = \frac{R}{Z} = \frac{R}{\sqrt{R^2 + (\omega L)^2}}$$

We consider an *RL* circuit in the following Example.

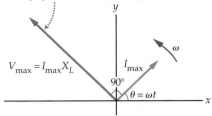

Voltage phasor leads current phasor by 90°.

▲ **FIGURE 24–16 Phasor diagram for an ac inductor circuit**
The maximum value of the inductor's voltage is $I_{max}X_L$, and its angle is 90° ahead (counterclockwise) of the current.

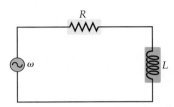

▲ **FIGURE 24–17 An alternating-current *RL* circuit**
An ac *RL* circuit consists of a generator, a resistor, and an inductor connected in series.

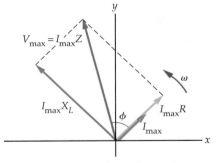

▲ **FIGURE 24–18 Phasor diagram for an *RL* circuit**
The maximum current in an *RL* circuit is $I_{max} = V_{max}/Z$, where V_{max} is the maximum voltage of the ac generator and $Z = \sqrt{R^2 + X_L^2}$ is the impedance of the circuit. The angle between the maximum voltage phasor and the current phasor is ϕ. Note that the voltage leads the current.

EXAMPLE 24–5 *RL* CIRCUIT

A 0.380-H inductor and a 225-Ω resistor are connected in series to an ac generator with an rms voltage of 30.0 V and a frequency of 60.0 Hz. Find the rms values of **(a)** the current in the circuit, **(b)** the voltage across the resistor, and **(c)** the voltage across the inductor.

PICTURE THE PROBLEM
Our sketch shows a 60.0-Hz generator connected in series with a 225-Ω resistor and a 0.380-H inductor. Because of the series connection, the same current flows through each of the circuit elements.

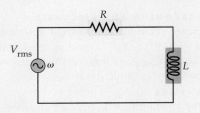

STRATEGY

a. The rms current in the circuit is $I_{rms} = V_{rms}/Z$, where the impedance is $Z = \sqrt{R^2 + (\omega L)^2}$.

CONTINUED ON NEXT PAGE

CONTINUED FROM PREVIOUS PAGE

b. The rms voltage across the resistor is $V_{rms,R} = I_{rms}R$.

c. The rms voltage across the inductor is $V_{rms,L} = I_{rms}X_L = I_{rms}\,\omega L$.

SOLUTION

Part (a)

1. Calculate the impedance of the circuit:

$$Z = \sqrt{R^2 + X_L^2} = \sqrt{R^2 + (\omega L)^2}$$
$$= \sqrt{(225\ \Omega)^2 + [2\pi(60.0\ \text{s}^{-1})(0.380\ \text{H})]^2} = 267\ \Omega$$

2. Use Z to find the rms current:

$$I_{rms} = \frac{V_{rms}}{Z} = \frac{30.0\ \text{V}}{267\ \Omega} = 0.112\ \text{A}$$

Part (b)

3. Multiply I_{rms} by the resistance, R, to find the rms voltage across the resistor:

$$V_{rms,R} = I_{rms}R = (0.112\ \text{A})(225\ \Omega) = 25.2\ \text{V}$$

Part (c)

4. Multiply I_{rms} by the inductive reactance, X_L, to find the rms voltage across the inductor:

$$V_{rms,L} = I_{rms}X_L = I_{rms}\omega L$$
$$= (0.112\ \text{A})2\pi(60.0\ \text{s}^{-1})(0.380\ \text{H}) = 16.0\ \text{V}$$

INSIGHT

As with the RC circuit, the individual voltages *do not* add up to the generator voltage. However, the generator rms voltage *is* equal to $\sqrt{V_{rms,R}^2 + V_{rms,L}^2}$.

PRACTICE PROBLEM

If the frequency in this circuit is increased, do you expect the rms current to increase, decrease, or stay the same? What is the current if the frequency is increased to 125 Hz?   [**Answer:** The current will decrease with frequency. At 125 Hz the current is 0.0803 A.]

Some related homework problems: Problem 32, Problem 33

The preceding Practice Problem shows that the current in an RL circuit decreases with increasing frequency. Conversely, as we saw in Example 24–3, an RC circuit has the opposite behavior. These results are summarized and extended in **Figure 24–19**, where we show the current in RL and RC circuits as a function of frequency. It is assumed in this plot that both circuits have the same resistance, R. The horizontal line at the top of the plot indicates the current that would flow if the circuits contained *only* the resistor.

CONCEPTUAL CHECKPOINT 24–3   COMPARE BRIGHTNESS: INDUCTOR CIRCUIT

Shown in the sketch below are two circuits with identical ac generators and lightbulbs. Circuit 2 differs from circuit 1 by the addition of an inductor in series with the light. Does the lightbulb in circuit 2 shine **(a)** more brightly, **(b)** less brightly, or **(c)** with the same intensity as that in circuit 1?

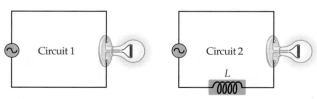

REASONING AND DISCUSSION

Since circuit 2 has both a resistance and an inductive reactance, its impedance, $Z = \sqrt{R^2 + X_L^2}$, is greater than that of circuit 1. Therefore, the current in circuit 2, $I_{rms} = V_{rms}/Z$, is less than the current in circuit 1, and hence the average power dissipated in lightbulb 2 is less.

ANSWER

(b) The lightbulb in circuit 2 shines less brightly.

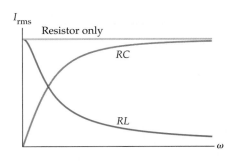

▶ **FIGURE 24–19 RMS currents in *RL* and *RC* circuits as a function of frequency**

A circuit with only a resistor has the same rms current regardless of frequency. In an *RC* circuit the current is low where the capacitive reactance is high (low frequency), and in the *RL* circuit the current is low where the inductive reactance is high (high frequency).

The basic principle illustrated in Conceptual Checkpoint 24–3 is used commercially in light dimmers. When you rotate the knob on a light dimmer in one direction or the other, adjusting the light's intensity to the desired level, what you are actually doing is moving an iron rod into or out of the coils of an inductor. This changes both the inductance of the inductor and the intensity of the light. For example, if the inductance is increased—by moving the iron rod deeper within the inductor's coil—the current in the circuit is decreased and the light dims. The advantage of dimming a light in this way is that no energy is dissipated by the inductor. In contrast, if one were to dim a light by placing a resistor in the circuit, the reduction in light would occur at the expense of wasted energy in the form of heat in the resistor. This needless dissipation of energy is avoided with the inductive light dimmer.

REAL-WORLD PHYSICS

Light dimmers

24–5 *RLC* Circuits

After having considered *RC* and *RL* circuits separately, we now consider circuits with all three elements, *R*, *L*, and *C*. In particular, if *R*, *L*, and *C* are connected in series, as in **Figure 24–20**, the resulting circuit is referred to as an ***RLC* circuit.** We now consider the behavior of such a circuit, using our previous results as a guide.

Phasor Diagram

As one might expect, a useful way to analyze the behavior of an *RLC* circuit is with the assistance of a phasor diagram. Thus, the phasor diagram for the circuit in Figure 24–20 is shown in **Figure 24–21**. Note that in addition to the current phasor, we show three separate voltage phasors corresponding to the resistor, inductor, and capacitor. The phasor diagram also shows that the voltage of the resistor is in phase with the current, the voltage of the inductor is 90° ahead of the current, and the voltage of the capacitor is 90° behind the current.

To find the total voltage in the system we must, as usual, perform a vector sum of the three voltage phasors. This process can be simplified if we first sum the inductor and capacitor phasors, which point in opposite directions along the same line. In the case shown in Figure 24–21 we assume for specificity that X_L is greater than X_C, so that the sum of these two voltage phasors is $I_{max}X_L - I_{max}X_C$. Combining this result with the phasor for the resistor voltage, and applying the Pythagorean theorem, we obtain the total maximum voltage:

$$V_{max} = \sqrt{(I_{max}R)^2 + (I_{max}X_L - I_{max}X_C)^2} = I_{max}\sqrt{R^2 + (X_L - X_C)^2} = I_{max}Z$$

The impedance of the circuit is thus defined as follows:

Impedance of an *RLC* Circuit

$$Z = \sqrt{R^2 + (X_L - X_C)^2} = \sqrt{R^2 + \left(\omega L - \frac{1}{\omega C}\right)^2} \qquad 24\text{–}16$$

SI unit: ohm, Ω

Note that this result for the impedance contains the expressions given for the *RC* and *RL* circuits as special cases. For example, in the *RC* circuit the inductance is zero; hence, $X_L = 0$, and the impedance becomes

$$Z = \sqrt{R^2 + X_C{}^2}$$

This relation is identical with the result given for the *RC* circuit in Equation 24–11.

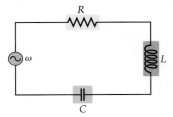

▲ **FIGURE 24–20 An alternating-current *RLC* circuit**

An ac *RLC* circuit consists of a generator, a resistor, an inductor, and a capacitor connected in series. The voltage may lead or lag the current, depending on the frequency of the generator and the values of *L* and *C*.

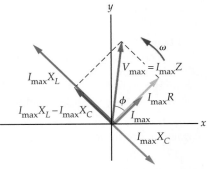

▲ **FIGURE 24–21 Phasor diagram for a typical *RLC* circuit**

In the case shown here, we assume that X_L is greater than X_C. All the results are the same for the opposite case, X_C greater than X_L, except that the phase angle ϕ changes sign, as can be seen in Equation 24–17.

In addition to the magnitude of the total voltage, we are also interested in the phase angle ϕ between it and the current. From the phasor diagram it is clear that this angle is given by

$$\tan \phi = \frac{I_{max}(X_L - X_C)}{I_{max}R} = \frac{X_L - X_C}{R} \qquad \text{24–17}$$

In particular, if X_L is greater than X_C, as in Figure 24–21, then ϕ is positive and the voltage leads the current. On the other hand, if X_C is greater than X_L it follows that ϕ is negative, so the voltage lags the current. In the special case $X_C = X_L$ the phase angle is zero, and the current and voltage are in phase. There is special significance to this last case, as we shall see in greater detail in the next section.

Finally, the power factor, $\cos \phi$, can also be obtained from the phasor diagram. Referring again to Figure 24–21, we see that

$$\cos \phi = \frac{R}{Z}$$

Note that this is precisely the result given earlier for RC and RL circuits.

EXAMPLE 24–6 DRAW YOUR PHASORS!

An ac generator with a frequency of 60.0 Hz and an rms voltage of 120.0 V is connected in series with a 175-Ω resistor, a 90.0-mH inductor, and a 15.0-μF capacitor. Draw the appropriate phasor diagram for this system and calculate the phase angle, ϕ.

PICTURE THE PROBLEM
Our sketch shows the RLC circuit described in the problem statement. In particular, the 60.0-Hz generator is connected in series with a 175-Ω resistor, a 90.0-mH inductor, and a 15.0-μF capacitor.

STRATEGY
To draw an appropriate phasor diagram we must first determine the values of X_C and X_L. These values determine immediately whether the voltage leads the current or lags it.

The precise value of the phase angle between the current and voltage is given by $\tan \phi = (X_L - X_C)/R$.

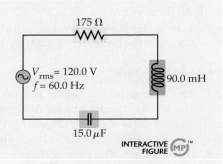

SOLUTION

1. Calculate the capacitive and inductive reactances:

$$X_C = \frac{1}{\omega C} = \frac{1}{2\pi(60.0 \text{ s}^{-1})(15.0 \times 10^{-6} \text{ F})} = 177 \text{ }\Omega$$

$$X_L = \omega L = 2\pi(60.0 \text{ s}^{-1})(90.0 \times 10^{-3} \text{ H}) = 33.9 \text{ }\Omega$$

2. Use $\tan \phi = (X_L - X_C)/R$ to find the phase angle, ϕ:

$$\phi = \tan^{-1}\left(\frac{X_L - X_C}{R}\right)$$

$$= \tan^{-1}\left(\frac{33.9 \text{ }\Omega - 177 \text{ }\Omega}{175 \text{ }\Omega}\right) = -39.3°$$

INSIGHT
We can now draw the phasor diagram for this circuit. First, the fact that X_C is greater than X_L means that the voltage of the circuit lags the current and that the phase angle is negative. In fact, we know that the phase angle has a magnitude of 39.3°. It follows that the phasor diagram for this circuit is as shown in the diagram.

Note that the lengths of the phasors $I_{max}R$, $I_{max}X_C$, and $I_{max}X_L$ in this diagram are drawn in proportion to the values of R, X_C, and X_L, respectively.

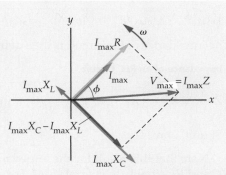

PRACTICE PROBLEM
Find the impedance and rms current for this circuit. [**Answer:** $Z = 226 \text{ }\Omega$, $I_{rms} = V_{rms}/Z = 0.531$ A]

Some related homework problems: Problem 52, Problem 53

The phase angle can also be obtained from the power factor: $\cos \phi = R/Z = (175\ \Omega)/(226\ \Omega) = 0.774$. Notice, however, that $\cos^{-1}(0.774) = \pm 39.3°$. Thus, because of the symmetry of cosine, the power factor determines only the magnitude of ϕ, not its sign. The sign can be obtained from the phasor diagram, of course, but using $\phi = \tan^{-1}(X_L - X_C)/R$ yields both the magnitude and sign of ϕ.

ACTIVE EXAMPLE 24–2   FIND THE INDUCTOR VOLTAGE

The circuit elements shown in the sketch are connected to an ac generator at points A and B. If the rms voltage of the generator is 41 V and its frequency is 75 Hz, what is the rms voltage across the inductor?

A   42 Ω   150 mH   35 μF   B

SOLUTION   *(Test your understanding by performing the calculations indicated in each step.)*

1. Calculate the capacitive reactance: $\qquad\qquad\qquad\qquad X_C = 61\ \Omega$

2. Calculate the inductive reactance: $\qquad\qquad\qquad\qquad X_L = 71\ \Omega$

3. Use Equation 24–16 to determine the impedance of the circuit: $\quad Z = 43\ \Omega$

4. Divide V_{rms} by Z to find the rms current: $\qquad\qquad\qquad I_{rms} = 0.95\ A$

5. Multiply I_{rms} by X_L to find the rms voltage of the inductor: $\quad V_{rms,L} = 67\ V$

INSIGHT
Note that the voltage across individual components in a circuit can be larger than the applied voltage. In this case the applied voltage is 41 V, whereas the voltage across the inductor is 67 V.

YOUR TURN
Find the rms voltage across **(a)** the resistor and **(b)** the capacitor.

(Answers to **Your Turn** *problems are given in the back of the book.)*

Table 24–1 summarizes the various results for ac circuit elements (R, L, and C) and their combinations (RC, RL, RLC).

TABLE 24–1 Properties of AC Circuit Elements and Their Combinations

Circuit element	Impedance, Z	Average power, P_{av}	Phase angle, ϕ	Phasor diagram
⌇W⌇	$Z = R$	$P_{av} = I_{rms}^2 R = V_{rms}^2/R$	$\phi = 0°$	Figure 24–3
⊣⊢	$Z = X_C = \dfrac{1}{\omega C}$	$P_{av} = 0$	$\phi = -90°$	Figure 24–8
⌁	$Z = X_L = \omega L$	$P_{av} = 0$	$\phi = +90°$	Figure 24–16
⌇W⌇ ⊣⊢	$Z = \sqrt{R^2 + X_C^2} = \sqrt{R^2 + \left(\dfrac{1}{\omega C}\right)^2}$	$P_{av} = I_{rms}V_{rms}\cos\phi$	$-90° < \phi < 0°$	Figure 24–10
⌇W⌇ ⌁	$Z = \sqrt{R^2 + X_L^2} = \sqrt{R^2 + (\omega L)^2}$	$P_{av} = I_{rms}V_{rms}\cos\phi$	$0° < \phi < 90°$	Figure 24–18
⌇W⌇ ⌁ ⊣⊢	$Z = \sqrt{R^2 + (X_L - X_C)^2}$ $= \sqrt{R^2 + \left(\omega L - \dfrac{1}{\omega C}\right)^2}$	$P_{av} = I_{rms}V_{rms}\cos\phi$	$-90° < \phi < 0°$ ($X_C > X_L$) $0° < \phi < 90°$ ($X_C < X_L$)	Figure 24–21

Large and Small Frequencies

In the limit of very large or small frequencies, the behavior of capacitors and inductors is quite simple, allowing us to investigate more complex circuits

(a) Original circuit   **(b)** High-frequency limit   **(c)** Low-frequency limit

▲ **FIGURE 24–22 High-frequency and low-frequency limits of an ac circuit**
(a) A complex circuit containing resistance, inductance, and capacitance. Although this is not a simple *RLC* circuit, we can still obtain useful results about the circuit in the limits of high and low frequencies. **(b)** The high-frequency limit, in which the inductor is essentially an open circuit, and the capacitor behaves like an ideal wire. **(c)** The low-frequency limit. In this case, the inductor is like an ideal wire, and the capacitor acts like an open circuit.

containing *R*, *L*, and *C*. For example, in the limit of large frequency the reactance of an inductor becomes very large, whereas that of a capacitor becomes vanishingly small. This means that a capacitor acts like a segment of ideal wire, with no resistance, and an inductor behaves like a very large resistor with practically no current—essentially, an open circuit.

By applying these observations to the circuit in **Figure 24–22 (a)**, for example, we can predict the current supplied by the generator at high frequencies. First, we replace the capacitor with an ideal wire and the inductor with an open circuit. This results in the simplified circuit shown in **Figure 24–22 (b)**. Clearly, the current in this circuit is 0.50 A; hence, we expect the current in the original circuit to approach this value as the frequency is increased.

In the opposite extreme of very small frequency we obtain the behavior that would be expected if the ac generator were replaced with a battery. Specifically, the reactance of an inductor vanishes as the frequency goes to zero, whereas that of a capacitor becomes extremely large. Thus, the roles of the inductor and capacitor are now reversed; it is the inductor that acts like an ideal wire, and the capacitor that behaves like an open circuit. For small frequencies, then, the circuit in Figure 24–22 (a) will behave the same as the circuit shown in **Figure 24–22 (c)**. The current in this circuit, and in the original circuit at low frequency, is 1.0 A.

EXAMPLE 24–7 FIND *R*

The circuit shown in the sketch is connected to an ac generator with an rms voltage of 120 V. What value must *R* have if the rms current in this circuit is to approach 1.0 A at high frequency?

PICTURE THE PROBLEM
The top diagram shows the original circuit with its various elements. The high-frequency behavior of this circuit is indicated in the bottom diagram, where the inductor has been replaced with an open circuit, and the capacitors have been replaced with ideal wires.

STRATEGY
The high-frequency circuit has only one remaining path through which current can flow. On this path the resistors with resistance *R* and 100 Ω are in series. Therefore, the total resistance of the circuit is $R_{\text{total}} = R + 100\ \Omega$. Finally, the rms current is $I_{\text{rms}} = V_{\text{rms}}/R_{\text{total}}$. Setting I_{rms} equal to 1.0 A gives us the value of *R*.

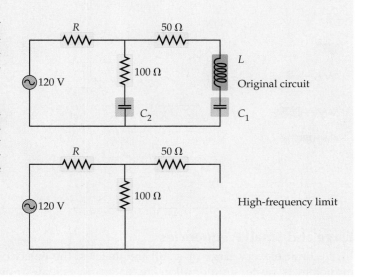

SOLUTION

1. Calculate the total resistance of the high-frequency circuit:

$$R_{total} = R + 100 \ \Omega$$

2. Write an expression for the rms current in the circuit:

$$I_{rms} = \frac{V_{rms}}{R_{total}} = \frac{V_{rms}}{R + 100 \ \Omega}$$

3. Solve for the resistance R:

$$R = \frac{V_{rms}}{I_{rms}} - 100 \ \Omega = \frac{120 \ V}{1.0 \ A} - 100 \ \Omega = 20 \ \Omega$$

INSIGHT

Note that no values are given for the capacitances and the inductance. At high enough frequencies the precise values of these quantities are unimportant.

PRACTICE PROBLEM

What is the rms current in this circuit in the limit of small frequency?   [**Answer:** The current approaches zero.]

Some related homework problems: Problem 50, Problem 51

24–6 Resonance in Electric Circuits

Many physical systems have natural frequencies of oscillation. For example, we saw in Chapter 13 that a child on a swing oscillates about the vertical with a definite, characteristic frequency determined by the length of the swing and the acceleration of gravity. Similarly, an object attached to a spring oscillates about its equilibrium position with a frequency determined by the "stiffness" of the spring and the mass of the object. Certain electric circuits have analogous behavior—their electric currents oscillate with certain characteristic frequencies. In this section we consider some examples of "oscillating" electric circuits.

LC Circuits

Perhaps the simplest circuit that displays an oscillating electric current is an **LC circuit** with no generator; that is, one that consists of nothing more than an inductor and a capacitor. Suppose, for example, that at $t = 0$ a charged capacitor has just been connected to an inductor and that there is no current in the circuit, as shown in **Figure 24–23 (a)**. Since the capacitor is charged, it has a voltage, $V = Q/C$, which causes a current to begin flowing through the inductor, as in **Figure 24–23 (b)**. Soon all the charge drains from the capacitor and its voltage drops to zero, but the current continues to flow because an inductor resists changes in its current. In fact, the current continues flowing until the capacitor becomes charged enough in the *opposite* direction to stop the current, as in **Figure 24–23 (c)**. At this point, the current begins to flow back the way it came, and the same sequence of events occurs again, leading to a *series of oscillations* in the current.

In the ideal case, the oscillations can continue forever, since neither an inductor nor a capacitor dissipates energy. The situation is completely analogous to a mass oscillating on a spring with no friction, as we indicate in Figure 24–23. At $t = 0$ the capacitor has a charge of magnitude Q on its plates, which means it stores the energy $U_C = Q^2/2C$. This situation is analogous to a spring being compressed a distance x and storing the potential energy $U = \frac{1}{2}kx^2$. At a later time the charge on the capacitor is zero, so it no longer stores any energy. The energy is not lost, however. Instead, it is now in the inductor, which carries a current I and stores the energy $U_L = \frac{1}{2}LI^2 = U_C$. In the mass–spring system, this corresponds to the mass being at the equilibrium position of the spring. At this time all the system's energy is the kinetic energy of the mass, $K = \frac{1}{2}mv^2 = U$, and none is stored in the spring. As the current continues to flow, it charges the capacitor with the opposite polarity until it reaches the magnitude Q and stores the same energy, U_C, as at $t = 0$. In the mass–spring system, this corresponds to the spring being stretched by the same distance x that it was originally compressed, which again stores all the initial energy as potential energy.

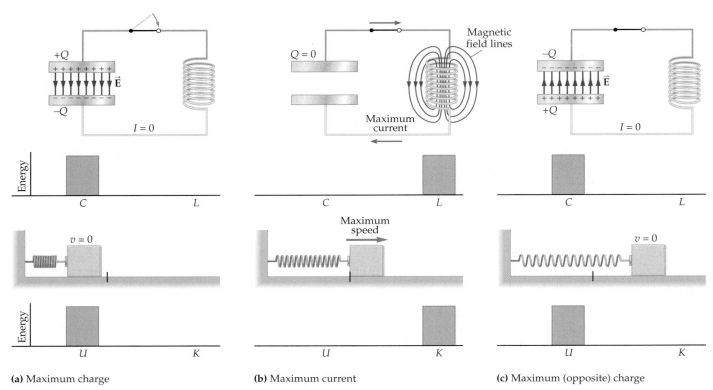

(a) Maximum charge   (b) Maximum current   (c) Maximum (opposite) charge

▲ **FIGURE 24–23 Oscillations in an *LC* circuit with no generator**
The current oscillations in an *LC* circuit are analogous to the oscillations of a mass on a spring. **(a)** At the moment the switch is closed, all the energy in the circuit is stored in the electric field of the charged capacitor. This is analogous to a mass at rest against a compressed spring, where all the energy of the system is stored in the spring. **(b)** A quarter of the way through a cycle, the capacitor is uncharged and the current in the inductor is a maximum. At this time, all the energy in the circuit is stored in the magnetic field of the inductor. In the mass–spring analog, the spring is at equilibrium and the mass has its maximum speed. All the system's energy is now in the form of kinetic energy. **(c)** After half a cycle, the capacitor is fully charged in the opposite direction and holds all the system's energy. This corresponds to a fully extended spring (with all the system's energy) and the mass at rest.

Thus, we see that a close analogy exists between a capacitor and a spring, and between an inductor and a mass. In addition, the charge on the capacitor is analogous to the displacement of the spring, and the current in the inductor is analogous to the speed of the mass. Thus, for example, the energy stored in the inductor, $\frac{1}{2}LI^2$, corresponds precisely to the kinetic energy of the mass, $\frac{1}{2}mv^2$. Comparing the potential energy of a spring, $\frac{1}{2}kx^2$, and the energy stored in a capacitor, $Q^2/2C$, we see that the stiffness of a spring is analogous to $1/C$. This makes sense, because a capacitor with a large capacitance, C, can store large quantities of charge with ease, just as a spring with a small force constant (if C is large, then $k = 1/C$ is small) can be stretched quite easily.

In the mass–spring system the natural angular frequency of oscillation is determined by the characteristics of the system. In particular, recall from Section 13–8 that

$$f_0 = \frac{1}{2\pi}\sqrt{\frac{k}{m}} \quad \text{or} \quad \omega_0 = 2\pi f_0 = \sqrt{\frac{k}{m}}$$

The natural frequency of the *LC* circuit can be determined by noting that the rms voltage across the capacitor C in Figure 24–23 must be the same as the rms voltage across the inductor L. This condition can be written as follows:

$$V_{\text{rms},C} = V_{\text{rms},L}$$

$$I_{\text{rms}}X_C = I_{\text{rms}}X_L$$

$$I_{\text{rms}}\left(\frac{1}{\omega_0 C}\right) = I_{\text{rms}}(\omega_0 L)$$

Solving for ω_0, we find

Natural Frequency of an LC Circuit

$$\omega_0 = \frac{1}{\sqrt{LC}} = 2\pi f \qquad\qquad 24\text{–}18$$

SI unit: s^{-1}

Note again the analogy between this result and that for a mass on a spring: if we make the following replacements, $m \rightarrow L$ and $k \rightarrow 1/C$, in the mass–spring result we find the expected LC result:

$$\omega_0 = \sqrt{\frac{k}{m}} = \sqrt{\frac{1}{LC}}$$

We summarize the mass–spring/LC circuit analogies in Table 24–2.

TABLE 24–2 Analogies Between a Mass on a Spring and an *LC* Circuit

Mass–spring system		LC circuit	
Position	x	Charge	q
Velocity	$v = \Delta x/\Delta t$	Current	$I = \Delta q/\Delta t$
Mass	m	Inductance	L
Force constant	k	Inverse capacitance	$1/C$
Natural Frequency	$\omega_0 = \sqrt{k/m}$	Natural frequency	$\omega_0 = \sqrt{1/LC}$

EXERCISE 24–6

It is desired to tune the natural frequency of an LC circuit to match the 88.5-MHz broadcast signal of an FM radio station. If a 1.50-μH inductor is to be used in the circuit, what capacitance is required?

SOLUTION

Solving $\omega_0 = 1/\sqrt{LC}$ for the capacitance, we find

$$C = \frac{1}{\omega_0{}^2 L} = \frac{1}{[2\pi(88.5 \times 10^6\ s^{-1})]^2(1.50 \times 10^{-6}\ H)} = 2.16 \times 10^{-12}\ F$$

Resonance

Whenever a physical system has a natural frequency, we can expect to find resonance when it is driven near that frequency. In a mass–spring system, for example, if we move the top end of the spring up and down with a frequency near the natural frequency, the displacement of the mass can become quite large. Similarly, if we push a person on a swing at the right frequency, the amplitude of motion will increase. These are examples of resonance in mechanical systems.

To drive an electric circuit, we can connect it to an ac generator. As we adjust the frequency of the generator, the current in the circuit will be a maximum at the natural frequency of the circuit. For example, consider the circuit shown in Figure 24–20. Here an ac generator drives a circuit containing an inductor, a capacitor, and a resistor. As we have already seen, the inductor and capacitor together establish a natural frequency $\omega_0 = 1/\sqrt{LC}$, and the resistor provides for energy dissipation.

The phasor diagram for this circuit is like the one shown in Figure 24–21. Recall that the maximum current in this circuit is

$$I_{max} = \frac{V_{max}}{Z}$$

Thus, the smaller the impedance, the larger the current. Hence, to obtain the largest possible current, we must have the smallest Z. Recall that the impedance is given by

$$Z = \sqrt{R^2 + (X_L - X_C)^2}$$

Writing Z in terms of frequency, we have

$$Z = \sqrt{R^2 + \left(\omega L - \frac{1}{\omega C}\right)^2}$$ 24–19

This expression for Z is plotted in **Figure 24–24 (a)**. Notice that the smallest value of the impedance is $Z = R$, and that this value is attained precisely at the frequency where $X_L = X_C$. We can see this mathematically by setting $X_L = X_C$ in the expression $Z = \sqrt{R^2 + (X_L - X_C)^2}$, which yields $Z = \sqrt{R^2 + 0} = R$, as expected. The frequency at which $X_L = X_C$ is the frequency for which $\omega L = 1/\omega C$. This frequency is $\omega = 1/\sqrt{LC} = \omega_0$, the *natural frequency* found in Equation 24–18 for LC circuits.

PROBLEM-SOLVING NOTE

A Circuit at Resonance

The resonance frequency of an RLC circuit depends only on the inductance, L, and the capacitance, C; that is, $\omega_0 = 1/\sqrt{LC}$. On the other hand, the impedance at resonance depends only on the resistance, R; that is, $Z = R$.

▶ **FIGURE 24–24** **Resonance in an**
***RLC* circuit**

(a) The impedance, Z, of an RLC circuit varies as a function of frequency. The minimum value of Z—which corresponds to the largest current—occurs at the resonance frequency $\omega_0 = 1/\sqrt{LC}$, where $X_C = X_L$. At this frequency $Z = R$. **(b)** Typical resonance peaks for an RLC circuit. The location of the peak is independent of resistance, but the resonance effect becomes smaller and more spread out with increasing R.

REAL-WORLD PHYSICS

Tuning a radio or television

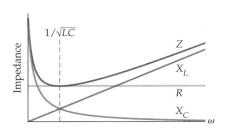

(a) Impedance is a minimum
($Z = R$) at resonance

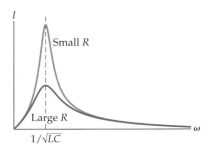

(b) Resonance curves for the current

Figure 24–24 (b) shows typical plots of the current in an RLC circuit. Note that the current peaks at the resonance frequency. Note also that increasing the resistance, although it reduces the maximum current in the circuit, does not change this frequency. It does, however, make the resonance peak flatter and broader. As a result, the resonance effect occurs over a wide range of frequencies and gives only a small increase in current. If the resistance is small, however, the peak is high and sharp. In this case, the resonance effect yields a large current that is restricted to a very small range of frequencies.

Radio and television tuners use low-resistance RLC circuits so they can pick up one station at a frequency f_1 without also picking up a second station at a nearby frequency f_2. In a typical radio tuner, for example, turning the knob of the tuner rotates one set of capacitor plates between a second set of plates, effectively changing both the plate separation and the plate area. This changes the capacitance in the circuit and the frequency at which it resonates. If the resonance peak is high and sharp—as occurs with low resistance—only the station broadcasting at the resonance frequency will be picked up and amplified. Other stations at nearby frequencies will produce such small currents in the tuning circuit that they will be undetectable.

Metal detectors also use resonance in RLC circuits, although in this case it is the inductance that changes rather than the capacitance. For example, when you walk through a metal detector at an airport, you are actually walking through a large coil of wire—an inductor. Metal objects on your person cause the inductance of the coil to increase slightly. This increase in inductance results in a small decrease in the resonance frequency of the RLC circuit to which the coil is connected. If the resonance peak is sharp and high, even a slight change in frequency results in a large change in current. It is this change in current that sets off the detector, indicating the presence of metal.

REAL-WORLD PHYSICS

Metal detectors

CONCEPTUAL CHECKPOINT 24–4   PHASE OF THE VOLTAGE

An RLC circuit is driven at its resonance frequency. Is its voltage **(a)** ahead of, **(b)** behind, or **(c)** in phase with the current?

REASONING AND DISCUSSION

At resonance the capacitive and inductive reactances are equal, which means that the voltage across the capacitor is equal in magnitude and opposite in direction to the voltage across the inductor. As a result, the net voltage in the system is simply the voltage across the resistor, which is in phase with the current.

ANSWER

(c) The voltage and current are in phase.

EXAMPLE 24–8   A CIRCUIT IN RESONANCE

An ac generator with an rms voltage of 25 V is connected in series to a 10.0-Ω resistor, a 53-mH inductor, and a 65-μF capacitor. Find **(a)** the resonance frequency of the circuit and **(b)** the rms current at resonance. In addition, sketch the phasor diagram at resonance.

PICTURE THE PROBLEM

As mentioned in the previous Conceptual Checkpoint, at resonance the magnitude of the voltage across the inductor is equal to the magnitude of the voltage across the capacitor. Since the phasors corresponding to these voltages point in opposite directions, however, they cancel. This leads to a net voltage phasor that is simply $I_{max}R$ in phase with the current. Finally, note that the lengths of the phasors $I_{max}R$, $I_{max}X_C$, and $I_{max}X_L$ are drawn in proportion to the values of R, X_C, and X_L, respectively.

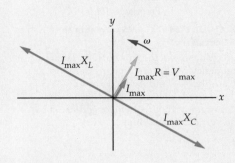

STRATEGY

a. We find the resonance frequency by substituting numerical values into $\omega_0 = 1/\sqrt{LC} = 2\pi f_0$.

b. At resonance the impedance of the circuit is simply the resistance; that is, $Z = R$. Thus, the rms current in the circuit is $I_{rms} = V_{rms}/Z = V_{rms}/R$.

SOLUTION

Part (a)

1. Calculate the resonance frequency:

$$\omega_0 = \frac{1}{\sqrt{LC}} = 2\pi f_0$$

$$f_0 = \frac{1}{2\pi\sqrt{LC}}$$

$$= \frac{1}{2\pi\sqrt{(53 \times 10^{-3}\,\text{H})(65 \times 10^{-6}\,\text{F})}} = 86\,\text{Hz}$$

Part (b)

2. Determine the impedance at resonance:

$$Z = R = 10.0\,\Omega$$

3. Divide the rms voltage by the impedance to find the rms current:

$$I_{rms} = \frac{V_{rms}}{Z} = \frac{25\,\text{V}}{10.0\,\Omega} = 2.5\,\text{A}$$

INSIGHT

If the frequency of this generator is increased above resonance, the inductive reactance, X_L, will be larger than the capacitive reactance, X_C, and hence the voltage will lead the current. If the frequency is lowered below resonance the voltage will lag the current.

PRACTICE PROBLEM

What is the magnitude of the rms voltage across the capacitor?   [**Answer:** 71 V. Note again that the voltage across a given circuit element can be much larger than the applied voltage.]

Some related homework problems: Problem 67, Problem 69

ACTIVE EXAMPLE 24–3   FIND THE OFF-RESONANCE CURRENT

Referring to the system in the previous Example, find the rms current when the generator operates at a frequency that is twice the resonance frequency.

SOLUTION   *(Test your understanding by performing the calculations indicated in each step.)*

1. Calculate the inductive reactance at the frequency $f = 2(86\,\text{Hz})$:

   $X_L = 57\,\Omega$

2. Calculate the capacitive reactance at the frequency $f = 2(86\,\text{Hz})$:

   $X_C = 14\,\Omega$

CONTINUED ON NEXT PAGE

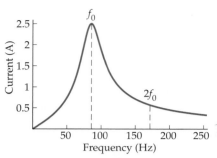

▲ **FIGURE 24–25 RMS current in an RLC circuit**
This plot shows the rms current versus frequency for the circuit considered in Example 24–8 and Active Example 24–3. The currents determined in these Examples occur at the frequencies indicated by the vertical dashed lines.

CONTINUED FROM PREVIOUS PAGE

3. Use these results plus the resistance R to find the impedance, Z: $Z = 44 \ \Omega$

4. Divide the rms voltage by the impedance to find the rms current: $I_{rms} = 0.57$ A

INSIGHT

As expected, the inductive reactance is greater than the capacitive reactance at this frequency. In addition, note that the rms current has been reduced by roughly a factor of 5 compared with its value in Example 24–8.

YOUR TURN

At what two frequencies is the current in this circuit equal to 1.5 A?

*(Answers to **Your Turn** problems are given in the back of the book.)*

The rms current as a function of frequency for the system considered in the previous Example and Active Example is shown in **Figure 24–25**. Notice that the two currents just calculated, at $f_0 = 86$ Hz and $f = 2f_0 = 2(86 \text{ Hz}) = 172$ Hz, are indicated on the graph.

THE BIG PICTURE   **PUTTING PHYSICS IN CONTEXT**

LOOKING BACK	LOOKING AHEAD
The concepts of current, resistance, capacitance, and inductance (Chapters 21 and 23) are used throughout this chapter. We also make extensive use of electrical power (Chapter 21). Alternating-current circuits have many similarities with oscillating mechanical systems. In particular, this chapter presents detailed connections between the behavior of a mass on a spring (Chapter 13) and the behavior of an *RLC* circuit.	An ac generator appears again in Chapter 25, where we show how the alternating current in an electric circuit can produce electromagnetic waves. Electromagnetic waves can be detected with an LC circuit whose frequency is matched to the frequency of the electromagnetic wave. This is discussed in Section 25–1.

CHAPTER SUMMARY

24–1 ALTERNATING VOLTAGES AND CURRENTS

An ac generator produces a voltage that varies with time as

$$V = V_{max} \sin \omega t \qquad \text{24–1}$$

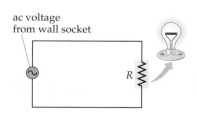

ac voltage from wall socket

Phasors
A phasor is a rotating vector representing a voltage or a current in an ac circuit. Phasors rotate counterclockwise about the origin in the x-y plane with an angular speed ω. The y component of a phasor gives the instantaneous value of that quantity.

rms Values
The rms, or root mean square, of a quantity x is the square root of the average value of x^2. For any quantity x that varies with time as a sine or a cosine, the rms value is

$$x_{rms} = \frac{1}{\sqrt{2}} x_{max} \qquad \text{24–4}$$

Standard dc formulas like $P = I^2 R$ can be converted to ac average formulas by using rms values. For example, $P_{av} = I_{rms}^2 R$.

24–2 CAPACITORS IN AC CIRCUITS

A capacitor in an ac circuit has a current that depends on the frequency and is out of phase with the voltage.

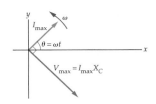

Capacitive Reactance
The rms current in a capacitor is

$$I_{rms} = \frac{V_{rms}}{X_C} \qquad 24\text{–}8$$

where X_C is the capacitive reactance. It is defined as follows:

$$X_C = \frac{1}{\omega C} \qquad 24\text{–}9$$

Phase Relation Between Voltage and Current in a Capacitor
The voltage across a capacitor lags the current by 90°.

Phasor Diagram for Capacitors
In a phasor diagram, the voltage of a capacitor is drawn at an angle that is 90° clockwise from the direction of the current phasor.

24–3 *RC* CIRCUITS

To analyze an *RC* circuit, the 90° phase difference between the resistor and capacitor voltages must be taken into account.

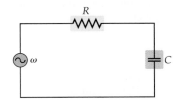

Impedance
The impedance, Z, of an *RC* circuit is analogous to the resistance in a simple resistor circuit and takes into account both the resistance, R, and the capacitive reactance, X_C:

$$Z = \sqrt{R^2 + X_C{}^2} = \sqrt{R^2 + \left(\frac{1}{\omega C}\right)^2} \qquad 24\text{–}11$$

Voltage and Current
The rms voltage and current in an *RC* circuit are related by

$$V_{rms} = I_{rms}\sqrt{R^2 + X_C{}^2} = I_{rms}Z$$

Power Factor
If the phase angle between the current and voltage in an *RC* circuit is ϕ, the average power consumed by the circuit is

$$P_{av} = I_{rms}V_{rms}\cos\phi \qquad 24\text{–}13$$

where $\cos\phi$ is referred to as the power factor.

24–4 INDUCTORS IN AC CIRCUITS

Inductors in ac circuits have frequency-dependent currents, and voltages that are out of phase with the current.

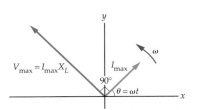

Inductive Reactance
The inductive reactance is

$$X_L = \omega L \qquad 24\text{–}14$$

The rms current in an inductor is

$$I_{rms} = \frac{V_{rms}}{X_L}$$

Phasor Diagram for Inductors
The voltage across an inductor leads the current by 90°. Thus, in a phasor diagram, the voltage phasor for an inductor is rotated 90° counterclockwise from the current phasor.

RL Circuits
The impedance of an *RL* circuit is

$$Z = \sqrt{R^2 + X_L{}^2} = \sqrt{R^2 + (\omega L)^2} \qquad 24\text{–}15$$

24–5 *RLC* CIRCUITS

An *RLC* circuit consists of a resistor, *R*, an inductor, *L*, and a capacitor, *C*, connected in series to an ac generator.

Impedance

The impedance of an *RLC* circuit is

$$Z = \sqrt{R^2 + (X_L - X_C)^2} = \sqrt{R^2 + \left(\omega L - \frac{1}{\omega C}\right)^2} \qquad \text{24–16}$$

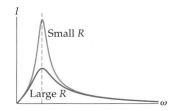

Large and Small Frequencies

In the limit of large frequencies, inductors behave like open circuits and capacitors are like ideal wires. For very small frequencies, the behaviors are reversed; inductors are like ideal wires, and capacitors act like open circuits.

24–6 RESONANCE IN ELECTRIC CIRCUITS

Electric circuits can have natural frequencies of oscillation, just like a pendulum or a mass on a spring.

LC Circuits

Circuits containing only an inductor and a capacitor have a natural frequency given by

$$\omega_0 = \frac{1}{\sqrt{LC}} = 2\pi f_0 \qquad \text{24–18}$$

Resonance

An *RLC* circuit connected to an ac generator has maximum current at the frequency $\omega = 1/\sqrt{LC}$. This effect is referred to as resonance.

PROBLEM-SOLVING SUMMARY

Type of Problem	Relevant Physical Concepts	Related Examples
Find the average power dissipated in a resistor.	To find the average power dissipated in a resistor, simply replace I with I_{rms} in $P = I^2R$, or replace V with V_{rms} in $P = V^2/R$.	Example 24–1
Calculate the rms voltage across a capacitor in an ac circuit.	In analogy with $V = IR$ for a resistor, the rms voltage across a capacitor is $V_{rms} = I_{rms}X_C$, where $X_C = 1/\omega C$.	Example 24–2
Find the impedance and rms voltage or rms current in an *RC* circuit.	The impedance in an *RC* circuit is $Z = \sqrt{R^2 + X_C{}^2} = \sqrt{R^2 + (1/\omega C)^2}$. The voltage and current in an *RC* circuit are related by the relation $V_{rms} = I_{rms}Z$.	Example 24–3 Active Example 24–1
Find the average power consumed by an ac circuit.	In an ac circuit, the average power consumed is $P_{av} = I_{rms}V_{rms} \cos \phi$, where the power factor $(\cos \phi)$ is defined as follows: $\cos \phi = R/Z$.	Example 24–4
Find the impedance and rms voltage or rms current in an *RL* circuit.	The impedance in an *RL* circuit is $Z = \sqrt{R^2 + X_L{}^2} = \sqrt{R^2 + (\omega L)^2}$. The voltage and current in an *RL* circuit are related by $V_{rms} = I_{rms}Z$.	Example 24–5
Find the impedance and rms voltage or rms current in an *RLC* circuit.	The impedance in an *RLC* circuit is $Z = \sqrt{R^2 + (X_L - X_C)^2} = \sqrt{R^2 + \left(\omega L - \frac{1}{\omega C}\right)^2}$. The voltage and current in an *RLC* circuit are related by the relation $V_{rms} = I_{rms}Z$.	Example 24–6 Active Example 24–2
Determine the resonance frequency and impedance of an *RLC* circuit.	In a circuit with a capacitance C and an inductance L, the resonance frequency is $\omega_0 = 1/\sqrt{LC}$. At resonance, the impedance of an *RLC* circuit is simply equal to its resistance; that is, $Z = R$.	Example 24–8 Active Example 24–3

CONCEPTUAL QUESTIONS

For instructor-assigned homework, go to www.masteringphysics.com

(Answers to odd-numbered Conceptual Questions can be found in the back of the book.)

1. How can the rms voltage of an ac circuit be nonzero when its average value is zero? Explain.

2. Why is the current in an ac circuit not always in phase with its voltage?

3. Does an *LC* circuit consume any power? Explain.

4. An *LC* circuit is driven at a frequency higher than its resonance frequency. What can be said about the phase angle, ϕ, for this circuit?

5. An *LC* circuit is driven at a frequency lower than its resonance frequency. What can be said about the phase angle, ϕ, for this circuit?

6. In Conceptual Checkpoint 24–3 we considered an ac circuit consisting of a lightbulb in series with an inductor. The effect of the inductor was to cause the bulb to shine less brightly.

Would the same be true in a direct-current (dc) circuit? Explain.

7. How do the resistance, capacitive reactance, and inductive reactance change when the frequency in a circuit is increased?

8. In the analogy between an *RLC* circuit and a mass on a spring, what is the analog of the current in the circuit? Explain.

9. In the analogy between an *RLC* circuit and a mass on a spring, the mass is analogous to the inductance, and the spring constant is analogous to the inverse of the capacitance. Explain.

10. Two *RLC* circuits have different values of *L* and *C*. Is it possible for these two circuits to have the same resonance frequency? Explain.

11. Can an *RLC* circuit have the same impedance at two different frequencies? Explain.

PROBLEMS AND CONCEPTUAL EXERCISES

Note: Answers to odd-numbered Problems and Conceptual Exercises can be found in the back of the book. **IP** *denotes an integrated problem, with both conceptual and numerical parts;* **BIO** *identifies problems of biological or medical interest;* **CE** *indicates a conceptual exercise.* **Predict/Explain** *problems ask for two responses:* **(a)** *your prediction of a physical outcome, and* **(b)** *the best explanation among three provided. On all problems, red bullets (•, ••, •••) are used to indicate the level of difficulty.*

SECTION 24–1 ALTERNATING VOLTAGES AND CURRENTS

1. • An ac generator produces a peak voltage of 55 V. What is the rms voltage of this generator?

2. • **European Electricity** In many European homes the rms voltage available from a wall socket is 240 V. What is the maximum voltage in this case?

3. • An rms voltage of 120 V produces a maximum current of 2.1 A in a certain resistor. Find the resistance of this resistor.

4. • The rms current in an ac circuit with a resistance of 150 Ω is 0.85 A. What are **(a)** the average and **(b)** the maximum power consumed by this circuit?

5. • A 3.33-kΩ resistor is connected to a generator with a maximum voltage of 141 V. Find **(a)** the average and **(b)** the maximum power delivered to this circuit.

6. •• A "75-watt" lightbulb uses an average power of 75 W when connected to an rms voltage of 120 V. **(a)** What is the resistance of the lightbulb? **(b)** What is the maximum current in the bulb? **(c)** What is the maximum power used by the bulb at any given instant of time?

7. ••• **Square-Wave Voltage I** The relationship $V_{rms} = V_{max}/\sqrt{2}$ is valid only for voltages that vary sinusoidally. Find the relationship between V_{rms} and V_{max} for the "square-wave" voltage shown in **Figure 24–26**.

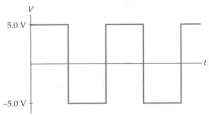

▲ **FIGURE 24–26** Problem 7

SECTION 24–2 CAPACITORS IN AC CIRCUITS

8. • The reactance of a capacitor is 65 Ω at a frequency of 57 Hz. What is its capacitance?

9. • The capacitive reactance of a capacitor at 60.0 Hz is 105 Ω. At what frequency is its capacitive reactance 72.5 Ω?

10. • A 105-μF capacitor is connected to an ac generator with an rms voltage of 20.0 V and a frequency of 100.0 Hz. What is the rms current in this circuit?

11. • The rms voltage across a 0.010-μF capacitor is 1.8 V at a frequency of 52 Hz. What are **(a)** the rms and **(b)** the maximum current through the capacitor?

12. •• An ac generator with a frequency of 30.0 Hz and an rms voltage of 12.0 V is connected to a 45.5-μF capacitor. **(a)** What is the maximum current in this circuit? **(b)** What is the current in the circuit when the voltage across the capacitor is 5.25 V and increasing? **(c)** What is the current in the circuit when the voltage across the capacitor is 5.25 V and decreasing?

13. •• The maximum current in a 22-μF capacitor connected to an ac generator with a frequency of 120 Hz is 0.15 A. **(a)** What is the maximum voltage of the generator? **(b)** What is the voltage across the capacitor when the current in the circuit is 0.10 A and increasing? **(c)** What is the voltage across the capacitor when the current in the circuit is 0.10 A and decreasing?

14. •• **IP** An rms voltage of 20.5 V with a frequency of 1.00 kHz is applied to a 0.395-μF capacitor. **(a)** What is the rms current in this circuit? **(b)** By what factor does the current change if the frequency of the voltage is doubled? **(c)** Calculate the current for a frequency of 2.00 kHz.

15. •• A circuit consists of a 1.00-kHz generator and a capacitor. When the rms voltage of the generator is 0.500 V, the rms current in the circuit is 0.430 mA. **(a)** What is the reactance of the capacitor at 1.00 kHz? **(b)** What is the capacitance of the capacitor? **(c)** If the rms voltage is maintained at 0.500 V, what is the rms current at 2.00 kHz? At 10.0 kHz?

16. •• **IP** A capacitor has an rms current of 21 mA at a frequency of 60.0 Hz when the rms voltage across it is 14 V. **(a)** What is the capacitance of this capacitor? **(b)** If the frequency is increased, will the current in the capacitor increase, decrease, or stay the same? Explain. **(c)** Find the rms current in this capacitor at a frequency of 410 Hz.

17. •• A 0.22-μF capacitor is connected to an ac generator with an rms voltage of 12 V. For what range of frequencies will the rms current in the circuit be less than 1.0 mA?

18. •• At what frequency will a generator with an rms voltage of 504 V produce an rms current of 7.50 mA in a 0.0150-μF capacitor?

SECTION 24–3 *RC* CIRCUITS

19. • Find the impedance of a 60.0-Hz circuit with a 45.5-Ω resistor connected in series with a 95.0-μF capacitor.

20. • An ac generator with a frequency of 105 Hz and an rms voltage of 22.5 V is connected in series with a 10.0-kΩ resistor and a 0.250-μF capacitor. What is the rms current in this circuit?

21. • The rms current in an *RC* circuit is 0.72 A. The capacitor in this circuit has a capacitance of 13 μF and the ac generator has a frequency of 150 Hz and an rms voltage of 95 V. What is the resistance in this circuit?

22. •• A 65.0-Hz generator with an rms voltage of 135 V is connected in series to a 3.35-kΩ resistor and a 1.50-μF capacitor. Find **(a)** the rms current in the circuit and **(b)** the phase angle, ϕ, between the current and the voltage.

23. •• **(a)** At what frequency must the circuit in Problem 22 be operated for the current to lead the voltage by 23.0°? **(b)** Using the frequency found in part (a), find the average power consumed by this circuit.

24. •• **(a)** Sketch the phasor diagram for an ac circuit with a 105-Ω resistor in series with a 32.2-μF capacitor. The frequency of the generator is 60.0 Hz. **(b)** If the rms voltage of the generator is 120 V, what is the average power consumed by the circuit?

25. •• Find the power factor for an *RC* circuit connected to a 70.0-Hz generator with an rms voltage of 155 V. The values of *R* and *C* in this circuit are 105 Ω and 82.4 μF, respectively.

26. •• **IP** **(a)** Determine the power factor for an *RC* circuit with $R = 4.0$ kΩ and $C = 0.35$ μF that is connected to an ac generator with an rms voltage of 24 V and a frequency of 150 Hz. **(b)** Will the power factor for this circuit increase, decrease, or stay the same if the frequency of the generator is increased? Explain.

27. ••• **Square-Wave Voltage II** The "square-wave" voltage shown in **Figure 24–27** is applied to an *RC* circuit. Sketch the shape of the instantaneous voltage across the capacitor, assuming the time constant of the circuit is equal to the period of the applied voltage.

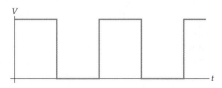

▲ **FIGURE 24–27** Problems 27, 42, and 98

SECTION 24–4 INDUCTORS IN AC CIRCUITS

28. • **CE Predict/Explain** When a long copper wire of finite resistance is connected to an ac generator, as shown in **Figure 24–28 (a)**, a certain amount of current flows through the wire. The wire is

now wound into a coil of many loops and reconnected to the generator, as indicated in **Figure 24–28 (b)**. **(a)** Is the current supplied to the coil greater than, less than, or the same as the current supplied to the uncoiled wire? **(b)** Choose the *best explanation* from among the following:

  I. More current flows in the circuit because the coiled wire is an inductor, and inductors tend to keep the current flowing in an ac circuit.

 II. The current supplied to the circuit is the same because the wire is the same. Simply wrapping the wire in a coil changes nothing.

III. Less current is supplied to the circuit because the coiled wire acts as an inductor, which increases the impedance of the circuit.

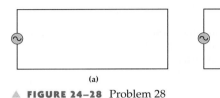

(a)   (b)

▲ **FIGURE 24–28** Problem 28

29. • An inductor has a reactance of 56.5 Ω at 75.0 Hz. What is its reactance at 60.0 Hz?

30. • What is the rms current in a 77.5-mH inductor when it is connected to a 60.0-Hz generator with an rms voltage of 115 V?

31. • What rms voltage is required to produce an rms current of 2.1 A in a 66-mH inductor at a frequency of 25 Hz?

32. •• A 525-Ω resistor and a 295-mH inductor are connected in series with an ac generator with an rms voltage of 20.0 V and a frequency of 60.0 Hz. What is the rms current in this circuit?

33. •• The rms current in an *RL* circuit is 0.26 A when it is connected to an ac generator with a frequency of 60.0 Hz and an rms voltage of 25 V. **(a)** Given that the inductor has an inductance of 145 mH, what is the resistance of the resistor? **(b)** Find the rms voltage across the resistor. **(c)** Find the rms voltage across the inductor. **(d)** Use your results from parts (b) and (c) to show that $\sqrt{V_{\mathrm{rms},R}^2 + V_{\mathrm{rms},L}^2}$ is equal to 25 V.

34. •• An ac generator with a frequency of 1.34 kHz and an rms voltage of 24.2 V is connected in series with a 2.00-kΩ resistor and a 315-mH inductor. **(a)** What is the power factor for this circuit? **(b)** What is the average power consumed by this circuit?

35. •• **IP** An rms voltage of 22.2 V with a frequency of 1.00 kHz is applied to a 0.290-mH inductor. **(a)** What is the rms current in this circuit? **(b)** By what factor does the current change if the frequency of the voltage is doubled? **(c)** Calculate the current for a frequency of 2.00 kHz.

36. •• A 0.22-μH inductor is connected to an ac generator with an rms voltage of 12 V. For what range of frequencies will the rms current in the circuit be less than 1.0 mA?

37. •• The phase angle in a certain *RL* circuit is 76° at a frequency of 60.0 Hz. If $R = 2.7$ Ω for this circuit, what is the value of the inductance, *L*?

38. •• An *RL* circuit consists of a resistor $R = 68$ Ω, an inductor, $L = 31$ mH, and an ac generator with an rms voltage of 120 V. **(a)** At what frequency will the rms current in this circuit be 1.5 A? For this frequency, what are **(b)** the rms voltage across the resistor, $V_{\mathrm{rms},R}$, and **(c)** the rms voltage across the inductor, $V_{\mathrm{rms},L}$? **(d)** Show that $V_{\mathrm{rms},R} + V_{\mathrm{rms},L} > 120$ V, but that $\sqrt{V_{\mathrm{rms},R}^2 + V_{\mathrm{rms},L}^2} = 120$ V.

39. •• **(a)** Sketch the phasor diagram for an ac circuit with a 105-Ω resistor in series with a 22.5-mH inductor. The frequency of the generator is 60.0 Hz. **(b)** If the rms voltage of the generator is 120 V, what is the average power consumed by the circuit?

40. •• **IP** In Problem 37, does the phase angle increase, decrease, or stay the same when the frequency is increased? Verify your answer by calculating the phase angle at 70.0 Hz.

41. •• A large air conditioner has a resistance of 7.0 Ω and an inductive reactance of 15 Ω. If the air conditioner is powered by a 60.0-Hz generator with an rms voltage of 240 V, find **(a)** the impedance of the air conditioner, **(b)** its rms current, and **(c)** the average power consumed by the air conditioner.

42. ••• **Square-Wave Voltage III** The "square-wave" voltage shown in Figure 24–27 is applied to an *RL* circuit. Sketch the shape of the instantaneous voltage across the inductor, assuming the time constant of the circuit is much less than the period of the applied voltage.

SECTION 24–5 *RLC* CIRCUITS

43. • **CE** An inductor and a capacitor are to be connected to a generator. Will the generator supply more current at *high* frequency if the inductor and capacitor are connected in series or in parallel? Explain.

44. • **CE** An inductor and a capacitor are to be connected to a generator. Will the generator supply more current at *low* frequency if the inductor and capacitor are connected in series or in parallel? Explain.

45. • **CE Predict/Explain (a)** When the ac generator in Figure 24–29 operates at high frequency, is the rms current in the circuit greater than, less than, or the same as when the generator operates at low frequency? **(b)** Choose the *best explanation* from among the following:

I. The current is the same because at high frequency the inductor is like an open circuit, and at low frequency the capacitor is like an open circuit. In either case the resistance of the circuit is *R*.

II. Less current flows at high frequency because in that limit the inductor acts like an open circuit, allowing no current to flow.

III. More current flows at high frequency because in that limit the capacitor acts like an ideal wire of zero resistance.

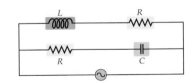

▲ **FIGURE 24–29** Problems 45 and 75

46. • **CE Predict/Explain (a)** When the ac generator in Figure 24–30 operates at high frequency, is the rms current in the circuit greater than, less than, or the same as when the generator operates at low frequency? **(b)** Choose the *best explanation* from among the following:

I. The current at high frequency is greater because the higher the frequency the more charge that flows through a circuit.

II. Less current flows at high frequency because in that limit the inductor is like an open circuit and current has only one path to flow through.

III. The inductor has zero resistance, and therefore the resistance of the circuit is the same at all frequencies. As a result the current is the same at all frequencies.

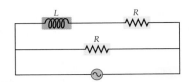

▲ **FIGURE 24–30** Problems 46, 50, and 76

47. • **CE Predict/Explain (a)** When the ac generator in **Figure 24–31** operates at high frequency, is the rms current in the circuit greater than, less than, or the same as when the generator operates at low frequency? **(b)** Choose the *best explanation* from among the following:

I. The capacitor has no resistance, and therefore the resistance of the circuit is the same at all frequencies. As a result the current is the same at all frequencies.

II. Less current flows at high frequency because in that limit the capacitor is like an open circuit and current has only one path to flow through.

III. More current flows at high frequency because in that limit the capacitor is like a short circuit and current has two parallel paths to flow through.

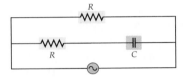

▲ **FIGURE 24–31** Problems 47, 51, 75, and 76

48. • Find the rms voltage across each element in an *RLC* circuit with $R = 9.9$ kΩ, $C = 0.15$ μF, and $L = 25$ mH. The generator supplies an rms voltage of 115 V at a frequency of 60.0 Hz.

49. • What is the impedance of a 1.50-kΩ resistor, a 105-mH inductor, and a 12.8-μF capacitor connected in series with a 60.0-Hz ac generator?

50. • Consider the circuit shown in Figure 24–30. The ac generator in this circuit has an rms voltage of 65 V. Given that $R = 15$ Ω and $L = 0.22$ mH, find the rms current in this circuit in the limit of **(a)** high frequency and **(b)** low frequency.

51. • Consider the circuit shown in Figure 24–31. The ac generator in this circuit has an rms voltage of 75 V. Given that $R = 15$ Ω and $C = 41$ μF, find the rms current in this circuit in the limit of **(a)** high frequency and **(b)** low frequency.

52. •• What is the phase angle in an *RLC* circuit with $R = 9.9$ kΩ, $C = 1.5$ μF, and $L = 250$ mH? The generator supplies an rms voltage of 115 V at a frequency of 60.0 Hz.

53. •• **IP** An *RLC* circuit has a resistance of 105 Ω, an inductance of 85.0 mH, and a capacitance of 13.2 μF. **(a)** What is the power factor for this circuit when it is connected to a 125-Hz ac generator? **(b)** Will the power factor increase, decrease, or stay the same if the resistance is increased? Explain. **(c)** Calculate the power factor for a resistance of 525 Ω.

54. •• An ac voltmeter, which displays the rms voltage between the two points touched by its leads, is used to measure voltages in the circuit shown in **Figure 24–32**. In this circuit, the ac generator has an rms voltage of 6.00 V and a frequency of 30.0 kHz. The inductance in the circuit is 0.300 mH, the capacitance is 0.100 μF, and the resistance is 2.50 Ω. What is the reading on a voltmeter when it is connected to points **(a)** A and B, **(b)** B and C, **(c)** A and C, and **(d)** A and D?

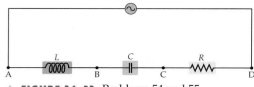

▲ **FIGURE 24–32** Problems 54 and 55

55. •• **IP** Consider the ac circuit shown in Figure 24–32, where we assume that the values of R, L, and C are the same as in the previous problem, and that the rms voltage of the generator is still 6.00 V. The frequency of the generator, however, is doubled to 60.0 kHz. Calculate the rms voltage across **(a)** the resistor, R, **(b)** the inductor, L, and **(c)** the capacitor, C. **(d)** Do you expect the sum of the rms voltages in parts (a), (b), and (c) to be greater than, less than, or equal to 6.00 V? Explain.

56. •• **(a)** Sketch the phasor diagram for an ac circuit with a 105-Ω resistor in series with a 22.5-mH inductor and a 32.2-μF capacitor. The frequency of the generator is 60.0 Hz. **(b)** If the rms voltage of the generator is 120 V, what is the average power consumed by the circuit?

57. •• A generator connected to an RLC circuit has an rms voltage of 120 V and an rms current of 34 mA. If the resistance in the circuit is 3.3 kΩ and the capacitive reactance is 6.6 kΩ, what is the inductive reactance of the circuit?

58. ••• **Manufacturing Plant Power** A manufacturing plant uses 2.22 kW of electric power provided by a 60.0-Hz ac generator with an rms voltage of 485 V. The plant uses this power to run a number of high-inductance electric motors. The plant's total resistance is $R = 25.0\ \Omega$ and its inductive reactance is $X_L = 45.0\ \Omega$. **(a)** What is the total impedance of the plant? **(b)** What is the plant's power factor? **(c)** What is the rms current used by the plant? **(d)** What capacitance, connected in series with the power line, will increase the plant's power factor to unity? **(e)** If the power factor is unity, how much current is needed to provide the 2.22 kW of power needed by the plant? Compare your answer with the current found in part (c). (Because power-line losses are proportional to the square of the current, a utility company will charge an industrial user with a low power factor a higher rate per kWh than a company with a power factor close to unity.)

SECTION 24–6 RESONANCE IN ELECTRIC CIRCUITS

59. • **CE** A capacitor and an inductor connected in series have a period of oscillation given by T. At the time $t = 0$ the capacitor has its maximum charge. In terms of T, what is the first time after $t = 0$ that **(a)** the current in the circuit has its maximum value and **(b)** the energy stored in the electric field is a maximum?

60. • **CE Predict/Explain** In an RLC circuit a second capacitor is added in *series* to the capacitor already present. **(a)** Does the resonance frequency increase, decrease, or stay the same? **(b)** Choose the *best explanation* from among the following:
   I. The resonance frequency stays the same because it depends only on the resistance in the circuit.
   II. Adding a capacitor in series increases the equivalent capacitance, and this decreases the resonance frequency.
   III. Adding a capacitor in series decreases the equivalent capacitance, and this increases the resonance frequency.

61. • **CE** In an RLC circuit a second capacitor is added in *parallel* to the capacitor already present. Does the resonance frequency increase, decrease, or stay the same? Explain.

62. • An RLC circuit has a resonance frequency of 2.4 kHz. If the capacitance is 47 μF, what is the inductance?

63. • At resonance, the rms current in an RLC circuit is 2.8 A. If the rms voltage of the generator is 120 V, what is the resistance, R?

64. •• **CE** The resistance in an RLC circuit is doubled. **(a)** Does the resonance frequency increase, decrease, or stay the same? Explain. **(b)** Does the maximum current in the circuit increase, decrease, or stay the same? Explain.

65. •• **CE** The voltage in a sinusoidally driven RLC circuit leads the current. **(a)** If we want to bring this circuit into resonance by changing the frequency of the generator, should the frequency be increased or decreased? Explain. **(b)** If we want to bring this circuit into resonance by changing the inductance instead, should the inductance be increased or decreased? Explain.

66. •• A 115-Ω resistor, a 67.6-mH inductor, and a 189-μF capacitor are connected in series to an ac generator. **(a)** At what frequency will the current in the circuit be a maximum? **(b)** At what frequency will the impedance of the circuit be a minimum?

67. •• **IP** An ac generator of variable frequency is connected to an RLC circuit with $R = 12\ \Omega$, $L = 0.15$ mH, and $C = 0.20$ mF. At a frequency of 1.0 kHz, the rms current in the circuit is larger than desired. Should the frequency of the generator be increased or decreased to reduce the current? Explain.

68. •• **(a)** Find the frequency at which a 33-μF capacitor has the same reactance as a 33-mH inductor. **(b)** What is the resonance frequency of an LC circuit made with this inductor and capacitor?

69. •• Consider an RLC circuit with $R = 105\ \Omega$, $L = 518$ mH, and $C = 0.200\ \mu$F. **(a)** At what frequency is this circuit in resonance? **(b)** Find the impedance of this circuit if the frequency has the value found in part (a), but the capacitance is increased to $0.220\ \mu$F. **(c)** What is the power factor for the situation described in part (b)?

70. •• **IP** An RLC circuit has a resonance frequency of 155 Hz. **(a)** If both L and C are doubled, does the resonance frequency increase, decrease, or stay the same? Explain. **(b)** Find the resonance frequency when L and C are doubled.

71. •• An RLC circuit has a capacitance of 0.29 μF. **(a)** What inductance will produce a resonance frequency of 95 MHz? **(b)** It is desired that the impedance at resonance be one-fifth the impedance at 11 kHz. What value of R should be used to obtain this result?

GENERAL PROBLEMS

72. • **CE BIO Persistence of Vision** Although an incandescent lightbulb appears to shine with constant intensity, this is an artifact of the eye's persistence of vision. In fact, the intensity of a bulb's light rises and falls with time due to the alternating current used in household circuits. If you could perceive these oscillations, would you see the light attain maximum brightness 60 or 120 times per second? Explain.

73. • **CE** An inductor in an LC circuit has a maximum current of 2.4 A and a maximum energy of 36 mJ. When the current in the inductor is 1.2 A, what is the energy stored in the capacitor?

74. • **CE** An RLC circuit is driven at its resonance frequency. Is its impedance greater than, less than, or equal to R? Explain.

75. • **CE Predict/Explain** Suppose the circuits shown in Figures 24–29 and 24–31 are connected to identical batteries, rather than to ac generators. **(a)** Assuming the value of R is the same in the two circuits, is the current in Figure 24–29 greater than, less

than, or the same as the current in Figure 24–31? **(b)** Choose the *best explanation* from among the following:

   I. The circuits have the same current because the capacitor acts like an open circuit and the inductor acts like a short circuit.

   II. The current in Figure 24–29 is larger because it has more circuit elements, each of which can carry current.

   III. The current in Figure 24–31 is larger because it has fewer circuit elements, meaning less resistance to current flow.

76. • **CE** Suppose the circuits shown in Figures 24–30 and 24–31 are connected to identical batteries, rather than to ac generators. Assuming the value of R is the same in the two circuits, is the current in Figure 24–30 greater than, less than, or the same as the current in Figure 24–31? Explain.

77. • **CE Predict/Explain** Consider a circuit consisting of a light-bulb and a capacitor, as shown in circuit 2 of Conceptual Checkpoint 24–2. **(a)** If the frequency of the generator is increased, does the intensity of the lightbulb increase, decrease, or stay the same? **(b)** Choose the *best explanation* from among the following:

   I. As the frequency increases it becomes harder to force current through the capacitor, and therefore the intensity of the lightbulb decreases.

   II. The intensity of the lightbulb increases because as the frequency becomes higher the capacitor acts more like a short circuit, allowing more current to flow.

   III. The intensity of the lightbulb is independent of frequency because the circuit contains a capacitor but not an inductor.

78. • **CE** Consider a circuit consisting of a lightbulb and an inductor, as shown in Conceptual Checkpoint 24–3. If the frequency of the generator is increased, does the intensity of the lightbulb increase, decrease, or stay the same? Explain.

79. • A 4.40-μF and an 8.80-μF capacitor are connected in *parallel* to a 60.0-Hz generator operating with an rms voltage of 115 V. What is the rms current supplied by the generator?

80. • A 4.40-μF and an 8.80-μF capacitor are connected in *series* to a 60.0-Hz generator operating with an rms voltage of 115 V. What is the rms current supplied by the generator?

81. • A 10.0-μF capacitor and a 30.0-μF capacitor are connected in parallel to an ac generator with a frequency of 60.0 Hz. What is the capacitive reactance of this pair of capacitors?

82. •• **CE** A generator drives an *RLC* circuit with the voltage V shown in **Figure 24–33**. The corresponding current I is also shown in the figure. **(a)** Is the inductive reactance of this circuit greater than, less than, or equal to its capacitive reactance? Explain. **(b)** Is the frequency of this generator greater than, less than, or equal to the resonance frequency of the circuit? Explain.

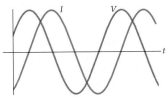

▲ **FIGURE 24–33** Problem 82

83. •• **IP** Consider the *RLC* circuit shown in Example 24–6, and the corresponding phasor diagram given in the Insight. **(a)** On the basis of the phasor diagram, can you conclude that the resonance frequency of this circuit is greater than, less than, or equal to 60.0 Hz? Explain. **(b)** Calculate the resonance frequency for

this circuit. **(c)** The impedance of this circuit at 60.0 Hz is 226 Ω. What is the impedance at resonance?

84. •• **IP** When a certain resistor is connected to an ac generator with a maximum voltage of 15 V, the average power dissipated in the resistor is 22 W. **(a)** What is the resistance of the resistor? **(b)** What is the rms current in the circuit? **(c)** We know that $P_{av} = I_{rms}^2 R$, and hence it seems that reducing the resistance should reduce the average power. On the other hand, we also know that $P_{av} = V_{rms}^2/R$, which suggests that reducing R increases P_{av}. Which conclusion is correct? Explain.

85. •• A 9.5-Hz generator is connected to a capacitor. If the current in the generator has its maximum value at $t = 0$, what is the earliest possible time that the voltage across the capacitor is a maximum?

86. •• The voltage across an inductor reaches its maximum value 25 ms before the current supplied by the generator reaches *its* maximum value. What is the lowest possible frequency at which the generator operates?

87. •• Find the average power consumed by an *RC* circuit connected to a 60.0-Hz generator with an rms voltage of 122 V. The values of R and C in this circuit are 3.30 kΩ and 2.75 μF, respectively.

88. •• A 1.15-$k\Omega$ resistor and a 505-mH inductor are connected in series to a 1250-Hz generator with an rms voltage of 14.2 V. **(a)** What is the rms current in the circuit? **(b)** What capacitance must be inserted in series with the resistor and inductor to reduce the rms current to half the value found in part (a)?

89. •• **IP** *RLC Phasor* The phasor diagram for an *RLC* circuit is shown in **Figure 24–34**. **(a)** If the resistance in this circuit is 525 Ω, what is the impedance? **(b)** If the frequency in this circuit is increased, will the impedance increase, decrease, or stay the same? Explain.

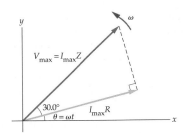

▲ **FIGURE 24–34** Problems 89, 90, and 91

90. •• **IP** Figure 24–34 shows the phasor diagram for an *RLC* circuit in which the impedance is 337 Ω. **(a)** What is the resistance, R, in this circuit? **(b)** Is this circuit driven at a frequency that is greater than, less than, or equal to the resonance frequency of the circuit? Explain.

91. •• **IP** An *RLC* circuit has a resistance $R = 25\ \Omega$ and an inductance $L = 160$ mH, and is connected to an ac generator with a frequency of 55 Hz. The phasor diagram for this circuit is shown in Figure 24–34. Find **(a)** the impedance, Z, and **(b)** the capacitance, C, for this circuit. **(c)** If the value of C is decreased, will the impedance of the circuit increase, decrease, or stay the same? Explain.

92. •• **IP Black-Box Experiment** You are given a sealed box with two electrical terminals. The box contains a 5.00-Ω resistor in series with either an inductor or a capacitor. When you attach an ac generator with an rms voltage of 0.750 V to the terminals of the box, you find that the current increases with increasing frequency. **(a)** Does the box contain an inductor or a capacitor?

Explain. **(b)** When the frequency of the generator is 25.0 kHz, the rms current is 87.2 mA. What is the capacitance or inductance of the unknown component in the box?

93. •• **IP** A circuit is constructed by connecting a 1.00-kΩ resistor, a 252-μF capacitor, and a 515-mH inductor in series. **(a)** What is the highest frequency at which the impedance of this circuit is equal to 2.00 kΩ? **(b)** To reduce the impedance of this circuit, should the frequency be increased or decreased from its value in part (a)? Explain.

94. •• An *RLC* circuit with $R = 25.0\ \Omega$, $L = 325$ mH, and $C = 45.2\ \mu$F is connected to an ac generator with an rms voltage of 24 V. Determine the average power delivered to this circuit when the frequency of the generator is **(a)** equal to the resonance frequency, **(b)** twice the resonance frequency, and **(c)** half the resonance frequency.

95. •• **A Light-Dimmer Circuit** The intensity of a lightbulb with a resistance of 120 Ω is controlled by connecting it in series with an inductor whose inductance can be varied from $L = 0$ to $L = L_{max}$. This "light dimmer" circuit is connected to an ac generator with a frequency of 60.0 Hz and an rms voltage of 110 V. **(a)** What is the average power dissipated in the lightbulb when $L = 0$? **(b)** The inductor is now adjusted so that $L = L_{max}$. In this case, the average power dissipated in the lightbulb is one-fourth the value found in part (a). What is the value of L_{max}?

96. ••• An electric motor with a resistance of 15 Ω and an inductance of 53 mH is connected to a 60.0-Hz ac generator. **(a)** What is the power factor for this circuit? **(b)** In order to increase the power factor of this circuit to 0.80, a capacitor is connected in series with the motor and inductor. Find the required value of the capacitance.

97. ••• **IP Tuning a Radio** A radio tuning circuit contains an *RLC* circuit with $R = 5.0\ \Omega$ and $L = 2.8\ \mu$H. **(a)** What capacitance is needed to produce a resonance frequency of 85 MHz? **(b)** If the capacitance is increased above the value found in part (a), will the impedance increase, decrease, or stay the same? Explain. **(c)** Find the impedance of the circuit at resonance. **(d)** Find the impedance of the circuit when the capacitance is 1% higher than the value found in part (a).

98. ••• If the maximum voltage in the square wave shown in Figure 24–27 is V_{max}, what are **(a)** the average voltage, V_{av}, and **(b)** the rms voltage, V_{rms}?

99. ••• An ac generator supplies an rms voltage of 5.00 V to an *RC* circuit. At a frequency of 20.0 kHz the rms current in the circuit is 45.0 mA; at a frequency of 25.0 kHz the rms current is 50.0 mA. What are the values of R and C in this circuit?

100. ••• An ac generator supplies an rms voltage of 5.00 V to an *RL* circuit. At a frequency of 20.0 kHz the rms current in the circuit is 45.0 mA; at a frequency of 25.0 kHz the rms current is 40.0 mA. What are the values of R and L in this circuit?

101. ••• An *RC* circuit consists of a resistor $R = 32\ \Omega$, a capacitor $C = 25\ \mu$F, and an ac generator with an rms voltage of 120 V. **(a)** At what frequency will the rms current in this circuit be 2.9 A? For this frequency, what are **(b)** the rms voltage across the resistor, $V_{rms,R}$, and **(c)** the rms voltage across the capacitor, $V_{rms,C}$? **(d)** Show that $V_{rms,R} + V_{rms,C} > 120$ V, but that $\sqrt{V_{rms,R}^2 + V_{rms,C}^2} = 120$ V.

PASSAGE PROBLEMS

Playing a Theremin

As mentioned in Chapter 20, a theremin is a musical instrument you play without touching. You may not have heard of a theremin before, but you have certainly heard one being played—in fact, theremins are responsible for the eerie background music on many science fiction movies, such as *The Day the Earth Stood Still*. A theremin-like instrument also plays a prominent part in the Beach Boys hit song "Good Vibrations."

So how *do* you play a theremin if you don't actually touch it? Well, a theremin has two antennas that extend from the main body of the instrument, a horizontal loop antenna to control the volume, and a vertical linear antenna to control the pitch. Varying the distance between your hand and the vertical antenna, for example, adjusts the capacitance of the *RLC* circuit to which that antenna is attached. Changing the capacitance changes the resonant frequency and that, in turn, changes the frequency of the note played by the instrument.

To produce notes spanning a full five octaves, the theremin employs a clever mechanism referred to as the heterodyne principle. Rather than produce the audio frequencies directly, the theremin uses two radio frequency oscillators, one at a fixed frequency, the other with a frequency controlled by the thereminist. The beat frequency between these two radio frequencies, which is in the audio frequency range, is what you actually hear. This mechanism is essentially the same as that used in FM (frequency-modulated) radio.

102. • Suppose a theremin uses an oscillator with a fixed frequency of 90.1 MHz and an *RLC* circuit with $R = 1.5\ \Omega$, $L = 2.08\ \mu$H, and $C = 1.50$ pF. What is the beat frequency of these two oscillators? (Audio frequencies range from about 20 Hz to 20,000 Hz.)

   A. 3740 Hz   **B.** 5100 Hz

   C. 4760 Hz   **D.** 9000 Hz

103. • If the thereminist moves one of her fingers and increases the capacitance of the system slightly, does the beat frequency increase, decrease, or stay the same?

104. • Find the new beat frequency if the thereminist increases the capacitance by 0.100% over its value in Problem 102. All other quantities stay the same.

   A. 761 Hz   **B.** 41,300 Hz

   C. 41,900 Hz   **D.** 86,300 Hz

105. • What is the rms current in the theremin's *RLC* circuit (Problem 102) if it is attached to an ac generator with an rms voltage of 25.0 V and a frequency of 90.0 MHz?

   A. 2.14 mA   **B.** 3.46 mA

   C. 8.06 A   **D.** 16.7 A

INTERACTIVE PROBLEMS

106. •• **IP Referring to Example 24–6** Suppose we would like to change the phase angle for this circuit to $\phi = -25.0°$, and that we would like to accomplish this by changing the resistor to a value other than 175 Ω. The inductor is still 90.0 mH, the capacitor is 15.0 μF, the rms voltage is 120.0 V, and the ac frequency is 60.0 Hz. **(a)** Should the resistance be increased or decreased? Explain. **(b)** Find the resistance that gives the desired phase angle. **(c)** What is the rms current in the circuit with the resistance found in part (b)?

107. •• **IP Referring to Example 24–6** You plan to change the frequency of the generator in this circuit to produce a phase angle of smaller magnitude. The resistor is still 175 Ω, the inductor is 90.0 mH, the capacitor is 15.0 μF, and the rms voltage is 120.0 V. **(a)** Should you increase or decrease the frequency? Explain. **(b)** Find the frequency that gives a phase angle of $-22.5°$. **(c)** What is the rms current in the circuit at the frequency found in part (b)?

25 Electromagnetic Waves

Most people think they know exactly what the world looks like—all you have to do, after all, is open your eyes and look. There's more to it than that, however. We all know, for example, that the simple yellow flower shown here (left) is a daisy. But is that really what a daisy looks like? To a bee—with its ability to see ultraviolet light invisible to us—the very same daisy is a flower with three concentric circles forming a "bull's-eye" centered on the nectar (right). Such "nectar guides" are a common feature of flowers as viewed by bees, but are generally invisible to humans. This chapter explores the nature and properties of electromagnetic radiation—the kind we know as visible light, and many other kinds as well.

E lectricity and magnetism can seem very different in many ways, but they are actually intimately related—in fact, electric and magnetic fields can be considered as different aspects of the same thing, like the two sides of a coin. For example, we have seen that an electric current produces a magnetic field, and a changing magnetic field produces an electric field. Because of fundamental connections like these, we refer to the phenomena of electricity and magnetism together as **electromagnetism.**

In this chapter we consider one of the most significant manifestations of electromagnetism; namely, that electric and magnetic fields can work together to create traveling waves called **electromagnetic waves.** As we shall see, these waves are responsible for everything from radio and TV signals, to the visible light we see all around us, to the X-rays that reveal our internal structure, and much more. In addition, the prediction, discovery, and technological development of electromagnetic waves are a fascinating success story in the history of science.

25–1 The Production of Electromagnetic Waves

Electromagnetic waves were predicted, and their properties were studied theoretically, decades before they were first produced with electric circuits in the lab. The prediction came from Scottish physicist James Clerk Maxwell (1831–1879), who, in 1864, hypothesized that since a changing magnetic field produces an electric field (Faraday's law) a changing electric field should similarly produce a magnetic field. In effect, Maxwell suggested a sort of "symmetry" between electric and magnetic fields.

Maxwell followed up on his suggestion by working out its mathematical consequences. Among these was that electric and magnetic fields, acting together, could produce an *electromagnetic wave* that travels with the speed of light. As a result, he proposed that visible light—which had previously been thought of as a completely separate phenomenon from electricity and magnetism—was, in fact, an electromagnetic wave. His theory also implied that electromagnetic waves would not be limited to visible light and that it should be possible to produce them with oscillating electric circuits similar to those studied in the previous chapter.

The first production and observation of electromagnetic waves in the lab were carried out by the German physicist Heinrich Hertz (1857–1894) in 1887. Hertz used what was basically an *LC* circuit to generate an alternating current and found that energy could be transferred from this circuit to a similar circuit several meters away. He was able to show, in addition, that the energy transfer exhibited such standard wave phenomena as reflection, refraction, interference, diffraction, and polarization. There could be no doubt that waves were produced by the first circuit and that they propagated across the room to the second circuit. Even more significantly, he was able to show that the speed of the waves was roughly the speed of light, as predicted by Maxwell.

It took only a few years for Hertz's experimental apparatus to be refined and improved to the point where it could be used in practical applications. The first to do so was Guglielmo Marconi (1874–1937), who immediately recognized the implications of the electromagnetic-wave experiments—namely, that waves could be used for communications, eliminating the wires necessary for telegraphy. He patented his first system in 1896 and gained worldwide attention when, in 1901, he received a radio signal in St. John's, Newfoundland, that had been sent from Cornwall, England. When Maxwell died, electromagnetic waves were still just a theory; twenty years later, they were revolutionizing communications.

To gain an understanding of electromagnetic waves, consider the simple electric circuit shown in **Figure 25–1**. Here we show an ac generator of period T connected to the center of an antenna, which is basically a long, straight wire with a break in the middle. Suppose at time $t = 0$ the generator gives the upper segment of the antenna a maximum positive charge and the lower segment a maximum negative charge, as shown in Figure 25–1 (a). A positive test charge placed on the x axis at point P will experience a downward force; hence, the electric field there is downward. A short time later, when the charge on the antenna is reduced in magnitude, the electric field at P also has a smaller magnitude. We show this result in Figure 25–1 (b).

More importantly, Figure 25–1 (b) also shows that the electric field produced at time $t = 0$ has not vanished, nor has it simply been replaced with the new, reduced-magnitude field. Instead, the original field has *moved farther away from the antenna*, to point Q. The reason that the reduction in charge on the antenna is felt at point P *before* it is felt at point Q is simply that it takes a finite time for this change in charge to be felt at a distance. This is analogous to the fact that a person near a lightning strike hears the thunder before a person who is half a mile away, or that a wave pulse on a string takes a finite time to move from one end of the string to the other.

After the generator has completed one-quarter of a cycle, at time $t = \frac{1}{4}T$, the antenna is uncharged and the field vanishes, as in Figure 25–1 (c). Still later the charges on the antenna segments change sign, giving rise to an electric field that points upward, as we see in Figures 25–1 (d) and (e). The field vanishes again after

▲ Electromagnetic waves are produced by (and detected as) oscillating electric currents in a wire or similar conducting element. The actual antenna is often much smaller than is commonly imagined—the bowl-shaped structures that we tend to think of as antennas, such as these microwave relay dishes, serve to focus the transmitted beam in a particular direction or concentrate the received signal on the actual detector.

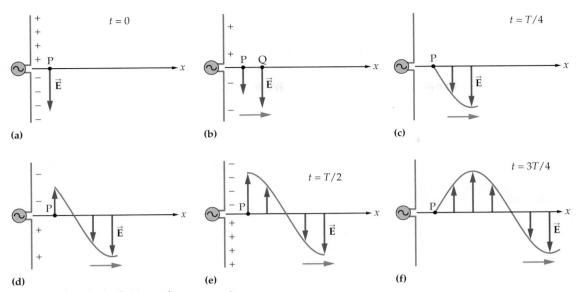

(a) $t = 0$

(b)

(c) $t = T/4$

(d)

(e) $t = T/2$

(f) $t = 3T/4$

▲ **FIGURE 25–1 Producing an electromagnetic wave**
A traveling electromagnetic wave produced by an ac generator attached to an antenna. **(a)** At $t = 0$ the electric field at point P is downward. **(b)** A short time later, the electric field at P is still downward, but now with a reduced magnitude. Note that the field created at $t = 0$ has moved to point Q. **(c)** After one-quarter of a cycle, at $t = \frac{1}{4}T$, the electric field at P vanishes. **(d)** The charge on the antenna has reversed polarity now, and the electric field at P points upward. **(e)** When the oscillator has completed half a cycle, $t = \frac{1}{2}T$, the field at point P is upward and of maximum magnitude. **(f)** At $t = \frac{3}{4}T$ the field at P vanishes again. The fields produced at earlier times continue to move away from the antenna.

three-quarters of a cycle, at $t = \frac{3}{4}T$, as shown in Figure 25–1 (f). Immediately after this time, the electric field begins to point downward once more. The net result is a wavelike electric field moving steadily away from the antenna. To summarize:

> The electric field produced by an antenna connected to an ac generator propagates away from the antenna, analogous to a wave on a string moving away from your hand as you wiggle it up and down.

This is really only half of the electromagnetic wave, however; the other half is a similar wave in the magnetic field. To see this, consider **Figure 25–2**, where we show the current in the antenna flowing upward at a time when the upper segment is positive. Pointing the thumb of the right hand in the direction of the current, and curling the fingers around the wire, as specified in the magnetic field RHR, we see that $\vec{\mathbf{B}}$ points into the page at the same time that $\vec{\mathbf{E}}$ points downward. It follows, then, that $\vec{\mathbf{E}}$ and $\vec{\mathbf{B}}$ are at right angles to each other. A more detailed analysis shows that $\vec{\mathbf{E}}$ and $\vec{\mathbf{B}}$ are perpendicular to each other at all times, and that they are also in phase; that is, when the magnitude of $\vec{\mathbf{E}}$ is at its maximum, so is the magnitude of $\vec{\mathbf{B}}$.

Combining the preceding results, we can represent the electric and magnetic fields in an electromagnetic wave as shown in **Figure 25–3**. Notice that not only are

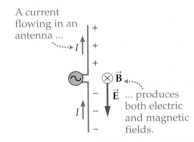

A current flowing in an antenna ...   ... produces both electric and magnetic fields.

▲ **FIGURE 25–2 Field directions in an electromagnetic wave**
At a time when the electric field produced by the antenna points downward, the magnetic field points into the page. In general, the electric and magnetic fields in an electromagnetic wave are always at right angles to each other.

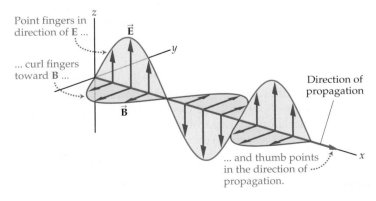

Point fingers in direction of E ...

... curl fingers toward B ...

... and thumb points in the direction of propagation.

Direction of propagation

◀ **FIGURE 25–3 The right-hand rule applied to an electromagnetic wave**
An electromagnetic wave propagating in the positive x direction. Note that $\vec{\mathbf{E}}$ and $\vec{\mathbf{B}}$ are perpendicular to each other and in phase. The direction of propagation is given by the thumb of the right hand, after pointing the fingers in the direction of $\vec{\mathbf{E}}$ and curling them toward $\vec{\mathbf{B}}$.

$\vec{\textbf{E}}$ and $\vec{\textbf{B}}$ perpendicular to each other, they are also perpendicular to the direction of propagation; hence, electromagnetic waves are **transverse** waves. (See Section 14–1 for a comparison of various types of waves.) Finally, the direction of propagation is given by another right-hand rule:

Direction of Propagation for Electromagnetic Waves

Point the fingers of your right hand in the direction of $\vec{\textbf{E}}$, curl your fingers toward $\vec{\textbf{B}}$, and your thumb will point in the direction of propagation.

This rule is consistent with the direction of propagation shown in Figure 25–3.

CONCEPTUAL CHECKPOINT 25–1 DIRECTION OF MAGNETIC FIELD

An electromagnetic wave propagates in the positive y direction, as shown in the sketch. If the electric field at the origin is in the positive z direction, is the magnetic field at the origin in **(a)** the positive x direction, **(b)** the negative x direction, or **(c)** the negative y direction?

REASONING AND DISCUSSION
Pointing the fingers of the right hand in the positive z direction (the direction of $\vec{\textbf{E}}$), we see that in order for the thumb to point in the direction of propagation (the positive y direction) the fingers must be curled toward the positive x direction. Therefore, $\vec{\textbf{B}}$ points in the positive x direction.

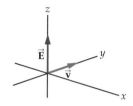

ANSWER
(a) $\vec{\textbf{B}}$ is in the positive x direction.

▶ **FIGURE 25–4 Receiving radio waves**
Basic elements of a tuning circuit used to receive radio waves. First, an incoming wave sets up an alternating current in the antenna. Next, the resonance frequency of the LC circuit is adjusted to match the frequency of the radio wave, resulting in a relatively large current in the circuit. This current is then fed into an amplifier to further increase the signal.

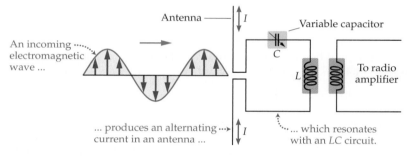

Electromagnetic waves can be detected in much the same way they are generated. Suppose, for instance, that an electromagnetic wave moves to the right, as in **Figure 25–4**. As the wave continues to move, its electric field exerts a force on electrons in the antenna that is alternately up and down, resulting in an alternating current. Thus the electromagnetic field makes the antenna behave much like an ac generator. If the antenna is connected to an LC circuit, as indicated in the figure, the resulting current can be relatively large if the resonant frequency of the circuit matches the frequency of the wave. This is the basic principle behind radio and television tuners. In fact, when you turn the tuning knob on a radio, you are actually changing the capacitance or the inductance in an LC circuit and, therefore, changing the resonance frequency.

Finally, though we have discussed the production of electromagnetic waves by means of an electric circuit and an antenna, this is certainly not the only way such waves can be generated. In fact, any time an electric charge is accelerated, it will radiate:

Accelerated charges radiate electromagnetic waves.

This condition applies no matter what the cause of the acceleration. In addition, the intensity of radiated electromagnetic waves depends on the orientation of the acceleration relative to the viewer. For example, viewing the antenna perpendicular to its length, so that the charges accelerate at right angles to the line of sight, results in maximum intensity, as illustrated in **Figure 25–5**. Conversely, viewing the antenna straight down from above, in the same direction as the acceleration, results in zero intensity.

REAL-WORLD PHYSICS
Radio and television communications

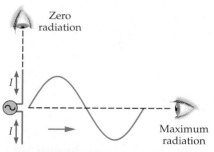

▲ **FIGURE 25–5 Electromagnetic waves and the line of sight**
Electromagnetic radiation is greatest when charges accelerate at right angles to the line of sight. Zero radiation is observed when the charges accelerate along the line of sight. These observations apply to electromagnetic waves of all frequencies.

25–2 The Propagation of Electromagnetic Waves

A sound wave or a wave on a string requires a medium through which it can propagate. For example, when the air is pumped out of a jar containing a ringing bell, its sound can no longer be heard. In contrast, we can still *see* that the bell is ringing. Thus, light can propagate through a vacuum, as can all other types of electromagnetic waves, such as radio waves and microwaves. In fact, electromagnetic waves travel through a vacuum with the maximum speed that *any* form of energy can have, as we discuss in detail in Chapter 29.

The Speed of Light

All electromagnetic waves travel through a vacuum with precisely the same speed, c. Since light is the form of electromagnetic wave most familiar to us, we refer to c as the *speed of light in a vacuum*. The approximate value of this speed is as follows:

Speed of Light in a Vacuum	
$c = 3.00 \times 10^8 \text{ m/s}$	25–1

This is a large speed, corresponding to about 186,000 mi/s. Put another way, a beam of light could travel around the world about seven times in a single second. In air the speed of light is slightly less than it is in a vacuum, and in denser materials, such as glass or water, the speed of light is reduced to about two-thirds of its vacuum value.

EXERCISE 25–1

The distance between Earth and the Sun is 1.50×10^{11} m. How long does it take for light to cover this distance?

SOLUTION

Recalling that speed is distance divided by time, it follows that the time t to cover a distance d is $t = d/v$. Using $v = c$, we find

$$t = \frac{d}{c} = \frac{1.50 \times 10^{11} \text{ m}}{3.00 \times 10^8 \text{ m/s}} = 500 \text{ s}$$

Noting that 500 s is $8\frac{1}{3}$ min, we say that Earth is about 8 light-minutes from the Sun.

Because the speed of light is so large, its value is somewhat difficult to determine. The first scientific attempt to measure the speed of light was made by Galileo (1564–1642), who used two lanterns for the experiment. Galileo opened the shutters of one lantern, and an assistant a considerable distance away was instructed to open the shutter on the second lantern as soon as he observed the light from Galileo's lantern. Galileo then attempted to measure the time that elapsed before he saw the light from his assistant's lantern. Since there was no perceptible time lag, beyond the normal human reaction time, Galileo could conclude only that the speed of light must be very great indeed.

The first to give a finite, numerical value to the speed of light was the Danish astronomer Ole Romer (1644–1710), though he did not set out to measure the speed of light at all. Romer was measuring the times at which the moons of Jupiter disappeared behind the planet, and he noticed that these eclipses occurred earlier when Earth was closer to Jupiter and later when Earth was farther away from Jupiter. This difference is illustrated in **Figure 25–6**. From the results of Exercise 25–1, we know that light requires about 16 minutes to travel from one side of Earth's orbit to the other, and this is roughly the discrepancy in eclipse times observed by Romer. In 1676 he announced a value for the speed of light of 2.25×10^8 m/s.

The first laboratory measurement of the speed of light was performed by the French scientist Armand Fizeau (1819–1896). The basic elements of his experiment, shown in **Figure 25–7**, are a mirror and a rotating, notched wheel. Light passing through one notch travels to a mirror a considerable distance away, is reflected

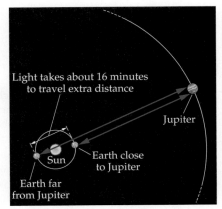

Light takes about 16 minutes to travel extra distance

Jupiter

Sun

Earth close to Jupiter

Earth far from Jupiter

▲ **FIGURE 25–6 Using Jupiter to determine the speed of light**

When the Earth is at its greatest distance from Jupiter, light takes about 16 minutes longer to travel between them. This time lag allowed Ole Romer to estimate the speed of light.

▶ **FIGURE 25–7 Fizeau's experiment to measure the speed of light**
If the time required for light to travel to the far mirror and back is equal to the time it takes the wheel to rotate from one notch to the next, light will pass through the wheel and on to the observer.

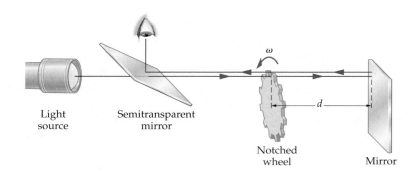

back, and then, if the rotational speed of the wheel is adjusted properly, passes through the *next notch* in the wheel. By measuring the rotational speed of the wheel and the distance from the wheel to the mirror, Fizeau was able to obtain a value of 3.13×10^8 m/s for the speed of light.

Today, experiments to measure the speed of light have been refined to such a degree that we now use it to *define* the meter, as was mentioned in Chapter 1. Thus, by definition, the speed of light in a vacuum is

$$c = 299\ 792\ 458 \text{ m/s}$$

For most routine calculations, however, the value $c = 3.00 \times 10^8$ m/s is adequate.

Maxwell's theoretical description of electromagnetic waves allowed him to obtain a simple expression for c in terms of previously known physical quantities. In particular, he found that c could be written as follows:

$$c = \frac{1}{\sqrt{\varepsilon_0 \mu_0}} \qquad \text{25–2}$$

Recall that $\varepsilon_0 = 8.85 \times 10^{-12}$ C^2/(N·m^2) occurs in the expression for the electric field due to a point charge; in fact, ε_0 determines the strength of the electric field. The constant $\mu_0 = 4\pi \times 10^{-7}$ T·m/A plays an equivalent role for the magnetic field. Thus, Maxwell was able to show that these two constants, which were determined by electrostatic and magnetostatic measurements, also combine to yield the speed of light—again demonstrating the symmetrical role that electric and magnetic fields play in electromagnetic waves. Substituting the values for ε_0 and μ_0 we find

$$c = \frac{1}{\sqrt{(8.85 \times 10^{-12}\ \text{C}^2/(\text{N·m}^2))(4\pi \times 10^{-7}\ \text{T·m/A})}} = 3.00 \times 10^8 \text{ m/s}$$

Clearly, Maxwell's theoretical expression agrees with experiment.

EXAMPLE 25–1 FIZEAU'S RESULTS

Consider a Fizeau experiment in which the wheel has 450 notches and rotates with a speed of 35 rev/s. Light passing through one notch travels to the mirror and back just in time to pass through the next notch. If the distance from the wheel to the mirror is 9500 m, what is the speed of light obtained by this measurement?

PICTURE THE PROBLEM
Our sketch shows an experimental setup similar to Fizeau's. The notched wheel is 9500 m from a mirror and spins with an angular speed of 35 rev/s. We show a few notches in our sketch, which represent the 450 notches on the actual wheel.

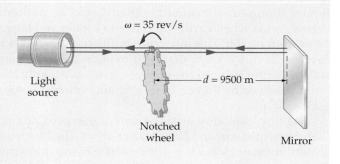

STRATEGY
The speed of light is the distance traveled, $2d$, divided by the time required, Δt. To find the time, we note that the wheel rotates from one notch to the next during this time; that is, it rotates through an angle $\Delta \theta = (1/450)$ rev. Knowing the rotational speed, ω, of the wheel, we can find the time using the relation $\Delta \theta = \omega\, \Delta t$ (Section 10–1).

SOLUTION

1. Find the time required for the wheel to rotate from one notch to the next:
$$\Delta t = \frac{\Delta \theta}{\omega} = \frac{(1/450)\text{ rev}}{35\text{ rev/s}} = 6.3 \times 10^{-5}\text{ s}$$

2. Divide the time into the distance to find the speed of light:
$$c = \frac{2d}{\Delta t} = \frac{2(9500\text{ m})}{6.3 \times 10^{-5}\text{ s}} = 3.0 \times 10^{8}\text{ m/s}$$

INSIGHT
Note that even with a rather large distance for the round trip, the travel time of the light is small, only 0.063 millisecond. This illustrates the great difficulty experimentalists faced in attempting to make an accurate measurement of c.

PRACTICE PROBLEM
If the wheel has 430 notches, what rotational speed is required for the return beam of light to pass through the next notch? [**Answer:** $\omega = 37$ rev/s]

Some related homework problems: Problem 16, Problem 17

Although the speed of light is enormous by earthly standards, it is useful to look at it from an astronomical perspective. Imagine, for example, that you could shrink the solar system to fit onto a football field, with the Sun at one end zone and Pluto at the other. On this scale, Earth would be a grain of sand located at the 2.5-yard line from the Sun, and light would take 8 min to cover that distance. To travel to Pluto, at the other end of the field, light would require about 5.5 hr. Thus, on this scale, the speed of light is like the crawl of a small caterpillar. When one recalls that the solar system is but a speck on the outskirts of the Milky Way galaxy, and that the nearest major galaxy to our own—the Andromeda galaxy—is about 2.2 million light-years away, the speed of light doesn't appear so great after all.

The Doppler Effect

In Section 14–5 we discussed the Doppler effect for sound waves—the familiar increase or decrease in frequency as a source of sound approaches or recedes. A similar Doppler effect applies to electromagnetic waves. There are two fundamental differences, however. First, sound waves require a medium through which to travel, whereas light can propagate across a vacuum. Second, the speed of sound can be different for different observers. For example, an observer approaching a source of sound measures an increased speed of sound, whereas an observer detecting sound from a moving source measures the usual speed of sound. For

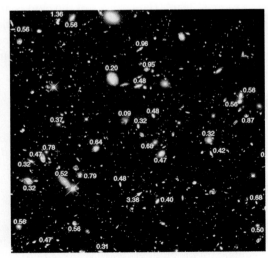

▲ Even traveling at 300 million meters per second, light from the Andromeda galaxy (left) takes over 2 million years to reach us. Yet Andromeda is one of our nearest cosmic neighbors. A sense of the true vastness of the universe is provided by the image known as the Hubble Deep Field (right). This long-exposure photograph, taken from orbit by the Hubble Space Telescope, shows over 1600 galaxies when examined closely. Most of them exhibit a Doppler red shift—that is, their light is shifted to lower frequencies by the Doppler effect, indicating that they are receding from Earth as the universe expands. The red shifts marked on the photo correspond to distances ranging from about 1.3 billion light-years to over 13 billion light-years (nearly 10^{23} miles).

this reason, the Doppler effect with sound is different for a moving observer than it is for a moving source (see Figure 14–18 for a direct comparison). In contrast, the speed of electromagnetic waves is *independent* of the motion of the source and observer, as we shall see in Chapter 29. Therefore, there is just one Doppler effect for electromagnetic waves, and it depends only on the *relative speed* between the observer and source.

For source speeds u that are small compared with the speed of light, the observed frequency f' from a source with frequency f is

$$f' = f\left(1 \pm \frac{u}{c}\right) \qquad 25\text{--}3$$

PROBLEM-SOLVING NOTE

Evaluating the Doppler Shift

Since everyday objects generally move with speeds much less than the speed of light, the Doppler-shifted frequency differs little from the original frequency. To see the Doppler effect more clearly, it is often useful to calculate the difference in frequency, $f' - f$, rather than the Doppler-shifted frequency itself.

Note that u in this expression is a speed and hence is always positive. The appropriate sign in front of the term u/c is chosen for a given situation—the plus sign applies to a source that is approaching the observer, the minus sign to a receding source. In addition, u is a *relative* speed between the source and the observer, both of which may be moving. For example, if an observer is moving in the positive x direction with a speed of 5 m/s, and a source ahead of the observer is moving in the positive x direction with a speed of 12 m/s, the relative speed is $u = 12$ m/s $- 5$ m/s $= 7$ m/s. Since the distance between the observer and source is increasing with time in this case, we would choose the minus sign in Equation 25–3.

EXERCISE 25–2

An FM radio station broadcasts at a frequency of 88.5 MHz. If you drive your car toward the station at 32.0 m/s, what change in frequency do you observe?

SOLUTION

We can find the change in frequency, $f' - f$, using Equation 25–3:

$$f' - f = f\frac{u}{c} = (88.5 \times 10^6 \text{ Hz})\frac{32.0 \text{ m/s}}{3.00 \times 10^8 \text{ m/s}} = 9.44 \text{ Hz}$$

Thus, the frequency changes by only 9.44 Hz = 0.00000944 MHz.

REAL-WORLD PHYSICS

Doppler radar

Common applications of the Doppler effect include the radar units used to measure the speed of automobiles, and the Doppler radar that is used to monitor the weather. In Doppler radar, electromagnetic waves are sent out into the atmosphere and are reflected back to the receiver. The change in frequency of the reflected beam relative to the outgoing beam provides a way of measuring the speed of the clouds and precipitation that reflected the beam. Thus, Doppler radar gives more information than just where a rainstorm is located; it also tells how it is moving. Measurements of this type are particularly important for airports, where information regarding areas of possible wind shear can be crucial for safety.

EXAMPLE 25–2 NEXRAD

REAL-WORLD PHYSICS

The Doppler weather radar used by the National Weather Service is referred to as Nexrad, which stands for next-generation radar. Nexrad commonly operates at a frequency of 2.7 GHz. If a Nexrad wave reflects from an approaching weather system moving with a speed of 28 m/s, find the difference in frequency between the outgoing and returning waves.

PICTURE THE PROBLEM

Our sketch shows the outgoing radar wave, the incoming weather system, and the returning radar wave. The speed of the weather system relative to the radar station is $u = 28$ m/s, and the frequency of the outgoing wave is $f = 2.7$ GHz.

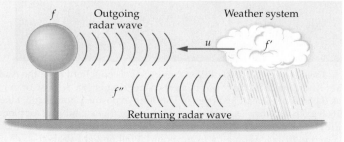

STRATEGY

Two Doppler effects are involved in this system. First, the outgoing wave is seen to have a frequency $f' = f(1 + u/c)$ by the weather system, since it is moving toward the source. The waves reflected by the weather system, then, have the frequency f'. Since the weather system acts like a moving source of radar with frequency f', an observer at the radar facility detects a frequency $f'' = f'(1 + u/c)$. Thus, given u and f, we can calculate the difference, $f'' - f$.

SOLUTION

1. Use $f' = f(1 + u/c)$ to calculate the difference in frequency, $f' - f$:

$$f' - f = fu/c = \frac{(2.7 \times 10^9 \text{ Hz})(28 \text{ m/s})}{3.00 \times 10^8 \text{ m/s}} = 250 \text{ Hz}$$

2. Now, use $f'' = f'(1 + u/c)$ to find the difference between f'' and f':

$$f'' - f' = f'u/c$$

3. Use the results of Step 1 to replace f' with $f + 250$ Hz:

$$f'' - (f + 250 \text{ Hz}) = (f + 250 \text{ Hz})u/c$$

4. Solve for the frequency difference, $f'' - f$:

$$f'' - f = (f + 250 \text{ Hz})u/c + 250 \text{ Hz}$$

$$= \frac{(2.7 \times 10^9 \text{ Hz} + 250 \text{ Hz})(28 \text{ m/s})}{3.00 \times 10^8 \text{ m/s}} + 250 \text{ Hz}$$

$$= 500 \text{ Hz}$$

INSIGHT

Notice that we focus on the *difference* in frequency between the very large numbers $f = 2{,}700{,}000{,}000$ Hz and $f'' = 2{,}700{,}000{,}500$ Hz. Clearly, it is more convenient to simply write $f'' - f = 500$ Hz.

In addition, note that the two Doppler shifts in this problem are analogous to the two Doppler shifts we found for the case of a train approaching a tunnel in Example 14–6.

PRACTICE PROBLEM

Find the difference in frequency if the weather system is receding with a speed of 21 m/s.   [**Answer:** $f'' - f = -380$ Hz]

Some related homework problems: Problem 23, Problem 25

25–3 The Electromagnetic Spectrum

When white light passes through a prism it spreads out into a rainbow of colors, with red on one end and violet on the other. All these various colors of light are electromagnetic waves, of course; they differ only in their frequency and, hence, their wavelength. The relationship between frequency and wavelength for any wave with a speed v is simply $v = f\lambda$, as was shown in Section 14–11. Because all electromagnetic waves in a vacuum have the same speed, c, it follows that f and λ are related as follows:

$$c = f\lambda \qquad\qquad 25\text{–}4$$

Thus, as the frequency of an electromagnetic wave increases, its wavelength decreases.

In the following Example we calculate the frequency of red and violet light, given the corresponding wavelengths. Notice that the wavelengths are given in units of nanometers (nm), where 1 nm $= 10^{-9}$ m. Occasionally the wavelength of light is given in terms of a non-SI unit referred to as the *angstrom* (Å), defined as follows: $1\text{Å} = 10^{-10}$ m.

EXAMPLE 25–3   ROSES ARE RED, VIOLETS ARE VIOLET

Find the frequency of red light, with a wavelength of 700.0 nm, and violet light, with a wavelength of 400.0 nm.

PICTURE THE PROBLEM

The visible electromagnetic spectrum, along with representative wavelengths, is shown in our diagram. In addition to the wavelengths of 700.0 nm for red light and 400.0 nm for violet light, we include 600.0 nm for yellowish orange light and 500.0 nm for greenish blue light.

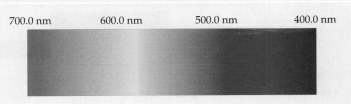

700.0 nm   600.0 nm   500.0 nm   400.0 nm

CONTINUED ON NEXT PAGE

CONTINUED FROM PREVIOUS PAGE

STRATEGY
We obtain the frequency by rearranging $c = f\lambda$ to yield $f = c/\lambda$.

SOLUTION

1. Substitute $\lambda = 700.0$ nm for red light:
$$f = \frac{c}{\lambda} = \frac{3.00 \times 10^8 \text{ m/s}}{700.0 \times 10^{-9} \text{ m}} = 4.29 \times 10^{14} \text{ Hz}$$

2. Substitute $\lambda = 400.0$ nm for violet light:
$$f = \frac{c}{\lambda} = \frac{3.00 \times 10^8 \text{ m/s}}{400.0 \times 10^{-9} \text{ m}} = 7.50 \times 10^{14} \text{ Hz}$$

INSIGHT
The frequency of visible light is extremely large. In fact, even for the relatively low frequency of red light, it takes only 2.33×10^{-15} s to complete one cycle. The *range* of visible frequencies is relatively small, however, when compared with other portions of the electromagnetic spectrum.

PRACTICE PROBLEM
What is the wavelength of light with a frequency of 5.25×10^{14} Hz? **[Answer: 571 nm]**

Some related homework problems: Problem 27, Problem 28

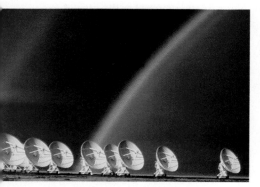

▲ Since the development of the first radiotelescopes in the 1950s, the radio portion of the electromagnetic spectrum has provided astronomers with a valuable new window on the universe. These antennas are part of the Very Large Array (VLA), located in San Augustin, New Mexico.

In principle, the frequency of an electromagnetic wave can have any positive value, and this full range of frequencies is known as the **electromagnetic spectrum.** Certain bands of the spectrum are given special names, as indicated in **Figure 25–8**. For example, we have just seen that visible light occupies a relatively narrow band of frequencies from 4.29×10^{14} Hz to 7.50×10^{14} Hz. In what follows, we discuss the various regions of the electromagnetic spectrum of most relevance to humans and our technology, in order of increasing frequency.

Radio Waves

($f \sim 10^6$ Hz to 10^9 Hz, $\lambda \sim$ 300 m to 0.3 m) The lowest-frequency electromagnetic waves of practical importance are *radio* and *television waves* in the frequency range of roughly 10^6 Hz to 10^9 Hz. Waves in this frequency range are produced in a variety of ways. For example, molecules and accelerated electrons in space give off radio waves, which radio astronomers detect with large dish receivers. Radio waves are also produced as a piece of adhesive tape is slowly peeled from a surface, as you can confirm by holding a transistor radio near the tape and listening for pops and snaps coming from the speaker. Most commonly, the radio waves we pick up with our radios and televisions are produced by alternating currents in metal antennas.

Microwaves

($f \sim 10^9$ Hz to 10^{12} Hz, $\lambda \sim$ 300 mm to 0.3 mm) Electromagnetic radiation with frequencies from 10^9 Hz to about 10^{12} Hz are referred to as *microwaves*. Waves in this frequency range are used to carry long-distance telephone conversations, as well as to cook our food. Microwaves, with wavelengths of about 1 mm to 30 cm,

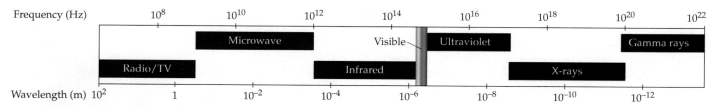

▲ **FIGURE 25–8 The electromagnetic spectrum**
Note that the visible portion of the spectrum is relatively narrow. The boundaries between various bands of the spectrum are not sharp but, instead, are somewhat arbitrary.

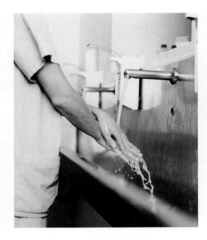

◀ We use infrared rays all the time, even though they are invisible to us. If you change the channel with a remote control, your signal is sent by an infrared ray; if you move your hand in front of a no-touch water faucet, an infrared ray detects the motion. In contrast, snakes called pit vipers can actually "see" infrared rays with the "pit" organs located just in front of their eyes. What must a remote control look like to one of these creatures?

are the highest-frequency electromagnetic waves that can be produced by electronic circuitry.

Infrared Waves

($f \sim 10^{12}$ **Hz to** 4.3×10^{14} **Hz,** $\lambda \sim$ **0.3 mm to 700 nm**) Electromagnetic waves with frequencies just below that of red light—roughly 10^{12} Hz to 4.3×10^{14} Hz— are known as *infrared* rays. These waves can be felt as heat on our skin but cannot be seen with our eyes. Many creatures, including various types of pit vipers, have specialized infrared receptors that allow them to "see" the infrared rays given off by a warm-blooded prey animal, even in total darkness. Infrared rays are often generated by the rotations and vibrations of molecules. In turn, when infrared rays are absorbed by an object, its molecules rotate and vibrate more vigorously, resulting in an increase in the object's temperature. Finally, many remote controls—for items ranging from TVs to DVD players to gas fireplaces—operate on a beam of infrared light with a wavelength of about 1000 nm. This infrared light is so close to the visible spectrum and so low in intensity that it cannot be felt as heat.

REAL-WORLD PHYSICS: BIO

Infrared receptors in pit vipers

Visible Light

($f \sim 4.3 \times 10^{14}$ **Hz to** 7.5×10^{14} **Hz,** $\lambda \sim$ **700 nm to 400 nm**) The portion of the electromagnetic spectrum most familiar to us is the spectrum of visible light, represented by the full range of colors seen in a rainbow. Each of the different colors, as perceived by our eyes and nervous system, is nothing more than an electromagnetic wave with a different frequency. Waves in this frequency range (4.3×10^{14} to 7.5×10^{14} Hz) are produced primarily by electrons changing their positions within an atom, as we discuss in detail in Chapter 31.

◀ Photographs made with infrared radiation are often called thermograms, since most infrared wavelengths can be felt as heat by the human skin. Thermograms provide a useful remote sensing technique for measuring temperature. In the photo on the left, the areas of the cat's head that are warmest (pink) and coolest (blue) can be clearly identified. In the photo on the right, an infrared satellite image of the Atlantic Ocean off the coast of North America, warmer colors are used to indicate higher sea surface temperatures. The swirling red streak running from lower left toward the upper right is the Gulf Stream.

**Biological effects of
ultraviolet light**

▲ Most ultraviolet radiation cannot penetrate Earth's atmosphere, but ultraviolet astronomical photographs, such as the one shown here, have been taken by cameras in orbit. The bright spots in this image of a spiral galaxy are areas of intense star formation, populated by hot young stars that radiate heavily in the ultraviolet.

Irradiated food

▲ The use of radiation to preserve food can be quite effective, but the technique is still controversial. Both boxes of strawberries shown here were stored for about 2 weeks at refrigerator temperature. Before storage, the box at right was irradiated to kill microorganisms and mold spores.

Ultraviolet Light

($f \sim 7.5 \times 10^{14}$ Hz to 10^{17} Hz, $\lambda \sim 400$ nm to 3 nm) When electromagnetic waves have frequencies just above that of violet light—from about 7.5×10^{14} Hz to 10^{17} Hz—they are called *ultraviolet* or *UV rays*. Although these rays are invisible, they often make their presence known by causing suntans with moderate exposure. More prolonged or intense exposure to UV rays can have harmful consequences, including an increased probability of developing a skin cancer. Fortunately, most of the UV radiation that reaches Earth from the Sun is absorbed in the upper atmosphere by ozone (O_3) and other molecules. A significant reduction in the ozone concentration in the stratosphere could result in an unwelcome increase of UV radiation on Earth's surface.

X-Rays

($f \sim 10^{17}$ Hz to 10^{20} Hz, $\lambda \sim 3$ nm to 0.003 nm) As the frequency of electromagnetic waves is raised even higher, into the range between about 10^{17} Hz to 10^{20} Hz, we reach the part of the spectrum known as *X-rays*. Typically, the X-rays used in medicine are generated by the rapid deceleration of high-speed electrons projected against a metal target, as we show in Section 31–17. These energetic rays, which are only weakly absorbed by the skin and soft tissues, pass through our bodies rather freely, except when they encounter bones, teeth, or other relatively dense material. This property makes X-rays most valuable for medical diagnosis, research, and treatment. Still, X-rays can cause damage to human tissue, and it is desirable to reduce unnecessary exposure to these rays as much as possible.

Gamma Rays

($f \sim 10^{20}$ Hz and higher, $\lambda \sim 0.003$ nm and smaller) Finally, electromagnetic waves with frequencies above about 10^{20} Hz are generally referred to as *gamma* (γ) *rays*. These rays, which are even more energetic than X-rays, are often produced as neutrons and protons rearrange themselves within a nucleus, or when a particle collides with its antiparticle, and the two annihilate each other. These processes are discussed in detail in Chapter 32. Gamma rays are also highly penetrating and destructive to living cells. It is for this reason that they are used to kill cancer cells and, more recently, microorganisms in food. Irradiated food, however, is a concept that has yet to become popular with the general public, even though NASA has irradiated astronauts' food since the 1960s. If you happen to see irradiated food in the grocery store, you will know that it has been exposed to γ rays from cobalt-60 for 20 to 30 minutes.

Notice that the visible part of the electromagnetic spectrum, so important to life on Earth, is actually the smallest of the frequency bands we have named. This accounts for the fact that a rainbow produces only a narrow band of color in the sky—if the visible band were wider, the rainbow would be wider as well. It should be remembered, however, that there is nothing particularly special about the visible band; in fact, it is even species dependent. For example, some bees and butterflies can see ultraviolet light, and, as mentioned previously, certain snakes can form images from infrared radiation.

One of the main factors in determining the visible range of frequencies is Earth's atmosphere. For example, if one examines the transparency of the atmosphere as a function of frequency, it is found that there is a relatively narrow range of frequencies for which the atmosphere is highly transparent. As eyes evolved in living systems on Earth, they could have evolved to be sensitive to various different frequency ranges. It so happens, however, that the range of frequencies that most animal eyes can detect matches nicely with the range of frequencies that the atmosphere allows to reach Earth's surface. This is a nice example of natural adaptation.

25-4 Energy and Momentum in Electromagnetic Waves

All waves transmit energy, and electromagnetic waves are no exception. When you walk outside on a sunny day, for example, you feel warmth where the sunlight strikes your skin. The energy creating this warm sensation originated in the Sun, and it has just completed a 93-million-mile trip when it reaches your body. In fact, the energy necessary for most of the life on Earth is transported here across the vacuum of space by electromagnetic waves traveling at the speed of light.

That electromagnetic waves carry energy with them is no surprise when you recall that they are composed of electric and magnetic fields, each of which has an associated energy density. For example, in Section 20–6 we showed that the energy density of an electric field of magnitude E is

$$u_E = \tfrac{1}{2}\varepsilon_0 E^2$$

Similarly, we showed in Section 23–9 that the energy density of a magnetic field of magnitude B is

$$u_B = \frac{1}{2\,\mu_0}B^2$$

It follows that the total energy density, u, of an electromagnetic wave is simply

$$u = u_E + u_B = \tfrac{1}{2}\varepsilon_0 E^2 + \frac{1}{2\,\mu_0}B^2 \qquad\qquad 25\text{--}5$$

As expected, both E and B contribute to the total energy carried by a wave. Not only that, but it can be shown that the electric and magnetic energy densities in an electromagnetic wave are, in fact, equal to each other—again demonstrating the symmetrical role played by the electric and magnetic fields. Thus, the total energy density of an electromagnetic field can be written in the following equivalent forms:

$$u = \tfrac{1}{2}\varepsilon_0 E^2 + \frac{1}{2\,\mu_0}B^2 = \varepsilon_0 E^2 = \frac{1}{\mu_0}B^2 \qquad\qquad 25\text{--}6$$

Since E and B vary sinusoidally with time, as indicated in Figure 25–3, it follows that their average values are zero—just like the current or voltage in an ac circuit. Therefore, to find the average energy density of an electromagnetic wave, we must use the rms values of E and B:

$$u_{av} = \tfrac{1}{2}\varepsilon_0 E_{rms}^2 + \frac{1}{2\,\mu_0}B_{rms}^2 = \varepsilon_0 E_{rms}^2 = \frac{1}{\mu_0}B_{rms}^2 \qquad\qquad 25\text{--}7$$

Recall that the rms value of a sinusoidally varying quantity x is related to its maximum value as $x_{rms} = x_{max}/\sqrt{2}$. Thus, for the electric and magnetic fields in an electromagnetic wave we have

$$E_{rms} = \frac{E_{max}}{\sqrt{2}}$$

$$B_{rms} = \frac{B_{max}}{\sqrt{2}} \qquad\qquad 25\text{--}8$$

The fact that the electric and magnetic energy densities are equal in an electromagnetic wave has a further interesting consequence. Setting u_E equal to u_B we obtain

$$\tfrac{1}{2}\varepsilon_0 E^2 = \frac{1}{2\,\mu_0}B^2$$

Rearranging slightly, and taking the square root, we find

$$E = \frac{1}{\sqrt{\varepsilon_0\mu_0}}B$$

Finally, from the fact that the speed of light is given by the relation $c = 1/\sqrt{\varepsilon_0\mu_0}$, it follows that

$$E = cB \qquad\qquad 25\text{–}9$$

Thus, not only does an electromagnetic wave have both an electric and a magnetic field, the fields must also have the specific ratio $E/B = c$.

EXERCISE 25–3

At a given instant of time the electric field in a beam of sunlight has a magnitude of 510 N/C. What is the magnitude of the magnetic field at this instant?

SOLUTION
Using $E = cB$, or $B = E/c$, we find

$$B = \frac{E}{c} = \frac{510\ \text{N/C}}{3.00 \times 10^8\ \text{m/s}} = 1.7 \times 10^{-6}\ \text{T}$$

Note that the units are consistent, since $1\ \text{T} = 1\ \text{N}/(\text{C}\cdot\text{m/s})$.

The amount of energy a wave delivers to a unit area in a unit time is referred to as its **intensity,** I. (Equivalently, since power is energy per time, the intensity of a wave is the power per unit area.) Imagine, for example, an electromagnetic wave of area A moving in the positive x direction, as in **Figure 25–9**. In the time Δt the wave moves through a distance $c\Delta t$; hence, all the energy in the volume $\Delta V = A(c\Delta t)$ is deposited on the area A in this time. Because energy is equal to the energy density times the volume, it follows that the energy in the volume ΔV is $\Delta U = u\Delta V$. Therefore, the intensity of the wave (energy per area per time) is

$$I = \frac{\Delta U}{A\Delta t} = \frac{u(Ac\Delta t)}{A\Delta t} = uc$$

Averaged over time, the intensity is $I_{av} = u_{av}c$. In terms of the electric and magnetic fields, we have

$$I = uc = \tfrac{1}{2}c\varepsilon_0 E^2 + \frac{1}{2\mu_0}cB^2 = c\varepsilon_0 E^2 = \frac{c}{\mu_0}B^2 \qquad 25\text{–}10$$

As before, to calculate an average intensity, we must replace E and B with their rms values.

Notice that the intensity is proportional to the square of the fields. This is analogous to the case of simple harmonic motion, where the energy of oscillation is proportional to the square of the amplitude, as we found in Section 14–15.

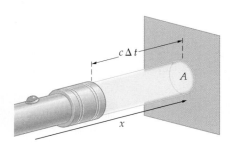

▲ **FIGURE 25–9 The energy in a beam of light**
A beam of light of cross-sectional area A shines on a surface. All the light energy contained in the volume $\Delta V = A(c\Delta t)$ strikes the surface in the time Δt.

EXAMPLE 25–4 LIGHTBULB FIELDS

A garage is illuminated by a single incandescent lightbulb dangling from a wire. If the bulb radiates light uniformly in all directions, and consumes an average electrical power of 50.0 W, what are **(a)** the average intensity of the light and **(b)** the rms values of E and B at a distance of 1.00 m from the bulb? (Assume that 5.00% of the electrical power consumed by the bulb is converted to light.)

PICTURE THE PROBLEM
The physical situation is pictured in the sketch. We assume that all the power radiated by the bulb passes uniformly through the area of a sphere of radius $r = 1.00$ m centered on the bulb.

STRATEGY

a. Recall that intensity is power per unit area; therefore, $I_{av} = P_{av}/A$. In this case the area is $A = 4\pi r^2$, the surface area of a sphere of radius r. The average power of the light is 5.00% of 50.0 W.

b. Once we know the intensity, I_{av}, we obtain the fields using Equation 25–10. Since the intensity is an average value, the corresponding fields are rms values.

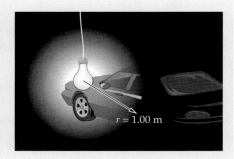

$r = 1.00$ m

SOLUTION

Part (a)

1. Calculate the average intensity at the surface of the sphere of radius $r = 1.00$ m:

$$I_{av} = \frac{P_{av}}{A} = \frac{(50.0 \text{ W})(0.0500)}{4\pi(1.00 \text{ m})^2} = 0.199 \text{ W/m}^2$$

Part (b)

2. Use $I_{av} = c\varepsilon_0 E_{rms}^2$ to find E_{rms}:

$$I_{av} = c\varepsilon_0 E_{rms}^2$$

$$E_{rms} = \sqrt{\frac{I_{av}}{c\varepsilon_0}} = \sqrt{\frac{0.199 \text{ W/m}^2}{(3.00 \times 10^8 \text{ m/s})(8.85 \times 10^{-12} \text{ C}^2/\text{N} \cdot \text{m}^2)}}$$

$$= 8.66 \text{ N/C}$$

3. Use $I_{av} = cB_{rms}^2/\mu_0$ to find B_{rms}:

$$I_{av} = cB_{rms}^2/\mu_0$$

$$B_{rms} = \sqrt{\frac{\mu_0 I_{av}}{c}} = \sqrt{\frac{(4\pi \times 10^{-7} \text{ N} \cdot \text{s}^2/\text{C}^2)(0.199 \text{ W/m}^2)}{(3.00 \times 10^8 \text{ m/s})}}$$

$$= 2.89 \times 10^{-8} \text{ T}$$

4. As a check, use the relation $E = cB$ and the results from Steps 2 and 3 to calculate the speed of light, c:

$$c = \frac{E_{rms}}{B_{rms}} = \frac{8.66 \text{ N/C}}{2.89 \times 10^{-8} \text{ T}} = 3.00 \times 10^8 \text{ m/s}$$

INSIGHT

Notice that the electric field in the light from the bulb has a magnitude ($8.66 \text{ N/C} = 8.66 \text{ V/m}$) that is relatively easy to measure. The magnetic field, however, is so small that a direct measurement would be difficult. This is common in electromagnetic waves because $B = E/c$, and c is such a large number.

PRACTICE PROBLEM

If the distance from the bulb is doubled, does E_{rms} increase, decrease, or stay the same? Check your answer by calculating E_{rms} at $r = 2.00$ m.   [**Answer:** E_{rms} decreases with increasing distance from the bulb. For $r = 2.00$ m, we find $E_{rms} = \frac{1}{2}(8.66 \text{ N/C}) = 4.33 \text{ N/C}$.]

Some related homework problems: Problem 51, Problem 62

ACTIVE EXAMPLE 25–1   **FIND E AND B IN A BEAM OF LIGHT**

A small laser emits a cylindrical beam of light 1.00 mm in diameter with an average power of 5.00 mW. Find the maximum values of E and B in this beam.

SOLUTION   *(Test your understanding by performing the calculations indicated in each step.)*

1. Divide the average power of the beam by its area to find the intensity:

   $I_{av} = 6370 \text{ W/m}^2$

2. Use $I_{av} = c\varepsilon_0 E_{rms}^2$ to find the rms electric field:

   $E_{rms} = 1550 \text{ N/C}$

3. Multiply E_{rms} by $\sqrt{2}$ to find E_{max}:

   $E_{max} = 2190 \text{ N/C}$

4. Divide E_{max} by c to find B_{max}:

   $B_{max} = 7.30 \times 10^{-6} \text{ T}$

INSIGHT

To find the cross-sectional area of the cylindrical beam in Step 1 we used $A = \pi d^2/4$.

YOUR TURN

Suppose the beam spreads out to twice its initial diameter. By what factor do E_{max} and B_{max} change?

*(Answers to **Your Turn** problems are given in the back of the book.)*

Finally, an electromagnetic wave also carries momentum. In fact, it can be shown that if a total energy U is absorbed by a given area, the momentum, p, that it receives is

$$p = \frac{U}{c}$$

25–11

For an electromagnetic wave absorbed by an area A, the total energy received in the time Δt is $U = u_{av}Ac\Delta t$; hence, the momentum, Δp, received in this time is

$$\Delta p = \frac{u_{av}Ac\Delta t}{c} = \frac{I_{av}A\Delta t}{c}$$

Since the average force is $F_{av} = \Delta p/\Delta t = u_{av}A = I_{av}A/c$, it follows that the average pressure (force per area) is simply

$$\text{pressure}_{av} = \frac{I_{av}}{c} \qquad\qquad 25\text{--}12$$

The pressure exerted by light is commonly referred to as **radiation pressure.**

CONCEPTUAL CHECKPOINT 25–2   MOMENTUM TRANSFER

When an electromagnetic wave carrying an energy U is absorbed by an object, the momentum the object receives is $p = U/c$. If, instead, the object reflects the wave, is the momentum the object receives **(a)** more than, **(b)** less than, or **(c)** the same as when it absorbs the wave?

REASONING AND DISCUSSION
When the object reflects the wave, it must supply not only enough momentum to stop the wave, $p = U/c$, but an equivalent amount of momentum to send the wave back in the opposite direction. Thus, the momentum the object receives is $p = 2U/c$. This situation is completely analogous to the momentum transfer of a ball that either sticks to a wall or bounces back the way it came.

ANSWER
(a) The object receives twice as much momentum.

Radiation pressure is a very real effect, though in everyday situations it is too small to be noticed. In principle, turning on a flashlight should give the user a "kick," like the recoil from firing a gun. In practice, the effect is much too small to be felt. In the next Exercise, we calculate the radiation pressure due to sunlight to get a feel for the magnitudes involved.

EXERCISE 25–4

On a sunny day, the average intensity of sunlight on Earth's surface is about 1.00×10^3 W/m^2. Find **(a)** the average radiation pressure due to the sunlight and **(b)** the average force exerted by the light on a 1.00-m $\times$ 2.50-m beach towel. For part (b), assume that the towel absorbs all the light that falls on it.

SOLUTION

a. Divide the average intensity by the speed of light to obtain the average pressure:

$$\text{pressure}_{av} = \frac{I_{av}}{c} = \frac{1.00 \times 10^3 \text{ W/m}^2}{3.00 \times 10^8 \text{ m/s}} = 3.33 \times 10^{-6} \text{ N/m}^2$$

b. Multiply pressure times area to find the force exerted on the towel:

$$F_{av} = \text{pressure}_{av}A = (3.33 \times 10^{-6} \text{ N/m}^2)(1.00 \text{ m} \times 2.50 \text{ m}) = 8.33 \times 10^{-6} \text{ N}$$

Since a newton is about a quarter of a pound, it follows that the force exerted on the beach towel by sunlight is incredibly small.

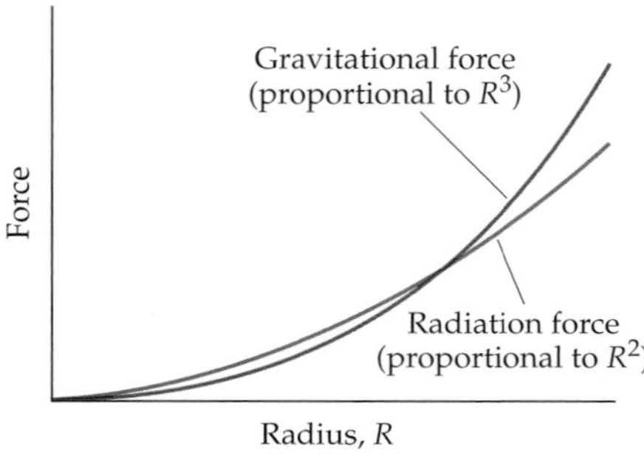

▲ **FIGURE 25–10 Radiation versus gravitational forces**
The forces exerted on a small particle of radius R by gravity and radiation pressure. For large R the gravitational force dominates; for small R the radiation force is larger.

Even though the radiation force on a beach towel is negligible, the effects of radiation pressure can be significant on particles that are very small. Consider a small object of radius R drifting through space somewhere in our solar system. The gravitational attraction it feels from the Sun depends on its mass. Since mass is proportional to volume, and volume is proportional to the cube of the particle's radius, the gravitational force on the particle varies as R^3. On the other hand, the radiation pressure is exerted over the area of the object, and the area is proportional to R^2. As R becomes smaller, the radiation pressure, with its R^2 dependence, decreases less rapidly than the gravitational force, with its R^3 dependence, as illustrated in **Figure 25–10**. Thus, for small enough particles, the radiation pressure from sunlight is actually more important than the gravitational force. It is for this reason that dust particles given off by a comet are "blown" away by sunlight, giving the comet a long tail that streams away from the sun. Some imaginative thinkers have envisaged a "sailing" ship designed to travel through the cosmos using this light-pressure "wind."

Finally, it has been discovered that the energy carried by light comes in discrete units, as if there were "particles" of energy in a light beam. So, in many ways, light behaves just like any other wave, showing the effects of diffraction and interference, whereas in other ways it acts like a particle. Therefore, light, and all electromagnetic waves, have a "dual" nature, exhibiting both wave and particle properties. As we shall see when we consider quantum physics in Chapter 30, this *wave–particle duality* plays a fundamental role in our understanding of modern physics.

25–5 Polarization

When looking into the blue sky of a crystal-clear day, we see light that is fairly uniform—as long as we refrain from looking too close to the Sun. However, for some animals, like the common honeybee, the light in the sky is far from uniform. The reason is that honeybees are sensitive to the **polarization** of light, an ability that aids in their navigation from hive to flower and back.

To understand what is meant by the polarization of light, or any other electromagnetic wave, consider the electromagnetic waves pictured in **Figure 25–11**. Each of these waves has an electric field that points along a single line. For example, the electric field in Figure 25–11 (a) points in either the positive or negative *z* direction. We say, then, that this wave is **linearly polarized** in the *z* direction. A wave of this sort might be produced by a straight-wire antenna oriented along the *z* axis. Similarly, the direction of polarization for the wave in Figure 25–11 (b) is in the *y-z* plane at an angle of 60° relative to the *y* direction. In general, *the polarization of an electromagnetic wave refers to the direction of its electric field.*

A convenient way to represent the polarization of a beam of light is shown in **Figure 25–12**. In part (a) of this figure, we indicate light that is polarized in the vertical direction. Not all light is polarized, however. In part (b) of Figure 25–12 we show light that is a combination of many waves with polarizations in different, random directions. Such light is said to be **unpolarized.** A common incandescent lightbulb produces unpolarized light because each atom in the heated filament sends out light of random polarization. Similarly, the light from the Sun is unpolarized.

Passing Light Through Polarizers

A beam of unpolarized light can be polarized in a number of ways, including by passing it through a **polarizer.** To be specific, a polarizer is a material that is

▲ Like many comets, comet Hale-Bopp, which passed relatively close to Earth in 1997, developed two tails as it approached the Sun. The straighter, blue tail in this photograph is gas boiling off the comet's head as volatile material is vaporized by the Sun's heat. The curved, whiter tail consists mostly of dust particles, blown away by the pressure of sunlight.

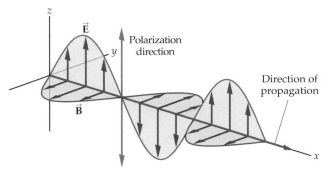

(a) Polarization in *z* direction

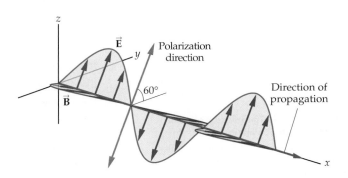

(b) Polarization 60° to *y* direction

▲ **FIGURE 25–11 Polarization of electromagnetic waves**
The polarization of an electromagnetic wave is the direction along which its electric field vector, $\vec{E}$, points. The cases shown illustrate **(a)** polarization in the *z* direction and **(b)** polarization in the *y-z* plane, at an angle of 60° with respect to the *y* axis.

▶ **FIGURE 25–12 Polarized versus unpolarized light**
A beam of light that is **(a)** polarized in the vertical direction and **(b)** unpolarized.

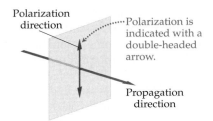
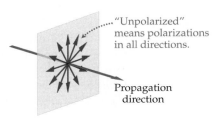

(a) Vertically polarized light

(b) Unpolarized light

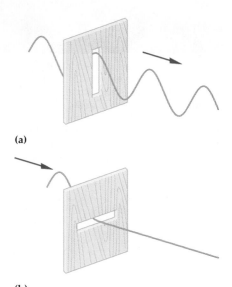

(a)

(b)

▲ **FIGURE 25–13 A mechanical analog of a polarizer**
(a) The polarization of the wave is in the same direction as the polarizer; hence, the wave passes through unaffected. **(b)** The polarization of the wave is at right angles to the direction of the polarizer. In this case the wave is absorbed.

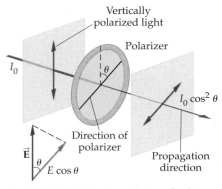

▲ **FIGURE 25–14 Transmission of polarized light through a polarizer**
A polarized beam of light, with intensity I_0, encounters a polarizer oriented at an angle θ relative to the polarization direction. The intensity of light transmitted through the polarizer is $I = I_0 \cos^2 \theta$. After passing through the polarizer, the light is polarized in the same direction as the polarizer.

composed of long, thin, electrically conductive molecules oriented in a specific direction. When a beam of light strikes a polarizer, it is readily absorbed if its electric field is parallel to the molecules; light whose electric field is perpendicular to the molecules passes through the material with little absorption. As a result, the light that passes through a polarizer is preferentially polarized along a specific direction. Common examples of a polarizer are the well-known Polaroid sheets used to make Polaroid sunglasses.

A simple mechanical analog of a polarizer is shown in **Figure 25–13**. Here we see a wave that displaces a string in the vertical direction as it propagates toward a slit cut into a block of wood. If the slit is oriented vertically, as in Figure 25–13 (a), the wave passes through unhindered. Conversely, when the slit is oriented horizontally it stops the wave, as indicated in Figure 25–13 (b). A polarizer performs a similar function on a beam of light.

We now consider what happens when a beam of light polarized in one direction encounters a polarizer oriented in a different direction. The situation is illustrated in **Figure 25–14**, where we see light with a vertical polarization and intensity I_0 passing through a polarizer with its preferred direction—its **transmission axis**—at an angle θ to the vertical. As shown in the figure, the component of $\vec{\mathbf{E}}$ along the transmission axis is $E \cos \theta$. Recalling that the intensity of light is proportional to the electric field squared, we see that the intensity, I, of the transmitted beam is reduced by the factor $\cos^2 \theta$. Therefore,

Law of Malus

$$I = I_0 \cos^2 \theta \qquad\qquad 25\text{–}13$$

This result is known as the **law of Malus,** after the French engineer Etienne-Louis Malus (1775–1812). Notice that the intensity is unchanged if $\theta = 0$, and is zero if $\theta = 90°$.

EXERCISE 25–5

Vertically polarized light with an intensity of 515 W/m² passes through a polarizer oriented at an angle θ to the vertical. Find the transmitted intensity of the light for **(a)** $\theta = 10.0°$, **(b)** $\theta = 45.0°$, and **(c)** $\theta = 90.0°$.

SOLUTION

Applying $I = I_0 \cos^2 \theta$ we obtain

    a. $I = (515 \text{ W/m}^2) \cos^2 10.0° = 499 \text{ W/m}^2$
    b. $I = (515 \text{ W/m}^2) \cos^2 45.0° = 258 \text{ W/m}^2$
    c. $I = (515 \text{ W/m}^2) \cos^2 90.0° = 0$

Just as important as the change in intensity is what happens to the *polarization* of the transmitted light:

> The transmitted beam of light is no longer polarized in its original direction; it is now polarized in the direction of the polarizer.

This effect also is illustrated in Figure 25–14. Thus, a polarizer changes both the intensity *and* the polarization of a beam of light.

Figure 25–15 shows an unpolarized beam of light with an intensity I_0 encountering a polarizer. In this case, there is no single angle θ; instead, to obtain the transmitted intensity, we must average $\cos^2 \theta$ over all angles. This has already been done in Section 24–1, where we considered rms values in ac circuits. As was shown there, the average of $\cos^2 \theta$ is one-half; thus, the transmitted intensity for an unpolarized beam with an intensity of I_0 is

Transmitted Intensity for an Unpolarized Beam

$$I = \tfrac{1}{2} I_0 \qquad\qquad 25\text{--}14$$

A common type of polarization experiment is shown in **Figure 25–16**. An unpolarized beam is first passed through a polarizer to give the light a specified polarization. The light next passes through a second polarizer, referred to as the **analyzer,** whose transmission axis is at an angle θ relative to the polarizer. The orientation of the analyzer can be adjusted to give a beam of light of variable intensity and polarization. We consider a situation of this type in the next Example.

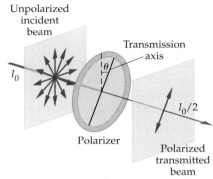

▲ **FIGURE 25–15** **Transmission of unpolarized light through a polarizer**
When an unpolarized beam of intensity I_0 passes through a polarizer, the transmitted beam has an intensity of $\frac{1}{2} I_0$ and is polarized in the direction of the polarizer.

◀ **FIGURE 25–16** **A polarizer and an analyzer**
An unpolarized beam of intensity I_0 is polarized in the vertical direction by a polarizer with a vertical transmission axis. Next, it passes through another polarizer, the analyzer, whose transmission axis is at an angle θ relative to the transmission axis of the polarizer. The final intensity of the beam is $I = \frac{1}{2} I_0 \cos^2 \theta$.

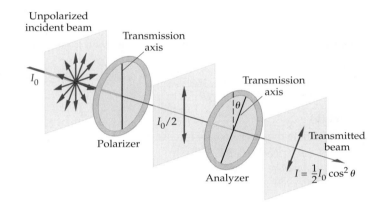

EXAMPLE 25–5   ANALYZE THIS

In the polarization experiment shown in our sketch, the final intensity of the beam is $0.200\, I_0$. What is the angle θ between the transmission axes of the analyzer and polarizer?

PICTURE THE PROBLEM
The experimental setup is shown in the sketch. As indicated, the intensity of the unpolarized incident beam, I_0, is reduced to $\frac{1}{2} I_0$ after passing through the first polarizer. The second polarizer reduces the intensity further, to $0.200\, I_0$.

STRATEGY
It is clear from the sketch that the analyzer reduces the intensity of the light by a factor of $1/2.50$; that is, the intensity is reduced from $I_0/2$ to $(1/2.50)(I_0/2) = I_0/5.00 = 0.200\, I_0$. Thus we must find the angle θ that gives this reduction. Recalling that an analyzer reduces intensity according to the relation $I = I_0 \cos^2 \theta$, we set $\cos^2 \theta = 1/2.50$ and solve for θ.

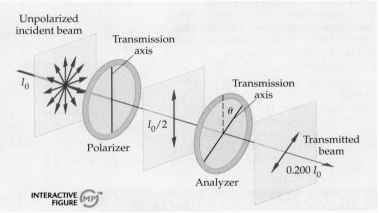

INTERACTIVE **MP** ™
FIGURE

SOLUTION

1. Set $\cos^2 \theta$ equal to $1/2.50$ and solve for $\cos \theta$:

$$\cos^2 \theta = \frac{1}{2.50}$$

$$\cos \theta = \frac{1}{\sqrt{2.50}}$$

2. Solve for the angle θ:

$$\theta = \cos^{-1}\!\left(\frac{1}{\sqrt{2.50}}\right) = 50.8°$$

CONTINUED ON NEXT PAGE

CONTINUED FROM PREVIOUS PAGE

INSIGHT

Since the analyzer absorbs part of the light as the beam passes through, it also absorbs energy. Therefore, in principle, the analyzer would experience a slight heating. As always, energy must be conserved.

PRACTICE PROBLEM

If the angle θ is increased slightly, does the final intensity increase, decrease, or stay the same? Check your answer by finding the final intensity for 60.0°.   [**Answer:** The final intensity decreases. For $\theta = 60.0°$, the final intensity is $I = 0.125\, I_0$.]

Some related homework problems: Problem 74, Problem 76, Problem 80

PROBLEM-SOLVING NOTE

Transmission Through Polarizing Filters

The intensity of light transmitted through a pair of polarizing filters depends only on the relative angle θ between the filters; it is independent of whether the filters are rotated clockwise or counterclockwise relative to each other.

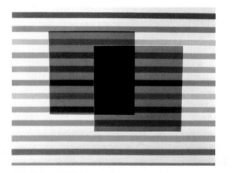

▲ **FIGURE 25–17 Overlapping polarizing sheets**

When unpolarized light strikes a single layer of polarizing material, half of the light is transmitted. This is true regardless of how the transmission axis is oriented. But no light at all can pass through a pair of polarizing filters with axes at right angles (crossed polarizers).

ACTIVE EXAMPLE 25–2 FIND THE TRANSMITTED INTENSITY

Calculate the transmitted intensity for the following two cases: **(a)** A vertically polarized beam of intensity I_0 passes through a polarizer with its transmission axis at 60° to the vertical. **(b)** A vertically polarized beam of intensity I_0 passes through two polarizers, the first with its transmission axis at 30° to the vertical, and the second with its transmission axis rotated an additional 30° to the vertical. (*Note:* In both cases, the final beam is polarized at 60° to the vertical.)

SOLUTION *(Test your understanding by performing the calculations indicated in each step.)*

Part (a)

1. Calculate the final intensity using $I = I_0 \cos^2 \theta$, with $\theta = 60°$: $\qquad\qquad I = \frac{1}{4} I_0$

Part (b)

2. Calculate the intermediate intensity using $I = I_0 \cos^2 \theta$, with $\theta = 30°$: $\qquad\qquad I = \frac{3}{4} I_0$

3. Calculate the final intensity using $I = I_0 \cos^2 \theta$, with $\theta = 30°$ again: $\qquad\qquad I = \frac{9}{16} I_0$

INSIGHT

Even though the direction of polarization is rotated by a total of 60° in each case, the final intensity is more than twice as great when two polarizers are used instead of just one. In general, the more polarizers that are used—and, hence, the more smoothly the direction of polarization changes—the greater the final intensity.

YOUR TURN

Calculate the transmitted intensity when three polarizers are used, the first at 20.0° to the vertical, the second at 40.0° to the vertical, and the third at 60.0° to the vertical. Compare your result with $\frac{9}{16} I_0 = 0.563\, I_0$ found with two polarizers.

*(Answers to **Your Turn** problems are given in the back of the book.)*

Polarizers with transmission axes at right angles to one another are referred to as "crossed polarizers." The transmission through a pair of crossed polarizers is zero, since $\theta = 90°$ in Malus's law. Crossed polarizers are illustrated in **Figure 25–17** and are referred to in the following Conceptual Checkpoint.

CONCEPTUAL CHECKPOINT 25–3 LIGHT TRANSMISSION

Consider a set of three polarizers. Polarizer 1 has a vertical transmission axis, and polarizer 3 has a horizontal transmission axis. Taken together, polarizers 1 and 3 are a pair of crossed polarizers. Polarizer 2, with a transmission angle at 45° to the vertical, is placed between polarizers 1 and 3, as shown in the sketch. A beam of unpolarized light shines on polarizer 1 from the left. Is the transmission of light through the three polarizers **(a)** zero or **(b)** nonzero?

REASONING AND DISCUSSION

Since polarizers 1 and 3 are still crossed, it might seem that no light can be transmitted. When one recalls, however, that a polarizer causes a beam to have a polarization in the direction of its transmission axis, it becomes clear that transmission is indeed possible.

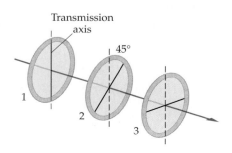

To be specific, some of the light that passes through polarizer 1 will also pass through polarizer 2, since the angle between their transmission axes is less than 90°. After passing through polarizer 2, the light is polarized at 45° to the vertical. As a result, some of this light can pass through polarizer 3 because, again, the angle between the polarization direction and the transmission axis is less than 90°.

In fact, the intensity of the incident light is reduced by a factor of 2 when it passes through polarizer 1, by a factor of 2 when it passes through polarizer 2 (since $\cos^2 45° = \frac{1}{2}$), and again by a factor of 2 when it passes through polarizer 3. The final intensity, then, is one-eighth the incident intensity.

ANSWER
(b) The transmission is nonzero.

There are many practical uses for crossed polarizers. For example, engineers often construct a plastic replica of a building, bridge, or similar structure to study the relative amounts of stress in its various parts in a technique known as *photoelastic stress analysis*. Dentists use the same technique to study stresses in teeth, and doctors use stress analysis when they design prosthetic joints. In this application, the plastic replica plays the role of polarizer 2 in Conceptual Checkpoint 25–3. In particular, in those regions of the structure where the stress is high, the plastic acts to rotate the plane of polarization and—like polarizer 2 in the Conceptual Checkpoint—allows light to pass through the system. By examining such models with crossed polarizers, engineers can gain valuable insight into the safety of the structures they plan to build, dentists can determine where a tooth is likely to break, and doctors can see where an artificial hip joint needs to be strengthened.

Another use of crossed polarizers is in the operation of a liquid-crystal display (LCD). There are basically three essential elements to each active area of an LCD—two crossed polarizers and a thin cell that holds a fluid composed of long, thin molecules known as a **liquid crystal.** The liquid crystal is selected for its ability to rotate the direction of polarization, and the thickness of the liquid-crystal cell is adjusted to give a rotation of 90°. Thus, in its "off" state, the liquid crystal rotates the direction of polarization and light passes through the crossed polarizers, as illustrated in **Figure 25–18 (a)**. In this state, the LCD is transparent—it allows ambient light to enter the display, reflect off the back, and exit the display, giving it the characteristic light background. When a voltage is applied to the liquid crystal, it no longer rotates the direction of polarization, and light is no longer transmitted through that area of the display. Thus, in the "on" state, shown in **Figure 25–18 (b)**, an area of the LCD appears black, which is how the black characters are

REAL-WORLD PHYSICS
Photoelastic stress analysis

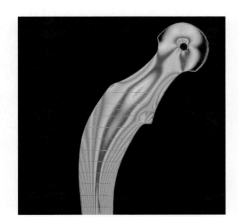

▲ In photoelastic stress analysis, a plastic model of an object being studied is placed between crossed polarizers. In this case, the object is a prosthetic hip joint. If the polarization of the light is unchanged by the plastic, it will not pass through the second polarizer. In areas where the plastic is stressed, however, it rotates the plane of polarization, allowing some of the light to pass through.

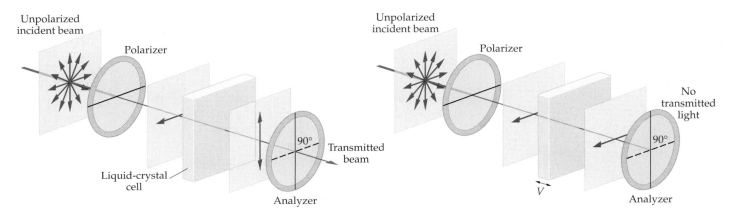

(a) Off (transmitted light gives bright background)   **(b) On** (dark characters formed where no light is transmitted)

▲ **FIGURE 25–18 Basic operation of a liquid-crystal display (LCD)**
(a) In the "off" state, the liquid crystal in the cell rotates the polarization of the light by 90°, allowing light to pass through the display. This produces the light background of the display. **(b)** In the "on" state, a voltage V is applied to the liquid-crystal cell, which means that the polarization of the light is no longer rotated. As a result, no light is transmitted and this element of the display appears black. This is how the dark characters are created on the display.

formed on the light background. Since very little energy is required to give the voltage necessary to turn a liquid-crystal cell "on," the LCD is very energy efficient. In addition, the LCD uses light already present in the environment; it does not need to produce its own light, as do some displays.

Finally, there are many organic compounds that are capable of rotating the polarization direction of a beam of light. Such compounds, which include sugar, turpentine, and tartaric acid, are said to be **optically active.** When a solution containing optically active molecules is placed between crossed polarizers, the amount of light that passes through the system gives a direct measure of the concentration of the active molecules in the solution.

Polarization by Scattering and Reflection

When unpolarized light is scattered from atoms or molecules, as in the atmosphere, or reflected by a solid or liquid surface, the light can acquire some degree of polarization. The basic reason for this is illustrated in **Figure 25–19**. In this case, we consider a vertically polarized beam of light that encounters a molecule. The light causes electrons in the molecule to oscillate in the vertical direction, that is, in the direction of the light's electric field. As we know, an accelerating charge radiates—thus, the molecule radiates as if it were a small vertical antenna, and this radiation is what we observe as scattered light. We also know, however, that an antenna gives off maximum radiation in the direction perpendicular to its length, and no radiation at all along its axis. Therefore, an observer at point A or B in Figure 25–19 sees maximum scattered radiation, whereas an observer at point C sees no radiation at all.

If we apply this same idea to an initially unpolarized beam of light scattering from a molecule, we find the situation pictured in **Figure 25–20**. In this case, electrons in the molecule oscillate in all directions within the plane of polarization; hence, an observer in the forward direction, at point A, sees scattered light of all polarizations—that is, unpolarized light. An observer at point B, however, sees no radiation from electrons oscillating horizontally, only from those oscillating vertically. Hence this observer sees vertically polarized light. An observer at an intermediate angle sees an intermediate amount of polarization.

This mechanism produces polarization in the light coming from the sky. In particular, maximum polarization is observed in a direction at right angles to the Sun, as can be seen in **Figure 25–21**. Thus, to a creature that is sensitive to the polarization of light, like a bee or certain species of birds, the light from the sky varies

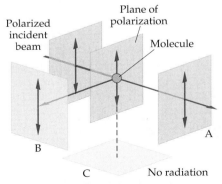

▲ **FIGURE 25–19 Vertically polarized light scattering from a molecule**
The molecule radiates scattered light, much like a small antenna. At points A and B strong scattered rays are observed; at point C no radiation is seen.

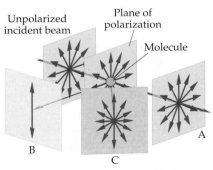

▲ **FIGURE 25–20 Unpolarized light scattering from a molecule**
In the forward direction, A, the scattered light is unpolarized. At right angles to the initial beam of light, B, the scattered light is polarized. Along other directions the light is only partially polarized.

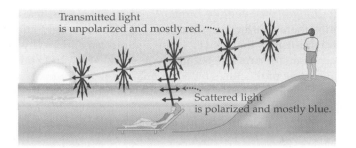

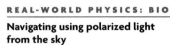

◀ **FIGURE 25–21 The effects of scattering on sunlight**
The scattering of sunlight by the atmosphere produces polarized light for an observer looking at right angles to the direction of the Sun. This observer also sees more blue light than red. An observer looking toward the Sun sees unpolarized light that contains more red light than blue.

REAL-WORLD PHYSICS: BIO

Navigating using polarized light from the sky

with the direction relative to the Sun. Experiments show that such creatures can use polarized light as an aid in navigation.

Another point of interest follows from the general observation that the amount of light scattered from a molecule is greatest when the wavelength of the light is comparable to the size of the molecule. The molecules in the atmosphere are generally much smaller than the wavelength of visible light, but blue light, with its relatively short wavelength, scatters more effectively than red light, with

◀ As it is scattered by air molecules and dust particles in the atmosphere, sunlight becomes polarized. The shorter wavelengths at the blue end of the spectrum are scattered most effectively, creating our familiar blue skies. Sunsets are red (top) because much of the blue light has been scattered as it passes through the atmosphere on the way to our eyes. The polarization of light from the sky can be demonstrated with a pair of photographs like those at left—one taken without a polarizing filter (left center), the other with a polarizer (left bottom). A smooth surface such as a lake (right) can also act as a polarizing filter. That is why the sky is darker and the clouds more easily seen in the reflected image than in the direct view.

its longer wavelength. Similarly, microscopic particles of dust in the upper atmosphere scatter the short-wavelength blue light most effectively. This basic mechanism is the answer to the age-old question, Why is the sky blue? Similarly, a sunset appears red because you are viewing the Sun through the atmosphere, and most of the Sun's blue light has been scattered off in other directions. In fact, the blue light that is missing from your sunset—giving it a red color—is the blue light of someone else's blue sky, as indicated in Figure 25–21.

Polarization also occurs when light reflects from a smooth surface, like the top of a table or the surface of a calm lake. A typical reflection situation, with unpolarized light from the Sun reflecting from the surface of a lake, is shown in **Figure 25–22.** When the light encounters molecules in the water, their electrons oscillate in the plane of polarization. For an observer at point A, however, only oscillations at right angles to the line of sight give rise to radiation. As a result, the reflected light from the lake is polarized horizontally. Polaroid sunglasses take advantage of this effect by using sheets of Polaroid material with a vertical transmission axis. With this orientation, the horizontally polarized reflected light—the glare—is not transmitted.

▶ **FIGURE 25–22 Polarization by reflection**
Because the observer sees no radiation from electrons moving along the line of sight, the radiation that is observed is polarized horizontally. Polaroid sunglasses with a vertical transmission axis reduce this kind of reflected glare.

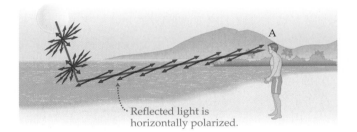

Reflected light is
horizontally polarized.

This presents a potential problem for the makers of digital watches and electronic devices with LCD displays. Suppose the person using such a device is wearing Polaroid sunglasses—not an unusual circumstance. As we saw in Figure 25–18, the light emerging from an LCD display is linearly polarized. If the polarization direction is vertical, the light will pass through the sunglasses and the display can be read as usual. On the other hand, if the polarization direction of the display is horizontal, it will appear completely black through a pair of Polaroid sunglasses—no light will pass through at all. The same effect can be seen by observing an LCD display through a pair of Polaroid sunglasses and slowly rotating the glasses or the display through 90°. Try this experiment sometime with a computer display or a digital watch, and you will see the result shown in the accompanying photographs. Clearly, it would not be wise to make an LCD display with a horizontal polarization direction—a person wearing Polaroid sunglasses would think the display was broken.

▶ To eliminate the glare from reflected light, which is horizontally polarized, the lenses of Polaroid sunglasses have a vertical transmission axis. LCD displays are usually constructed so that the polarized light that comes from them is also vertically polarized. Thus a person wearing Polaroid glasses can view an LCD display (left). If the pair of glasses or the LCD device is rotated 90°, the display becomes invisible (right).

THE BIG PICTURE PUTTING PHYSICS IN CONTEXT

LOOKING BACK

Electric and magnetic fields (from Chapters 19 and 22, respectively) are used throughout this chapter. In fact, we see in Sections 25–1 and 25–2 that $\vec{E}$ and $\vec{B}$ fields are intimately linked, and that together they can produce light waves, radio waves, infrared waves, and other forms of electromagnetic radiation.

The Doppler effect, which we studied for the case of sound in Chapter 14, is applied to light in Section 25–2. We find that the situation is simpler for light than it was for sound in that there is only a single Doppler effect for light, as opposed to two distinct cases with sound.

In Section 25–4 we study the energy, momentum, and intensity associated with electromagnetic waves. These are the same concepts introduced in Chapter 8 (energy), Chapter 9 (momentum), and Chapter 14 (intensity), respectively.

LOOKING AHEAD

The straight-line propagation of light presented in Sections 25–1 and 25–2 is the basis for the geometrical optics to be presented in Chapters 26 and 27.

The speed of light, $c = 3.00 \times 10^8$ m/s, applies to all forms of electromagnetic radiation propagating through a vacuum. When electromagnetic radiation propagates through a medium like glass, however, its speed is reduced. This leads to a change in direction of propagation referred to as refraction, as we shall see in Section 26–5.

Light is a fascinating phenomenon, with characteristics of both waves and particles. In this chapter we focused on the wavelike properties of light, and we will expand on this topic in Chapter 28. The particle-like aspects of light will be presented in Chapter 30, when we study quantum physics.

CHAPTER SUMMARY

25–1 THE PRODUCTION OF ELECTROMAGNETIC WAVES

Electromagnetic waves are traveling waves of oscillating electric and magnetic fields.

$\vec{E}$ and $\vec{B}$

The electric and magnetic fields in an electromagnetic wave are perpendicular to each other and to the direction of propagation. They are also in phase.

Direction of Propagation

To find the direction of propagation of an electromagnetic wave, point the fingers of your right hand in the direction of $\vec{E}$, then curl them toward $\vec{B}$. Your thumb will be pointing in the direction of propagation.

Generation of Electromagnetic Waves

Any accelerated charge will radiate energy in the form of electromagnetic waves.

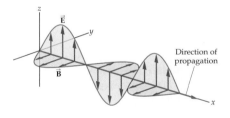

25–2 THE PROPAGATION OF ELECTROMAGNETIC WAVES

Electromagnetic waves can travel through a vacuum, and all electromagnetic waves in a vacuum have precisely the same speed, $c = 3.00 \times 10^8$ m/s. This is referred to as the *speed of light in a vacuum*.

Doppler Effect

Electromagnetic radiation experiences a Doppler effect that is analogous to that observed in sound waves. For relative speeds, u, that are small compared with the speed of light, c, the Doppler effect shifts the frequency of an electromagnetic wave as follows:

$$f' = f\left(1 \pm \frac{u}{c}\right) \qquad \text{25–3}$$

where the plus sign is used when source and receiver are approaching, the minus sign when they are receding.

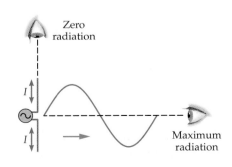

25–3 THE ELECTROMAGNETIC SPECTRUM

Electromagnetic waves can have any frequency from zero to infinity. The entire range of waves with different frequencies is referred to as the electromagnetic spectrum. Several "bands" of frequency are given special names.

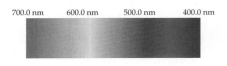

Frequency Bands in the Electromagnetic Spectrum

Radio/TV, 10^6 Hz to 10^9 Hz; microwaves, 10^9 Hz to 10^{12} Hz; infrared, 10^{12} Hz to 10^{14} Hz; visible, 4.29×10^{14} Hz to 7.50×10^{14} Hz; ultraviolet, 10^{15} Hz to 10^{17} Hz; X-rays, 10^{17} Hz to 10^{20} Hz; γ rays, above 10^{20} Hz.

Frequency–Wavelength Relationship

The frequency, f, and wavelength, λ, of all electromagnetic waves in a vacuum are related as follows:

$$c = f\lambda \qquad\qquad 25\text{–}4$$

25–4 ENERGY AND MOMENTUM IN ELECTROMAGNETIC WAVES

Electromagnetic waves carry both energy and momentum, shared equally between the electric and magnetic fields.

Energy Density

The energy density, u, of an electromagnetic wave can be written in several equivalent forms:

$$u = \tfrac{1}{2}\varepsilon_0 E^2 + \frac{1}{2\mu_0}B^2 = \varepsilon_0 E^2 = \frac{1}{\mu_0}B^2 \qquad\qquad 25\text{–}6$$

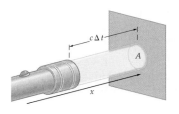

Ratio of Electric and Magnetic Fields

In an electromagnetic field, the magnitudes E and B are related as follows:

$$E = cB \qquad\qquad 25\text{–}9$$

Intensity

The intensity, I, of an electromagnetic field is the power per area. It can be expressed in the following forms:

$$I = uc = \tfrac{1}{2}c\varepsilon_0 E^2 + \frac{1}{2\mu_0}cB^2 = c\varepsilon_0 E^2 = \frac{c}{\mu_0}B^2 \qquad\qquad 25\text{–}10$$

Momentum

An electromagnetic wave that delivers an energy U to an object transfers the momentum

$$p = \frac{U}{c} \qquad\qquad 25\text{–}11$$

Radiation Pressure

If an electromagnetic wave shines on an object with an average intensity I_{av}, the average pressure exerted on the object by the radiation is

$$\text{pressure}_{av} = \frac{I_{av}}{c} \qquad\qquad 25\text{–}12$$

25–5 POLARIZATION

The polarization of a beam of light is the direction along which its electric field points. An unpolarized beam has components with polarization in random directions.

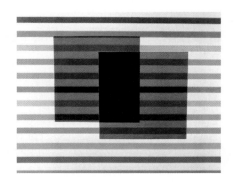

Polarizer

A polarizer transmits only light whose electric field has a component in the direction of the polarizer's transmission axis.

Law of Malus

If light with intensity I_0 encounters a polarizer with a transmission axis at a direction θ relative to its polarization, the transmitted intensity, I, is

$$I = I_0 \cos^2 \theta \qquad\qquad 25\text{–}13$$

Unpolarized Light Passing Through a Polarizer

When an unpolarized beam of light of intensity I_0 passes through a polarizer, the transmitted intensity, I, is reduced by half:

$$I = \tfrac{1}{2}I_0 \qquad\qquad 25\text{–}14$$

Polarization by Scattering

Light scattered by the atmosphere is polarized when viewed at right angles to the Sun.

Polarization by Reflection

When light reflects from a horizontal surface, like a tabletop or the surface of a lake, it is partially polarized in the horizontal direction.

PROBLEM-SOLVING SUMMARY

Type of Problem	Relevant Physical Concepts	Related Examples
Find the difference in frequency produced by the Doppler effect.	If an object moves with speed u and emits electromagnetic waves of speed c and frequency f, the Doppler-shifted frequency is $f' = f(1 \pm u/c)$. In this expression, the plus sign corresponds to a source approaching the observer; the minus sign corresponds to the source receding from the observer.	Example 25–2
Relate the frequency, f, wavelength, λ, and speed of an electromagnetic wave.	Electromagnetic waves in a vacuum always travel at the speed of light, c, regardless of their frequency. If either the frequency or wavelength of a wave is known, the other quantity can be obtained from the general relation $c = \lambda f$.	Example 25–3
Relate the intensity of an electromagnetic wave to its electric and magnetic fields.	An electromagnetic wave always consists of both electric and magnetic fields. The intensity of a wave, I (energy per area per time), can be related to either field as follows: $I = c\varepsilon_0 E^2 = cB^2/\mu_0$.	Example 25–4 Active Example 25–1
Find the average radiation pressure exerted by an electromagnetic wave.	When electromagnetic waves fall on an object, they exert a pressure known as the radiation pressure. The average radiation pressure, P_{av}, is directly proportional to the average intensity of the wave; that is, $P_{av} = I_{av}/c$.	Exercise 25–4
Find the transmitted intensity of unpolarized light passing through a polarizing filter.	When unpolarized light of intensity I_0 passes through a polarizing filter, its intensity is cut in half; that is, $I = \frac{1}{2}I_0$.	Example 25–5
Find the transmitted intensity of polarized light passing through a polarizing filter.	When polarized light of intensity I_0 passes through a polarizing filter, its transmitted intensity is $I = I_0 \cos^2 \theta$, where θ is the angle between the polarization direction of the light and the polarization direction of the filter.	Example 25–5 Active Example 25–2

CONCEPTUAL QUESTIONS

For instructor-assigned homework, go to www.masteringphysics.com

(Answers to odd-numbered Conceptual Questions can be found in the back of the book.)

1. Explain why the "invisible man" would be unable to see.

2. The magnitude of the Doppler effect tells how rapidly a weather system is moving. What determines whether the system is approaching or receding?

3. Explain why radiation pressure is more significant on a grain of dust in interplanetary space when the grain is very small.

4. While wearing your Polaroid sunglasses at the beach, you notice that they reduce the glare from the water better when you are sitting upright than when you are lying on your side. Explain.

5. You want to check the time while wearing your Polaroid sunglasses. If you hold your forearm horizontally, you can read the time easily. If you hold your forearm vertically, however, so that you are looking at your watch sideways, you notice that the display is black. Explain.

6. **BIO Polarization and the Ground Spider** The ground spider *Drassodes cupreus*, like many spiders, has several pairs of eyes. It has been discovered that one of these pairs of eyes acts as a set of polarization filters, with one eye's polarization direction oriented at 90° to the other eye's polarization direction. In addition,

experiments show that the spider uses these eyes to aid in navigating to and from its burrow. Explain how such eyes might aid navigation.

7. The electromagnetic waves we pick up on our radios are typically polarized. In contrast, the indoor light we see every day is typically unpolarized. Explain.

8. You are given a sheet of Polaroid material. Describe how to determine the direction of its transmission axis if none is indicated on the sheet.

9. Can sound waves be polarized? Explain.

10. At a garage sale you find a pair of "Polaroid" sunglasses priced to sell. You are not sure, however, if the glasses are truly Polaroid, or if they simply have tinted lenses. How can you tell which is the case? Explain.

11. **3-D Movies** Modern-day 3-D movies are produced by projecting two different images onto the screen, with polarization directions that are at 90° relative to one another. Viewers must wear headsets with polarizing filters to experience the 3-D effect. Explain how this works.

PROBLEMS AND CONCEPTUAL EXERCISES

Note: Answers to odd-numbered Problems and Conceptual Exercises can be found in the back of the book. **IP** *denotes an integrated problem, with both conceptual and numerical parts;* **BIO** *identifies problems of biological or medical interest;* **CE** *indicates a conceptual exercise.* **Predict/Explain** *problems ask for two responses:* **(a)** *your prediction of a physical outcome, and* **(b)** *the best explanation among three provided. On all problems, red bullets* (•, ••, •••) *are used to indicate the level of difficulty.*

SECTION 25–1 THE PRODUCTION OF ELECTROMAGNETIC WAVES

1. • **CE** If the electric field in an electromagnetic wave is increasing in magnitude at a particular time, is the magnitude of the magnetic field at the same time increasing or decreasing? Explain.

2. • The electric field of an electromagnetic wave points in the positive y direction. At the same time, the magnetic field of this wave points in the positive z direction. In what direction is the wave traveling?

3. • An electric charge on the x axis oscillates sinusoidally about the origin. A distant observer is located at a point on the $+y$ axis. **(a)** In what direction will the electric field oscillate at the observer's location? **(b)** In what direction will the magnetic field oscillate at the observer's location? **(c)** In what direction will the electromagnetic wave propagate at the observer's location?

4. • An electric charge on the z axis oscillates sinusoidally about the origin. A distant observer is located at a point on the $+y$ axis. **(a)** In what direction will the electric field oscillate at the observer's location? **(b)** In what direction will the magnetic field oscillate at the observer's location? **(c)** In what direction will the electromagnetic wave propagate at the observer's location?

5. •• Give the direction (N, S, E, W, up, or down) of the missing quantity for each of the four electromagnetic waves listed in Table 25–1.

TABLE 25–1 Problem 5

Direction of propagation	Direction of electric field	Direction of magnetic field
N	W	**(a)**
N	**(b)**	W
up	S	**(c)**
(d)	down	S

6. •• Give the direction ($\pm x$, $\pm y$, $\pm z$) of the missing quantity for each of the four electromagnetic waves listed in Table 25–2.

TABLE 25–2 Problem 6

Direction of propagation	Direction of electric field	Direction of magnetic field
$+x$	$+y$	**(a)**
$+x$	**(b)**	$+y$
$-y$	$+z$	**(c)**
(d)	$+z$	$+y$

7. •• **IP** At a particular instant of time, a light beam traveling in the positive z direction has an electric field given by $\vec{E} = (6.22 \text{ N/C})\hat{x} + (2.87 \text{ N/C})\hat{y}$. The magnetic field in the beam has a magnitude of 2.28×10^{-8} T at the same time. **(a)** Does the magnetic field at this time have a z component that is positive, negative, or zero? Explain. **(b)** Write $\vec{B}$ in terms of unit vectors.

8. •• **IP** A light beam traveling in the negative z direction has a magnetic field $\vec{B} = (3.02 \times 10^{-9} \text{ T})\hat{x} + (-5.28 \times 10^{-9} \text{ T})\hat{y}$ at a given instant of time. The electric field in the beam has a magnitude of 1.82 N/C at the same time. **(a)** Does the electric field at this time have a z component that is positive, negative, or zero? Explain. **(b)** Write $\vec{E}$ in terms of unit vectors.

SECTION 25–2 THE PROPAGATION OF ELECTROMAGNETIC WAVES

9. • **CE** Three electromagnetic waves have electric and magnetic fields pointing in the directions shown in **Figure 25–23**. For each of the three cases, state whether the wave propagates in the $+x$, $-x$, $+y$, $-y$, $+z$, or $-z$ direction.

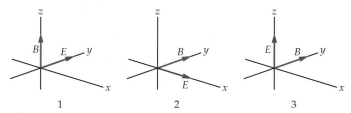

▲ **FIGURE 25–23** Problem 9

10. • The light-year (ly) is a unit of distance commonly used in astronomy. It is defined as the distance traveled by light in a vacuum in one year. **(a)** Express 1 ly in km. **(b)** Express the speed of light, c, in units of ly per year. **(c)** Express the speed of light in feet per nanosecond.

11. • Alpha Centauri, the closest star to the sun, is 4.3 ly away. How far is this in meters?

12. • **Mars Rover** When the Mars rover was deployed on the surface of Mars in July 1997, radio signals took about 12 min to travel from Earth to the rover. How far was Mars from Earth at that time?

13. •• A distant star is traveling directly away from Earth with a speed of 37,500 km/s. By what factor are the wavelengths in this star's spectrum changed?

14. •• **IP** A distant star is traveling directly toward Earth with a speed of 37,500 km/s. **(a)** When the wavelengths in this star's spectrum are measured on Earth, are they greater or less than the wavelengths we would find if the star were at rest relative to us? Explain. **(b)** By what fraction are the wavelengths in this star's spectrum shifted?

15. •• **IP** The frequency of light reaching Earth from a particular galaxy is 12% lower than the frequency the light had when it was emitted. **(a)** Is this galaxy moving toward or away from Earth? Explain. **(b)** What is the speed of this galaxy relative to the Earth? Give your answer as a fraction of the speed of light.

16. •• **Measuring the Speed of Light** Galileo attempted to measure the speed of light by measuring the time elapsed between his opening a lantern and his seeing the light return from his assistant's lantern. The experiment is illustrated in **Figure 25–24**. What distance, d, must separate Galileo and his assistant in order for the human reaction time, $\Delta t = 0.2$ s, to introduce no more than a 15% error in the speed of light?

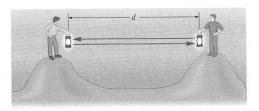

▲ **FIGURE 25–24** Problem 16

17. •• **Measuring the Speed of Light: Michelson** In 1926, Albert Michelson measured the speed of light with a technique similar to that used by Fizeau. Michelson used an eight-sided mirror rotating at 528 rev/s in place of the toothed wheel, as illustrated in **Figure 25–25**. The distance from the rotating mirror to a distant reflector was 35.5 km. If the light completed the 71.0-km round trip in the time it took the mirror to complete one-eighth of a revolution, what is the speed of light?

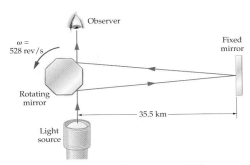

▲ **FIGURE 25–25** Problems 17 and 92

18. •• **Communicating with the *Voyager* Spacecraft** When the *Voyager I* and *Voyager II* spacecraft were exploring the outer planets, NASA flight controllers had to plan the crafts' moves well in advance. How many seconds elapse between the time a command is sent from Earth and the time the command is received by *Voyager* at Neptune? Assume the distance from Earth to Neptune is 4.5×10^{12} m.

19. •• A father and his daughter are interested in the same baseball game. The father sits next to his radio at home and listens to the game; his daughter attends the game and sits in the outfield bleachers. In the bottom of the ninth inning a home run is hit. If the father's radio is 132 km from the radio station, and the daughter is 115 m from home plate, who hears the home run first? (Assume that there is no time delay between the baseball being hit and its sound being broadcast by the radio station. In addition, let the speed of sound in the stadium be 343 m/s.)

20. •• **IP** **(a)** How fast would a motorist have to be traveling for a yellow ($\lambda = 590$ nm) traffic light to appear green ($\lambda = 550$ nm) because of the Doppler shift? **(b)** Should the motorist be traveling toward or away from the traffic light to see this effect? Explain.

21. •• Most of the galaxies in the universe are observed to be moving away from Earth. Suppose a particular galaxy emits orange light with a frequency of 5.000×10^{14} Hz. If the galaxy is receding from Earth with a speed of 3325 km/s, what is the frequency of the light when it reaches Earth?

22. •• Two starships, the *Enterprise* and the *Constitution,* are approaching each other head-on from a great distance. The separation between them is decreasing at a rate of 782.5 km/s. The *Enterprise* sends a laser signal toward the *Constitution*. If the *Constitution* observes a wavelength $\lambda = 670.3$ nm, what wavelength was emitted by the *Enterprise*?

23. •• Baseball scouts often use a radar gun to measure the speed of a pitch. One particular model of radar gun emits a microwave signal at a frequency of 10.525 GHz. What will be the increase in frequency if these waves are reflected from a 90.0-mi/h fastball headed straight toward the gun? (*Note:* 1 mi/h = 0.447 m/s)

24. •• A state highway patrol car radar unit uses a frequency of 8.00×10^9 Hz. What frequency difference will the unit detect from a car receding at a speed of 44.5 m/s from a stationary patrol car?

25. •• Consider a spiral galaxy that is moving directly away from Earth with a speed $V = 3.600 \times 10^5$ m/s at its center, as shown in **Figure 25–26**. The galaxy is also rotating about its center, so that points in its spiral arms are moving with a speed $v = 6.400 \times 10^5$ m/s relative to the center. If light with a frequency of 8.230×10^{14} Hz is emitted in both arms of the galaxy, what frequency is detected by astronomers observing the arm that is moving **(a)** toward and **(b)** away from Earth? (Measurements of this type are used to map out the speed of various regions in distant, rotating galaxies.)

▲ **FIGURE 25–26** Problems 25 and 93

26. ••• **IP** A highway patrolman sends a 24.150-GHz radar beam toward a speeding car. The reflected wave is lower in frequency by 4.04 kHz. **(a)** Is the car moving toward or away from the radar gun? Explain. **(b)** What is the speed of the car? [*Hint:* For small values of x, the following approximation may be used: $(1 + x)^2 \approx 1 + 2x$.]

SECTION 25–3 THE ELECTROMAGNETIC SPECTRUM

27. • **BIO Dental X-rays** X-rays produced in the dentist's office typically have a wavelength of 0.30 nm. What is the frequency of these rays?

28. • Find the frequency of blue light with a wavelength of 460 nm.

29. • Yellow light has a wavelength $\lambda = 590$ nm. How many of these waves would span the 1.0-mm thickness of a dime?

30. • How many red wavelengths ($\lambda = 705$ nm) tall are you?

31. • A cell phone transmits at a frequency of 1.75×10^8 Hz. What is the wavelength of the electromagnetic wave used by this phone?

32. • **BIO Human Radiation** Under normal conditions, humans radiate electromagnetic waves with a wavelength of about 9.0 microns. **(a)** What is the frequency of these waves? **(b)** To what portion of the electromagnetic spectrum do these waves belong?

33. • **BIO UV Radiation** Ultraviolet light is typically divided into three categories. UV-A, with wavelengths between 400 nm and 320 nm, has been linked with malignant melanomas. UV-B radiation, which is the primary cause of sunburn and other skin cancers, has wavelengths between 320 nm and 280 nm. Finally, the region known as UV-C extends to wavelengths of 100 nm. **(a)** Find the range of frequencies for UV-B radiation. **(b)** In which of these three categories does radiation with a frequency of 7.9×10^{14} Hz belong?

34. • **Communicating with a Submarine** Normal radiofrequency waves cannot penetrate more than a few meters below the surface of the ocean. One method of communicating with submerged submarines uses very low frequency (VLF) radio waves. What is the wavelength (in air) of a 10.0-kHz VLF radio wave?

35. •• **IP** When an electromagnetic wave travels from one medium to another with a different speed of propagation, the frequency of the wave remains the same. Its wavelength, however, changes. **(a)** If the wave speed decreases, does the wavelength increase or decrease? Explain. **(b)** Consider a case where the wave speed decreases from c to $\frac{3}{4}c$. By what factor does the wavelength change?

36. •• **IP (a)** Which color of light has the higher frequency, red or violet? **(b)** Calculate the frequency of blue light with a wavelength of 470 nm, and red light with a wavelength of 680 nm.

37. •• ULF (ultra low frequency) electromagnetic waves, produced in the depths of outer space, have been observed with wavelengths in excess of 29 million kilometers. What is the period of such a wave?

38. •• A television is tuned to a station broadcasting at a frequency of 6.60×10^7 Hz. For best reception, the rabbit-ear antenna used by the TV should be adjusted to have a tip-to-tip length equal to half a wavelength of the broadcast signal. Find the optimum length of the antenna.

39. •• An AM radio station's antenna is constructed to be $\lambda/4$ tall, where λ is the wavelength of the radio waves. How tall should the antenna be for a station broadcasting at a frequency of 810 kHz?

40. •• As you drive by an AM radio station, you notice a sign saying that its antenna is 112 m high. If this height represents one quarter-wavelength of its signal, what is the frequency of the station?

41. •• Find the difference in wavelength ($\lambda_1 - \lambda_2$) for each of the following pairs of radio waves: **(a)** $f_1 = 50$ kHz and $f_2 = 52$ kHz, **(b)** $f_1 = 500$ kHz and $f_2 = 502$ kHz.

42. •• Find the difference in frequency ($f_1 - f_2$) for each of the following pairs of radio waves: **(a)** $\lambda_1 = 300.0$ m and $\lambda_2 = 300.5$ m, **(b)** $\lambda_1 = 30.0$ m and $\lambda_2 = 30.5$ m.

SECTION 25–4   ENERGY AND MOMENTUM IN ELECTROMAGNETIC WAVES

43. • **CE** If the rms value of the electric field in an electromagnetic wave is doubled, **(a)** by what factor does the rms value of the magnetic field change? **(b)** By what factor does the average intensity of the wave change?

44. • **CE** The radiation pressure exerted by beam of light 1 is half the radiation pressure of beam of light 2. If the rms electric field of beam 1 has the value E_0, what is the rms electric field in beam 2?

45. • The maximum magnitude of the electric field in an electromagnetic wave is 0.0400 V/m. What is the maximum magnitude of the magnetic field in this wave?

46. • What is the rms value of the electric field in a sinusoidal electromagnetic wave that has a maximum electric field of 88 V/m?

47. •• The magnetic field in an electromagnetic wave has a peak value given by $B = 3.7\ \mu T$. For this wave, find **(a)** the peak electric field strength, **(b)** the peak intensity, and **(c)** the average intensity.

48. •• What is the maximum value of the electric field in an electromagnetic wave whose maximum intensity is 5.00 W/m²?

49. •• What is the maximum value of the electric field in an electromagnetic wave whose average intensity is 5.00 W/m²?

50. •• **IP** Electromagnetic wave 1 has a maximum electric field of $E_0 = 52$ V/m, and electromagnetic wave 2 has a maximum magnetic field of $B_0 = 1.5\ \mu T$. **(a)** Which wave has the greater intensity? **(b)** Calculate the intensity of each wave.

51. •• A 65-kW radio station broadcasts its signal uniformly in all directions. **(a)** What is the average intensity of its signal at a distance of 250 m from the antenna? **(b)** What is the average intensity of its signal at a distance of 2500 m from the antenna?

52. •• At what distance will a 45-W lightbulb have the same apparent brightness as a 120-W bulb viewed from a distance of 25 m? (Assume that both bulbs convert electrical power to light with the same efficiency, and radiate light uniformly in all directions.)

53. •• What is the ratio of the sunlight intensity reaching Pluto compared with the sunlight intensity reaching Earth? (On average, Pluto is 39 times as far from the Sun as is Earth.)

54. •• **IP** In the following, assume that lightbulbs radiate uniformly in all directions and that 5.0% of their power is converted to light. **(a)** Find the average intensity of light at a point 2.0 m from a 120-W red lightbulb ($\lambda = 710$ nm). **(b)** Is the average intensity 2.0 m from a 120-W blue lightbulb ($\lambda = 480$ nm) greater than, less than, or the same as the intensity found in part (a)? Explain. **(c)** Calculate the average intensity for part (b).

55. •• A 5.0-mW laser produces a narrow beam of light. How much energy is contained in a 1.0-m length of its beam?

56. •• What length of a 5.0-mW laser's beam will contain 9.5 mJ of energy?

57. •• **Sunlight Intensity** After filtering through the atmosphere, the Sun's radiation illuminates Earth's surface with an average intensity of 1.0 kW/m². Assuming this radiation strikes the 15-m × 45-m black, flat roof of a building at normal incidence, calculate the average force the radiation exerts on the roof.

58. •• **IP (a)** Find the electric and magnetic field amplitudes in an electromagnetic wave that has an average energy density of 1.0 J/m³. **(b)** By what factor must the field amplitudes be increased if the average energy density is to be doubled to 2.0 J/m³?

59. •• **Lasers for Fusion** Two of the most powerful lasers in the world are used in nuclear fusion experiments. The NOVA laser produces 40.0 kJ of energy in a pulse that lasts 2.50 ns, and the NIF laser (under construction) will produce a 10.0-ns pulse with 3.00 MJ of energy. **(a)** Which laser produces more energy in each pulse? **(b)** Which laser produces the greater average power during each pulse? **(c)** If the beam diameters are the same, which laser produces the greater average intensity?

60. •• **BIO** You are standing 2.5 m from a 150-W lightbulb. **(a)** If the pupil of your eye is a circle 5.0 mm in diameter, how much energy enters your eye per second? (Assume that 5.0% of the lightbulb's power is converted to light.) **(b)** Repeat part (a) for the case of a 1.0-mm-diameter laser beam with a power of 0.50 mW.

61. •• **BIO Laser Safety** A 0.75-mW laser emits a narrow beam of light that enters the eye, as shown in **Figure 25–27.** **(a)** How much energy is absorbed by the eye in 0.2 s? **(b)** The eye focuses this beam to a tiny spot on the retina, perhaps 5.0 μm in diameter. What is the average intensity of light (in W/cm²) at this spot? **(c)** Damage to the retina can occur if the average intensity of light exceeds 1.0×10^{-2} W/cm². By what factor has the intensity of this laser beam exceeded the safe value?

$d = 5.0 \, \mu m$

▲ **FIGURE 25–27** Problems 61 and 94

62. •• Find the rms electric and magnetic fields at a point 2.50 m from a lightbulb that radiates 75.0 W of light uniformly in all directions.

63. •• A 0.50-mW laser produces a beam of light with a diameter of 1.5 mm. **(a)** What is the average intensity of this beam? **(b)** At what distance does a 150-W lightbulb have the same average intensity as that found for the laser beam in part (a)? (Assume that 5.0% of the bulb's power is converted to light.)

64. •• A laser emits a cylindrical beam of light 2.4 mm in diameter. If the average power of the laser is 2.8 mW, what is the rms value of the electric field in the laser beam?

65. •• **(a)** If the laser in Problem 64 shines its light on a perfectly absorbing surface, how much energy does the surface receive in 12 s? **(b)** What is the radiation pressure exerted by the beam?

66. ••• **BIO Laser Surgery** Each pulse produced by an argon–fluoride excimer laser used in PRK and LASIK ophthalmic surgery lasts only 10.0 ns but delivers an energy of 2.50 mJ. **(a)** What is the power produced during each pulse? **(b)** If the beam has a diameter of 0.850 mm, what is the average intensity of the beam during each pulse? **(c)** If the laser emits 55 pulses per second, what is the average power it generates?

67. ••• A pulsed laser produces brief bursts of light. One such laser emits pulses that carry 0.350 J of energy but last only 225 fs. **(a)** What is the average power during one of these pulses? **(b)** Assuming the energy is emitted in a cylindrical beam of light 2.00 mm in diameter, calculate the average intensity of this laser beam. **(c)** What is the rms electric field in this wave?

SECTION 25–5 POLARIZATION

68. • **CE Predict/Explain** Consider the two polarization experiments shown in **Figure 25–28**. **(a)** If the incident light is unpolarized, is the transmitted intensity in case A greater than, less than, or the same as the transmitted intensity in case B? **(b)** Choose the *best explanation* from among the following:

   I. The transmitted intensity is the same in either case; the first polarizer lets through one-half the incident intensity, and the second polarizer is at an angle θ relative to the first.

   II. Case A has a smaller transmitted intensity than case B because the first polarizer is at an angle θ relative to the incident beam.

   III. Case B has a smaller transmitted intensity than case A because the direction of polarization is rotated by an angle θ in the clockwise direction in case B.

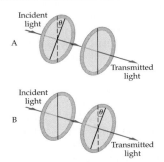

▲ **FIGURE 25–28** Problems 68, 69, and 70

69. • **CE Predict/Explain** Consider the two polarization experiments shown in Figure 25–28. **(a)** If the incident light is polarized in the horizontal direction, is the transmitted intensity in case A greater than, less than, or the same as the transmitted intensity in case B? **(b)** Choose the *best explanation* from among the following:

   I. The two cases have the same transmitted intensity because the angle between the polarizers is θ in each case.

   II. The transmitted intensity is greater in case B because all of the initial beam gets through the first polarizer.

   III. The transmitted intensity in case B is smaller than in case A; in fact, the transmitted intensity in case B is zero because the first polarizer is oriented vertically.

70. • **CE** Suppose linearly polarized light is incident on the polarization experiments shown in Figure 25–28. In what direction, relative to the vertical, must the incident light be polarized if the transmitted intensity is to be the same in both experiments? Explain.

71. • **CE** An incident beam of light with an intensity I_0 passes through a polarizing filter whose transmission axis is at an angle θ to the vertical. As the angle is changed from $\theta = 0$ to $\theta = 90°$, the intensity as a function of angle is given by one of the curves in **Figure 25–29**. Give the color of the curve corresponding to an incident beam that is **(a)** unpolarized, **(b)** vertically polarized, and **(c)** horizontally polarized.

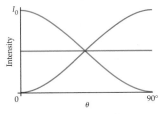

▲ **FIGURE 25–29** Problem 71

72. • Vertically polarized light with an intensity of 0.55 W/m^2 passes through a polarizer whose transmission axis is at an angle of 65.0° with the vertical. What is the intensity of the transmitted light?

73. • A person riding in a boat observes that the sunlight reflected by the water is polarized parallel to the surface of the water. The person is wearing polarized sunglasses with the polarization axis vertical. If the wearer leans at an angle of 21.5° to the vertical, what fraction of the reflected light intensity will pass through the sunglasses?

74. •• Unpolarized light passes through two polarizers whose transmission axes are at an angle of 30.0° with respect to each other. What fraction of the incident intensity is transmitted through the polarizers?

75. •• In Problem 74, what should be the angle between the transmission axes of the polarizers if it is desired that one-tenth of the incident intensity be transmitted?

76. •• Unpolarized light with intensity I_0 falls on a polarizing filter whose transmission axis is vertical. The axis of a second polarizing filter makes an angle of θ with the vertical. Plot a graph that shows the intensity of light transmitted by the second filter (expressed as a fraction of I_0) as a function of θ. Your graph should cover the range $\theta = 0°$ to $\theta = 360°$.

77. •• **IP** A beam of vertically polarized light encounters two polarizing filters, as shown in **Figure 25–30**. **(a)** Rank the three cases, A, B, and C, in order of increasing transmitted intensity. Indicate ties where appropriate. **(b)** Calculate the transmitted intensity for each of the cases in Figure 25–30, assuming that the incident intensity is 37.0 W/m². Verify that your numerical results agree with the rankings in part (a).

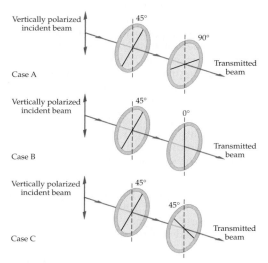

▲ **FIGURE 25–30** Problems 77 and 78

78. •• **IP** Repeat Problem 77, this time assuming that the polarizers to the left in Figure 25–30 are at an angle of 22.5° to the vertical rather than 45°. The incident intensity is again 37.0 W/m².

79. •• **IP BIO Optical Activity** Optically active molecules have the property of rotating the direction of polarization of linearly polarized light. Many biologically important molecules have this property, some causing a counterclockwise rotation (negative rotation angle), others causing a clockwise rotation (positive rotation angle). For example, a 5.00 gram per 100 mL solution of *l*-leucine causes a rotation of −0.550°; the same concentration of *d*-glutamic acid causes a rotation of 0.620°. **(a)** If placed between crossed polarizers, which of these solutions transmits the greater intensity? Explain. **(b)** Find the transmitted intensity for each of these solutions when placed between crossed polarizers. The incident beam is unpolarized and has an intensity of 12.5 W/m².

80. •• A helium–neon laser emits a beam of unpolarized light that passes through three Polaroid filters, as shown in **Figure 25–31**. The intensity of the laser beam is I_0. **(a)** What is the intensity of the beam at point A? **(b)** What is the intensity of the beam at point B? **(c)** What is the intensity of the beam at point C? **(d)** If filter 2 is removed, what is the intensity of the beam at point C?

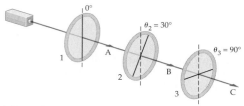

▲ **FIGURE 25–31** Problems 80, 81, 100, and 101

81. ••• Referring to Figure 25–31, suppose that filter 3 is at a general angle θ with the vertical, rather than the angle 90°. **(a)** Find an expression for the transmitted intensity as a function of θ. **(b)** Plot your result from part (a), and determine the maximum transmitted intensity. **(c)** At what angle θ does maximum transmission occur?

GENERAL PROBLEMS

82. • **CE** Suppose the magnitude of the electric field in an electromagnetic wave is doubled. **(a)** By what factor does the magnitude of the magnetic field change? **(b)** By what factor does the maximum intensity of the wave change?

83. • **CE** If "sailors" of the future use radiation pressure to propel their ships, should the surfaces of their sails be absorbing or reflecting? Explain.

84. • Sunlight at the surface of Earth has an average intensity of about 1.00×10^3 W/m². Find the rms values of the electric and magnetic fields in the sunlight.

85. • **BIO** A typical medical X-ray has a frequency of 1.50×10^{19} Hz. What is the wavelength of such an X-ray?

86. • How many hydrogen atoms, 0.10 nm in diameter, must be placed end to end to fit into one wavelength of 410-nm violet light?

87. • **BIO Radiofrequency Ablation** In radiofrequency (RF) ablation, a small needle is inserted into a cancerous tumor. When radiofrequency oscillating currents are sent into the needle, ions in the neighboring tissue respond by vibrating rapidly, causing local heating to temperatures as high as 100°C. This kills the cancerous cells and, because of the small size of the needle, relatively few of the surrounding healthy cells. A typical RF ablation treatment uses a frequency of 750 kHz. What is the wavelength that such radio waves would have in a vacuum?

88. •• **CE Figure 25–32** shows four polarization experiments in which unpolarized incident light passes through two polarizing filters with different orientations. Rank the four cases in order of increasing amount of transmitted light. Indicate ties where appropriate.

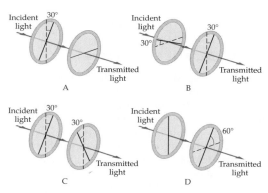

▲ **FIGURE 25–32** Problem 88

89. •• **IP (a)** What minimum intensity must a laser beam have if it is to levitate a tiny black (100% absorbing) sphere of radius $r = 0.5 \ \mu$m and mass $= 1.6 \times 10^{-15}$ kg? Comment on the feasibility of such levitation. **(b)** If the radius of the sphere is doubled but its mass remains the same, will the minimum intensity be greater than, less than, or equal to the value found in part (a)? Explain. **(c)** Find the minimum intensity for the situation described in part (b).

90. •• The *Apollo 11* **Reflector** One of the experiments placed on the Moon's surface by *Apollo 11* astronauts was a reflector that is used to measure the Earth–Moon distance with high accuracy. A laser beam on Earth is bounced off the reflector, and its round-trip travel time is recorded. If the travel time can be measured to within an accuracy of 0.030 ns, what is the uncertainty in the Earth–Moon distance?

91. •• The H_β line of the hydrogen atom's spectrum has a normal wavelength $\lambda_\beta = 486$ nm. This same line is observed in the spectrum of a distant quasar, but lengthened by 20.0 nm. What is the speed of the quasar relative to Earth, assuming it is moving along our line of sight?

92. •• **IP** Suppose the distance to the fixed mirror in Figure 25–25 is decreased to 20.5 km. **(a)** Should the angular speed of the rotating mirror be increased or decreased to ensure that the experiment works as described in Problem 17? **(b)** Find the required angular speed, assuming the speed of light is 3.00×10^8 m/s.

93. •• **IP** Suppose the speed of the galaxy in Problem 25 is increased by a factor of 10; that is, $V = 3.600 \times 10^6$ m/s. The speed of the arms, v, and the frequency of the light remain the same. **(a)** Does the arm near the top of Figure 25–26 show a red shift (toward lower frequency) or a blue shift (toward higher frequency)? Does the lower arm show a red or a blue shift? Explain. What frequency is detected by astronomers observing **(b)** the upper arm and **(c)** the lower arm?

94. •• **IP BIO** Consider the physical situation illustrated in Figure 25–27. **(a)** Is E_{rms} in the incident laser beam greater than, less than, or the same as E_{rms} where the beam hits the retina? Explain. **(b)** If the intensity of the beam at the retina is equal to the damage threshold, 1.0×10^{-2} W/cm^2, what is the value of E_{rms} at that location? **(c)** If the diameter of the spot on the retina is reduced by a factor of 2, by what factor does the intensity increase? By what factor does E_{rms} increase?

95. •• **BIO Polaroid Vision in a Spider** Experiments show that the ground spider *Drassodes cupreus* uses one of its several pairs of eyes as a polarization detector. In fact, the two eyes in this pair have polarization directions that are at right angles to one another. Suppose linearly polarized light with an intensity of 825 W/m^2 shines from the sky onto the spider, and that the intensity transmitted by one of the polarizing eyes is 232 W/m^2. **(a)** For this eye, what is the angle between the polarization direction of the eye and the polarization direction of the incident light? **(b)** What is the intensity transmitted by the other polarizing eye?

96. •• A state highway patrol car radar unit uses a frequency of 9.00×10^9 Hz. What frequency difference will the unit detect from a car approaching a parked patrol car with a speed of 35.0 m/s?

97. •• What is the ratio of the sunlight intensity reaching Mercury compared with the sunlight intensity reaching Earth? (On average, Mercury's distance from the Sun is 0.39 that of Earth's.)

98. •• What area is needed for a solar collector to absorb 45.0 kW of power from the Sun's radiation if the collector is 75.0% efficient? (At the surface of Earth, sunlight has an average intensity of 1.00×10^3 W/m^2.)

99. •• **BIO Near-Infrared Brain Scans** Light in the near-infrared (close to visible red) can penetrate surprisingly far through human tissue, a fact that is being used to "illuminate" the interior of the brain in a noninvasive technique known as near-infrared spectroscopy (NIRS). In this procedure, illustrated in **Figure 25–33**, an optical fiber carrying a beam of infrared laser light with a power of 1.5 mW and a cross-sectional diameter of 1.2 mm is placed against the skull. Some of the light enters the brain, where it scatters from hemoglobin in the blood. The scattered light is picked up by a detector and analyzed by a computer. **(a)** According to the **Beer–Lambert law**, the intensity of light, I, decreases with penetration distance, d, as $I = I_0 e^{-\mu d}$, where I_0 is the initial intensity of the beam and $\mu = 4.7$ cm^{-1} for a typical case. Find the intensity of the laser beam after it penetrates through 3.0 cm of tissue. **(b)** Find the electric field of the initial light beam.

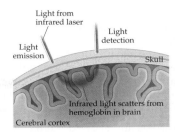

▲ **FIGURE 25–33** The basic elements of a near-infrared brain scan. (Problem 99)

A patient undergoing a near-infrared brain scan. (Problem 99)

100. •• Three polarizers are arranged as shown in Figure 25–31. If the incident beam of light is unpolarized and has an intensity of 1.60 W/m^2, find the transmitted intensity **(a)** when $\theta_2 = 25.0°$ and $\theta_3 = 50.0°$, and **(b)** when $\theta_2 = 50.0°$ and $\theta_3 = 25.0°$.

101. •• Repeat Problem 100, this time assuming an incident beam that is vertically polarized. The intensity of the incident beam is 1.60 W/m^2.

102. •• A lightbulb emits light uniformly in all directions. If the rms electric field of this light is 16.0 N/C at a distance of 1.35 m from the bulb, what is the average total power radiated by the bulb?

103. •• **IP** A beam of light is a mixture of unpolarized light with intensity I_u and linearly polarized light with intensity I_p. The polarization direction for the polarized light is vertical. When this mixed beam of light is passed through a polarizer that is vertical, the transmitted intensity is 16.8 W/m^2; when the polarizer is at an angle of 55.0° with the vertical, the transmitted intensity is 8.68 W/m^2. **(a)** Is I_u greater than, less than, or equal to I_p? Explain. **(b)** Calculate I_u and I_p.

104. ••• **BIO** As mentioned in Problem 95, one pair of eyes in a particular species of ground spider has polarization directions that are at right angles to one another. Suppose that linearly polarized light is incident on such a spider. **(a)** Prove that the transmitted intensity of one eye plus the transmitted intensity from the other eye is equal to the incident intensity. **(b)** If the transmitted intensities for the two eyes are 163 W/m^2 and 662 W/m^2, through what angle must the spider rotate to make the transmitted intensities equal to one another?

105. ••• A typical home may require a total of 2.00×10^3 kWh of energy per month. Suppose you would like to obtain this energy from sunlight, which has an average daylight intensity of 1.00×10^3 W/m^2. Assuming that sunlight is available 8.0 h per day, 25 d per month (accounting for cloudy days), and that you have a way to store energy from your collector when the Sun isn't shining, determine the smallest collector size that will provide the needed energy, given a conversion efficiency of 25%.

106. ••• At the top of Earth's atmosphere, sunlight has an average intensity of 1360 W/m^2. If the average distance from Earth to the Sun is 1.50×10^{11} m, at what rate does the Sun radiate energy?

107. ••• **IP** A typical laser used in introductory physics laboratories produces a continuous beam of light about 1.0 mm in diameter. The average power of such a laser is 0.75 mW. What are **(a)** the average intensity, **(b)** the peak intensity, and **(c)** the average energy density of this beam? **(d)** If the beam is reflected from a mirror, what is the maximum force the laser beam can exert on it? **(e)** Describe the orientation of the laser beam relative to the mirror for the case of maximum force.

108. ••• Four polarizers are set up so that the transmission axis of each successive polarizer is rotated clockwise by an angle θ relative to the previous polarizer. Find the angle θ for which unpolarized light is transmitted through these four polarizers with its intensity reduced by a factor of 25.

109. ••• **BIO Optical Activity of Sugar** The sugar concentration in a solution (e.g., in a urine specimen) can be measured conveniently by using the *optical activity* of sugar and other asymmetric molecules. In general, an optically active molecule, like sugar, will rotate the plane of polarization through an angle that is proportional to the thickness of the sample and to the concentration of the molecule. To measure the concentration of a given solution, a sample of known thickness is placed between two polarizing filters that are at right angles to each other, as shown in **Figure 25–34**. The intensity of light transmitted through the two filters can be compared with a calibration chart to determine the concentration. **(a)** What percentage of the incident (unpolarized) light will pass through the first filter? **(b)** If no sample is present, what percentage of the initial light will pass through the second filter? **(c)** When a particular sample is placed between the two filters, the intensity of light emerging from the second filter is 40.0% of the incident intensity. Through what angle did the sample rotate the plane of polarization? **(d)** A second sample has half the sugar concentration of the first sample. Find the intensity of light emerging from the second filter in this case.

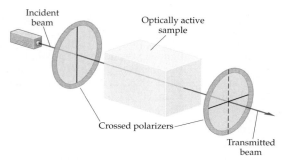

Incident beam

Optically active sample

Crossed polarizers

Transmitted beam

▲ **FIGURE 25–34** Problem 109

PASSAGE PROBLEMS

Visible-Light Curing in Dentistry

An essential part of modern dentistry is visible-light curing (VLC), a procedure that hardens the restorative materials used in fillings, veneers, and other applications. These "curing lights" work by activating molecules known as photoinitiators within the restorative materials. The photoinitiators, in turn, start a process of polymerization that causes monomers to link together to form a tough, solid polymer network. Thus, with VLC a dentist can apply and shape soft restorative materials as desired, shine a bright light on the result as shown in **Figure 25–35**, and in 20 seconds have a completely hardened—or cured—final product.

The most common photoinitiator is camphoroquinone (CPQ). To cure CPQ in the least time, one should illuminate it with light having a wavelength of approximately 465 nm. Many VLC units use a halogen light, but there are some draw-

backs to this approach. First, the filament in a halogen light is heated to a temperature of about 3000 K, which can cause heat degradation of components in the curing unit itself. Second, less than 1% of the energy given off by a halogen bulb is visible light, so a halogen bulb must have a high power rating to produce the desired light intensity.

More recently, VLC units have begun to use LEDs as their light source. These lights stay cool, emit their energy output as visible light at the desired wavelength, and provide light with an intensity as high as 1000 mW/cm², which is about 10 times the intensity of sunlight on the surface of the Earth.

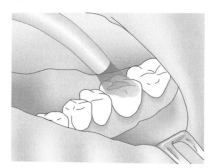

▲ **FIGURE 25–35** An intense beam of light cures, or hardens, the restorative material used to fill a cavity. (Problems 110, 111, 112, and 113)

110. • What is the color of the light that is most effective at activating the photoinitiator CPQ?

   A. red   **B.** yellow

   C. green   **D.** blue

111. • What is the frequency of the light that is most effective at activating a CPQ molecule?

   A. 140 Hz   **B.** 1.00×10^{14} Hz

   C. 6.45×10^{14} Hz   **D.** 1.55×10^{15} Hz

112. • Suppose a VLC unit uses an LED that produces light with an average intensity of 975 mW/cm². What is the rms value of the electric field in this beam of light?

   A. 606 N/C   **B.** 1920 N/C

   C. 3.67×10^6 N/C   **D.** 3.32×10^7 N/C

113. • How much radiation pressure does the beam of light in Problem 112 exert on a tooth, assuming the tooth absorbs all the light?

   A. 3.25×10^{-5} N/m²   **B.** 5.70×10^{-3} N/m²

   C. 3.67×10^6 N/m²   **D.** 1.10×10^{15} N/m²

INTERACTIVE PROBLEMS

114. •• **IP Referring to Example 25–5** Suppose the incident beam of light is linearly polarized in the same direction θ as the transmission axis of the analyzer. The transmission axis of the polarizer remains vertical. **(a)** What value must θ have if the transmitted intensity is to be $0.200\, I_0$? **(b)** If θ is increased from the value found in part (a), does the transmitted intensity increase, decrease, or stay the same? Explain.

115. •• **Referring to Example 25–5** Suppose the incident beam of light is linearly polarized in the vertical direction. In addition, the transmission axis of the analyzer is at an angle of 80.0° to the vertical. What angle should the transmission axis of the polarizer make with the vertical if the transmitted intensity is to be a maximum?

26 Geometrical Optics

We all learn early in life that light travels in straight lines—but we also learn, when we first look into a mirror or through a glass of water, that these lines can take some odd bends and turns. The mirrors shown here, for example, present a distorted view of their surroundings, simply because light striking their surface is reflected onto new directions. As we shall see in this chapter, optical devices that change the direction of light—such as mirrors and lenses—follow simple geometrical laws that account for their ability to form images. By analyzing the direction of light rays before and after they encounter a mirror or lens, we can determine the precise location, size, and orientation of the image it produces.

W hen you look into a mirror, or through a pair of binoculars, you see images of various objects in your surroundings. These images are formed by mirrors or lenses that redirect light coming from the objects. In the case of mirrors, light is *reflected* onto a new path. In lenses the speed of light is reduced, resulting in a change of direction referred to as *refraction*. By changing the direction in which light moves, both mirrors and lenses can be used to create images of various sizes and orientations.

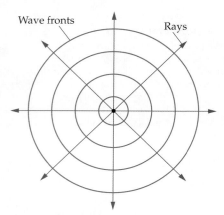

▲ **FIGURE 26–1 Wave fronts and rays**

In this case, the wave fronts indicate the crests of water waves. The rays indicate the local direction of motion.

26–1 The Reflection of Light

Perhaps the simplest way to change the direction of light is by reflection from a shiny surface. To understand this process in detail, it is convenient to describe light in terms of "wave fronts" and "rays." As we shall see later in this chapter, these concepts are equally useful in understanding the behavior of lenses.

Wave Fronts and Rays

Consider the waves created by a rock dropped into a still pool of water. As we know, these waves form concentric outward-moving circles. A simplified representation of this system is given in **Figure 26–1**, where the circles indicate the crests of the waves. We refer to these circles as **wave fronts.** In addition, the radial motion of the waves is indicated by the outward-pointing arrows, referred to as **rays.** Notice that the rays are always at right angles to the wave fronts.

A similar situation applies to electromagnetic waves radiated by a small source, as illustrated in **Figure 26–2 (a)**. In this case, however, the waves move outward in three dimensions, giving rise to spherical wave fronts. As expected, **spherical waves** such as these have rays that point radially outward.

In **Figure 26–2 (b)** we show that as one moves farther from the source of spherical waves the wave fronts become increasingly flat and the rays more nearly parallel. In the limit of increasing distance, the wave fronts approach perfectly flat planes. Such **plane waves,** with their planar wave fronts and parallel rays, are useful idealizations for investigating the properties of mirrors and lenses.

▶ **FIGURE 26–2 Spherical and planar wave fronts**

(a) As spherical waves move farther from their source, the wave fronts become increasingly flat. **(b)** In the limit of large distances the wave fronts are planes, and the corresponding rays are parallel to one another.

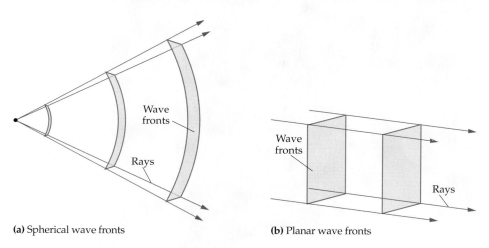

(a) Spherical wave fronts

(b) Planar wave fronts

Finally, we usually simplify our representation of light beams even further by omitting the wave fronts and plotting only one, or a few, rays. For example, in **Figure 26–3** both the incident plane wave and the reflected plane wave are shown as single rays pointing in the direction of propagation. The direction of these rays is considered next.

The Law of Reflection

To characterize the behavior of light as it reflects from a mirror or other shiny object, we begin by drawing the normal, which is simply a line perpendicular to the reflecting surface at the point of incidence. Relative to the normal, the incident ray strikes the surface at the angle θ_i, the *angle of incidence,* as shown in Figure 26–3. Similarly, the *angle of reflection,* θ_r, is the angle the reflected ray makes with the normal. The relationship between these two angles is very simple—they are equal:

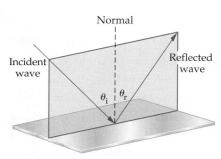

▲ **FIGURE 26–3 Reflection from a smooth surface**

In this simplified representation, the incident and reflected waves are indicated by single rays pointing in the direction of propagation. Notice that the angle of reflection, θ_r, is equal to the angle of incidence, θ_i. In addition, the incident ray, reflected ray, and the normal all lie in the same plane.

Law of Reflection

$$\theta_r = \theta_i$$

26–1

Note, in addition, that the incident ray, the normal, and the reflected ray all lie in the same plane, as is also clear from Figure 26–3.

(a) Specular reflection

(b) Diffuse reflection

The reflection of light from a smooth shiny surface, as in **Figure 26–4 (a)**, is referred to as **specular reflection.** Notice that all the reflected light moves in the same direction. In contrast, if a surface is rough, as in **Figure 26–4 (b)**, the reflected light is sent out in a variety of directions, giving rise to **diffuse reflection.** For example, when the surface of a road is wet, the water creates a smooth surface, and headlights reflecting from the road undergo specular reflection. As a result, the reflected light goes in a single direction, giving rise to an intense glare. When the same road is dry its surface is microscopically rough; hence, light is reflected in many directions and glare is not observed. The law of reflection is obeyed in either case, of course; it is the surface that is different, not the underlying physics.

A novel variation on specular versus diffuse reflection occurs in a new type of electronic chip known as a Digital Micromirror Device (DMD). These small devices consist of as many as 1.3 million microscopic plane mirrors, each of which, though smaller than the diameter of a human hair, can be oriented independently in response to electrical signals. The reflection from each micromirror is specular, and if all 1.3 million mirrors are oriented in the same direction, the DMD acts like a small plane mirror. Conversely, if the mirrors are oriented randomly, the reflection from the DMD is diffuse. When a DMD is used to project a movie, as will soon be the case in certain theaters, each micromirror will play the role of a single pixel in the projected image. In such a system, the light directed onto the DMD cycles rapidly from red to green to blue, and each of the mirrors reflects only the appropriate colors for that pixel onto the screen (**Figure 26–5**). The result is a projected image of great brilliance and vividness that eliminates the need for film.

▲ When rays of light are reflected from a smooth, flat surface, both the incident and reflected rays make identical angles with the normal. Consequently, parallel rays are reflected in the same direction, as shown in Figure 26–4 (a). The result is specular reflection, which produces a "mirror image" of the source of the rays.

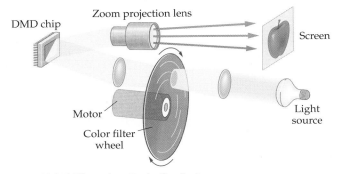

DMD chip

Zoom projection lens

Screen

Motor

Color filter wheel

Light source

▲ **FIGURE 26–5 A Digital Micromirror Projection System**
A digital projection system based on micromirrors reflects incoming light onto a distant screen. Each micromirror produces one pixel of the final image. The color and intensity of a given pixel are determined by the amount of red, green, and blue light the corresponding micromirror reflects to the screen.

REAL-WORLD PHYSICS

Micromirror devices and digital movie projection

26–2 Forming Images with a Plane Mirror

We are all familiar with looking at ourselves in a mirror. If the mirror is perfectly flat—that is, a **plane mirror**—we see an upright image of ourselves as far behind the mirror as we are in front. In addition, the image is reversed right to left; for example, if we raise our right hand, the mirror image raises its left hand. Lettering read in a mirror is reversed right to left as well. This is the reason ambulances and other emergency vehicles use mirror-image writing on the front of their vehicles—when viewed in the rear-view mirror of a car, the writing looks normal. In this section we use the law of reflection to derive these and other results.

Before we delve into the details of mirror images, however, let's pause for a moment to consider the process by which physical objects produce images in our

▲ An ant's leg provides a sense of scale in this photo of an array of Digital Micromirror Device (DMD) mirrors. Each mirror has an area of 16 μm^2 and can pivot 10° in either direction about a diagonal axis.

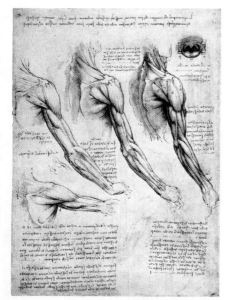

eyes. First, any object that is near you at this moment is bathed in light coming at it from all directions. As this object reflects the incoming light back into the room, every point on it acts like an omnidirectional source of light. When we view the object, the light coming from any given point enters our eyes and is focused to a point on our retina. This is the case *for every point* that we can see on the object— each point on the object is detected by a different point on the retina. This results in a one-to-one mapping between the physical object and the image on the retina.

The formation of a mirror image occurs in a similar manner, except that the light from an object reflects from a mirror before it enters our eyes. This is illustrated in **Figure 26–6 (a)**, where we show an object—a small flower—placed in front of a plane mirror. Rays of light leaving the top of the flower at point P are shown reflecting from the mirror and entering the eye of an observer. To the observer, it appears that the rays are emanating from point P′ *behind the mirror*.

We can show that the image is the same distance behind the mirror as the object is in front. Consider **Figure 26–6 (b)**, where we indicate the distance of the object from the mirror by d_o, and the distance from the image to the mirror by d_i. One ray from the top of the flower is shown reflecting from the mirror and entering the observer's eye. We also show the extension of the reflected ray back to the image. By the law of reflection, if the angle of incidence at point A is θ, the angle of reflection is also θ. Therefore, the straight line from the observer to the image cuts across the normal line with an angle θ on either side, as shown. Clearly, then, the angles indicated by ϕ must also be equal to one another. Combining these results, we see that triangle PAQ shares a side and two adjacent angles with triangle P′AQ; hence, the two triangles are equal. It follows that the distance d_o is equal to the distance d_i, as expected.

▶ **FIGURE 26–6 Locating a mirror image**
(a) Rays of light from point P at the top of the flower appear to originate from point P′ behind the mirror. **(b)** Construction showing that the distance from the object to the mirror is the same as the distance from the image to the mirror.

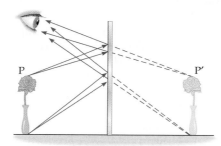

(a) Image formed by a plane mirror

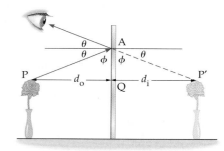

(b) Object distance (d_o) equals image distance (d_i)

In the following Example, we once again apply the law of reflection to the flower and its mirror image.

EXAMPLE 26–1 REFLECTING ON A FLOWER

An observer is at table level, a distance d to the left of a flower of height h. The flower itself is a distance d to the left of a mirror, as shown in the sketch. Note that a ray of light propagating from the top of the flower to the observer's eye reflects from the mirror at a height y above the table. Find y in terms of the height of the flower, h.

PICTURE THE PROBLEM
The physical situation is shown in the diagram, along with a ray from the top of the flower to the eye of the observer. The point where the ray hits the mirror is a height y above the table. The flower is a distance d to the left of the mirror, and its image is a distance d to the right of the mirror. Finally, the observer's eye is a distance $2d$ to the left of the mirror.

STRATEGY
To find y, we must use the fact that the angle of reflection is equal to the angle of incidence. In the diagram we indicate these two angles by the symbol θ. It follows, then, that $\tan\theta$ obtained from the small yellow triangle must be equal to $\tan\theta$ obtained from the larger green triangle. Setting these expressions for $\tan\theta$ equal to one another yields a relation that can be solved for y.

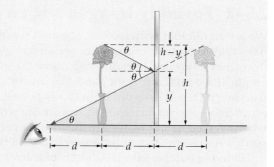

SOLUTION

1. Write an expression for tan θ using the yellow triangle:

$$\tan \theta = \frac{h - y}{d}$$

2. Now, write a similar expression for tan θ using the green triangle:

$$\tan \theta = \frac{y}{2d}$$

3. Set the two expressions for tan θ equal to one another:

$$\frac{h - y}{d} = \frac{y}{2d}$$

4. Rearrange the preceding equation and solve for *y*:

$$h - y = \frac{y}{2}$$

$$y = \frac{2}{3}h$$

INSIGHT

Note that the observer will see the entire image of the flower in a section of mirror that is only two-thirds the height of the flower.

PRACTICE PROBLEM

If the observer moves farther from the base of the flower, does the point of reflection move upward, downward, or stay at the same location? As a check on your answer, calculate the point of reflection for the case where the distance between the observer and the base of the flower is $2d$.   [**Answer:** The point of reflection moves upward. In this case, we find $y = \frac{3}{4}h$.]

Some related homework problems: Problem 5, Problem 7

To summarize, the basic features of reflection by a plane mirror are the following:

Properties of Mirror Images Produced by Plane Mirrors

- A mirror image is upright, but appears reversed right to left.
- A mirror image appears to be the same distance behind the mirror that the object is in front of the mirror.
- A mirror image is the same size as the object.

CONCEPTUAL CHECKPOINT 26–1   HEIGHT OF MIRROR

To save expenses, you would like to buy the shortest mirror that will allow you to see your entire body. Should the mirror be **(a)** half your height, **(b)** two-thirds your height, or **(c)** equal to your height?

REASONING AND DISCUSSION

First, to see your feet, the mirror must extend from your eyes downward to a point halfway between your eyes and feet, as shown in the sketch.

Similarly, the mirror must extend upward from your eyes half the distance to the top of your head. Altogether, then, the mirror must have a height equal to half the distance from your eyes to your feet plus half the distance from your eyes to the top of your head—that is, half your total height.

ANSWER

(a) The mirror needs to be only half your height.

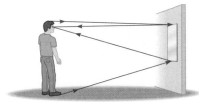

An application of mirror images that has been used in military aircraft for many years, and is now beginning to appear in commercial automobiles, is the *heads-up display*. In the case of a car, a small illuminated display screen is recessed in the dashboard, out of direct sight of the driver. The screen shows important information, like the speed of the car, in mirror image. The driver sees the information not by looking directly at the screen—which is hidden from view—but by looking at its reflection in the windshield. Thus, while still looking straight ahead (heads up), the driver can see both the road and the reading of the speedometer.

A similar device is used in theaters to provide subtitles for the hearing impaired. In this case, a transparent plastic screen is mounted on the arm of a person's chair. This screen is adjusted in such a way that the person can look through it to see the movie, and at the same time see the reflection of a screen in the back of the theater that gives the subtitles in mirror-image form.

▲ In this heads-up display in an airplane cockpit, important flight information is reflected on a transparent screen in the windshield, enabling the pilot to view the data without diverting attention from the scene ahead.

EXAMPLE 26–2 TWO-DIMENSIONAL CORNER REFLECTOR

Two mirrors are placed at right angles, as shown in the diagram. An incident ray of light makes an angle of 30° with the *x* axis and reflects from the lower mirror. Find the angle the outgoing ray makes with the *y* axis after it reflects once from each mirror.

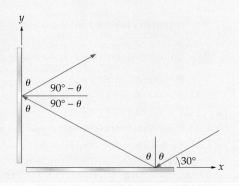

PICTURE THE PROBLEM

The physical system is shown in the diagram. Notice that the normal for the first reflection is vertical, whereas it is horizontal for the second reflection. It follows that the angle of incidence for the first reflection is $\theta = 90° - 30° = 60°$. Similarly, the angle of incidence for the second reflection is $90° - \theta = 30°$.

SOLUTION

We apply the law of reflection to each of the two reflections. For the first reflection, the angle of incidence is $\theta = 60°$; hence, the reflected ray also makes a 60° angle with the vertical.

At the second reflection, the normal is horizontal; hence, the angle of incidence is $90° - \theta = 30°$. After the second reflection, the outgoing ray is 30° above the horizontal and hence it makes an angle $\theta = 60°$ with respect to the *y* axis, as shown.

INSIGHT

Note that the outgoing ray travels parallel to the incoming ray, but in exactly the opposite direction. This is true regardless of the value of the initial angle θ.

PRACTICE PROBLEM

If the incoming ray hits the horizontal mirror farther to the right, is the angle the outgoing ray makes with the vertical greater than θ, equal to θ, or less than θ? **[Answer:** The angle of the outgoing ray is still equal to θ. The distance between the incoming and outgoing rays would be increased, however.**]**

Some related homework problems: Problem 8, Problem 16

REAL-WORLD PHYSICS

Corner reflectors and the Earth–Moon distance

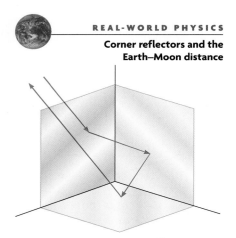

▲ **FIGURE 26–7 A corner reflector**

A three-dimensional corner reflector constructed from three plane mirrors at right angles to one another. A ray entering the reflector is sent back in the direction from which it came.

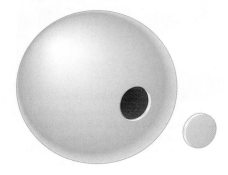

If three plane mirrors are joined at right angles, as in **Figure 26–7**, the result is a **corner reflector.** A corner reflector behaves in three dimensions the same as two mirrors at right angles in two dimensions; namely, a ray incident on the corner reflector is sent back in the same direction from which it came. This type of behavior has led to many useful applications for corner reflectors.

One of the more interesting applications involves the only Apollo experiments still returning data from the Moon. On *Apollos 11, 14*, and *15* the astronauts placed retroreflector arrays consisting of 100 corner reflectors on the lunar surface. Scientists at observatories on Earth can send a laser beam to the appropriate location on the Moon, where the retroreflector sends it back to its source. By measuring the round-trip travel time of the light, it is possible to determine the Earth–Moon distance to an accuracy of about 3 cm out of a total distance of roughly 385,000 km! Over the years, these measurements have revealed that the Moon is moving away from Earth at the rate of 3.8 cm per year.

26–3 Spherical Mirrors

We now consider another type of mirror that is encountered frequently in everyday life—the spherical mirror. These mirrors get their name from the fact that they have the same shape as a section of a sphere, as is shown in **Figure 26–8**, where a portion of a spherical shell of radius *R* is cut away from the rest of the shell. If the outside of this spherical section is a reflecting surface, the result is a **convex** mirror; if the inside surface is reflecting, we have a **concave** mirror.

Convex and concave spherical mirrors are illustrated in **Figure 26–9**, where we also indicate the **center of curvature,** *C*, and the **principal axis.** The center of

◀ **FIGURE 26–8 Spherical mirrors**

A spherical mirror has the same shape as a section of a sphere. If the outside surface of this section is reflecting, the mirror is convex. If the inside surface reflects, the mirror is concave.

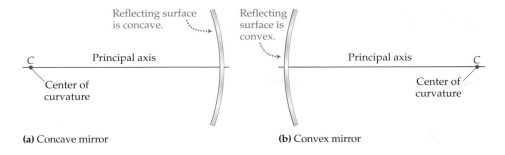

(a) Concave mirror   **(b)** Convex mirror

▲ **FIGURE 26–9** **Concave and convex mirrors**
(a) A concave mirror and its center of curvature, C, which is on the same side as the reflecting surface. **(b)** A convex mirror and its center of curvature, C. In this case C is on the opposite side of the mirror from its reflecting surface.

curvature is the center of the sphere with radius R of which the mirror is a section, and the principal axis is a straight line drawn through the center of curvature and the midpoint of the mirror. Note that the principal axis intersects the mirror at right angles.

To investigate the behavior of spherical mirrors, suppose that a beam of light is directed toward either a convex or a concave mirror along its principal axis. For example, several parallel rays of light are approaching a convex mirror in **Figure 26–10**. After reflecting from the mirror, the rays diverge as if they originated from a single point behind the mirror called the **focal point,** F. (This is strictly true only for light rays close to the principal axis, as we point out in greater detail later in this section. All of our results for spherical mirrors assume rays close to the axis.)

Now, let's find the **focal length;** that is, the distance from the surface of the mirror to the focal point. This can be done with the aid of **Figure 26–11**, which shows a single ray reflecting from the mirror. The first thing to notice about this diagram is that a straight line drawn through the center of curvature always intersects the mirror at right angles; hence, the line through C is the normal to the surface at the point of incidence, A. Since the incoming ray is parallel to the principal axis, it follows that the angle of incidence, θ, is equal to the angle FCA. Next, the law of reflection states that the angle of reflection must equal θ, which means that the angle CAF is also θ. We see, then, that CAF is an isosceles triangle, with the sides CF and FA of equal length. Finally, for small angles θ, the length CF is approximately equal to half the length $CA = R$; that is, $CF \sim \frac{1}{2}R$. Therefore, to this same approximation, the distance FB is also $\frac{1}{2}R$.

Thus, when considering a convex mirror of radius R, we will always use the following result for the focal length:

Focal Length for a Convex Mirror of Radius R

$$f = -\tfrac{1}{2}R \qquad\qquad 26\text{–}2$$

SI unit: m

The minus sign in this expression is used to indicate that the focal point lies behind the mirror. This is part of a general sign convention for mirrors that will be discussed in detail in the next section.

The situation is similar for a concave mirror. First, rays parallel to the principal axis are reflected by the mirror and brought together at a focal point, F, as shown in **Figure 26–12**. The same type of analysis used for the convex mirror can be applied to the concave mirror as well, with the result that the focal point is a distance $\frac{1}{2}R$ in front of the mirror. Thus, for a concave mirror,

Focal Length for a Concave Mirror of Radius R

$$f = \tfrac{1}{2}R \qquad\qquad 26\text{–}3$$

SI unit: m

This time f is positive, since the focal point is in front of the mirror. Note that in this case the rays of light actually pass through the focal point, in contrast to the behavior of the convex mirror.

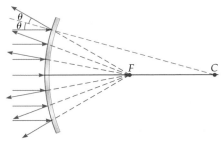

▲ **FIGURE 26–10** **Parallel rays on a convex mirror**
When parallel rays of light reflect from a convex mirror, they diverge as if originating from a focal point halfway between the surface of the mirror and its center of curvature.

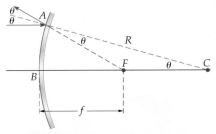

▲ **FIGURE 26–11** **Ray diagram for a convex mirror**
Ray diagram used to locate the focal point, F.

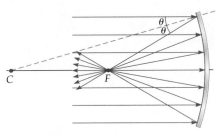

▲ **FIGURE 26–12** **Parallel rays on a concave mirror**
Parallel rays reflecting from a concave mirror pass through the focal point F, which is halfway between the surface of the mirror and the center of curvature.

Suppose you would like to use the Sun to start a fire in the wilderness. Which type of mirror, concave or convex, would work best?

REASONING AND DISCUSSION
First, note that rays from the Sun are essentially parallel, since it is at such a great distance. Therefore, if a convex mirror is used, the situation is like that shown in Figure 26–10, in which the rays are spread out after reflection. Conversely, a concave mirror will bring the rays together at a point, as in Figure 26–12. Clearly, by focusing the sunlight at one point, the concave mirror will stand a better chance of starting a fire.

ANSWER
The concave mirror is the one to use.

A final point regards the approximation made earlier in this section; namely, that the angle θ is small. This is equivalent to saying that the distance between the principal axis of the mirror and the incoming rays is much less than the radius of curvature of the mirror, R. When rays are displaced from the axis by distances comparable to the radius R, as in **Figure 26–13 (a)**, the result is that they do not all pass through the focal point—the farther a ray is from the axis, the more it misses the focal point. In a case like this, the mirror will produce a blurred image, an effect known as **spherical aberration.** This effect can be reduced to undetectable levels by restricting the incoming rays to only those near the axis. These axis-hugging rays are referred to as **paraxial rays.**

REAL-WORLD PHYSICS
Parabolic mirrors

▶ **FIGURE 26–13 Spherical aberration and the parabolic mirror**

(a) Rays far from the axis of a spherical mirror are not reflected through its focal point. The result is a blurred image referred to as spherical aberration. **(b)** A mirror with a parabolic cross section brings all parallel rays to a focus at one point, regardless of how far they are from the axis of the mirror.

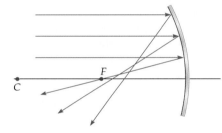 

(a) A spherical mirror blurs the focus   **(b)** A parabolic mirror has a single focal point

Another way to eliminate spherical aberration is to construct a mirror with a *parabolic* cross section, as shown in **Figure 26–13 (b)**. One of the key properties of a parabola is that rays parallel to its axis are reflected through the same point F regardless of their distance from the axis. Thus a parabolic mirror produces a sharp image from all the rays coming into it. For this reason, astronomical mirrors, like that of the Hale telescope on Mount Palomar, California, are polished to a parabolic shape to give the greatest possible light-gathering ability and the sharpest possible images.

The same principle works in reverse as well. For example, if a source of light is placed at the focal point of a parabolic mirror, as at point F in Figure 26–13 (b), the mirror will redirect the light into an intense, unidirectional beam that can be aimed in a precise direction. Applications of this effect include flashlights, car headlights, and the giant arc lights that sweep across the sky to announce a grand opening.

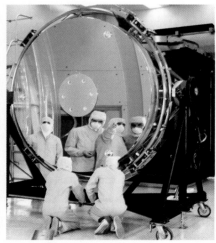

▲ The mirror of the Hubble Space Telescope (HST) was to have a perfectly parabolic shape. Unfortunately, owing to a mistake in the grinding of the mirror, the images produced by the HST were marred by spherical aberration. This necessitated a repair mission to the orbiting telescope, during which astronauts installed corrective optics to compensate for the defect.

26–4 Ray Tracing and the Mirror Equation

The images formed by a spherical mirror can be more varied than those produced by a plane mirror. In a plane mirror the image is always upright, the same size as the object, and the same distance from the mirror as the object. In the case of spherical mirrors, the image can be either upright or inverted, larger or smaller than the object, and closer or farther from the mirror than the object.

To find the orientation, size, and location of an image in a spherical mirror, we use two techniques. The first, referred to as **ray tracing,** gives the orientation of the image as well as qualitative information on its location and size. If drawn carefully to scale, a ray diagram can also give quantitative results. The second method, using a relation referred to as the **mirror equation,** provides precise quantitative information without the need for accurate scale drawings. Both methods are presented in this section.

Ray Tracing

The basic idea behind ray tracing is to follow the path of representative rays of light as they reflect from a mirror and form an image. This was done for the plane mirror in Section 26–2. Ray tracing for spherical mirrors is a straightforward extension of the same basic techniques.

There are three rays with simple behavior that are used most often in ray tracing with spherical mirrors. These rays are illustrated in **Figure 26–14** for a concave mirror, and in **Figure 26–15** for a convex mirror. We start with the parallel ray (*P* ray), which, as its name implies, is parallel to the principal axis of the mirror. As we know from the previous section, a parallel ray is reflected through the focal point of a concave mirror, as shown by the purple ray in Figure 26–14. Similarly, a *P* ray reflects from a convex mirror along a line that *extends back* through the focal point, as with the purple ray in Figure 26–15.

Next, a ray that passes through the focal point of a concave mirror is reflected parallel to the axis, as indicated by the green ray in Figure 26–14. Thus, in a sense, a focal-point ray (*F* ray) is the reverse of a *P* ray. In general, any ray is equally valid in either the forward or reverse direction. The corresponding *F* ray for a convex mirror is shown in green in Figure 26–15.

Finally, note that any straight line extending from the center of curvature intersects the mirror at right angles. Thus, a ray moving along such a path is reflected back along the same path. Center-of-curvature rays (*C* rays) are illustrated in red in Figures 26–14 and 26–15.

To see how these rays can be used to obtain an image, consider the convex mirror shown in **Figure 26–16**. In front of the mirror is an object, represented symbolically by the red arrow. Also indicated in the figure are the three rays described above. Note that these rays diverge from the mirror as if they had originated from the tip of the orange arrow behind the mirror. This arrow is the image of the object; in fact, since no light passes through the image, we call it a **virtual image.** As we can see from the diagram, the virtual image is upright, smaller than the object, and located between the mirror and the focal point *F*.

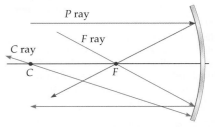

▲ **FIGURE 26–14** **Principal rays used in ray tracing for a concave mirror**
The parallel ray (*P* ray) reflects through the focal point. The focal ray (*F* ray) reflects parallel to the axis, and the center-of-curvature ray (*C* ray) reflects back along its incoming path.

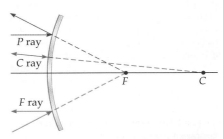

▲ **FIGURE 26–15** **Principal rays used in ray tracing for a convex mirror**
The parallel ray (*P* ray) reflects along a direction that extends back to the focal point. Similarly, a focal ray (*F* ray) moves toward the focal point until it reflects from the mirror, after which it moves parallel to the axis. A center-of-curvature ray (*C* ray) is directed toward the center of curvature. It reflects from the mirror back along its incoming path.

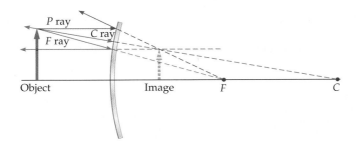

◀ **FIGURE 26–16** **Image formation with a convex mirror**
Ray diagram showing an image formed by a convex mirror. The three outgoing rays (*P*, *F*, and *C*) extend back to a single point at the top of the image.

Even though we drew three rays in Figure 26–16, any two would have given the intersection point at the tip of the virtual image. This is commonly the case with ray diagrams. When possible, it is useful to draw all three rays as a check on your results.

Finally, in the limit that the object is very close to a convex mirror, the mirror is essentially flat and behaves like a plane mirror. Thus the virtual image will be about the same distance behind the mirror that the image is in front, and about the same size as the object. Conversely, if the object is far from the mirror, the image is very small and practically at the focal point. These limits are illustrated in **Figure 26–17.**

▶ **FIGURE 26–17 Image size and location in a convex mirror**
(a) When an object is close to a convex mirror, the image is practically the same size and distance from the mirror. **(b)** In the limit that the object is very far from the mirror, the image is small and close to the focal point.

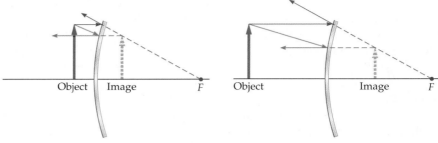

(a) An object close to a convex mirror   **(b)** An object far from a convex mirror

PROBLEM-SOLVING NOTE

Using Rays to Locate the Image of a Spherical Mirror

To find the top of an image, draw the three rays, *P, F,* and *C,* from the top of the object. The rays will either intersect at the top of the image (real image) or extend backward to the top of the image (virtual image). When drawing the rays, remember that a *P* ray is *p*arallel to the axis of the mirror, the *C* ray goes either through or toward the mirror's *c*enter of curvature, and the *F* ray goes either through or toward the *f*ocal point of the mirror.

Next we consider the image formed by a concave mirror. In particular, we examine the following three cases: (i) The object is farther from the mirror than the center of curvature; (ii) the object is between the center of curvature and the focal point; and (iii) the object is between the mirror and the focal point. The *F* and *P* rays for case (i) are drawn in **Figure 26–18 (a)**, showing that the image is inverted, closer to the mirror, and smaller than the object. In addition, because light rays pass through the image, we call it a **real image.** The corresponding ray diagram for case (ii) is shown in **Figure 26–18 (b)**. Note that the image is again real and inverted, but it is now farther from the mirror and larger than the object—as the object moves closer to the focal point, an observer will see a large inverted image of the object far in front of the mirror. The *C* ray was not useful in these cases and therefore was omitted from the diagrams.

▶ **FIGURE 26–18 Image formation with a concave mirror**
Ray diagrams showing the image formed by a concave mirror when the object is **(a)** beyond the center of curvature and **(b)** between the center of curvature and the focal point.

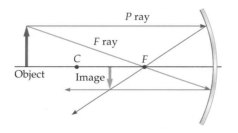

(a) An object beyond *C*   **(b)** An object between *C* and *F*

The final case, in which the object is between the mirror and the focal point, is considered in the next Example.

EXAMPLE 26–3 IMAGE FORMATION

Use a ray diagram to find the location, size, and orientation of the image formed by a concave mirror when the object is between the mirror and the focal point.

PICTURE THE PROBLEM
The physical system is shown in the diagram, along with the three principal rays. Notice that after reflection these three rays diverge from one another, just as if they had originated at a point behind the mirror.

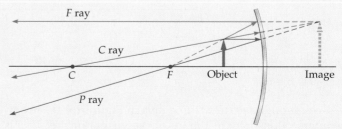

INTERACTIVE FIGURE (MP)™

SOLUTION
The three outgoing rays (*P, F,* and *C*) extend back to form an image behind the mirror that is upright (same orientation as the object) and enlarged. We now discuss these three rays one at a time:

***P* ray:** The *P* ray is the most straightforward of the three. It starts parallel to the axis, then reflects through the focal point.

***F* ray:** The *F* ray does not go through the focal point, as is usually the case. Instead, it starts on a line that *extends back* to the focal point, then reflects parallel to the axis.

C **ray:** The *C* ray starts at the top of the object, contacts the mirror at right angles, then reflects back along its initial path and through the center of curvature.

INSIGHT

Makeup mirrors are concave mirrors with fairly large focal lengths. The person applying makeup is between the mirror and its focal point, as is the object in this Example, and therefore sees an upright and enlarged image, as desired.

PRACTICE PROBLEM

If the object in the diagram is moved closer to the mirror, does the image increase or decrease in size?   [**Answer:** As the object moves closer, the mirror behaves more like a plane mirror. Thus, the image becomes smaller, so that it is closer in size to the object.]

Some related homework problems: Problem 27, Problem 30

CONCEPTUAL CHECKPOINT 26–3   REARVIEW MIRRORS

The passenger-side rearview mirrors in newer cars often have warning labels that read, OBJECTS IN MIRROR ARE CLOSER THAN THEY APPEAR. Are these rearview mirrors concave or convex?

REASONING AND DISCUSSION

Objects in the mirror are closer than they appear because the mirror produces an image that is reduced in size, which makes the object look as if it is farther away. In addition, we know that the rearview mirror always gives an upright image, no matter how close or far away the object. The mirror that always produces upright and reduced images is the convex mirror.

ANSWER

The mirrors are convex.

The imaging characteristics of convex and concave mirrors are summarized in Table 26–1.

TABLE 26–1 Imaging Characteristics of Convex and Concave Spherical Mirrors

CONVEX MIRROR Object location	Image orientation	Image size	Image type
Arbitrary	Upright	Reduced	Virtual

CONCAVE MIRROR Object location	Image orientation	Image size	Image type
Beyond *C*	Inverted	Reduced	Real
C	Inverted	Same as object	Real
Between *F* and *C*	Inverted	Enlarged	Real
Just beyond *F*	Inverted	Approaching infinity	Real
Just inside *F*	Upright	Approaching infinity	Virtual
Between mirror and *F*	Upright	Enlarged	Virtual

We now show how to obtain results about the images formed by mirrors in a quantitative manner.

The Mirror Equation

The mirror equation is a precise mathematical relationship between the object distance and the image distance for a given mirror. To obtain this relation, we use the ray diagrams shown in **Figure 26–19**. Note that the distance from the mirror to the object is d_o, the distance from the mirror to the image is d_i, and the distance from the mirror to the center of curvature is R. In addition, the height of the object is h_o, and the height of the image is h_i. Since the image is inverted, its height is negative; thus $-h_i$ is positive.

The ray in Figure 26–19 (a) hits the mirror at its midpoint, where the principal axis is the normal to the mirror. As a result, the ray reflects at an angle θ below

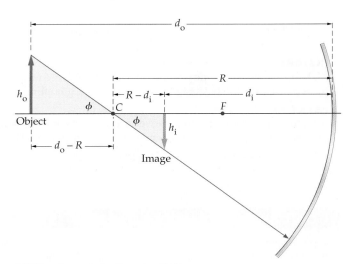

(a) Triangles to derive Equation 26–4

(b) Triangles to derive Equation 26–5

▲ **FIGURE 26–19 Ray diagrams used to derive the mirror equation**
(a) The two similar triangles in this case are used to obtain Equation 26–4. **(b)** These similar triangles yield Equation 26–5.

the principal axis that is equal to its incident angle θ above the axis. Therefore, the two green triangles in this diagram are similar, from which it follows that $h_o/d_o = (-h_i)/d_i$, or

$$\frac{h_o}{-h_i} = \frac{d_o}{d_i} \qquad \text{26–4}$$

Next, Figure 26–19 (b) shows the C ray for this mirror. From the figure it is clear that the two yellow triangles in this diagram are also similar, since they are both right triangles and share the common angle ϕ. Thus, $h_o/(d_o - R) = (-h_i)/(R - d_i)$, or

$$\frac{h_o}{-h_i} = \frac{d_o - R}{R - d_i} \qquad \text{26–5}$$

Setting these two expressions for $h_o/(-h_i)$ equal gives us

$$\frac{d_o}{d_i} = \frac{d_o - R}{R - d_i} \quad \text{or} \quad 1 = \frac{1 - \dfrac{R}{d_o}}{\dfrac{R}{d_i} - 1}$$

Rearranging, we find $R/d_o + R/d_i = 2$. Finally, dividing by R and recalling that $f = \frac{1}{2}R$ for a concave mirror, we get the **mirror equation:**

The Mirror Equation

$$\frac{1}{d_o} + \frac{1}{d_i} = \frac{1}{f} \qquad \text{26–6}$$

We apply the mirror equation to a simple system in the following Exercise.

EXERCISE 26–1

The concave side of a spoon has a focal length of 5.00 cm. Find the image distance for this "mirror" when the object distance is **(a)** 25.0 cm, **(b)** 9.00 cm, and **(c)** 2.00 cm. These three cases correspond to Figure 26–18 (a), Figure 26–18 (b), and Example 26–3, respectively.

SOLUTION
Solving Equation 26–6 for d_i, we obtain the following results:

a. $d_i = \dfrac{d_o f}{d_o - f} = \dfrac{(25.0 \text{ cm})(5.00 \text{ cm})}{25.0 \text{ cm} - 5.00 \text{ cm}} = 6.25 \text{ cm}$

b. $d_i = \dfrac{d_o f}{d_o - f} = \dfrac{(9.00 \text{ cm})(5.00 \text{ cm})}{9.00 \text{ cm} - 5.00 \text{ cm}} = 11.3 \text{ cm}$

c. $d_i = \dfrac{d_o f}{d_o - f} = \dfrac{(2.00 \text{ cm})(5.00 \text{ cm})}{2.00 \text{ cm} - 5.00 \text{ cm}} = -3.33 \text{ cm}$

Note that the image distance in Exercise 26–1 is negative when the object is closer to the mirror than the focal point [$d_o < f$, as in part (c)]. We know from Example 26–3 that this is also the case where the image is *behind* the mirror. Thus, the *sign* of the image distance indicates the *side* of the mirror on which the image is located. As long as we are discussing signs, it should be noted that the mirror equation applies equally well to a *convex* mirror, as long as we recall that the focal length in this case is *negative*. The following Exercise calculates the image distance for the convex mirror shown in Figure 26–16.

EXERCISE 26–2

The convex mirror in Figure 26–16 has a 20.0-cm radius of curvature. Find the image distance for this mirror when the object distance is 6.33 cm, as it is in Figure 26–16.

SOLUTION

Recalling that $f = -\frac{1}{2}R$ for a convex mirror, we find $f = -10.0$ cm and

$$d_i = \frac{d_o f}{d_o - f} = \frac{(6.33 \text{ cm})(-10.0 \text{ cm})}{6.33 \text{ cm} - (-10.0 \text{ cm})} = -3.88 \text{ cm}$$

This result agrees with the image shown in Figure 26–16.

Next, we consider the height of an image, which is given by the relation in Equation 26–4. Solving this equation for h_i we find

$$h_i = -\left(\frac{d_i}{d_o}\right)h_o \qquad\qquad 26\text{–}7$$

The ratio of the height of the image to the height of the object is defined as the **magnification,** *m*; that is, $m = h_i/h_o$. From Equation 26–7, we see that

Magnification, m

$$m = \frac{h_i}{h_o} = -\frac{d_i}{d_o} \qquad\qquad 26\text{–}8$$

The sign of the magnification gives the orientation of the image. For example, if both d_o and d_i are positive, as in Figure 26–18, the magnification is negative and the image is inverted. Conversely, if the image is behind the mirror, so that d_i is negative, the magnification is positive, and the image is upright. An example of this case is shown in Example 26–3 and Figure 26–16. Finally, the magnitude of the magnification gives the factor by which the size of the image is increased or decreased compared with the object. In the special case of an image with the same size and orientation as the object, as in a plane mirror, the magnification is one.

The sign conventions for mirrors are summarized below:

Focal Length
f is positive for concave mirrors.
f is negative for convex mirrors.

Magnification
m is positive for upright images.
m is negative for inverted images.

Image Distance
d_i is positive for images in front of a mirror (real images).
d_i is negative for images behind a mirror (virtual images).

PROBLEM-SOLVING NOTE

Applying the Mirror Equation

To use the mirror equation correctly, be careful to use the appropriate signs for all the known quantities. The final answer will also have a sign, which gives additional information about the system.

Object Distance

d_o is positive for objects in front of a mirror (real objects).

d_o is negative for objects behind a mirror (virtual objects).

The case of a negative object distance—that is, a virtual object—can occur when the image from one mirror serves as the object for another mirror or lens. For example, if mirror 1 produces an image that is behind mirror 2, we say that the image of mirror 1 is a virtual object for mirror 2. Such situations will be considered in the next chapter.

We now apply the mirror and magnification equations to specific Examples.

EXAMPLE 26–4 CHECKING IT TWICE

After leaving some presents under the tree, Santa notices his image in a shiny, spherical Christmas ornament. The ornament is 8.50 cm in diameter and 1.10 m away from Santa. Curious to know the location and size of his image, Santa consults a book on physics. Knowing that Santa likes to "check it twice," what results should he obtain, assuming his height is 1.75 m?

PICTURE THE PROBLEM

The physical situation is illustrated in the sketch, along with the image Santa sees in the ornament. Because the spherical ornament is a convex mirror, the image it forms is upright and reduced.

STRATEGY

The ornament is a convex mirror of radius $R = \frac{1}{2}(8.50 \text{ cm}) = 4.25$ cm. We can find the location of the image, then, by using the mirror equation with the focal length given by $f = -\frac{1}{2}R$.

Once we have determined the location of the image, we can find its size using $h_i = -(d_i/d_o)h_o = mh_o$, where $h_o = 1.75$ m.

SOLUTION

1. Calculate the focal length for the ornament:

$$f = -\frac{1}{2}R = -\frac{1}{2}(4.25 \text{ cm}) = -0.0213 \text{ m}$$

2. Use the mirror equation to find the image distance, d_i:

$$\frac{1}{d_i} = \frac{1}{f} - \frac{1}{d_o} = \frac{1}{(-0.0213 \text{ m})} - \frac{1}{1.10 \text{ m}} = -47.9 \text{ m}^{-1}$$

$$d_i = \frac{1}{(-47.9 \text{ m}^{-1})} = -0.0209 \text{ m}$$

3. Determine the magnification of the image using $m = -(d_i/d_o)$:

$$m = -\left(\frac{d_i}{d_o}\right) = -\left(\frac{-0.0209 \text{ m}}{1.10 \text{ m}}\right) = 0.0190$$

4. Find the image height with $h_i = mh_o$:

$$h_i = mh_o = (0.0190)(1.75 \text{ m}) = 0.0333 \text{ m}$$

INSIGHT

Thus, Santa's image is 2.09 cm behind the surface of the ornament—about halfway between the surface and the center—and 3.33 cm high. Note that his image fits on the surface of the ornament with room to spare, and therefore Santa can see the reflection of his entire body.

PRACTICE PROBLEM

How far from the ornament must Santa stand if his image is to be 1.75 cm behind its surface?   [**Answer:** 9.81 cm]

Some related homework problems: Problem 31, Problem 34, Problem 39

ACTIVE EXAMPLE 26–1   SAY AHHH: FIND THE MAGNIFICATION OF THE TOOTH

A dentist uses a small mirror attached to a thin rod to examine one of your teeth. When the tooth is 1.20 cm in front of the mirror, the image it forms is 9.25 cm behind the mirror. Find **(a)** the focal length of the mirror and **(b)** the magnification of the image.

SOLUTION *(Test your understanding by performing the calculations indicated in each step.)*

1. Use the information given to identify the object and image distances: $d_o = 1.20$ cm, $d_i = -9.25$ cm

Part (a)

2. Use the mirror equation to calculate the focal length:

$f = 1.38 \text{ cm}$

Part (b)

3. Use $m = -(d_i/d_o)$ to find the magnification:

$m = 7.71$

INSIGHT

Because the focal length is positive, and the magnification is greater than one, we conclude that the dentist's mirror is concave. Note that it will make the tooth look 7.71 times larger than life. A convex mirror, in contrast, always produces a magnification less than one.

YOUR TURN

If the mirror is moved closer to the tooth, will the magnification of the tooth increase or decrease? Find the magnification if the distance from the tooth to the mirror is reduced to 1.00 cm.

*(Answers to **Your Turn** problems are given in the back of the book.)*

Finally, recall that a spherical mirror behaves like a plane mirror when the object is close to the mirror. But just what exactly do we mean by *close*? Well, in this case, *close* means that the object distance should be small in comparison with the radius of curvature. For example, if the radius of curvature were to go to infinity, the mirror would behave like a plane mirror for all object distances. Simply put, a sphere with $R \rightarrow \infty$ has a surface that is essentially as flat as a plane. Letting $f = \frac{1}{2}R$ go to infinity in the mirror equation yields

$$\frac{1}{d_o} + \frac{1}{d_i} = \frac{1}{f} \rightarrow 0$$

Therefore, in this limit $d_i = -d_o$, as expected for a plane mirror.

26–5 The Refraction of Light

When a wave propagates from a medium in which its speed is v_1 to another in which its speed is $v_2 \neq v_1$, it will, in general, change its direction of motion. This phenomenon is called **refraction.** To understand the cause of refraction, consider the behavior of a marching band as it moves from a solid section of ground to an area where the ground is muddy, as indicated in **Figure 26–20**. In this analogy, the rows of the marching band correspond to the wave fronts of a traveling wave, and the solid and muddy sections of the ground represent media in which the wave speed is different.

If the speed of the marchers on solid ground is v_1, then after a time Δt they have advanced a distance $v_1 \Delta t$. On the other hand, if the marchers in the mud move at the reduced speed v_2, they advance only a distance $v_2 \Delta t$ in the same time. This causes a bend in the line of marchers, as can be seen in Figure 26–20, and therefore a change in the direction of motion.

Figure 26–21 shows a simplified version of Figure 26–20, with three "wave fronts" taking the place of the marchers. Also shown is a ray drawn at right angles to the wave fronts, and a normal to the interface between the two types of ground. The angle of incidence is θ_1, and from the figure we can see that this is also the angle between the incoming wave front and the interface. The outgoing wave front and its ray are characterized by a different angle, θ_2. From the geometry of the figure, we see that the green and brown triangles share a common side, AB. Therefore,

$$\sin\theta_1 = \frac{v_1\Delta t}{\text{AB}} \quad \text{and} \quad \sin\theta_2 = \frac{v_2\Delta t}{\text{AB}}$$

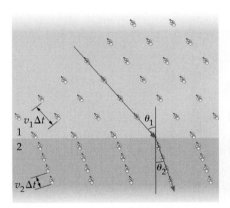

▲ **FIGURE 26–20 An analogy for refraction**

As a marching band moves from an area where the ground is solid to one where it is soft and muddy, the direction of motion changes.

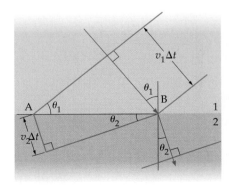

▲ **FIGURE 26–21 The basic mechanism of refraction**

Refraction is the bending of wave fronts and a change in direction of propagation due to a change in speed.

TABLE 26–2 Index of Refraction for Common Substances

Substance	Index of refraction, n
SOLIDS	
Diamond	2.42
Flint glass	1.66
Crown glass	1.52
Fused quartz (glass)	1.46
Ice	1.31
LIQUIDS	
Benzene	1.50
Ethyl alcohol	1.36
Water	1.33
GASES	
Carbon dioxide	1.00045
Air	1.000293

▲ The water in this tank contains a fluorescent dye, making it easier to see the refraction of the beam as it passes from the air into the water.

Eliminating the common factor $\Delta t / AB$ yields

$$\frac{\sin \theta_1}{v_1} = \frac{\sin \theta_2}{v_2} \qquad 26\text{–}9$$

Thus, we see that the direction of propagation is directly related to the speed of propagation.

The speed of light, in turn, depends on the medium through which it travels. For example, we know that in a vacuum the speed of light is $c = 3.00 \times 10^8$ m/s. When light propagates through water, however, its speed is reduced by a factor of 1.33. In general, the speed of light in a given medium, v, is determined by the medium's **index of refraction**, n, defined as follows:

Definition of the Index of Refraction, n

$$v = \frac{c}{n} \qquad 26\text{–}10$$

Representative values of the index of refraction for a variety of media are given in Table 26–2.

EXERCISE 26–3

How long does it take for light to travel 2.50 m in water?

SOLUTION

The speed of light in water is c/n, where $n = 1.33$. Therefore, the time required to cover 2.50 m is

$$t = \frac{d}{v} = \frac{d}{(c/n)} = \frac{2.50 \text{ m}}{\left(\frac{3.00 \times 10^8 \text{ m/s}}{1.33}\right)} = 1.11 \times 10^{-8} \text{ s}$$

Returning to the direction of propagation, let's suppose light has the speed $v_1 = c/n_1$ in one medium and $v_2 = c/n_2$ in a second medium. The direction of propagation in these two media is related by Equation 26–9:

$$\frac{\sin \theta_1}{(c/n_1)} = \frac{\sin \theta_2}{(c/n_2)}$$

Elimination of the common factor c yields the following relation, known as Snell's law:

Snell's Law

$$n_1 \sin \theta_1 = n_2 \sin \theta_2 \qquad 26\text{–}11$$

A typical application of Snell's law is given in the following Exercise.

EXERCISE 26–4

A beam of light in air enters **(a)** water ($n = 1.33$) or **(b)** diamond ($n = 2.42$) at an angle of 60.0° relative to the normal. Find the angle of refraction for each case.

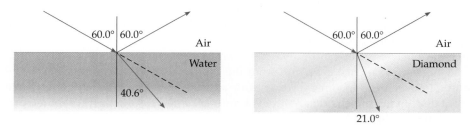

SOLUTION

Since the beam starts in air, we refer to Table 26–2 and set $n_1 = 1.000293$, or simply $n_1 = 1.00$ to three significant figures.

a. With $n_2 = 1.33$ we find

$$\theta_2 = \sin^{-1}\!\left(\frac{1.00}{n_2}\sin\theta_1\right) = \sin^{-1}\!\left(\frac{1.00}{1.33}\sin 60.0°\right) = 40.6°$$

b. Setting $n_2 = 2.42$, we get

$$\theta_2 = \sin^{-1}\!\left(\frac{1.00}{n_2}\sin\theta_1\right) = \sin^{-1}\!\left(\frac{1.00}{2.42}\sin 60.0°\right) = 21.0°$$

From the preceding Exercise we can see that the greater the difference in index of refraction between two different materials, the greater the difference in direction of propagation. In addition, light is bent closer to the normal in the medium where its speed is less. Of course, the opposite is true when light passes into a medium in which its speed is greater, as can be seen by reversing the incident and refracted rays. The qualitative features of refraction are as follows:

- When a ray of light enters a medium where the index of refraction is increased, and hence the speed of the light is *decreased*, the ray is bent *toward* the normal.
- When a ray of light enters a medium where the index of refraction is decreased, and hence the speed of the light is *increased*, the ray is bent *away* from the normal.
- There is no change in direction of propagation if there is no change in index of refraction. The greater the change in index of refraction, the greater the change in propagation direction.
- If a ray of light goes from one medium to another along the normal, it is undeflected, regardless of the index of refraction.

The last property listed follows directly from Snell's law: If θ_1 is zero, then $0 = n_2 \sin\theta_2$, which means that $\theta_2 = 0$. Refraction is explored further in the following Example.

PROBLEM-SOLVING NOTE

Applying Snell's Law

To apply Snell's law correctly, recall that the two angles in $n_1 \sin\theta_1 = n_2 \sin\theta_2$ are always measured relative to the normal at the interface. Also, note that each angle is associated with its corresponding index of refraction. For example, the angle θ_1 is the angle to the normal in the substance with an index of refraction equal to n_1.

EXAMPLE 26–5   SITTING ON A DOCK OF THE BAY

One night, while on vacation in the Caribbean, you walk to the end of a dock and, for no particular reason, shine your laser pointer into the water. When you shine the beam of light on the water a horizontal distance of 2.4 m from the dock, you see a glint of light from a shiny object on the sandy bottom—perhaps a gold doubloon. If the pointer is 1.8 m above the surface of the water, and the water is 5.5 m deep, what is the horizontal distance from the end of the dock to the shiny object?

PICTURE THE PROBLEM

The person standing at the end of the dock and the shiny object on the bottom are shown in the sketch. Note also that all the known distances are indicated, along with the angle of incidence, θ_1, and the angle of refraction, θ_2. Finally, the appropriate indices of refraction from Table 26–2 are given as well.

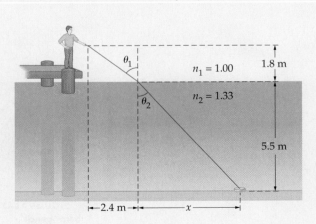

STRATEGY

We can use Snell's law and basic trigonometry to find the horizontal distance to the shiny object.

First, the information given in the problem determines the angle of incidence, θ_1. In particular, we can see from the sketch that $\tan\theta_1 = (2.4 \text{ m})/(1.8 \text{ m})$.

Second, Snell's law, $(n_1 \sin\theta_1 = n_2 \sin\theta_2)$, gives the angle of refraction, θ_2.

Finally, the sketch shows that the horizontal distance to the shiny object is 2.4 m + x. We can find the distance x from θ_2, since $\tan\theta_2 = x/(5.5 \text{ m})$.

CONTINUED ON NEXT PAGE

CONTINUED FROM PREVIOUS PAGE

SOLUTION

1. Find the angle of incidence from the information given in the problem:

$$\theta_1 = \tan^{-1}\left(\frac{2.4\ m}{1.8\ m}\right) = 53°$$

2. Use Snell's law to calculate the angle of refraction:

$$\theta_2 = \sin^{-1}\left(\frac{n_1}{n_2}\sin\theta_1\right) = \sin^{-1}\left[\left(\frac{1.00}{1.33}\right)\sin 53°\right] = 37°$$

3. Calculate x using $\tan\theta_2 = x/(5.5\ m)$:

$$\tan\theta_2 = \frac{x}{5.5\ m}$$
$$x = (5.5\ m)\tan\theta_2 = (5.5\ m)\tan 37° = 4.1\ m$$

4. Add 2.4 m to x to find the total horizontal distance to the shiny object:

$$\text{distance} = 2.4\ m + x = 2.4\ m + 4.1\ m = 6.5\ m$$

INSIGHT

Notice that if the water were to be removed, the incident beam of light would continue along its original path, characterized by the angle θ_1, and overshoot the hoped-for doubloon by a significant distance. Similarly, if you were to simply stand at the end of the dock and look out into the water at a glint of gold on the bottom, you would be looking in a direction (again characterized by θ_1) that is too high—the gold would be below your line of sight and closer to you than you think.

PRACTICE PROBLEM

If the index of refraction for water were 1.35 instead of 1.33, would the doubloon be farther from the dock, nearer the dock, or the same distance calculated above? Check your answer by calculating the distance with $n_2 = 1.35$.   [**Answer:** The doubloon would be closer to the dock. The horizontal distance in this case would be 6.4 m.]

Some related homework problems: Problem 52, Problem 56, Problem 60

REAL-WORLD PHYSICS

Apparent depth

REAL-WORLD PHYSICS

Mirages

Refraction is responsible for a number of common "optical illusions." For example, we all know that a pencil placed in a glass of water appears to be bent, though it is still perfectly straight. The cause of this illusion is shown in **Figure 26–22**, where we see that rays leaving the water bend away from the normal and make the pencil appear to be above its actual position. This is an example of what is known as *apparent depth*, in which an object appears to be closer to the water's surface than it really is. The relation between the true depth and apparent depth is considered in Problem 61.

Similarly, refraction can cause a mirage, which makes hot, dry ground in the distance appear to be covered with water. Basically, hot air near the surface is less dense—and hence has a smaller index of refraction—than the cooler air higher

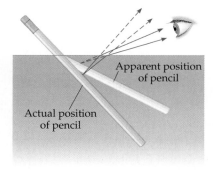

▲ **FIGURE 26–22 Refraction and the "bent" pencil**

Refraction causes a pencil to appear bent when placed in water. Note that rays leaving the water are bent away from the normal and hence extend back to a point that is higher than the actual position of the pencil.

▲ One of the most common mirages, often seen in hot weather, makes a stretch of road look like the surface of a lake. The blue color that so resembles water to our eyes is actually an image of the sky, refracted by the hot, low-density air above the road.

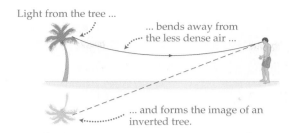

Light from the tree ...

... bends away from the less dense air ...

... and forms the image of an inverted tree.

A mirage is produced when light bends upward due to the low index of refraction of heated air near the ground.

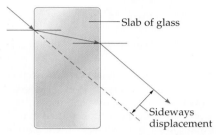

Slab of glass

Sideways displacement

▲ **FIGURE 26–24 Light propagating through a glass slab**
When a ray of light passes through a glass slab, it first refracts toward the normal, then away from the normal. The net result is that the ray continues in its original direction but is displaced sideways by a finite distance.

up. Thus, as light propagates toward the ground, it bends away from the normal until, eventually, it travels upward and enters the eye of an observer, as indicated in **Figure 26–23**. What appears to be a reflecting pool of water in the distance, then, is actually an image of the sky.

Finally, if light passes through a refracting slab, like the sheet of glass in **Figure 26–24**, it undergoes two refractions—one at each surface of the slab. The first refraction bends the light rays closer to the normal, and the second refraction bends the rays away from the normal. As can be seen in the figure, the two changes in direction cancel, so that the final direction of the light is the same as its original direction. The light has been displaced slightly, however, by an amount proportional to the thickness of the slab. The displacement distance is calculated in Problem 119.

CONCEPTUAL CHECKPOINT 26–4   REFRACTION IN A PRISM

A horizontal ray of light encounters a prism, as shown in the first diagram. After passing through the prism, is the ray **(a)** deflected upward, **(b)** still horizontal, or **(c)** deflected downward?

REASONING AND DISCUSSION
When the ray enters the prism, it is bent toward the normal, which deflects it *downward*, as shown in the second diagram. When it leaves through the opposite side of the prism, it is bent away from the normal. Because the sides of a prism are angled in opposite directions, however, bending away from the normal in the second refraction also causes a *downward* deflection.

The net result, then, is a downward deflection of the ray.

ANSWER
(c) The ray deflects downward.

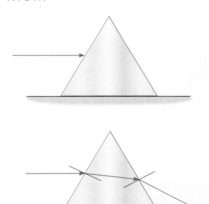

We will use the results of this Conceptual Checkpoint in the next section when we investigate the behavior of lenses. For now, we turn to two additional phenomena associated with refraction.

Total Internal Reflection

Figure 26–25 (a) shows a ray of light in water encountering a water–air interface. In such a case, it is observed that part of the light is reflected back into the water at the interface—as from the surface of a mirror—while the rest emerges into the air along a direction that is bent away from the normal according to Snell's law. If the angle of incidence is increased, as in **Figure 26–25 (b)**, the angle of refraction increases as well. At some critical angle of incidence, θ_c, the refracted beam no longer enters the air but instead is directed parallel to the water–air interface (**Figure 26–25 (c)**). In this case, the angle of refraction is 90°. For angles of incidence greater than the critical angle, as in **Figure 26–25 (d)**, it is observed that all the light is reflected back into the water. This phenomenon is referred to as **total internal reflection.**

We can find the critical angle for total internal reflection by setting $\theta_2 = 90°$ and applying Snell's law:

$$n_1 \sin \theta_c = n_2 \sin 90° = n_2$$

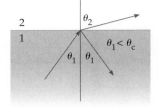

(a) Small angle of incidence

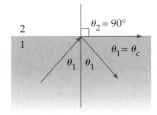

(b) Larger angle of incidence

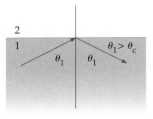

(c) Refracted beam parallel to interface

(d) Total internal reflection

▲ **FIGURE 26–25 Total internal reflection**
Total internal reflection can occur when light propagates from a region with a high index of refraction to one with a lower index of refraction. Part **(a)** shows an incident ray in medium 1 encountering the interface with medium 2, which has a lower index of refraction. A portion of the ray is transmitted to region 2 at the angle θ_2, given by Snell's law, and a portion of the ray is reflected back into medium 1 at the angle of incidence, θ_1. The sum of the intensities of the refracted and reflected rays equals the intensity of the incident ray. In part **(b)** the angle of incidence has been increased, and the refracted beam makes a smaller angle with the interface. When the angle of incidence equals the critical angle, θ_c, as in part **(c)**, the refracted ray propagates parallel to the interface. For incident angles greater than θ_c there is no refracted ray at all, as shown in part **(d)**, and all of the incident intensity goes into the reflected ray.

Therefore, the critical angle is given by the following relation:

Critical Angle for Total Internal Reflection, θ_c

$$\sin \theta_c = \frac{n_2}{n_1} \qquad\qquad 26\text{–}12$$

PROBLEM-SOLVING NOTE

Total Internal Reflection

Remember that total internal reflection can occur only on the side of an interface between two different substances that has the greater index of refraction.

Since $\sin \theta$ is always less than or equal to 1, the index of refraction, n_1, must be larger than the index of refraction n_2 if the Equation 26–12 is to give a physical solution. Thus, total internal reflection can occur only when light in one medium encounters an interface with another medium in which the speed of light is greater. For example, light moving from water to air can undergo total internal reflection, as shown in Figure 26–25, but light moving from air to water cannot.

EXAMPLE 26–6 LIGHT TOTALLY REFLECTED

Find the critical angle for light traveling from glass ($n = 1.50$) to **(a)** air ($n = 1.00$) and **(b)** water ($n = 1.33$).

PICTURE THE PROBLEM
Our sketch shows the two cases considered in the problem. Note that in each case the incident medium is glass; therefore $n_1 = 1.50$. For part (a) it follows that $n_2 = 1.00$, and for part (b) $n_2 = 1.33$.

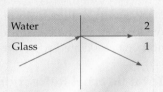

STRATEGY
The critical angle, θ_c, is defined in Equation 26–12. It is straightforward to obtain θ_c by simply substituting the appropriate indices of refraction for each case in the relation $\theta_c = \sin^{-1}(n_2/n_1)$.

SOLUTION

Part (a)

1. Solve Equation 26–12 for θ_c, using $n_1 = 1.50$ and $n_2 = 1.00$:
$$\theta_c = \sin^{-1}\left(\frac{n_2}{n_1}\right) = \sin^{-1}\left(\frac{1.00}{1.50}\right) = 41.8°$$

Part (b)

2. Solve Equation 26–12 for θ_c, using $n_1 = 1.50$ and $n_2 = 1.33$:
$$\theta_c = \sin^{-1}\left(\frac{n_2}{n_1}\right) = \sin^{-1}\left(\frac{1.33}{1.50}\right) = 62.5°$$

INSIGHT
Note that in the case of water the two indices of refraction are closer in value; hence, light escapes from glass to water over a wider range of incident angles (0° to 62.5°) than from glass to air (0° to 41.8°). In general, if the indices of refraction of two media are close in value, only light rays with large angles of incidence will undergo total internal reflection.

PRACTICE PROBLEM

Suppose the incident ray is in a different type of glass, with a glass–air critical angle of 40.0°. Is the index of refraction of this glass more than or less than 1.50? Verify your answer with a calculation.   [**Answer:** The index of refraction is greater than 1.50. The calculated value is 1.56.]

Some related homework problems: Problem 62, Problem 63, Problem 64

Total internal reflection is frequently put to practical use. For example, many binoculars contain a set of prisms—referred to as *Porro prisms*—that use total internal reflection to "fold" a relatively long light path into the short length of the binoculars, as shown in **Figure 26–26**. This allows the user to hold with ease a relatively short device that has the same optical behavior as a set of long, unwieldy telescopes. Thus, the characteristic zigzag shape of a binocular is not a fashion statement, but a reflection of its internal optical construction.

Optical fibers are another important application of total internal reflection. These thin fibers are generally composed of a glass or plastic core with a high index of refraction surrounded by an outer coating, or cladding, with a low index of refraction. Light is introduced into the core of the fiber at one end. It then propagates along the fiber in a zigzag path, undergoing one total internal reflection after another, as indicated in **Figure 26–27**. The core is so transparent that even after light propagates through a 1-km length of fiber, the amount of absorption is roughly the same as if the light had simply passed through a glass window. In addition, the total internal reflections allow the fiber to go around corners, and even to be tied into knots, and still deliver the light to the other end.

The ability of optical fibers to convey light along curved paths has been put to good use in various fields of medicine. In particular, devices known as *endoscopes* allow physicians to examine the interior of the body by snaking a flexible tube containing optical fibers into the part of the body to be examined. For example, a type of endoscope called the bronchoscope can be inserted into the nose or throat, threaded through the bronchial tubes, and eventually placed in the lungs. There, the bronchoscope delivers light through one set of fibers and returns light to the physician through another set of fibers. In some cases, the bronchoscope can even be used to retrieve small samples from the lung for further analysis. Similarly, the colonoscope can be used to examine the colon, making it one of the most important weapons in the fight against colon cancer.

Finally, optical fibers are important components of telecommunication systems. Not only are they small, light, and flexible, but they are also immune to the type of electrical interference that can degrade information carried on copper wires. Even more important, however, is that optical fibers can carry thousands of times more information than an electric current in a wire. For example, it takes

Porro prisms in binoculars

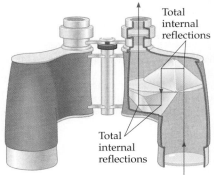

▲ **FIGURE 26–26 Porro prisms**
Prisms are used to "fold" the light path within a pair of binoculars. This makes the binoculars easier to handle.

Optical fibers and endoscopes

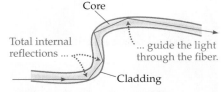

▲ **FIGURE 26–27 An optical fiber**
An optical fiber channels light along its core by a series of total internal reflections between the core and the cladding.

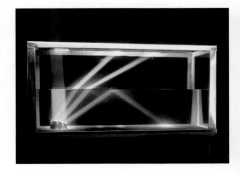

◀ At left, a beam of light enters a tank of water from above and is reflected by mirrors oriented at different angles. Most of the light in the first two beams passes through the water–air interface, undergoing refraction as it leaves the water. Only a small portion of the light is reflected back down into the tank. (The weak beams are hard to see, but the spots they make on the bottom of the tank are clearly visible.) The third beam, however, strikes the interface at an angle of incidence greater than the critical angle. As a result, all of the beam is reflected, as if the surface of the water were a mirror. This phenomenon of total internal reflection makes it possible to "pipe" light through tiny optical fibers such as the one shown at right.

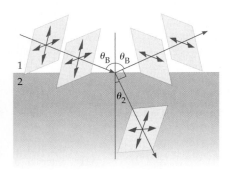

▲ **FIGURE 26–28 Brewster's angle**
When light is incident at Brewster's angle, θ_B, the reflected and refracted rays are perpendicular to each other. In addition, the reflected light is completely polarized parallel to the reflecting surface.

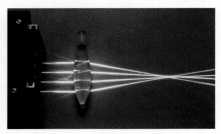

▲ The paths of light rays through a concave (diverging) lens and a convex (converging) lens.

▶ **FIGURE 26–29 A variety of converging and diverging lenses**
Converging and diverging lenses come in a variety of shapes. Generally speaking, converging lenses are thicker in the middle than at the edges, and diverging lenses are thinner in the middle. We will use double convex and double concave lenses when we present examples of converging and diverging lenses, respectively.

only a single optical fiber to transmit several television programs and tens of thousands of telephone conversations, all at the same time.

Total Polarization

As explained in Section 25–5, light reflected from a nonmetallic surface is generally polarized to some degree. For example, the light reflected from the surface of a lake is preferentially polarized in the horizontal direction. The polarization of reflected light is complete for one special angle of incidence, **Brewster's angle,** θ_B, defined as follows:

> Reflected light is completely polarized when the reflected and refracted beams are at right angles to one another. The direction of polarization is parallel to the reflecting surface.

This situation is illustrated in **Figure 26–28**. Brewster's angle is named for its discoverer, and the inventor of the kaleidoscope, the Scottish physicist Sir David Brewster (1781–1868).

To calculate Brewster's angle, we begin by applying Snell's law with an incident angle equal to θ_B:

$$n_1 \sin \theta_B = n_2 \sin \theta_2$$

Next, we note from Figure 26–28 that $\theta_B + 90° + \theta_2 = 180°$; that is, $\theta_B + \theta_2 = 90°$, or $\theta_2 = 90° - \theta_B$. Therefore, a standard trigonometric identity (Appendix A) leads to the following relation: $\sin \theta_2 = \sin(90° - \theta_B) = \cos \theta_B$. Combining this result with the preceding equation, we obtain

Brewster's Angle, θ_B

$$\tan \theta_B = \frac{n_2}{n_1} \qquad\qquad\qquad 26\text{–}13$$

We calculate Brewster's angle for a typical situation in the next Exercise.

EXERCISE 26–5

Find Brewster's angle for light reflected from the top of a glass ($n = 1.50$) coffee table.

SOLUTION
Letting $n_1 = 1.00$ and $n_2 = 1.50$, we find

$$\theta_B = \tan^{-1}\!\left(\frac{n_2}{n_1}\right) = 56.3°$$

26–6 Ray Tracing for Lenses

As we have seen, a ray of light can be redirected as it passes from one medium to another. A device that takes advantage of this effect, and uses it to focus light and form images, is a **lens.** Typically, a lens is a thin piece of glass or other transparent substance that can be characterized by the effect it has on light. In particular, converging lenses take parallel rays of light and bring them together at a focus; diverging lenses cause parallel rays to spread out as if diverging from a point source. A variety of converging and diverging lenses are illustrated in **Figure 26–29**,

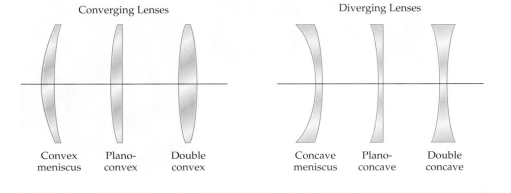

Converging Lenses			Diverging Lenses		
Convex meniscus	Plano-convex	Double convex	Concave meniscus	Plano-concave	Double concave

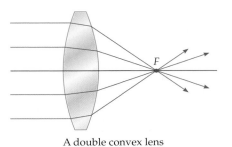
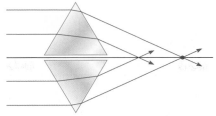

A double convex lens   Two prisms stacked "back to back"

◀ **FIGURE 26–30 A convex lens compared with a pair of prisms**

The behavior of a convex lens is similar to that of two prisms placed back to back. In both cases, light parallel to the axis is made to converge. Note that the lens, because of its curved shape, brings light to a focus at the focal point, *F*.

▲ Drops of dew can serve as double convex lenses, producing tiny, inverted images of objects beyond their focal points.

though we consider only the most basic types here—namely, the double concave (or simply concave) and the double convex (or simply convex).

First, consider a convex lens, as shown in **Figure 26–30**. To see qualitatively why such a lens is converging, note that it is similar to two prisms placed back to back. Recalling the bending of light described for a prism in Conceptual Checkpoint 26–4, we expect parallel rays of light to be brought together. In fact, lenses are shaped so that they bring parallel light to a focus at a **focal point,** *F*, along the center line, or axis, of the lens, as indicated in the figure.

Similarly, a concave lens is qualitatively the same as two prisms brought together point to point, as we see in **Figure 26–31**. In this case, parallel rays are bent away from the axis of the lens. When the diverging rays from a lens are extended back, they appear to originate at a focal point *F* on the axis of the lens.

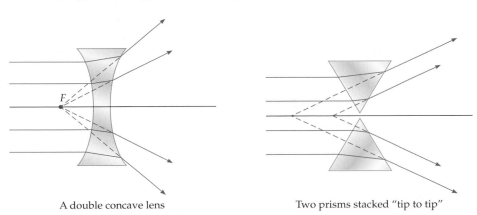

A double concave lens   Two prisms stacked "tip to tip"

▲ **FIGURE 26–31 A concave lens compared with a pair of prisms**
A concave lens and two prisms placed point to point have similar behavior. In both cases, parallel light is made to diverge.

To determine the type of image formed by a convex or concave lens, we can use ray tracing, as we did with mirrors. The three principal rays for lenses are shown in **Figures 26–32** and **26–33**. Their properties are as follows:

- The *P* ray—or parallel ray—approaches the lens parallel to its axis. The *P* ray is bent so that it passes through the focal point of a convex lens or extrapolates back to the focal point on the same side of a concave lens.

- The *F* ray (focal-point ray) on a convex lens is drawn through the focal point and on to the lens, as pictured in Figure 26–32. The lens bends the ray parallel to the axis—basically the reverse of a *P* ray. For a concave lens, the *F* ray is drawn toward the focal point *on the other side* of the lens, as in Figure 26–33. Before it gets there, however, it passes through the lens and is bent parallel to the axis.

- The midpoint ray (*M* ray) goes through the middle of the lens, which is basically like a thin slab of glass. For ideal lenses, which are infinitely thin, the *M* ray continues in its original direction with negligible displacement after passing through the lens.

To illustrate ray tracing, we start with the concave lens shown in **Figure 26–34**. Notice that the three rays originating from the top of the object extend backward

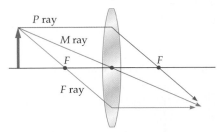

▲ **FIGURE 26–32 The three principal rays used for ray tracing with convex lenses**

The image formed by a convex lens can be found by using the rays shown here. The *P* ray propagates parallel to the principal axis until it encounters the lens, where it is refracted to pass through the focal point on the far side of the lens. The *F* ray passes through the focal point on the near side of the lens, then leaves the lens parallel to the principal axis. The *M* ray passes through the middle of the lens with no deflection. Note that in each case we consider the lens to be of negligible thickness (thin-lens approximation). Therefore, the *P* and *F* rays are depicted as undergoing refraction at the center of the lens, rather than at its front and back surfaces, and the *M* ray has zero displacement. This convention is adopted in all subsequent figures.

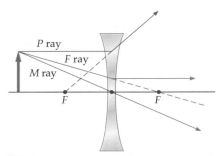

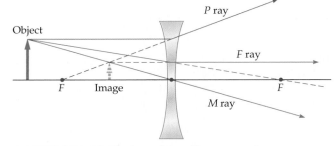

▲ **FIGURE 26–33** The three principal rays used for ray tracing with concave lenses

▲ **FIGURE 26–34** The image formed by a concave lens
Ray tracing can be used to find the image produced by a concave lens. Note that the *P*, *F*, and *M* rays all extend back to the top of the virtual image, which is upright and reduced in size.

to a single point on the left side of the lens—to an observer on the right side of the lens this point is the top of the image. Our ray diagram also shows that the image is upright and reduced in size. In addition, the image is virtual, since it is on the same side of the lens as the object. These are general features of the image formed by a concave lens.

The behavior of a convex lens is more interesting in that the type of image it forms depends on the location of the object. For example, if the object is placed beyond the focal point, as in **Figure 26–35 (a)**, the image is on the opposite side of the lens and light passes through it—it is a real image. Notice as well that the image is inverted. If the object is placed between the lens and the focal point, the result, shown in **Figure 26–35 (b)**, is an image that is virtual (on the same side as the object) and upright.

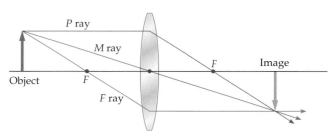

(a) Object beyond focal point *F*

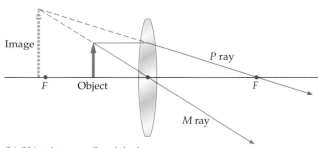

(b) Object between *F* and the lens

▲ **FIGURE 26–35** Ray tracing for a convex lens
(a) The object is beyond the focal point. The image in this case is real and inverted. **(b)** The object is between the lens and the focal point. In this case the image is virtual, upright, and enlarged.

The imaging characteristics of concave and convex lenses are summarized in Table 26–3. Compare with Table 26–1, which presents the same information for mirrors.

TABLE 26–3 Imaging Characteristics of Concave and Convex Lenses

CONCAVE LENS Object location	Image orientation	Image size	Image type
Arbitrary	Upright	Reduced	Virtual
CONVEX LENS Object location	Image orientation	Image size	Image type
Beyond *F*	Inverted	Reduced or enlarged	Real
Just beyond *F*	Inverted	Approaching infinity	Real
Just inside *F*	Upright	Approaching infinity	Virtual
Between lens and *F*	Upright	Enlarged	Virtual

Finally, the location of the focal point depends on the index of refraction of the lens as well as that of the surrounding medium. This effect is considered in the next Conceptual Checkpoint.

CONCEPTUAL CHECKPOINT 26–5 A LENS IN WATER

The lens shown in the diagram below (left) is generally used in air. If it is placed in water instead, does its focal length **(a)** increase, **(b)** decrease, or **(c)** stay the same?

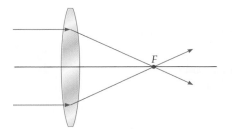 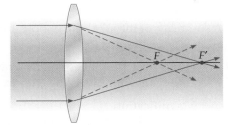

REASONING AND DISCUSSION
In water, the difference in index of refraction between the lens and its surroundings is less than when it is in air. Therefore, recalling the discussion following Exercise 26–4, we conclude that light is bent less by the lens when it is in water, as illustrated in the second diagram (right).

As a result, the focal length of the lens is increased.

This explains why our vision is so affected by immersing our eyes in water—the focusing ability of the eye is greatly altered by the water, as we can see from the diagram above. On the other hand, if we wear goggles, so that our eyes are still in contact with air, our vision is normal.

ANSWER
(a) The focal length increases.

REAL-WORLD PHYSICS: BIO
Underwater vision

26–7 The Thin-Lens Equation

To calculate the precise location and size of the image formed by a lens, we use an equation that is analogous to the mirror equation. This equation can be derived by referring to **Figure 26–36**, which shows the image produced by a convex lens, along with the P and M rays that locate the image.

First, note that the P ray creates two similar triangles on the right side of the lens in Figure 26–36 (a). Since the triangles are similar, it follows that

$$\frac{h_o}{f} = \frac{-h_i}{d_i - f}$$

26–14

In this expression, f is the **focal length**—that is, the distance from the lens to the focal point, F—and we use $-h_i$ on the right side of the equation, since h_i is negative for an inverted image. Next, the M ray forms another pair of similar triangles, shown in Figure 26–36 (b), from which we obtain the following:

$$\frac{h_o}{d_o} = \frac{-h_i}{d_i}$$

26–15

Combining these two relations, we obtain the thin-lens equation:

Thin-Lens Equation

$$\frac{1}{d_o} + \frac{1}{d_i} = \frac{1}{f}$$

26–16

Finally, the magnification, m, of the image is defined in the same way as for a mirror:

$$h_i = mh_o$$

26–17

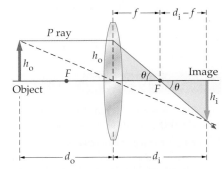

(a) Triangles to derive Equation 26–14

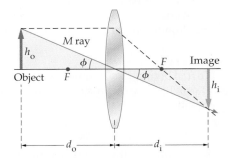

(b) Triangles to derive Equation 26–15

▲ FIGURE 26–36 Ray diagrams used to derive the thin-lens equation
(a) The two similar triangles in this case are used to obtain Equation 26–14.
(b) These similar triangles yield Equation 26–15.

Rearranging Equation 26–15, we find that $h_i = -(d_i/d_o)h_o$. Therefore, the magnification for a lens, just as for a mirror, is

Magnification, m

$$m = -\frac{d_i}{d_o}$$

26–18

As before, the sign of the magnification indicates the orientation of the image, and the magnitude gives the amount by which its size is enlarged or reduced compared with the object.

A summary of the sign conventions for lenses is as follows:

Focal Length
f is positive for converging (convex) lenses.
f is negative for diverging (concave) lenses.

Magnification
m is positive for upright images (same orientation as object).
m is negative for inverted images (opposite orientation of object).

Image Distance
d_i is positive for real images (images on the opposite side of the lens from the object).
d_i is negative for virtual images (images on the same side of the lens as the object).

Object Distance
d_o is positive for real objects (from which light diverges).
d_o is negative for virtual objects (toward which light converges).

Examples of virtual objects will be presented in Chapter 27.

We now apply the thin-lens equation and the definition of magnification to typical lens systems.

PROBLEM-SOLVING NOTE

Applying the Thin-Lens Equation

To use the thin-lens equation correctly, be careful to use the appropriate signs for all the known quantities. The final answer will also have a sign, which gives additional information about the system.

EXAMPLE 26–7 OBJECT DISTANCE AND FOCAL LENGTH

A lens produces a real image that is twice as large as the object and is located 15 cm from the lens. Find **(a)** the object distance and **(b)** the focal length of the lens.

PICTURE THE PROBLEM

Because the image is *real*, the lens must be convex, and the object must be outside the focal point, as we indicate in our sketch. [Compare with Figure 26–35 (a).] Note that the image is inverted, which means that the magnification is actually -2. In addition, the distance to the real image is given as $d_i = 15$ cm.

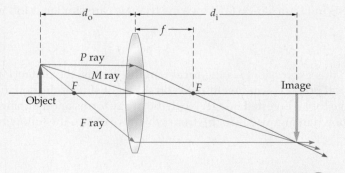

STRATEGY

To find both d_o and f requires two independent relations. One is provided by the magnification, the other by the thin-lens equation.

a. We can use the magnification, $m = -d_i/d_o$, to find the object distance, d_o. As noted before, $m = -2$ in this case.

b. We now use the values of d_i and d_o in the thin-lens equation, $1/d_o + 1/d_i = 1/f$, to find the focal length, f.

INTERACTIVE FIGURE (MP)™

SOLUTION

Part (a)

1. Use $m = -d_i/d_o$ to find the object distance, d_o:

$$m = -\frac{d_i}{d_o} = -2 \quad \text{or} \quad d_o = -\frac{d_i}{m} = -\left(\frac{15\ \text{cm}}{-2}\right) = 7.5\ \text{cm}$$

Part (b)

2. Use the thin-lens equation to find $1/f$:

$$\frac{1}{f} = \frac{1}{d_o} + \frac{1}{d_i} = \frac{1}{7.5 \text{ cm}} + \frac{1}{15 \text{ cm}} = \frac{1}{5.0 \text{ cm}}$$

3. Invert $1/f$ to find the focal length, f:

$$f = \left(\frac{1}{5.0 \text{ cm}}\right)^{-1} = 5.0 \text{ cm}$$

INSIGHT

As expected for a convex lens, the focal length is positive. In addition, note that the object distance is greater than the focal length, in agreement with both Figure 26–35 (a) and our sketch for this Example. Finally, the magnification produced by this lens is not always −2. In fact, it depends on the precise location of the object, as we see in the following Practice Problem.

PRACTICE PROBLEM

Suppose we would like to have a magnification of −3 using the same lens. **(a)** Should the object be moved closer to the lens or farther from it? **(b)** Calculate the object and image distances for this case.   [**Answer: (a)** The object should be moved closer to the lens. This moves the image farther from the lens and makes it larger. **(b)** $d_o = 6.67$ cm and $d_i = 3d_o = 20.0$ cm]

Some related homework problems: Problem 78, Problem 81, Problem 114

ACTIVE EXAMPLE 26–2 **FIND THE MAGNIFICATION**

An object is placed 12 cm in front of a diverging lens with a focal length of −7.9 cm. Find **(a)** the image distance and **(b)** the magnification.

SOLUTION *(Test your understanding by performing the calculations indicated in each step.)*

a. Use the thin-lens equation to find the image distance, d_i: $d_i = -4.8$ cm

b. Use $m = -d_i/d_o$ to find the magnification: $m = 0.40$

INSIGHT

Since the image distance is negative, it follows that the image is virtual and, hence, on the same side of the lens as the object, as expected for a concave (diverging) lens. In addition, the fact that the magnification is positive means that the image is upright. These numerical values correspond to the system illustrated in Figure 26–34.

YOUR TURN

If we would like a larger magnification, should the object be moved closer to the lens or farther from it? Calculate the object and image distances that give a magnification of 0.75.

*(Answers to **Your Turn** problems are given in the back of the book.)*

26–8 Dispersion and the Rainbow

As discussed earlier in this chapter, different materials, such as air, water, and glass, have different indices of refraction. There is more to the story, however. The index of refraction for a given material also depends on the frequency of the light—in general, the higher the frequency, the higher the index of refraction. This means, for example, that violet light—with its high frequency and large index of refraction—bends more when refracted by a given material than does red light. The result is that white light, with its mixture of frequencies, is spread out by refraction, so that different colors travel in different directions. This "spreading out" of light according to color is known as **dispersion.**

EXAMPLE 26–8 **PRISMATICS**

A flint-glass prism is made in the shape of a 30°-60°-90° triangle, as shown in the diagram. Red and violet light are incident on the prism at right angles to its vertical side. Given that the index of refraction of flint glass is 1.66 for red light and 1.70 for violet light, find the angle each ray makes with the horizontal when it emerges from the prism.

CONTINUED ON NEXT PAGE

CONTINUED FROM PREVIOUS PAGE

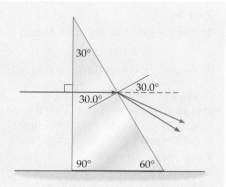

PICTURE THE PROBLEM

The prism and the red and violet rays are shown in our sketch. Note that the angle of incidence on the vertical side of the prism is 0°; hence, the angle of refraction is also 0° for both rays. On the slanted side of the prism, the rays have an angle of incidence equal to 30.0°. Their angles of refraction are different, however.

STRATEGY

To find the final angle for each ray, we apply Snell's law with the appropriate index of refraction. Note, however, that the angle of refraction is measured relative to the normal, which itself is 30.0° above the horizontal. Therefore, the angle each ray makes with the horizontal is the angle of refraction minus 30.0°.

SOLUTION

1. Apply Snell's law with $n_1 = 1.66$, $\theta_1 = 30.0°$, and $n_2 = 1.00$ to find the angle of refraction, θ_2, for red light:

$$n_1 \sin \theta_1 = n_2 \sin \theta_2$$
$$\theta_2 = \sin^{-1}\left(\frac{n_1}{n_2} \sin \theta_1\right) = \sin^{-1}\left(\frac{1.66}{1.00} \sin 30.0°\right) = 56.1°$$

2. Subtract 30.0° to find the angle relative to the horizontal:

$$56.1° - 30.0° = 26.1°$$

3. Next, apply Snell's law with $n_1 = 1.70$, $\theta_1 = 30.0°$, and $n_2 = 1.00$ to find the angle of refraction, θ_2, for violet light:

$$\theta_2 = \sin^{-1}\left(\frac{n_1}{n_2} \sin \theta_1\right) = \sin^{-1}\left(\frac{1.70}{1.00} \sin 30.0°\right) = 58.2°$$

4. Subtract 30.0° to find the angle relative to the horizontal:

$$58.2° - 30.0° = 28.2°$$

INSIGHT

This is the reason for the familiar "rainbow" of colors seen with a prism. It is also the cause of a common defect of simple lenses—referred to as chromatic aberration—in which different colors focus at different points. In the next chapter, we show how two or more lenses in combination can be used to correct this problem.

PRACTICE PROBLEM

If green light emerges from the prism at an angle of 27.0° below the horizontal, what is the index of refraction for this color of light?   [**Answer:** $n = 1.68$]

Some related homework problems: Problem 91, Problem 92

REAL-WORLD PHYSICS

The rainbow

Perhaps the most famous and striking example of dispersion is provided by the rainbow, which is caused by the dispersion of light in droplets of rain. The physical situation is illustrated in **Figure 26–37**, which shows a single drop of rain and an incident beam of sunlight. When sunlight enters the drop, it is separated into its red and violet components by dispersion, as shown. The light then reflects from the back of the drop and finally refracts and undergoes additional dispersion as it leaves the drop.

Note that the final direction of the light is almost opposite to its incident direction, falling short by only 40° to 42°, depending on the color of the light. To be specific, violet light—which is bent the most by refraction—changes its direction of propagation by 320°, so that it is 40° away from moving in the direction of the Sun. Red light, which is not bent as much as violet light, changes its direction by only 318° and hence moves in a direction that is 42° away from the Sun.

To see how the rainbow is formed in the sky, imagine standing with your back to the setting Sun, looking toward an area where rain is falling. Consider a single drop as it falls toward the ground. This drop, like all the other drops in the area, is sending out light of all colors in different directions. When the drop is at an angle of 42° above the horizontal, we see the red light coming from it, as indicated in **Figure 26–38**. As the drop continues to fall, its angle above the horizontal decreases. Eventually it reaches a height where its angle with the horizontal is 40°, at which point the violet light from the drop reaches our eye. In between, the drop has sent all the other colors of the rainbow to us for our enjoyment.

▲ A rainbow over Isaac Newton's childhood home in the manor house of Woolsthorpe, near Grantham, Lincolnshire, England. (Note the apple tree near the right side of the house.)

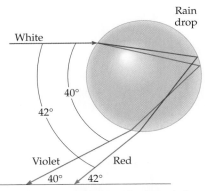

▲ **FIGURE 26–37 Dispersion in a raindrop**
White light entering a raindrop is spread out by dispersion into its various color components—like red and violet. (The angles shown in this figure are exaggerated for clarity.) This is the basic mechanism responsible for the formation of a rainbow. Rays of light that reflect twice inside the raindrop before exiting produce the "secondary" rainbow, which is above the normal, or "primary," rainbow. The sequence of colors in a secondary rainbow is reversed from that in the primary rainbow. A faint secondary bow can be seen in the photograph on the previous page.

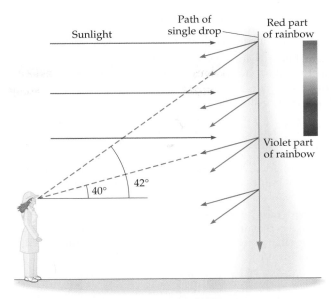

▲ **FIGURE 26–38 How rainbows are produced**
As a single drop of rain falls toward the ground, it sends all the colors of the rainbow to an observer. Note that the top of the rainbow is red; the bottom is violet. (The angles in this figure have been exaggerated for clarity.)

THE BIG PICTURE PUTTING PHYSICS IN CONTEXT

LOOKING BACK

The concepts of light rays and wave fronts relate directly to the discussion of light propagation in Chapter 25.

We make extensive use of trigonometric functions in Section 26–5, where the refraction of light is presented. Note especially the calculations in Example 26–5. Trigonometric functions were first introduced in Chapter 3 when we discussed vector components.

LOOKING AHEAD

The material on dispersion and the rainbow (Section 26–8) relates directly to chromatic aberration, one of the lens aberrations discussed in Section 27–6.

The index of refraction plays an important role in Chapter 28, when we study interference in thin films. As we shall see, the wavelength of light depends on the index of refraction of the medium in which it propagates. The wavelength, in turn, is needed to determine whether the light experiences constructive or destructive interference.

CHAPTER SUMMARY

26–1 THE REFLECTION OF LIGHT

The direction of light can be changed by reflecting it from a shiny surface.

Wave Fronts
A wave front is a surface on which the phase of a wave is constant.

Rays
The direction of wave propagation is indicated by rays, which are always at right angles to wave fronts.

Law of Reflection
The law of reflection states that the angle of reflection, θ_r, is equal to the angle of incidence, θ_i; that is, $\theta_r = \theta_i$.

Specular/Diffuse Reflection
A smooth surface reflects light in a single direction (specular reflection). A rough surface reflects light in many directions (diffuse reflection).

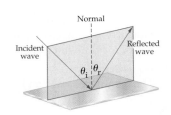

26–2 FORMING IMAGES WITH A PLANE MIRROR

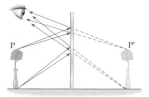

The image formed by a plane mirror has the following characteristics:

- The image is upright, but appears reversed right to left.
- The image appears to be the same distance behind the mirror that the object is in front of the mirror.
- The image is the same size as the object.

26–3 SPHERICAL MIRRORS

A spherical mirror has a spherical reflecting surface. A convex spherical mirror has a reflecting surface that bulges outward. A concave spherical mirror has a hollowed reflecting surface.

Focal Point and Focal Length for a Convex Mirror

A convex mirror reflects rays that are parallel to its principal axis so that they diverge, as if they had originated from a focal point, F, behind the mirror. The focal length of a convex mirror is

$$f = -\tfrac{1}{2}R \qquad \qquad 26\text{–}2$$

Focal Point and Focal Length for a Concave Mirror

A concave mirror reflects rays that are parallel to its principal axis so that they pass through a point known as the focal point, F. The distance from the surface of the mirror to the focal point is the focal length, f. The focal length for a mirror with a radius of curvature R is

$$f = \tfrac{1}{2}R \qquad \qquad 26\text{–}3$$

Spherical Aberration and Paraxial Rays

Paraxial rays are rays that are close to the principal axis of a mirror. Rays that are farther from the axis produce a blurred effect known as spherical aberration.

26–4 RAY TRACING AND THE MIRROR EQUATION

The location, size, and orientation of an image produced by a mirror can be found qualitatively using a ray diagram or quantitatively using a relation known as the mirror equation.

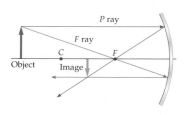

Ray Tracing

Ray tracing involves drawing two or three of the rays that have particularly simple behavior. These rays originate at a point on the object and intersect at the corresponding point on the image.

Real/Virtual Images

An image is said to be real if light passes through the apparent position of the image itself; it is virtual if light does not pass through the image.

Mirror Equation

The mirror equation relates the object distance, d_o, image distance, d_i, and focal length, f:

$$\frac{1}{d_o} + \frac{1}{d_i} = \frac{1}{f} \qquad \qquad 26\text{–}6$$

The focal length is positive for a concave mirror, negative for a convex mirror. Similarly, the image distance is positive for an image in front of the mirror and negative for an image behind the mirror.

Magnification

The magnification of an image is

$$m = -\frac{d_i}{d_o} \qquad \qquad 26\text{–}8$$

26–5 THE REFRACTION OF LIGHT

Refraction is the change in direction of light due to a change in its speed.

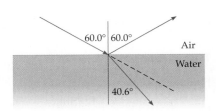

Index of Refraction
The index of refraction, n, quantifies how much a medium slows the speed of light. In particular, the speed of light in a medium is

$$v = \frac{c}{n} \qquad 26\text{--}10$$

Snell's Law
Snell's law relates the index of refraction and angle of incidence in one medium (n_1 , θ_1) to the index of refraction and angle of refraction in another medium (n_2 , θ_2):

$$n_1 \sin \theta_1 = n_2 \sin \theta_2 \qquad 26\text{--}11$$

Qualitative Properties of Refraction
Refracted light is bent closer to the normal in a medium where its speed is reduced and away from the normal in a medium where its speed is increased.

Total Internal Reflection
When light in a medium in which its speed is relatively low encounters a medium in which its speed is greater, the light will be totally reflected back into its original medium if its angle of incidence exceeds the critical angle, θ_c, given by

$$\sin \theta_c = \frac{n_2}{n_1} \qquad 26\text{--}12$$

Total Polarization
Reflected light is totally polarized parallel to the surface when the reflected and refracted rays are at right angles. This condition occurs at Brewster's angle, θ_B, given by

$$\tan \theta_B = \frac{n_2}{n_1} \qquad 26\text{--}13$$

26–6 RAY TRACING FOR LENSES

As with mirrors, ray tracing is a convenient way to determine the qualitative features of an image formed by a lens.

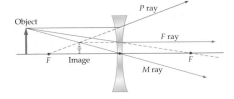

Lens
A lens is an object that uses refraction to bend light and form images.

26–7 THE THIN-LENS EQUATION

The precise location of an image formed by a lens can be obtained using the thin-lens equation, which has the same form as the mirror equation.

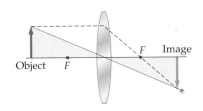

Thin-Lens Equation
The thin-lens equation relates the object distance, d_o, the image distance, d_i, and the focal length, f, for a lens:

$$\frac{1}{d_o} + \frac{1}{d_i} = \frac{1}{f} \qquad 26\text{--}16$$

Magnification
The magnification, m, of an image formed by a lens is given by the same expression used for the images produced by mirrors:

$$m = -\frac{d_i}{d_o} \qquad 26\text{--}18$$

Sign Conventions
The focal length is positive for a converging lens, negative for a diverging lens. The magnification is positive for an upright image, negative for an inverted image. The image distance is positive when the image is on the opposite side of a lens from the object, negative when it is on the same side of the lens.

26–8 DISPERSION AND THE RAINBOW

The index of refraction depends on frequency—generally, the higher the frequency, the higher the index of refraction. This difference in index of refraction causes light of different colors to be refracted in different directions (dispersion). Rainbows are caused by dispersion within raindrops.

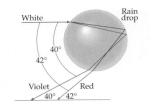

PROBLEM-SOLVING SUMMARY

Type of Problem	Relevant Physical Concepts	Related Examples
Find the approximate location and size of the image produced by a spherical mirror.	Begin by drawing the three principal rays, P, F, and C, from the top of the object. The intersection of these rays gives the location of the top of the image.	Example 26–3
Determine precise values for the location and size of the image produced by a spherical mirror.	The mirror equation, $1/d_o + 1/d_i = 1/f$, relates the object distance, d_o, the image distance, d_i, and the focal length, f. The magnification of the image is given by $m = -(d_i/d_o)$.	Example 26–4 Active Example 26–1
Relate the angles of incidence and refraction.	The angles of incidence and refraction as a beam of light passes from one medium to another are related by Snell's law: $n_1 \sin \theta_1 = n_2 \sin \theta_2$.	Example 26–5
Find the critical angle for total internal reflection.	A light beam in a medium with an index of refraction n_1 may undergo total internal reflection when it encounters a second medium with an index of refraction $n_2 < n_1$. The critical angle at which total internal reflection begins is given by the relation $\sin \theta_c = n_2/n_1$.	Example 26–6
Determine precise values for the location and size of the image produced by a convex or concave lens.	The thin-lens equation, $1/d_o + 1/d_i = 1/f$, relates the object distance, d_o, the image distance, d_i, and the focal length, f. The magnification of the image is given by $m = -(d_i/d_o)$.	Example 26–7 Active Example 26–2

CONCEPTUAL QUESTIONS

For instructor-assigned homework, go to www.masteringphysics.com

(Answers to odd-numbered Conceptual Questions can be found in the back of the book.)

1. Two plane mirrors meet at right angles at the origin, as indicated in **Figure 26–39**. Suppose an L-shaped object has the position and orientation labeled A. Draw the location and orientation of *all* the images of object A formed by the two mirrors.

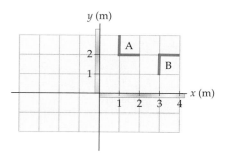

▲ **FIGURE 26–39** Conceptual Questions 1 and 2

2. Two plane mirrors meet at right angles at the origin, as indicated in Figure 26–39. Suppose an L-shaped object has the position and orientation labeled B. Draw the location and orientation of *all* the images of object B formed by the two mirrors.

3. What is the radius of curvature of a plane mirror? What is its focal length? Explain.

4. Dish receivers for satellite TV always use the concave side of the dish, never the convex side. Explain.

5. Suppose you would like to start a fire by focusing sunlight onto a piece of paper. In Conceptual Checkpoint 26–2 we saw that a concave mirror would be better than a convex mirror for this purpose. At what distance from the mirror should the paper be held for best results?

6. When light propagates from one medium to another, does it always bend toward the normal? Explain.

7. A swimmer at point B in **Figure 26–40** needs help. Two lifeguards depart simultaneously from their tower at point A, but they follow different paths. Although both lifeguards run with equal speed on the sand and swim with equal speed in the water, the lifeguard who follows the longer path, ACB, arrives at point B before the lifeguard who follows the shorter, straight-line path from A to B. Explain.

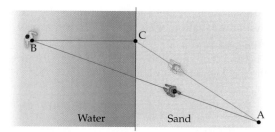

▲ **FIGURE 26–40** Conceptual Question 7

8. When you observe a mirage on a hot day, what are you actually seeing when you gaze at the "pool of water" in the distance?

9. Explain the difference between a virtual and a real image.

10. Sitting on a deserted beach one evening, you watch as the last bit of the Sun approaches the horizon. Just before the Sun disappears from sight, is the top of the Sun actually above or below the horizon? That is, if Earth's atmosphere could be instantly removed just before the Sun disappeared, would the Sun still be visible, or would it be below the horizon? Explain.

11. A large, empty coffee mug sits on a table. From your vantage point the bottom of the mug is not visible. When the mug is filled with water, however, you *can* see the bottom of the mug. Explain.

12. **The Disappearing Eyedropper** The accompanying photograph shows eyedroppers partially immersed in oil (left) and water (right). Explain why the dropper is invisible in the oil.

What happened to the dropper?
(Conceptual Questions 12 and 13)

13. The Invisible Man In the H. G. Wells novel *The Invisible Man*, a person becomes invisible by altering his index of refraction to match that of air. This is the idea behind the disappearing eyedropper in Conceptual Question 12. If the invisible man could actually do this, would he be able to see? Explain.

14. What's the Secret? The top of **Figure 26–41** shows the words SECRET CODE written in different colors. If you place a cylindrical rod of glass or plastic just above the words, you find that SECRET appears inverted, but CODE does not. Explain.

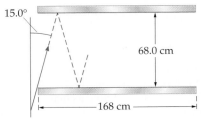

▲ **FIGURE 26–41** Conceptual Question 14

PROBLEMS AND CONCEPTUAL EXERCISES

Note: Answers to odd-numbered Problems and Conceptual Exercises can be found in the back of the book. **IP** *denotes an integrated problem, with both conceptual and numerical parts;* **BIO** *identifies problems of biological or medical interest;* **CE** *indicates a conceptual exercise.* **Predict/Explain** *problems ask for two responses:* **(a)** *your prediction of a physical outcome, and* **(b)** *the best explanation among three provided. On all problems, red bullets (•, ••, •••) are used to indicate the level of difficulty.*

(The outside medium is assumed to be air, with an index of refraction of 1.00, unless specifically stated otherwise.)

SECTION 26–1 THE REFLECTION OF LIGHT

1. • A laser beam is reflected by a plane mirror. It is observed that the angle between the incident and reflected beams is 28°. If the mirror is now rotated so that the angle of incidence increases by 5.0°, what is the new angle between the incident and reflected beams?

2. •• The reflecting surfaces of two mirrors form a vertex with an angle of 120°. If a ray of light strikes mirror 1 with an angle of incidence of 55°, find the angle of reflection of the ray when it leaves mirror 2.

3. •• A ray of light reflects from a plane mirror with an angle of incidence of 37°. If the mirror is rotated by an angle θ, through what angle is the reflected ray rotated?

4. •• **IP** A small vertical mirror hangs on the wall, 1.40 m above the floor. Sunlight strikes the mirror, and the reflected beam forms a spot on the floor 2.50 m from the wall. Later in the day, you notice that the spot has moved to a point 3.75 m from the wall. **(a)** Were your two observations made in the morning or in the afternoon? Explain. **(b)** What was the change in the Sun's angle of elevation between your two observations?

5. •• Sunlight enters a room at an angle of 32° above the horizontal and reflects from a small mirror lying flat on the floor. The reflected light forms a spot on a wall that is 2.0 m behind the mirror, as shown in **Figure 26–42**. If you now place a pencil under the edge of the mirror nearer the wall, tilting it upward by 5.0°, how much higher on the wall (Δy) is the spot?

6. •• You stand 1.50 m in front of a wall and gaze downward at a small vertical mirror mounted on it. In this mirror you can see the reflection of your shoes. If your eyes are 1.85 m above your feet, through what angle should the mirror be tilted for you to see your eyes reflected in the mirror? (The location of the mirror remains the same, only its angle to the vertical is changed.)

7. •• **IP** Standing 2.3 m in front of a small vertical mirror, you see the reflection of your belt buckle, which is 0.72 m below your eyes. **(a)** What is the vertical location of the mirror relative to the level of your eyes? **(b)** What angle do your eyes make with the horizontal when you look at the buckle? **(c)** If you now move backward until you are 6.0 m from the mirror, will you still see the buckle, or will you see a point on your body that is above or below the buckle? Explain.

8. •• How many times does the light beam shown in **Figure 26–43** reflect from **(a)** the top and **(b)** the bottom mirror?

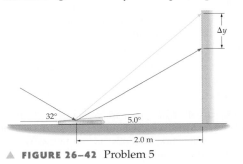

15.0°

68.0 cm

168 cm

▲ **FIGURE 26–43** Problems 8 and 102

SECTION 26–2 FORMING IMAGES WITH A PLANE MIRROR

9. • **CE** If you view a clock in a mirror, do the hands rotate clockwise or counterclockwise?

Δy

32° 5.0°

2.0 m

▲ **FIGURE 26–42** Problem 5

Which way do the hands go? (Problem 9)

10. • A 12.5-foot-long, nearsighted python is stretched out perpendicular to a plane mirror, admiring its reflected image. If the greatest distance to which the snake can see clearly is 26.0 ft, how close must its head be to the mirror for it to see a clear image of its tail?

11. •• **(a)** How rapidly does the distance between you and your mirror image decrease if you walk directly toward a mirror with a speed of 2.6 m/s? **(b)** Repeat part (a) for the case in which you walk toward a mirror but at an angle of 38° to its normal.

12. •• You are 1.9 m tall and stand 3.2 m from a plane mirror that extends vertically upward from the floor. On the floor 1.5 m in front of the mirror is a small table, 0.80 m high. What is the minimum height the mirror must have for you to be able to see the top of the table in the mirror?

13. •• The rear window in a car is approximately a rectangle, 1.3 m wide and 0.30 m high. The inside rearview mirror is 0.50 m from the driver's eyes, and 1.50 m from the rear window. What are the minimum dimensions for the rearview mirror if the driver is to be able to see the entire width and height of the rear window in the mirror without moving her head?

14. •• **IP** You hold a small plane mirror 0.50 m in front of your eyes, as shown in **Figure 26–44** (not to scale). The mirror is 0.32 cm high, and in it you see the image of a tall building behind you. **(a)** If the building is 95 m behind you, what vertical height of the building, H, can be seen in the mirror at any one time? **(b)** If you move the mirror closer to your eyes, does your answer to part (a) increase, decrease, or stay the same? Explain.

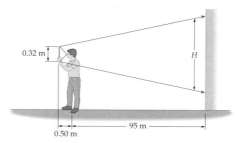

▲ **FIGURE 26–44** Problems 14 and 113

15. •• Two rays of light converge toward each other, as shown in **Figure 26–45**, forming an angle of 27°. Before they intersect, however, they are reflected from a circular plane mirror with a diameter of 11 cm. If the mirror can be moved horizontally to

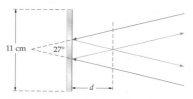

▲ **FIGURE 26–45** Problem 15

the left or right, what is the greatest possible distance d from the mirror to the point where the reflected rays meet?

16. •• For a corner reflector to be effective, its surfaces must be precisely perpendicular. Suppose the surfaces of a corner reflector left on the Moon's surface by the Apollo astronauts formed a 90.001° angle with each other. If a laser beam is bounced back to Earth from this reflector, how far (in kilometers) from its starting point will the reflected beam strike Earth? For simplicity, assume the beam reflects from only two sides of the reflector, and that it strikes the first surface at precisely 45°.

SECTION 26–3 SPHERICAL MIRRORS

17. • **CE** Astronomers often use large mirrors in their telescopes to gather as much light as possible from faint distant objects. Should the mirror in their telescopes be concave or convex? Explain.

18. • A section of a sphere has a radius of curvature of 0.86 m. If this section is painted with a reflective coating on both sides, what is the focal length of **(a)** the convex side and **(b)** the concave side?

19. • A mirrored-glass gazing globe in a garden is 31.9 cm in diameter. What is the focal length of the globe?

20. • Sunlight reflects from a concave piece of broken glass, converging to a point 15 cm from the glass. What is the radius of curvature of the glass?

SECTION 26–4 RAY TRACING AND THE MIRROR EQUATION

21. • **CE** You hold a shiny tablespoon at arm's length and look at the back side of the spoon. **(a)** Is the image you see of yourself upright or inverted? **(b)** Is the image enlarged or reduced? **(c)** Is the image real or virtual?

22. • **CE** You hold a shiny tablespoon at arm's length and look at the front side of the spoon. **(a)** Is the image you see of yourself upright or inverted? **(b)** Is the image enlarged or reduced? **(c)** Is the image real or virtual?

23. • **CE** An object is placed in front of a convex mirror whose radius of curvature is R. What is the greatest distance behind the mirror that the image can be formed?

24. • **CE** An object is placed to the left of a concave mirror, beyond its focal point. In which direction will the image move when the object is moved farther to the left?

25. • **CE** An object is placed to the left of a convex mirror. In which direction will the image move when the object is moved farther to the left?

26. • A small object is located 30.0 cm in front of a concave mirror with a radius of curvature of 40.0 cm. Where will the image be formed?

27. • Use ray diagrams to show whether the image formed by a convex mirror increases or decreases in size as an object is brought closer to the mirror's surface.

28. • An object with a height of 46 cm is placed 2.4 m in front of a concave mirror with a focal length of 0.50 m. **(a)** Determine the approximate location and size of the image using a ray diagram. **(b)** Is the image upright or inverted?

29. • Find the location and magnification of the image produced by the mirror in Problem 28 using the mirror and magnification equations.

30. • An object with a height of 46 cm is placed 2.4 m in front of a convex mirror with a focal length of −0.50 m. **(a)** Determine the approximate location and size of the image using a ray diagram. **(b)** Is the image upright or inverted?

31. • Find the location and magnification of the image produced by the mirror in Problem 30 using the mirror and magnification equations.

32. •• During a daytime football game you notice that a player's reflective helmet forms an image of the Sun 4.8 cm behind the surface of the helmet. What is the radius of curvature of the helmet, assuming it to be roughly spherical?

33. •• **IP** A magician wishes to create the illusion of a 2.74-m-tall elephant. He plans to do this by forming a virtual image of a 50.0-cm-tall model elephant with the help of a spherical mirror. **(a)** Should the mirror be concave or convex? **(b)** If the model must be placed 3.00 m from the mirror, what radius of curvature is needed? **(c)** How far from the mirror will the image be formed?

34. •• A person 1.7 m tall stands 0.66 m from a reflecting globe in a garden. **(a)** If the diameter of the globe is 18 cm, where is the image of the person, relative to the surface of the globe? **(b)** How large is the person's image?

35. •• Shaving/makeup mirrors typically have one flat and one concave (magnifying) surface. You find that you can project a magnified image of a lightbulb onto the wall of your bathroom if you hold the mirror 1.8 m from the bulb and 3.5 m from the wall. **(a)** What is the magnification of the image? **(b)** Is the image erect or inverted? **(c)** What is the focal length of the mirror?

36. •• **The Hale Telescope** The 200-inch-diameter concave mirror of the Hale telescope on Mount Palomar has a focal length of 16.9 m. An astronomer stands 20.0 m in front of this mirror. **(a)** Where is her image located? Is it in front of or behind the mirror? **(b)** Is her image real or virtual? How do you know? **(c)** What is the magnification of her image?

37. •• A concave mirror produces a virtual image that is three times as tall as the object. **(a)** If the object is 28 cm in front of the mirror, what is the image distance? **(b)** What is the focal length of this mirror?

38. •• A concave mirror produces a real image that is three times as large as the object. **(a)** If the object is 22 cm in front of the mirror, what is the image distance? **(b)** What is the focal length of this mirror?

39. •• The virtual image produced by a convex mirror is one-quarter the size of the object. **(a)** If the object is 36 cm in front of the mirror, what is the image distance? **(b)** What is the focal length of this mirror?

40. •• **IP** A 5.7-ft-tall shopper in a department store is 17 ft from a convex security mirror. The shopper notices that his image in the mirror appears to be only 6.4 in. tall. **(a)** Is the shopper's image upright or inverted? Explain. **(b)** What is the mirror's radius of curvature?

41. •• You view a nearby tree in a concave mirror. The inverted image of the tree is 3.8 cm high and is located 7.0 cm in front of the mirror. If the tree is 23 m from the mirror, what is its height?

42. ••• A shaving/makeup mirror produces an erect image that is magnified by a factor of 2.2 when your face is 25 cm from the mirror. What is the mirror's radius of curvature?

43. ••• A concave mirror with a focal length of 36 cm produces an image whose distance from the mirror is one-third the object distance. Find the object and image distances.

SECTION 26-5 THE REFRACTION OF LIGHT

44. • **CE Predict/Explain** When a ray of light enters a glass lens surrounded by air, it slows down. **(a)** As it leaves the glass, does its speed increase, decrease, or stay the same? **(b)** Choose the *best explanation* from among the following:

   I. Its speed increases because the ray is now propagating in a medium with a smaller index of refraction.

   II. The speed decreases because the speed of light decreases whenever light moves from one medium to another.

   III. The speed will stay the same because the speed of light is a universal constant.

45. • **CE Samurai Fishing** A humorous scene in Akira Kurosawa's classic film *The Seven Samurai* shows the young samurai Kikuchiyo wading into a small stream and plucking a fish from it for his dinner. **(a)** As Kikuchiyo looks through the water to the fish, does he see it in the general direction of point 1 or point 2 in **Figure 26–46**? **(b)** If the fish looks up at Kikuchiyo, does it see Kikuchiyo's head in the general direction of point 3 or point 4?

▲ **FIGURE 26–46** Problem 45

46. • **CE** When color A and color B are sent through a prism, color A is bent more than color B. Which color travels more rapidly in the prism? Explain.

47. • **CE Day Versus Night** **(a)** Imagine for a moment that the Earth has no atmosphere. Over the period of a year, is the number of daylight hours at your home greater than, less than, or equal to the number of nighttime hours? **(b)** Repeat part (a), only this time take into account the Earth's atmosphere.

48. • **CE Predict/Explain** A kitchen has twin side-by-side sinks. One sink is filled with water, the other is empty. **(a)** Does the sink with water appear to be deeper, shallower, or the same depth as the empty sink? **(b)** Choose the *best explanation* from among the following:

   I. The sink with water appears deeper because you have to look through the water to see the bottom.

   II. Water bends the light, making an object under the water appear to be closer to the surface. Thus the water-filled sink appears shallower.

   III. The sinks are identical, and therefore have the same depth. This doesn't change by putting water in one of them.

49. • **CE** A light beam undergoes total internal reflection at the interface between medium A, in which it propagates, and medium B, on the other side of the interface. Which medium has the greater index of refraction? Explain.

50. • Light travels a distance of 0.960 m in 4.00 ns in a given substance. What is the index of refraction of this substance?

51. • Find the ratio of the speed of light in water to the speed of light in diamond.

52. • **Ptolemy's Optics** One of the many works published by the Greek astronomer Ptolemy (A.D. ca. 100–170) was *Optics*. In this book Ptolemy reports the results of refraction experiments he conducted by observing light passing from air into water. His results are as follows: angle of incidence = 10.0°, angle of refraction = 8.00°; angle of incidence = 20.0°, angle of refraction = 15.5°. Find the percentage error in the calculated index of refraction of water for each of Ptolemy's measurements.

53. • Light enters a container of benzene at an angle of 43° to the normal; the refracted beam makes an angle of 27° with the normal. Calculate the index of refraction of benzene.

54. • The angle of refraction of a ray of light traveling into an ice cube from air is 38°. Find the angle of incidence.

55. • **IP (a)** Referring to Problem 54, suppose the ice melts, but the angle of refraction remains the same. Is the corresponding angle of incidence greater than, less than, or the same as it was for ice? Explain. **(b)** Calculate the angle of incidence for part (a).

56. •• A submerged scuba diver looks up toward the calm surface of a freshwater lake and notes that the Sun appears to be 35° from the vertical. The diver's friend is standing on the shore of the lake. At what angle above the horizon does the friend see the sun?

57. •• A pond with a total depth (ice + water) of 3.25 m is covered by a transparent layer of ice, with a thickness of 0.38 m. Find the time required for light to travel vertically from the surface of the ice to the bottom of the pond.

58. •• Light is refracted as it travels from a point A in medium 1 to a point B in medium 2. If the index of refraction is 1.33 in medium 1 and 1.51 in medium 2, how long does it take light to go from A to B, assuming it travels 331 cm in medium 1 and 151 cm in medium 2?

59. •• You have a semicircular disk of glass with an index of refraction of $n = 1.52$. Find the incident angle θ for which the beam of light in **Figure 26–47** will hit the indicated point on the screen.

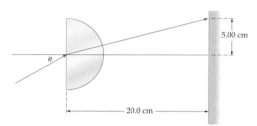

5.00 cm

θ

20.0 cm

▲ **FIGURE 26–47** Problems 59 and 66

60. •• The observer in **Figure 26–48** is positioned so that the far edge of the bottom of the empty glass (not to scale) is just visible. When the glass is filled to the top with water, the center of the bottom of the glass is just visible to the observer. Find the height, H, of the glass, given that its width is $W = 6.2$ cm.

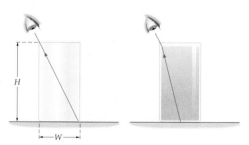

H

W

▲ **FIGURE 26–48** Problem 60

61. •• A coin is lying at the bottom of a pool of water that is 6.5 feet deep. Viewed from directly above the coin, how far below the surface of the water does the coin appear to be? (The coin is assumed to be small in diameter; therefore, we can use the small-angle approximations $\sin \theta \cong \tan \theta \cong \theta$.)

62. •• A ray of light enters the long side of a 45°-90°-45° prism and undergoes two total internal reflections, as indicated in **Figure 26–49**. The result is a reversal in the ray's direction of propagation. Find the minimum value of the prism's index of refraction, n, for these internal reflections to be total.

90°

45°   45°

▲ **FIGURE 26–49** Problems 62 and 63

63. •• When the prism in Problem 62 is immersed in a fluid with an index of refraction of 1.21, the internal reflections shown in Figure 26–49 are still total. The reflections are no longer total, however, when the prism is immersed in a fluid with $n = 1.43$. Use this information to set upper and lower limits on the possible values of the prism's index of refraction.

64. •• **IP** A glass paperweight with an index of refraction n rests on a desk, as shown in **Figure 26–50**. An incident ray of light enters the horizontal top surface of the paperweight at an angle $\theta = 77°$ to the vertical. **(a)** Find the minimum value of n for which the internal reflection on the vertical surface of the paperweight is total. **(b)** If θ is decreased, is the minimum value of n increased or decreased? Explain.

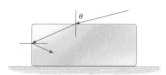

θ

▲ **FIGURE 26–50** Problems 64 and 65

65. •• **IP** Suppose the glass paperweight in Figure 26–50 has an index of refraction $n = 1.38$. **(a)** Find the value of θ for which the reflection on the vertical surface of the paperweight exactly satisfies the condition for total internal reflection. **(b)** If θ is increased, is the reflection at the vertical surface still total? Explain.

66. •• **IP** Consider the physical system shown in Figure 26–47 and described in Problem 59. **(a)** If the index of refraction of the glass is increased, will the desired value of θ increase or decrease? Explain. **(b)** Find the value of θ for the case of flint glass ($n = 1.66$).

67. •• While studying physics at the library late one night, you notice the image of the desk lamp reflected from the varnished tabletop. When you turn your Polaroid sunglasses sideways, the reflected image disappears. If this occurs when the angle between the incident and reflected rays is 110°, what is the index of refraction of the varnish?

68. ••• A horizontal beam of light enters a 45°-90°-45° prism at the center of its long side, as shown in **Figure 26–51**. The emerging ray moves in a direction that is 34° below the horizontal. What is the index of refraction of this prism?

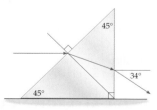

45°

34°

45°

▲ **FIGURE 26–51** Problems 68 and 123

69. ••• A laser beam enters one of the sloping faces of the equilateral glass prism ($n = 1.42$) in **Figure 26–52** and refracts through the prism. Within the prism the light travels horizontally. What is the angle θ between the direction of the incident ray and the direction of the outgoing ray?

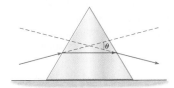

▲ **FIGURE 26–52** Problems 69 and 92

SECTION 26–6   RAY TRACING FOR LENSES

70. • **(a)** Use a ray diagram to determine the approximate location of the image produced by a concave lens when the object is at a distance $\frac{1}{2}|f|$ from the lens. **(b)** Is the image upright or inverted? **(c)** Is the image real or virtual? Explain.

71. • **(a)** Use a ray diagram to determine the approximate location of the image produced by a concave lens when the object is at a distance $2|f|$ from the lens. **(b)** Is the image upright or inverted? **(c)** Is the image real or virtual? Explain.

72. • An object is a distance $f/2$ from a convex lens. **(a)** Use a ray diagram to find the approximate location of the image. **(b)** Is the image upright or inverted? **(c)** Is the image real or virtual? Explain.

73. • An object is a distance $2f$ from a convex lens. **(a)** Use a ray diagram to find the approximate location of the image. **(b)** Is the image upright or inverted? **(c)** Is the image real or virtual? Explain.

74. •• Two lenses that are 35 cm apart are used to form an image, as shown in **Figure 26–53**. Lens 1 is converging and has a focal length $f_1 = 14$ cm; lens 2 is diverging and has a focal length $f_2 = -7.0$ cm. The object is placed 24 cm to the left of lens 1. **(a)** Use a ray diagram to find the approximate location of the image. **(b)** Is the image upright or inverted? **(c)** Is the image real or virtual? Explain.

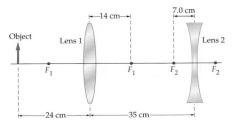

▲ **FIGURE 26–53** Problems 74 and 83

75. •• Two lenses that are 35 cm apart are used to form an image, as shown in **Figure 26–54**. Lens 1 is diverging and has a focal length $f_1 = -7.0$ cm; lens 2 is converging and has a focal length $f_2 = 14$ cm. The object is placed 24 cm to the left of lens 1. **(a)** Use a ray diagram to find the approximate location of the image. **(b)** Is the image upright or inverted? **(c)** Is the image real or virtual? Explain.

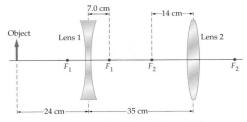

▲ **FIGURE 26–54** Problems 75 and 84

SECTION 26–7   THE THIN-LENS EQUATION

76. • A convex lens is held over a piece of paper outdoors on a sunny day. When the paper is held 26 cm below the lens, the sunlight is focused on the paper and the paper ignites. What is the focal length of the lens?

77. • A concave lens has a focal length of -32 cm. Find the image distance and magnification that result when an object is placed 29 cm in front of the lens.

78. • When an object is located 46 cm to the left of a lens, the image is formed 17 cm to the right of the lens. What is the focal length of the lens?

79. • An object with a height of 2.54 cm is placed 36.3 mm to the left of a lens with a focal length of 35.0 mm. **(a)** Where is the image located? **(b)** What is the height of the image?

80. •• A lens for a 35-mm camera has a focal length given by $f = 55$ mm. **(a)** How close to the film should the lens be placed to form a sharp image of an object that is 5.0 m away? **(b)** What is the magnification of the image on the film?

81. •• **IP** An object is located to the left of a convex lens whose focal length is 34 cm. The magnification produced by the lens is $m = 3.0$. **(a)** To increase the magnification to 4.0, should the object be moved closer to the lens or farther away? Explain. **(b)** Calculate the distance through which the object should be moved.

82. •• **IP** You have two lenses at your disposal, one with a focal length $f_1 = +40.0$ cm, the other with a focal length $f_2 = -40.0$ cm. **(a)** Which of these two lenses would you use to project an image of a lightbulb onto a wall that is far away? **(b)** If you want to produce an image of the bulb that is enlarged by a factor of 2.00, how far from the wall should the lens be placed?

83. •• **(a)** Determine the distance from lens 1 to the final image for the system shown in Figure 26–53. **(b)** What is the magnification of this image?

84. •• **(a)** Determine the distance from lens 1 to the final image for the system shown in Figure 26–54. **(b)** What is the magnification of this image?

85. •• **IP** An object is located to the left of a concave lens whose focal length is -34 cm. The magnification produced by the lens is $m = \frac{1}{3}$. **(a)** To decrease the magnification to $m = \frac{1}{4}$, should the object be moved closer to the lens or farther away? **(b)** Calculate the distance through which the object should be moved.

86. •• **IP BIO** Albert is nearsighted, and without his eyeglasses he can focus only on objects less than 2.2 m away. **(a)** Are Albert's eyeglasses concave or convex? Explain. **(b)** To correct Albert's nearsightedness, his eyeglasses must produce a virtual, upright image at a distance of 2.2 m when viewing an infinitely distant object. What is the focal length of Albert's eyeglasses?

87. •• A small insect viewed through a convex lens is 1.4 cm from the lens and appears twice its actual size. What is the focal length of the lens?

88. ••• **IP** A friend tells you that when he takes off his eyeglasses and holds them 23 cm above a printed page the image of the print is erect but reduced to 0.67 of its actual size. **(a)** Is the image real or virtual? How do you know? **(b)** What is the focal length of your friend's glasses? **(c)** Are the lenses in the glasses concave or convex? Explain.

89. ••• **IP** A friend tells you that when she takes off her eyeglasses and holds them 23 cm above a printed page the image of the print is erect but enlarged to 1.5 times its actual size. **(a)** Is the image real or virtual? How do you know? **(b)** What is the focal length of your friend's glasses? **(c)** Are the lenses in the glasses concave or convex? Explain.

SECTION 26–8 DISPERSION AND THE RAINBOW

90. • **CE Predict/Explain** You take a picture of a rainbow with an infrared camera, and your friend takes a picture at the same time with visible light. **(a)** Is the height of the rainbow in the infrared picture greater than, less than, or the same as the height of the rainbow in the visible-light picture? **(b)** Choose the *best explanation* from among the following:
   I. The height will be greater because the top of a rainbow is red, and so infrared light would be even higher.
   II. The height will be less because infrared light is below the visible spectrum.
   III. A rainbow is the same whether seen in visible light or infrared; therefore the height is the same.

91. •• The index of refraction for red light in a certain liquid is 1.320; the index of refraction for violet light in the same liquid is 1.332. Find the dispersion $(\theta_v - \theta_r)$ for red and violet light when both are incident on the flat surface of the liquid at an angle of 45.00° to the normal.

92. •• A horizontal incident beam consisting of white light passes through an equilateral prism, like the one shown in Figure 26–52. What is the dispersion $(\theta_v - \theta_r)$ of the outgoing beam if the prism's index of refraction is $n_v = 1.505$ for violet light and $n_r = 1.421$ for red light?

93. •• The focal length of a lens is inversely proportional to the quantity $(n - 1)$, where n is the index of refraction of the lens material. The value of n, however, depends on the wavelength of the light that passes through the lens. For example, one type of flint glass has an index of refraction of $n_r = 1.572$ for red light and $n_v = 1.605$ in violet light. Now, suppose a white object is placed 24.00 cm in front of a lens made from this type of glass. If the red light reflected from this object produces a sharp image 55.00 cm from the lens, where will the violet image be found?

GENERAL PROBLEMS

94. • **CE Jurassic Park** A *T. rex* chases the heroes of Steven Spielberg's *Jurassic Park* as they desperately try to escape in their Jeep. The *T. rex* is closing in fast, as they can see in the outside rearview mirror. Near the bottom of the mirror they also see the following helpful message: OBJECTS IN THE MIRROR ARE CLOSER THAN THEY APPEAR. Is this mirror concave or convex? Explain.

95. • **CE** The receiver for a dish antenna is placed in front of the concave surface of the dish. If the radius of curvature of the dish is R, how far in front of the dish should the receiver be placed? Explain.

96. • **CE Predict/Explain** If a lens is immersed in water, its focal length changes, as discussed in Conceptual Checkpoint 26–5. **(a)** If a spherical mirror is immersed in water, does its focal length increase, decrease, or stay the same? **(b)** Choose the *best explanation* from among the following:
   I. The focal length will increase because the water will cause more bending of light.
   II. Water will refract the light. This, combined with the reflection due to the mirror, will result in a decreased focal length.
   III. The focal length stays the same because it depends on the fact that the angle of incidence is equal to the angle of reflection for a mirror. This is unaffected by the presence of the water.

97. • **CE Predict/Explain** A glass slab surrounded by air causes a sideways displacement in a beam of light. **(a)** If the slab is now placed in water, does the displacement it causes increase, decrease, or stay the same? **(b)** Choose the *best explanation* from among the following:

   I. The displacement of the beam increases because of the increased refraction due to the water.
   II. The displacement of the beam is decreased because with water surrounding the slab there is a smaller difference in index of refraction between the slab and its surroundings.
   III. The displacement stays the same because it is determined only by the properties of the slab; in particular, the material it is made of and its thickness.

98. • **CE** Referring to Conceptual Question 12, suppose the same type of glass used in an eyedropper is made into a convex lens with a focal length f. If this lens is immersed in the oil of the bottle on the left in the photo, will its focal length be 0, $f/2$, $2f$, or ∞? (*Hint:* See Conceptual Checkpoint 26–5.)

99. • **CE** Two identical containers are filled with different transparent liquids. The container with liquid A appears to have a greater depth than the container with liquid B. Which liquid has the greater index of refraction? Explain.

100. •• **CE** Is the image you see in a three-dimensional corner reflector upright or inverted?

101. •• **CE Inverse Lenses** Suppose we mold a hollow piece of plastic into the shape of a double concave lens. The "lens" is watertight, and its interior is filled with air. We now place this lens in water and shine a beam of light on it. **(a)** Does the lens converge or diverge the beam of light? Explain. **(b)** If our hollow lens is double convex instead, does it converge or diverge a beam of light when immersed in water? Explain.

102. •• **IP** Suppose the separation between the two mirrors in Figure 26–43 is increased by moving the top mirror upward. **(a)** Will this affect the number of reflections made by the beam of light? If so, how? **(b)** What is the total number of reflections made by the beam of light when the separation between the mirrors is 145 cm?

103. •• Standing 2.0 m in front of a small vertical mirror you see the reflection of your belt buckle, which is 0.70 m below your eyes. If you remain 2.0 m from the mirror but climb onto a stool, how high must the stool be to allow you to see your knees in the mirror? Assume that your knees are 1.2 m below your eyes.

104. •• **IP Apparent Size of Floats in a Termometro Lentos** The Galileo thermometer, or Termometro Lentos (slow thermometer in Italian), consists of a vertical, cylindrical flask containing a fluid and several glass floats of different color. The floats all have the same dimensions, but they appear to differ in size depending on their location within the cylinder. **(a)** Does a float near the front surface of the cylinder (the surface closest to you) appear to be larger or smaller than a float near the back surface? **(b)** Figure 26–55 shows a ray diagram for a float near the front surface of the cylinder. Draw a ray diagram for a float at the center of the cylinder, and show that the change in apparent size agrees with your answer to part (a).

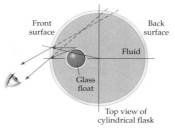

▲ **FIGURE 26–55** Problem 104

105. •• **(a)** Find the two locations where an object can be placed in front of a concave mirror with a radius of curvature of 39 cm such that its image is twice its size. **(b)** In each of these cases, state whether the image is real or virtual, upright or inverted.

106. •• A convex mirror with a focal length of −85 cm is used to give a truck driver a view behind the vehicle. **(a)** If a person who is 1.7 m tall stands 2.2 m from the mirror, where is the person's image located? **(b)** Is the image upright or inverted? **(c)** What is the size of the image?

107. •• **IP** The three laser beams shown in **Figure 26–56** meet at a point at the back of a solid, transparent sphere. **(a)** What is the index of refraction of the sphere? **(b)** Is there a finite index of refraction that will make the three beams come to a focus at the center of the sphere? If your answer is yes, give the required index of refraction; if your answer is no, explain why not.

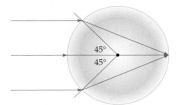

▲ **FIGURE 26–56** Problem 107

108. •• The speed of light in substance A is x times greater than the speed of light in substance B. Find the ratio n_A/n_B in terms of x.

109. •• **IP** A film of oil, with an index of refraction of 1.48 and a thickness of 1.50 cm, floats on a pool of water, as shown in **Figure 26–57**. A beam of light is incident on the oil at an angle of 60.0° to the vertical. **(a)** Find the angle θ the light beam makes with the vertical as it travels through the water. **(b)** How does your answer to part (a) depend on the thickness of the oil film? Explain.

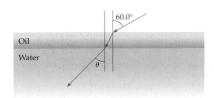

▲ **FIGURE 26–57** Problems 109, 110, and 111

110. •• **IP** Consider the physical system shown in Figure 26–57. For this problem we assume that the angle of incidence at the air–oil interface can be varied from 0° to 90°. **(a)** What is the maximum possible value for θ, the angle of refraction in the water? **(b)** If an oil with a larger index of refraction is used, does your answer to part (a) increase or decrease? Explain.

111. •• **IP** Consider the physical system shown in Figure 26–57, only this time let the direction of the light rays be reversed. **(a)** Find the angle of incidence θ at the water–oil interface such that the condition for total internal reflection at the oil–air surface is exactly satisfied. **(b)** If θ is decreased, is the reflection at the oil–air interface still total? Explain.

112. •• **Figure 26–58** shows a ray of light entering one end of an optical fiber at an angle of incidence $\theta_i = 50.0°$. The index of refraction of the fiber is 1.62. **(a)** Find the angle θ the ray makes with the normal when it reaches the curved surface of the fiber. **(b)** Show that the internal reflection from the curved surface is total.

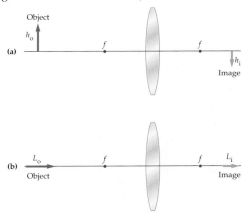

▲ **FIGURE 26–58** Problems 112 and 120

113. •• Suppose the person's eyes in Figure 26–44 are 1.6 m above the ground and that the small plane mirror can be moved up or down. **(a)** Find the height of the bottom of the mirror such that the lowest point the person can see on the building is 19.6 m above the ground. **(b)** With the mirror held at the height found in part (a), what is the highest point on the building the person can see?

114. •• An arrow 2.00 cm long is located 75.0 cm from a lens that has a focal length $f = 30.0$ cm. **(a)** If the arrow is perpendicular to the principal axis of the lens, as in **Figure 26–59 (a)**, what is its lateral magnification, defined as h_i/h_o? **(b)** Suppose, instead, that the arrow lies *along* the principal axis, extending from 74.0 cm to 76.0 cm from the lens, as indicated in **Figure 26–59 (b)**. What is the longitudinal magnification of the arrow, defined as L_i/L_o? (*Hint*: Use the thin-lens equation to locate the image of each end of the arrow.)

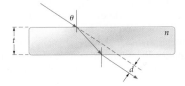

▲ **FIGURE 26–59** Problems 114 and 115

115. •• Repeat Problem 114, this time for a diverging lens with a focal length $f = -30.0$ cm.

116. ••• A convex lens with $f_1 = 20.0$ cm is mounted 40.0 cm to the left of a concave lens. When an object is placed 30.0 cm to the left of the convex lens, a real image is formed 60.0 cm to the right of the concave lens. What is the focal length f_2 of the concave lens?

117. ••• Two thin lenses, with focal lengths f_1 and f_2, are placed in contact. What is the effective focal length of the double lens?

118. ••• When an object is placed a distance d_o in front of a curved mirror, the resulting image has a magnification m. Find an expression for the focal length of the mirror, f, in terms of d_o and m.

119. ••• **A Slab of Glass** Give a symbolic expression for the sideways displacement d of a light ray passing through the slab of glass shown in **Figure 26–60**. The thickness of the glass is t, its index of refraction is n, and the angle of incidence is θ.

▲ **FIGURE 26–60** Problem 119

120. ••• Referring to Figure 26–58, show that the internal reflection from the curved surface of the fiber is always total for any incident angle θ_i, provided the index of refraction of the fiber exceeds $\sqrt{2}$.

121. ••• **Least Time** A beam of light propagates from point A in medium 1 to point B in medium 2, as shown in **Figure 26–61**. The index of refraction is different in these two media; therefore, the light follows a refracted path that obeys Snell's law. **(a)** Calculate the time required for light to travel from A to B along the refracted path. **(b)** Compare the time found in part (a) with the time it takes for light to travel from A to B along a straight-line path. (Note that the time on the straight-line path is longer than the time on the refracted path. In general, the shortest time between two points in different media is along the path given by Snell's law.)

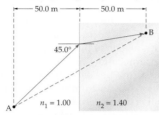

▲ **FIGURE 26–61** Problem 121

122. ••• The ray of light shown in **Figure 26–62** passes from medium 1 to medium 2 to medium 3. The index of refraction in medium 1 is n_1, in medium 2 it is $n_2 > n_1$, and in medium 3 it is $n_3 > n_2$. Show that medium 2 can be ignored when calculating the angle of refraction in medium 3; that is, show that $n_1 \sin \theta_1 = n_3 \sin \theta_3$.

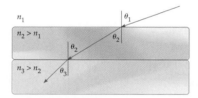

▲ **FIGURE 26–62** Problem 122

123. ••• **IP** A beam of light enters the sloping side of a 45°-90°-45° glass prism with an index of refraction $n = 1.66$. The situation is similar to that shown in Figure 26–51, except that the angle of incidence of the incoming beam can be varied. **(a)** Find the angle of incidence for which the reflection on the vertical side of the prism exactly satisfies the condition for total internal reflection. **(b)** If the angle of incidence is increased, is the reflection at the vertical surface still total? Explain. **(c)** What is the minimum value of n such that a horizontal beam like that in Figure 26–51 undergoes total internal reflection at the vertical side of the prism?

PASSAGE PROBLEMS

The Focal Length of a Lens

A number of factors play a role in determining the focal length of a lens. First and foremost is the shape of the lens. As a general rule, a lens that is thicker in the middle will converge light, a lens that is thinner in the middle will diverge light.

Another important factor is the index of refraction of the lens material, n_{lens}. For example, imagine comparing two lenses with identical shapes but made of different materials. The lens with the larger index of refraction bends light more, bringing it to a focus in a shorter distance. As a result, a larger index of refraction implies a smaller focal length. In fact, the focal length of a lens surrounded by air ($n = 1$) is given by the **lens maker's formula:**

$$\frac{1}{f_{\text{in air}}} = (n_{\text{lens}} - 1)\left(\frac{1}{R_1} - \frac{1}{R_2}\right)$$

In this expression R_1 and R_2 are the radii of curvature of the front and back surfaces of the lens, respectively. For given values of R_1 and R_2—that is, for a given shape—the focal length of the lens becomes smaller as the index of refraction increases.

A lens is not always surrounded by air, however. More generally, the fluid in which the lens is immersed may have an index of refraction given by n_{fluid}. In this case, the focal length is given by

$$\frac{1}{f_{\text{in fluid}}} = \left(\frac{n_{\text{lens}} - n_{\text{fluid}}}{n_{\text{fluid}}}\right)\left(\frac{1}{R_1} - \frac{1}{R_2}\right)$$

It follows, then, that the focal lengths of a lens surrounded by air or by a general fluid are related by

$$f_{\text{in fluid}} = \left[\frac{(n_{\text{lens}} - 1)n_{\text{fluid}}}{n_{\text{lens}} - n_{\text{fluid}}}\right]f_{\text{in air}}$$

This relation shows that the surrounding fluid can change the magnitude of the focal length, or even cause it to become infinite. The fluid can also change the sign of the focal length, which determines whether the lens is diverging or converging.

124. • A converging lens with a focal length in air of $f = +5.25$ cm is made from ice. What is the focal length of this lens if it is immersed in benzene? (Refer to Table 26–2.)

   A. −20.7 cm   **B.** −18.1 cm

   C. −12.8 cm   **D.** −11.2 cm

125. • A diverging lens with $f = -12.5$ cm is made from ice. What is the focal length of this lens if it is immersed in ethyl alcohol? (Refer to Table 26–2.)

   A. 102 cm   **B.** 105 cm

   C. 118 cm   **D.** 122 cm

126. • Calculate the focal length of a lens in water, given that the index of refraction of the lens is $n_{\text{lens}} = 1.52$ and its focal length in air is 25.0 cm. (Refer to Table 26–2.)

   A. 57.8 cm   **B.** 66.0 cm

   C. 91.0 cm   **D.** 104 cm

127. • Suppose a lens is made from fused quartz (glass), and that its focal length in air is −7.75 cm. What is the focal length of this lens if it is immersed in benzene? (Refer to Table 26–2.)

   A. −130 cm   **B.** 134 cm

   C. 141 cm   **D.** −145 cm

INTERACTIVE PROBLEMS

128. •• **Referring to Example 26–3** Suppose the radius of curvature of the mirror is 5.0 cm. **(a)** Find the object distance that gives an upright image with a magnification of 1.5. **(b)** Find the object distance that gives an inverted image with a magnification of −1.5.

129. •• **IP Referring to Example 26–3** An object is 4.5 cm in front of the mirror. **(a)** What radius of curvature must the mirror have if the image is to be 2.2 cm in front of the mirror? **(b)** What is the magnification of the image? **(c)** If the object is moved closer to the mirror, does the magnification of the image increase in magnitude, decrease in magnitude, or stay the same?

130. •• **Referring to Example 26–7 (a)** What object distance is required to give an image with a magnification of +2.0? Assume that the focal length of the lens is +5.0 cm. **(b)** What is the location of the image in this case?

131. •• **IP Referring to Example 26–7** Suppose the convex lens is replaced with a concave lens with a focal length of −5.0 cm. **(a)** Where must the object be placed to form an image with a magnification of 0.50? **(b)** What is the location of the image in this case? **(c)** If we now move the object closer to the lens, does the magnification of the image increase, decrease, or stay the same?

27 Optical Instruments

Large modern astronomical telescopes are seldom used for visual observations. More often, the light they capture is directed to a scientific instrument, such as a camera, spectrograph, or semiconducting device. When a human observer is involved, it's often within the telescope rather than behind it. The astronomer in this photo is actually perched above the 200-inch mirror of the great Hale reflecting telescope on Mount Palomar, at the "prime focus" of the telescope. This chapter explores how mirrors and lenses are used to create a variety of optical devices, from telescopes that let us scan the cosmos, to microscopes that give us access to the world of the very small—and even to eyeglasses that help us read a newspaper.

Human vision can be aided in a number of ways by the practical application of optics. For example, a pair of glasses or contact lenses can correct a person's faulty eyesight to produce normal vision. Optics can also extend the abilities of human sight by magnifying very small objects to a visible size or by making distant objects seem near at hand. In this chapter, after discussing the optics of a normal eye and the closely related behavior of a camera, we consider a variety of optical instruments designed to either correct or extend our abilities to see the natural world.

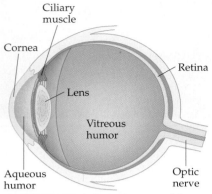

▲ **FIGURE 27–1 Basic elements of the human eye**

Light enters the eye through the cornea and the lens. It is focused onto the retina by the ciliary muscles, which change the shape of the lens.

REAL-WORLD PHYSICS: BIO

Optical properties of the human eye

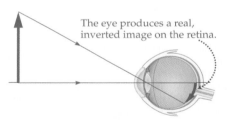

▲ **FIGURE 27–2 Image production in the eye**

27–1 The Human Eye and the Camera

The Human Eye

The eye is a marvelously sensitive and versatile optical instrument, allowing us to observe objects as distant as the stars or as close as the book in our hands. Perhaps more amazing is the fact that the eye can accomplish all this even though its basic structure is that of a spherical bag of water 2.5 cm in diameter. The slight differences that set the eye apart from a bag of water make all the difference in terms of optical performance, however.

The fundamental elements of an eye are illustrated in **Figure 27–1**. Basically, light enters the eye through the transparent outer coating of the eye, the *cornea*, and then passes through the *aqueous humor*, the adjustable *lens*, and the jellylike *vitreous humor* before reaching the light-sensitive *retina* at the back of the eye, as illustrated in **Figure 27–2**. The retina is covered with millions of small structures known as *rods* and *cones*, which, when stimulated by light, send electrical impulses along the *optic nerve* to the brain. How the nerve impulses are processed, so that we interpret the upside-down image on the retina as a right-side-up object, is another story altogether; here we concentrate on the optical properties of the eye.

To begin, we note that most of the refraction needed to produce an image occurs at the cornea, as light first enters the eye. The reason is that the difference in index of refraction is greater at the air–cornea interface than at any interface within the eye. Specifically, the index of refraction of air is $n = 1.00$, whereas that of the cornea is about $n = 1.38$, just slightly greater than the index of refraction of water ($n = 1.33$). When light passes from the cornea to the aqueous humor, the index of refraction changes from $n = 1.38$ to $n = 1.33$. Next, light encounters the lens ($n = 1.40$) and then the vitreous humor ($n = 1.34$) before arriving at the retina.

In the end, the lens accounts for only about a quarter of the total refraction produced by the eye—but it is a crucial contribution nonetheless. By altering the shape of the lens with the *ciliary muscles*, we are able to change the precise amount of refraction the lens produces, which, in turn, changes its focal length. Specifically, when we view a distant object, our ciliary muscles are *relaxed*, as shown in **Figure 27–3 (a)**, allowing the lens to be relatively flat. As a result, it causes little refraction and its focal length is at its greatest. When we view a nearby object, the lens must shorten its focal length and cause more refraction, as shown in **Figure 27–3 (b)**. Thus, the ciliary muscles *tense* to give the lens a greater curvature. The process of changing the shape of the lens, and hence adjusting its focal length, is referred to as **accommodation.** Producing the proper accommodation is no easy feat for a newborn but is automatic for an adult.

The fact that the ciliary muscles must be tensed to focus on nearby objects means that our eyes can "tire" from muscular strain. That is why it is beneficial to pause occasionally from reading and to look off into the distance. Viewing distant

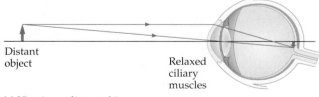

(a) Viewing a distant object

▶ **FIGURE 27–3 Accommodation in the human eye**

(a) When the eye is viewing a distant object, the ciliary muscles are relaxed and the focal length of the lens is at its greatest. **(b)** When the eye is focusing on a near object, the ciliary muscles are tensed, changing the shape and reducing the focal length of the lens.

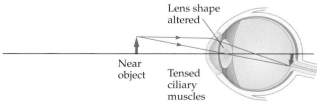

(b) Viewing a near object

objects allows the ciliary muscles to relax, thus reducing the strain on our eyes to a minimum.

The lens can be distorted only so much, however; hence, there is a limit to how close the eye can focus. The shortest distance at which a sharp focus can be obtained is the **near point**—anything closer will appear fuzzy no matter how hard we try to focus on it. For young people the near-point distance, N, is typically about 25 cm, but it increases with age. Persons 40 years of age may experience a near point that is 40 cm from the eye, and in later years the near point may move to 500 cm or more. The extension of the near point to greater distances with age is referred to as *presbyopia* (or "short arm" syndrome) and is due to the lens becoming less flexible. Thus, as one ages it is not uncommon to have to move a piece of paper away from the eyes in order to focus, and eventually reading glasses may be necessary.

At the other end of the scale, the **far point** is the greatest distance an object can be from the eye and still be in focus. Since we can focus on the Moon and stars, it is clear that the normal far point is essentially infinity.

Finally, the amount of light that reaches the retina is controlled by a colored diaphragm called the *iris*. As the iris expands or contracts it adjusts the size of the *pupil*, the opening through which light enters the eye. In bright light the pupil closes down to about 1 mm in diameter. On the darkest nights, the dark-adapted pupil can open up to a diameter of about 7.0 mm.

▲ The pigmented iris of the human eye responds automatically to changing levels of illumination, dilating the pupil in dim light and contracting it in bright light.

EXAMPLE 27–1 JOURNEY TO NEAR POINT

The near-point distance of a given eye is $N = 25$ cm. Treating the eye as if it were a single thin lens a distance 2.5 cm from the retina, find the focal length of the lens when it is focused on an object **(a)** at the near point and **(b)** at infinity. (Typical values for the effective lens–retina distance range from 1.7 cm to 2.5 cm.)

PICTURE THE PROBLEM
The eye, and the simplified thin-lens equivalent, are shown in the sketch. Note that the horizontal axis is broken in order to bring the object and eye into the same sketch. The object and image distances are indicated as well.

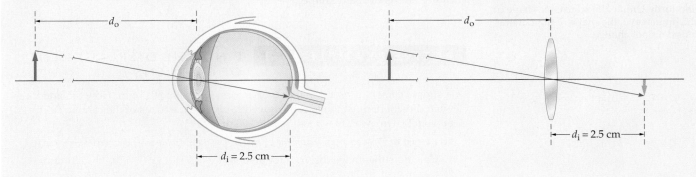

STRATEGY
The focal length can be found using the thin-lens equation, $1/d_o + 1/d_i = 1/f$. The image distance is $d_i = 2.5$ cm for both (a) and (b). For part (a) the object distance is $d_o = 25$ cm; for part (b) the object distance is $d_o = \infty$.

SOLUTION

Part (a)

1. Substitute $d_o = 25$ cm and $d_i = 2.5$ cm into the thin-lens equation and solve for f:

$$\frac{1}{f} = \frac{1}{d_o} + \frac{1}{d_i} = \frac{1}{25\text{ cm}} + \frac{1}{2.5\text{ cm}} = 0.44\text{ cm}^{-1}$$

$$f = \frac{1}{0.44\text{ cm}^{-1}} = 2.3\text{ cm}$$

Part (b)

2. Substitute $d_o = \infty$ and $d_i = 2.5$ cm into the thin-lens equation and solve for f:

$$\frac{1}{f} = \frac{1}{d_o} + \frac{1}{d_i} = \frac{1}{\infty} + \frac{1}{2.5\text{ cm}} = \frac{1}{2.5\text{ cm}}$$

$$f = 2.5\text{ cm}$$

CONTINUED ON NEXT PAGE

CONTINUED FROM PREVIOUS PAGE

INSIGHT

Note that the effective focal length of the eye changes by only about 2 mm in changing focus from the near point to infinity. Thus, if the shape of the eye is changed even slightly—so that the distance from the lens to the retina is increased or decreased by a couple millimeters—the eye will no longer be able to function properly. We return to this point in the next section when we discuss near- and farsightedness.

PRACTICE PROBLEM

At what object distance is the effective focal length of the eye equal to 2.4 cm?   [**Answer:** $d_o = 60$ cm]

Some related homework problems: Problem 4, Problem 5

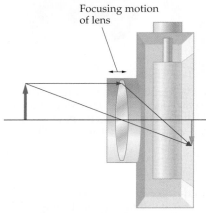

Focusing motion of lens

▲ **FIGURE 27–4 Basic elements of a camera**

A camera forms a real, inverted image on photographic film or an electronic sensor, and is focused by moving the lens back and forth. Unlike the adjustable shape of the human eye, the shape of the camera lens does not change.

Perhaps you have noticed that when you are looking at a light-colored background, or a clear sky, one or more "spots" may float across your field of vision. Many people have these "floaters." In fact, some people are occasionally fooled by one of these spots into thinking a fly is buzzing about their head. For this reason, these floaters are called *muscae volitantes*, which means, literally, "flying flies." Muscae volitantes are caused by cells, cell fragments, or other small impurities suspended either in the vitreous humor of the eye or in the lens itself. These are usually harmless but can be symptoms of a detached retina.

The Camera

A simple camera, such as the one illustrated in **Figure 27–4**, operates in much the same way as the eye. In particular, the lens of the camera forms a real, inverted image on a light-sensitive material—which in this case is either photographic film or, more commonly these days, the charge-coupled device (CCD) in a digital camera. The focusing mechanism is different, however. To focus a digital camera at different distances the lens is moved either toward or away from the CCD. Thus, the eye focuses by changing the shape of a stationary lens; the camera focuses by moving a lens of fixed shape.

ACTIVE EXAMPLE 27–1 FIND THE DISPLACEMENT OF THE LENS

A simple camera uses a thin lens with a focal length of 50.0 mm. How far, and in what direction, must the lens be moved to change the focus of the camera from a person 20.0 m away to a person only 3.00 m away?

SOLUTION *(Test your understanding by performing the calculations indicated in each step.)*

1. Calculate the image distance for an object distance of 20.0 m: $d_{i1} = 5.01$ cm

2. Calculate the image distance for an object distance of 3.00 m: $d_{i2} = 5.08$ cm

3. Find the difference in image distance: 0.07 cm = 0.7 mm

INSIGHT

Since the image distance is greater for the person at 3.00 m, it follows that the lens must be moved *away* from the film by 0.7 mm to change the focus the desired amount. Note how little displacement is required.

YOUR TURN

Suppose the lens is moved an additional 0.7 mm away from the film. At what distance is the camera focused now?

*(Answers to **Your Turn** problems are given in the back of the book.)*

REAL-WORLD PHYSICS

Speed and aperture settings on a camera

The aperture of a camera is analogous to the pupil of an eye, and like the pupil, its size can be adjusted—the greater the size, the more light that is available to make an image. Photographers often characterize the size of the aperture with a dimensionless quantity called the **f-number,** which is defined as follows:

$$f\text{-number} = \frac{\text{focal length}}{\text{diameter of aperture}} = \frac{f}{D} \qquad 27\text{–}1$$

Notice that the larger the diameter of the aperture, the smaller the f-number; in fact, $D = f/(f\text{-number})$.

Professional-grade cameras have aperture settings indicated by a sequence of f-numbers such as the following:

$$2, \quad 2.8, \quad 4, \quad 5.6, \quad 8, \quad 11, \quad 16$$

For example, a camera lens with a focal length of 50.0 mm and an aperture setting of 4 has an aperture diameter of $D = (50.0 \text{ mm})/4 = 12.5$ mm. This is often referred to by photographers as an $f/4$ setting for the aperture. Similarly, turning the aperture ring to the $f/2$ setting opens the aperture to a diameter of $(50.0 \text{ mm})/2 = 25.0$ mm. Since the aperture is a circular opening, the area through which light enters the camera, $A = \pi D^2/4$, varies as the square of the aperture diameter D and thus inversely as the square of the f-number.

The amount of light that falls on the photographic film or CCD in a camera is also controlled by the shutter speed. A shutter speed of 1/500, for example, means that the shutter is open for only 1/500 of a second—very effective for "freezing" the motion of a high-speed object. Changing the shutter speed to 1/250 doubles the time the shutter is open and, hence, doubles the light received by the film. Typical camera shutter speeds are 1/1000, 1/500, 1/250, 1/125, and so on. (On professional-grade cameras, these speeds are indicated on a dial as 1000, 500, 250, 125 to save space.)

Suppose, for example, that a photograph receives the proper amount of light when the shutter speed is 1/500 and the aperture f-number is 4. If the photographer decides to take a second shot with a shutter speed of 1/250, what f-number is required to maintain the correct exposure? Since the slower shutter speed doubles the light entering the camera, the area of the aperture must be halved to compensate. To halve the area, the diameter must be *reduced* by a factor of $\sqrt{2} \approx 1.4$, which means the f-number must be *increased* by a factor of 1.4. Hence, the photographer must change the aperture setting to $4 \times (1.4) = 5.6$. Note that each of the f-numbers listed earlier is larger than the previous one by a factor of roughly 1.4, leading to a factor of 2 difference in the light received by the film.

27–2 Lenses in Combination and Corrective Optics

Whereas a normal eye can provide sharp images for objects over a wide range of distances, some eyes have difficulty focusing on distant objects, and others are unable to focus as close as the normal near point. Problems such as these can be corrected with glasses or contact lenses used in combination with the eye's lens. In general, a combination of lenses can have beneficial properties not possible with a single lens, as we shall see throughout the rest of this chapter. First, in this section, we discuss how to analyze a system with more than one lens; we then apply these results to correcting near- and farsightedness.

The basic operating principle for a system consisting of more than one lens is the following:

> The *image* produced by one lens acts as the *object* for the next lens.
> This is true regardless of whether the image produced by the first lens
> is real or virtual, or whether it is in front of or behind the second lens.

This principle finds many applications, since most optical instruments—like cameras, microscopes, and telescopes—use a number of lenses to produce the desired results.

As an example of a lens system, consider the two lenses shown in **Figure 27–5 (a)**. An object is 20.0 cm to the left of the convex lens, which is 50.0 cm from a concave lens to its right. Given that the focal lengths of the convex and concave lenses are

(a) Original object

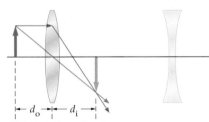

(b) Image from first lens

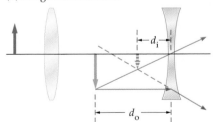

(c) Image from second lens

▲ **FIGURE 27–5 A two-lens system**
In this system, a convex and a concave lens are separated by 50.0 cm. **(a)** An object is placed 20.0 cm to the left of the convex lens, whose focal length is 10.0 cm. **(b)** The image formed by the convex lens is 20.0 cm to its right. This image is the object for the concave lens. **(c)** The object for this lens is 30.0 cm to its left. Because the focal length of this lens is −12.5 cm, it forms an image 8.82 cm to its left. This is the final image of the system.

10.0 cm and −12.5 cm, respectively, we would like to find the location and orientation of the image produced by the two lenses acting together.

The first step is illustrated in **Figure 27–5 (b)**, where we determine the image formed by the convex lens using a ray diagram. The precise location of "image 1" is given by the thin-lens equation:

$$\frac{1}{d_o} + \frac{1}{d_i} = \frac{1}{f} \quad \text{or} \quad \frac{1}{20.0 \text{ cm}} + \frac{1}{d_i} = \frac{1}{10.0 \text{ cm}}$$

$$d_i = 20.0 \text{ cm}$$

Note that image 1 is inverted and 20.0 cm to the right of the convex lens.

The next step is to note that image 1 is 50.0 cm − 20.0 cm = 30.0 cm to the left of the concave lens. Considering image 1 to be the object for the concave lens, we obtain the ray diagram shown in **Figure 27–5 (c)**. The thin-lens equation, with $d_o = 30.0$ cm and $f = -12.5$ cm, yields the following image distance:

$$\frac{1}{d_o} + \frac{1}{d_i} = \frac{1}{f} \quad \text{or} \quad \frac{1}{30.0 \text{ cm}} + \frac{1}{d_i} = \frac{1}{-12.5 \text{ cm}}$$

$$d_i = -8.82 \text{ cm}$$

Thus the final image of the two-lens system is 8.82 cm to the left of the concave lens. The orientation and size of the final image can be found as follows:

> The total magnification produced by a lens system is equal to the *product* of the magnifications produced by each lens individually.

Using Equation 26–18, $m = -d_i/d_o$, we find that the first lens produces magnification $m_1 = -(20.0 \text{ cm})/(20.0 \text{ cm}) = -1$. Similarly, the second lens causes a magnification of $m_2 = -(-8.82 \text{ cm})/(30.0 \text{ cm}) = 0.294$. The total magnification of the system, then, is $m = m_1 m_2 = -0.294$, showing that the final image is inverted and reduced in size by a factor of 0.294 compared with the original object. This value is in agreement with the results shown in the ray diagrams of Figure 27–5.

The following Active Example considers the effect of reversing the order of the two lenses.

PROBLEM-SOLVING NOTE

Finding the Image for a Multilens System

To find the final image of a multilens system, consider the lenses one at a time. In particular, find the image for the first lens, then use that image as the object for the next lens, and so on.

ACTIVE EXAMPLE 27–2   FIND THE FINAL IMAGE

Find the location and orientation of the final image produced by the system shown in Figure 27–5 if the positions of the two lenses are reversed.

SOLUTION *(Test your understanding by performing the calculations indicated in each step.)*

1. Use the thin-lens equation to find the image distance for the concave lens, with $d_{o1} = 20.0$ cm: $d_{i1} = -7.69$ cm

2. Find the object distance for the convex lens: $d_{o2} = 57.7$ cm

3. Use this object distance and the thin-lens equation to find the location of the final image: $d_{i2} = 12.1$ cm

INSIGHT

The final image is 12.1 cm to the right of the convex lens. The magnification produced by the first lens is $m_1 = -d_{i1}/d_{o1} = 0.385$, and that produced by the second lens is $m_2 = -d_{i2}/d_{o2} = -0.210$. Hence, the total magnification is $m = m_1 m_2 = -0.0809$, showing that the final image is reduced in size by a factor of 0.0809 and inverted. Notice how different the results are when we simply reverse the order of the lenses—just like looking through the wrong end of a pair of binoculars.

YOUR TURN

Consider again the system shown in Figure 27–5. If we replace the concave lens with a concave mirror whose focal length is $f = 12.5$ cm, what are the location and magnification of the final image?

*(Answers to **Your Turn** problems are given in the back of the book.)*

We now show how a two-lens system (the eye plus an external lens) can correct abnormal vision.

Nearsightedness

When a person with normal vision relaxes the ciliary muscles of the eye, an object at infinity is in focus. In a nearsighted (myopic) person, however, a totally relaxed eye focuses only out to a finite distance from the eye—the far point. Thus a person with this condition is said to be nearsighted because objects near the eye can be focused, whereas objects beyond the far point are fuzzy.

The problem in this situation is that the eye converges the light coming into it in too short a distance—in other words, the focal length of the eye is less than the distance from the lens to the retina. This condition is illustrated in **Figure 27–6**, where we see that an object at infinity forms an image in front of the retina, because of the elongation of the eye. The effect need not be large; as we saw in the previous section, an elongation of only a millimeter or two is enough to cause a problem.

To correct this condition, we need to "undo" some of the excess convergence produced by the eye, so that images again fall on the retina. This can be done by placing a *diverging* lens in front of the eye. Specifically, consider an object at infinity—which would ordinarily appear blurry to a nearsighted person. If a concave lens with the proper focal length produces a virtual image of this object at the nearsighted person's far point, as in **Figure 27–7**, the person's relaxed eye can now focus on the object. We consider this situation in the next Example.

PROBLEM-SOLVING NOTE

Correcting Nearsightedness

To correct for nearsightedness, one must use a lens that produces an image at the person's far-point distance when the object is at infinity. Note that the far point is closer to the lens in a pair of glasses than it is to the eye, since the glasses are a finite distance in front of the eyes.

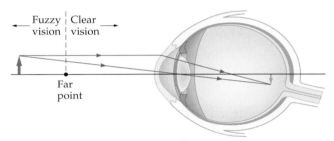

▲ **FIGURE 27–6** **Eye shape and nearsightedness**
An eye that is elongated can cause nearsightedness. In this case, an object at infinity comes to a focus in front of the retina.

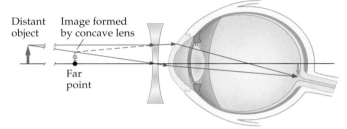

▲ **FIGURE 27–7** **Correcting nearsightedness**
A diverging lens in front of the eye can correct for nearsightedness. The concave lens focuses light from an object beyond the far point to produce an image that is at the far point. The eye can now focus on the image of the object.

EXAMPLE 27–2 EXTENDED VISION

 REAL-WORLD PHYSICS: BIO A nearsighted person has a far point that is 323 cm from her eye. If the lens in a pair of glasses is 2.00 cm in front of this person's eye, what focal length must it have to allow her to focus on distant objects?

PICTURE THE PROBLEM

In our sketch, we show an object at infinity and the corresponding image produced by the concave lens. As is usual with a concave lens, the image is upright, reduced, and on the same side of the lens as the object. In addition, the image is placed at the person's far point, where the eye can focus on it.

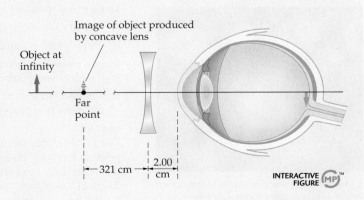

STRATEGY

We can find the focal length of the lens using the thin-lens equation, $1/d_o + 1/d_i = 1/f$.

In this case, $d_o = \infty$, since the object is infinitely far away.

As for the image, it must be 323 cm from the eye, which is 323 cm − 2.00 cm = 321 cm in front of the lens. Thus, the image distance is $d_i = -321$ cm, where the minus sign is required because the image is on the same side of the lens as the object.

With these values for d_o and d_i, it is straightforward to find the focal length, f.

CONTINUED ON NEXT PAGE

CONTINUED FROM PREVIOUS PAGE

SOLUTION

1. Substitute $d_o = \infty$ and $d_i = -321$ cm in the thin-lens equation:

$$\frac{1}{d_o} + \frac{1}{d_i} = \frac{1}{f} = \frac{1}{\infty} + \frac{1}{-321 \text{ cm}}$$

2. Solve for the focal length, f:

$$f = -321 \text{ cm}$$

INSIGHT

With a lens of this focal length, the person can focus on distant objects with a relaxed eye. In addition, note that the focal length is equal to the image distance, $f = d_i$. This is always the case when the object distance is infinite.

PRACTICE PROBLEM

If these glasses are used to view an object 525 cm from the eye, how far from the eye is the image produced by the concave lens? [**Answer:** The image is 201 cm in front of the eye.]

Some related homework problems: Problem 31, Problem 33, Problem 34

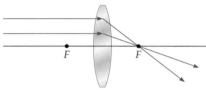

(a) Large refractive power

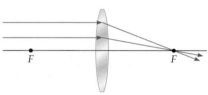

(b) Small refractive power

▲ **FIGURE 27–8 Refractive power**
Light is bent (refracted) more by a lens with a short focal length than one with a long focal length. Therefore, lens **(a)** has a greater refractive power than lens **(b)**.

The ability of a lens to refract light—its **refractive power**—is related to its focal length. For example, the shorter the focal length, the more strongly a lens refracts light, as indicated in **Figure 27–8**. Thus refractive power depends inversely on the focal length. By definition, then, we say that the refractive power of a lens is $1/f$, where f is measured in meters:

Refractive Power

$$\text{refractive power} = \frac{1}{f}$$

SI unit: diopter $= \text{m}^{-1}$

27–2

Lenses are typically characterized by optometrists in terms of diopters rather than in terms of focal length.

As an example of the meaning of diopters, a lens with a refractive power of 10.0 diopters has a focal length of $1/(10.0 \text{ m}^{-1}) = 10.0$ cm (a converging lens), and a lens with a refractive power of -10.0 diopters has a focal length of -10.0 cm (a diverging lens). In Example 27–2, the lens required to correct nearsightedness had a refractive power of $1/(-3.21 \text{ m}) = -0.312$ diopter.

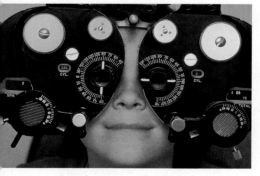

▲ In order to prescribe the right corrective lenses, it is necessary to measure the refractive properties of the patient's eyes. This device, known as a *phoropter*, allows the optometrist to see how lenses with various optical characteristics affect the patient's vision.

ACTIVE EXAMPLE 27–3 FIND THE REFRACTIVE POWER

A person has a far point that is 5.50 m from his eyes. If this person is to wear glasses that are 2.00 cm from his eyes, what refractive power, in diopters, must his lenses have?

SOLUTION *(Test your understanding by performing the calculations indicated in each step.)*

1. Identify the object distance: $\quad d_o = \infty$
2. Identify the image distance: $\quad d_i = -548$ cm
3. Use the thin-lens equation to calculate the focal length of the lenses: $\quad f = -548$ cm
4. Convert f to meters and invert to find the refractive power: $\quad$ refractive power $= -0.182$ diopter

INSIGHT

The fact that this person's far point is farther away (closer to infinity) than the far point in Example 27–2 means that the lenses do not have to be as "strong" to correct the vision. As a result, the magnitude of the refractive power in this case (0.182 diopter) is less than the corresponding magnitude (0.312 diopter) in Example 27–2.

YOUR TURN

A person wears glasses with a refractive power of -0.500 diopter. If the glasses are 2.00 cm in front of the eyes, what is the distance from the eyes to the far point?

*(Answers to **Your Turn** problems are given in the back of the book.)*

Today, in addition to glasses and contact lenses, a number of medical procedures are available to correct nearsightedness. Some of these procedures involve using laser beams to reshape the cornea of the eye. These techniques are discussed in Chapter 31, where we consider lasers in detail. Here we present two alternative procedures that change the shape of a cornea by mechanical means.

Perhaps the simplest such technique is the implantation of either an *intracorneal ring* or an *Intact* in the cornea of an eye. An Intact, in particular, consists of two small, clear crescents made of the same material used in contact lenses. These crescents are slipped into "tunnels" that are cut into the cornea. When the crescents are in place they tend to stretch the cornea outward and flatten its surface. This flattening increases the focal length of the eye and corrects for nearsightedness by -1.0 to -3.0 diopters. If necessary, the Intacts can be removed and replaced with others that cause more or less flattening of the cornea. Intracorneal rings are similar, except that they consist of a single ring rather than the two crescents used in Intacts.

A more complex method involves making radial incisions in the cornea with a highly precise diamond blade that cuts to a specified depth. This method is referred to as *radial keratotomy* or RK (**Figure 27–9**). The cuts allow the peripheral parts of the cornea to bulge outward, which, in turn, causes the central portion to flatten. As with an Intact, the flattening of the cornea corrects for nearsightedness.

Farsightedness

A person who is farsighted (hyperopic) can see clearly beyond a certain distance—the near point—but cannot focus on closer objects. Basically, the vision of a farsighted person differs from that of a person with normal vision by having a near point that is much farther from the eye than the usual 25 cm. As a result, a farsighted person is typically unable to read clearly, since a book would have to be held at such a great distance to come into focus.

Farsightedness can be caused by an eyeball that is shorter than normal, as illustrated in **Figure 27–10**, or by a lens that becomes sufficiently stiff with age that it can no longer take on the shape required to focus on nearby objects. In such cases, rays from an object inside the near point are brought to a focus behind the retina. Thus the focal length of the farsighted eye is too large—stated another way, the farsighted eye does not converge the incoming light strongly enough to focus on the retina.

This problem can be corrected by "preconverging" the light—that is, by using a converging lens in front of the eye to add to its insufficient convergence. For example, suppose an object is inside a person's near point, as in **Figure 27–11**. If a converging lens placed in front of the eye can produce an image that is far away—that is, beyond the near point—the farsighted person can view the object with ease. Such a system is considered in the next Example.

REAL-WORLD PHYSICS: BIO
Intracorneal rings and Intacts

REAL-WORLD PHYSICS: BIO
Radial keratotomy

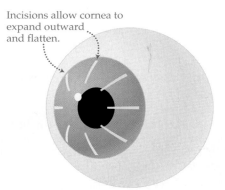

Incisions allow cornea to expand outward and flatten.

▲ **FIGURE 27–9 Radial keratotomy**
In radial keratotomy, a series of radial incisions is made around the periphery of the cornea. This allows the cornea to expand outward, resulting in a flattening of the cornea's central region. The reduced curvature of the cornea increases the focal length of the eye, allowing it to focus on distant objects.

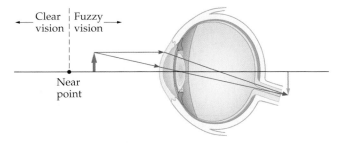

▲ **FIGURE 27–10 Eye shape and farsightedness**
An eye that is shorter than normal can cause farsightedness. Note that an object inside the near point comes to a focus behind the retina.

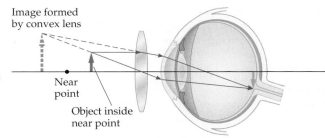

▲ **FIGURE 27–11 Correcting farsightedness**
A converging lens in front of the eye can correct for farsightedness. The convex lens focuses light from an object inside the near point to produce an image that is beyond the near point. The eye can now focus on the image of the object.

EXAMPLE 27–3 HIS VISION IS A FAR SIGHT BETTER

REAL-WORLD PHYSICS: BIO

A farsighted person wears glasses that enable him to read a book held at a distance of 25.0 cm from his eyes, even though his near-point distance is 57.0 cm. If his glasses are at a distance of 2.00 cm from his eyes, find the focal length and refractive power required of his lenses to place the image of the book at the near point.

PICTURE THE PROBLEM

The physical situation is shown in our sketch. Notice that we use a convex lens, and that the image produced by the lens is upright and farther from the eye than the book. These features are in agreement with the qualitative characteristics shown in Figure 27–11. In our case, however, the image of the book is exactly at the near point.

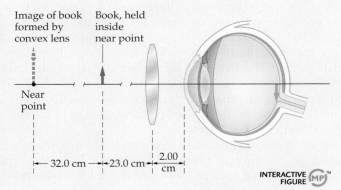

STRATEGY

The focal length of the lens can be found using the thin-lens equation. First, the object distance is $d_o = 23.0$ cm, taking into account the 2.00 cm between the glasses and the eye. Similarly, the desired image distance is $d_i = -55.0$ cm. Note that the minus sign is required on the image distance, since the image is on the same side of the lens as the object.

SOLUTION

1. Use the thin-lens equation to find the focal length, f:

$$\frac{1}{f} = \frac{1}{d_o} + \frac{1}{d_i} = \frac{1}{23.0 \text{ cm}} + \frac{1}{-55.0 \text{ cm}} = 0.0253 \text{ cm}^{-1}$$

$$f = \frac{1}{0.0253 \text{ cm}^{-1}} = 39.5 \text{ cm}$$

2. The refractive power is $1/f$, with f measured in meters:

$$\text{refractive power} = \frac{1}{f} = \frac{1}{0.395 \text{ m}} = 2.53 \text{ diopters}$$

INSIGHT

Notice that the book is between the lens and its focal point. As a result, the image is upright [see Figure 26–35 (b)], as desired for reading glasses. In addition, the image is virtual, as was also the case for the concave lens in Example 27–2. Thus, even though a "virtual image" can sound like one that isn't real or of practical importance, you are viewing a virtual image every time you look through a pair of glasses.

PRACTICE PROBLEM

A second person has a near-point distance that is greater than 57.0 cm. Is the refractive power of the lenses required for this person greater than or less than 2.53 diopters? To check your answer, calculate the refractive strength for a near-point distance of 67.0 cm. [**Answer:** The refractive power must be greater than 2.53 diopters. We find 2.81 diopters for a near-point distance of 67.0 cm.]

Some related homework problems: Problem 30, Problem 32

CONCEPTUAL CHECKPOINT 27–1 EYEGLASSES TO START A FIRE

Bill and Ted are on an excellent camping trip when they decide to start a fire by focusing sunlight with a pair of eyeglasses. If Bill is nearsighted and Ted is farsighted, should they use **(a)** Bill's glasses or **(b)** Ted's glasses?

REASONING AND DISCUSSION

To focus the parallel rays of light from the Sun to a point requires a converging lens. As we have seen, nearsightedness is corrected with a diverging lens; farsightedness is corrected with a converging lens. Ted's eyeglasses are converging; therefore, they should be the ones used to start a fire.

ANSWER

(b) Ted's eyeglasses would be the more excellent choice.

To consider the effect of a pair of contact lenses, we simply take into account that contacts are placed directly against the eye. This means that the eye–object distance is the same as the lens–object distance. We make use of this fact in the following Active Example.

ACTIVE EXAMPLE 27-4 CONTACT: FIND THE FOCAL LENGTH OF CONTACT LENSES

Find the focal length of a pair of contact lenses that will allow a person with a near-point distance of 145 cm to read a newspaper held 25.1 cm from the eyes.

SOLUTION *(Test your understanding by performing the calculations indicated in each step.)*

1. Identify the object distance: $d_o = 25.1$ cm
2. Identify the image distance: $d_i = -145$ cm
3. Use the thin-lens equation to calculate the focal length: $f = 30.4$ cm

INSIGHT

Note that the image distance is negative, since the image is on the same side of the lens as the object.

YOUR TURN

Suppose a second person has a near-point distance of 205 cm. Is the focal length of the second person's contacts greater than or less than the focal length of the first person's contacts? Find the focal length of the second person's contacts.

(Answers to **Your Turn** *problems are given in the back of the book.)*

PROBLEM-SOLVING NOTE

Correcting Farsightedness

To correct for farsightedness, one must use a lens that produces an image at the person's near-point distance when the object is closer than the near point. Note that the distance from the object to the lens in a pair of glasses is less than the distance from the object to the eye.

PROBLEM-SOLVING NOTE

Finding the Focal Length of Contact Lenses

Contact lenses are so named because they are in direct contact with the eye. Therefore, the object distance for a contact lens is the same as the distance from the eye to the object. Similarly, the near point is the same distance from both the contact lens and the eye.

REAL-WORLD PHYSICS: BIO

Keratometers

An important step in assuring that contact lenses fit a patient's eye properly is to measure the radius of curvature of the cornea. This is accomplished with a device known as a *keratometer*. To begin the procedure, a brightly lit object is brought near the eye. The light from the object reflects from the front surface of the cornea, just as light reflects from a convex, spherical mirror. In the next step of the procedure, the keratometer measures the magnification of the mirror image produced by the cornea. Finally, a straightforward application of the mirror equation determines the radius of curvature. A specific example of a keratometer in use is given in Problem 90.

Another eye condition that requires corrective optics is **astigmatism.** In most cases, astigmatism is due to an irregular curvature of the cornea, with a greater curvature in one direction than in another. For example, the curvature in the vertical plane may be greater than the curvature in the horizontal plane. As a result, if the eye is focused for light coming into it from one direction, it will be out of focus for light arriving from a different direction. Almost everyone has some degree of astigmatism, usually quite mild, but in serious cases it can cause distorted or blurry vision at all distances.

27–3 The Magnifying Glass

A **magnifying glass** is nothing more than a simple convex lens. Working together with the eye as part of a two-lens system, a *magnifier* can make objects appear to be many times larger than their actual size. As we shall see, the magnifying glass works by moving the near point closer to the eye—much as a converging lens corrects farsightedness. Basically, the magnifier allows an object to be viewed from a reduced distance, and this is what makes it appear larger.

To be more specific about the apparent size of an object, we consider the angle it subtends on the retina—after all, the more area an image takes up on the retina, the larger it will seem to us. For example, suppose an object of height h_o is a distance d_o from the eye, as in **Figure 27–12 (a).** The image formed by this object subtends an angle θ on the retina of the eye, as shown in the figure. If the angle is small, as is often the case, it can be approximated by the tangent of the angle, since $\tan \theta \approx \theta$ for small angles. Referring to the figure, we see that the angle θ is approximately

$$\theta \approx \frac{h_o}{d_o}$$

▶ **FIGURE 27–12 Angular size and distance**
The angular size of an object depends on its distance from the eye, even though its height, h_o, remains the same.

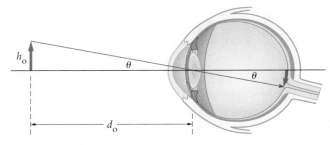

(a) Small angular size

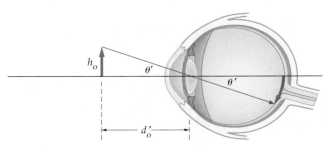

(b) Same object, larger angular size

If the object is now moved closer to the eye, to a distance d'_o, the angle subtended by the image is larger. Using the small-angle approximation again, and referring to **Figure 27–12 (b)**, we find that the new angle is

$$\theta' \approx \frac{h_o}{d'_o} > \theta$$

Thus, moving an object closer to the eye increases its apparent size, since its image covers a larger portion of the retina. There is a limit, however, to how close an object can come to the unaided eye and still be in focus—the near point. This is where a magnifier comes into play.

Suppose, then, that you would like to see as much detail as possible on a small object—a feather, perhaps, or a flower petal. With the unaided eye you can bring the object to the near point, a distance N from the eye, as in **Figure 27–13 (a)**. If the height of the object is h_o, the angular size of the object on the retina is approximately

$$\theta = \frac{h_o}{N}$$

Now, consider placing a convex lens of focal length $f < N$ just in front of the eye, as shown in **Figure 27–13 (b)**. If the object is brought to the focal point of this lens, its image will be infinitely far from the eye, where it can be viewed in focus with ease. As we can see from the figure, the angular size of the image is approximately

$$\theta' = \frac{h_o}{f}$$

Note that θ' is greater than θ—since f is less than N—and hence the object appears larger. The factor by which the object is enlarged, referred to as the **angular magnification,** M, is defined as follows:

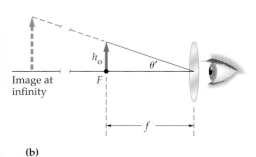

(a)

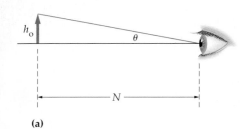

Image at infinity

(b)

▲ **FIGURE 27–13 How a simple magnifier works**
(a) An object viewed with the unaided eye at the near point of the eye subtends an angle $\theta \approx h_o/N$. **(b)** With a magnifier, the object can be viewed from the distance f, which is less than N. As a result, the angular size of the object is $\theta' \approx h_o/f$, which is greater than θ.

Angular Magnification, *M*

$$M = \frac{\theta'}{\theta}$$

27–3

SI unit: dimensionless

Using the angles θ and θ' obtained previously, we find

$$M = \frac{h_o/f}{h_o/N} = \frac{N}{f} \qquad\qquad 27\text{--}4$$

A typical situation is considered in the following Exercise.

EXERCISE 27–1

A person with a near-point distance of 30 cm examines a stamp with a magnifying glass. If the magnifying glass produces an angular magnification of 6, what is its focal length?

SOLUTION
Solving $M = N/f$ for the focal length we find

$$f = \frac{N}{M} = \frac{30\text{ cm}}{6} = 5\text{ cm}$$

Note that an object examined with a magnifier appears larger precisely because it is closer to the eye. In fact, the angular size of the object in Figure 27–13 (b) is h_o/f regardless of whether the magnifier is present or not. If the magnifier is not present, however, the object is out of focus, and then its increased angular size is of no practical value. The magnifier is beneficial in that it brings the object into focus at this close distance.

In addition, since the image produced by the magnifier is at infinity, the rays entering the eye are parallel. This means that a person can view the image with a completely relaxed eye. If the same object is viewed with the unaided eye at the near point, the ciliary muscles are fully tensed, causing eye strain. Thus, not only does the magnifier enlarge the object, it also makes it more comfortable to view.

Now, of course, it is also possible to produce an enlarged image with a convex lens without holding the lens close to the eye. In fact, if the lens is held at some distance from the eye, and just under a focal length from the object to be viewed, we see an enlarged image. There are two distinct disadvantages to holding a magnifier far from the eye, however. First, it is difficult to hold the lens motionless at a distance, as opposed to bracing it against the face. Second, when the lens is held at a distance, only a small portion of the object can be viewed at any given time. When the lens is held close up, however, the entire object can be viewed at once.

As we shall see in the next two sections, the magnifier plays an important role in both the microscope and the telescope.

▲ A magnifying glass produces an enlarged image when held near an object. The image is upright and virtual.

CONCEPTUAL CHECKPOINT 27–2   USING A MAGNIFIER

Person 1, with a near-point distance of 25 cm, and person 2, with a near-point distance of 50 cm, both use the same magnifying glass. Does **(a)** person 1 or **(b)** person 2 benefit more from using the magnifier?

REASONING AND DISCUSSION
The person with the greater near-point distance cannot see an object as closely with the unaided eye as can the person with the smaller near-point distance. With the magnifier, however, both people can view an object from the same close-in distance. Hence, the person with the larger near-point distance benefits more from the magnifier.

ANSWER
(b) Person 2, with the 50-cm near-point distance, benefits more.

It is possible to obtain a magnification that is slightly greater than the value N/f given in Equation 27–4. After all, this magnification is for an image formed at infinity; if the image were closer to the eye, it would appear larger. The closest the image can be—and still be in focus—is the near point. In the next Example we compare the magnifications that result when the image is at infinity and at the near point.

EXAMPLE 27–4	COMPARING MAGNIFICATIONS

A biologist with a near-point distance of $N = 26$ cm examines an insect wing through a magnifying glass whose focal length is 4.3 cm. Find the angular magnification when the image produced by the magnifier is **(a)** at infinity and **(b)** at the near point.

PICTURE THE PROBLEM

Our sketch shows the situation in which the image formed by the magnifying glass is at the near point. To produce an image at this point, the object must be closer to the magnifier than its focal point. Therefore, the object appears larger in this case than it does in the case where the object is at the focal length and the image is at infinity.

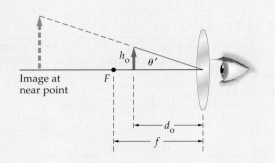

STRATEGY

a. The magnification when the image is at infinity is $M = N/f$, as given in Equation 27–4.

b. The angular size of the object in this case is $\theta' \approx h_o/d_o$, where d_o is the object distance that places the image at the near point. Thus the magnification is $M = \theta'/\theta = (h_o/d_o)/(h_o/N) = N/d_o$. Note that this result differs from the magnification in part (a) only in that f has been replaced with d_o. All that remains is to calculate d_o using the thin-lens equation.

SOLUTION

Part (a)

1. Substitute $N = 26$ cm and $f = 4.3$ cm into $M = N/f$:

$$M = \frac{N}{f} = \frac{26 \text{ cm}}{4.3 \text{ cm}} = 6.0$$

Part (b)

2. Use the thin-lens equation to find d_o, given $f = 4.3$ cm and $d_i = -26$ cm:

$$\frac{1}{d_o} = \frac{1}{f} - \frac{1}{d_i} = \frac{1}{4.3 \text{ cm}} - \frac{1}{-26 \text{ cm}} = 0.27 \text{ cm}^{-1}$$

$$d_o = \frac{1}{0.27 \text{ cm}^{-1}} = 3.7 \text{ cm}$$

3. Calculate the magnification using $M = N/d_o$:

$$M = \frac{N}{d_o} = \frac{26 \text{ cm}}{3.7 \text{ cm}} = 7.0$$

INSIGHT

We see that moving the object closer to the lens, which brings the image in from infinity to the near point, results in an increase in magnification of 1.0—from 6.0 to 7.0.

PRACTICE PROBLEM

What is the magnification if the object is placed 4.0 cm in front of the magnifying glass?   [**Answer:** $M = N/d_o = 6.5$]

Some related homework problems: Problem 50, Problem 52

Thus we see from Example 27–4 that the magnification of a magnifying glass with the object at a distance d_o is $M = N/d_o$, where N is the near-point distance of the observer. In the special case of an image at infinity, the object distance is the focal length, and $M = N/f$. It can be shown (see Problem 106) that the greatest magnification—which occurs when the image is at the near point—is $M = 1 + N/f$. The magnification results are summarized here:

$$M = \frac{N}{f} \quad \text{(image at infinity)}$$

$$M = 1 + \frac{N}{f} \quad \text{(image at near point)}$$

27–5

Note that Example 27–4 is simply a special case of these general expressions.

The early microscopes produced by Antonie van Leeuwenhoek (1632–1723) were, in fact, simply powerful magnifying glasses using a single lens mounted in a hole in a flat metal plate. The object to be examined with the microscope was placed on the head of a small, movable pin placed just below the lens. The observer would then hold the top surface of the lens close to the eye for the most stable, wide-angle view. Leeuwenhoek's instruments were capable of magnifying objects by as much as 275 times, enough that he could make the first detailed microscopic descriptions of single-celled animals, red blood cells, plant cells, and much more. Microscopes in common use today have more than one lens, as we describe in the next section.

27–4 The Compound Microscope

Although a magnifying glass is a useful device, higher magnifications and improved optical quality can be obtained with a **microscope.** The simplest microscope consists of two converging lenses fixed at either end of a tube. Such an instrument, illustrated in **Figure 27–14**, is sometimes referred to as a *compound microscope.*

The basic optical elements of a microscope are the **objective** and the **eyepiece.** The objective is a converging lens with a relatively short focal length that is placed near the object to be viewed. It forms a real, inverted, and enlarged image, as shown in **Figure 27–15**. The precise location of the image is adjusted when the microscope is focused by moving the objective up or down. This image serves as the object for the second lens in the microscope—the eyepiece. In fact, the eyepiece is simply a magnifier that views the image of the objective, giving it an additional enlargement.

The final magnification of the microscope, then, is simply the product of the magnification of the objective and the magnification of the eyepiece. For example, a microscope might have a 10× eyepiece (meaning it magnifies 10 times) and a 50× objective. When these two lenses are used together, the magnification of the microscope is 500×.

In a typical situation, the object to be examined is placed only a small distance beyond the focal point of the objective, which means that $d_o \approx f_{\text{objective}}$. The magnification produced by the objective is given by Equation 26–18:

$$m_{\text{objective}} = -\frac{d_i}{d_o} \approx -\frac{d_i}{f_{\text{objective}}}$$

Next, the image formed by the objective is essentially at the focal point of the eyepiece. This means that the eyepiece forms a virtual image at infinity that the observer can view with a relaxed eye. The angular magnification of the eyepiece is given by Equation 27–4:

$$M_{\text{eyepiece}} = \frac{N}{f_{\text{eyepiece}}}$$

Multiplying these magnifications, we find the total magnification of the microscope:

$$M_{\text{total}} = m_{\text{objective}}M_{\text{eyepiece}} = \left(-\frac{d_i}{f_{\text{objective}}}\right)\left(\frac{N}{f_{\text{eyepiece}}}\right)$$

$$= -\frac{d_i N}{f_{\text{objective}} f_{\text{eyepiece}}} \qquad \text{27–6}$$

The minus sign indicates that the image is inverted.

Typical numerical values are considered in the following Example.

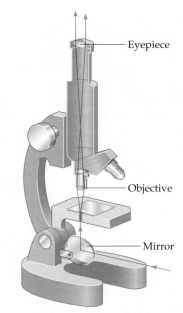

▲ **FIGURE 27–14 Basic elements of a compound microscope**

A compound microscope consists of two lenses—an objective and an eyepiece—fixed at either end of a movable tube.

PROBLEM-SOLVING NOTE

Lens Placement in a Microscope

In a working microscope, the distance between the lenses is greater than the sum of their focal lengths.

▶ **FIGURE 27–15 The operation of a compound microscope**

In a compound microscope the object is placed just outside the focal point of the objective. The resulting enlarged image is then enlarged further by the eyepiece, which is basically a magnifying glass.

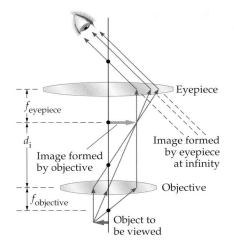

EXAMPLE 27–5 A MICROSCOPIC VIEW

In biology class, a student with a near-point distance of $N = 25$ cm uses a microscope to view an amoeba. If the objective has a focal length of 1.0 cm, the eyepiece has a focal length of 2.5 cm, and the amoeba is 1.1 cm from the objective, what is the magnification produced by the microscope?

CONTINUED ON NEXT PAGE

CONTINUED FROM PREVIOUS PAGE

PICTURE THE PROBLEM
The lenses of the microscope are shown in the sketch. Note that the distance from the objective to the object is d_o and the distance from the objective to its image is d_i.

STRATEGY
The magnification of the microscope is given by Equation 27–6. The only unknown in this expression is the image distance, d_i. We can find this by using the thin-lens equation.

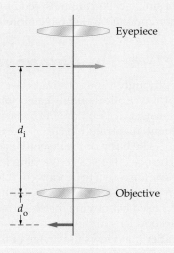

SOLUTION

1. Use the thin-lens equation to find the image distance, d_i:

$$\frac{1}{d_i} = \frac{1}{f} - \frac{1}{d_o} = \frac{1}{1.0 \text{ cm}} - \frac{1}{1.1 \text{ cm}} = 0.091 \text{ cm}^{-1}$$

$$d_i = \frac{1}{0.091 \text{ cm}^{-1}} = 11 \text{ cm}$$

2. Use Equation 27–6 to find the magnification of the microscope:

$$M_{total} = -\frac{d_i N}{f_{objective} f_{eyepiece}} = -\frac{(11 \text{ cm})(25 \text{ cm})}{(1.0 \text{ cm})(2.5 \text{ cm})} = -110$$

INSIGHT
Thus the amoeba appears 110 times larger and is inverted. If the amoeba is to be viewed with a relaxed eye, the image formed by the objective should be at the focal point of the eyepiece, which will then form an image at infinity. Therefore, the length of the tube containing the objective and eyepiece is 11 cm + 2.5 cm = 13.5 cm in this case.

PRACTICE PROBLEM
If the focal length of the eyepiece is increased, does the magnitude of the magnification increase or decrease? Check your response by calculating the magnification when the focal length of the eyepiece is 3.5 cm. [**Answer:** The magnification is reduced in magnitude. Its new value is −79.]

Some related homework problems: Problem 58, Problem 61

27–5 Telescopes

A telescope is similar in many respects to a microscope—both use two converging lenses to give a magnified image of an object with small angular size. In the case of a microscope the object itself is small and close at hand; in the case of the telescope the object may be as large as a galaxy, but its angular size can be very small because of its great distance. The major difference between these instruments is that the telescope must deal with an object that is essentially infinitely far away.

For this reason, it is clear that the light entering the objective of a telescope from a distant object is focused at the focal point of the objective, as shown in **Figure 27–16**. If the image formed by the objective has a height h_i, the angular size of the object is approximately

$$\theta = -\frac{h_i}{f_{objective}}$$

Note that the minus sign is included because h_i is negative for the inverted image formed by the objective.

As in the microscope, the image of the objective is the object for the eyepiece, which is basically a magnifier. Thus, if the image of the objective is placed

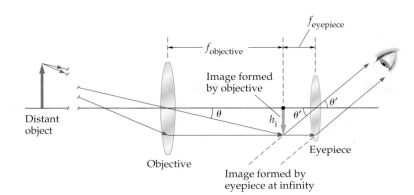

◀ **FIGURE 27–16 Basic elements of a telescope**
A telescope focuses light from a distant object at its focal point. The image of the objective is placed at the focal point of the eyepiece to produce an enlarged image that can be viewed with a relaxed eye.

at the focal point of the eyepiece, it will form an image that is at infinity, as indicated in Figure 27–16. In this configuration, the observer can view the final image of the telescope with a completely relaxed eye. The angular size of the image formed by the eyepiece, θ', is shown in Figure 27–16. Clearly, this angle is approximately

PROBLEM-SOLVING NOTE

Lens Placement in a Telescope

In a working telescope, the distance between the lenses is approximately equal to the sum of their focal lengths. In addition, the focal length of the eyepiece is significantly less than that of the objective.

$$\theta' = -\frac{h_i}{f_{eyepiece}}$$

To find the total angular magnification of the telescope, we take the ratio of θ' to θ:

$$M_{total} = \frac{\theta'}{\theta} = \frac{f_{objective}}{f_{eyepiece}} \qquad 27\text{--}7$$

Thus, for example, a telescope with an objective whose focal length is 1500 mm and an eyepiece whose focal length is 10.0 mm produces an angular magnification of 150.

CONCEPTUAL CHECKPOINT 27–3 COMPARING TELESCOPES

Two telescopes have identical eyepieces, but telescope A is twice as long as telescope B. Is the magnification of telescope A **(a)** greater than the magnification of telescope B, **(b)** less than the magnification of telescope B, or **(c)** is there no way to tell?

REASONING AND DISCUSSION
Note in Figure 27–16 that the total length of the telescope is $f_{objective}$ plus $f_{eyepiece}$. Thus, because the scopes have identical eyepieces, it follows that telescope A must have a greater objective focal length than telescope B. We know, then, from $M_{total} = f_{objective}/f_{eyepiece}$, that the magnification of telescope A is greater than the magnification of telescope B.

ANSWER
(a) Telescope A has a greater magnification than telescope B.

Telescopes using two or more lenses, as in Figure 27–16, are referred to as *refractors*. In fact, the first telescopes constructed for astronomical purposes, made by Galileo starting in 1609, were refractors. Galileo's telescopes differed from the telescopes in common use today, however, in that they used a diverging lens for the eyepiece. We consider this type of telescope in Problems 75, 86, and 94. By the end of 1609, Galileo had produced a telescope whose angular magnification was 20. This was more than enough to enable him to see—for the first time in human history—mountains on the Moon, stars in the Milky Way, the phases of Venus, and moons orbiting Jupiter. As a result of his telescopic observations, Galileo became a firm believer in the Copernican model of the solar system.

In the next Example we consider a standard refractor with two converging lenses. In particular, we show how the length of such a telescope is related to the focal lengths of its objective and eyepiece.

EXAMPLE 27–6 CONSIDERING THE TELESCOPE AT LENGTH

A telescope has a magnification of 40.0 and a length of 1230 mm. What are the focal lengths of the objective and eyepiece?

PICTURE THE PROBLEM
Our sketch shows that the overall length of the telescope, L, is equal to the sum of the focal lengths of the objective and the eyepiece. In addition, note that the objective forms an inverted image of the distant object, which the eyepiece enlarges. The final image seen by the observer, then, is inverted.

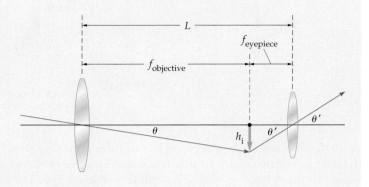

STRATEGY
We can determine the two unknowns, $f_{objective}$ and $f_{eyepiece}$, from the two independent bits of information given in the problem statement:

The magnification, $M_{total} = f_{objective}/f_{eyepiece}$, is equal to 40.0.

The total length, $L = f_{objective} + f_{eyepiece}$, is equal to 1230 mm.

Combining this information gives us the two focal lengths.

SOLUTION

1. Use the magnification equation to write $f_{objective}$ in terms of $f_{eyepiece}$:

$$M_{total} = \frac{f_{objective}}{f_{eyepiece}} = 40.0$$
$$f_{objective} = 40.0 f_{eyepiece}$$

2. Substitute $f_{objective} = 40.0 f_{eyepiece}$ into the expression for the total length of the telescope:

$$L = f_{objective} + f_{eyepiece}$$
$$= 40.0 f_{eyepiece} + f_{eyepiece} = 41.0 f_{eyepiece} = 1230 \text{ mm}$$

3. Divide 1230 mm by 41.0 to find the focal length of the eyepiece:

$$f_{eyepiece} = \frac{1230 \text{ mm}}{41.0} = 30.0 \text{ mm}$$

4. Multiply $f_{eyepiece}$ by 40.0 to find the focal length of the objective:

$$f_{objective} = 40.0 f_{eyepiece} = 40.0(30.0 \text{ mm}) = 1200 \text{ mm}$$

INSIGHT
The focal lengths found in this example are typical of those used in many popular amateur telescopes. A telescope with a magnification of 40.0 can easily show the moons of Jupiter and the rings of Saturn.

PRACTICE PROBLEM
A telescope 1820 mm long has an objective with a focal length of 1780 mm. Find the focal length of the eyepiece and magnification of the telescope. [**Answer:** $f_{eyepiece} = 40.0$ mm, $M_{total} = 44.5$]

Some related homework problems: Problem 68, Problem 69, Problem 78

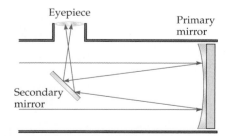

▲ **FIGURE 27–17 A Newtonian reflecting telescope**
The large primary mirror collects incoming light and reflects it off a small secondary mirror into the eyepiece.

Because telescopes are typically used to view objects that are very dim, it is desirable to have an objective with as large a diameter, or aperture, as possible. If the diameter is doubled, for example, the light gathered by a telescope increases by the same factor as the area of its objective: $2^2 = 4$. In addition, a larger aperture results in a higher-resolution image, as we shall see in the next chapter.

For a refractor, a large aperture means a very large and heavy piece of glass. In fact, the world's largest refractor, the Yerkes refractor at Williams Bay, Wisconsin, has an objective that is only 1 m across. If refractors were made much larger, the objective lens would sag and distort under its own weight.

The telescopes with the largest apertures are reflectors, one example of which is illustrated in **Figure 27–17**. Invented by Isaac Newton in 1671, and referred to as a Newtonian reflector, this type of telescope uses a mirror in place of an objective lens. As with the refractor, the mirror forms an image which the eyepiece then magnifies. Since a mirror can be much thinner and lighter than a lens, and can be supported all over its back surface instead of just around the edges as with a lens, it has many advantages for a large scope. In fact, the largest telescopes in the world are the twin Keck reflecting telescopes atop Hawaii's Mauna Kea. These telescopes have hexagonal objective mirrors that are 10 m across, which means

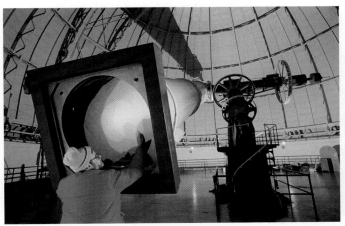

▲ At left, the 1-m objective lens of the Yerkes refractor—the largest refracting telescope ever constructed—being dusted to remove spiders. If lenses were made much larger than this, they would sag and distort under their own weight. At right, one of the twin 10-m Keck reflecting telescopes atop Mauna Kea on the Big Island of Hawaii, where the air is cold, thin, and dry—ideal conditions for astronomical observation. Clearly, the use of interchangeable objectives, commonly found in microscopes, is not feasible for large telescopes. Instead, magnification is varied by using eyepieces with different focal lengths. It is rare for a person to actually look through one of these telescopes, however—instead, they are usually fitted with sophisticated cameras, spectrographs, or other instruments.

that each gathers 100 times as much light as the Yerkes refractor. When used together, the Keck telescopes constitute the world's largest pair of binoculars!

27–6 Lens Aberrations

An ideal lens brings all parallel rays of light that strike it together at a single focal point. Real lenses, however, never quite live up to the ideal. Instead, a real lens blurs the focal point into a small but finite region of space. This, in turn, blurs the image it forms. The deviation of a lens from ideal behavior is referred to as an **aberration.**

One example is **spherical aberration,** which is related to the shape of a lens. **Figure 27–18** shows parallel rays of light passing through a lens with spherical aberration; note that the rays do not meet at a single focal point. This is completely analogous to the spherical aberration present in mirrors with a spherical cross section, as shown in Figure 26–13. To prevent spherical aberration, a lens must be ground and polished to a very precise nonspherical shape.

Another common aberration in lenses is **chromatic aberration,** which is due to the unavoidable dispersion present in any refracting material. Just as a prism splits white light into a spectrum of colors, each color bent by a different amount, so too does a lens bend light of different colors by different amounts as shown in **Figure 27–19**. The result is that white light passing through a lens does not focus to

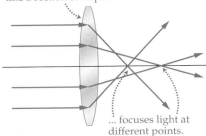

A lens whose surface is shaped like a section of a sphere ...

... focuses light at different points.

▲ **FIGURE 27–18** **Spherical aberration**
In a lens with spherical aberration, light striking the lens at different locations comes together at different focal points.

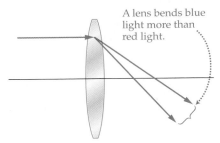

A lens bends blue light more than red light.

▲ **FIGURE 27–19** **Chromatic aberration in a converging lens**
Chromatic aberration is caused by the fact that blue light bends more than red light as it passes through a lens. As a result, different colors have different focal points.

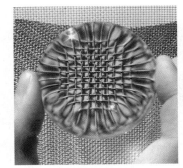

▲ Chromatic aberration typical of a simple lens.

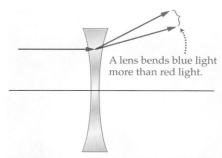

A lens bends blue light more than red light.

▲ **FIGURE 27–20   Chromatic aberration in a diverging lens**
A diverging lens also bends blue light more than red light, though the rays are bent in the opposite direction compared with the bending by a converging lens.

▶ **FIGURE 27–21   Correcting for chromatic aberration**
An achromatic doublet is a combination of a converging and a diverging lens made of different types of glass. In the converging lens the blue light is bent toward the axis of the lens more than the red light. In the diverging lens the opposite is the case. The net result is that the red and blue light pass through the same focal point.

a single point. This is why you sometimes see a fringe of color around an image seen through a simple lens.

Chromatic aberration can be corrected by combining two or more lenses to form a compound lens. For example, suppose a convex lens made from glass A bends red and blue light as shown in Figure 27–19. Note that the blue light is bent more, as expected. A second lens, concave this time, is made of glass B. This lens also bends blue light more than red light, but since it is a diverging lens, it bends both the red and blue light in the opposite direction compared with the convex lens, as we see in **Figure 27–20.** If the amount of bending of blue light compared with red light is different for glasses A and B, it is possible to construct a convex and a concave lens in such a way that the opposite directions in which they bend red and blue light can be made to cancel while still causing the light to come to a focus. This combination is illustrated in **Figure 27–21.** Such a lens is said to be **achromatic.** Three lenses connected similarly can produce even better correction for chromatic aberration. Lenses of this type are referred to as **apochromatic.**

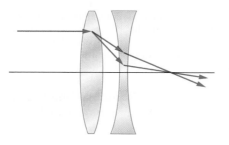

Two lenses canceling chromatic aberration

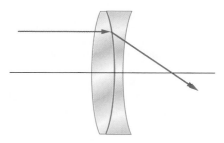

An achromatic doublet

Many lenses in optical instruments are not single lenses at all but, instead, are compound achromatic lenses. For example, the "lens" in a 35-mm camera may actually contain five or more individual lenses—some of them converging, some diverging, some made of one type of glass, some made of another. A zoom lens for a camera not only must correct for chromatic aberration, but must also be able to change its focal length. Thus, it is not uncommon for such a lens to contain as many as twelve individual lenses.

THE BIG PICTURE   PUTTING PHYSICS IN CONTEXT

LOOKING BACK	LOOKING AHEAD
Ray diagrams from Chapter 26 are used throughout this chapter to determine the location, orientation, and size of an image. We also extend ray diagrams to cases involving more than a single lens.	Light has been treated as straight-line rays in both Chapters 26 and 27. In Chapter 28, however, we consider the wave properties of light, and show that rays—while useful—do not tell the whole story.
Dispersion, which was discussed in Chapter 26 in relation to rainbows, appears again in this chapter when we consider chromatic aberration in Section 27–6.	Ray diagrams showing light reflected from a mirror are used in relativity (Chapter 29). There we analyze a "light clock," and show that a moving clock runs at a slower rate than a clock at rest.

CHAPTER SUMMARY

27–1 THE HUMAN EYE AND THE CAMERA

The human eye forms a real, but inverted, image on the retina; a camera forms a real, but inverted, image on light-sensitive material.

Focusing the Eye
The eye is focused by the ciliary muscles, which change the shape of the lens. This process is referred to as accommodation.

Focusing a Camera

A camera is focused by moving the lens closer to or farther away from the light-sensitive material. The shape of the lens is unchanged.

Near Point

The near point is the closest distance to which the eye can focus. A typical value for the near-point distance, N, is 25 cm.

Far Point

The far point is the greatest distance at which the eye can focus. In a normal eye the far point is infinity.

f-Number

The *f*-number of a lens relates the diameter of the aperture, D, to the focal length, f, as follows:

$$f\text{-number} = \frac{\text{focal length}}{\text{diameter of aperture}} = \frac{f}{D} \qquad 27\text{--}1$$

27–2 LENSES IN COMBINATION AND CORRECTIVE OPTICS

The basic idea used in analyzing systems with more than one lens is the following: The *image* produced by one lens acts as the *object* for the next lens.

In addition, the total magnification of a system of lenses is given by the product of the magnifications produced by each lens individually.

Nearsightedness

Nearsightedness is a condition in which clear vision is restricted to a region relatively close to the eye—in other words, the far point is not at infinity but instead is a finite distance from the eye. This condition can be corrected with diverging lenses placed in front of the eyes.

Farsightedness

A person who is farsighted can see clearly only at relatively large distances from the eye—that is, the person's near point is much farther from the eye than the more typical values of 25 to 40 cm. Farsightedness can be corrected by placing converging lenses in front of the eyes.

Refractive Power and Diopters

The refractive power of a lens refers to its ability to bend light and is measured in diopters. It is defined as follows:

$$\text{refractive power} = \frac{1}{f} \qquad 27\text{--}2$$

In this expression, the focal length, f, must be measured in meters.

The greater the magnitude of the refractive power, the more strongly the lens bends light. A positive refractive power indicates a converging lens; a negative refractive power indicates a diverging lens.

27–3 THE MAGNIFYING GLASS

A magnifying glass is simply a converging lens. It works by allowing an object to be viewed at a distance less than the near-point distance. Since the distance is reduced, the angular size is increased.

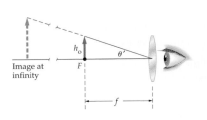

Magnification

A person with a near-point distance N using a magnifying glass with a focal length f experiences the following magnifications, M:

$$M = \frac{N}{f} \qquad \text{(image at infinity)}$$

$$M = 1 + \frac{N}{f} \quad \text{(image at near point)} \qquad 27\text{--}5$$

27–4 THE COMPOUND MICROSCOPE

A compound microscope uses two lenses in combination—an objective and an eyepiece—to produce a magnified image.

The object to be viewed is placed just outside the focal length of the objective. The image formed by the objective is then viewed by the eyepiece, giving additional magnification.

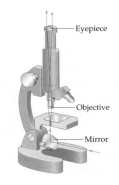

Magnification

The magnification produced by a compound microscope is

$$M_{\text{total}} = -\frac{d_i N}{f_{\text{objective}} f_{\text{eyepiece}}}$$ 27–6

where N is the near-point distance and d_i is the image distance for the objective. The magnifications quoted for microscopes assume a near-point distance of 25 cm.

27–5 TELESCOPES

A telescope provides magnified views of distant objects using two lenses. The objective lens focuses the incoming light at its focal point; the eyepiece magnifies the image formed by the objective.

Magnification

The total magnification produced by a telescope is

$$M_{\text{total}} = \frac{f_{\text{objective}}}{f_{\text{eyepiece}}}$$ 27–7

Length of a Telescope

The length, L, of a telescope is the sum of the focal lengths of its two lenses:

$$L = f_{\text{objective}} + f_{\text{eyepiece}}$$

Reflecting Telescopes

A reflecting telescope uses a mirror in place of an objective lens. The largest telescopes are reflectors, as are many amateur telescopes.

27–6 LENS ABERRATIONS

Any deviation of a lens from ideal behavior is referred to as an aberration.

Spherical Aberration

In spherical aberration, parallel rays of light passing through a lens fail to go through a single focal point. Spherical aberration is related to the shape of a lens.

Chromatic Aberration

Chromatic aberration results from dispersion within a refracting material. It causes different colors to focus at different points. Achromatic lenses correct for chromatic aberration by combining two lenses with different refractive properties.

PROBLEM-SOLVING SUMMARY

Type of Problem	Relevant Physical Concepts	Related Examples
Find the effective focal length of an eye when focused on a given object.	Use the thin-lens equation with an image distance of 1.7 cm to 2.5 cm to take into account the distance from the lens to the retina.	Example 27–1
Determine the optical behavior of a multilens system.	Apply the thin-lens equation to each lens one at a time. Let the image of one lens be the object for the next lens.	Active Example 27–2
Find the focal length of corrective glasses.	In the case of a nearsighted person, make sure the image is no farther away from the eye than the far point. For a farsighted person, make sure the image is no closer to the eye than the near point.	Examples 27–2, 27–3 Active Examples 27–3, 27–4

CONCEPTUAL QUESTIONS

For instructor-assigned homework, go to www.masteringphysics.com

(Answers to odd-numbered Conceptual Questions can be found in the back of the book.)

1. Why is it restful to your eyes to gaze off into the distance?

2. If a lens is cut in half through a plane perpendicular to its surface, does it show only half an image?

3. If your near-point distance is N, how close can you stand to a mirror and still be able to focus on your image?

4. When you open your eyes underwater, everything looks blurry. Can this be thought of as an extreme case of nearsightedness or farsightedness? Explain.

5. Would you benefit more from a magnifying glass if your near-point distance is 25 cm or if it is 15 cm? Explain.

6. When you use a simple magnifying glass, does it matter whether you hold the object to be examined closer to the lens than its focal length or farther away? Explain.

7. Is the final image produced by a telescope real or virtual? Explain.

8. Does chromatic aberration occur in mirrors? Explain.

PROBLEMS AND CONCEPTUAL EXERCISES

Note: Answers to odd-numbered Problems and Conceptual Exercises can be found in the back of the book. **IP** *denotes an integrated problem, with both conceptual and numerical parts;* **BIO** *identifies problems of biological or medical interest;* **CE** *indicates a conceptual exercise.* **Predict/Explain** *problems ask for two responses:* **(a)** *your prediction of a physical outcome, and* **(b)** *the best explanation among three provided. On all problems, red bullets (•, ••, •••) are used to indicate the level of difficulty.*

(The outside medium is assumed to be air, with an index of refraction of 1.00, unless specifically stated otherwise.)

SECTION 27–1   THE HUMAN EYE AND THE CAMERA

1. • **CE Predict/Explain BIO Octopus Eyes** To focus its eyes, an octopus does not change the shape of its lens, as is the case in humans. Instead, an octopus moves its rigid lens back and forth, as in a camera. This changes the distance from the lens to the retina and brings an object into focus. **(a)** If an object moves closer to an octopus, must the octopus move its lens closer to or farther from its retina to keep the object in focus? **(b)** Choose the *best explanation* from among the following:
   I. The lens must move closer to the retina—that is, farther away from the object—to compensate for the object moving closer to the eye.
   II. When the object moves closer to the eye, the image produced by the lens will be farther behind the lens; therefore, the lens must move farther from the retina.

2. • Your friend is 1.9 m tall. **(a)** When she stands 3.2 m from you, what is the height of her image formed on the retina of your eye? (Consider the eye to consist of a thin lens 2.5 cm from the retina.) **(b)** What is the height of her image when she is 4.2 m from you?

3. • Which forms the larger image on the retina of your eye: a 43-ft tree seen from a distance of 210 ft, or a 12-in. flower viewed from a distance of 2.0 ft?

4. • Approximating the eye as a single thin lens 2.60 cm from the retina, find the eye's near-point distance if the smallest focal length the eye can produce is 2.20 cm.

5. • Referring to Problem 4, what is the focal length of the eye when it is focused on an object at a distance of **(a)** 285 cm and **(b)** 28.5 cm?

6. • Four camera lenses have the following focal lengths and f-numbers:

Lens	Focal length (mm)	f-number
A	150	$f/1.2$
B	150	$f/5.6$
C	35	$f/1.2$
D	35	$f/5.6$

Rank these lenses in order of increasing aperture diameter. Indicate ties where appropriate.

7. • **BIO** The focal length of the human eye is approximately 1.7 cm. **(a)** What is the f-number for the human eye in bright light, when the pupil diameter is 2.0 mm? **(b)** What is the f-number in dim light, when the pupil diameter has expanded to 7.0 mm?

8. • **IP** A camera with a 55-mm-focal-length lens has aperture settings of 2.8, 4, 8, 11, and 16. **(a)** Which setting has the largest aperture diameter? **(b)** Calculate the five possible aperture diameters for this camera.

9. • The actual frame size of "35-mm" film is 24 mm × 36 mm. You want to take a photograph of your friend, who is 1.9 m tall. Your camera has a 55-mm-focal-length lens. How far from the camera should your friend stand in order to produce a 36-mm-tall image on the film?

10. • To completely fill a frame of "35-mm" film, the image produced by a camera must be 36 mm high. If a camera has a focal length of 150 mm, how far away must a 2.0-m-tall person stand to produce an image that fills the frame?

11. •• You are taking a photograph of a poster on the wall of your dorm room, so you can't back away any farther than 3.0 m to take the shot. The poster is 0.80 m wide and 1.2 m tall, and you want the image to fit in the 24-mm × 36-mm frame of the film in your camera. What is the longest focal length lens that will work?

12. •• A photograph is properly exposed when the aperture is set to $f/8$ and the shutter speed is 125. Find the approximate shutter speed needed to give the same exposure if the aperture is changed to $f/2.4$.

13. •• You are taking pictures of the beach at sunset. Just before the Sun sets, a shutter speed of $f/11$ produces a properly exposed picture. Shortly after the Sun sets, however, your light meter indicates that the scene is only one-quarter as bright as before. **(a)** If you don't change the aperture, what approximate shutter speed is needed for your second shot? **(b)** If, instead, you keep the shutter speed at $1/100$ s, what approximate f-stop will be needed for the second shot?

14. •• **IP** You are taking a photograph of a horse race. A shutter speed of 125 at $f/5.6$ produces a properly exposed image, but the running horses give a blurred image. Your camera has f-stops of 2, 2.8, 4, 5.6, 8, 11, and 16. **(a)** To use the shortest possible exposure time (i.e., highest shutter speed), which f-stop should you use? **(b)** What is the shortest exposure time you can use and *still* get a properly exposed image?

15. •• **The Hale Telescope** The 200-in. (5.08-m) diameter mirror of the Hale telescope on Mount Palomar has a focal length $f = 16.9$ m. **(a)** When the detector is placed at the focal point of the mirror (the "prime focus"), what is the f-ratio for this telescope? **(b)** The coudé focus arrangement uses additional mirrors to bend the light path and increase the effective focal length to 155.4 m. What is the f-ratio of the telescope when the coudé focus is being used? (*Coudé* is French for "elbow," since the light path is "bent like an elbow." This arrangement is useful when the light needs to be focused onto a distant instrument.)

SECTION 27–2   LENSES IN COMBINATION AND CORRECTIVE OPTICS

16. • **CE Predict/Explain** Two professors are stranded on a deserted island. Both wear glasses, though one is nearsighted and the other is farsighted. **(a)** Which person's glasses should be used to focus the rays of the Sun and start a fire? **(b)** Choose the *best explanation* from among the following:
   I. A nearsighted person can focus close, so that person's glasses should be used to focus the sunlight on a piece of moss at a distance of a couple inches.
   II. A farsighted person can't focus close, so the glasses to correct that person's vision are converging. A converging lens is what you need to concentrate the rays of the Sun.

17. • **CE** A clerk at the local grocery store wears glasses that make her eyes look larger than they actually are. Is the clerk nearsighted or farsighted? Explain.

18. • **CE** The umpire at a baseball game wears glasses that make his eyes look smaller than they actually are. Is the umpire near-sighted or farsighted? Explain.

19. • Construct a ray diagram for Active Example 27–2.

20. • The cornea of a normal human eye has an optical power of +43.0 diopters. What is its focal length?

21. • A myopic student is shaving without his glasses. If his eyes have a far point of 1.6 m, what is the greatest distance he can stand from the mirror and still see his image clearly?

22. • An eyeglass prescription calls for a lens with an optical power of +2.7 diopters. What is the focal length of this lens?

23. •• Two thin lenses, with $f_1 = +25.0$ cm and $f_2 = -42.5$ cm, are placed in contact. What is the focal length of this combination?

24. •• Two thin lenses have refractive powers of +4.00 diopters and −2.35 diopters. What is the refractive power of the two if they are placed in contact? (Note that these are the same two lenses described in the previous problem.)

25. •• Two concave lenses, each with $f = -12$ cm, are separated by 6.0 cm. An object is placed 24 cm in front of one of the lenses. Find (a) the location and (b) the magnification of the final image produced by this lens combination.

26. •• **IP BIO** The focal length of a relaxed human eye is approximately 1.7 cm. When we focus our eyes on a close-up object, we can change the refractive power of the eye by about 16 diopters. (a) Does the refractive power of our eyes increase or decrease by 16 diopters when we focus closely? Explain. (b) Calculate the focal length of the eye when we focus closely.

27. •• **IP BIO Diopter Change in Diving Cormorants** Double-crested cormorants (*Phalacrocorax auritus*) are extraordinary birds—they can focus on objects in the air, just like we can, but they can also focus underwater as they pursue their prey. To do so, they have one of the largest *accommodation ranges* in nature—that is, they can change the focal length of their eyes by amounts that are greater than is possible in other animals. When a cormorant plunges into the ocean to catch a fish, it can change the refractive power of its eyes by about 45 diopters, as compared to only 16 diopters of change possible in the human eye. (a) Should this change of 45 diopters be an increase or a decrease? Explain. (b) If the focal length of the cormorant's eyes is 4.2 mm before it enters the water, what is the focal length after the refractive power changes by 45 diopters?

28. •• A converging lens of focal length 8.000 cm is 20.0 cm to the left of a diverging lens of focal length −6.00 cm. A coin is placed 12.0 cm to the left of the converging lens. Find (a) the location and (b) the magnification of the coin's final image.

29. •• Repeat Problem 28, this time with the coin placed 18.0 cm to the right of the diverging lens.

30. •• Find the focal length of contact lenses that would allow a farsighted person with a near-point distance of 176 cm to read a book at a distance of 10.1 cm.

31. •• Find the focal length of contact lenses that would allow a nearsighted person with a 135-cm far point to focus on the stars at night.

32. •• What focal length should a pair of contact lenses have if they are to correct the vision of a person with a near point of 56 cm?

33. •• A nearsighted person wears contacts with a focal length of −8.5 cm. If this person's far-point distance with her contacts is 8.5 m, what is her uncorrected far-point distance?

34. •• Without his glasses, Isaac can see objects clearly only if they are less than 4.5 m from his eyes. What focal length glasses worn 2.1 cm from his eyes will allow Isaac to see distant objects clearly?

35. •• A person whose near-point distance is 49 cm wears a pair of glasses that are 2.0 cm from her eyes. With the aid of these glasses, she can now focus on objects 25 cm away from her eyes. Find the focal length and refractive power of her glasses.

36. •• A pair of eyeglasses is designed to allow a person with a far-point distance of 2.50 m to read a road sign at a distance of 25.0 m. Find the focal length required of these glasses if they are to be worn (a) 2.00 cm or (b) 1.00 cm from the eyes.

37. •• **IP** Your favorite aunt can read a newspaper only if it is within 15.0 cm of her eyes. (a) Is your aunt nearsighted or farsighted? Explain. (b) Should your aunt wear glasses that are converging or diverging to improve her vision? Explain. (c) How many diopters of refractive power must her glasses have if they are worn 2.00 cm from the eyes and allow her to read a newspaper at a distance of 25.0 cm?

38. •• **IP** The relaxed eyes of a patient have a refractive power of 48.5 diopters. (a) Is this patient nearsighted or farsighted? Explain. (b) If this patient is nearsighted, find the far point. If this person is farsighted, find the near point. (For the purposes of this problem, treat the eye as a single-lens system, with the retina 2.40 cm from the lens.)

39. •• **IP** You are comfortably reading a book at a distance of 24 cm. (a) What is the refractive power of your eyes? (b) Does the refractive power of your eyes increase or decrease when you move the book farther away? Explain. (For the purposes of this problem, treat the eye as a single-lens system, with the retina 2.40 cm from the lens.)

40. •• Without glasses, your Uncle Albert can see things clearly only if they are between 25 cm and 170 cm from his eyes. (a) What power eyeglass lens will correct your uncle's myopia? Assume the lenses will sit 2.0 cm from his eyes. (b) What is your uncle's near point when wearing these glasses?

41. •• A 2.05-cm-tall object is placed 30.0 cm to the left of a converging lens with a focal length $f_1 = 20.5$ cm. A diverging lens, with a focal length $f_2 = -42.5$ cm, is placed 30.0 cm to the right of the first lens. How tall is the final image of the object?

42. •• A simple camera telephoto lens consists of two lenses. The objective lens has a focal length $f_1 = +39.0$ cm. Precisely 36.0 cm behind this lens is a concave lens with a focal length $f_2 = -10.0$ cm. The object to be photographed is 4.00 m in front of the objective lens. (a) How far behind the concave lens should the film be placed? (b) What is the linear magnification of this lens combination?

43. •• **IP** With unaided vision, a librarian can focus only on objects that lie at distances between 5.0 m and 0.50 m. (a) Which type of lens (converging or diverging) is needed to correct his near-sightedness? Explain. (b) Which type of lens will correct his far-sightedness? Explain. (c) Find the refractive power needed for each part of the bifocal eyeglass lenses that will give the librarian normal visual acuity from 25 cm out to infinity. (Assume the lenses rest 2.0 cm from his eyes.)

44. •• **IP** With unaided vision, a physician can focus only on objects that lie at distances between 5.0 m and 0.50 m. (a) Which type of lens (converging or diverging) is needed to correct her nearsightedness? Explain. (b) Which type of lens will correct her farsightedness? Explain. (c) Find the refractive power needed for each part of the bifocal contact lenses that will give the physician normal visual acuity from 25 cm out to infinity.

45. •• A person's prescription for her new bifocal glasses calls for a refractive power of -0.445 diopter in the distance-vision part, and a power of $+1.85$ diopters in the close-vision part. What are the near and far points of this person's uncorrected vision? Assume the glasses are 2.00 cm from the person's eyes, and that the person's near-point distance is 25.0 cm when wearing the glasses.

46. •• A person's prescription for his new bifocal eyeglasses calls for a refractive power of -0.0625 diopter in the distance-vision part and a power of $+1.05$ diopters in the close-vision part. Assuming the glasses rest 2.00 cm from his eyes and that the corrected near-point distance is 25.0 cm, determine the near and far points of this person's uncorrected vision.

47. ••• Two lenses, with $f_1 = +20.0$ cm and $f_2 = +30.0$ cm, are placed on the x axis, as shown in **Figure 27–22**. An object is fixed 50.0 cm to the left of lens 1, and lens 2 is a variable distance x to the right of lens 1. Find the lateral magnification and location of the final image relative to lens 2 for the following cases: **(a)** $x = 115$ cm; **(b)** $x = 30.0$ cm; **(c)** $x = 0$. **(d)** Show that your result for part (c) agrees with the relation for the effective focal length of two lenses in contact, $1/f_{eff} = 1/f_1 + 1/f_2$.

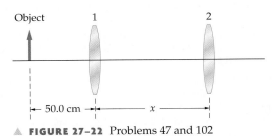

▲ **FIGURE 27–22** Problems 47 and 102

48. ••• A converging lens with a focal length of 4.0 cm is to the left of a second identical lens. When a feather is placed 12 cm to the left of the first lens, the final image is the same size and orientation as the feather itself. What is the separation between the lenses?

SECTION 27–3   THE MAGNIFYING GLASS

49. • The Moon is 3476 km in diameter and orbits the Earth at an average distance of 384,400 km. **(a)** What is the angular size of the Moon as seen from Earth? **(b)** A penny is 19 mm in diameter. How far from your eye should the penny be held to produce the same angular diameter as the Moon?

50. •• A magnifying glass is a single convex lens with a focal length of $f = +14.0$ cm. **(a)** What is the angular magnification when this lens forms a (virtual) image at $-\infty$? How far from the object should the lens be held? **(b)** What is the angular magnification when this lens forms a (virtual) image at the person's near point (assumed to be 25 cm)? How far from the object should the lens be held in this case?

51. •• **IP** A student has two lenses, one of focal length $f_1 = 5.0$ cm and the other with focal length $f_2 = 13$ cm. **(a)** When used as a simple magnifier, which of these lenses can produce the greater magnification? Explain. **(b)** Find the maximum magnification produced by each of these lenses.

52. •• A beetle 4.73 mm long is examined with a simple magnifier of focal length $f = 10.1$ cm. If the observer's eye is relaxed while using the magnifier, and has a near-point distance of 25.0 cm, what is the apparent length of the beetle?

53. •• To engrave wishes of good luck on a watch, an engraver uses a magnifier whose focal length is 8.65 cm. If the image formed by the magnifier is at the engraver's near point of 25.6 cm, find **(a)** the distance between the watch and the magnifier and

(b) the angular magnification of the engraving. Assume the magnifying glass is directly in front of the engraver's eyes.

54. •• A jeweler examines a diamond with a magnifying glass. If the near-point distance of the jeweler is 20.8 cm, and the focal length of the magnifying glass is 7.50 cm, find the angular magnification when the diamond is held at the focal point of the magnifier. Assume the magnifying glass is directly in front of the jeweler's eyes.

55. •• In Problem 54, find the angular magnification when the diamond is held 5.59 cm from the magnifying glass.

56. ••• A person with a near-point distance of 25 cm finds that a magnifying glass gives an angular magnification that is 1.5 times larger when the image of the magnifier is at the near point than when the image is at infinity. What is the focal length of the magnifying glass?

SECTION 27–4   THE COMPOUND MICROSCOPE

57. • **CE** You have two lenses: lens 1 with a focal length of 0.45 cm and lens 2 with a focal length of 1.9 cm. If you construct a microscope with these lenses, which one should you use as the objective? Explain.

58. • A compound microscope has an objective lens with a focal length of 2.2 cm and an eyepiece with a focal length of 5.4 cm. If the image produced by the objective is 12 cm from the objective, what magnification does this microscope produce?

59. • **BIO** A typical red blood cell subtends an angle of only 1.9×10^{-5} rad when viewed at a person's near-point distance of 25 cm. Suppose a red blood cell is examined with a compound microscope in which the objective and eyepiece are separated by a distance of 12.0 cm. Given that the focal length of the eyepiece is 2.7 cm, and the focal length of the objective is 0.49 cm, find the magnitude of the angle subtended by the red blood cell when viewed through this microscope.

60. •• The medium-power objective lens in a laboratory microscope has a focal length $f_{objective} = 4.00$ mm. **(a)** If this lens produces a lateral magnification of -40.0, what is its "working distance"; that is, what is the distance from the object to the objective lens? **(b)** What is the focal length of an eyepiece lens that will provide an overall magnification of 125?

61. •• A compound microscope has the objective and eyepiece mounted in a tube that is 18.0 cm long. The focal length of the eyepiece is 2.62 cm, and the near-point distance of the person using the microscope is 25.0 cm. If the person can view the image produced by the microscope with a completely relaxed eye, and the magnification is -4525, what is the focal length of the objective?

62. •• In Problem 61, what is the distance between the objective lens and the object to be examined?

63. •• The barrel of a compound microscope is 15 cm in length. The specimen will be mounted 1.0 cm from the objective, and the eyepiece has a 5.0-cm focal length. Determine the focal length of the objective lens.

64. •• A compound microscope uses a 75.0-mm lens as the objective and a 2.0-cm lens as the eyepiece. The specimen will be mounted 122 mm from the objective. Determine **(a)** the barrel length and **(b)** the total magnification produced by the microscope.

65. ••• The "tube length" of a microscope is defined to be the difference between the (objective) image distance and objective focal length: $L = d_i - f_{objective}$. Many microscopes are standardized to a tube length of $L = 160$ mm. Consider such a microscope whose objective lens has a focal length $f_{objective} = 7.50$ mm.

(a) How far from the object should this lens be placed? **(b)** What focal length eyepiece would give an overall magnification of −55? **(c)** What focal length eyepiece would give an overall magnification of −110?

SECTION 27–5 TELESCOPES

66. • **CE** Two telescopes of different length produce the same angular magnification. Is the focal length of the long telescope's eyepiece greater than or less than the focal length of the short telescope's eyepiece? Explain.

67. • **CE** To construct a telescope, you are given a lens with a focal length of 32 mm and a lens with a focal length of 1600 mm. **(a)** On the basis of focal length alone, which lens should be the objective and which the eyepiece? Explain. **(b)** What magnification would this telescope produce?

68. • A grade school student plans to build a 35-power telescope as a science fair project. She starts with a magnifying glass with a focal length of 5.0 cm as the eyepiece. What focal length is needed for her objective lens?

69. • A 55-power refracting telescope has an eyepiece with a focal length of 5.0 cm. How long is the telescope?

70. • An amateur astronomer wants to build a small refracting telescope. The only lenses available to him have focal lengths of 5.00 cm, 10.0 cm, 20.0 cm, and 30.0 cm. **(a)** What is the greatest magnification that can be obtained using two of these lenses? **(b)** How long is the telescope with the greatest magnification?

71. • A pirate sights a distant ship with a spyglass that gives an angular magnification of 22. If the focal length of the eyepiece is 11 mm, what is the focal length of the objective?

72. •• A telescope has lenses with focal lengths $f_1 = +30.0$ cm and $f_2 = +5.0$ cm. **(a)** What distance between the two lenses will allow the telescope to focus on an infinitely distant object and produce an infinitely distant image? **(b)** What distance between the lenses will allow the telescope to focus on an object that is 5.0 m away and to produce an infinitely distant image?

73. •• Jason has a 25-power telescope whose objective lens has a focal length of 120 cm. To make his sister appear smaller than normal, he turns the telescope around and looks through the objective lens. What is the angular magnification of his sister when viewed through the "wrong" end of the telescope?

74. •• **Roughing It with Science** A professor shipwrecked on Hooligan's Island decides to build a telescope from his eyeglasses and some coconut shells. Fortunately, the professor's eyes require different prescriptions, with the left lens having a power of +5.0 diopters and the right lens having a power of +2.0 diopters. **(a)** Which lens should he use as the objective? **(b)** What is the angular magnification of the professor's telescope?

75. •• **Galileo's Telescope** Galileo's first telescope used a convex objective lens with a focal length $f = 1.7$ m and a concave eyepiece, as shown in **Figure 27–23**. When this telescope is focused on an infinitely distant object, and produces an infinitely distant image, its angular magnification is +3.0. **(a)** What is the focal length of the eyepiece? **(b)** How far apart are the two lenses?

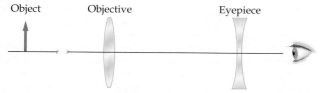

▲ **FIGURE 27–23** Problems 75, 86, and 94

76. •• The Moon has an angular size of 0.50° when viewed with unaided vision from Earth. Suppose the Moon is viewed through a telescope with an objective whose focal length is 53 cm and an eyepiece whose focal length is 25 mm. What is the angular size of the Moon as seen through this telescope?

77. •• In Problem 76, an eyepiece is selected to give the Moon an angular size of 15°. What is the focal length of this eyepiece?

78. •• A telescope is 275 mm long and has an objective lens with a focal length of 257 mm. **(a)** What is the focal length of the eyepiece? **(b)** What is the magnification of this telescope?

GENERAL PROBLEMS

79. • **CE Predict/Explain BIO Intracorneal Ring** An intracorneal ring is a small plastic device implanted in a person's cornea to change its curvature. By changing the shape of the cornea, the intracorneal ring can correct a person's vision. **(a)** If a person is nearsighted, should the ring increase or decrease the cornea's curvature? **(b)** Choose the *best explanation* from among the following:
    I. The intracorneal ring should increase the curvature of the cornea so that it bends light more. This will allow it to focus on light coming from far away.
    II. The intracorneal ring should decrease the curvature of the cornea so it's flatter and bends light less. This will allow parallel rays from far away to be focused.

80. • **CE BIO** The lens in a normal human eye, with aqueous humor on one side and vitreous humor on the other side, has a refractive power of 15 diopters. Suppose a lens is removed from an eye and surrounded by air. In this case, is its refractive power greater than, less than, or equal to 15 diopters? Explain.

81. • **CE** An optical system consists of two lenses, one with a focal length of 0.50 cm and the other with a focal length of 2.3 cm. If the separation between the lenses is 12 cm, is the instrument a microscope or a telescope? Explain.

82. • **CE** An optical system consists of two lenses, one with a focal length of 50 cm and the other with a focal length of 2.5 cm. If the separation between the lenses is 52.5 cm, is the instrument a microscope or a telescope? Explain.

83. • **CE Predict/Explain BIO Treating Cataracts** When the lens in a person's eye becomes clouded by a cataract, the lens can be removed with a process called phacoemulsification and replaced with a man-made intraocular lens. The intraocular lens restores clear vision, but its focal length cannot be changed to allow the user to focus at different distances. In most cases, the intraocular lens is adjusted for viewing of distant objects, and corrective glasses are worn when viewing nearby objects. **(a)** Should the refractive power of the corrective glasses be positive or negative? **(b)** Choose the *best explanation* from among the following:
    I. The refractive power should be positive—converging—because the intraocular lens will make the person farsighted.
    II. A negative refractive power is required to bring the focal point of the intraocular lens in from infinity to a finite value.

84. •• **IP** The greatest refractive power a patient's eyes can produce is 44.1 diopters. **(a)** Is this patient nearsighted or farsighted? Explain. **(b)** If this patient is nearsighted, find the far point. If this person is farsighted, find the near point. (For the purposes of this problem, treat the eye as a single-lens system, with the retina 2.40 cm from the lens.)

85. •• **IP** You are observing a rare species of bird in a distant tree with your unaided eyes. **(a)** What is the refractive power of your eyes? **(b)** Does the refractive power of your eyes increase

or decrease when you shift your view to the guidebook in your hands? Explain. (For the purposes of this problem, treat the eye as a single-lens system, with the retina 2.40 cm from the lens.)

86. •• Galileo's original telescope (Figure 27–23) used a convex objective and a concave eyepiece. Use a ray diagram to show that this telescope produces an upright image when a distant object is being viewed. Assume that the eyepiece is to the right of the object and that the right-hand focal point of the eyepiece is just to the left of the objective's right-hand focal point. In addition, assume that the focal length of the eyepiece has a magnitude that is about one-quarter the focal length of the objective.

87. •• **IP** For each of the following cases, use a ray diagram to show that the angular sizes of the image and the object are identical if both angles are measured *from the center of the lens*. **(a)** A convex lens with the object outside the focal length. **(b)** A convex lens with the object inside the focal length. **(c)** A concave lens with the object outside the focal length. **(d)** Given that the angular size does not change, how does a simple magnifier work? Explain.

88. •• **IP** You have two lenses, with focal lengths $f_1 = +2.60$ cm and $f_2 = +20.4$ cm. **(a)** How would you arrange these lenses to form a magnified image of the Moon? **(b)** What is the maximum angular magnification these lenses could produce? **(c)** How would you arrange the same two lenses to form a magnified image of an insect? **(d)** If you use the magnifier of part (c) to view an insect, what is the angular magnification when the insect is held 2.90 cm from the objective lens?

89. •• **BIO** The eye is actually a multiple-lens system, but we can approximate it with a single-lens system for most of our purposes. When the eye is focused on a distant object, the optical power of the equivalent single lens is +41.4 diopters. **(a)** What is the effective focal length of the eye? **(b)** How far in front of the retina is this "equivalent lens" located?

90. •• **BIO Fitting Contact Lenses with a Keratometer** When a patient is being fitted with contact lenses, the curvature of the patient's cornea is measured with an instrument known as a keratometer. A lighted object is held near the eye, and the keratometer measures the magnification of the image formed by reflection from the front of the cornea. If an object is held 10.0 cm in front of a patient's eye, and the reflected image is magnified by a factor of 0.035, what is the radius of curvature of the patient's cornea?

91. •• **Pricey Stamp** A rare 1918 "Jenny" stamp, depicting a misprinted, upside-down Curtiss JN-4 "Jenny" airplane, sold at auction for $525,000. A collector uses a simple magnifying glass to examine the "Jenny," obtaining a linear magnification of 2.5 when the stamp is held 2.76 cm from the lens. What is the focal length of the magnifying glass?

92. •• **IP** A person needs glasses with a refractive power of −1.35 diopters to be able to focus on distant objects. **(a)** Is this person nearsighted or farsighted? Explain. **(b)** What is this person's (unaided) far point?

93. •• **IP BIO A Big Eye** The largest eye ever to exist on Earth belonged to an extinct species of ichthyosaur, *Temnodontosaurus platyodon*. This creature had an eye that was 26.4 cm in diameter. It is estimated that this ichthyosaur also had a relatively large pupil, giving it an effective aperture setting of about $f/1.1$. **(a)** Assuming its pupil was one-third the diameter of the eye, what was the approximate focal length of the ichthyosaur's eye? **(b)** When the ichthyosaur narrowed its pupil in bright light, did its f-number increase or decrease? Explain.

94. •• Consider a Galilean telescope, as illustrated in Figure 27–23, constructed from two lenses with focal lengths of 75.6 cm and −18.0 mm. **(a)** What is the distance between these lenses if an infinitely distant object is to produce an infinitely distant image? **(b)** What is the angular magnification when the lenses are separated by the distance calculated in part (a)?

95. •• A converging lens forms a virtual object 12 cm to the right of a second lens that has a refracting power of 3.75 diopter. **(a)** Where is the image? **(b)** Is the image real or virtual? Explain.

96. •• A farsighted person uses glasses with a refractive power of 3.6 diopters. The glasses are worn 2.5 cm from his eyes. What is this person's near point when not wearing glasses?

97. ••• **Landing on an Aircraft Carrier** The Long-Range Lineup System (LRLS) used to ensure safe landings on aircraft carriers consists of a series of Fresnel lenses of different colors. Each lens focuses light in a different, specific direction, and hence which light a pilot sees on approach determines whether the plane is above, below, or on the proper landing path. The basic idea behind a Fresnel lens, which has the same optical properties as an ordinary lens, is shown in **Figure 27–24**, along with a photo of the LRLS. Suppose an object (a lightbulb in this case) is 17.1 cm behind a Fresnel lens, and that the corresponding image is a distance $d_i = d$ in front of the lens. If the object is moved to a distance of 12.0 cm behind the lens, the image distance doubles to $d_i = 2d$. In the LRLS, it is desired to have the image of the lightbulb at infinity. What object distance will give this result for this particular lens?

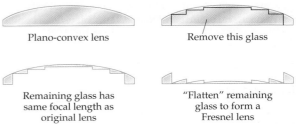

Plano-convex lens   Remove this glass

Remaining glass has   "Flatten" remaining
same focal length as   glass to form a
original lens   Fresnel lens

(a)

(b)

▲ **FIGURE 27–24 Fresnel lenses and the Long-Range Lineup System**

(a) A lens causes light to refract at its surface; therefore, the interior glass can be removed without changing its optical properties. This produces a Fresnel lens, which is much lighter than the original lens. **(b)** If an airplane is on the correct approach path, the pilot will see an amber light, called the "meatball," in line with the row of blue lights.

98. ••• When using a telescope to photograph a faint astronomical object, you need to maximize the amount of light energy that falls on each square millimeter of the image on the film. For a given telescope and object, the total light that falls on the film is proportional to the length of the exposure, so a long exposure will reveal fainter objects than a short exposure. Show that for a given length of exposure, the brightness of the image is inversely proportional to the *square* of the f-number of the telescope system.

99. ••• A Cassegrain astronomical telescope uses two mirrors to form the image. The larger (concave) objective mirror has a focal length $f_1 = +50.0$ cm. A small convex secondary mirror is mounted 43.0 cm in front of the primary. As shown in **Figure 27–25**, light is reflected from the secondary through a hole in the center of the primary, thereby forming a real image 8.00 cm behind the primary mirror. What is the radius of curvature of the secondary mirror?

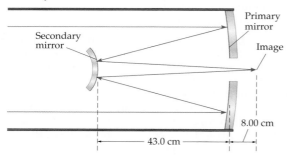

▲ **FIGURE 27–25** Problem 99

100. ••• **IP** A convex lens ($f = 20.0$ cm) is placed 10.0 cm in front of a plane mirror. A matchstick is placed 25.0 cm in front of the lens, as shown in **Figure 27–26**. **(a)** If you look through the lens toward the mirror, where will you see the image of the matchstick? **(b)** Is the image real or virtual? Explain. **(c)** What is the magnification of the image? **(d)** Is the image upright or inverted?

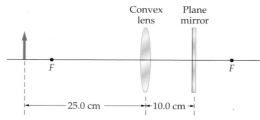

▲ **FIGURE 27–26** Problems 100 and 101

101. ••• Repeat Problem 100 for the case where the converging lens is replaced with a diverging lens with $f = -20.0$ cm. Everything else in the problem remains the same.

102. ••• Repeat Problem 47 for the case where lens 1 is replaced with a diverging lens with $f_1 = -20.0$ cm. Everything else in the problem remains the same.

103. ••• The diameter of a collimated laser beam can be expanded or reduced by using two converging lenses, with focal lengths f_1 and f_2, mounted a distance $f_1 + f_2$ from each other, as shown in **Figure 27–27**. What is the ratio of the two beam diameters, (d_1/d_2), expressed in terms of the focal lengths?

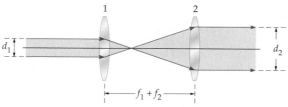

▲ **FIGURE 27–27** Problem 103

104. ••• Consider three lenses with focal lengths of 25.0 cm, -15.0 cm, and 11.0 cm positioned on the x axis at $x = 0$, $x = 0.400$ m, and $x = 0.500$ m, respectively. An object is at $x = -122$ cm. Find **(a)** the location and **(b)** the orientation and magnification of the final image produced by this lens system.

105. ••• Because a concave lens cannot form a real image of a real object, it is difficult to measure its focal length precisely. One method uses a second, convex, lens to produce a virtual object for the concave lens. Under the proper conditions, the concave lens will form a real image of the virtual object! A student conducting a laboratory project on concave lenses makes the following observations: When a lamp is placed 42.0 cm to the left of a particular convex lens, a real (inverted) image is formed 37.5 cm to the right of the lens. The lamp and convex lens are kept in place while a concave lens is mounted 15.0 cm to the right of the convex lens. A real image of the lamp is now formed 35.0 cm to the right of the concave lens. What is the focal length of each lens?

106. ••• A person with a near-point distance N uses a magnifying glass with a focal length f. Show that the greatest magnification that can be achieved with this magnifier is $M = 1 + N/f$.

PASSAGE PROBLEMS

BIO Cataracts and Intraocular Lenses

A cataract is an opacity or "cloudiness" that develops in the lens of an eye. The result can be serious degradation of vision, or even blindness. In fact, cataracts are the leading cause of blindness worldwide, and in the United States 60% of the population between the ages of 65 and 74 have cataracts to some extent. Cataracts can be caused by prolonged exposure to electromagnetic radiation of almost any form, including microwaves,

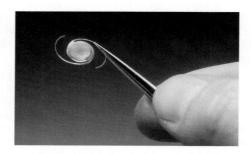

(a) A spring-loaded, nonadaptive intraocular lens

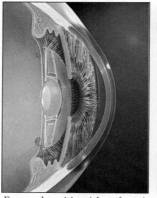

Forward position (close focus)   Rear position (focus at infinity)

(b) An adaptive intraocular lens in action

▲ **FIGURE 27–28** **Intraocular Lenses**

In cases of severe cataract, an intraocular lens (IOL) can be implanted in place of the eye's natural lens. An adaptive IOL makes use of the same muscles that change he shape of the natural lens, although the lens moves forward and back rather than changing shape.

ultraviolet rays, and infrared rays. For example, cataracts are unusually common among airline pilots, who encounter intense UV exposure at high altitude, and glassblowers, who are exposed to infrared radiation for long periods of time.

Cataracts are generally treated by removing the affected lens with a technique referred to as phacoemulsification. After the natural lens is removed, it is replaced with a man-made, intraocular lens, or IOL. In many cases, the IOL is rigid; neither its focal length nor location can be changed. These lenses are designed to allow the eye to see clearly at infinity, but corrective glasses or contacts must be worn for close vision. More recently, adaptive IOLs have been developed that flex when the focusing muscles of the eye contract, thus allowing a degree of accommodation. This is illustrated in **Figure 27–28**, where we see the IOL move forward to focus on a close object. Notice that the focal length of the adaptive IOL is fixed, just as with a normal IOL, but the eye muscles can change its location—the same as in a camera when it focuses.

107. • A patient receives a rigid IOL whose focus cannot be changed—it is designed to provide clear vision of objects at infinity. The patient will use corrective contacts to allow for close vision. Should the refractive power of the corrective contacts be positive or negative?

108. • Referring to the previous problem, find the refractive power of contacts that will allow the patient to focus on a book at a distance of 23.0 cm.

  A. 0.0435 diopter   **B.** 0.230 diopter

  C. 4.35 diopters   **D.** 8.70 diopters

109. • Suppose a flexible, adaptive IOL has a focal length of 3.00 cm. How far forward must the IOL move to change the focus of the eye from an object at infinity to an object at a distance of 50.0 cm?

  A. 1.9 mm   **B.** 2.8 mm

  C. 3.1 mm   **D.** 3.2 mm

INTERACTIVE PROBLEMS

110. •• **IP Referring to Example 27–2** Suppose a person's eyeglasses have a focal length of -301 cm, are 2.00 cm in front of the eyes, and allow the person to focus on distant objects. **(a)** Is this person's far point greater than or less than 323 cm, which is the far point for glasses the same distance from the eyes and with a focal length of -321 cm? Explain. **(b)** Find the far point for this person.

111. •• **IP Referring to Example 27–2** In Example 27–2, a person has a far-point distance of 323 cm. If this person wears glasses 2.00 cm in front of the eyes with a focal length of -321 cm, distant objects can be brought into focus. Suppose a second person's far point is 353 cm. **(a)** Is the magnitude of the focal length of the eyeglasses that allow this person to focus on distant objects greater than or less than 321 cm? Assume the glasses are 2.00 cm in front of the eyes. **(b)** Find the required focal length for the second person's eyeglasses.

112. •• **IP Referring to Example 27–3** Suppose a person's eyeglasses have a refractive power of 2.75 diopters and that they allow the person to focus on an object that is just 25.0 cm from the eye. The glasses are 2.00 cm in front of the eyes. **(a)** Is this person's near point greater than or less than 57.0 cm, which is the near-point distance when the glasses have a refractive power of 2.53 diopters? Explain. **(b)** Find the near point for this person.

113. •• **IP Referring to Example 27–3** Suppose a person's near-point distance is 67.0 cm. **(a)** Is the refractive power of the eyeglasses that allow this person to focus on an object just 25.0 cm from the eye greater than or less than 2.53 diopters, which is the refractive power when the near-point distance is 57.0 cm? The glasses are worn 2.00 cm in front of the eyes. **(b)** Find the required refractive power for this person's eyeglasses.

28 Physical Optics: Interference and Diffraction

Most of the colors that we see every day can be understood in terms of absorption and reflection—an old-fashioned fire engine absorbs green light and reflects red light, a grassy lawn absorbs red light and reflects green light, and so on. Yet the distinctive iridescent hues of this blue morpho butterfly are created by different physical mechanisms. These mechanisms, known as interference and diffraction, cannot be understood solely in terms of geometrical optics. Instead, they have their origin in the wave nature of light, the subject of this chapter.

In our discussion of optics to this point we have treated light in terms of "rays" that propagate in straight lines. This description is valid in a wide variety of circumstances, as we have seen, though it ignores the fact that light is an electromagnetic *wave*. There are situations, however, in which the ray model fails and the wave nature of light is of central importance. For example, any

time we deal with objects or sources that have characteristic dimensions comparable to the wavelength of light, we shall find new effects not predicted by ray optics. In this chapter we show that the wave properties of light are responsible for such phenomena as the operation of a CD, the appearance of images on a television screen, and the brilliant iridescent colors of a butterfly's wing.

28–1 Superposition and Interference

One of the fundamental aspects of wave behavior is **superposition,** in which the net displacement caused by a combination of waves is the algebraic sum of the displacements caused by each wave individually. If waves add to cause a larger displacement, we say that they interfere **constructively;** if the net displacement is reduced, the interference is **destructive.** Examples of wave superposition on a string are given in Figures 14–19 and 14–20. Because light also exhibits wave behavior, with propagating electric and magnetic fields, it too can show interference effects and a resulting increase or decrease in brightness.

Interference is noticeable, however, only if certain conditions are met. In the case of light, for example, the light should be **monochromatic;** that is, it should have a single color and hence a single frequency. In addition, if two or more sources of light are to show interference, they must maintain a constant phase relationship with one another—that is, they must be **coherent.** Sources whose relative phases vary randomly with time show no discernible interference patterns and are said to be **incoherent.** Incoherent light sources include incandescent and fluorescent lights. In contrast, lasers emit light that is both monochromatic and coherent.

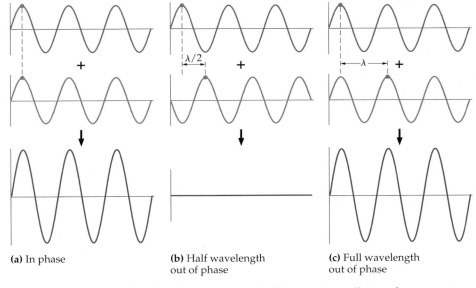

(a) In phase (b) Half wavelength out of phase (c) Full wavelength out of phase

◀ **FIGURE 28–1 Constructive and destructive interference**

(a) Waves that are in phase add to give a larger amplitude. This is constructive interference. **(b)** Waves that are half a wavelength out of phase interfere destructively. If the individual waves have equal amplitudes, as here, their sum will have zero amplitude. **(c)** When waves are one wavelength out of phase, the result is again constructive interference, exactly the same as when the waves are in phase.

The basic principle that determines whether waves will interfere constructively or destructively is their phase relative to one another. For example, if two waves have zero phase difference—that is, the waves are in phase—they add constructively, and the net result is an increased amplitude, as **Figure 28–1 (a)** indicates. If waves of equal amplitude are 180° out of phase, however, the net result is zero amplitude and destructive interference, as indicated in **Figure 28–1 (b).** Notice that a *180° difference in phase* corresponds to waves being out of step by *half a wavelength.* Similarly, if the phase difference between two waves is 360°—which corresponds to one full wavelength—the interference is again constructive, as in **Figure 28–1 (c).**

Finally, Figure 28–1 shows the blue wave shifted to the right, meaning it is *ahead* of the red wave by half a wavelength in part (b), and *ahead* by a full wavelength in part (c). The same results are obtained, however, if the blue wave is *behind* the red wave. Thus, for example, a phase difference of −180° produces destructive interference, the same as in Figure 28–1 (b), and a phase difference of −360° produces constructive interference, just as in Figure 28–1 (c).

Let's apply the preceding observations to a system consisting of two radio antennas radiating electromagnetic waves of frequency f and wavelength λ, as in **Figure 28–2.** If the antennas are connected to the same transmitter, they emit waves that are in phase and coherent. When waves from each antenna reach point P_0, they have traveled the same distance—that is, the same number of wavelengths—and hence they are still in phase. As a result, point P_0 experiences constructive

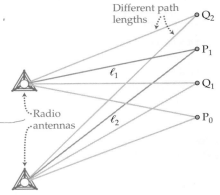

▲ **FIGURE 28–2 Two radio antennas transmitting the same signal**

At point P_0, midway between the antennas, the waves travel the same distance, and hence they interfere constructively. At point P_1 the distance ℓ_2 is greater than the distance ℓ_1 by one wavelength; thus, P_1 is also a point of constructive interference. At Q_1 the distance ℓ_2 is greater than the distance ℓ_1 by half a wavelength, and the waves interfere destructively at that point.

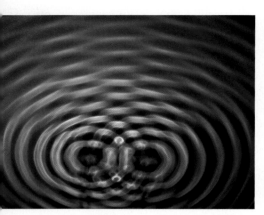

▲ Two sets of water waves, radiating outward in circular patterns from point sources, create an interference pattern where they overlap.

interference and the radio signal is strong. This location corresponds to the situation shown in Figure 28–1 (a).

To reach point P_1, the waves from the two antennas must travel different distances, ℓ_1 and ℓ_2, as indicated in Figure 28–2. If the difference in these distances is one wavelength, $\ell_2 - \ell_1 = \lambda$, the waves are 360° out of phase, and point P_1 also has constructive interference. This location corresponds to Figure 28–1 (c). Similarly, if the difference in path length to a point P_m is an integer $m = 0, 1, 2, \ldots$ times the wavelength, $\ell_2 - \ell_1 = m\lambda$, that point will also be a location of constructive interference.

Finally, the difference in path length to point Q_1 is half a wavelength, $\ell_2 - \ell_1 = \lambda/2$, and hence the waves cancel there, just as in Figure 28–1 (b). Destructive interference also occurs at Q_2, where the difference in path length is one and a half wavelengths, $\ell_2 - \ell_1 = 3\lambda/2$. In general, for $m = 1, 2, 3, \ldots$, destructive interference occurs at points Q_m where the difference in path lengths is $\ell_2 - \ell_1 = \left(m - \frac{1}{2}\right)\lambda$.

To summarize, *constructive interference* satisfies the conditions

$$\ell_2 - \ell_1 = m\lambda \qquad m = 0, 1, 2, \ldots \qquad \textit{(constructive interference)}$$

Similarly, *destructive interference* satisfies the conditions

$$\ell_2 - \ell_1 = \left(m - \tfrac{1}{2}\right)\lambda \quad m = 1, 2, 3, \ldots \qquad \textit{(destructive interference)}$$

These conditions are applied to a specific physical system in the following Example.

EXAMPLE 28–1 TWO MAY NOT BE BETTER THAN ONE

Two friends tune their radios to the same frequency and pick up a signal transmitted simultaneously by a pair of antennas. The friend who is equidistant from the antennas, at P_0, receives a strong signal. The friend at point Q_1 receives a very weak signal. Find the wavelength of the radio waves if $d = 7.50$ km, $L = 14.0$ km, and $y = 1.88$ km. Assume that Q_1 is the first point of minimum signal as one moves away from P_0 in the y direction.

PICTURE THE PROBLEM
Our sketch shows the radio antennas and the two locations mentioned in the problem statement. Notice that the radio antennas are separated by a distance $d = 7.50$ km in the y direction, and that the points P_0 and Q_1 have a y-direction separation of $y = 1.88$ km. The distance to P_0 and Q_1 in the x direction is $L = 14.0$ km.

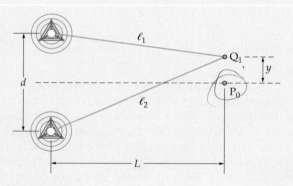

STRATEGY
Since point Q_1 is the first *minimum* in the y direction from the *maximum* at point P_0, we know that the path difference, $\ell_2 - \ell_1$, is half a wavelength. Thus we can determine λ by calculating the lengths ℓ_2 and ℓ_1 and setting their difference equal to $\lambda/2$.

SOLUTION

1. Calculate the path length ℓ_1:

$$\ell_1 = \sqrt{L^2 + \left(\frac{d}{2} - y\right)^2}$$

$$= \sqrt{(14.0 \text{ km})^2 + \left(\frac{7.50 \text{ km}}{2} - 1.88 \text{ km}\right)^2} = 14.1 \text{ km}$$

2. Calculate the path length ℓ_2:

$$\ell_2 = \sqrt{L^2 + \left(\frac{d}{2} + y\right)^2}$$

$$= \sqrt{(14.0 \text{ km})^2 + \left(\frac{7.50 \text{ km}}{2} + 1.88 \text{ km}\right)^2} = 15.1 \text{ km}$$

3. Set $\ell_2 - \ell_1$ equal to $\lambda/2$ and solve for the wavelength:

$$\ell_2 - \ell_1 = \tfrac{1}{2}\lambda$$

$$\lambda = 2(\ell_2 - \ell_1) = 2(15.1 \text{ km} - 14.1 \text{ km}) = 2.0 \text{ km}$$

INSIGHT
We see that radio waves are rather large. In fact, the distance from one crest to the next (the wavelength) for these waves is about 1.2 miles. Recalling that radio waves travel at the speed of light, the corresponding frequency is $f = c/\lambda = 150 \text{ kHz}$.

PRACTICE PROBLEM
Suppose the wavelength broadcast by these antennas is changed, and that the y distance between P_0 and Q_1 (first minimum) increases as a result. Is the new wavelength greater than or less than 2.0 km? Find the wavelength for the case $y = 2.91$ km.
[**Answer:** Greater than 2.0 km; $\lambda = 3.0$ km]

Some related homework problems: Problem 1, Problem 3, Problem 4

28-2 Young's Two-Slit Experiment

We now consider a classic physics experiment that not only demonstrates the wave nature of light, but also allows one to determine the wavelength of a beam of light, just as we did for the radio waves in Example 28–1. The experiment was first performed in 1801 by the English physician and physicist Thomas Young (1773–1829), whose medical background contributed to his studies of vision, and whose love of languages made him a key figure in deciphering the Rosetta Stone. The experiment, in simplest form, consists of a beam of monochromatic light that passes through a small slit in a screen and then illuminates two slits, S_1 and S_2, in a second screen. After the light passes through the two slits, it shines on a distant screen, as **Figure 28–3** shows, where an interference pattern of bright and dark "fringes" is observed.

In this "two-slit experiment" the slit in the first screen serves only to produce a small source of light that prevents the interference pattern on the distant screen from becoming smeared out. The key elements in the experiment are the two slits in the second screen. Since they are equidistant from the single slit, as shown in Figure 28–3, the light passing through them has the same phase. Thus the two slits act as monochromatic, coherent sources of light—analogous to the two radio antennas in Example 28–1—as needed to produce interference.

If light were composed of small particles, or "corpuscles" as Newton referred to them, they would simply pass straight through the two slits and illuminate the distant screen directly behind each slit. If light is a wave, on the other hand, each slit acts as the source of new waves, analogous to water waves passing through two small openings. This is referred to as **Huygens's principle** and is illustrated in **Figure 28–4**. Notice that light radiates away from the slits in all forward directions—not just in the direction of the incoming light. The result is that light is spread out over a large area on the distant screen; it is not localized in small regions directly behind the slits. Thus, an experiment like this can readily distinguish between the two models of light.

Of key importance is the fact that the illumination of the distant screen is not only spread out, but also consists of alternating bright and dark fringes, as indicated in Figure 28–3. These fringes are the direct result of constructive and destructive interference. For example, the central bright fringe is midway between the two slits; hence, the path lengths for light from the slits are equal. Because light coming from the slits is in phase, it follows that the light is also in phase at the midpoint, giving rise to constructive interference—just like at point P_0 in Figure 28–2.

The next bright fringe occurs when the difference in path length from the two slits is equal to one wavelength of light, as with point P_1 in Figure 28–2. In most experimental situations the distance to the screen is much greater than the separation d of the slits; hence, the light travels to a point on the screen along approximately parallel paths, as indicated in **Figure 28–5**. Therefore, the path difference, $\Delta\ell$, is

$$\Delta\ell = d\sin\theta$$

As a result, the bright fringe closest to the midpoint occurs at the angle θ given by the condition $d\sin\theta = \lambda$, or $\sin\theta = \lambda/d$. In general, a bright fringe occurs whenever

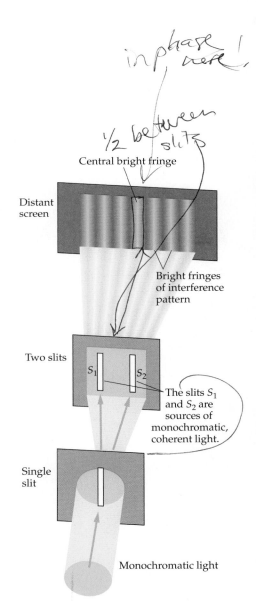

▲ **FIGURE 28-3 Young's two-slit experiment**
The first screen produces a small source of light that illuminates the two slits, S_1 and S_2. After passing through these slits, the light spreads out into an interference pattern of alternating bright and dark fringes on a distant screen.

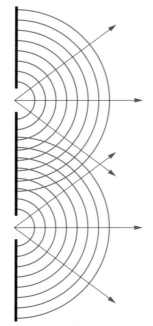

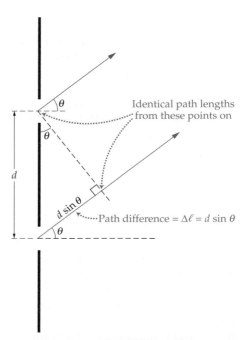

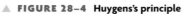

▲ **FIGURE 28–4 Huygens's principle**
According to Huygens's principle, each of the two slits in Young's experiment acts as a source of light waves propagating outward in all forward directions. It follows that light from the two sources can overlap, resulting in an interference pattern.

▲ **FIGURE 28–5 Path difference in the two-slit experiment**
Light propagating from two slits to a distant screen along parallel paths; note that the paths make an angle θ relative to the normal to the slits. The difference in path length is $\Delta \ell = d \sin \theta$, where d is the slit separation.

the path difference, $\Delta \ell = d \sin \theta$, is equal to $m\lambda$, where $m = 0, \pm 1, \pm 2, \ldots$. Therefore, we find that bright fringes satisfy the following conditions:

Conditions for Bright Fringes (Constructive Interference) in a Two-Slit Experiment

$$d \sin \theta = m\lambda \qquad m = 0, \pm 1, \pm 2, \ldots \qquad \text{28–1}$$

Note that the $m = 0$ fringe occurs at $\theta = 0$, which corresponds to the central fringe. In addition, positive values of m indicate fringes above the central bright fringe; negative values indicate fringes below the central bright fringe.

Between the bright fringes we find dark fringes, where destructive interference occurs. The condition for a dark fringe is that the difference in path lengths be $\pm \lambda/2, \pm 3\lambda/2, \pm 5\lambda/2, \ldots$. Notice that the dark fringes are analogous to the points $Q_1, Q_2, \ldots$ in Figure 28–2. It follows that the angles corresponding to dark fringes are given by the following conditions:

Conditions for Dark Fringes (Destructive Interference) in a Two-Slit Experiment

$$d \sin \theta = \left(m - \tfrac{1}{2}\right)\lambda \quad m = 1, 2, 3, \ldots \qquad \text{(above central bright fringe)}$$

$$d \sin \theta = \left(m + \tfrac{1}{2}\right)\lambda \quad m = -1, -2, -3, \ldots \text{ (below central bright fringe)}$$

28–2

Clearly, $m = +1$ corresponds to a path difference of $\Delta \ell = \lambda/2$, the value $m = +2$ corresponds to a path difference of $\Delta \ell = 3\lambda/2$, and so on. These dark fringes are above the central bright fringe. Similarly, $m = -1$ corresponds to a path difference of $-\lambda/2$, and $m = -2$ corresponds to a path difference of $-3\lambda/2$. These dark fringes are below the central bright fringe. **Figure 28–6** shows the numbering systems for both bright and dark fringes. (All problems in this text refer to fringes above the central bright fringe, so only the first set of conditions in Equation 28–2 will be needed.)

Since $\sin \theta$ is less than or equal to 1, it follows from Equation 28–1 that d must be greater than or equal to λ to show the $m = \pm 1$ bright fringes. In a typical

▲ An interference pattern created by monochromatic laser light passing through two slits.

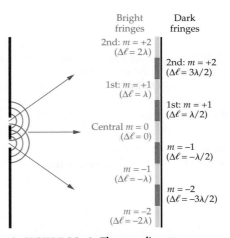

▲ **FIGURE 28–6 The two-slit pattern**
Numbering systems for bright and dark fringes.

situation d may be 100 times larger than λ, which means that the angle to the first dark or bright fringe will be roughly half a degree. If d is too much larger than λ, however, the angle between successive minima and maxima is so small that they tend to merge together, making the interference pattern difficult to discern.

EXERCISE 28–1

Red light ($\lambda = 752$ nm) passes through a pair of slits with a separation of 6.20×10^{-5} m. Find the angles corresponding to **(a)** the first bright fringe and **(b)** the second dark fringe above the central bright fringe.

SOLUTION

a. Referring to Figure 28–6 we find that $m = +1$ corresponds to the first bright fringe above the central bright fringe; hence,

$$\theta = \sin^{-1}\left(m\frac{\lambda}{d}\right) = \sin^{-1}\left[(1)\frac{7.52 \times 10^{-7}\text{ m}}{6.20 \times 10^{-5}\text{ m}}\right] = 0.695°$$

b. In this case $m = +2$, therefore

$$\theta = \sin^{-1}\left[\left(m - \tfrac{1}{2}\right)\frac{\lambda}{d}\right] = \sin^{-1}\left[\left(2 - \tfrac{1}{2}\right)\frac{7.52 \times 10^{-7}\text{ m}}{6.20 \times 10^{-5}\text{ m}}\right] = 1.04°$$

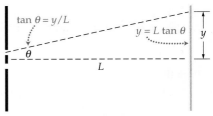

▲ **FIGURE 28–7 Linear distance in an interference pattern**
If light propagates at an angle θ relative to the normal to the slits, it is displaced a linear distance $y = L \tan \theta$ on the distant screen.

A convenient way to characterize the location of interference fringes is in terms of their linear distance from the central fringe, as indicated in **Figure 28–7**. If the distance to the screen is L—and L is much greater than the slit separation d—it follows that the linear distance y is given by the following expression:

Linear Distance from Central Fringe

$$y = L \tan \theta \qquad \qquad 28\text{–}3$$

In the next Example we show how a measurement of the linear distance between fringes can determine the wavelength of light.

PROBLEM-SOLVING NOTE

Angular Versus Linear Position of Fringes

The angle at which a bright or dark fringe occurs is determined by the wavelength of the light and the separation of the slits. The linear position of a fringe on a screen is determined by the distance from the slits to the screen.

EXAMPLE 28–2 BLUE LIGHT SPECIAL

Two slits with a separation of 8.5×10^{-5} m create an interference pattern on a screen 2.3 m away. **(a)** If the tenth bright fringe above the central fringe is a linear distance of 12 cm above the central fringe, what is the wavelength of light used in the experiment? **(b)** What is the linear distance from the central bright fringe to the tenth dark fringe above it?

CONTINUED ON NEXT PAGE

CONTINUED FROM PREVIOUS PAGE

PICTURE THE PROBLEM
Our sketch shows that the first bright fringe above the central fringe corresponds to $m = +1$ in Equation 28–1, the second bright fringe corresponds to $m = +2$, and so on. Therefore, $m = +10$ in Equation 28–1 gives the position of the tenth bright fringe. Similarly, the first dark fringe corresponds to $m = +1$, and so the tenth dark fringe is given by $m = +10$ in the top equation of Equation 28–2.

Finally, we note that the separation of the slits is $d = 8.5 \times 10^{-5}$ m, the distance to the screen is $L = 2.3$ m, and the vertical distance to the $m = +10$ bright fringe is $y = 12$ cm $= 0.12$ m.

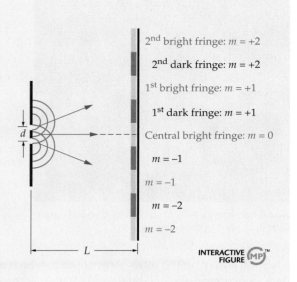

2nd bright fringe: $m = +2$

2nd dark fringe: $m = +2$

1st bright fringe: $m = +1$

1st dark fringe: $m = +1$

Central bright fringe: $m = 0$

$m = -1$

$m = -1$

$m = -2$

$m = -2$

STRATEGY

a. To find the wavelength, we first determine the angle to the tenth fringe using $y = L \tan \theta$ (Equation 28–3). Once we know θ, we use the condition for bright fringes to determine the wavelength. That is, we set $m = +10$ in Equation 28–1 ($d \sin \theta = m\lambda$) and solve for λ.

b. We use $d \sin \theta = (m - \frac{1}{2})\lambda$ (Equation 28–2) with $m = +10$ and λ from part (a) to determine the angle θ. Next, we substitute θ in $y = L \tan \theta$ to find the linear distance.

SOLUTION

Part (a)

1. Calculate the angle to the tenth bright fringe:

$$y = L \tan \theta$$
$$\theta = \tan^{-1}\left(\frac{y}{L}\right) = \tan^{-1}\left(\frac{0.12 \text{ m}}{2.3 \text{ m}}\right) = 3.0°$$

2. Use $\sin \theta = m\lambda/d$ to find the wavelength:

$$\lambda = \frac{d}{m} \sin \theta = \left(\frac{8.5 \times 10^{-5} \text{ m}}{10}\right) \sin(3.0°)$$
$$= 4.4 \times 10^{-7} \text{ m} = 440 \text{ nm}$$

Part (b)

3. Find the angle corresponding to the tenth dark fringe:

$$d \sin \theta = \left(m - \tfrac{1}{2}\right)\lambda$$
$$\theta = \sin^{-1}\left[\left(m - \tfrac{1}{2}\right)\lambda/d\right]$$
$$= \sin^{-1}\left[\left(10 - \tfrac{1}{2}\right)(4.4 \times 10^{-7} \text{ m})/(8.5 \times 10^{-5} \text{ m})\right] = 2.8°$$

4. Use $y = L \tan \theta$ to find the linear distance:

$$y = L \tan \theta = (2.3 \text{ m}) \tan(2.8°) = 0.11 \text{ m}$$

INSIGHT
Note that we have expressed the wavelength of the light in nanometers, a common unit of measure for light waves. Referring to the electromagnetic spectrum shown in Example 25–3, we see that light with a wavelength of 440 nm is dark blue.

PRACTICE PROBLEM
(a) If the wavelength of light used in this experiment is increased, does the linear distance to the tenth bright fringe above the central fringe increase, decrease, or stay the same? **(b)** Check your reasoning by calculating the linear distance to the tenth bright fringe for a wavelength of 550 nm.   [**Answer: (a)** Increase; **(b)** $y = 0.15$ m > 0.12 m]

Some related homework problems: Problem 17, Problem 23, Problem 25

Finally, we consider the effect of changing the medium through which the light propagates in a two-slit experiment.

CONCEPTUAL CHECKPOINT 28–1 FRINGE SPACING

A two-slit experiment is performed in the air. Later, the same apparatus is immersed in water and the experiment is repeated. When the apparatus is in water, are the interference fringes **(a)** more closely spaced, **(b)** more widely spaced, or **(c)** spaced the same as when the apparatus was in air?

REASONING AND DISCUSSION

The angles corresponding to bright fringes are related to the wavelength by the equation $d \sin \theta = m\lambda$. From this relation it is clear that if λ is increased, the angle θ (and hence the spacing between fringes) also increases; if λ is decreased, the angle θ decreases. Thus the behavior of the two-slit experiment in water depends on how the wavelength of light changes in water.

Recall that when light goes from air ($n = 1.00$) to a medium in which the index of refraction is $n > 1$, the speed of propagation decreases by the factor n:

$$v = \frac{c}{n}$$

The frequency of light, f, is unchanged throughout as it goes from one medium to another. Therefore, the fact that the speed $v = \lambda f$ decreases by a factor n means that the wavelength λ decreases by the same factor. Hence, if the wavelength of light is λ when $n = 1$, its wavelength in a medium with an index of refraction $n > 1$ is

$$\lambda_n = \frac{\lambda}{n} \qquad\qquad 28\text{–}4$$

As a result, the wavelength of light is less in water than in air, and therefore, the interference fringes are more closely spaced when the experiment is performed in water.

ANSWER

(a) The fringes are more closely spaced.

The relation $\lambda_n = \lambda/n$ plays an important role in the interference observed in thin films, as we shall see in the next section.

28–3 Interference in Reflected Waves

Waves that reflect from objects at different locations can interfere with one another, just like the light from two different slits. In fact, interference due to reflected waves is observed in many everyday circumstances, as we show next. Before we can understand the physics behind this type of interference, however, we must note that reflected waves change their phase in two completely different ways. First, the phase changes in proportion to the distance the waves travel—just as with light in the two-slit experiment. For example, the phase of a wave that travels half a wavelength changes by 180°, and the phase of a wave that travels one wavelength changes by 360°. Second, the phase of a reflected wave can change as a result of the reflection process itself. We begin by considering the latter type of phase change.

Phase Changes Due to Reflection

Phase changes due to reflection have been discussed before in connection with waves on a string in Chapter 14. In particular, we observed at that time that a wave on a string reflects differently depending on whether the end of the string is tied to a solid support, as in Figure 14–7, or is free to move up and down, as in Figure 14–8. Specifically, the wave with a loose end is reflected back exactly as it approached the end; that is, there is no phase change. Conversely, a wave on a string that is tied down is inverted when reflected. This is equivalent to changing the phase of the wave by 180°, or half a wavelength.

Since light is a wave, it undergoes an analogous phase change on reflection. As indicated in **Figure 28–8 (a)**, a light wave that encounters a region with a lower index of refraction is reflected with no phase change, like a wave on a string whose end is free to move. In contrast, when light encounters a region with a larger index of refraction, as in **Figure 28–8 (b)**, it is reflected with a phase change of half a wavelength, like a wave on a string whose end is fixed. This half-wavelength phase change also applies to light reflected from a solid surface, such as a mirror.

▶ **FIGURE 28–8 Phase change with reflection**
(a) An electromagnetic wave reflects with no phase change when it encounters a medium with a lower index of refraction. **(b)** An electromagnetic wave reflects with a half-wavelength (180°) phase change when it encounters a medium with a larger index of refraction.

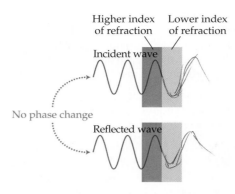

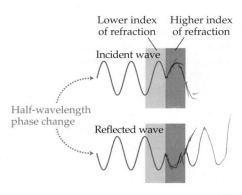

(a) Reflection from lower index of refraction

(b) Reflection from higher index of refraction

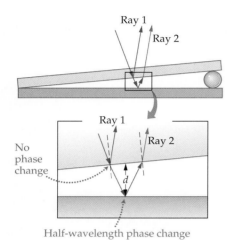

▲ **FIGURE 28–9 An air wedge**
In an air wedge, interference occurs between the light reflected from the bottom surface of the top plate of glass (ray 1) and light reflected from the top surface of the bottom plate of glass (ray 2). (The two rays are shown widely separated for clarity; in reality, they should be almost on top of one another.)

PROBLEM-SOLVING NOTE

Interference of Reflected Light

When determining whether reflected light interferes constructively or destructively, it is essential to take into account the phase changes that can occur under reflection.

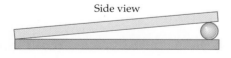

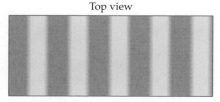

▲ **FIGURE 28–10 Interference fringes in an air wedge**
The interference fringes in an air wedge are regularly spaced, as shown in the top view of the wedge.

To summarize:

> There is no phase change when light reflects from a region with a lower index of refraction.

> There is a half-wavelength phase change when light reflects from a region with a higher index of refraction, or from a solid surface.

We now apply these observations to the case of an air wedge.

Air Wedge

An interesting example of reflection interference is provided by two plates of glass that touch at one end and have a small separation at the other, as shown in **Figure 28–9**. Note that the air between the plates occupies a thin, wedge-shaped region; hence, this type of arrangement is referred to as an **air wedge.**

The predominant interference effect in this system is between light reflected from the bottom surface of the top glass plate and light reflected from the upper surface of the lower plate, since these surfaces are physically so close together. Consider, for example, the two rays illustrated in Figure 28–9. Ray 1 reflects at the glass-to-air interface; it experiences no phase change. Ray 2 travels a distance d through the air ($n = 1.00$), reflects from the air-to-glass interface, then travels essentially the same distance d in the opposite direction before rejoining ray 1 (the rays in Figure 28–9 are shown widely separated for clarity). Since the reflection from air to glass results in a 180° phase change—the same phase change as if the wave had traveled half a wavelength—the *effective* path length of ray 2 is

$$\Delta \ell_{\text{eff}} = d + \tfrac{1}{2}\lambda + d = \tfrac{1}{2}\lambda + 2d$$

If the effective path length is an integer number of wavelengths, $\lambda/2 + 2d = m\lambda$, rays 1 and 2 will interfere constructively. Dividing by the wavelength, we have the following condition for **constructive interference:**

$$\frac{1}{2} + \frac{2d}{\lambda} = m \qquad m = 1, 2, 3, \ldots \qquad \text{28–5}$$

Similarly, if the effective path length of ray 2 is an odd half integer there will be **destructive interference:**

$$\frac{1}{2} + \frac{2d}{\lambda} = m + \frac{1}{2} \qquad m = 0, 1, 2, \ldots \qquad \text{28–6}$$

Since the distance between the plates, d, increases linearly with the distance from the point where the glass plates touch, it follows that the dark and bright interference fringes are evenly spaced, as shown in **Figure 28–10**.

CONCEPTUAL CHECKPOINT 28–2 DARK OR BRIGHT FRINGE?

Is the point where the glass plates touch in an air wedge **(a)** a dark fringe or **(b)** a bright fringe?

REASONING AND DISCUSSION

At the point where the glass plates touch, the separation d is zero. Setting d equal to zero in Equation 28–5 gives $\frac{1}{2} = m$, which can never be satisfied with an integer value of m. As a result, we conclude that the point of contact is not a bright fringe.

Considering Equation 28–6 in the limit of $d = 0$ yields $\frac{1}{2} = m + \frac{1}{2}$, which is satisfied by $m = 0$. Therefore, the point of contact of the glass plates is the first dark fringe in the system.

We can understand the origin of the dark fringe at the point of contact by recalling that ray 1 undergoes no phase change on reflection, whereas ray 2 experiences a 180° phase change due to reflection. It follows, then, that when the path difference, $2d$, approaches zero, rays 1 and 2 will cancel with destructive interference.

ANSWER

(a) A dark fringe occurs where the two glass plates touch.

The next Example uses the number of fringes observed in an air wedge to calculate the thickness of a human hair.

EXAMPLE 28–3 SPLITTING HAIRS

An air wedge is formed by placing a human hair between two glass plates on one end, and allowing them to touch on the other end. When this wedge is illuminated with red light ($\lambda = 771$ nm), it is observed to have 179 bright fringes. How thick is the hair?

PICTURE THE PROBLEM

The air wedge used in this experiment is shown in the sketch. Note that the separation of the plates on the end with the hair is equal to the thickness of the hair, t. It is at this point that the 179th bright fringe is observed.

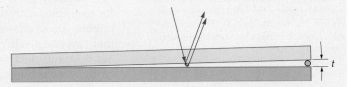

STRATEGY

Recall that the condition for bright fringes is $\frac{1}{2} + 2d/\lambda = m$, and that $m = 1$ corresponds to the first bright fringe, $m = 2$ corresponds to the second bright fringe, and so on. Clearly, then, the 179th bright fringe is given by $m = 179$. Substituting this value for m in $\frac{1}{2} + 2d/\lambda = m$, and setting the plate separation, d, equal to the thickness of the hair, t, allows us to solve for t.

SOLUTION

1. Solve the bright-fringe condition, $\frac{1}{2} + 2d/\lambda = m$, for the plate separation, d:

$$\frac{1}{2} + \frac{2d}{\lambda} = m$$

$$d = \frac{\lambda}{2}\left(m - \tfrac{1}{2}\right)$$

2. Set $m = 179$, and solve for the hair thickness, $t = d$:

$$t = \frac{\lambda}{2}\left(m - \tfrac{1}{2}\right)$$

$$= \frac{(771 \times 10^{-9}\text{ m})}{2}\left(179 - \tfrac{1}{2}\right)$$

$$= 6.88 \times 10^{-5}\text{ m} = 68.8\ \mu\text{m}$$

INSIGHT

Thus, a human hair has a diameter of roughly 70 micrometers. Note that to measure the thickness of a hair we have used a "ruler" with units that are comparable to the distance to be measured. In this case, the hair has a thickness about 100 times larger than the wavelength of the light that we used.

PRACTICE PROBLEM

If a thicker hair is used in this experiment, will the number of bright fringes increase, decrease, or stay the same? How many bright fringes will be observed if the hair has a thickness of 80.0 μm?   [**Answer:** Increase; 208 fringes]

Some related homework problems: Problem 37, Problem 42

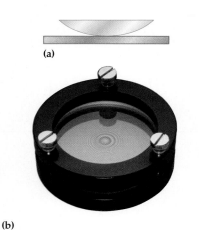

(a)

(b)

▲ **FIGURE 28–11   A system for generating Newton's rings**

(a) A variation on the air wedge is produced by placing a piece of glass with a spherical cross section on top of a plate of glass. **(b)** A top view of the system shown in part (a). The circular interference fringes are referred to as Newton's rings.

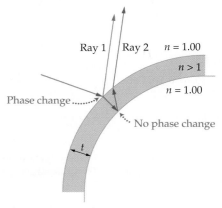

▲ **FIGURE 28–12   Interference in thin films**
The phase of ray 1 changes by half a wavelength due to reflection; the phase of ray 2 changes by $2t/\lambda_n$, where λ_n is the wavelength of light within the thin film.

Clearly, the location of interference fringes is very sensitive to even extremely small changes in plate separation. This can be illustrated dramatically by simply pressing down lightly on the upper plate with a finger. Even though the finger causes no visible change in the glass plate, the fringes are observed to move. By measuring the displacement of the fringes it is possible to calculate the tiny deflection, or bend, the finger caused in the plate. Devices using this type of mechanism are frequently used to show small displacements that would be completely invisible to the naked eye.

Newton's Rings

A system similar to an air wedge, but with a slightly different geometry, is shown in **Figure 28–11 (a)**. In this case, the upper glass plate is replaced by a curved piece of glass with a spherical cross section. Still, the mechanism producing interference is the same as before. The result is a series of circular interference fringes, as shown in **Figure 28–11 (b)**, referred to as **Newton's rings.**

Notice that the fringes become more closely spaced as one moves farther from the center of the pattern. The reason is that the curved surface of the upper piece of glass moves away from the lower plate at a progressively faster rate. As a result, the horizontal distance required to go from one fringe to the next becomes less as one moves away from the center.

Newton's rings can be used to test the shape of a lens. Imperfections in the ring pattern indicate slight distortions in the lens. As in the case of an air wedge, even a very small change in shape can cause a significant displacement of the interference fringes.

Thin Films

Perhaps the most commonly observed examples of interference are provided by thin films, such as those found in soap bubbles and oil slicks. We are all familiar, for example, with the swirling patterns of color that are seen on the surface of a bubble. These colors are the result of the constructive and destructive interference that can occur when white light reflects from a thin film. In particular, some colors undergo destructive interference and are *eliminated* from the incident light, while others colors are *enhanced* by constructive interference.

To determine the conditions for constructive and destructive interference in a thin film, consider **Figure 28–12**. Here we show a thin film of thickness t and index of refraction $n > 1$, with air ($n = 1.00$) on either side. To analyze this system, we proceed in much the same way as we did for the air wedge earlier in this section. Specifically, we focus on the phase change of rays 1 and 2, taking into account phase changes due to both reflection and path-length difference.

First, note that ray 1 reflects from the air-to-film interface; hence, its phase changes by half a wavelength. The effective path length for ray 1, then, is

$$\ell_{\mathrm{eff},1} = \tfrac{1}{2}\lambda$$

Dividing by the wavelength, λ, gives the phase change of ray 1 in terms of the wavelength:

$$\frac{\ell_{\mathrm{eff},1}}{\lambda} = \frac{1}{2} \qquad\qquad 28\text{–}7$$

Recall that if two rays of light differ in phase by half a wavelength, the result is destructive interference.

Next, ray 2 reflects from the film-to-air interface; hence, it experiences no phase change due to reflection. It does, however, have a phase change as a result of traveling an extra distance $2t$ through the film. Thus the effective path length for ray 2 is

$$\ell_{\mathrm{eff},2} = 2t$$

To put this path length in terms of the wavelength, we must recall that if the wavelength of light in a vacuum is λ_{vacuum}, its wavelength in a medium with an index of refraction n is

$$\lambda_n = \frac{\lambda_{\text{vacuum}}}{n}$$

Dividing the path length by λ_n gives us the phase change of ray 2 in terms of the wavelength within the film:

$$\frac{\ell_{\text{eff},2}}{\lambda_n} = \frac{2t}{\lambda_n} = \frac{2nt}{\lambda_{\text{vacuum}}} \qquad 28\text{–}8$$

Finally, we can calculate the difference in phase changes for rays 1 and 2 using the preceding results:

$$\text{difference in phase changes} = \frac{2nt}{\lambda_{\text{vacuum}}} - \frac{1}{2} \qquad 28\text{–}9$$

Note that in the limit of zero film thickness, $t = 0$, the phase difference is $-\frac{1}{2}$. This corresponds to destructive interference, since moving a wave back half a wavelength is equivalent to moving it ahead half a wavelength. In general, then, destructive interference occurs if any of the following conditions are satisfied:

$$\frac{2nt}{\lambda_{\text{vacuum}}} - \frac{1}{2} = -\frac{1}{2}, \frac{1}{2}, \frac{3}{2}, \cdots$$

Adding $\frac{1}{2}$ to each side yields our final result for **destructive interference:**

$$\frac{2nt}{\lambda_{\text{vacuum}}} = m \qquad m = 0, 1, 2, \ldots \qquad 28\text{–}10$$

Similarly, if the phase difference between the rays is equal to an integer, m, the result is **constructive interference:**

$$\frac{2nt}{\lambda_{\text{vacuum}}} - \frac{1}{2} = m \qquad m = 0, 1, 2, \ldots \qquad 28\text{–}11$$

These conditions are applied in the next Example.

▲ The swirling colors typical of soap bubbles are created by interference, both destructive and constructive, which eliminates certain wavelengths from the reflected light while enhancing others. Which colors are removed or enhanced at a given point depends on the precise thickness of the film in that region.

EXAMPLE 28–4   RED LIGHT SPECIAL

A beam consisting of red light ($\lambda_{\text{vacuum}} = 662$ nm) and blue light ($\lambda_{\text{vacuum}} = 465$ nm) is directed at right angles onto a thin soap film. If the film has an index of refraction $n = 1.33$ and is suspended in air ($n = 1.00$), find the smallest nonzero thickness for which it appears red in reflected light.

PICTURE THE PROBLEM
Our sketch shows a soap film of thickness t and index of refraction $n = 1.33$ suspended in air ($n = 1.00$). We consider each of the colors in the incoming beam of light separately, with one blue ray and one red ray. If the film is to look red in reflected light, it follows that the reflected blue light must cancel due to destructive interference, as indicated.

STRATEGY
As mentioned above, the desired thickness of the film is such that blue light satisfies one of the conditions for destructive interference given in Equation 28–10. These conditions are $2nt/\lambda_{\text{vacuum}} = m$, where m is equal to 0, 1, 2, and so on. Because the case $m = 0$ corresponds to zero thickness, $t = 0$, it follows that the smallest nonzero thickness is given by $m = 1$.

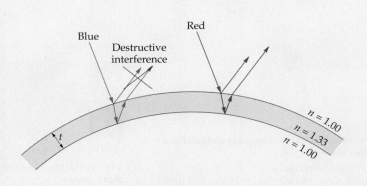

CONTINUED ON NEXT PAGE

CONTINUED FROM PREVIOUS PAGE

SOLUTION

1. Solve $2nt/\lambda_{\text{vacuum}} = m$ for the thickness t:

$$\frac{2nt}{\lambda_{\text{vacuum}}} = m \quad \text{or} \quad t = \frac{m\lambda_{\text{vacuum}}}{2n}$$

2. Calculate the thickness using $m = 1$:

$$t = \frac{m\lambda_{\text{vacuum}}}{2n} = \frac{(1)(465 \text{ nm})}{2(1.33)} = 175 \text{ nm}$$

INSIGHT

Although blue light is canceled at this thickness, red light is not. In fact, repeating the preceding calculation shows that red light is not canceled until the thickness of the film is 249 nm.

PRACTICE PROBLEM

What is the smallest thickness of film that gives *constructive* interference for the blue light in this system?   [**Answer:** $t = 87.4$ nm]

Some related homework problems: Problem 32, Problem 38

The connection between the thickness of a film and the color it shows in reflected light is illustrated in **Figure 28–13**. Notice that in thicker regions of the film, the longer wavelengths of light interfere destructively, resulting in reflected light that is bluish. In the limit of zero thickness the condition for destructive interference, $2nt/\lambda_{\text{vacuum}} = m$, is satisfied for all wavelengths with m set equal to zero. Hence, an extremely thin film appears dark in reflected light—it is essentially one large dark fringe.

Another example of a thin-film interference is provided by a film floating on a liquid or coating a solid surface. For example, **Figure 28–14** shows a thin film floating on water. If the index of refraction of the film is greater than that of water, the situation in terms of interference is essentially the same as for a thin film suspended in air. In particular, the ray reflected from the top surface of the film undergoes a phase change of half a wavelength; the ray reflected from the bottom of the film has no phase change due to reflection. Thus the interference conditions given in Equation 28–10 and 28–11 still apply.

On the other hand, suppose a thin film has an index of refraction that is greater than 1.00 but less than the index of refraction of the material on which it is supported, as in **Figure 28–15**. In a case like this, there is a reflection phase change at both the upper and the lower surfaces of the film. The condition for destructive interference for this type of system is discussed in the following Active Example.

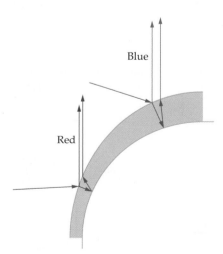

▲ **FIGURE 28–13 Thickness and color in a thin film**
Thicker portions of thin film appear blue, since the long-wavelength red light experiences destructive interference. Thinner regions appear red because the short-wavelength blue light interferes destructively.

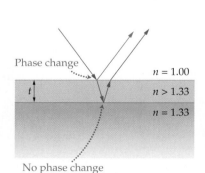

▲ **FIGURE 28–14 A thin film with one phase change**
If the index of refraction of the film is greater than that of the water, the situation in terms of phase changes is essentially the same as for a thin film suspended in air.

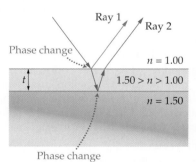

▲ **FIGURE 28–15 A thin film with two phase changes**
A thin film is applied to a material with a relatively large index of refraction. If the index of refraction of the film is less than that of the material that supports it, there will be a phase change for reflections from both surfaces of the film. Films of this type are often used in nonreflective coatings.

ACTIVE EXAMPLE 28–1   NONREFLECTIVE COATING: FIND THE THICKNESS

Camera lenses ($n = 1.52$) are often coated with a thin film of magnesium fluoride ($n = 1.38$). These "nonreflective coatings" use destructive interference to reduce unwanted reflections. Find the condition for destructive interference in this case, and calculate the minimum thickness required to give destructive interference for light in the middle of the visible spectrum (yellow-green light, $\lambda_{\text{vacuum}} = 565$ nm).

SOLUTION *(Test your understanding by performing the calculations indicated in each step.)*

1. Give the phase change (in units of the wavelength) for ray 1 in Figure 28–15: $\frac{1}{2}$

2. Give the phase change (in units of the wavelength) for ray 2 in Figure 28–15: $\frac{1}{2} + 2t/\lambda_n$

3. Set the difference in phase changes equal to one-half: $2t/\lambda_n = \frac{1}{2}$

4. Solve for the thickness: $t = \lambda_n/4$

5. Substitute numerical values: $t = 102$ nm

INSIGHT

Because the thickness of the film should be a quarter of a wavelength, these non-reflective films are often referred to as quarter-wave coatings.

YOUR TURN

Suppose a different coating material with a larger index of refraction is used on this lens. Is the desired minimum thickness greater than or less than it was with magnesium fluoride? Calculate the new minimum thickness, assuming the index of refraction for the coating material is $n = 1.45$.

*(Answers to **Your Turn** problems are given in the back of the book.)*

▲ The lenses of binoculars and cameras often have a blue, purple, or amber tint, the product of their antireflection coating. The coating is a thin film that reduces reflection from the lens surfaces by destructive interference.

REAL-WORLD PHYSICS

Nonreflective coating

Interference in CDs

Destructive interference plays a crucial role in the operation of a CD. The basic idea behind these devices is that information is encoded in the form of a series of "bumps" on an otherwise smooth reflecting surface. A laser beam directed onto the surface is reflected back to a detector, and as the intensity of the reflected beam varies due to the bumps, the information on the CD is decoded—similar to using dots and dashes to send information in Morse code.

Imagine the laser beam illuminating a small area on the surface of a CD. As a bump moves into the beam, as in **Figure 28–16**, there are two components to the reflected beam—one from the top of the bump, the other from the bottom. If these two beams are out of phase by half a wavelength, there will be destructive interference and the detector will receive a weak signal. When the beam is reflected solely from the top of the bump, the detector again receives a strong signal, since there is no longer an interference effect. As the bump moves out of the beam we again have the condition for destructive interference, and the detector signal again falls. Thus the bumps give the detector a series of "on" and "off" signals that can be converted to sound, pictures, or other types of information.

▲ This microscopic view of the surface of a CD reveals the pattern of tiny bumps that encodes the information. From the back, these areas take the form of indentations—hence they are commonly known as "pits," even though to the laser beam scanning the CD they appear as raised areas.

REAL-WORLD PHYSICS

Reading the information on a CD

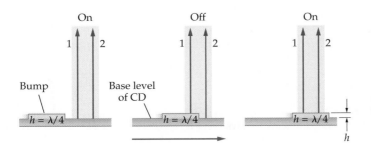

◀ **FIGURE 28–16 Reading information on a CD**

As a "bump" on a CD moves through the laser beam, the detector receives a weak "off" signal when the bump enters and leaves the beam. When the beam reflects entirely from the top of a bump, or from the base level of the CD, the detector receives a strong "on" signal.

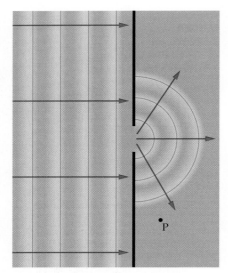

▲ FIGURE 28–17 Diffraction of water waves

As water waves pass through an opening, they diffract, or change direction. Thus an observer at point P detects waves even though this point is not on a line with the original direction of the waves and the opening. All waves exhibit similar diffraction behavior.

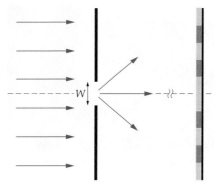

▲ FIGURE 28–18 Single-slit diffraction

When light of wavelength λ passes through a slit of width W, a "diffraction pattern" of bright and dark fringes is formed.

▶ FIGURE 28–19 Locating the first dark fringe in single-slit diffraction

The location of the first dark fringe in single-slit diffraction can be determined by considering pairs of waves radiating from the top and bottom half of a slit. **(a)** A wave pair originating at points 1 and 1' has a path difference of $(W/2) \sin \theta$. These waves interfere destructively if the path difference is equal to half a wavelength. **(b)** The rest of the light coming from the slit can be considered to consist of additional wave pairs, like 2 and 2', 3 and 3', and so on. Each wave pair has the same path difference.

Let's find the necessary height h of a bump if the red light of a ruby laser, with a wavelength of 694 nm, is to have destructive interference. Figure 28–16 shows the situation, in which ray 1 reflects from the top of a bump, and ray 2 reflects from the base level of the CD. The path difference between the rays is $2h$; thus, to have destructive interference, $2h$ must be equal to half a wavelength, $2h = \lambda/2$, or

$$h = \frac{\lambda}{4}$$

In the case of a ruby laser, the required height of a bump is about 174 nm.

28–4 Diffraction

If light is indeed a wave, it must exhibit behavior similar to that displayed by the water waves in **Figure 28–17**. Note that the waves are initially traveling directly to the right. After passing through the gap in the barrier, however, they spread out and travel in all possible forward directions, in accordance with Huygens's principle. Thus, an observer at point P detects waves even though she is not on a direct line with the initial waves and the gap. In general, waves always bend, or **diffract,** when they pass by a barrier or through an opening.

A familiar example of diffraction is the observation that you can hear a person talking even when that person is out of sight around a corner. The sound waves from the person bend around the corner, just like the water waves in Figure 28–17. It might seem, then, that light cannot be a wave, since it does not bend around a corner along with the sound. There is a significant difference between sound and light waves, however; namely, their wavelengths differ by many orders of magnitude—from a meter or so for sound to about 10^{-7} m for light. As we shall see, the angle through which a wave bends is greater the larger the wavelength of the wave; hence, diffraction effects in light are typically small compared with those in sound waves and water waves.

To investigate the diffraction of light we start by considering the behavior of a beam of light as it passes through a single, narrow slit in a screen.

Single-Slit Diffraction

Consider monochromatic light of wavelength λ passing through a narrow slit of width W, as shown in **Figure 28–18**. After passing through the slit the light shines on a distant screen, which we assume to be much farther from the slit than the width W. According to Huygens's principle, each point within the slit can be considered as a source of new waves that radiate toward the screen. The interference of these sources with one another generates a diffraction pattern.

We can understand the origin of a single-slit diffraction pattern by referring to **Figure 28–19**, where we show light propagating to the screen from various points in

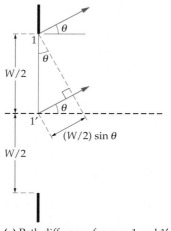

(a) Path difference for rays 1 and 1'

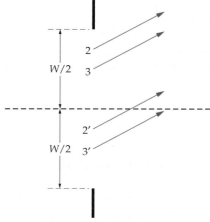

(b) Same path difference for all other pairs of rays

the slit. For example, consider waves from points 1 and 1′ propagating to the screen at an angle θ relative to the initial direction of the light. Since the screen is distant, the waves from 1 and 1′ travel on approximately parallel paths to reach the screen. From the figure, then, it is clear that the path difference for these waves is $(W/2) \sin \theta$. Similarly, the same path difference applies to the "wave pairs" from points 2 and 2′, the wave pairs from points 3 and 3′, and so on through all points in the slit.

In the forward direction, $\theta = 0°$, the path difference is zero; $(W/2) \sin 0° = 0$. As a result, all wave pairs interfere constructively, giving maximum intensity at the center of the diffraction pattern. However, if θ is increased until the path difference is half a wavelength, $(W/2) \sin \theta = \lambda/2$, each wave pair experiences destructive interference. Thus, the *first minimum*, or dark fringe, in the diffraction pattern occurs at the angle given by

$$W \sin \theta = \lambda$$

To find the second dark fringe, imagine dividing the slit into four regions, as illustrated in **Figure 28–20**. Within the upper two regions, we perform the same wave-pair construction described above; the same is done with the lower two regions. In this case, the path difference between wave pairs 1 and 1′ is $(W/4) \sin \theta$. As before, destructive interference first occurs when the path difference is $\lambda/2$. Solving for $\sin \theta$ in this case gives us the condition for the *second dark fringe:*

$$W \sin \theta = 2\lambda$$

The next dark fringe can be found by dividing the slit into six regions, with a path difference between wave pairs of $(W/6) \sin \theta$. In this case, the condition for destructive interference is $W \sin \theta = 3\lambda$. In general, then, dark fringes satisfy the following conditions:

Conditions for Dark Fringes in Single-Slit Interference

$$W \sin \theta = m\lambda \qquad m = \pm 1, \pm 2, \pm 3, \ldots \qquad \text{28–12}$$

Note that by including both positive and negative values for m, we have taken into account the symmetry of the diffraction pattern about its midpoint.

EXERCISE 28–2

Monochromatic light passes through a slit of width 1.2×10^{-5} m. If the first dark fringe of the resulting diffraction pattern is at angle $\theta = 3.25°$, what is the wavelength of the light?

SOLUTION
Solving Equation 28–12 for the wavelength gives us

$$\lambda = \frac{W \sin \theta}{m} = \frac{(1.2 \times 10^{-5} \text{ m}) \sin(3.25°)}{1} = 680 \text{ nm}$$

Notice that we use $m = 1$, since this is the *first* dark fringe.

The bright fringes in a diffraction pattern consist of the central fringe plus additional bright fringes approximately halfway between successive dark fringes. Note, in addition, that the central fringe is about twice as wide as the other bright fringes. In the small-angle approximation ($\sin \theta \sim \theta$), the central fringe extends from $\theta = \lambda/W$ to $\theta = -\lambda/W$, and hence its width is given by the following:

$$\text{approximate angular width of central fringe} = 2\frac{\lambda}{W} \qquad \text{28–13}$$

Thus we see that the wavelength λ plays a crucial role in diffraction patterns and that λ/W gives the characteristic angle of "bending" produced by diffraction. In the following Conceptual Checkpoint we consider the role played by the width of the slit.

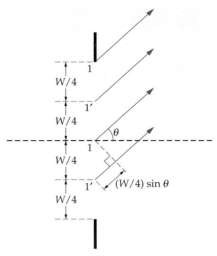

▲ **FIGURE 28–20 Locating the second dark fringe in single-slit diffraction**
To find the second dark fringe in a single-slit diffraction pattern, we divide the slit into four regions and consider wave pairs that originate from points separated by a distance $W/4$. Destructive interference occurs when the path difference, $(W/4) \sin \theta$, is half a wavelength.

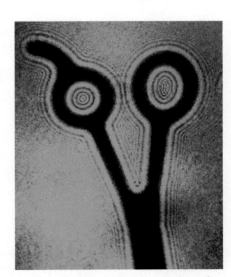

▲ Upon close inspection, the shadow of a sharp edge is seen to consist of numerous fringes produced by diffraction.

CONCEPTUAL CHECKPOINT 28–3 WIDTH OF CENTRAL BRIGHT FRINGE

If the width of the slit through which light passes is reduced, does the central bright fringe **(a)** become wider, **(b)** become narrower, or **(c)** remain the same size?

REASONING AND DISCUSSION
It might seem that making the slit narrower will cause the diffraction pattern to be narrower as well. Recall, however, that the diffraction pattern is produced by waves propagating from all parts of the slit. If the slit is wide, the incoming wave passes through with little deflection. If it is small, on the other hand, it acts like a point source, and light is radiated over a broad range of angles. Therefore, the smaller slit produces a wider central fringe.

This result is also confirmed by considering Equation 28–13, where we see that a smaller value of W results in a wider central fringe.

ANSWER
(a) The central bright fringe is wider.

As mentioned earlier, waves diffract whenever they encounter some sort of barrier or opening. It follows, then, that the shadow cast by an object is really not as sharp as it may seem. On closer examination, as shown in the photo on the previous page, the shadow of an object such as a pair of scissors actually consists of a tightly spaced series of diffraction fringes. Thus, shadows are not the sharp boundaries implied by geometrical optics but, instead, are smeared out on a small length scale. Under ordinary conditions the diffraction pattern in a shadow is not readily visible. However, a similar diffraction pattern can be observed by simply holding two fingers close together before your eyes—try it.

EXAMPLE 28–5 EXPLORING THE DARK SIDE

Light with a wavelength of 511 nm forms a diffraction pattern after passing through a single slit of width 2.20×10^{-6} m. Find the angle associated with **(a)** the first and **(b)** the second dark fringe above the central bright fringe.

PICTURE THE PROBLEM
In our sketch we identify the first and second dark fringes above the central bright fringe. Note that the first dark fringe corresponds to $m = 1$, and the second corresponds to $m = 2$. Finally, the width of the slit is $W = 2.20 \times 10^{-6}$ m.

STRATEGY
We can find the desired angles by using the condition for dark fringes, $W \sin \theta = m\lambda$ (Equation 28–12). As mentioned above, we use $m = 1$ for part (a), and $m = 2$ for part (b). The values of λ and W are given in the problem statement.

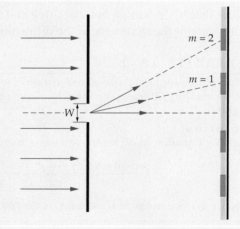

INTERACTIVE FIGURE (MP)™

SOLUTION

Part (a)
1. Solve for θ using $m = 1$:
$$\theta = \sin^{-1}\left(\frac{m\lambda}{W}\right) = \sin^{-1}\left[\frac{(1)(511 \times 10^{-9} \text{ m})}{2.20 \times 10^{-6} \text{ m}}\right] = 13.4°$$

Part (b)
1. Solve for θ using $m = 2$:
$$\theta = \sin^{-1}\left(\frac{m\lambda}{W}\right) = \sin^{-1}\left[\frac{(2)(511 \times 10^{-9} \text{ m})}{2.20 \times 10^{-6} \text{ m}}\right] = 27.7°$$

INSIGHT
Notice that the angle to the second dark fringe is *not* simply twice the angle to the first dark fringe. This is because the angle θ depends on the sine function, which is not linear. If you look at a plot of the sine function, as in Figure 13–15, you will see that $\sin \theta$ is slightly less than θ for the angles considered here. Therefore, to double the value of $\sin \theta$ you must increase θ by slightly more than a factor of 2.

PRACTICE PROBLEM

Suppose the wavelength of light in this experiment is changed to give the first dark fringe at an angle greater than 13.4°. Is the required wavelength greater than or less than 511 nm? Check your answer by calculating the wavelength required to give the first dark fringe at $\theta = 15.0°$. [**Answer:** The wavelength must be larger; $\lambda = 569$ nm.]

Some related homework problems: Problem 44, Problem 45, Problem 50

ACTIVE EXAMPLE 28–2   FIND THE LINEAR DISTANCE

In a single-slit experiment, light passes through the slit and forms a diffraction pattern on a screen 2.31 m away. If the wavelength of light is 632 nm, and the width of the slit is 4.20×10^{-5} m, find the linear distance on the screen from the center of the diffraction pattern to the first dark fringe.

SOLUTION *(Test your understanding by performing the calculations indicated in each step.)*

1. Find the angle to the first dark fringe: $\qquad\qquad\qquad \theta = 0.862°$
2. Use $y = L \tan \theta$ to find the linear distance: $\qquad\quad y = 3.48$ cm

INSIGHT

Note that the linear distance is found using Equation 28–3, just as for the two-slit experiment.

YOUR TURN

Find the linear distance from the center of the pattern to the second dark fringe.

(Answers to **Your Turn** *problems are given in the back of the book.)*

▲ The paradoxical bright dot at the center of a circular shadow, known as Poisson's bright spot, is a convincing proof of the wave nature of light.

The photo to the right shows a particularly interesting diffraction phenomenon: the shadow produced by a penny. As expected, the edges of the shadow show diffraction fringes, but of even greater interest is the bright point of light seen in the center of the shadow. This is referred to as "Poisson's bright spot," after the French scientist Siméon D. Poisson (1781–1840) who predicted its existence. It should be noted that Poisson did not believe that light is a wave. In fact, he used his prediction of a bright spot to show that the wave model of light was absurd and must be rejected—after all, how could a bright spot occur in the darkest part of a shadow? When experiments soon after his prediction showed conclusively that the bright spot does exist, however, the wave model of light gained almost universal acceptance. We know today that light has both wave and particle properties, a fact referred to as the *wave–particle duality*. This will be discussed in detail in Chapter 30.

28–5 Resolution

Diffraction affects, and ultimately limits, the way we see the world. There is a difference, for example, in how sharply we can see a scene depending on the size of our pupil—in general, the larger the pupil, the sharper the vision. For example, an eagle, which has pupils that are even larger than ours, can see a small creature on the ground with greater acuity than is possible for a person. Similarly, a camera or a telescope with a large aperture can "see" with greater detail than the human eye. The sharpness of vision—in particular, the ability to visually separate closely spaced objects—is referred to as **resolution.**

The way in which diffraction affects resolution can be seen by considering the diffraction pattern created by a circular aperture—such as the pupil of an eye. Just as with the diffraction pattern of a slit, a circular aperture creates a pattern of alternating bright and dark regions. The difference, as one might expect, is that the diffraction pattern of a circular opening is circular in shape, as we see in **Figure 28–21**.

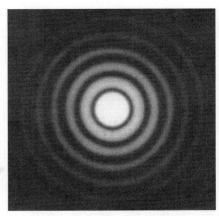

▲ **FIGURE 28–21 Diffraction from a circular opening**

Light passing through a circular aperture creates a circular diffraction pattern of alternating bright and dark regions.

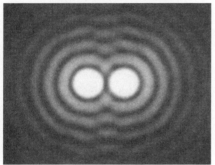

▲ **FIGURE 28–22 Resolving two point sources: Rayleigh's criterion**

If the angular separation between two sources is not great enough (top), their diffraction patterns overlap to the point where they appear as a single elongated source. With greater angular separation (bottom), the individual sources can be distinguished as separate.

In a slit pattern, the first dark fringe occurs at the angle given by $\sin \theta = \lambda/W$, where W is the width of the slit. A circular aperture of diameter D produces a central bright spot and a dark fringe at an angle θ from the center line given by the following expression:

First Dark Fringe for the Diffraction Pattern of a Circular Opening

$$\sin \theta = 1.22 \frac{\lambda}{D}$$

28–14

So we see that the change in geometry from a slit of width W to a circular opening of diameter D is reflected in the replacement of $(1)\lambda/W$ with $(1.22)\lambda/D$.

The physical significance of this result is that even if you focus perfectly on a point source of light, it will form a circular image of finite size on the retina. This blurs the image, replacing a point with a small circle. Thus, the diffraction of light through the pupil limits the resolution of your eye. In addition, it should be noted that the wavelength in Equation 28–14 refers to the wavelength in the medium in which the diffraction pattern is observed. For example, the wavelength that should be used in considering the eye is $\lambda_n = \lambda/n$, where n is the eye's average index of refraction (approximately $n = 1.36$).

Diffraction-induced smearing also makes it difficult to visually separate objects that are close to one another. In particular, if two closely spaced sources of light are smeared by diffraction, the circles they produce may overlap, making it difficult to tell if there are two sources or only one. The condition that is used to determine whether two sources can be visually separated is called **Rayleigh's criterion:**

> If the first dark fringe of one circular diffraction pattern passes through the center of a second diffraction pattern, the two sources responsible for the patterns will appear to be a single source.

This condition is illustrated in **Figure 28–22.**

To put Rayleigh's criterion in quantitative terms, note that for small angles (as is usually the case) the location of the first dark fringe is given by $\theta = 1.22\lambda/D$. Therefore, *two objects can be seen as separate only if their angular separation is greater than the following minimum:*

Rayleigh's Criterion

$$\theta_{\min} = 1.22 \frac{\lambda}{D}$$

28–15

Note that the larger the diameter D of the aperture, the smaller the angular separation that can be resolved and, hence, the greater the resolution.

EXERCISE 28–3

Find $\theta_{\min}$ for yellow light (551 nm) and an aperture diameter of 5.00 mm.

SOLUTION

Substituting into Equation 28–15 we find

$$\theta_{\min} = 1.22 \frac{\lambda}{D} = 1.22 \left(\frac{551 \times 10^{-9} \text{ m}}{5.00 \times 10^{-3} \text{ m}} \right) = 0.000134 \text{ rad} = 0.00770°$$

Because our pupils have diameters of about 5.00 mm, the small value of $\theta_{\min}$ indicates that human vision has the potential for relatively high resolution.

PROBLEM-SOLVING NOTE

Angular Resolution and the Index of Refraction

When applying the condition $\theta_{\min} = 1.22\lambda/D$, it is important to remember that λ refers to the wavelength in the region between the aperture and the screen (retina, film, etc.) on which the diffraction pattern is observed. If the index of refraction in this region is n, the wavelength is reduced from λ to λ/n. (See Conceptual Checkpoint 28–1.)

An example of diffraction-limited resolution is illustrated in the accompanying photos. In the first photo we see a brilliant light in the distance that may be the single headlight of an approaching motorcycle or the unresolved image of two headlights on a car. If we are seeing the headlights of a car, the angular separation between them will increase as the car approaches. When the angular separation

▲ Resolving the headlights of an approaching car. If the headlights were true point sources and the atmosphere perfectly transparent (or absent), the individual headlights could be distinguished at a much greater distance, as Example 28–6 shows.

exceeds $1.22\lambda/D$, as in the second photo, we are able to distinguish the two headlights as separate sources of light. As the car continues to approach, its individual headlights become increasingly distinct, as shown in the third photo.

The distance at which the two headlights can be resolved increases as the size of the aperture increases. This dependence is considered in detail in the following Example.

EXAMPLE 28–6   MOTORCYCLE OR CAR?

The linear distance separating the headlights of a car is 1.1 m. Assuming light of 460 nm, a pupil diameter of 5.0 mm, and an average index of refraction for the eye of 1.36, find the maximum distance at which the headlights can be distinguished as two separate sources of light.

PICTURE THE PROBLEM
Our sketch shows the car a distance L from the observer, with the headlights separated by a linear distance $y = 1.1$ m. The condition for the headlights to be resolved is that their angular separation be at least θ_{min}.

STRATEGY
To find the maximum distance, we must first determine the minimum angular separation, which is given by $\theta_{min} = 1.22\lambda_n/D$. In this expression, $\lambda_n = \lambda/1.36$.

Once we know θ_{min}, we can find the distance L using the trigonometric relation $\tan\theta_{min} = y/L$, which follows directly from our sketch.

SOLUTION

1. Find the minimum angular separation for the headlights to be resolved:

$$\theta_{min} = 1.22\frac{\lambda/n}{D}$$

$$= 1.22\left[\frac{(460\times10^{-9}\text{ m})/1.36}{5.0\times10^{-3}\text{ m}}\right] = 8.3\times10^{-5}\text{ rad}$$

2. Solve the relation $\tan\theta_{min} = y/L$ for the distance L:

$$L = \frac{y}{\tan\theta_{min}}$$

3. Substitute numerical values to find L:

$$L = \frac{y}{\tan\theta_{min}} = \frac{1.1\text{ m}}{\tan(8.3\times10^{-5}\text{ rad})} = 13{,}000\text{ m}$$

INSIGHT
Thus, the car must be about 8 mi away before the headlights appear to merge. This, of course, is the ideal case. In the real world, the finite size of the headlights and the blurring effects of the atmosphere greatly reduce the maximum distance.

PRACTICE PROBLEM
If the pupil diameter for an eagle is 6.2 mm, from what distance can it resolve the car's headlights under ideal conditions? [**Answer:** About 9.9 mi]

Some related homework problems: Problem 62, Problem 82

ACTIVE EXAMPLE 28-3 RESOLVING IDA AND DACTYL

The asteroid Ida is orbited by its own small "moon" called Dactyl. If the separation between these two asteroids is 2.5 km, what is the maximum distance at which the Hubble Space Telescope (aperture diameter = 2.4 m) can still resolve them with 550-nm light?

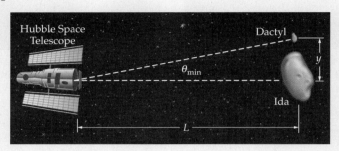

SOLUTION *(Test your understanding by performing the calculations indicated in each step.)*

1. Calculate the minimum angular separation for the asteroids to be resolved:

$$\theta_{min} = 2.8 \times 10^{-7} \text{ rad}$$

2. Express L in terms of y and θ_{min}:

$$L = y/\tan(\theta_{min})$$

3. Substitute numerical values to find L:

$$L = 8.9 \times 10^9 \text{ m} = 5.5 \times 10^6 \text{ mi}$$

INSIGHT

If the asteroids are farther away than this distance, the Hubble Space Telescope will not be able to image them as two separate objects. In fact, the asteroids are close enough to be viewed separately, as we see in the photo on page 390.

YOUR TURN

If the aperture diameter of the telescope is increased, does the maximum resolution distance increase or decrease? Calculate the maximum resolution distance for an aperture diameter of 3.0 m.

*(Answers to **Your Turn** problems are given in the back of the book.)*

▲ **FIGURE 28-23** Pointillism
Paul Signac's *The Mills at Owerschie* (1905), an example of pointillism.

REAL-WORLD PHYSICS: BIO
Pointillism and painting

REAL-WORLD PHYSICS: BIO
Color television image

▲ **FIGURE 28-24** Pixels on a television screen
A typical pixel on the screen of a color television consists of three closely spaced color spots: one red, one blue, and one green. These are the only colors the television actually produces.

We conclude by considering two interesting examples of diffraction-limited resolution. First, in the artistic style known as *pointillism*, the artist applies paint to a canvas in the form of small dots of color. When viewed from a distance the individual dots cannot be resolved and the painting appears to be painted with continuous colors. An example is given in **Figure 28-23**. If this painting is viewed from a distance of more than a few meters, the color dots blend into one another.

The second example is the formation of a picture on the screen of a television. Although a television can show all the colors of the rainbow, it in fact produces only three colors—red, green, and blue, the so-called additive primaries. These three colors are grouped together tightly to form the "pixels" on the screen, as **Figure 28-24** shows. From a distance the three individual color spots can no longer be distinguished, and the eye sees the net effect of the three colors combined. Since any color can be created with the proper amounts of the three primary colors—red, green, and blue—the television screen can reproduce any picture.

To see this effect in action, try the following: look for a region on a television screen or a computer monitor where the picture is yellow. Since yellow is created by mixing red and green light equally, you will see on close examination (perhaps with the aid of a magnifying glass) that pixels in the yellow region of the screen have both the red and green dots illuminated, but the blue dots are dark. As you slowly move away from the screen, note that the red and green dots merge, leaving the brain with the sensation of a yellow light, even though there are no yellow color dots on the screen.

28–6 Diffraction Gratings

As we saw earlier in this chapter, a screen with one or two slits can produce striking patterns of interference fringes. It is natural to wonder what interference effects may be produced if the number of slits is increased significantly. In general, we refer to a system with a large number of slits as a **diffraction grating.** There are many ways of producing a grating; for example, one might use a diamond stylus to cut thousands of slits in the aluminum coating on a sheet of glass. Alternatively, one might photoreduce an image of parallel lines onto a strip of film. In some cases it is possible to produce gratings with as many as 40,000 slits—or "lines," as they are often called—per centimeter.

The interference pattern formed by a diffraction grating consists of a series of sharp, widely spaced bright fringes—called *principal maxima*—separated by relatively dark regions with a number of weak secondary maxima, as indicated in **Figure 28–25** for the case of five slits. In the limit of a large number of slits, the principal maxima become more sharply peaked, and the secondary maxima become insignificant. As one might expect, the angle at which a principal maximum occurs depends on the wavelength of light that passes through the grating. In this way, a grating acts much like a prism—sending the various colors of white light off in different directions. A grating, however, can spread the light out over a wider range of angles than a prism.

To determine the angles at which principal maxima are found, consider a grating with a large number of slits, each separated from the next by a distance d, as shown in **Figure 28–26**. A beam of light with wavelength λ is incident on the grating from the left and is diffracted onto a distant screen. At an angle θ to the incident direction the path difference between successive slits is $d \sin \theta$, as Figure 28–26 shows. Therefore, constructive interference, and a principal maximum, occurs when the path difference is an integral number of wavelengths, λ:

Constructive Interference in a Diffraction Grating

$$d \sin \theta = m\lambda \qquad m = 0, \pm 1, \pm 2, \ldots \qquad 28\text{–}16$$

Notice that the angle θ becomes larger as d is made smaller. In particular, if a grating has more lines per centimeter (smaller d), light passing through it will be spread out over a larger range of angles.

EXERCISE 28–4

Find the slit spacing necessary for 450-nm light to have a first-order ($m = 1$) principal maximum at 15°.

SOLUTION

Solving Equation 28–16 for d we obtain

$$d = \frac{m\lambda}{\sin \theta} = \frac{(1)(450 \times 10^{-9} \text{ m})}{\sin 15°} = 1.7 \times 10^{-6} \text{ m}$$

A grating is often characterized in terms of its number of lines per unit length, N. For example, a particular grating might have 2250 lines per centimeter. The corresponding slit separation, d, is simply the inverse of the number of lines per length. In this case the slit separation is $d = 1/N = 1/(2250 \text{ cm}^{-1}) = 4.44 \times 10^{-4} \text{ cm} = 4.44 \times 10^{-6} \text{ m}$.

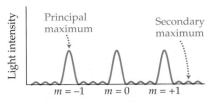

▲ **FIGURE 28–25 Diffraction pattern for five slits**

The interference pattern produced by a diffraction grating with five slits. The large "principal" maxima are sharper than the maxima in the two-slit apparatus. The small "secondary" maxima are negligible compared with the principal maxima.

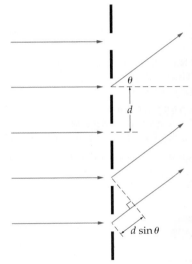

▲ **FIGURE 28–26 Path-length difference in a diffraction grating**

A simple diffraction grating consists of a number of slits with a spacing d. The difference in path length for rays from neighboring slits is $d \sin \theta$.

EXAMPLE 28–7 A SECOND-ORDER MAXIMUM

When 546-nm light passes through a particular diffraction grating, a second-order principal maximum is observed at an angle of 16.0°. How many lines per centimeter does this grating have?

CONTINUED ON NEXT PAGE

CONTINUED FROM PREVIOUS PAGE

PICTURE THE PROBLEM
Our sketch shows the first few principal maxima on either side of the center of the diffraction pattern. The second-order maximum ($m = 2$) is the second maximum above the central peak, and it occurs at an angle of 16.0°.

STRATEGY
First, we can use $\sin\theta = m\lambda/d$ to find the necessary spacing d, given that m, λ, and θ are specified in the problem statement. Next, the number of lines per centimeter, N, is simply the inverse of the spacing d; that is, $N = 1/d$.

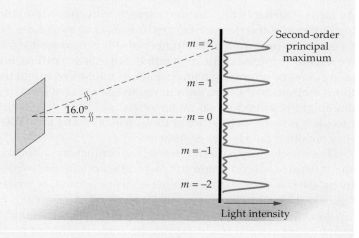

SOLUTION

1. Calculate the distance d between slits:
$$d = \frac{m\lambda}{\sin\theta} = \frac{(2)(546 \times 10^{-9}\ \text{m})}{\sin 16.0°} = 3.96 \times 10^{-6}\ \text{m}$$

2. Take the inverse of d to find the number of lines per meter:
$$N = \frac{1}{d} = \frac{1}{3.96 \times 10^{-6}\ \text{m}} = 2.53 \times 10^{5}\ \text{m}^{-1}$$

3. Convert to lines per centimeter:
$$N = 2.53 \times 10^{5}\ \text{m}^{-1}\left(\frac{1\ \text{m}}{100\ \text{cm}}\right) = 2530\ \text{cm}^{-1}$$

INSIGHT
Thus, this diffraction grating must have 2530 lines per centimeter. Though this is a lot of lines to pack into a distance of one centimeter, it is common to find this many lines or more in a typical diffraction grating.

PRACTICE PROBLEM
If the grating is ruled with more lines per centimeter, does the angle to the second-order maximum increase, decrease, or stay the same? Check your answer by calculating the angle for a grating with 3530 lines per centimeter.   [**Answer:** Increase; $\theta = 22.7°$]

Some related homework problems: Problem 63, Problem 64, Problem 74

Acousto-optic modulation

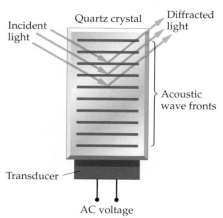

▲ **FIGURE 28–27 An acousto-optical modulator**
Acousto-optical modulators use sound waves, and the density variations they produce, to diffract light. The angle at which the light diffracts can be controlled by the frequency of the sound.

Of the many ways to produce a diffraction grating, one of the more novel is by *acousto-optic modulation* (AOM). In this technique, light is diffracted not by a series of slits but by a series of high-density wave fronts produced by a sound wave propagating through a solid or a liquid. For example, in the AOM device shown in **Figure 28–27**, sound waves propagate through a quartz crystal, producing a series of closely spaced, parallel wave fronts. An incoming beam of light diffracts from these wave fronts, giving rise to an intense outgoing beam. If the sound is turned off, however, the incoming light passes through the crystal without being deflected. Thus, simply turning the sound on or off causes the diffracted beam to be switched on or off, whereas changing the frequency of the sound can change the angle of the diffracted beam. Many laser printers use AOMs to control the laser beam responsible for "drawing" the desired image on a light-sensitive surface.

X-ray Diffraction

There is another type of diffraction grating that is not made by clever applications of technology but occurs naturally—the crystal. The key characteristic of a crystal is that it has a regular, repeating structure. In particular, crystals generally consist of regularly spaced planes of atoms or ions; these planes, just like the wave fronts in an AOM, can diffract an incoming beam of electromagnetic radiation.

For a diffraction grating to be effective, however, the wavelength of the radiation, λ, must be comparable to the spacing, d, in Equation 28–16. In a typical crystal, the spacing between atomic planes is roughly an angstrom; that is, $d \sim 10^{-10}$ m ~ 0.1 nm. Notice that this distance is much less than the wavelength of visible light, which is roughly 400 to 700 nm. Therefore, visible light will not produce useful diffraction effects from a crystal. However, if we consider the full electromagnetic spectrum, as presented in Figure 25–8, we see that wavelengths of 0.1 nm fall within the X-ray portion of the spectrum. Indeed, X-rays produce vivid diffraction patterns when sent through crystals (**Figure 28–28**).

Today, *X-ray diffraction* is a valuable scientific tool. First, the angle at which principal maxima occur in an X-ray diffraction pattern can determine the precise distance between various planes of atoms in a particular crystal. Second, the symmetry of the pattern determines the type of crystal structure. More sophisticated analysis of X-ray diffraction patterns—in particular, the angles and intensities of diffraction maxima—can be used to help determine the structures of even large organic molecules. In fact, it was in part through examination of X-ray diffraction patterns that J. D. Watson and F. H. C. Crick were able to deduce the double-helix structure of DNA in 1953.

Grating Spectroscopes

As noted earlier in this section, a grating can produce a wide separation in the various colors contained in a beam of light. This phenomenon is used as a means of measuring the corresponding wavelengths with an instrument known as a *grating spectroscope*. As we see in **Figure 28–29**, light entering a grating spectroscope is diffracted as it passes through a grating. Next, the angle of diffraction of a given color is determined by a small telescope mounted on a rotating base. Finally, application of Equation 28–16 allows one to determine the wavelength of the light to great precision. Devices of this type have played a key role in elucidating the expanding nature of the universe, as we show in the following Active Example.

ACTIVE EXAMPLE 28–4   FIND THE WAVELENGTHS

A grating spectroscope with a line separation of 4.600×10^{-6} m is used to analyze light from the distant quasar (quasi-stellar object) designated 3C 273. With this instrument it is determined that light from the quasar exhibits a Doppler "red shift" (Section 25–2), indicating motion away from Earth. For example, some of the light given off by hydrogen atoms in the lab has a principal maximum at an angle of 6.067°. When this same hydrogen light is analyzed from the quasar, it is found to have a principal maximum at an angle of 7.030°. Find the wavelength in the lab, and the red-shifted wavelength from the quasar.

SOLUTION *(Test your understanding by performing the calculations indicated in each step.)*

1. Use Equation 28–16 to solve for λ: $\lambda = d \sin \theta$
2. Substitute $\theta = 6.067°$: $\lambda = 486.2$ nm
3. Substitute $\theta = 7.030°$: $\lambda = 563.0$ nm

INSIGHT
Thus we find that the wavelength from the quasar is 15.8% longer than the wavelength in the lab. If we combine this information with the Doppler effect for light (Section 25–2), we can determine the speed at which the quasar is receding from Earth.

YOUR TURN
Find the recession speed of the quasar.

(Answers to **Your Turn** *problems are given in the back of the book.)*

Reflection Gratings

Yet another way to produce a diffraction grating is to inscribe lines on a reflecting surface, with the regions between the lines acting as coherent sources of light. An everyday example of this type of *reflection grating* is a CD. As we saw earlier in this chapter, the information on a CD is encoded in the form of a series of bumps, and these bumps spiral around the CD, creating a tightly spaced set of lines. When a beam of monochromatic light is incident on a CD, as in **Figure 28–30**, a series of

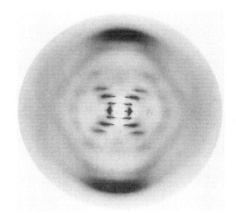

▲ **FIGURE 28–28** **X-ray diffraction**
X-ray diffraction pattern produced by DNA. A photo like this one helped Watson and Crick deduce the double-helix structure of the DNA molecule in 1953.

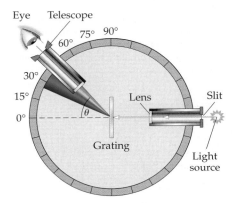

▲ **FIGURE 28–29** **A grating spectroscope**
Because gratings can spread light out over a wider angle than prisms, they are used in most modern spectroscopes.

REAL-WORLD PHYSICS
Measuring the red shift of a quasar

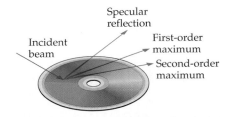

▲ **FIGURE 28–30** **Reflection from a CD**
When a laser beam shines on a CD, a number of reflected beams are observed. The most intense of these is the specular beam, whose angle of reflection is equal to its angle of incidence—the same as if the CD were a plane mirror. Additional reflected beams are observed at angles corresponding to the principal maxima of the grating.

reflected beams is created. One beam reflects at an angle equal to the incident angle—this is referred to as the *specular beam*, since it is the beam that would be expected from a smooth plane mirror. In addition, a number of other reflected beams are observed, each corresponding to a different principal maximum of the grating.

If white light is reflected from a CD, different colors in the incident light are reflected at different angles. This is why light reflected from a CD shows the colors of the rainbow. A similar effect occurs in light reflected from feathers, or from the wing of a butterfly like the blue morpho shown on page 976. For example, a microscopic examination of a butterfly wing shows that it consists of a multitude of plates, much like shingles on a roof. On each of these plates is a series of closely spaced ridges. These ridges act like the grooves on a CD, producing reflected light of different color in different directions. This type of coloration is referred to as **iridescence.** The next time you are able to examine an iridescent object, notice how the color changes as you change the angle from which it is viewed.

REAL-WORLD PHYSICS: BIO

Iridescence in nature

▶ Diffraction can occur when light falls on any surface having grooves with a spacing comparable to the wavelength of the light. Many common surfaces can act as diffraction gratings, including artificially created ones, such as the CDs at left, and natural ones, such as the fly's eye at right. If the incident light is white, comprising a range of different wavelengths, diffraction gives rise to a rainbow effect.

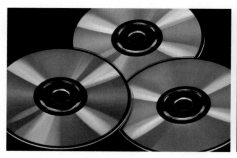

THE BIG PICTURE PUTTING PHYSICS IN CONTEXT

LOOKING BACK

Superposition and interference in light waves (Section 28–1) are completely analogous to superposition and interference in sound waves, as discussed in Chapter 14.

Phase changes due to the reflection of light from an interface (Section 28–3) is just like the phase changes seen when a wave on a string reflects from an end that is either tied down or free to move. See in particular Section 14–2.

LOOKING AHEAD

One of the key experiments related to relativity was the measurement of the speed of light in different directions by Michelson and Morley (Chapter 29). Their experiment was based on observing interference fringes, just like those seen in Young's two-slit experiment (Section 28–2).

The Heisenberg uncertainty principle that is so fundamental to quantum physics can be understood as analogous to the diffraction of light, as we shall see in Section 30–6.

CHAPTER SUMMARY

28–1 SUPERPOSITION AND INTERFERENCE

The simple addition of two or more waves to give a resultant wave is referred to as superposition. When waves are superposed, the result may be a wave of greater amplitude (constructive interference) or of reduced amplitude (destructive interference).

Monochromatic Light
Monochromatic light consists of waves with a single frequency and, hence, a single color.

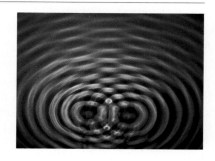

Coherent/Incoherent Light

Light waves that maintain a constant phase relationship with one another are referred to as coherent. Light waves in which the relative phases vary randomly with time are said to be incoherent.

28–2 YOUNG'S TWO-SLIT EXPERIMENT

Interference effects in light are shown clearly in Young's two-slit experiment, in which light passing through two slits forms bright and dark interference "fringes."

Conditions for Bright Fringes

Bright fringes in a two-slit experiment occur at angles θ given by the following relation:

$$d \sin \theta = m\lambda \qquad m = 0, \pm1, \pm2, \dots \qquad \text{28–1}$$

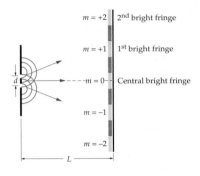

In this expression, λ is the wavelength of the light and d is the separation of the slits. The various values of the integer m correspond to different bright fringes.

Conditions for Dark Fringes

The locations of dark fringes in a two-slit experiment are given by the following:

$$d \sin \theta = \left(m - \tfrac{1}{2}\right)\lambda \quad m = 1, 2, 3, \dots \qquad \text{(above central bright fringe)}$$
$$d \sin \theta = \left(m + \tfrac{1}{2}\right)\lambda \quad m = -1, -2, -3, \dots \text{(below central bright fringe)}$$

28–2

Linear Distance

If the screen on which the interference pattern is projected in a two-slit experiment is a distance L from the slits, the linear distance to a given bright or dark fringe is $y = L \tan \theta$.

28–3 INTERFERENCE IN REFLECTED WAVES

Light waves reflected from different locations can interfere, just like light from the slits in a two-slit experiment.

Phase Changes Due to Reflection

No phase change occurs when light is reflected from a region with a lower index of refraction, whereas a 180° (half-wavelength) phase change occurs when light reflects from a region with a higher index of refraction, or from a solid surface.

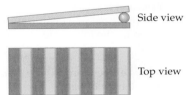

Side view

Top view

Air Wedge

Two plates of glass that touch on one end and have a small separation on the other end form an air wedge. When light of wavelength λ shines on an air wedge, bright fringes occur when the separation between the plates, d, is such that

$$\frac{1}{2} + \frac{2d}{\lambda} = m \qquad m = 1, 2, 3, \dots \qquad \text{28–5}$$

Similarly, dark fringes occur when the following conditions are satisfied:

$$\frac{1}{2} + \frac{2d}{\lambda} = m + \frac{1}{2} \qquad m = 0, 1, 2, \dots \qquad \text{28–6}$$

Newton's Rings

When a piece of glass with a spherical cross section is placed on a flat sheet of glass, the resulting interference fringes form a set of concentric circles known as Newton's rings.

Thin Films

Thin films, like those in a soap bubble, can produce colors in reflected light by eliminating other colors with destructive interference.

28–4 DIFFRACTION

When a wave encounters an obstacle, or passes through an opening, it changes direction. This phenomenon is referred to as diffraction.

Single-Slit Diffraction

When monochromatic light of wavelength λ passes through a single slit of width W, it forms a diffraction pattern of alternating bright and dark fringes.

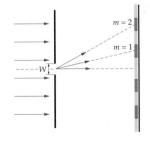

Condition for Dark Fringes

The condition that determines the location of dark fringes in single-slit diffraction is

$$W \sin \theta = m\lambda \qquad m = \pm1, \pm2, \pm3, \dots \qquad \text{28–12}$$

Bright Fringes

Bright fringes are located approximately halfway between successive dark fringes. In addition, the central bright fringe is approximately twice as wide as the other bright fringes.

28–5 RESOLUTION

Resolution refers to the ability of a visual system, like the eye or a camera, to distinguish closely spaced objects.

First Dark Fringe

A circular aperture of diameter D produces a circular diffraction pattern in which the first dark fringe occurs at the angle θ given by the following condition:

$$\sin \theta = 1.22 \frac{\lambda}{D} \qquad \text{28–14}$$

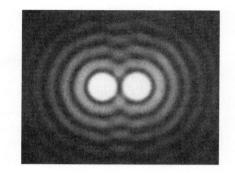

Rayleigh's Criterion: Qualitative Statement

Rayleigh's criterion states that two objects become blurred together when the first dark fringe of one object's diffraction pattern passes through the center of the other object's diffraction pattern.

Rayleigh's Criterion: Quantitative Statement

In quantitative terms, Rayleigh's criterion states that if the angular separation between two objects is less than a certain minimum, $\theta_{\text{min}} = 1.22\lambda/D$, they will appear to be a single object.

28–6 DIFFRACTION GRATINGS

A diffraction grating is a large number of slits through which a beam of light can pass.

Principal Maxima

The principal maxima produced by a diffraction grating occur at the angles given by the following conditions:

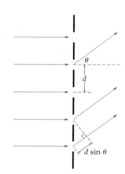

$$d \sin \theta = m\lambda \qquad m = 0, \pm 1, \pm 2, \dots \qquad \text{28–16}$$

In this expression, d is the distance between successive slits and λ is the wavelength of light.

Number of Lines per Centimeter

Diffraction gratings are often characterized by the number of lines, or slits, they have per centimeter. If the number of lines per centimeter is N, the spacing between slits is $d = 1/N$, where d is measured in centimeters.

Reflection Gratings

Diffraction gratings can also be constructed from a reflecting surface with a large number of reflecting lines, like a CD or a butterfly wing.

Iridescence

When white light shines on a reflecting grating, different colors in the light are reflected at different angles. The color effects produced in this way are referred to as iridescence.

PROBLEM-SOLVING SUMMARY

Type of Problem	Relevant Physical Concepts	Related Examples
Determine the angular and linear positions of fringes in a two-slit experiment.	The angular positions are determined by constructive and destructive interference, with the results given in Equations 28–1 and 28–2. The linear distance is given by simple geometry, as in Equation 28–3.	Example 28–2
Determine whether reflection results in constructive or destructive interference.	When reflection is involved, the condition for constructive versus destructive interference involves both the difference in path length and the phase change that may result from reflection.	Examples 28–3, 28–4 Active Example 28–1
Find the angle corresponding to dark fringes in a diffraction pattern.	A single slit produces a dark fringe when a ray from the top of the slit follows a path that is an integer number of half wavelengths longer than the path followed by a ray starting at the center of the slit. The result is the condition given in Equation 28–12.	Example 28–5 Active Example 28–2

Determine whether two nearby objects can be resolved.	To be resolved, nearby objects must have an angular separation that is at least $1.22 \lambda/D$, where D is the diameter of the aperture and λ is the wavelength of the light.	Example 28–6 Active Example 28–3
Find the location of the principal maxima produced by a diffraction grating.	The principal maxima in a diffraction grating pattern are at the same angular positions as the bright fringes in a two-slit experiment, as given in Equation 28–1.	Example 28–7 Active Example 28–4

CONCEPTUAL QUESTIONS

For instructor-assigned homework, go to www.masteringphysics.com

(Answers to odd-numbered Conceptual Questions can be found in the back of the book.)

1. When two light waves interfere destructively, what happens to their energy?

2. What happens to the two-slit interference pattern if the separation between the slits is less than the wavelength of light?

3. If a radio station broadcasts its signal through two different antennas simultaneously, does this guarantee that the signal you receive will be stronger than from a single antenna? Explain.

4. How would you expect the interference pattern of a two-slit experiment to change if white light is used instead of monochromatic light?

5. Suppose a sheet of glass is placed in front of one of the slits in a two-slit experiment. If the thickness of the glass is such that the light reaching the two slits is 180° out of phase, how does this affect the interference pattern?

6. Describe the changes that would be observed in the two-slit interference pattern if the entire experiment were to be submerged in water.

7. Explain why the central spot in Newton's rings is dark.

8. Two identical sheets of glass are coated with films of different materials but equal thickness. The colors seen in reflected light from the two films are different. Give a reason that can account for this observation.

9. Spy cameras use lenses with very large apertures. Why are large apertures advantageous in such applications?

10. A cat's eye has a pupil that is elongated in the vertical direction. How does the resolution of a cat's eye differ in the horizontal and vertical directions?

11. Which portion of the soap film in the accompanying photograph is thinnest? Explain.

Conceptual Question 11

12. The color of an iridescent object, like a butterfly wing or a feather, appears to be different when viewed from different directions. The color of a painted surface appears the same from all viewing angles. Explain the difference.

PROBLEMS AND CONCEPTUAL EXERCISES

Note: Answers to odd-numbered Problems and Conceptual Exercises can be found in the back of the book. **IP** *denotes an integrated problem, with both conceptual and numerical parts;* **BIO** *identifies problems of biological or medical interest;* **CE** *indicates a conceptual exercise.* **Predict/Explain** *problems ask for two responses:* **(a)** *your prediction of a physical outcome, and* **(b)** *the best explanation among three provided. On all problems, red bullets (•, ••, •••) are used to indicate the level of difficulty.*

(In all problems involving sound waves, take the speed of sound to be 343 m/s.)

SECTION 28–1 SUPERPOSITION AND INTERFERENCE

1. • Two sources emit waves that are coherent, in phase, and have wavelengths of 26.0 m. Do the waves interfere constructively or destructively at an observation point 78.0 m from one source and 143 m from the other source?

2. • Repeat Problem 1 for observation points that are **(a)** 91.0 m and 221 m and **(b)** 44.0 m and 135 m from the two sources.

3. •• Two sources emit waves that are in phase with each other. What is the longest wavelength that will give constructive interference at an observation point 161 m from one source and 295 m from the other source?

4. •• A person driving at 17 m/s crosses the line connecting two radio transmitters at right angles, as shown in **Figure 28–31**. The transmitters emit identical signals in phase with each other, which the driver receives on the car radio. When the car is at

▲ **FIGURE 28–31** Problems 4 and 11

point A, the radio picks up a maximum net signal. **(a)** What is the longest possible wavelength of the radio waves? **(b)** How long after the car passes point A does the radio experience a minimum in the net signal? Assume that the wavelength has the value found in part (a).

5. •• Two students in a dorm room listen to a pure tone produced by two loudspeakers that are in phase. Students A and B in **Figure 28–32** hear a maximum sound. What is the lowest possible frequency of the loudspeakers?

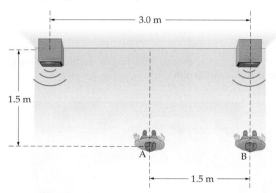

▲ **FIGURE 28–32** Problems 5 and 6

6. •• If the loudspeakers in Problem 5 are 180° out of phase, determine whether a 185-Hz tone heard at location B is a maximum or a minimum.

7. •• A microphone is located on the line connecting two speakers that are 0.845 m apart and oscillating in phase. The microphone is 2.55 m from the midpoint of the two speakers. What are the lowest two frequencies that produce an interference maximum at the microphone's location?

8. •• A microphone is located on the line connecting two speakers that are 0.845 m apart and oscillating 180° out of phase. The microphone is 2.25 m from the midpoint of the two speakers. What are the lowest two frequencies that produce an interference maximum at the microphone's location?

9. •• Moe, Larry, and Curly stand in a line with a spacing of 1.00 m. Larry is 3.00 m in front of a pair of stereo speakers 0.800 m apart, as shown in **Figure 28–33**. The speakers produce a single-frequency tone, vibrating in phase with each other. What are the two lowest frequencies that allow Larry to hear a loud tone while Moe and Curly hear very little?

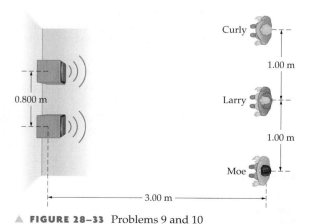

▲ **FIGURE 28–33** Problems 9 and 10

10. •• **IP** In Figure 28–33 the two speakers emit sound that is 180° out of phase and of a single frequency, *f*. **(a)** Does Larry hear a

sound intensity that is a maximum or a minimum? Does your answer depend on the frequency of the sound? Explain. **(b)** Find the lowest two frequencies that produce a maximum sound intensity at the positions of Moe and Curly.

11. •• **IP** Suppose the car radio in Problem 4 picks up a minimum net signal at point A. **(a)** What is the largest possible value for the wavelength of the radio waves? **(b)** If the radio transmitters use a wavelength that is half the value found in part (a), will the car radio pick up a net signal at point A that is a maximum or a minimum? Explain. **(c)** What is the second largest wavelength that will result in a minimum signal at point A?

SECTION 28–2 YOUNG'S TWO-SLIT EXPERIMENT

12. • **CE** Consider a two-slit interference pattern, with monochromatic light of wavelength λ. What is the path difference $\Delta \ell$ for **(a)** the fourth bright fringe and **(b)** the third dark fringe above the central bright fringe? Give your answers in terms of the wavelength of the light.

13. • **CE (a)** Does the path-length difference $\Delta \ell$ increase or decrease as you move from one bright fringe of a two-slit experiment to the next bright fringe farther out? **(b)** What is $\Delta \ell$ in terms of the wavelength λ of the light?

14. • **CE Predict/Explain** A two-slit experiment with red light produces a set of bright fringes. **(a)** Will the spacing between the fringes increase, decrease, or stay the same if the color of the light is changed to blue? **(b)** Choose the *best explanation* from among the following:

   I. The spacing between the fringes will increase because blue light has a greater frequency than red light.
   II. The fringe spacing decreases because blue light has a shorter wavelength than red light.
   III. Only the wave property of light is important in producing the fringes, not the color of the light. Therefore the spacing stays the same.

15. • **CE** A two-slit experiment with blue light produces a set of bright fringes. Will the spacing between the fringes increase, decrease, or stay the same if **(a)** the separation of the slits is decreased, or **(b)** the experiment is immersed in water?

16. • Laser light with a wavelength $\lambda = 670$ nm illuminates a pair of slits at normal incidence. What slit separation will produce first-order maxima at angles of $\pm 35°$ from the incident direction?

17. • Monochromatic light passes through two slits separated by a distance of 0.0334 mm. If the angle to the third maximum above the central fringe is 3.21°, what is the wavelength of the light?

18. • In Young's two-slit experiment, the first dark fringe above the central bright fringe occurs at an angle of 0.31°. What is the ratio of the slit separation, *d*, to the wavelength of the light, λ?

19. •• **IP** A two-slit experiment with slits separated by 48.0×10^{-5} m produces a second-order maximum at an angle of 0.0990°. **(a)** Find the wavelength of the light used in this experiment. **(b)** If the slit separation is increased but the second-order maximum stays at the same angle, does the wavelength increase, decrease, or stay the same? Explain. **(c)** Calculate the wavelength for a slit separation of 68.0×10^{-5} m.

20. •• A two-slit pattern is viewed on a screen 1.00 m from the slits. If the two third-order minima are 22.0 cm apart, what is the width (in cm) of the central bright fringe?

21. •• Light from a He–Ne laser ($\lambda = 632.8$ nm) strikes a pair of slits at normal incidence, forming a double-slit interference pattern on a screen located 1.40 m from the slits. **Figure 28–34**

shows the interference pattern observed on the screen. What is the slit separation?

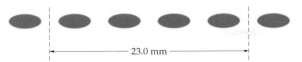

▲ **FIGURE 28–34** Problems 21 and 24

22. •• Light with a wavelength of 546 nm passes through two slits and forms an interference pattern on a screen 8.75 m away. If the linear distance on the screen from the central fringe to the first bright fringe above it is 5.36 cm, what is the separation of the slits?

23. •• A set of parallel slits for optical interference can be made by holding two razor blades together (carefully!) and scratching a pair of lines on a glass microscope slide that has been painted black. When monochromatic light strikes these slits at normal incidence, an interference pattern is formed on a distant screen. The thickness of each razor blade used to make the slits is 0.230 mm, and the screen is 2.50 m from the slits. If the center-to-center separation of the fringes is 7.15 mm, what is the wavelength of the light?

24. •• **IP** Suppose the interference pattern shown in Figure 28–34 is produced by monochromatic light passing through two slits, with a separation of 135 μm, and onto a screen 1.20 m away. **(a)** What is the wavelength of the light? **(b)** If the frequency of this light is increased, will the bright spots of the pattern move closer together or farther apart? Explain.

25. •• A physics instructor wants to produce a double-slit interference pattern large enough for her class to see. For the size of the room, she decides that the distance between successive bright fringes on the screen should be at least 2.50 cm. If the slits have a separation $d = 0.0220$ mm, what is the minimum distance from the slits to the screen when 632.8-nm light from a He–Ne laser is used?

26. •• **IP** When green light ($\lambda = 505$ nm) passes through a pair of double slits, the interference pattern shown in **Figure 28–35 (a)** is observed. When light of a different color passes through the same pair of slits, the pattern shown in **Figure 28–35 (b)** is observed. **(a)** Is the wavelength of the second color longer or shorter than 505 nm? Explain. **(b)** Find the wavelength of the second color. (Assume that the angles involved are small enough to set sin $\theta \approx$ tan θ.)

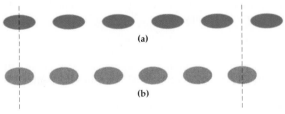

(a)

(b)

▲ **FIGURE 28–35** Problems 26 and 27

27. •• **IP** The interference pattern shown in Figure 28–35 (a) is produced by green light with a wavelength of $\lambda = 505$ nm passing through two slits with a separation of 127 μm. After passing through the slits, the light forms a pattern of bright and dark spots on a screen located 1.25 m from the slits. **(a)** What is the distance between the two vertical, dashed lines in Figure 28–35 (a)? **(b)** If it is desired to produce a more tightly packed interference pattern, like the one shown in Figure 28–35 (b), should the frequency of the light be increased or decreased? Explain.

SECTION 28–3 INTERFERENCE IN REFLECTED WAVES

28. • **CE** Figure 28–36 shows four different cases where light of wavelength λ reflects from both the top and the bottom of a thin film of thickness d. The indices of refraction of the film and the media above and below it are indicated in the figure. For which of the cases will the two reflected rays undergo constructive interference if **(a)** $d = \lambda/4$ or **(b)** $d = \lambda/2$?

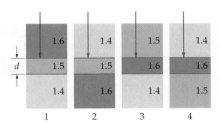

▲ **FIGURE 28–36** Problem 28

29. • **CE** The oil film floating on water in the accompanying photo appears dark near the edges, where it is thinnest. Is the index of refraction of the oil greater than or less than that of the water? Explain.

Light reflected from a film of oil. (Problem 29)

30. • A soap bubble with walls 401 nm thick floats in air. If this bubble is illuminated perpendicularly with sunlight, what wavelength (and color) will be absent in the reflected light? Assume that the index of refraction of the soap film is 1.33. (Refer to Example 25–3 for the connection between wavelength and color.)

31. • A soap film ($n = 1.33$) is 825 nm thick. White light strikes the film at normal incidence. What visible wavelengths will be constructively reflected if the film is surrounded by air on both sides? (Refer to Example 25–3 for the range of visible wavelengths.)

32. • White light is incident on a soap film ($n = 1.30$) in air. The reflected light looks bluish because the red light ($\lambda = 670$ nm) is absent in the reflection. What is the minimum thickness of the soap film?

33. • A 742-nm-thick soap film ($n_{film} = 1.33$) rests on a glass plate ($n_{glass} = 1.52$). White light strikes the film at normal incidence. What visible wavelengths will be constructively reflected from the film? (Refer to Example 25–3 for the range of visible wavelengths.)

34. • An oil film ($n = 1.38$) floats on a water puddle. You notice that green light ($\lambda = 521$ nm) is absent in the reflection. What is the minimum thickness of the oil film?

35. •• A radio broadcast antenna is 36.00 km from your house. Suppose an airplane is flying 2.230 km above the line connecting the broadcast antenna and your radio, and that waves reflected from the airplane travel 88.00 wavelengths farther than waves that travel directly from the antenna to your house. **(a)** Do you observe constructive or destructive interference between the direct and reflected waves? (*Hint:* Does a phase change occur when the waves are reflected?) **(b)** The situation just described occurs when the plane is above a point on the ground that is two-thirds of the way from the antenna to your house. What is the wavelength of the radio waves?

36. •• **IP Newton's Rings** Monochromatic light with $\lambda = 648$ nm shines down on a plano-convex lens lying on a piece of plate glass, as shown in **Figure 28–37**. When viewed from above, one sees a set of concentric dark and bright fringes, referred to as Newton's rings (See Figure 28–11 for a photo of Newton's rings.). **(a)** If the radius of the twelfth dark ring from the center is measured to be 1.56 cm, what is the radius of curvature, R, of the lens? **(b)** If light with a longer wavelength is used with this system, will the radius of the twelfth dark ring be greater than or less than 1.56 cm? Explain.

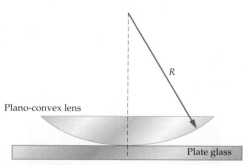

▲ **FIGURE 28–37** Problems 36 and 90

37. •• Light is incident from above on two plates of glass, separated on both ends by small wires of diameter $d = 0.600$ μm. Considering only interference between light reflected from the bottom surface of the upper plate and light reflected from the upper surface of the lower plate, state whether the following wavelengths give constructive or destructive interference: **(a)** $\lambda = 600.0$ nm; **(b)** $\lambda = 800.0$ nm; **(c)** $\lambda = 343.0$ nm.

38. •• **(a)** What is the minimum soap-film thickness ($n = 1.33$) in air that will produce constructive interference in reflection for red ($\lambda = 652$ nm) light? **(b)** Which visible wavelengths will destructively interfere when reflected from this film? (Refer to Example 25–3 for the range of visible wavelengths.)

39. •• **IP** A thin layer of magnesium fluoride ($n = 1.38$) is used to coat a flint-glass lens ($n = 1.61$). **(a)** What thickness should the magnesium fluoride film have if the reflection of 565-nm light is to be suppressed? Assume that the light is incident at right angles to the film. **(b)** If it is desired to suppress the reflection of light with a higher frequency, should the coating of magnesium fluoride be made thinner or thicker? Explain.

40. ••• White light is incident normally on a thin soap film ($n = 1.33$) suspended in air. **(a)** What are the two minimum thicknesses that will constructively reflect yellow ($\lambda = 590$ nm) light? **(b)** What are the two minimum thicknesses that will *destructively* reflect yellow ($\lambda = 590$ nm) light?

41. ••• A thin coating ($t = 340.0$ nm, $n = 1.480$) is placed on a glass lens. Which visible (400 nm $< \lambda <$ 700 nm) wavelength(s) will be absent in the reflected beam if **(a)** the glass has an index of refraction $n = 1.350$, and **(b)** the glass has an index of refraction $n = 1.675$?

42. ••• Two glass plates are separated by fine wires with diameters $d_1 = 0.0500$ mm and $d_2 = 0.0520$ mm, as indicated in **Figure 28–38**. The wires are parallel and separated by a distance of 7.00 cm. If monochromatic light with $\lambda = 589$ nm is incident

from above, what is the distance (in cm) between adjacent dark bands in the reflected light? (Consider interference only between light reflected from the bottom surface of the upper plate and light reflected from the upper surface of the lower plate.)

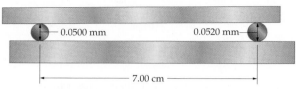

▲ **FIGURE 28–38** Problem 42

SECTION 28–4 DIFFRACTION

43. • **CE** A single-slit diffraction pattern is formed on a distant screen. Assuming the angles involved are small, by what factor will the width of the central bright spot on the screen change if **(a)** the wavelength is doubled, **(b)** the slit width is doubled, or **(c)** the distance from the slit to the screen is doubled?

44. • What width single slit will produce first-order diffraction minima at angles of $\pm23°$ from the central maximum with 690-nm light?

45. • Diffraction also occurs with sound waves. Consider 1300-Hz sound waves diffracted by a door that is 84 cm wide. What is the angle between the two first-order diffraction minima?

46. • Green light ($\lambda = 546$ nm) strikes a single slit at normal incidence. What width slit will produce a central maximum that is 2.50 cm wide on a screen 1.60 m from the slit?

47. • Light with a wavelength of 676 nm passes through a slit 7.64 μm wide and falls on a screen 1.85 m away. Find the linear distance on the screen from the central bright fringe to the first bright fringe above it.

48. • Repeat Problem 47, only this time find the distance on the screen from the central bright fringe to the third dark fringe above it.

49. •• **IP** A single slit is illuminated with 610-nm light, and the resulting diffraction pattern is viewed on a screen 2.3 m away. **(a)** If the linear distance between the first and second dark fringes of the pattern is 12 cm, what is the width of the slit? **(b)** If the slit is made wider, will the distance between the first and second dark fringes increase or decrease? Explain.

50. •• How many dark fringes will be produced on either side of the central maximum if green light ($\lambda = 553$ nm) is incident on a slit that is 8.00 μm wide?

51. •• **IP** The diffraction pattern shown in **Figure 28–39** is produced by passing He–Ne laser light ($\lambda = 632.8$ nm) through a single slit and viewing the pattern on a screen 1.50 m behind the slit. **(a)** What is the width of the slit? **(b)** If monochromatic yellow light with a wavelength of 591 nm is used with this slit instead, will the distance indicated in Figure 28–39 be greater than or less than 15.2 cm? Explain.

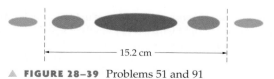

▲ **FIGURE 28–39** Problems 51 and 91

52. •• A screen is placed 1.00 m behind a single slit. The central maximum in the resulting diffraction pattern on the screen is 1.60 cm wide—that is, the two first-order diffraction minima

are separated by 1.60 cm. What is the distance between the two second-order minima?

SECTION 28–5   RESOLUTION

53. • **CE Predict/Explain** (a) In principle, do your eyes have greater resolution on a dark cloudy day or on a bright sunny day? (b) Choose the *best explanation* from among the following:
I. Your eyes have greater resolution on a cloudy day because your pupils are open wider to allow more light to enter the eye.
II. Your eyes have greater resolution on a sunny day because the bright light causes your pupil to narrow down to a smaller opening.

54. • **CE** Is resolution greater with blue light or red light, all other factors being equal? Explain.

55. • Two point sources of light are separated by 5.5 cm. As viewed through a 12-μm-diameter pinhole, what is the maximum distance from which they can be resolved (a) if red light ($\lambda = 690$ nm) is used, or (b) if violet light ($\lambda = 420$ nm) is used?

56. • A spy camera is said to be able to read the numbers on a car's license plate. If the numbers on the plate are 5.0 cm apart, and the spy satellite is at an altitude of 160 km, what must be the diameter of the camera's aperture? (Assume light with a wavelength of 550 nm.)

57. • **Splitting Binary Stars** As seen from Earth, the red dwarfs Krüger 60A and Krüger 60B form a binary star system with an angular separation of 2.5 arc seconds. What is the smallest diameter telescope that could theoretically resolve these stars using 550-nm light? (*Note:* 1 arc sec = 1/3600°)

58. • Find the minimum aperture diameter of a camera that can resolve detail on the ground the size of a person (2.0 m) from an SR-71 Blackbird airplane flying at an altitude of 27 km. (Assume light with a wavelength of 450 nm.)

59. • **The Resolution of Hubble** The Hubble Space Telescope (HST) orbits Earth at an altitude of 613 km. It has an objective mirror that is 2.4 m in diameter. If the HST were to look down on Earth's surface (rather than up at the stars), what is the minimum separation of two objects that could be resolved using 550-nm light? [*Note:* The HST is used only for astronomical work, but a (classified) number of similar telescopes are in orbit for spy purposes.]

60. •• A lens that is "optically perfect" is still limited by diffraction effects. Suppose a lens has a diameter of 120 mm and a focal length of 640 mm. (a) Find the angular width (that is, the angle from the bottom to the top) of the central maximum in the diffraction pattern formed by this lens when illuminated with 540-nm light. (b) What is the linear width (diameter) of the central maximum at the focal distance of the lens?

61. •• The resolution of a telescope is ultimately limited by the diameter of its objective lens or mirror. A typical amateur astronomer's telescope may have a 6.0-in.-diameter mirror. (a) What is the minimum angular separation (in arc seconds) of two stars that can be resolved with a 6.0-in. scope? (Take λ to be at the center of the visible spectrum, about 550 nm, and see Problem 57 for the definition of an arc second.) (b) What is the minimum distance (in km) between two points on the Moon's surface that can be resolved by a 6.0-in. scope? (*Note:* The average distance from Earth to the Moon is 384,400 km.)

62. •• Early cameras were little more than a box with a pinhole on the side opposite the film. (a) What angular resolution would you expect from a pinhole with a 0.50-mm diameter? (b) What is the greatest distance from the camera at which two point objects 15 cm apart can be resolved? (Assume light with a wavelength of 520 nm.)

SECTION 28–6   DIFFRACTION GRATINGS

63. • A grating has 787 lines per centimeter. Find the angles of the first three principal maxima above the central fringe when this grating is illuminated with 655-nm light.

64. • Suppose you want to produce a diffraction pattern with X-rays whose wavelength is 0.030 nm. If you use a diffraction grating, what separation between lines is needed to generate a pattern with the first maximum at an angle of 14°? (For comparison, a typical atom is a few tenths of a nanometer in diameter.)

65. • A diffraction grating has 2200 lines/cm. What is the angle between the first-order maxima for red light ($\lambda = 680$ nm) and blue light ($\lambda = 410$ nm)?

66. • A diffraction grating with 345 lines/mm is 1.00 m in front of a screen. What is the wavelength of light whose first-order maxima will be 16.4 cm from the central maximum on the screen?

67. • The yellow light from a helium discharge tube has a wavelength of 587.5 nm. When this light illuminates a certain diffraction grating it produces a first-order principal maximum at an angle of 1.250°. Calculate the number of lines per centimeter on the grating.

68. •• **IP** The second-order maximum produced by a diffraction grating with 560 lines per centimeter is at an angle of 3.1°. (a) What is the wavelength of the light that illuminates the grating? (b) If a grating with a larger number of lines per centimeter is used with this light, is the angle of the second-order maximum greater than or less than 3.1°? Explain.

69. •• White light strikes a grating with 7600 lines/cm at normal incidence. How many complete visible spectra will be formed on either side of the central maximum? (Refer to Example 25–3 for the range of visible wavelengths.)

70. •• White light strikes a diffraction grating (890 lines/mm) at normal incidence. What is the highest-order visible maximum that is formed? (Refer to Example 25–3 for the range of visible wavelengths.)

71. •• White light strikes a diffraction grating (760 lines/mm) at normal incidence. What is the longest wavelength that forms a second-order maximum?

72. •• A light source emits two distinct wavelengths [$\lambda_1 = 430$ nm (violet); $\lambda_2 = 630$ nm (orange)]. The light strikes a diffraction grating with 450 lines/mm at normal incidence. Identify the colors of the first eight interference maxima on either side of the central maximum.

73. •• A laser emits two wavelengths ($\lambda_1 = 420$ nm; $\lambda_2 = 630$ nm). When these two wavelengths strike a grating with 450 lines/mm, they produce maxima (in different orders) that coincide. (a) What is the order (m) of each of the two overlapping lines? (b) At what angle does this overlap occur?

74. •• **IP** When blue light with a wavelength of 465 nm illuminates a diffraction grating, it produces a first-order principal maximum but no second-order maximum. (a) Explain the absence of higher-order principal maxima. (b) What is the maximum spacing between lines on this grating?

75. •• Monochromatic light strikes a diffraction grating at normal incidence before illuminating a screen 2.10 m away. If the first-order maxima are separated by 1.53 m on the screen, what is the distance between the two second-order maxima?

76. ••• A diffraction grating with a slit separation d is illuminated by a beam of monochromatic light of wavelength λ. The diffracted beam is observed at an angle ϕ relative to the incident direction. If the plane of the grating bisects the angle between the incident and diffracted beams, show that the mth maximum will be observed at an angle that satisfies the relation $m\lambda = 2d \sin(\phi/2)$, with $m = 0, \pm 1, \pm 2, \ldots$.

GENERAL PROBLEMS

77. • **CE** Monochromatic light with a wavelength λ passes through a single slit of width W and forms a diffraction pattern of alternating bright and dark fringes. **(a)** If the width of the slit is decreased, do the dark fringes move outward or inward? Explain. **(b)** What width is necessary for the first dark fringe to move outward to infinity? Give your answer in terms of λ.

78. • **CE Predict/Explain (a)** If a thin liquid film floating on water has an index of refraction less than that of water, will the film appear bright or dark in reflected light as its thickness goes to zero? **(b)** Choose the *best explanation* from among the following:
  I. The film will appear bright because as the thickness of the film goes to zero the phase difference for reflected rays goes to zero.
  II. The film will appear dark because there is a phase change at both interfaces, and this will cause destructive interference of the reflected rays.

79. • **CE** If the index of refraction of an eye could be magically reduced, would the eye's resolution increase or decrease? Explain.

80. • **CE** In order to increase the resolution of a camera, should its f-number be increased or decreased? Explain.

81. • Diffraction effects often involve small angles, and we usually make the approximation $\sin \theta \cong \tan \theta$. To see how accurate this approximation is, complete the following table.

θ (deg)	θ (rad)	$\sin \theta$	$\tan \theta$	$\sin \theta / \tan \theta$
0.0100°				
1.00°				
5.00°				
10.0°				
20.0°				
30.0°				
40.0°				

82. •• When reading the printout from a laser printer, you are actually looking at an array of tiny dots. If the pupil of your eye is 4.3 mm in diameter when reading a page held 28 cm from your eye, what is the minimum separation of adjacent dots that can be resolved? (Assume light with a wavelength of 540 nm, and use 1.36 as the index of refraction for the interior of the eye.)

83. •• The headlights of a pickup truck are 1.32 m apart. What is the greatest distance at which these headlights can be resolved as separate points of light on a photograph taken with a camera whose aperture has a diameter of 12.5 mm? (Take $\lambda = 555$ nm.)

84. •• **Antireflection Coating** A glass lens ($n_{\text{glass}} = 1.52$) has an antireflection coating of MgF$_2$ ($n = 1.38$). **(a)** For 517-nm light, what minimum thickness of MgF$_2$ will cause the reflected rays R_2 and R_4 in **Figure 28–40** to interfere destructively, assuming normal incidence? **(b)** Interference will also occur between the forward-moving rays R_1 and R_3 in Figure 28–40. What minimum thickness of MgF$_2$ will cause these two rays to interfere constructively?

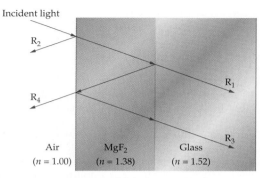

Incident light

Air ($n = 1.00$)   MgF$_2$ ($n = 1.38$)   Glass ($n = 1.52$)

▲ **FIGURE 28–40**   Problem 84

85. •• **IP** White light reflected at normal incidence from a soap bubble ($n = 1.33$) in air produces an interference maximum at $\lambda = 575$ nm but no interference minima in the visible spectrum. **(a)** Explain the absence of interference minima in the visible. **(b)** What are the possible thicknesses of the soap film? (Refer to Example 25–3 for the range of visible wavelengths.)

86. •• A thin film of oil ($n = 1.30$) floats on water ($n = 1.33$). When sunlight is incident at right angles to this film, the only colors that are enhanced by reflection are blue (458 nm) and red (687 nm). Estimate the thickness of the oil film.

87. •• The yellow light of sodium, with wavelengths of 588.99 nm and 589.59 nm, is normally incident on a grating with 494 lines/cm. Find the linear distance between the first-order maxima for these two wavelengths on a screen 2.55 m from the grating.

88. •• **IP** A thin soap film ($n = 1.33$) suspended in air has a uniform thickness. When white light strikes the film at normal incidence, violet light ($\lambda_V = 420$ nm) is constructively reflected. **(a)** If we would like green light ($\lambda_G = 560$ nm) to be constructively reflected, instead, should the film's thickness be increased or decreased? **(b)** Find the new thickness of the film. (Assume the film has the minimum thickness that can produce these reflections.)

89. •• **IP** A thin film of oil ($n = 1.40$) floats on water ($n = 1.33$). When sunlight is incident at right angles to this film, the only colors that are absent from the reflected light are blue (458 nm) and red (687 nm). Estimate the thickness of the oil film.

90. •• **IP** Sodium light, with a wavelength of $\lambda = 589$ nm, shines downward onto the system shown in Figure 28–37. When viewed from above, you see a series of concentric circles known as Newton's rings. **(a)** Do you expect a bright or a dark spot at the center of the pattern? Explain. **(b)** If the radius of curvature of the plano-convex lens is $R = 26.1$ m, what is the radius of the tenth-largest dark ring? (Only rings of nonzero radius will be counted as "rings.")

91. •• **IP** Figure 28–39 shows a single-slit diffraction pattern formed by light passing through a slit of width $W = 11.2\ \mu$m and illuminating a screen 0.855 m behind the slit. **(a)** What is the wavelength of the light? **(b)** If the width of the slit is decreased, will the distance indicated in Figure 28–39 be greater than or less than 15.2 cm? Explain.

92. •• **BIO Entoptic Halos** Images produced by structures within the eye (like lens fibers or cell fragments) are referred to as entoptic images. These images can sometimes take the form of "halos" around a bright light seen against a dark background. The halo in such a case is actually the bright outer rings of a circular diffraction pattern, like Figure 28–21, with the central bright spot not visible because it overlaps the direct image of the light. Find the diameter of the eye structure that causes a circular diffraction pattern with the first dark ring at an angle of 3.7° when viewed with monochromatic light of wavelength 630 nm. (Typical eye structures of this type have diameters on the order of 10 μm. Also, the index of refraction of the vitreous humor is 1.336.)

93. ••• White light is incident on a soap film ($n = 1.33$, thickness = 800.0 nm) suspended in air. If the incident light makes a 45° angle with the normal to the film, what visible wavelength(s) will be constructively reflected? (Refer to Example 25–3 for the range of visible wavelengths.)

94. ••• **IP** A system like that shown in Figure 28–26 consists of N slits, each transmitting light of intensity I_0. The light from each slit has the same phase and the same wavelength. The net intensity I observed at an angle θ due to all N slits is

$$I = I_0 \left[\frac{\sin(N\phi/2)}{\sin(\phi/2)} \right]^2$$

In this expression, $\phi = (2\pi d/\lambda) \sin \theta$, where λ is the wavelength of the light. **(a)** Show that the intensity in the limit $\theta \to 0$ is $I = N^2 I_0$. This is the maximum intensity of the interference pattern. **(b)** Show that the first points of zero intensity on either side of $\theta = 0$ occur at $\phi = 2\pi/N$ and $\phi = -2\pi/N$. **(c)** Does the central maximum ($\theta = 0$) of this pattern become narrower or broader as the number of slits is increased? Explain.

95. ••• Two plates of glass are separated on both ends by small wires of diameter d. Derive an expression for the condition for constructive interference when light of wavelength λ is incident normally on the plates. Consider only interference between waves reflected from the bottom of the top plate and the top of the bottom plate.

96. ••• A curved piece of glass with a radius of curvature R rests on a flat plate of glass. Light of wavelength λ is incident normally on this system. Considering only interference between waves reflected from the curved (lower) surface of glass and the top surface of the plate, show that the radius of the nth dark ring is

$$r_n = \sqrt{n\lambda R - n^2\lambda^2/4}$$

97. ••• **BIO** **The Resolution of the Eye** The resolution of the eye is ultimately limited by the pupil diameter. What is the smallest diameter spot the eye can produce on the retina if the pupil diameter is 4.25 mm? Assume light with a wavelength of $\lambda = 550$ nm. (*Note*: The distance from the pupil to the retina is 25.4 mm. In addition, the space between the pupil and the retina is filled with a fluid whose index of refraction is $n = 1.36$.)

PASSAGE PROBLEMS

Resolving Lines on an HDTV

The American Television Systems Committee (ATSC) sets the standards for high-definition television (HDTV). One of the approved HDTV formats is 1080p, which means 1080 horizontal lines scanned progressively (p)—that is, one line after another in sequence from top to bottom. Another standard is 1080i, which stands for 1080 lines interlaced (i). In this system it takes two scans of the screen to show a complete picture: the first scan shows the "even" horizontal lines, the second scan shows the "odd" horizontal lines. Interlacing was the norm for television displays until the 1970s, and is still used in most standard-definition TVs today. Progressive scanning became more popular with the advent of computer monitors, and is used today in LCD, DLP, and plasma HDTVs.

In addition, the ATSC sets the standard for the shape of displays. For example, it defines a "wide screen" to be one with a

16:9 ratio; that is, the width of the display is greater than the height by the factor 16/9. This ratio is just a little larger than the golden ratio, $\phi = \left(1 + \sqrt{5}\right)/2 = 1.618\ldots$, which is generally believed to be especially pleasing to the eye. Whatever the shape or definition of a TV, the ATSC specifies that it project 30 frames per second on a progressive display, or 60 fields per second on an interlace display, where each field is half the horizontal lines.

For the following problems, assume that 1080 horizontal lines are displayed on a television with a screen that is 15.7 inches high (32-inch diagonal), and that the light coming from the screen has a wavelength of 645 nm. Also, assume that the pupil of your eye has a diameter of 5.50 mm, and that the index of refraction of the interior of the eye is 1.36.

98. • What is the minimum angle your eye can resolve, according to the Rayleigh criterion and the above assumptions?

 A. 0.862×10^{-4} rad   B. 1.05×10^{-4} rad
 C. 1.43×10^{-4} rad   D. 1.95×10^{-4} rad

99. • What is the linear separation between horizontal lines on the screen?

 A. 0.0235 mm   B. 0.145 mm
 C. 0.369 mm   D. 0.926 mm

100. • What is the angular separation of the horizontal lines as viewed from a distance of 12.0 feet?

 A. 1.01×10^{-4} rad   B. 2.53×10^{-4} rad
 C. 2.56×10^{-4} rad   D. 12.1×10^{-4} rad

101. • According to the Rayleigh criterion, what is the closest you can be to the TV screen before resolving the individual horizontal lines? (In practice you can be considerably closer than this distance before resolving the lines.)

 A. 3.51 ft   B. 4.53 ft
 C. 11.5 ft   D. 14.0 ft

INTERACTIVE PROBLEMS

102. •• **IP** Referring to Example 28–2 Suppose we change the slit separation to a value other than 8.5×10^{-5} m, with the result that the linear distance to the tenth bright fringe above the central bright fringe increases from 12 cm to 18 cm. The screen is still 2.3 m from the slits, and the wavelength of the light is 440 nm. **(a)** Did we increase or decrease the slit separation? Explain. **(b)** Find the new slit separation.

103. •• **IP** Referring to Example 28–2 The wavelength of the light is changed to a value other than 440 nm, with the result that the linear distance to the *seventh* bright fringe above the central bright fringe is 12 cm. The screen is still 2.3 m from the slits, and the slit separation is 8.5×10^{-5} m. **(a)** Is the new wavelength longer or shorter than 440 nm? Explain. **(b)** Find the new wavelength.

104. •• **IP** Referring to Example 28–5 The light used in this experiment has a wavelength of 511 nm. **(a)** If the width of the slit is decreased, will the angle to the first dark fringe above the central bright fringe increase or decrease? Explain. **(b)** Find the angle to the first dark fringe if the reduced slit width is 1.50×10^{-6} m.

105. •• **IP** Referring to Example 28–5 The width of the slit in this experiment is 2.20×10^{-6} m. **(a)** If the frequency of the light is decreased, will the angle to the first dark fringe above the central bright fringe increase or decrease? Explain. **(b)** Find the angle to the first dark fringe if the reduced frequency is 5.22×10^{14} Hz.

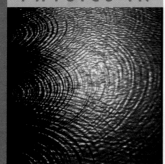

Waves and Particles: A Theme of Modern Physics

We usually think of waves and matter as distinct. In these pages we review what waves are and how we recognize them. Then we show that waves and matter are not so distinct after all.

❶ What is a wave?

A wave is a traveling disturbance. Some waves propagate through a medium, such as water or air, but some—such as light—can travel in a vacuum. Waves often carry energy, although the particles within a wave usually just oscillate back and forth as the wave passes by. The speed of a wave is determined by the properties of the medium through which the wave travels. As the following examples show, a "wave" can take many forms.

Kelvin-Helmholtz waves The ripple-shaped clouds in this photo mark a type of wave that forms in the shear zone between atmospheric layers moving at different velocities.

Kelvin-Helmholtz waves are also common in Jupiter's atmosphere, where intense wind bands shear against each other. The white waves to the left of the Great Red Spot are each larger than our moon.

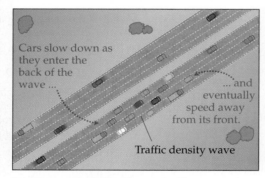

Density waves in traffic Clumpy traffic represents a type of *density wave*. At certain traffic densities, transient disturbances can set up long-lasting, self-sustaining density waves, with cars slowing as they approach the rear of the wave and eventually speeding away from its front. The wave itself may propagate in the direction opposite to the motion of the cars.

Galactic density waves The arms of a spiral galaxy are also density waves—zones in which the stars and gas that orbit the galactic center are unusually tightly packed. The arm propagates because its extra density strengthens its gravitational pull, speeding up the matter approaching its leading edge and slowing the matter leaving its trailing edge. The arms are bright because their high density ignites star formation.

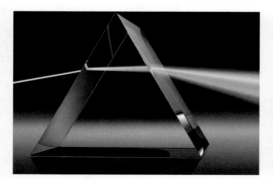

Electromagnetic waves As we learned in Chapter 25, light and other electromagnetic waves represent oscillating, mutually generating electric and magnetic fields. Electromagnetic waves can propagate in a vacuum—they do not require a material medium.

Gravitational waves Einstein's theory of general relativity implies that moving masses can produce waves in the fabric of spacetime. This artist's impression shows the pattern of gravitational waves (actually invisible) that would be produced by a pair of white dwarf stars orbiting each other.

② What properties do all waves share?

- All waves exhibit *superposition* and its consequence, *interference*. When two or more waves of the same type coincide in space, their effects are additive, as diagrammed at right for two waves moving in opposite directions on a string.

- All waves also exhibit *diffraction*: they bend around obstacles. Diffraction is a consequence of the way in which waves propagate and interfere.

If a phenomenon exhibits interference and diffraction, you know it is a wave.

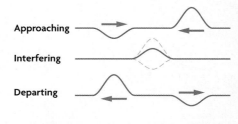

Approaching

Interfering

Departing

Interference and diffraction: Water waves
Water waves passing through slits in a breakwater diffract and interfere. The sand has formed a point at a node in the interference pattern.

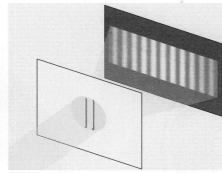

Interference and diffraction: Light This is the diffraction pattern formed when a wave (light in this case) passes through a double slit.

Standing wave in a vibrating cup of coffee
A standing wave will form whenever identical waves travel in opposite directions—as when a wave reflects back on itself.

③ Light consists of waves—or does it?

Light exhibits interference and diffraction, so clearly it's a wave.

However, if you shine light of very low intensity through a double slit onto a CCD chip, what you'll see at first is not a faint diffraction pattern, but instead a random pattern of sharp dots, as if tiny "light bullets" are striking the CCD. Over time, though, these dots accumulate to form the diffraction pattern that is diagnostic of a wave phenomenon!

As we'll explore in Chapter 30, light is both wavelike and particle-like—our everyday experience tells us that these phenomena are mutually exclusive, but in reality they are not.

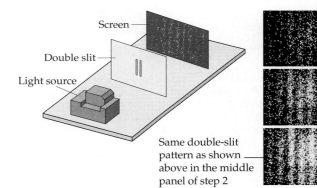

Screen

Double slit

Light source

Same double-slit pattern as shown above in the middle panel of step 2

④ Matter is made of particles—or is it?

Now shoot electrons through a double slit. Initially you see the random dots you expect if electrons act like little bullets—but over time the dots accumulate to form a classic diffraction pattern, meaning that the electrons must be waves! This pattern forms even if you fire the electrons through the double slit one at a time—meaning that *each individual electron* "knows" about both slits and interferes with itself.

Thus, electrons, like light, have both wave and particle properties. As we'll learn in Chapter 30, this finding applies to all particles—therefore, surprising as it may seem, solid objects also have wavelike properties.

The wavelengths of electrons and other particles are much shorter than those of light. That is why an electron microscope can see smaller details than a light microscope can. (Recall from Section 28-5 that resolution is limited by wavelength.)

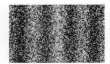

Two-slit diffraction pattern formed by electrons. This pattern appears even if the electrons pass through the device one by one.

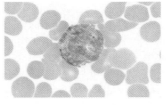

Light micrograph of blood cells

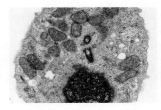

Electron micrograph showing structure in a single cell

29 Relativity

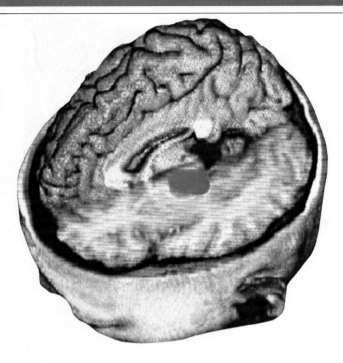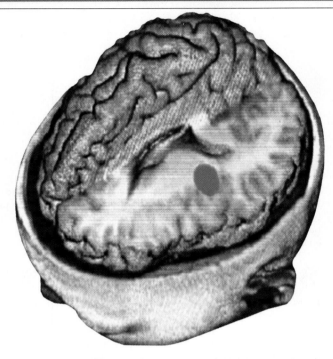

If someone reads this caption to you, a specific region of your brain is activated as you process the information. On the other hand, you may try to recall the words that were read to you after a period of time has elapsed, in which case a different region of your brain is activated as you retrieve the words from your memory. How do we know? The images shown here are the result of positron emission tomography (PET) scans of the brain, in which a radioactive tracer is injected into the bloodstream. In the image on the left, a person is listening to words being read, and the hippocampus region is activated, causing a higher concentration of radioactive blood flow in that area; in the image on the right, the person tries to recall the words, and the temporoparietal region is activated. What is really amazing about these images, however, is that they show areas where matter (electrons) and antimatter (positrons emitted by the radioactive tracer) are annihilating inside the brain and sending out bursts of energy. The energy released in these annihilations is given by the familiar equation ($E = mc^2$), developed by a 26-year-old patent clerk who would soon become the most famous scientist of our time. This chapter explores some of the major contributions of Albert Einstein, including his special and general theories of relativity, and shows how they have radically altered our views of space, time, matter, and energy.

What we refer to today as "modern physics" can be thought of as having started around the beginning of the twentieth century, when two fundamentally new ways of looking at nature were introduced. One of these developments was the introduction of the quantum hypothesis by Max Planck. This will be considered in detail in the next chapter. The other revolutionary development was Albert Einstein's theory of relativity.

The one thing these new theories had in common was that they showed Newton's laws to be incomplete. As it turns out, Newton's laws apply only to objects of macroscopic size and relatively low speeds. In this chapter we explore the surprising behavior associated with speeds approaching the speed of light. We find, in fact, that clocks run slow, metersticks become shorter, and objects become more massive as their speed increases. Although this may seem like science fiction rather than science fact, experiment shows that the predictions of Einstein's theory are indeed correct.

29–1 The Postulates of Special Relativity

At a time when some scientists thought physics was almost completely understood, with only minor details to be straightened out, physics was changed forever with the introduction of the **special theory of relativity.** Published in 1905 by Albert Einstein (1879–1955), a 26-year-old patent clerk (third class) in Berne, Switzerland, relativity fundamentally altered our understanding of such basic physical concepts as time, length, mass, and energy. It may come as a surprise, then, that Einstein's theory of relativity is based on just two simply stated postulates, and that algebra is all the mathematics required to work out its main results.

The postulates of special relativity put forward by Einstein can be stated as follows:

Equivalence of Physical Laws
The laws of physics are the same in all inertial frames of reference.

Constancy of the Speed of Light
The speed of light in a vacuum, $c = 3.00 \times 10^8 \, \text{m/s}$, is the same in all inertial frames of reference, independent of the motion of the source or the receiver.

All the consequences of relativity explored in this chapter follow as a direct result of these postulates.

The first postulate is certainly reasonable. Recall from Section 5–2 that an inertial frame of reference is one in which Newton's laws of motion are obeyed. Specifically, an object with no force acting on it has zero acceleration in all inertial frames. Einstein's first postulate simply extends this notion of an inertial frame to cover *all* the known laws of physics, including those dealing with thermodynamics, electricity, magnetism, and electromagnetic waves. For example, a mechanics experiment performed on the surface of the Earth (which is approximately an inertial frame) gives the same results as when the same experiment is carried out in an airplane moving with constant velocity. In addition, the behavior of heat, magnets, and electric circuits is the same in the airplane as on the ground, as indicated in **Figure 29–1**.

All inertial frames of reference move with constant velocity (that is, zero acceleration) relative to one another. Hence, the special theory of relativity is "special" in the sense that it restricts our considerations to frames with no acceleration. The more general case, in which accelerated motion is considered, is the subject of the *general* theory of relativity, which we discuss later in this chapter. In the case of Earth, the accelerations associated with its orbital and rotational motions are small enough to be ignored in most experiments. Thus, unless otherwise stated, we shall consider the Earth and objects moving with constant velocity relative to it to be inertial frames of reference.

The second postulate of relativity is less intuitive than the first. Specifically, it states that light travels with the same speed, c, regardless of whether the source or observer is in motion. To understand the implications of this assertion, consider for a moment the case of waves on water. In **Figure 29–2 (a)** we see an observer at

▲ Albert Einstein in his twenties.

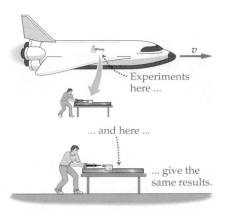

Experiments here ...

... and here ...

... give the same results.

▲ **FIGURE 29–1 Inertial frames of reference**

The two observers shown are in different inertial frames of reference. According to the first postulate of relativity, physical experiments will give identical laws of nature in the two frames.

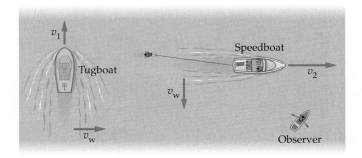

(a) Speed of water waves independent of speed of source

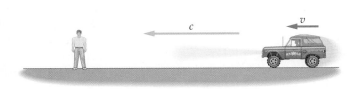

(b) Speed of light waves independent of speed of source

▲ **FIGURE 29–2 Wave speed versus source speed**

The speed of a wave is independent of the speed of the source that generates it. **(a)** Water waves produced by a slow-moving tugboat have the same speed as those produced by a high-powered speedboat. **(b)** The speed of a beam of light, c, is independent of the speed of its source.

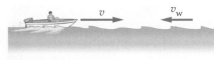

(a) High relative speed

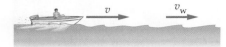

(b) Low relative speed

▲ **FIGURE 29–3 Wave speed versus observer speed**
The speed of a wave depends on the speed of the observer relative to the medium through which the wave propagates. **(a)** The water waves move relative to the observer with speed $v + v_w$. **(b)** In this case, the waves move relative to the observer with speed $v - v_w$.

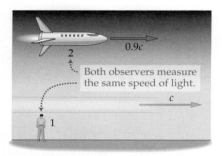

▲ **FIGURE 29–4 The speed of light for different observers**
A beam of light is moving to the right with a speed c relative to observer 1. Observer 2 is moving to the right with a speed of $0.9c$. Still, from the point of view of observer 2, the beam of light is moving to the right with a speed of c, in agreement with the second postulate of relativity.

rest relative to the water, and two moving sources generating waves. The waves produced by both the speedboat and the tugboat travel at the characteristic speed of water waves, v_w, once they are generated. Thus the observer sees a wave speed that is independent of the speed of the source—just as postulated for light, and shown in **Figure 29–2 (b)**.

On the other hand, suppose the observer is in motion with a speed v with respect to the water. If the observer is moving to the right, and water waves are moving to the left with a speed v_w, as in **Figure 29–3 (a)**, the waves move past the observer with a speed $v + v_w$. Similarly, if the water waves are moving to the right, as in **Figure 29–3 (b)**, the observer finds them to have a speed $v - v_w$. Clearly, the fact that the observer is in motion with respect to the medium through which the waves are traveling (water in this case) means that the speed of the water waves depends on the speed of the observer.

Before Einstein's theory of relativity, it was generally accepted that a similar situation would apply to light waves. In particular, light was thought to propagate through a hypothetical medium, referred to as the *luminiferous ether*, or the ether for short, that permeates all space. Since the Earth rotates about its axis with a speed of roughly 1000 mi/h at the equator, and orbits the Sun with a speed of about 67,000 mi/h, it follows that it must move relative to the ether. If this is the case, it should be possible to detect this motion by measuring differences in the speed of light propagating in different directions—just as in the case of water waves. Extremely precise experiments were carried out to this end by the American physicists A. A. Michelson (1852–1931) and E. W. Morley (1838–1923) from 1883 to 1887. They were unable to detect *any* difference in the speed of light. More recent and accurate experiments have come to precisely the same conclusion; namely, the second postulate of relativity is an accurate description of the way light behaves.

To see how counterintuitive the second postulate can be, consider the situation illustrated in **Figure 29–4**. In this case a ray of light is propagating to the right with a speed c relative to observer 1. A second observer is moving to the right as well, with a speed of $0.9c$. Although it seems natural to think that observer 2 should see the ray of light passing with a speed of only $0.1c$, this is not the case. Observer 2, like observers in all inertial frames of reference, sees the ray go by with the speed of light, c.

For the observations given in Figure 29–4 to be valid—that is, for both observers to measure the same speed of light—the behavior of space and time must differ from our everyday experience when speeds approach the speed of light. This is indeed the case, as we shall see in considerable detail in the next few sections. In everyday circumstances, however, the physics described by Newton's laws are perfectly adequate. In fact, Newton's laws are valid in the limit of very small speeds, whereas Einstein's theory of relativity gives correct results for all speeds from zero up to the speed of light.

Since all inertial observers measure the same speed for light, they are all equally correct in claiming that they are at rest. For example, observer 1 in Figure 29–4 may say that he is at rest and that observer 2 is moving to the right with a speed of $0.9c$. Observer 2, however, is equally justified in saying that she is at rest and that observer 1 is in motion with a speed of $0.9c$ to the left. From the point of view of relativity, both observers are equally correct. There is no absolute rest or absolute motion, only motion relative to something else.

Finally, note that it would not make sense for observer 2 in Figure 29–4 to have a speed greater than that of light. If this were the case, it would not be possible for the light ray to pass the observer, much less to pass with the speed c. Thus we conclude that *the ultimate speed in the universe is the speed of light in a vacuum*. In the next several sections of this chapter, we shall see several more ways of arriving at precisely the same conclusion.

29–2 The Relativity of Time and Time Dilation

We generally think of time as moving forward at a constant rate, as suggested by our everyday experience. This is simply not the case, however, when dealing with speeds approaching the speed of light. If you were to observe a spaceship moving past you with a speed of 0.5c, for example, you would notice that the clocks on the ship run slow compared with your clocks—even if they were identical in all other respects.

To calculate the difference between the rates of a moving clock and one at rest, consider the "light clock" shown in **Figure 29–5**. In this clock, a cycle begins when a burst of light is emitted from the light source S. The light then travels a distance d to a mirror, where it is reflected. It travels back a distance d to the detector D, and triggers the next burst of light. Each round trip of light can be thought of as one "tick" of the clock.

We begin by calculating the time interval between the ticks of this clock when it is at rest; that is, when its speed relative to the observer making the measurement is zero. Since the light covers a total distance 2d with a constant speed c, the time between ticks is simply

$$\Delta t_0 = \frac{2d}{c} \qquad\qquad 29\text{–}1$$

The subscript 0 indicates the clock is at rest ($v = 0$) when the measurement is made.

In contrast, consider the same light clock moving with a finite speed v, as in **Figure 29–6**. Notice that the light must now follow a zigzag path in order to complete a tick of the clock. Since this path is clearly longer than 2d, and the speed of the light is still the same—according to the second postulate of relativity—the time between ticks must be greater than Δt_0. With more time elapsing between ticks, the clock runs slow. We refer to this phenomenon as **time dilation,** because the time interval for one tick has been increased—or dilated—from Δt_0 for a clock at rest relative to an observer to $\Delta t > \Delta t_0$ for a clock in motion relative to an observer.

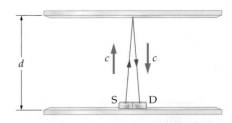

▲ **FIGURE 29–5 A stationary light clock**
Light emitted by the source S travels to a mirror a distance d away and is reflected back into the detector D. The time between emission and detection is one cycle, or one "tick," of the clock.

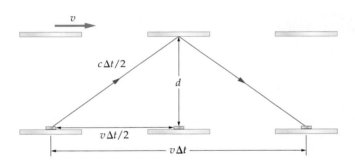

◀ **FIGURE 29–6 A moving light clock**
A moving light clock requires a time Δt to complete one cycle. Note that the light follows a zigzag path that is longer than 2d; hence, the time between ticks is greater for the moving clock than it is for the clock at rest.

To calculate the dilated time, Δt, notice that in the time $\Delta t/2$ the clock moves a horizontal distance $v\Delta t/2$, which is halfway to its position at the end of the tick. The distance traveled by the light in this time is $c\Delta t/2$, which is the hypotenuse of the right triangle shown in Figure 29–6. Applying the Pythagorean theorem to this triangle, we find the following relation:

$$\left(\frac{v\Delta t}{2}\right)^2 + d^2 = \left(\frac{c\Delta t}{2}\right)^2$$

Solving for the time Δt, we find

$$\Delta t = \frac{2d}{\sqrt{c^2 - v^2}} = \frac{2d}{c\sqrt{1 - v^2/c^2}}$$

Recalling that $\Delta t_0 = 2d/c$, we can relate the two time intervals as follows:

Time Dilation

$$\Delta t = \frac{\Delta t_0}{\sqrt{1 - v^2/c^2}} \qquad\qquad 29\text{–}2$$

SI unit: s

Notice that $\Delta t = \Delta t_0$ for $v = 0$, as expected. For speeds v that are greater than zero but less than c, the denominator in Equation 29–2 is less than 1. As a result, it

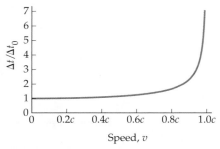

▲ FIGURE 29–7 Time dilation as a function of speed
As the speed of a clock increases, the time required for it to advance by 1 s increases slowly at first and then rapidly near the speed of light.

follows that Δt is greater than Δt_0. Finally, in the limit that the speed v approaches the speed of light, c, we observe that the denominator of Equation 29–2 vanishes, and the time interval Δt goes to infinity. This behavior is illustrated in **Figure 29–7**, where we show the ratio $\Delta t / \Delta t_0$ as a function of the speed v. The fact that Δt goes to infinity means that it takes an infinite amount of time for one tick—in other words, as v approaches the speed of light a clock slows to the point of stopping. Clearly, then, the speed of light provides a natural upper limit to the possible speed of an object.

EXERCISE 29–1

A spaceship carrying a light clock moves with a speed of $0.500c$ relative to an observer on Earth. According to this observer, how long does it take for the spaceship's clock to advance 1.00 s?

SOLUTION
Substituting $\Delta t_0 = 1.00$ s and $v = 0.500c$ in Equation 29–2, we obtain

$$\Delta t = \frac{\Delta t_0}{\sqrt{1 - v^2/c^2}} = \frac{1.00 \text{ s}}{\sqrt{1 - (0.500c)^2/c^2}} = \frac{1.00 \text{ s}}{\sqrt{1 - 0.25}} = 1.15 \text{ s}$$

Even at this high speed, the relativistic effect is relatively small—only about 15%.

Equation 29–2 applies to any type of clock, not just the light clock. If this were not the case—if different clocks ran at different rates when in motion with constant velocity—the first postulate of relativity would be violated.

CONCEPTUAL CHECKPOINT 29–1   THE RATE OF TIME

A clock moving with a finite speed v is observed to run slow. If the speed of light were twice as large as it actually is, would the factor by which the clock runs slow be **(a)** increased, **(b)** decreased, or **(c)** unchanged?

REASONING AND DISCUSSION
As Figure 29–7 shows, the factor by which time is dilated increases as the speed of a clock approaches the speed of light. If the speed of light were twice as large, the speed of a moving clock would be a smaller fraction of the speed of light; hence, the time dilation factor would be less.

Extending the preceding argument, it follows that in the limit of an infinite speed of light there would be no time dilation at all. This conclusion is verified by noting that in the limit $c \to \infty$ the denominator in Equation 29–2 is equal to 1, so that Δt and Δt_0 are equal. Since the speed of light is practically infinite in everyday terms, the relativistic time dilation effect is negligible for everyday objects.

ANSWER
(b) The factor by which the clock runs slow would be decreased.

PROBLEM-SOLVING NOTE

Identifying the Proper Time

The key to solving problems involving time dilation is to correctly identify the proper time, Δt_0. Simply put, the proper time is the time between events that occur at the same location.

Before we proceed, it is useful to introduce some of the terms commonly employed in relativity. First, an **event** is a physical occurrence that happens at a *specified location* at a *specified time*. In three dimensions, for example, we specify an event by giving the values of the coordinates x, y, and z as well as the time t. If two events occur at the same location but at different times, the time between the events is referred to as the **proper time:**

> The *proper time* is the amount of time separating two events that occur at the *same location*.

As an example, the proper time between ticks in a light clock is the time between the emission of light (event 1) and its detection (event 2) when the clock is at rest relative to the observer. Thus, Δt_0 in Equation 29–2 is the proper time, and Δt is the corresponding time when the clock moves relative to the observer with a speed v.

In Exercise 29–1 we calculated the relativistic effect of time dilation for the case of a clock moving with a speed equal to half the speed of light. In everyday circumstances, however, speeds are never so large. In fact, the greatest speed a human might reasonably attain today is the speed of the space shuttle in orbit. As we saw in Chapter 12, this speed is only about 7700 m/s, or 17,000 mi/h. Although this is a rather large speed, it is still only 1/39,000th the speed of light.

To find the time dilation in a case like this we cannot simply substitute $v = 7700$ m/s into Equation 29–2, since a typical calculator does not have enough decimal places to give the correct answer. You might want to try it yourself; your calculator will probably give the incorrect result that Δt is equal to Δt_0. To find the correct answer, we must use the binomial expansion (Appendix A) to reexpress Equation 29–2 as follows:

$$\Delta t = \frac{\Delta t_0}{\sqrt{1 - v^2/c^2}} \approx \Delta t_0 \left(1 + \frac{1}{2}\frac{v^2}{c^2} + \cdots \right)$$

Substituting the speed of the space shuttle and the speed of light, we find

$$\Delta t \approx \Delta t_0 \left[1 + \frac{1}{2}\left(\frac{7700 \text{ m/s}}{3.00 \times 10^8 \text{ m/s}}\right)^2\right] = \Delta t_0 (1 + 3.3 \times 10^{-10})$$

Thus, a clock on board the space shuttle runs slow by a factor of 1.00000000033. At this rate, it would take almost 100 years for the clock on the shuttle to lose 1 s compared with a clock on Earth.

Clearly, such small differences in time cannot be measured with an ordinary clock. In recent years, however, atomic clocks have been constructed that have an accuracy sufficient to put relativity to a direct test. The physicists J. C. Hafele and R. E. Keating conducted such a test in 1971 by placing one atomic clock on board a jet airplane and leaving an identical clock at rest in the laboratory. After flying the moving clock at high speed for many hours, the experimenters found it to have run slower than the clock left in the lab. The discrepancy in times agreed with the predictions of relativity. Today, if an atomic clock is transported from one location to another, the relativistic effects of time dilation must be taken into account; otherwise the clock will give a time that is behind the correct value.

Another aspect of time dilation is the fact that different observers disagree on *simultaneity*. For example, suppose observer 1 notes that two events at different locations occur at the same time. For observer 2, who is moving with a speed v relative to observer 1, these same two events are not simultaneous. Thus, relativity not only changes the rate at which time progresses for different observers, it also changes the amount of time that separates events.

Space Travel and Biological Aging

To this point, our discussion of time dilation has been applied solely to clocks. Clocks are not the only objects that show time dilation, however. In fact,

> relativistic time dilation applies equally to *all physical processes*, including chemical reactions and biological functions.

Thus an astronaut in a moving spaceship *ages more slowly* than one who remains on Earth and by precisely the same factor that a clock on the spaceship runs more slowly than one at rest. To the astronaut, however, time seems to progress as usual. The implications of time dilation for a high-speed trip to a nearby star are considered in the next Example.

EXAMPLE 29–1   BENNY AND JENNY—SEPARATED AT LAUNCH

Astronaut Benny travels to Vega, the fifth brightest star in the night sky, leaving his 35.0-year-old twin sister Jenny behind on Earth. Benny travels with a speed of 0.990c, and Vega is 25.3 light-years from Earth. **(a)** How long does the trip take from the point of view of Jenny? **(b)** How much has Benny aged when he arrives at Vega?

CONTINUED ON NEXT PAGE

CONTINUED FROM PREVIOUS PAGE

PICTURE THE PROBLEM

Our sketch shows the spacecraft as it travels in a straight line from Earth to Vega. The speed of the spacecraft relative to Earth is 99% of the speed of light, $v = 0.990c$. In addition, the distance to Vega is 25.3 light-years; that is, $d = 25.3$ ly. [*Note:* One light-year is the distance light travels with a speed c in one year. Specifically, 1 light-year = 1 ly = c(1 y).]

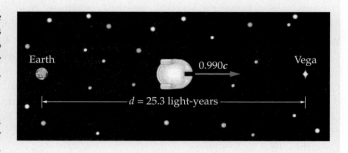

STRATEGY

The two events of interest in this problem are (1) leaving Earth and (2) arriving at Vega. For Jenny, these two events clearly occur at different locations. It follows that the time interval for her is Δt, and not the proper time, Δt_0.

For Benny, however, the two events occur at the same location—namely, just outside the spacecraft door. (In fact, from Benny's point of view the spacecraft is at rest, and the stars are in motion.) Therefore, the time interval measured by Benny is the proper time, Δt_0.

Finally, we note that Δt and Δt_0 are related by Equation 29–2; that is, $\Delta t = \Delta t_0 / \sqrt{1 - v^2/c^2}$.

SOLUTION

Part (a)

1. The spacecraft covers a distance d in a time Δt with a speed $v = 0.990c$. Use $v = d/\Delta t$ to solve for the time, Δt:

$$v = \frac{d}{\Delta t}$$

$$\Delta t = \frac{d}{v} = \frac{25.3 \text{ ly}}{0.990c} = \frac{25.3c(1\text{y})}{0.990c} = 25.6 \text{ y}$$

Part (b)

2. Rearrange $\Delta t = \Delta t_0 / \sqrt{1 - v^2/c^2}$ (Equation 29–2) to solve for the proper time, Δt_0:

$$\Delta t = \frac{\Delta t_0}{\sqrt{1 - v^2/c^2}}$$

$$\Delta t_0 = \Delta t \sqrt{1 - v^2/c^2}$$

3. Substitute $v = 0.990c$ and $\Delta t = 25.6$ y to find Δt_0:

$$\Delta t_0 = \Delta t \sqrt{1 - v^2/c^2}$$

$$= (25.6 \text{ y})\sqrt{1 - \frac{(0.990c)^2}{c^2}} = 3.61 \text{ y}$$

INSIGHT

Thus when Benny reaches Vega, he is only 38.6 y old, whereas Jenny, who stayed behind on Earth, is 60.6 y old.

From the point of view of Benny, the trip took 3.61 y at a speed of 0.990c. As a result, he would say that the distance covered in traveling to Vega was only (3.61 y)(0.990c) = 3.57 ly. We consider this result in greater detail in the next section.

PRACTICE PROBLEM

How fast must Benny travel if he wants to age only 2.00 y during his trip to Vega?   [**Answer:** $v = 0.997c$]

Some related homework problems: Problem 8, Problem 14

In the following Active Example, we determine the speed of an astronaut by considering the change in her heart rate.

ACTIVE EXAMPLE 29–1   **HEARTTHROB: FIND THE ASTRONAUT'S SPEED**

An astronaut traveling with a speed v relative to Earth takes her pulse and finds that her heart beats once every 0.850 s. Mission control on Earth, which monitors her heart activity, observes one heartbeat every 1.40 s. What is the astronaut's speed relative to Earth?

SOLUTION *(Test your understanding by performing the calculations indicated in each step.)*

1. Identify the proper time, Δt_0, and $\Delta t_0 = 0.850$ s
   the dilated time, Δt: $\Delta t = 1.40$ s

2. Solve the time-dilation expression, $\Delta t = \Delta t_0/\sqrt{1 - v^2/c^2}$, for the speed, v:

$$v = c\sqrt{1 - \Delta t_0^2/\Delta t^2}$$

3. Substitute numerical values:

$$v = 0.795c$$

YOUR TURN

What speed is required for mission control to measure a time between heartbeats that is $2(1.40 \text{ s}) = 2.80 \text{ s}$?

*(Answers to **Your Turn** problems are given in the back of the book.)*

The Decay of the Muon

A particularly interesting example of time dilation involves subatomic particles called *muons* that are created by cosmic radiation high in Earth's atmosphere. A muon is an unstable particle; in fact, a muon at rest exists for only about 2.20×10^{-6} s, on average, before it decays. Suppose, for example, that a muon is created at an altitude of 5.00 km above the surface of the Earth. If this muon travels toward the ground with a speed of $0.995c$ for 2.20×10^{-6} s, it will cover a distance of only 657 m before decaying. Thus, without time dilation, one would conclude that the muons produced at high altitude should not reach the surface of the Earth. It is found, however, that large numbers of muons do in fact reach the ground. The reason is that they age more slowly due to their motion—just like the astronaut traveling to Vega considered in Example 29–1. The next Example examines this time dilation effect.

EXAMPLE 29–2 THE LIFE AND TIMES OF A MUON

Consider muons traveling toward Earth with a speed of $0.995c$ from their point of creation at a height of 5.00 km. **(a)** Find the average lifetime of these muons, assuming that a muon at rest has an average lifetime of 2.20×10^{-6} s. **(b)** Calculate the average distance these muons can cover before decaying.

PICTURE THE PROBLEM

Our sketch shows a muon that has just been created at an altitude of 5.00 km. The muon is heading straight for the ground with a speed $v = 0.995c$. As a result of its high speed, the muon's "internal clock" runs slow, allowing it to live longer as seen by an observer on Earth.

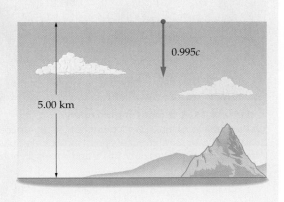

STRATEGY

The two events to be considered in this case are (1) the creation of a muon and (2) the decay of a muon. From the point of view of an observer on Earth, these two events occur at different locations. It follows that the corresponding time is Δt.

From the muon's point of view, it is at rest and the Earth is moving toward it at $0.995c$. Hence, for the muon, the two events occur at the same location, and the time between them is $\Delta t_0 = 2.20 \times 10^{-6}$ s. We can find the time Δt by using time dilation, $\Delta t = \Delta t_0/\sqrt{1 - v^2/c^2}$.

SOLUTION

Part (a)

1. Substitute $v = 0.995c$ and $\Delta t_0 = 2.20 \times 10^{-6}$ s into Equation 29–2:

$$\Delta t = \frac{\Delta t_0}{\sqrt{1 - v^2/c^2}}$$

$$= \frac{2.20 \times 10^{-6} \text{ s}}{\sqrt{1 - (0.995c)^2/c^2}} = 22.0 \times 10^{-6} \text{ s}$$

Part (b)

2. Multiply $v = 0.995c$ times the time 22.0×10^{-6} s to find the average distance covered:

$$d_{\text{av}} = (0.995c)(22.0 \times 10^{-6} \text{ s}) = 6570 \text{ m}$$

CONTINUED ON NEXT PAGE

CONTINUED FROM PREVIOUS PAGE

INSIGHT

Thus relativistic time dilation allows the muons to travel about 10 times farther (6570 m instead of 657 m) than would have been expected from nonrelativistic physics. As a result, muons are detected on Earth's surface.

PRACTICE PROBLEM

Muons produced at the laboratory of the European Council for Nuclear Research (CERN) in Geneva are accelerated to high speeds. At these speeds, the muons are observed to have a lifetime 30.00 times greater than the lifetime of muons at rest. What is the speed of the muons at CERN? [**Answer:** $v = 0.9994c$]

Some related homework problems: Problem 11, Problem 13, Problem 17

29–3 The Relativity of Length and Length Contraction

Just as time is altered for an observer moving with a speed close to the speed of light, so too is distance. For example, a meterstick moving with a speed of $0.5c$ would appear noticeably shorter than a meterstick at rest. As the speed of the meterstick approaches c, its length diminishes toward zero.

To see why lengths contract, and to calculate the amount of contraction, recall the example of the twins Benny and Jenny and the trip to Vega. From Jenny's point of view on Earth, Benny's trip took 25.6 y and covered a distance of $(0.990c)(25.6 \text{ y}) = 25.3 \text{ ly}$. From Benny's point of view, however, the trip took only 3.61 y. Since both twins agree on their relative velocity, it follows that as far as Benny is concerned, his trip covered a distance of only $(0.990c)(3.61 \text{ y}) = 3.57 \text{ ly}$. Thus, from the point of view of the astronaut, Earth and Vega move by at a speed of $0.990c$, and the distance between them is not 25.3 ly, but only 3.57 ly. This is an example of length contraction, as illustrated in **Figure 29–8**.

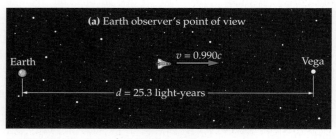

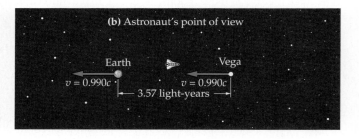

▲ **FIGURE 29–8 A relativistic trip to Vega**

(a) From the Earth observer's point of view the spaceship is traveling with a speed of $0.990c$, covering a distance of 25.3 ly in a time of 25.6 y. **(b)** From the astronaut's point of view the spaceship is at rest, and Earth and Vega are moving with a speed of $0.990c$. For the astronaut the trip takes only 3.61 y and covers a contracted distance of only 3.57 ly.

In general, we would like to determine the contracted length L of an object moving with a speed v. When the object is at rest ($v = 0$), we say that its length is the **proper length,** L_0:

> The proper length is the distance between two points as measured by an observer who is at rest with respect to them.

PROBLEM-SOLVING NOTE

Identifying the Proper Length

The key to solving problems involving length contraction is to correctly identify the proper length, L_0. Specifically, the proper length is the distance between two points that are at rest relative to the observer.

In the Benny and Jenny Example, Jenny is at rest with respect to Earth and Vega. As a result, the distance between them as measured by Jenny is the proper length; that is, $L_0 = 25.3 \text{ ly}$. The contracted length, $L = 3.57 \text{ ly}$, is measured by Benny.

As for the times measured for the trip, from Jenny's point of view the two events (event 1, departing Earth; event 2, arriving at Vega) occur at different locations. As a result, she measures the *dilated time*, $\Delta t = 25.6 \text{ y}$, even though she also measures the *proper length*, L_0. In contrast, Benny measures the *proper time*, $\Delta t_0 = 3.61 \text{ y}$, and the *contracted length*, L. Note that one must be careful to determine from the definitions given previously which observer measures the proper time and which observer measures the proper length—*it should never be assumed*, for example, that just because one observer measures the proper time that observer also measures the proper length.

We now use these observations to obtain a general expression relating L and L_0. To begin, note that both observers measure the same relative speed, v. For Jenny, the speed is $v = L_0/\Delta t$, and for Benny it is $v = L/\Delta t_0$. Setting these speeds equal we obtain

$$v = \frac{L_0}{\Delta t} = \frac{L}{\Delta t_0}$$

Solving for L in terms of L_0, we find $L = L_0(\Delta t_0/\Delta t)$. Finally, using Equation 29–2 to express Δt in terms of Δt_0 we obtain the following:

Length Contraction

$$L = L_0\sqrt{1 - \frac{v^2}{c^2}} \qquad\qquad 29\text{–}3$$

SI unit: m

Substituting the numerical values from the Benny and Jenny Example gives us $L = (25.3 \text{ ly})\sqrt{1 - (0.990c)^2/c^2} = 3.57 \text{ ly}$, as expected.

Note that if $v = 0$ in Equation 29–3, we find that $L = L_0$. As v approaches the speed of light, the contracted length L approaches zero. In general, the length of a moving object is always less than its proper length. **Figure 29–9** shows L as a function of speed v for a meterstick, where we see again that the speed of light is the ultimate speed possible.

PROBLEM-SOLVING NOTE

Measuring Proper Length and Proper Time

Keep in mind that the observer who measures the proper length is *not* necessarily the same observer who measures the proper time.

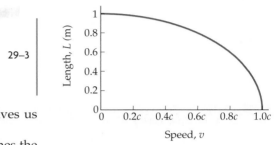

▲ **FIGURE 29–9 Length contraction**
The length of a meterstick as a function of its speed. Note that the length shrinks to zero in the limit that its speed approaches the speed of light.

EXAMPLE 29–3 HALF A METER

Find the speed for which the length of a meterstick is 0.500 m.

PICTURE THE PROBLEM
Our sketch shows a moving meterstick with a contracted length $L = 0.500$ m. The meterstick at rest has its proper length, $L_0 = 1.00$ m.

STRATEGY
We can find the desired speed by applying length contraction (Equation 29–3) to the moving meterstick. In particular, given L and L_0 we can solve $L = L_0\sqrt{1 - v^2/c^2}$ for the speed, v.

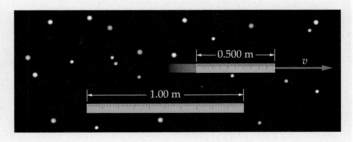

SOLUTION

1. Solve Equation 29–3 for the speed v:

$$L = L_0\sqrt{1 - \frac{v^2}{c^2}}$$

$$v = c\sqrt{1 - \frac{L^2}{L_0^2}}$$

2. Substitute numerical values:

$$v = c\sqrt{1 - \frac{L^2}{L_0^2}} = c\sqrt{1 - \frac{(0.500 \text{ m})^2}{(1.00 \text{ m})^2}} = 0.866c$$

INSIGHT
Note that a person traveling along with the moving meterstick would see it as having its proper length of 1.00 m. From this person's point of view, it is the *other* meterstick that is only 0.500 m long. Thus, lengths—like times—are relative, and depend on the observer making the measurement.

PRACTICE PROBLEM
Find the length of a meterstick that is moving with half the speed of light.   [**Answer:** $L = 0.866$ m]

Some related homework problems: Problem 24, Problem 29

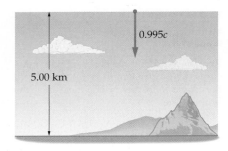

(a) Earth observer's point of view

(b) Muon's point of view

◀ **FIGURE 29–10 A muon reaches Earth's surface**
(a) From Earth's point of view a muon travels downward with a speed of 0.995c for a distance of 5.00 km. The muon can reach Earth's surface only if it ages slowly due to its motion.
(b) From the muon's point of view Earth is moving upward with a speed of 0.995c, and the distance to Earth's surface is only 499 m. From this point of view the muon reaches Earth's surface because the distance is contracted, not because the muon lives longer.

In **Figure 29–10** we illustrate the effects of length contraction for the case of a muon traveling toward Earth's surface. From the point of view of the Earth, Figure 29–10 (a), the muon travels a distance of 5.00 km at a speed of 0.995c. To cover this distance, the muon must live about 10 times longer than it would at rest. From the point of view of the muon, Figure 29–10 (b), the Earth is moving upward at a speed of 0.995c, <u>and the distance</u> the Earth must travel to reach the muon is only $L = (5.00 \text{ km}) \sqrt{1 - (0.995c)^2/c^2} = 499$ m. The Earth can easily cover this distance during the muon's resting lifetime of 2.20 μs.

One final comment: The length contraction calculated in Equation 29–3 *pertains only to lengths in the direction of relative motion.* Lengths at right angles to the relative motion are unaffected.

CONCEPTUAL CHECKPOINT 29–2 ANGLE OF REPOSE

An astronaut is resting on a bed inclined at an angle θ above the floor of a spaceship, as shown in the first sketch. From the point of view of an observer who sees the spaceship moving to the right with a speed approaching c, is the angle the bed makes with the floor **(a)** greater than, **(b)** less than, or **(c)** equal to the angle θ observed by the astronaut?

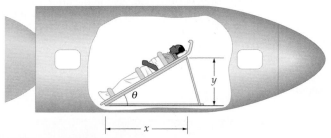

REASONING AND DISCUSSION
A person observing the spaceship moving with a speed v notices a contracted length, x', in the direction of motion, but an unchanged length, y, perpendicular to the direction of motion, as shown in the second sketch.

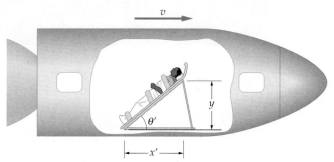

As a result of the contraction in just one direction, the bed is inclined at an angle greater than the angle θ measured by the astronaut.

ANSWER
(a) The angle will be greater than θ.

The fact that lengths in different directions contract differently has interesting effects on the way a rapidly moving object appears to the eye. Additional effects are related to the finite time it takes for light to propagate to the eye from different parts of an object. An example of the way an object would actually look if it moved past us at nearly the speed of light is shown in **Figure 29–11.**

(a) Streetcar at rest

(b) Streetcar at relativistic speed

◀ **FIGURE 29–11 Relativistic distortions**
If the streetcar depicted at rest in part **(a)** moved to the left at nearly the speed of light, it would look like the computer-simulated image in part **(b)**. The streetcar would appear compressed in the direction of its motion. Because the streetcar would move significantly in the time it took for light from it to reach you, however, you would also observe other odd effects—for example, you would be able to see its back surface even from a position directly alongside it.

29–4 The Relativistic Addition of Velocities

Suppose you are piloting a spaceship in deep space, moving toward an asteroid with a speed of 25 mi/h. To signal a colleague on the asteroid, you activate a beam of light on the nose of the ship and point it in the direction of motion. Since light in a vacuum travels with the same speed c relative to all inertial observers, the speed of the light beam relative to the asteroid is simply c, *not* $c + 25$ mi/h. Clearly, then, simple addition of velocities, which seems to work just fine for everyday speeds, no longer applies.

The correct way to add velocities, valid for all speeds from zero to the speed of light, was obtained by Einstein. Consider, for example, a one-dimensional system consisting of a spaceship (1), a probe (2), and an asteroid (3), as indicated in **Figure 29–12**. Suppose the spaceship moves toward the asteroid with the velocity v_{13}, where the subscripts mean "the velocity of object 1 relative to object 3." Since the spaceship moves to the right in Figure 29–12 (a), its velocity is positive ($v_{13} > 0$); if it moved to the left, its velocity would be negative ($v_{13} < 0$). Now imagine that the ship sends out a probe, whose velocity relative to the ship is v_{21}, as shown in Figure 29–12 (b). The question is, how do we add velocities correctly to get the velocity of the probe relative to the asteroid, v_{23}?

The "classical" answer is to simply add the velocities, as we did in Chapter 3; that is, $v_{23} = v_{21} + v_{13}$. Einstein showed that the correct result, taking into account relativity, contains an additional factor:

Relativistic Addition of Velocities

$$v_{23} = \frac{v_{21} + v_{13}}{1 + \dfrac{v_{21}v_{13}}{c^2}}$$

29–4

SI unit: m/s

Notice that if the speed of light were infinite, $c \rightarrow \infty$, the denominator would be $1 + v_{13}v_{21}/(\infty)^2 = 1$ and we would recover classical velocity addition. Thus, it is the *finite* speed of light that is responsible for relativistic effects.

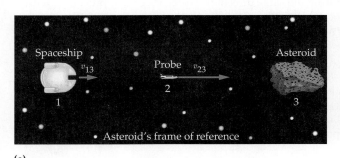

(a)

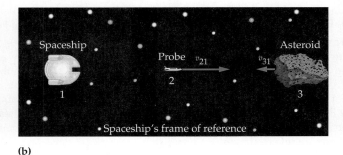

(b)

▲ **FIGURE 29–12 The velocity of a probe: Two frames of reference**
A spaceship approaches an asteroid and sends a probe to investigate it. **(a)** In the asteroid's frame of reference, where the asteroid is at rest, the spaceship approaches with the velocity v_{13} and the probe approaches with the velocity v_{23}. **(b)** In the spaceship's frame of reference, the probe is launched with the velocity v_{21} and the asteroid approaches with the velocity $v_{31} = -v_{13}$. The relativistic sum of the velocities v_{13} and v_{21}, using Equation 29–4, yields the probe's velocity relative to the asteroid, v_{23}.

PROBLEM-SOLVING NOTE

Adding Velocities

To add velocities correctly, it is important to identify each velocity in terms of the object or observer relative to which it is measured, just as for velocity addition in nonrelativistic physics (Section 3–6).

To return to the example at the beginning of this section, what if the probe is actually a beam of light, with the velocity $v_{21} = c$? In this case, relativistic velocity addition gives the speed of the probe (light) relative to the asteroid as

$$v_{23} = \frac{v_{21} + v_{13}}{1 + \frac{v_{21}v_{13}}{c^2}} = \frac{c + v_{13}}{1 + \frac{c\,v_{13}}{c^2}} = \frac{c\left(1 + \frac{v_{13}}{c}\right)}{1 + \frac{v_{13}}{c}} = c$$

Thus, as expected, both the observer in the spaceship (1) and the observer on the asteroid (3) measure the same velocity for the light beam, $v_{21} = v_{23} = c$, regardless of the velocity of the ship relative to the asteroid, v_{13}.

We've seen that Equation 29–4 gives the correct result for a beam of light, but how does it work when applied to speeds much smaller than the speed of light? As an example, suppose a person on a spaceship moving at 25 mi/h throws a ball with a speed of 15 mi/h in the direction of the asteroid. According to the classical addition of velocities, the velocity of the ball relative to the asteroid is 15 mi/h + 25 mi/h = 40 mi/h. Application of Equation 29–4 gives the following result:

$$v = 39.99999999999997 \text{ mi/h}$$

Thus, in any practical measurement, the velocity of the ball relative to the asteroid will be 40 mi/h. We conclude that the classical result, $v_{23} = v_{21} + v_{13}$, although not strictly correct, is appropriate for small speeds.

EXERCISE 29–2

Suppose the spaceship described previously is approaching an asteroid with a speed of $0.750c$. If the spaceship launches a probe toward the asteroid with a speed of $0.800c$ relative to the ship, what is the speed of the probe relative to the asteroid?

SOLUTION

Substituting $v_{13} = 0.750c$ and $v_{21} = 0.800c$ in Equation 29–4 we obtain

$$v_{23} = \frac{v_{21} + v_{13}}{1 + \frac{v_{21}v_{13}}{c^2}} = \frac{0.800c + 0.750c}{1 + \frac{(0.800c)(0.750c)}{c^2}} = 0.969c$$

As expected, the speed relative to the asteroid is less than c. Note, however, that classical velocity addition gives a strikingly different prediction for the speed relative to the asteroid; namely, $v = 0.800c + 0.750c = 1.550c$. Thus, velocity addition is one more way that experiments can verify the predictions of relativity.

In the following Example and Active Example, we apply relativistic velocity addition to a variety of physical systems.

EXAMPLE 29–4   GENERATION NEXT

At starbase Faraway Point, you observe two spacecraft approaching from the same direction. The *La Forge* is approaching with a speed of $0.906c$, and the *Picard* is approaching with a speed of $0.806c$. **(a)** Find the velocity of the *La Forge* relative to the *Picard*. **(b)** Find the velocity of the *La Forge* relative to the *Picard* if the *La Forge*'s direction of motion is reversed.

PICTURE THE PROBLEM

Our sketch shows the two spacecraft approaching the starbase along the same line, moving in the positive direction. The speed of the *La Forge* is $0.906c$, and the speed of the *Picard* is $0.806c$. In addition, note that we have numbered the objects in this system as follows: *Picard* (1); *La Forge* (2); Faraway Point (3). Finally, we show the *La Forge* ahead of the *Picard* in our sketch, but the final result is the same even if their positions are reversed.

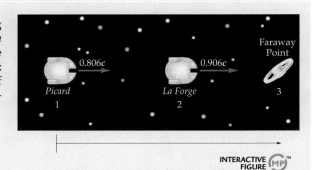

INTERACTIVE FIGURE (MP)™

STRATEGY

a. The key to solving a problem like this is to choose the velocities v_{13}, v_{21}, and v_{23} in a way that is consistent with our numbering scheme. In this case, we know that $v_{13} = 0.806c$ and $v_{23} = 0.906c$. The velocity we want to find, *La Forge* relative to *Picard*, is v_{21}. This can be determined by substituting the known quantities into Equation 29–4.

b. In this case the *La Forge* is traveling in the negative direction; therefore, $v_{23} = -0.906c$.

SOLUTION

Part (a)

1. Use relativistic velocity addition (Equation 29–4) and straightforward algebra to solve for v_{21}:

$$v_{23} = \frac{v_{21} + v_{13}}{1 + \dfrac{v_{21}v_{13}}{c^2}} \quad \text{or} \quad v_{21} = \frac{v_{23} - v_{13}}{1 - \dfrac{v_{23}v_{13}}{c^2}}$$

2. Substitute $v_{13} = 0.806c$ and $v_{23} = 0.906c$ to find v_{21}, the velocity of the *La Forge* relative to the *Picard*:

$$v_{21} = \frac{v_{23} - v_{13}}{1 - \dfrac{v_{23}v_{13}}{c^2}} = \frac{0.906c - 0.806c}{1 - \dfrac{(0.906c)(0.806c)}{c^2}} = 0.371c$$

Part (b)

3. Use the result from part (a), only this time with $v_{23} = -0.906c$:

$$v_{21} = \frac{v_{23} - v_{13}}{1 - \dfrac{v_{23}v_{13}}{c^2}} = \frac{(-0.906c) - 0.806c}{1 - \dfrac{(-0.906c)(0.806c)}{c^2}} = -0.989c$$

INSIGHT

A nonrelativistic calculation of the velocity of the *La Forge* relative to the *Picard* in part (a) gives $0.100c$, considerably less than the correct value of $0.371c$ obtained from relativistic velocity addition. In part (b), a nonrelativistic calculation gives a speed of $1.712c$, as opposed to the correct speed, $0.989c$, which is less than c. Also, note that the negative value of v_{21} means the *Picard* sees the *La Forge* moving to the left.

Finally, when we set up Equation 29–4, we arbitrarily chose the *Picard*, *La Forge*, and Faraway Point to be 1, 2, and 3, respectively. It is important to note, however, that any assignment of 1, 2, and 3 is equally valid and will yield precisely the same results.

PRACTICE PROBLEM

Find the velocity of the *La Forge* relative to the *Picard* if the *Picard*'s direction of motion is reversed but the *La Forge* is still moving toward Faraway Point.   [**Answer:** In this case $v_{13} = -0.806c$; hence, $v_{21} = 0.989c$. The *Picard* sees the *La Forge* moving to the right.]

Some related homework problems: Problem 40, Problem 41, Problem 42

ACTIVE EXAMPLE 29–2	FIND THE RELATIVISTIC LENGTH

As a spaceship approaches a distant planet with a speed of $0.445c$, it launches a probe toward the planet. The proper length of the probe is 10.0 m, and its length as measured by an observer on the spaceship is 7.50 m. What is the length of the probe as measured by an observer on the planet?

SOLUTION *(Test your understanding by performing the calculations indicated in each step.)*

1. Use length contraction, Equation 29–3, to find the velocity of the probe relative to the spaceship:
    $$v_{21} = 0.661c$$

2. Use relativistic velocity addition, Equation 29–4, to find the velocity of the probe relative to the planet:
    $$v_{13} = 0.445c$$
    $$v_{21} = 0.661c$$
    $$v_{23} = 0.855c$$

3. Use the velocity of the probe relative to the planet to calculate its contracted length:
    $$L = 5.19 \text{ m}$$

INSIGHT

Therefore, the probe has a length of 10.0 m when it is at rest relative to an observer. After it is launched from the spaceship, it has a length of 7.50 m relative to the ship, and a length of 5.19 m relative to the planet the ship is approaching.

YOUR TURN

Find the length of the probe relative to the planet if the spaceship moves *away* from the planet with a speed of $0.445c$ when it launches the probe.

*(Answers to **Your Turn** problems are given in the back of the book.)*

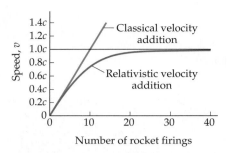

▲ FIGURE 29–13 Relativistic velocity addition

A comparison of classical velocity addition and relativistic velocity addition. Classically, a spacecraft that continues to fire its rockets attains a greater and greater speed. The correct, relativistic, result is that the spacecraft approaches the speed of light in the limit of an infinite number of rocket firings.

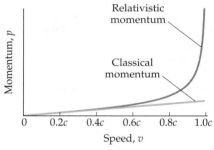

▲ FIGURE 29–14 Relativistic momentum

The magnitude of momentum as a function of speed. Classically, the momentum increases linearly with speed, as shown by the straight line. The correct relativistic momentum increases to infinity as the speed of light is approached.

To get a better feeling for relativistic velocity addition, consider a spacecraft, initially at rest, that increases its speed by $0.1c$ when it fires its rockets. At first, the speed of the spacecraft increases linearly with the number of times the rockets are fired, as indicated by the line labeled "Classical velocity addition" in **Figure 29–13**. As the spacecraft's speed approaches the speed of light, however, further rocket firings have less and less effect, as seen in the curve labeled "Relativistic velocity addition." In the limit of infinite time—and an infinite number of rocket firings—the speed of the spacecraft approaches c, without ever attaining that value.

29–5 Relativistic Momentum

The first postulate of relativity states that the laws of physics are the same for all observers in all inertial frames of reference. Among the most fundamental of these laws are the conservation of momentum and the conservation of energy for an isolated system. In this section we consider the relativistic expression for momentum, leaving energy conservation for the next section.

When one considers the unusual way that relativistic velocities add, as given in Equation 29–4, it comes as no surprise that the classical expression for momentum, $p = mv$, is not valid for all speeds. As an example, we saw in Chapter 9 that if a large mass with a speed v collides elastically with a small mass at rest, the small mass is given a speed $2v$. Clearly, this cannot happen if the speed of the large mass is greater than $0.5c$, since the small mass cannot have a speed greater than the speed of light. Thus the nonrelativistic relation $p = mv$ must be modified for speeds comparable to c.

A detailed analysis shows that the correct relativistic expression for the magnitude of momentum is the following:

Relativistic Momentum

$$p = \frac{mv}{\sqrt{1 - \dfrac{v^2}{c^2}}}$$

29–5

SI unit: $\text{kg} \cdot \text{m/s}$

As v approaches the speed of light, the relativistic momentum becomes significantly larger than the classical momentum, eventually diverging to infinity as $v \rightarrow c$, as shown in **Figure 29–14**. For low speeds the classical and relativistic results agree.

EXERCISE 29–3

Find **(a)** the classical and **(b)** the relativistic momentum of a 2.4-kg mass moving with a speed of $0.81c$.

SOLUTION

a. Evaluate $p = mv$:

$$p = mv = (2.4 \text{ kg})(0.81)(3.00 \times 10^8 \text{ m/s}) = 5.8 \times 10^8 \text{ kg} \cdot \text{m/s}$$

b. Evaluate $p = mv/\sqrt{1 - v^2/c^2}$:

$$p = \frac{mv}{\sqrt{1 - \dfrac{v^2}{c^2}}} = \frac{(2.4 \text{ kg})(0.81)(3.00 \times 10^8 \text{ m/s})}{\sqrt{1 - \dfrac{(0.81c)^2}{c^2}}} = 9.9 \times 10^8 \text{ kg} \cdot \text{m/s}$$

As expected, the relativistic momentum is larger in magnitude than the classical momentum.

Next we consider a system in which the relativistic momentum is conserved.

EXAMPLE 29–5 THE UNKNOWN MASS

A satellite, initially at rest in deep space, explodes into two pieces. One piece has a mass of 150 kg and moves away from the explosion with a speed of 0.76c. The other piece moves away from the explosion in the opposite direction with a speed of 0.88c. Find the mass of the second piece of the satellite.

PICTURE THE PROBLEM
The two pieces of the satellite and their speeds (0.76c and 0.88c) are indicated in the sketch. Note that the two pieces move in opposite directions, which means that they move away from one another along the same line.

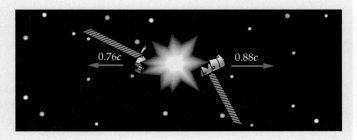

STRATEGY
The basic idea in this system is that because no external forces act on the satellite, its total momentum must be conserved. The initial momentum is zero; hence, the final momentum must be zero as well. This means that the pieces will move in opposite directions, as mentioned, and with momenta of equal magnitude.

Thus, we begin by calculating the magnitude of the momentum for the first piece of the satellite. Next, we set the momentum of the second piece equal to the same magnitude and solve for the mass.

SOLUTION

1. Calculate the magnitude of the momentum for the piece of the satellite with a mass $m_1 = 150$ kg and a speed $v_1 = 0.76c$:

$$p_1 = \frac{m_1 v_1}{\sqrt{1 - \frac{v_1^2}{c^2}}}$$

$$= \frac{(150 \text{ kg})(0.76)(3.00 \times 10^8 \text{ m/s})}{\sqrt{1 - \frac{(0.76c)^2}{c^2}}}$$

$$= 5.3 \times 10^{10} \text{ kg} \cdot \text{m/s}$$

2. Set the momentum of the second piece of the satellite equal to the momentum of the first piece:

$$p_2 = \frac{m_2 v_2}{\sqrt{1 - \frac{v_2^2}{c^2}}} = p_1$$

3. Solve the above relation for m_2, the mass of the second piece of the satellite, and substitute the numerical values $v_2 = 0.88c$ and $p_1 = 5.3 \times 10^{10}$ kg·m/s:

$$m_2 = \left(\frac{p_1}{v_2}\right)\sqrt{1 - \frac{v_2^2}{c^2}}$$

$$= \left[\frac{5.3 \times 10^{10} \text{ kg} \cdot \text{m/s}}{(0.88)(3.00 \times 10^8 \text{ m/s})}\right]\sqrt{1 - \frac{(0.88c)^2}{c^2}}$$

$$= 95 \text{ kg}$$

INSIGHT
In comparison, a classical calculation, with $p = mv$, gives the erroneous result that the mass of the second piece of the satellite is 130 kg. This overestimate of the mass is due to the omission of the factor $1/\sqrt{1 - v^2/c^2}$ in the classical momentum.

PRACTICE PROBLEM
If the mass of the second piece of the satellite had been 210 kg, what would its speed have been?   [**Answer:** $v_2 = 0.64c$]

Some related homework problems: Problem 49, Problem 50

Equation 29–5 is sometimes thought of in terms of a mass that increases with speed. Suppose, for example, that an object has the mass m_0 when it is at rest; that is, its **rest mass** is m_0. When the speed of the object is v, its momentum is

$$p = \frac{m_0 v}{\sqrt{1 - \frac{v^2}{c^2}}} = \left(\frac{m_0}{\sqrt{1 - \frac{v^2}{c^2}}}\right)v$$

If we now make the identification $m = m_0/\sqrt{1 - v^2/c^2}$ we can write the momentum as follows:

$$p = mv$$

Thus the classical expression for momentum can be used for all speeds if we simply interpret the mass as increasing with speed according to the expression

$$m = \frac{m_0}{\sqrt{1 - \dfrac{v^2}{c^2}}}$$

29–6

Note that m approaches infinity as $v \rightarrow c$. Hence a constant force acting on an object generates less and less acceleration, $a = F/m$, as the speed of light is approached. This gives one further way of seeing that the speed of light cannot be exceeded.

Though Equation 29–6 helps to show why an object of finite mass can never be accelerated to speeds exceeding the speed of light, it must be noted that the concept of a relativistic mass has its limitations. For example, the relativistic kinetic energy of an object is *not* obtained by simply replacing m in $\frac{1}{2}mv^2$ with the expression given in Equation 29–6, as we shall see in detail in the next section.

29–6 Relativistic Energy and $E = mc^2$

We have just seen that, from the point of view of momentum, an object's mass increases as its speed increases. Therefore, when work is done on an object, part of the work goes into increasing its speed, and part goes into increasing its mass. It follows, then, that *mass is another form of energy*. This result, like time dilation, was completely unanticipated before the introduction of the theory of relativity.

Consider, for example, an object whose mass while at rest is m_0. Einstein was able to show that when the object moves with a speed v, its **total energy**, E, is given by the following expression:

Relativistic Energy

$$E = \frac{m_0 c^2}{\sqrt{1 - \dfrac{v^2}{c^2}}} = mc^2$$

29–7

SI unit: J

This is Einstein's most famous result from relativity; that is, $E = mc^2$, where m is the relativistic mass given in Equation 29–6.

Note that the total energy, E, does not vanish when the speed goes to zero, as does the classical kinetic energy. Instead, the energy of an object at rest—its **rest energy, E_0**—is

Rest Energy

$$E_0 = m_0 c^2$$

29–8

SI unit: J

Because the speed of light is so large, it follows that the mass of an object times the *speed of light squared* is an enormous amount of energy, as illustrated in the following Exercise.

EXERCISE 29–4

Find the rest energy of a 0.12-kg apple.

SOLUTION

Substituting $m_0 = 0.12$ kg and $c = 3.00 \times 10^8$ m/s in Equation 29–8 we find

$$E_0 = m_0 c^2 = (0.12 \text{ kg})(3.00 \times 10^8 \text{ m/s})^2 = 1.1 \times 10^{16} \text{ J}$$

To put the result of Exercise 29–4 in context, let's compare it with the total yearly energy usage in the United States, which is about 10^{20} J. This means that if the rest energy of the apple could be converted entirely to usable forms of energy, it could supply the energy needs of the entire United States for about an hour.

Put another way, if the rest energy of the apple could be used to light a 100-W lightbulb, it would stay lit for about 10 million years.

This example illustrates the basic principle behind the operation of nuclear power plants, in which small decreases in mass—due to various nuclear reactions—are used to generate electrical energy. The type of reactions that power such plants are referred to as **fission reactions,** in which a large nucleus splits into smaller nuclei and neutrons, as described in Section 32–5. For example, the nucleus of a uranium-235 atom may decay into two smaller nuclei and a number of neutrons. Since the mass of the uranium nucleus is greater than the sum of the masses of the fragments of the decay, the reaction releases an enormous amount of energy. In fact, 1 lb of uranium can produce about 3×10^6 kWh of electrical energy, compared with the 1 kWh that can be produced by the combustion of 1 lb of coal.

The Sun is also powered by the conversion of mass to energy. In this case, however, the energy is released by **fusion reactions,** in which two very small nuclei combine to form a larger nucleus. The detailed reactions are presented in Section 32–6. In the following Example, we determine the amount of mass lost by the Sun per second.

REAL-WORLD PHYSICS

Nuclear power—converting mass to energy

REAL-WORLD PHYSICS

Converting mass to energy to power the Sun

 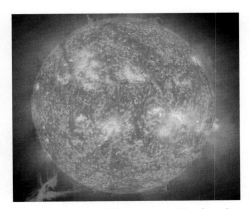

▲ A nuclear reactor (left) converts mass to energy by means of fission reactions, in which large nuclei (such as those of uranium or plutonium) are split into smaller fragments. This photo shows the blue glow referred to as Cherenkov radiation in a nuclear reactor at the Centre of Atomic Energy in Saclay, France. The Sun (right) and other stars are powered by fusion reactions, in which small nuclei (such as those of hydrogen) combine to form heavier ones (such as those of helium). Although this process is utilized in the hydrogen bomb, it has not yet been successfully harnessed as a practical source of power on Earth.

EXAMPLE 29–6 THE PRODIGAL SUN

Energy is radiated by the Sun at the rate of about 3.92×10^{26} W. Find the corresponding decrease in the Sun's mass for every second that it radiates.

PICTURE THE PROBLEM
Our sketch indicates energy radiated continuously by the Sun at the rate $P = 3.92 \times 10^{26}$ W $= 3.92 \times 10^{26}$ J/s. We label the energy given off in the time interval $\Delta t = 1.00$ s with ΔE. The corresponding decrease in mass is Δm.

STRATEGY
If the energy radiated by the Sun in 1.00 s is ΔE, the corresponding decrease in mass, according to the relation $E = mc^2$, is given by $\Delta m = \Delta E/c^2$.

To find ΔE, we simply recall (Chapter 7) that power is energy per time, $P = \Delta E/\Delta t$. Thus the energy radiated by the Sun in 1.00 s is $\Delta E = P\Delta t$, with $\Delta t = 1.00$ s.

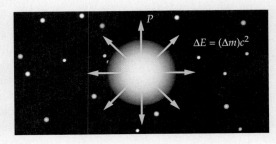

$\Delta E = (\Delta m)c^2$

SOLUTION

1. Calculate the energy radiated by the Sun in 1.00 s:

$\Delta E = P\Delta t = (3.92 \times 10^{26} \text{ J/s})(1.00 \text{ s}) = 3.92 \times 10^{26} \text{ J}$

2. Divide ΔE by the speed of light squared, c^2, to find the decrease in mass:

$\Delta m = \dfrac{\Delta E}{c^2} = \dfrac{3.92 \times 10^{26} \text{ J}}{(3.00 \times 10^8 \text{ m/s})^2} = 4.36 \times 10^9 \text{ kg}$

CONTINUED ON NEXT PAGE

INSIGHT

Thus the Sun loses a rather large amount of mass each second—in fact, roughly the equivalent of 2000 space shuttles. Since the Sun has a mass of 1.99×10^{30} kg, however, the mass it loses in 1500 y is only 10^{-10} of its total mass. Even after 1.5 billion years of radiating at its present rate, the Sun will lose a mere 0.01% of its mass. Clearly, the Sun will not evaporate into space anytime soon.

PRACTICE PROBLEM

Find the power radiated by a star whose mass decreases by the mass of the Moon (7.36×10^{22} kg) in half a million years (5.00×10^5 y). [**Answer:** $P = 4.21 \times 10^{26}$ W, slightly more than the power of the Sun.]

Some related homework problems: Problem 58, Problem 63, Problem 69

CONCEPTUAL CHECKPOINT 29–3   COMPARE THE MASS

When you compress a spring between your fingers, does its mass **(a)** increase, **(b)** decrease, or **(c)** stay the same?

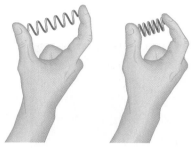

REASONING AND DISCUSSION

When the spring is compressed by an amount x, its energy is increased by the amount $\Delta E = \frac{1}{2}kx^2$, as we saw in Chapter 8. Since the energy of the spring has increased, its mass increases as well, by the amount $\Delta m = \Delta E/c^2$.

ANSWER

(a) The mass of the spring increases.

As one might expect, the increase in the mass of a compressed spring is generally too small to be measured. For example, if the energy of a spring increases by 1.00 J, its mass increases by only $\Delta m = (1.00 \text{ J})/c^2 = 1.11 \times 10^{-17}$ kg.

Matter and Antimatter

A particularly interesting aspect of the equivalence of mass and energy is the existence of **antimatter.** For every elementary particle known to exist, there is a corresponding antimatter particle that has precisely the same mass but exactly the opposite charge. For example, an *electron* has a mass $m_e = 9.11 \times 10^{-31}$ kg and a charge $-e = -1.60 \times 10^{-19}$ C; an *antielectron* has a mass of 9.11×10^{-31} kg and a charge equal to $+1.60 \times 10^{-19}$ C. Since an antielectron has a positive charge, it is generally referred to as a **positron.**

Antimatter is frequently created in accelerators, where particles collide at speeds approaching the speed of light. In fact, it is possible to create antiatoms in the lab made entirely of antimatter. An intriguing possibility is that the universe may actually contain entire antigalaxies of antimatter.

If this is indeed the case, one would have to be a bit careful about visiting such a galaxy, because particles of matter and antimatter have a rather interesting behavior when they meet—they **annihilate** one another. This situation is illustrated in **Figure 29–15**, where we show an electron and a positron coming into contact. The result is that the particles cease to exist, which satisfies charge conservation, since the net charge of the system is zero before and after the annihilation. As for energy conservation, the mass of the two particles is converted into two gamma rays, which are similar to X-rays only more energetic. Each of the gamma rays must have an energy that is at least $E = m_e c^2$. Thus, in matter-antimatter annihilation the particles vanish in a burst of radiation.

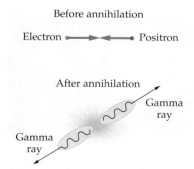

▲ **FIGURE 29–15 Electron-positron annihilation**

An electron and a positron annihilate when they come into contact. The result is the emission of two energetic gamma rays with no mass. The mass of the original particles has been converted into the energy of the gamma rays.

Electron-positron annihilation is the basis for the diagnostic imaging technique called positron emission tomography (PET), which is often used to examine biological processes within the brain, heart, and other organs. In a typical PET brain scan, for example, a patient is injected with glucose (the primary energy source for brain activity) that has been "tagged" with radioactive tracers. These tracers emit positrons in a nuclear reaction described in Section 32–7, and the resulting positrons, in turn, encounter electrons in the brain and undergo annihilation. The resulting gamma rays exit through the patient's skull and are monitored by the PET scanner (**Figure 29–16**), which converts them into false-color images showing the glucose metabolism levels within the brain. Thus, surprising as it may seem, this powerful diagnostic tool actually relies on the annihilation of matter and antimatter inside a person's brain.

The conversion between mass and energy can go the other way, as well. That is, an energetic gamma ray, which has zero mass, can be converted into a particle-antiparticle pair. For example, a gamma ray with an energy of at least $(2m_e)c^2$ can be converted into an electron and a positron; that is, the energy of the gamma ray can be converted into the rest energy of the two particles. Thus we can see that the equivalence of mass and energy, as given by the relation $E = mc^2$, has far-reaching implications.

Relativistic Kinetic Energy

When work is done on an isolated object, accelerating it from rest to a finite speed v, its total energy increases, as given by Equation 29–7. We refer to the *increase* in the object's energy as its *kinetic energy*. Thus the total energy, E, of an object is the sum of its rest energy, m_0c^2, and its kinetic energy, K. In particular,

$$E = \frac{m_0c^2}{\sqrt{1 - \dfrac{v^2}{c^2}}} = m_0c^2 + K$$

Solving for the kinetic energy, we find

Relativistic Kinetic Energy

$$K = \frac{m_0c^2}{\sqrt{1 - \dfrac{v^2}{c^2}}} - m_0c^2 \qquad\qquad 29\text{–}9$$

SI unit: J

As a check, note that the kinetic energy is zero when the speed is zero, as expected. A comparison between the relativistic and classical kinetic energies is presented in **Figure 29–17**.

Although the expression in Equation 29–9 looks nothing like the familiar classical kinetic energy, $\frac{1}{2}mv^2$, it does approach this value in the limit of small speeds. To see this, we can expand Equation 29–9 for small v using the binomial expansion given in Appendix A. The result is as follows:

$$K = \frac{m_0c^2}{\sqrt{1 - \dfrac{v^2}{c^2}}} - m_0c^2 = m_0c^2\left[\frac{1}{\sqrt{1 - \dfrac{v^2}{c^2}}}\right] - m_0c^2$$

$$= m_0c^2\left[1 + \frac{1}{2}\left(\frac{v^2}{c^2}\right) + \frac{3}{8}\left(\frac{v^2}{c^2}\right)^2 + \cdots\right] - m_0c^2$$

The second term in the square brackets is much smaller than the first term for everyday speeds. For example, if the speed v is equal to $0.00001c$ (a rather large everyday speed of roughly 6000 mi/h), the second term is only one ten-millionth of a percent of the first term. Later terms in the expansion are smaller still. Hence, for all practical purposes, the kinetic energy for low speeds is

$$K = m_0c^2\left[1 + \frac{1}{2}\left(\frac{v^2}{c^2}\right)\right] - m_0c^2 = m_0c^2 + \frac{1}{2}m_0v^2 - m_0c^2$$

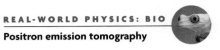
Positron emission tomography

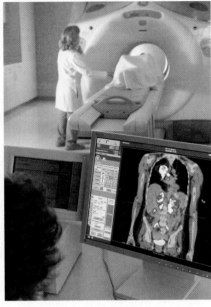

▲ **FIGURE 29–16 A PET scanner**
In a PET scan, positrons emitted by a radioactively labeled metabolite are annihilated when they collide with electrons. The mass of the two particles is converted to energy in the form of a pair of gamma rays, which are always emitted in diametrically opposite directions. A ring of detectors surrounding the patient records the radiation and uses it to construct an image. In this way it is possible to map the location of particular metabolic activities in the body.

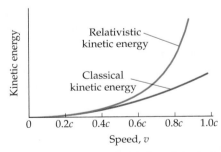

▲ **FIGURE 29–17 Relativistic and classical kinetic energies**
The relativistic kinetic energy (upper curve) goes to infinity as the speed of light is approached. The classical kinetic energy (lower curve) agrees with the relativistic result when the speed is small compared with the speed of light.

PROBLEM-SOLVING NOTE

Rest Mass

To correctly evaluate the relativistic kinetic energy, it is necessary to use the rest mass in Equation 29–9.

Canceling the rest energy, $m_0 c^2$, we find

$$K = \tfrac{1}{2} m_0 v^2$$

Note that the subscript 0 in this expression simply emphasizes the fact that the mass to be used is the rest mass.

We apply the relativistic kinetic energy in the next Example.

EXAMPLE 29–7 RELATIVISTIC KINETIC ENERGY

An observer watching a high-speed spaceship pass by notices that a clock on board runs slow by a factor of 1.50. **(a)** Find the speed of the clock relative to the observer. **(b)** If the rest mass of the clock is 0.320 kg, what is its kinetic energy?

PICTURE THE PROBLEM

Our sketch shows the high-speed spaceship moving by an observer on a distant asteroid. To the observer, the clock in the ship is running slow. In fact, 1.50 s elapse on the observer's clock in the time it takes the spaceship's clock to advance only 1.00 s.

STRATEGY

a. To begin, we can find the speed of the spaceship, v, by using time dilation, as given in Equation 29–2. In particular, we set $\Delta t = \Delta t_0 / \sqrt{1 - v^2/c^2} = 1.50 \, \Delta t_0$ and solve for v.

b. Now that we know both the speed and the rest mass ($m_0 = 0.320$ kg) of the clock, we can find its kinetic energy by applying Equation 29–9: $K = m_0 c^2 / \sqrt{1 - v^2/c^2} - m_0 c^2$.

SOLUTION

Part (a)

1. Use time dilation to solve for the speed, v:

$$\Delta t = \frac{\Delta t_0}{\sqrt{1 - \dfrac{v^2}{c^2}}}$$

$$v = c \sqrt{1 - \left(\frac{\Delta t_0}{\Delta t} \right)^2}$$

2. The dilated time, Δt, is greater than the proper time, Δt_0, by a factor of 1.50. Therefore, we substitute $\Delta t_0 / \Delta t = 1/1.50$ in the expression for v from Step 1:

$$v = c \sqrt{1 - \left(\frac{1}{1.50} \right)^2} = 0.745c$$

Part (b)

3. Substitute $v = 0.745c$ and $m_0 = 0.320$ kg into the expression for the relativistic kinetic energy (Equation 29–9):

$$K = \frac{m_0 c^2}{\sqrt{1 - \dfrac{v^2}{c^2}}} - m_0 c^2 = m_0 c^2 \left(\frac{1}{\sqrt{1 - \dfrac{v^2}{c^2}}} - 1 \right)$$

$$= m_0 c^2 \left(\frac{1}{\sqrt{1 - \dfrac{(0.745c)^2}{c^2}}} - 1 \right) = m_0 c^2 (1.50 - 1)$$

$$= (0.320 \text{ kg})(3.00 \times 10^8 \text{ m/s})^2 (1.50 - 1) = 1.44 \times 10^{16} \text{ J}$$

INSIGHT

In comparison, the classical kinetic energy at this speed is only $\tfrac{1}{2} m_0 v^2 = \tfrac{1}{2} m_0 (0.745c)^2 = m_0 c^2 (0.278) = 7.99 \times 10^{15} \text{ J} < m_0 c^2 (1.50 - 1) = 1.44 \times 10^{16} \text{ J}$. In fact, the classical kinetic energy is *always* less than the relativistic kinetic energy at any given speed, as we can see in Figure 29–17. This is because more work must be done to accelerate a particle that is becoming more massive with speed, as is the case with relativity.

PRACTICE PROBLEM

How fast must the clock be moving if its relativistic kinetic energy is to be 5.00×10^{16} J?   [**Answer:** $v = 0.931c$]

Some related homework problems: Problem 55, Problem 65, Problem 66

Once again, we see that the speed of light is the ultimate speed possible for an object of finite rest mass. As is clear from Figure 29–17, the kinetic energy of an object goes to infinity as its speed approaches c. Thus to accelerate an object to the speed of light would require an infinite amount of energy. Any finite amount of work will increase the speed only to a speed less than c.

29–7 The Relativistic Universe

It is a profound understatement to say that relativity has revolutionized our understanding of the universe. If we think back over the results presented in the last several sections—time dilation, length contraction, increasing mass, mass–energy equivalence—it is clear that relativity reveals to us a universe that is far richer and more varied in its behavior than was ever imagined before. In fact, it is often said that the universe is not only stranger than we imagine, but stranger than we *can* imagine.

By way of analogy, it is almost as if we had spent our lives on a small island on the equator. We would have no knowledge of snow or deserts or mountain ranges. Although our knowledge of Earth would be valid for our small island, we would have an incomplete picture of the world. Our situation with respect to relativity is similar. Before relativity, our knowledge of the physical universe seemed complete, with only minor details to be worked out. After all, Newton's laws and other fundamental principles of physics gave correct predictions for virtually everything we experienced. What Einstein revealed with his theory of relativity, however, was that we were seeing only a small part of the whole and that the behavior we experience at low speeds cannot be extended to high speeds. We weren't missing small details—like the mid-ocean islanders, we were missing most of the picture.

Now, it might seem that relativity plays no significant role in our daily lives, since we do not move at speeds approaching the speed of light. In many respects this observation is correct—relativity is not used to design better cars or airplanes, nor is it used to calculate the orbits needed to send astronauts to the Moon or to Mars. On the other hand, every major hospital has a particle accelerator in its basement to produce radioactive elements for various types of treatments. The accelerator brings particles to speeds very close to the speed of light; hence, relativistic effects cannot be ignored. In fact, for an accelerator to work properly, it must be constructed with relativistic effects taken into account. Similarly, GPS technology can provide accurate positional information only by taking into account relativistic effects—both those having to do with speed, as we have seen in the previous sections, as well as relativistic effects due to gravity, to be discussed in the next section. Thus, as these examples show, we now live in a world where relativity is truly an important part of our everyday lives.

29–8 General Relativity

Einstein's **general theory of relativity** applies to accelerated frames of reference and to gravitation. In fact, the theory provides a link between these two types of physical processes that leads to a new interpretation of gravity.

Consider two observers, both standing within closed elevators. Observer 1 is in an elevator that is at rest on the surface of the Earth, as **Figure 29–18** shows. If this observer drops or throws an object, it falls toward the floor of the elevator with an acceleration equal to the acceleration of gravity.

Observer 2 stands in an identical elevator located in deep space. If this elevator is at rest, or moving with a constant velocity, the observer experiences weightlessness within the elevator, as shown in **Figure 29–19**. If an object is released, it remains in place. Now suppose the elevator is given an upward acceleration equal to the acceleration of gravity, g, as indicated in **Figure 29–20**. An object that is released now remains at rest relative to the background stars while the floor of the elevator accelerates upward toward it with the acceleration g. Similarly, if observer 2 throws the ball horizontally it will follow a parabolic path to the floor, just

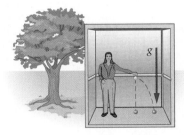

▲ **FIGURE 29–18 A frame of reference in a gravitational field**

An observer in an elevator at rest on Earth's surface. If the observer drops or throws a ball, it falls with a downward acceleration of g.

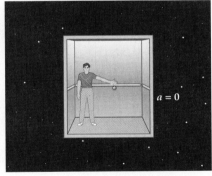

▲ **FIGURE 29–19 An inertial frame of reference with no gravitational field**

An observer in an elevator in deep space experiences weightlessness. If the observer releases an object, it remains at rest.

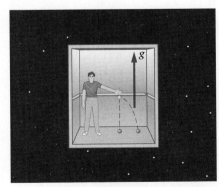

▲ **FIGURE 29–20 An accelerated frame of reference**

If the elevator in Figure 29–19 is given an upward acceleration of magnitude g, the observer in the elevator will note that objects that are dropped or thrown fall toward the floor of the elevator with an acceleration g, just as for the observer in Figure 29–18.

▶ **FIGURE 29–21** **A light experiment in two different frames of reference**
(a) In a nonaccelerating elevator, a beam of light travels on a straight line as it crosses the elevator. **(b)** In an accelerated elevator, the elevator moves upward as the light crosses the elevator; hence, the light strikes the opposite wall at a lower level. The path of the light in this case appears parabolic to the observer riding in the elevator.

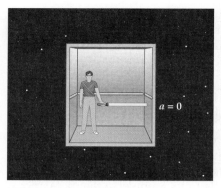

(a) Nonaccelerated elevator

(b) Accelerated elevator

▲ **FIGURE 29–22** **The principle of equivalence**

By the principle of equivalence, a beam of light in a gravitational field should follow a parabolic path, just as in the accelerated elevator in Figure 29–21. The amount of bending of the light's path has been exaggerated here for clarity.

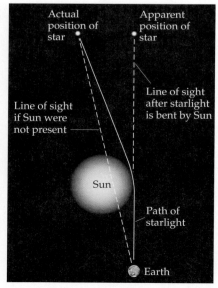

▲ **FIGURE 29–23** **Gravitational bending of light**

As light from a distant star passes close to the Sun, it is bent. The result is that an observer on Earth sees the star along a line of sight that is displaced away from the center of the Sun.

as for observer 1. In addition, the floor of the elevator exerts a force mg on the feet of observer 1 to give that observer (whose mass is m) an upward acceleration g.

We conclude, then, that when observer 2 conducts an experiment in his accelerating elevator, the results are the same as those obtained by observer 1 in her elevator at rest on Earth. Einstein extended these observations to a general principle, the **principle of equivalence:**

Principle of Equivalence
All physical experiments conducted in a uniform gravitational field and in an accelerated frame of reference give identical results.

Thus the two observers cannot tell, without looking outside the elevator, whether they are at rest in a gravitational field or in deep space in an accelerating elevator.

Now let's apply the principle of equivalence to a simple experiment involving light. If the observer in **Figure 29–21 (a)** shines a flashlight toward the opposite wall of the elevator (which has zero acceleration), the light strikes the wall at its initial height. If the same experiment is conducted in an accelerated elevator, as in **Figure 29–21 (b)**, the elevator is accelerating upward during the time the light travels across the elevator. Thus, by the time the light reaches the far wall, it strikes it at a lower level. In fact, the light has followed a parabolic path, just as one would expect for a ball that was projected horizontally.

Applying the principle of equivalence, Einstein concluded that a beam of light in a gravitational field must also bend downward, just as it does in an accelerated elevator; that is, *gravity bends light.* This phenomenon is illustrated in **Figure 29–22**, where the amount of bending has been exaggerated for clarity.

In order to put Einstein's prediction to the test, it is necessary to increase the amount of bending as much as possible to make it large enough to be measured. Thus, we need to use the strongest gravitational field available. In our solar system, the strongest gravitational field is provided by the Sun; hence, experiments were planned to look for the bending of light produced by the Sun.

To see what effect the Sun's gravitational field might have on light, consider the Sun and a ray of light from a distant star, as shown in **Figure 29–23**. As the light passes the Sun, it is bent, as indicated. So an observer on Earth must look in a direction that is farther from the Sun than the actual direction of the distant star—the Sun's gravitational field displaces the distant stars to apparent positions farther from the Sun. If we imagine the Sun moving in front of a background field of stars, as in **Figure 29–24**, the stars near the Sun are displaced outward. It is almost as if the Sun were a lens, slightly distorting the scene behind it.

Because the Sun is so bright, it is possible to carry out an experiment like that shown in Figures 29–23 and 29–24 only during a total eclipse of the Sun, when the Sun's light is blocked by the Moon. During the eclipse, photographs can be taken to show the positions of the background stars. Later, these photographs can be compared with photographs of the same star field taken 6 months later, when the Sun is on the other side of Earth. Comparing the photographs allows one to measure the displacement of the stars. This experiment was carried out during an expedition to

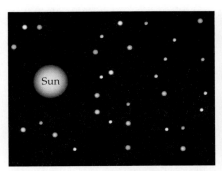  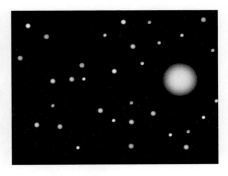

▲ **FIGURE 29–24 Bending of light near the Sun**
As the Sun moves across a starry background, the stars near it appear to be displaced outward, away from the center of the Sun. This is the effect that was used in the first experimental confirmation of general relativity.

Africa in 1919 by Sir Arthur Eddington. His results confirmed the predictions of the general theory of relativity and made Einstein a household name.

Since gravity can bend light, the more powerful the gravitational force, the more the bending, and the more dramatic the results. In Figure 12–18 we saw what can happen when a large galaxy or a cluster of galaxies, with its immense gravitational field, lies between us and a more distant galaxy. The intermediate galaxy can produce significant bending of light, resulting in multiple images of the distant galaxy that can form arcs or crosses. Some examples of such *gravitational lensing* are shown in **Figure 29–25**.

REAL-WORLD PHYSICS
Gravitational lensing

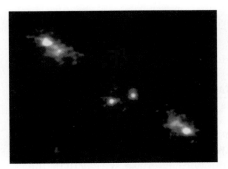 

▲ **FIGURE 29–25 Gravitational lensing**
The images created by gravitational lenses take a number of forms. Sometimes the light from a distant object is stretched out into an arc or even a complete ring. In other cases, such as that of the distant quasar at left, we may see a pair of images. (The quasar appears at the upper left and lower right. The lensing galaxy is not visible in the photo—the small dots at center are unrelated objects.) In still other instances, four images may be produced, as in the famous "Einstein cross" shown at right. The lensing galaxy at the center is some 400 million light-years from us, while the quasar whose multiple images surround it is about 20 times farther away.

An intense gravitational field can also be produced when a star burns up its nuclear fuel and collapses to a very small size. In such a case, the gravitational field can become strong enough to actually trap light; that is, to bend it around to the point that it cannot escape the star. Since such a star cannot emit light of its own, it is referred to as a **black hole.** Black holes, by definition, cannot be directly observed; however, their presence can be inferred by their gravitational effects on other bodies. It is also possible to detect the intense radiation emitted by ionized matter as it falls into a black hole. (Recall from Section 25–1 that accelerated charges produce electromagnetic radiation.) By these and other means, the existence of black holes in the centers of many galaxies has been firmly established. In fact, it is now thought that black holes may be relatively common in the universe.

REAL-WORLD PHYSICS
Black holes

Recall that in Chapter 12 we calculated the escape speed for Earth. The result was

$$v_e = \sqrt{\frac{2GM_E}{R_E}}$$

▲ Because they swallow up all light, black holes are themselves invisible. However, they can be detected indirectly in several ways. If one member of a binary star collapses to become a black hole, an accretion disk may form around it, as shown in the artist's conception at left. The accretion disk is a ring of matter wrenched from its companion, whirling around at ever greater speeds as it spirals into the black hole. The radiation emitted by this matter as it falls into the abyss is the signature of a black hole. The same mechanism, on a much vaster scale, probably accounts for the enormous radiation from active galaxies, such as M87 (right), a giant elliptical galaxy with an enormous jet of matter emanating from its nucleus. Black holes millions of times more massive than the Sun are thought to be present in the cores of such galaxies—and possibly in all galaxies.

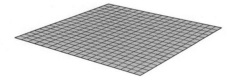

(a) Flat space, away from massive objects

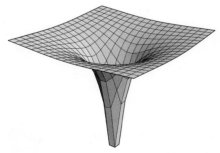

(b) Warped space, near a massive object

▲ **FIGURE 29–26 Warped space and black holes**

(a) Regions of space that are far from any large masses can be thought of as flat. In these regions, light propagates in straight lines. **(b)** Near a large concentrated mass, such as a black hole, space can be thought of as "warped." In these regions, the paths of light rays are bent.

Replacing M_E with M and R_E with R gives the escape speed for any spherical body of mass M and radius R. Now, if we set the escape speed equal to the speed of light, $v_e = c$, we find the following:

$$v_e = \sqrt{\frac{2GM}{R}} = c$$

Solving for R, we find that for an astronomical body to be a black hole, its radius must be no greater than

$$R = \frac{2GM}{c^2} \qquad\qquad 29\text{--}10$$

Although this calculation is too simplistic and not entirely correct, the final result given in Equation 29–10, known as the **Schwarzschild radius,** does indeed agree with the results of general relativity. In the following Exercise, we calculate the radius of a black hole for a specific case.

EXERCISE 29–5

To what radius must Earth be condensed for it to become a black hole?

SOLUTION

Substituting G, M_E, and c in Equation 29–10 we get

$$R = \frac{2GM_E}{c^2} = \frac{2(6.67 \times 10^{-11}\,\text{N}\cdot\text{m}^2/\text{kg}^2)(5.98 \times 10^{24}\,\text{kg})}{(3.00 \times 10^8\,\text{m/s})^2} = 8.86\,\text{mm}$$

Thus Earth would have to be reduced to roughly the size of a walnut for it to become a black hole.

A convenient way to visualize the effects of intense gravitational fields is to think of space as a sheet of rubber with a square array of grid lines, as shown in **Figure 29–26.** In the absence of mass, the sheet is flat, and the grid lines are straight, as in Figure 29–26 (a). In this case, a beam of light follows a straight-line path, parallel to a grid line. If a large mass is present, however, the sheet of rubber is deformed, as in Figure 29–26 (b), and light rays follow the curved paths of the grid lines. In cases where one large mass orbits another, as in **Figure 29–27,** the result is a series of ripples moving outward through the rubber sheet. These ripples represent **gravity waves,** one of the many intriguing predictions of general relativity.

When a gravity wave passes through a given region of space, it causes a local deformation of space, as Figure 29–27 suggests. Early attempts to detect gravity waves were based on measuring the distortion a gravity wave would produce in a large metal bar. Unfortunately, the sensitivity of these devices was too low to detect the weak waves that are thought to pass through the Earth all the time. Ironically, these detectors *were* sensitive enough to detect the relatively strong gravity waves that must have accompanied the 1987 supernova explosion in the nearby Large Magellanic Cloud (a small satellite galaxy of our Milky Way). As luck would have it, however, none of the detectors were operating at the time.

The next generation of gravity wave detectors is now nearing completion. These detectors, which go by the name of Laser Interferometer Gravitational Wave Observatory, or LIGO for short, should be sensitive enough to detect several gravitational wave events per year. One type of event that LIGO will be looking for is the final death spiral of a neutron star as it plunges into a black hole. The neutron star might orbit the black hole for 150,000 years, but only in the last 15 minutes of its life does it find fame, because during those few minutes its acceleration is great enough to produce gravity waves detectable by LIGO.

The basic operating principle of LIGO is to send a laser beam in two different directions along 4-km-long vacuum tubes. At the far end of each tube the beam is reflected back to its starting point. If the two beams travel equal distances, they will interfere constructively when reunited, just like the two rays that meet at the center of a two-slit interference pattern, as discussed in Section 28–2. If the path lengths along the two tunnels are slightly different, however, the beams will have at least partial destructive interference when they combine. The resulting difference in intensity is what LIGO will measure.

The connection with gravity waves is indicated in **Figure 29–28**. Here we see that as a gravity wave passes a LIGO facility, it will cause one tube to increase in length and the other tube to decrease in length. Hence, if the observatory can detect a small enough change in length, it will "see" the gravity wave. Just how much change in length is a gravity wave expected to produce? Incredibly, a typical strong gravity wave will change the length of a tube by less than the diameter of an atomic nucleus. Thus, the task for LIGO is to measure this small change in length in a tube that is 4 km long—no easy task, but the LIGO scientists are confident it can be done.

If gravity waves are indeed observed in the early years of the twenty-first century, as expected, it will open an entirely new window on the universe. So far, information about the universe comes to our telescopes by way of electromagnetic waves. We have certainly learned a lot about the universe in this way; in fact, whenever we observe in a different part of the electromagnetic spectrum, we find surprising new phenomena. One can only imagine what additional wonders will be observed when an entirely different spectrum—the gravity wave spectrum—is explored.

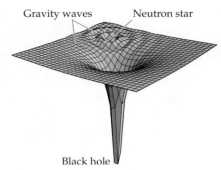

▲ **FIGURE 29–27 Gravity waves**
Gravity waves can be thought of as "ripples" in the warped space described in Figure 29–26. In the case illustrated here, a neutron star orbits a black hole. As a result of its acceleration, the neutron star emits gravity waves.

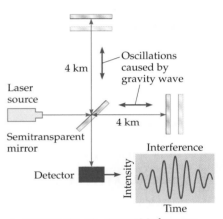

▲ **FIGURE 29–28 How LIGO detects a gravity wave**
When a gravity wave passes through the Earth, it "distorts" everything in its path. The idea behind LIGO is to detect these distortions using interference effects similar to those observed in Young's two-slit experiment (Chapter 28). In particular, LIGO sends laser light along two 4-km vacuum tubes oriented at right angles to one another. When a gravity wave passes, one arm will be lengthened and the other arm will be shortened. The resulting difference in path length changes the interference pattern, allowing us to "see" the gravity wave.

THE BIG PICTURE **PUTTING PHYSICS IN CONTEXT**

LOOKING BACK

The operation of a light clock, as described in Section 29–2, depends on the reflection of light from a mirror, which was studied in detail in Chapters 26 and 27.

The kinetic energy introduced originally in Chapter 7, $K = \frac{1}{2}mv^2$, is shown in this chapter to be valid only for speeds small compared to the speed of light. The generalized kinetic energy, valid for any speed, is presented in Section 29–6.

LOOKING AHEAD

Photons, which always travel at the speed of light when propagating through a vacuum, are the ultimate relativistic particles. We shall see in Chapter 30 that even though their rest mass is zero, they still have finite momentum.

The energy released in nuclear reactions is due to a conversion of mass to energy, in accordance with $E = mc^2$ (Section 29–6). This will be discussed in detail in Chapter 32.

CHAPTER SUMMARY

29–1 THE POSTULATES OF SPECIAL RELATIVITY

Einstein's theory of relativity is based on just two postulates.

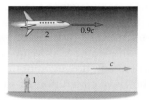

Equivalence of Physical Laws
The laws of physics are the same in all inertial frames of reference.

Constancy of the Speed of Light
The speed of light in a vacuum, $c = 3.00 \times 10^8$ m/s, is the same in all inertial frames of reference, independent of the motion of the source or the receiver.

29–2 THE RELATIVITY OF TIME AND TIME DILATION

Clocks that move relative to one another keep time at different rates. In particular, a moving clock runs slower than one that is at rest relative to a given observer.

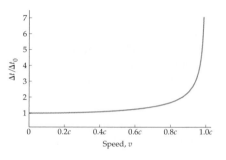

Proper Time
The proper time, Δt_0, is the amount of time separating two events that occur at the same location.

Time Dilation
If two events separated by a proper time Δt_0 occur in a frame of reference moving with a speed v relative to an observer, the dilated time measured by the observer, Δt, is given by

$$\Delta t = \frac{\Delta t_0}{\sqrt{1 - v^2/c^2}}$$ 29–2

Space Travel and Biological Aging
Time dilation applies equally to all physical processes, including chemical reactions and biological functions.

29–3 THE RELATIVITY OF LENGTH AND LENGTH CONTRACTION

The length of an object depends on its speed relative to a given observer.

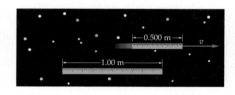

Proper Length
The proper length, L_0, is the distance between two points as measured by an observer who is at rest with respect to them.

Contracted Length
An object with a proper length L_0 moving with a speed v relative to an observer has a contracted length L given by:

$$L = L_0\sqrt{1 - \frac{v^2}{c^2}}$$ 29–3

Direction of Contraction
Lengths contract only in the direction of motion.

29–4 THE RELATIVISTIC ADDITION OF VELOCITIES

Simple velocity addition, $v = v_1 + v_2$, is valid only in the limit of very small speeds.

Relativistic Addition of Velocities
Suppose object 1 moves with a velocity v_{13} relative to object 3. If object 2 moves along the same straight line with a velocity v_{21} relative to object 1, the velocity of object 2 relative to object 3, v_{23}, is

$$v_{23} = \frac{v_{21} + v_{13}}{1 + \dfrac{v_{21}v_{13}}{c^2}}$$ 29–4

Ultimate Speed
If two velocities, v_1 and v_2, are less than the speed of light, c, then their relativistic sum, v, is also less than c. Thus it is not possible to increase the speed of an object from a value less than c to a value greater than c.

29–5 RELATIVISITIC MOMENTUM

The momentum of an object of mass m and speed v is

$$p = \frac{mv}{\sqrt{1 - \dfrac{v^2}{c^2}}} \qquad\qquad 29\text{--}5$$

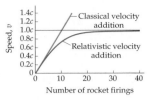

This expression is valid for all speeds between zero and the speed of light and reduces to $p = mv$ in the limit of small speeds.

29–6 RELATIVISITIC ENERGY AND $E = mc^2$

One of the most important results of relativity is that mass is another form of energy.

To put it another way, mass and energy are two different aspects of the same quantity.

Relativistic Energy
The total energy, E, of an object with rest mass m_0 and speed v is

$$E = \frac{m_0 c^2}{\sqrt{1 - \dfrac{v^2}{c^2}}} = mc^2 \qquad\qquad 29\text{--}7$$

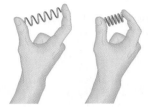

Rest Energy
When an object is at rest, its energy E_0 is

$$E_0 = m_0 c^2 \qquad\qquad 29\text{--}8$$

Relativistic Kinetic Energy
The relativistic kinetic energy, K, of an object of rest mass m_0 moving with a speed v is its total energy, E, minus its rest energy, E_0. In particular,

$$K = \frac{m_0 c^2}{\sqrt{1 - \dfrac{v^2}{c^2}}} - m_0 c^2 \qquad\qquad 29\text{--}9$$

29–8 GENERAL RELATIVITY

General relativity deals with accelerated frames of reference and with gravity.

Principle of Equivalence
One of the basic principles on which general relativity is founded is the following: All physical experiments conducted in a gravitational field and in an accelerated frame of reference give identical results.

Radius of a Black Hole
For an astronomical body of mass M and radius R to be a black hole, its radius must be less than or equal to the following value, known as the Schwarzschild radius:

$$R = \frac{2GM}{c^2} \qquad\qquad 29\text{--}10$$

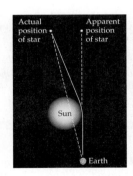

PROBLEM-SOLVING SUMMARY

Type of Problem	Relevant Physical Concepts	Related Examples
Find the time between two events as measured by different observers.	The time between events that occur at the same location is the proper time, Δt_0. The time measured by an observer with a relative speed v is $\Delta t = \Delta t_0 / \sqrt{1 - v^2/c^2}$.	Examples 29–1, 29–2 Active Example 29–1
Find the distance between two points as measured by different observers.	The distance between points that are at rest relative to the observer is the proper length, L_0. The distance measured by an observer with a relative speed v is $L = L_0 \sqrt{1 - v^2/c^2}$.	Example 29–3

Add velocities relativistically.	The correct rule for adding two velocities, v_{21} and v_{13}, to obtain the total velocity v_{23} is $v_{23} = (v_{21} + v_{13})/(1 + v_{21}v_{13}/c^2)$. In the limit of small velocities, this equation reduces to simple addition, as one would expect. For velocities comparable to the speed of light, simple addition fails, because the final velocity can never be greater than c.	Example 29–4 Active Example 29–2
Find the relativistic momentum.	The momentum of an object with a rest mass m_0 and speed v is $p = m_0 v/\sqrt{1 - v^2/c^2}$.	Example 29–5
Find the energy equivalence of a given mass.	The rest energy of an object with a rest mass m_0 is $E_0 = m_0 c^2$.	Example 29–6
Find the relativistic kinetic energy.	The kinetic energy of an object with a rest mass m_0 and speed v is $K = m_0 c^2/\sqrt{1 - v^2/c^2} - m_0 c^2$.	Example 29–7

CONCEPTUAL QUESTIONS

For instructor-assigned homework, go to www.masteringphysics.com

(Answers to odd-numbered Conceptual Questions can be found in the back of the book.)

1. Some distant galaxies are moving away from us at speeds greater than $0.5c$. What is the speed of the light received on Earth from these galaxies? Explain.

2. The speed of light in glass is less than c. Why is this not a violation of the second postulate of relativity?

3. How would velocities add if the speed of light were infinitely large? Justify your answer by considering Equation 29–4.

4. Describe some of the everyday consequences that would follow if the speed of light were 35 mi/h.

5. When we view a distant galaxy, we notice that the light coming from it has a longer wavelength (it is "red-shifted") than the

corresponding light here on Earth. Is this consistent with the postulate that all observers measure the same speed of light? Explain.

6. According to the theory of relativity, the maximum speed for any particle is the speed of light. Is there a similar restriction on the maximum energy of a particle? Is there a maximum momentum? Explain.

7. Give an argument that shows that an object of finite mass cannot be accelerated from rest to a speed greater than the speed of light in a vacuum.

PROBLEMS AND CONCEPTUAL EXERCISES

Note: Answers to odd-numbered Problems and Conceptual Exercises can be found in the back of the book. **IP** *denotes an integrated problem, with both conceptual and numerical parts;* **BIO** *identifies problems of biological or medical interest;* **CE** *indicates a conceptual exercise.* **Predict/Explain** *problems ask for two responses:* **(a)** *your prediction of a physical outcome, and* **(b)** *the best explanation among three provided. On all problems, red bullets (•, ••, •••) are used to indicate the level of difficulty. (For speeds, v, that are much less than the speed of light, c, the following expansion may be used:* $1/\sqrt{1 - v^2/c^2} \approx 1 + \frac{1}{2}(v^2/c^2)$.)

SECTION 29–1 THE POSTULATES OF SPECIAL RELATIVITY

1. • **CE Predict/Explain** You are in a spaceship, traveling directly away from the Moon with a speed of $0.9c$. A light signal is sent in your direction from the surface of the Moon. **(a)** As the signal passes your ship, do you measure its speed to be greater than, less than, or equal to $0.1c$? **(b)** Choose the *best explanation* from among the following:
   I. The speed you measure will be greater than $0.1c$; in fact, it will be c, since all observers in inertial frames measure the same speed of light.
   II. You will measure a speed less than $0.1c$ because of time dilation, which causes clocks to run slow.
   III. When you measure the speed you will find it to be $0.1c$, which is the difference between c and $0.9c$.

2. • Albert is piloting his spaceship, heading east with a speed of $0.90c$. Albert's ship sends a light beam in the forward (eastward) direction, which travels away from his ship at a speed c. Meanwhile, Isaac is piloting his ship in the westward direction, also at $0.90c$, toward Albert's ship. With what speed does Isaac see Albert's light beam pass his ship?

SECTION 29–2 THE RELATIVITY OF TIME AND TIME DILATION

3. • **CE** A street performer tosses a ball straight up into the air (event 1) and then catches it in his mouth (event 2). For each of the following observers, state whether the time they measure between these two events is the proper time or the dilated time: **(a)** the street performer; **(b)** a stationary observer on the other side of the street; **(c)** a person sitting at home watching the performance on TV; **(d)** a person observing the performance from a moving car.

4. • **CE Predict/Explain** A clock in a moving rocket is observed to run slow. **(a)** If the rocket reverses direction, does the clock run slow, fast, or at its normal rate? **(b)** Choose the *best explanation* from among the following:
   I. The clock will run slow, just as before. The rate of the clock depends only on relative speed, not on direction of motion.
   II. When the rocket reverses direction the rate of the clock reverses too, and this makes it run fast.
   III. Reversing the direction of the rocket undoes the the time dilation effect, and so the clock will now run at its normal rate.

5. • **CE Predict/Explain** Suppose you are a traveling salesman for SSC, the Spacely Sprockets Company. You travel on a spaceship that reaches speeds near the speed of light, and you are paid by the hour. **(a)** When you return to Earth after a sales trip, would you prefer to be paid according to the clock at Spacely Sprockets universal headquarters on Earth, according to the clock on the spaceship in which you travel, or would your pay be the same in either case? **(b)** Choose the *best explanation* from among the following:

    I. You want to be paid according to the clock on Earth, because the clock on the spaceship runs slow when it approaches the speed of light.

    II. Collect your pay according to the clock on the spaceship because according to you the clock on Earth has run slow.

    III. Your pay would be the same in either case because motion is relative, and all inertial observers will agree on the amount of time that has elapsed.

6. • A neon sign in front of a café flashes on and off once every 4.1 s, as measured by the head cook. How much time elapses between flashes of the sign as measured by an astronaut in a spaceship moving toward Earth with a speed of $0.84c$?

7. • A lighthouse sweeps its beam of light around in a circle once every 7.5 s. To an observer in a spaceship moving away from Earth, the beam of light completes one full circle every 15 s. What is the speed of the spaceship relative to Earth?

8. • Refer to Example 29–1. How much does Benny age if he travels to Vega with a speed of $0.9995c$?

9. • As a spaceship flies past with speed v, you observe that 1.0000 s elapses on the ship's clock in the same time that 1.0000 min elapses on Earth. How fast is the ship traveling, relative to the Earth? (Express your answer as a fraction of the speed of light.)

10. • Donovan Bailey set a world record for the 100-m dash on July 27, 1996. If observers on a spaceship moving with a speed of $0.7705c$ relative to Earth saw Donovan Bailey's run and measured his time to be 15.44 s, find the time that was recorded on Earth.

11. • Find the average distance (in the Earth's frame of reference) covered by the muons in Example 29–2 if their speed relative to Earth is $0.750c$.

12. •• **The Pi Meson** An elementary particle called a pi meson (or pion for short) has an average lifetime of 2.6×10^{-8} s when at rest. If a pion moves with a speed of $0.99c$ relative to Earth, find **(a)** the average lifetime of the pion as measured by an observer on Earth and **(b)** the average distance traveled by the pion as measured by the same observer. **(c)** How far would the pion have traveled relative to Earth if relativistic time dilation did not occur?

13. •• **The Σ^- Particle** The Σ^- is an exotic particle that has a lifetime (when at rest) of 0.15 ns. How fast would it have to travel in order for its lifetime, as measured by laboratory clocks, to be 0.25 ns?

14. •• **IP (a)** Is it possible for you to travel far enough and fast enough so that when you return from a trip, you are younger than your stay-at-home sister, who was born 5.0 y after you? **(b)** Suppose you fly on a rocket with a speed $v = 0.99c$ for 1 y, according to the ship's clocks and calendars. How much time elapses on Earth during your 1-y trip? **(c)** If you were 22 y old when you left home and your sister was 17, what are your ages when you return?

15. •• The radar antenna on a navy ship rotates with an angular speed of 0.29 rad/s. What is the angular speed of the antenna as measured by an observer moving away from the antenna with a speed of $0.82c$?

16. •• An observer moving toward Earth with a speed of $0.95c$ notices that it takes 5.0 min for a person to fill her car with gas. Suppose, instead, that the observer had been moving away from Earth with a speed of $0.80c$. How much time would the observer have measured for the car to be filled in this case?

17. •• **IP** An astronaut moving with a speed of $0.65c$ relative to Earth measures her heart rate to be 72 beats per minute. **(a)** When an Earth-based observer measures the astronaut's heart rate, is the result greater than, less than, or equal to 72 beats per minute? Explain. **(b)** Calculate the astronaut's heart rate as measured on Earth.

18. •• **BIO** Newly sprouted sunflowers can grow at the rate of 0.30 in. per day. One such sunflower is left on Earth, and an identical one is placed on a spacecraft that is traveling away from Earth with a speed of $0.94c$. How tall is the sunflower on the spacecraft when a person on Earth says his is 2.0 in. high?

19. •• An astronaut travels to Mars with a speed of 8350 m/s. After a month (30.0 d) of travel, as measured by clocks on Earth, how much difference is there between the Earth clock and the spaceship clock? Give your answer in seconds.

20. •• As measured in Earth's frame of reference, two planets are 424,000 km apart. A spaceship flies from one planet to the other with a constant velocity, and the clocks on the ship show that the trip lasts only 1.00 s. How fast is the ship traveling?

21. •• Captain Jean-Luc is piloting the USS *Enterprise XXIII* at a constant speed $v = 0.825c$. As the *Enterprise* passes the planet Vulcan, he notices that his watch and the Vulcan clocks both read 1:00 P.M. At 3:00 P.M., according to his watch, the *Enterprise* passes the planet Endor. If the Vulcan and Endor clocks are synchronized with each other, what time do the Endor clocks read when the *Enterprise* passes by?

22. ••• **IP** A plane flies with a constant velocity of 222 m/s. The clocks on the plane show that it takes exactly 2.00 h to travel a certain distance. **(a)** According to ground-based clocks, will the flight take slightly more or slightly less than 2.00 h? **(b)** Calculate how much longer or shorter than 2.00 h this flight will last, according to clocks on the ground.

SECTION 29–3 THE RELATIVITY OF LENGTH AND LENGTH CONTRACTION

23. • **CE** If the universal speed of light in a vacuum were larger than 3.00×10^8 m/s, would the effects of length contraction be greater or less than they are now? Explain.

24. • How fast does a 250-m spaceship move relative to an observer who measures the ship's length to be 150 m?

25. • Suppose the speed of light in a vacuum were only 25.0 mi/h. Find the length of a bicycle being ridden at a speed of 20.0 mi/h as measured by an observer sitting on a park bench, given that its proper length is 1.89 m.

26. • A rectangular painting is 124 cm wide and 80.5 cm high, as indicated in **Figure 29–29**. At what speed, v, must the painting move parallel to its width if it is to appear to be square?

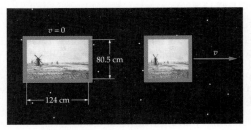

▲ **FIGURE 29–29** Problem 26

27. • The Linac portion of the Fermilab Tevatron contains a high-vacuum tube that is 64 m long, through which protons travel with an average speed $v = 0.65c$. How long is the Linac tube, as measured in the proton's frame of reference?

28. •• A cubical box is 0.75 m on a side. **(a)** What are the dimensions of the box as measured by an observer moving with a speed of $0.88c$ parallel to one of the edges of the box? **(b)** What is the volume of the box, as measured by this observer?

29. •• When parked, your car is 5.0 m long. Unfortunately, your garage is only 4.0 m long. **(a)** How fast would your car have to be moving for an observer on the ground to find your car shorter than your garage? **(b)** When you are driving at this speed, how long is your garage, as measured in the car's frame of reference?

30. •• An astronaut travels to a distant star with a speed of $0.55c$ relative to Earth. From the astronaut's point of view, the star is 7.5 ly from Earth. On the return trip, the astronaut travels with a speed of $0.89c$ relative to Earth. What is the distance covered on the return trip, as measured by the astronaut? Give your answer in light-years.

31. •• **IP** Laboratory measurements show that an electron traveled 3.50 cm in a time of 0.200 ns. **(a)** In the rest frame of the electron, did the lab travel a distance greater than or less than 3.50 cm? Explain. **(b)** What is the electron's speed? **(c)** In the electron's frame of reference, how far did the laboratory travel?

32. •• You and a friend travel through space in identical spaceships. Your friend informs you that he has made some length measurements and that his ship is 150 m long but that yours is only 120 m long. From your point of view, **(a)** how long is your friend's ship, **(b)** how long is your ship, and **(c)** what is the speed of your friend's ship relative to yours?

33. •• A ladder 5.0 m long leans against a wall inside a spaceship. From the point of view of a person on the ship, the base of the ladder is 3.0 m from the wall, and the top of the ladder is 4.0 m above the floor. The spaceship moves past the Earth with a speed of $0.90c$ in a direction parallel to the floor of the ship. Find the angle the ladder makes with the floor as seen by an observer on Earth.

34. ••• When traveling past an observer with a relative speed v, a rocket is measured to be 9.00 m long. When the rocket moves with a relative speed $2v$, its length is measured to be 5.00 m. **(a)** What is the speed v? **(b)** What is the proper length of the rocket?

35. ••• **IP** The starships *Picard* and *La Forge* are traveling in the same direction toward the Andromeda galaxy. The *Picard* moves with a speed of $0.90c$ relative to the *La Forge*. A person on the *La Forge* measures the length of the two ships and finds the same value. **(a)** If a person on the *Picard* also measures the lengths of the two ships, which of the following is observed: (i) the *Picard* is longer; (ii) the *La Forge* is longer; or (iii) both ships have the same length? Explain. **(b)** Calculate the ratio of the proper length of the *Picard* to the proper length of the *La Forge*.

SECTION 29–4 THE RELATIVISTIC ADDITION OF VELOCITIES

36. • A spaceship moving toward Earth with a speed of $0.90c$ launches a probe in the forward direction with a speed of $0.10c$ relative to the ship. Find the speed of the probe relative to Earth.

37. • Suppose the probe in Problem 36 is launched in the opposite direction to the motion of the spaceship. Find the speed of the probe relative to Earth in this case.

38. • A spaceship moving relative to an observer with a speed of $0.70c$ shines a beam of light in the forward direction, directly toward the observer. Use Equation 29–4 to calculate the speed of the beam of light relative to the observer.

39. • Suppose the speed of light is 35 mi/h. A paper girl riding a bicycle at 22 mi/h throws a rolled-up newspaper in the forward direction, as shown in **Figure 29–30**. If the paper is thrown with a speed of 19 mi/h relative to the bike, what is its speed, v, with respect to the ground?

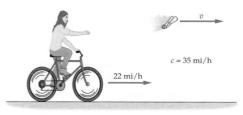

▲ **FIGURE 29–30** Problem 39

40. •• Two asteroids head straight for Earth from the same direction. Their speeds relative to Earth are $0.80c$ for asteroid 1 and $0.60c$ for asteroid 2. Find the speed of asteroid 1 relative to asteroid 2.

41. •• Two rocket ships approach Earth from opposite directions, each with a speed of $0.8c$ relative to Earth. What is the speed of one ship relative to the other?

42. •• A spaceship and an asteroid are moving in the same direction away from Earth with speeds of $0.77c$ and $0.41c$, respectively. What is the relative speed between the spaceship and the asteroid?

43. •• An electron moves to the right in a laboratory accelerator with a speed of $0.84c$. A second electron in a different accelerator moves to the left with a speed of $0.43c$ relative to the first electron. Find the speed of the second electron relative to the lab.

44. •• **IP** Two rocket ships are racing toward Earth, as shown in **Figure 29–31**. Ship A is in the lead, approaching the Earth at $0.80c$ and separating from ship B with a relative speed of $0.50c$. **(a)** As seen from Earth, what is the speed, v, of ship B? **(b)** If ship A increases its speed by $0.10c$ relative to the Earth, does the relative speed between ship A and ship B increase by $0.10c$, by more than $0.10c$, or by less than $0.10c$? Explain. **(c)** Find the relative speed between ships A and B for the situation described in part (b).

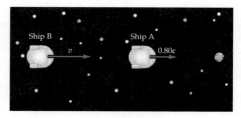

▲ **FIGURE 29–31** Problem 44

45. •• **IP** An inventor has proposed a device that will accelerate objects to speeds greater than c. He proposes to place the object to be accelerated on a conveyor belt whose speed is $0.80c$. Next, the entire system is to be placed on a second conveyor belt that also has a speed of $0.80c$, thus producing a final speed of $1.6c$. **(a)** Construction details aside, should you invest in this scheme? **(b)** What is the actual speed of the object relative to the ground?

SECTION 29–5 RELATIVISTIC MOMENTUM

46. • A 4.5×10^6-kg spaceship moves away from Earth with a speed of $0.75c$. What is the magnitude of the ship's **(a)** classical and **(b)** relativistic momentum?

47. • An asteroid with a mass of 8.2×10^{11} kg is observed to have a relativistic momentum of magnitude 7.74×10^{20} kg · m/s. What is the speed of the asteroid relative to the observer?

48. •• An object has a relativistic momentum that is 7.5 times greater than its classical momentum. What is its speed?

49. •• A football player with a mass of 88 kg and a speed of 2.0 m/s collides head-on with a player from the opposing team whose mass is 120 kg. The players stick together and are at rest after the collision. Find the speed of the second player, assuming the speed of light is 3.0 m/s.

50. •• In the previous problem, suppose the speed of the second player is 1.2 m/s. What is the speed of the players after the collision?

51. •• A space probe with a rest mass of 8.2×10^7 kg and a speed of $0.50c$ smashes into an asteroid at rest and becomes embedded within it. If the speed of the probe–asteroid system is $0.26c$ after the collision, what is the rest mass of the asteroid?

52. •• At what speed does the classical momentum, $p = mv$, give an error, when compared with the relativistic momentum, of **(a)** 1.00% and **(b)** 5.00%?

53. •• A proton has 1836 times the rest mass of an electron. At what speed will an electron have the same momentum as a proton moving at $0.0100c$?

SECTION 29–6   RELATIVISTIC ENERGY AND $E = mc^2$

54. • **CE** Particles A through D have the following rest energies and total energies:

Particle	Rest Energy	Total Energy
A	$6E$	$6E$
B	$2E$	$4E$
C	$4E$	$6E$
D	$3E$	$4E$

Rank these particles in order of increasing **(a)** rest mass, **(b)** kinetic energy, and **(c)** speed. Indicate ties where appropriate.

55. • Find the work that must be done on a proton to accelerate it from rest to a speed of $0.90c$.

56. • If a neutron moves with a speed of $0.99c$, what are its **(a)** total energy, **(b)** rest energy, and **(c)** kinetic energy?

57. • A spring with a force constant of 584 N/m is compressed a distance of 39 cm. Find the resulting increase in the spring's mass.

58. • When a certain capacitor is charged, its mass increases by 8.3×10^{-16} kg. How much energy is stored in the capacitor?

59. • What minimum energy must a gamma ray have to create an electron-antielectron pair?

60. • When a proton encounters an antiproton, the two particles annihilate each other, producing two gamma rays. Assuming the particles were at rest when they annihilated, find the energy of each of the two gamma rays produced. (*Note:* The rest energies of an antiproton and a proton are identical.)

61. • A rocket with a mass of 2.7×10^6 kg has a relativistic kinetic energy of 2.7×10^{23} J. How fast is the rocket moving?

62. •• An object has a total energy that is 5.5 times its rest energy. What is its speed?

63. •• A nuclear power plant produces an average of 1.0×10^3 MW of power during a year of operation. Find the corresponding change in mass of reactor fuel, assuming all of the energy released by the fuel can be converted directly to electrical energy. (In a practical reactor, only a relatively small fraction of the energy can be converted to electricity.)

64. •• A helium atom has a rest mass of $m_{\text{He}} = 4.002603$ u. When disassembled into its constituent particles (2 protons, 2 neutrons, 2 electrons), the well-separated individual particles have the following masses: $m_\text{p} = 1.007276$ u, $m_\text{n} = 1.008665$ u, $m_\text{e} = 0.000549$ u. How much work is required to completely disassemble a helium atom? (*Note:* 1 u of mass has a rest energy of 931.49 MeV.)

65. •• What is the percent difference between the classical kinetic energy, $K_{\text{cl}} = \frac{1}{2}m_0 v^2$, and the correct relativistic kinetic energy, $K = m_0 c^2 / \sqrt{1 - v^2/c^2} - m_0 c^2$, at a speed of **(a)** $0.10c$ and **(b)** $0.90c$?

66. •• A proton has 1836 times the rest mass of an electron. At what speed will an electron have the same kinetic energy as a proton moving at $0.0250c$?

67. •• **IP** Consider a baseball with a rest mass of 0.145 kg. **(a)** How much work is required to increase the speed of the baseball from 25.0 m/s to 35.0 m/s? **(b)** Is the work required to increase the speed of the baseball from 200,000,025 m/s to 200,000,035 m/s greater than, less than, or the same as the amount found in part (a)? Explain. **(c)** Calculate the work required for the increase in speed indicated in part (b).

68. •• **IP** A particle has a kinetic energy equal to its rest energy. **(a)** What is the speed of this particle? **(b)** If the kinetic energy of this particle is doubled, does its speed increase by a more than, less than, or exactly a factor of 2? Explain. **(c)** Calculate the speed of a particle whose kinetic energy is twice its rest energy.

69. ••• A lump of putty with a mass of 0.240 kg and a speed of $0.980c$ collides head-on and sticks to an identical lump of putty moving with the same speed. After the collision the system is at rest. What is the mass of the system after the collision?

SECTION 29–8   GENERAL RELATIVITY

70. • Find the radius to which the Sun must be compressed for it to become a black hole.

71. •• **The Black Hole in the Center of the Milky Way** Recent measurements show that the black hole at the center of the Milky Way galaxy, which is believed to coincide with the powerful radio source Sagittarius A*, is 2.6 million times more massive than the Sun; that is, $M = 5.2 \times 10^{36}$ kg. **(a)** What is the maximum radius of this black hole? **(b)** Find the acceleration of gravity at the Schwarzschild radius of this black hole, using the expression for R given in Equation 29–10. **(c)** How does your answer to part (b) change if the mass of the black hole is doubled? Explain.

GENERAL PROBLEMS

72. • **CE** Two observers are moving relative to one another. Which of the following quantities will they always measure to have the same value: **(a)** their relative speed; **(b)** the time between two events; **(c)** the length of an object; **(d)** the speed of light in a vacuum; **(e)** the speed of a third observer?

73. • **CE** You are standing next to a runway as an airplane lands. **(a)** If you and the pilot observe a clock in the cockpit, which of you measures the proper time? **(b)** If you and the pilot observe a large clock on the control tower, which of you measures the proper time? **(c)** Which of you measures the proper length of the airplane? **(d)** Which of you measures the proper length of the runway?

74. • **CE** Which clock runs slower relative to a clock on the North Pole: clock 1 on an airplane flying from New York to Los Angeles, or clock 2 on an airplane flying from Los Angeles to New York? Assume each plane has the same speed relative to the surface of the Earth. Explain.

75. • **CE** An apple drops from the bough of a tree to the ground. Is the mass of the apple near the top of its fall greater than, less than, or the same as its mass after it has landed? Explain.

76. • **CE Predict/Explain** Consider two apple pies that are identical in every respect, except that pie 1 is piping hot and pie 2 is at room temperature. **(a)** If identical forces are applied to the two pies, is the acceleration of pie 1 greater than, less than, or equal to the acceleration of pie 2? **(b)** Choose the *best explanation* from among the following:

  I. The acceleration of pie 1 is greater because the fact that it is hot means it has the greater energy.

  II. The fact that pie 1 is hot means it behaves as if it has more mass than pie 2, and therefore it has a smaller acceleration.

  III. The pies have the same acceleration regardless of their temperature because they have identical rest masses.

77. • **CE** Is the mass of a warm cup of tea greater than, less than, or the same as the mass of the same cup of tea when it has cooled? Explain.

78. • **CE Predict/Explain** An uncharged capacitor is charged by moving some electrons from one plate of the capacitor to the other plate. **(a)** Is the mass of the charged capacitor greater than, less than, or the same as the mass of the uncharged capacitor? **(b)** Choose the *best explanation* from among the following:

  I. The charged capacitor has more mass because it is storing energy within it, just like a compressed spring.

  II. The charged capacitor has less mass because some of its mass now appears as the energy of the electric field between its plates.

  III. The capacitor has the same mass whether it is charged or not because charging it only involves moving electrons from one plate to the other without changing the total number of electrons.

79. •• **Cosmic Rays** Protons in cosmic rays have been observed with kinetic energies as large as 1.0×10^{20} eV. **(a)** How fast are these protons moving? Give your answer as a fraction of the speed of light. **(b)** Show that the kinetic energy of a single one of these protons is much greater than the kinetic energy of a 15-mg ant walking with a speed of 8.8 mm/s.

80. •• An apple falls from a tree, landing on the ground 3.7 m below. How long is the apple in the air, as measured by an observer moving toward Earth with a speed of $0.89c$?

81. •• What is the momentum of a proton with 1.50×10^3 MeV of kinetic energy? (*Note*: The rest energy of a proton is 938 MeV.)

82. •• **IP** A container holding 2.00 moles of an ideal monatomic gas is heated at constant volume until the temperature of the gas increases by 112 F°. **(a)** Does the mass of the gas increase, decrease, or stay the same? Explain. **(b)** Calculate the change in mass of the gas, if any.

83. •• A ^{14}C nucleus, initially at rest, emits a beta particle. The beta particle is an electron with 156 keV of kinetic energy. **(a)** What is the speed of the beta particle? **(b)** What is the momentum of the beta particle? **(c)** What is the momentum of the nucleus after it emits the beta particle? **(d)** What is the speed of the nucleus after it emits the beta particle?

84. •• A clock at rest has a rectangular shape, with a width of 24 cm and a height of 12 cm. When this clock moves parallel to its width with a certain speed v its width and height are the same. Relative to a clock at rest, how long does it take for the moving clock to advance by 1.0 s?

85. •• A starship moving toward Earth with a speed of $0.75c$ launches a shuttle craft in the forward direction. The shuttle, which has a proper length of 12.5 m, is only 6.25 m long as viewed from Earth. What is the speed of the shuttle relative to the starship?

86. •• When a particle of charge q and momentum p enters a uniform magnetic field at right angles it follows a circular path of radius $R = p/qB$, as shown in **Figure 29–32**. What radius does this expression predict for a proton traveling with a speed $v = 0.99c$ through a magnetic field $B = 0.20$ T if you use (a) the nonrelativistic momentum ($p = mv$) or **(b)** the relativistic momentum ($p = mv/\sqrt{1 - v^2/c^2}$)?

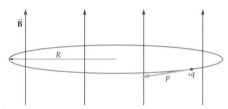

▲ **FIGURE 29–32** Problem 86

87. •• **IP** A starship moving away from Earth with a speed of $0.75c$ launches a shuttle craft in the reverse direction, that is, toward Earth. **(a)** If the speed of the shuttle relative to the starship is $0.40c$, and its proper length is 13 m, how long is the shuttle as measured by an observer on Earth? **(b)** If the shuttle had been launched in the forward direction instead, would its length as measured by an observer on Earth be greater than, less than, or the same as the length found in part (a)? Explain. **(c)** Calculate the length for the case described in part (b).

88. •• A 2.5-m titanium rod in a moving spacecraft is at an angle of 45° with respect to the direction of motion. The craft moves directly toward Earth at $0.98c$. As viewed from Earth, **(a)** how long is the rod and **(b)** what angle does the rod make with the direction of motion?

89. •• Electrons are accelerated from rest through a potential difference of 276,000 V. What is the final speed predicted **(a)** classically and **(b)** relativistically?

90. •• The rest energy, m_0c^2, of a particle with a kinetic energy K and a momentum p can be determined as follows:

$$m_0c^2 = \frac{(pc)^2 - K^2}{2K}$$

Suppose a pion (a subatomic particle) is observed to have a kinetic energy $K = 35.0$ MeV and a momentum $p = 5.61 \times 10^{-20}$ kg^3 m/s = 105 MeV/c. What is the rest energy of the pion? Give your answer in MeV.

91. •• A small star of mass m orbits a supermassive black hole of mass M. **(a)** Find the orbital speed of the star if its orbital radius is $2R$, where R is the Schwarzschild radius (Equation 29–10). **(b)** Repeat part (a) for an orbital radius equal to R.

92. •• **IP** Consider a "relativistic air track" on which two identical air carts undergo a completely inelastic collision. One cart is initially at rest; the other has an initial speed of $0.650c$. **(a)** In classical physics, the speed of the carts after the collision would be $0.325c$. Do you expect the final speed in this relativistic collision to be greater than or less than $0.325c$? Explain. **(b)** Use relativistic

momentum conservation to find the speed of the carts after they collide and stick together.

93. ••• **IP** In Conceptual Checkpoint 29–2 we considered an astronaut at rest on an inclined bed inside a moving spaceship. From the point of view of observer 1, on board the ship, the astronaut has a length L_0 and is inclined at an angle θ_0 above the floor. Observer 2 sees the spaceship moving to the right with a speed v. **(a)** Show that the length of the astronaut as measured by observer 2 is

$$L = L_0\sqrt{1 - \left(\frac{v^2}{c^2}\right)\cos^2\theta_0}$$

(b) Show that the angle θ the astronaut makes with the floor of the ship, as measured by observer 2, is given by

$$\tan\theta = \frac{\tan\theta_0}{\sqrt{1 - v^2/c^2}}$$

94. ••• A pulsar is a collapsed, rotating star that sends out a narrow beam of radiation, like the light from a lighthouse. With each revolution, we see a brief, intense pulse of radiation from the pulsar. Suppose a pulsar is receding directly away from Earth with a speed of $0.800c$, and the starship *Endeavor* is sent out toward the pulsar with a speed of $0.950c$ relative to Earth. If an observer on Earth finds that 153 pulses are emitted by the pulsar every second, at what rate does an observer on the *Endeavor* see pulses emitted?

95. ••• Show that the total energy of an object is related to its momentum by the relation $E^2 = p^2c^2 + (m_0c^2)^2$.

96. ••• Show that if $0 < v_1 < c$ and $0 < v_2 < c$ are two velocities pointing in the same direction, the relativistic sum of these velocities, v, is greater than v_1 and greater than v_2 but less than c. In particular, show that this is true even if v_1 and v_2 are greater than $0.5c$.

97. ••• Show that an object with momentum p and rest mass m_0 has a speed given by

$$v = \frac{c}{\sqrt{1 + (m_0c/p)^2}}$$

98. ••• **Decay of the Σ^- Particle** When at rest, the Σ^- particle has a lifetime of 0.15 ns before it decays into a neutron and a pion. One particular Σ^- particle is observed to travel 3.0 cm in the lab before decaying. What was its speed? (*Hint:* Its speed was not $\frac{2}{3}c$.)

PASSAGE PROBLEMS

Relativity in a TV Set

The first televisions used cathode-ray tubes, or CRTs, to form a picture. Even today, when plasma screens, liquid-crystal displays (LCD), and digital light processing (DLP) systems are increasingly popular, the CRT is still a reliable and inexpensive choice for TVs and computer monitors.

The basic idea behind a CRT is fairly simple: use a beam of electrons to "paint" a picture on a fluorescent screen. This is illustrated in **Figure 29–33**. First, a heated coil at the negative terminal of the tube (the cathode) produces electrons which are accelerated toward the positive terminal (the anode) to form a beam of electrons—the so-called "cathode ray." A series of horizontal and vertical deflecting plates then direct the beam to any desired spot on a fluorescent screen to produce a glowing dot that can be seen. Moving the glowing dot rapidly around the screen, and varying its intensity with the control grid, allows one to produce a glowing image of any desired object. The first televised image—a dollar sign—was transmitted by Philo T. Farnsworth in 1927, and television inventors have been seeing dollar signs ever since.

The interior of a CRT must be a very good vacuum, typically 10^{-7} of an atmosphere or less, to ensure electrons aren't

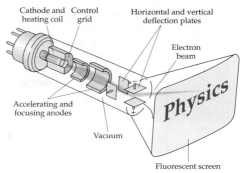

▲ **FIGURE 29–33** A cathode-ray tube. (Problems 99, 100, 101, and 102)

scattered by air molecules on their way to the screen. Electrons in a television set are accelerated through a potential difference of 25.0 kV, which is sufficient to give them speeds comparable to the speed of light. As a result, relativity must be used to accurately determine their behavior. Thus, even in something as commonplace as a TV set, Einstein's theory of relativity proves itself to be of great practical value.

99. • Find the speed of an electron accelerated through a voltage of 25.0 kV—*ignoring* relativity. Express your answer as a fraction times the speed of light. (Speeds over about $0.1c$ are generally regarded as relativistic.)

   A. $0.221c$ **B.** $0.281c$

   C. $0.312c$ **D.** $0.781c$

100. • When relativistic effects are included, do you expect the speed of the electrons to be greater than, less than, or the same as the result found in the previous problem?

101. • Find the speed of the electrons in Problem 99, this time using a correct relativistic calculation. As before, express your answer as a fraction times the speed of light.

   A. $0.301c$ **B.** $0.312c$

   C. $0.412c$ **D.** $0.953c$

102. • Suppose the accelerating voltage in Problem 99 is increased by a factor of 10. What is the correct relativistic speed of an electron in this case?

   A. $0.205c$ **B.** $0.672c$

   C. $0.740c$ **D.** $0.862c$

INTERACTIVE PROBLEMS

103. •• **Referring to Example 29–4** The *Picard* approaches starbase Faraway Point with a speed of $0.806c$, and the *La Forge* approaches the starbase with a speed of $0.906c$. Suppose the *Picard* now launches a probe toward the starbase. **(a)** What velocity must the probe have relative to the *Picard* if it is to be at rest relative to the *La Forge*? **(b)** What velocity must the probe have relative to the *Picard* if its velocity relative to the *La Forge* is to be $0.100c$? **(c)** For the situation described in part (b), what is the velocity of the probe relative to the Faraway Point starbase?

104. •• **Referring to Example 29–4** Faraway Point starbase launches a probe toward the approaching starships. The probe has a velocity relative to the *Picard* of $-0.906c$. The *Picard* approaches starbase Faraway Point with a speed of $0.806c$, and the *La Forge* approaches the starbase with a speed of $0.906c$. **(a)** What is the velocity of the probe relative to the *La Forge*? **(b)** What is the velocity of the probe relative to Faraway Point starbase?

30 Quantum Physics

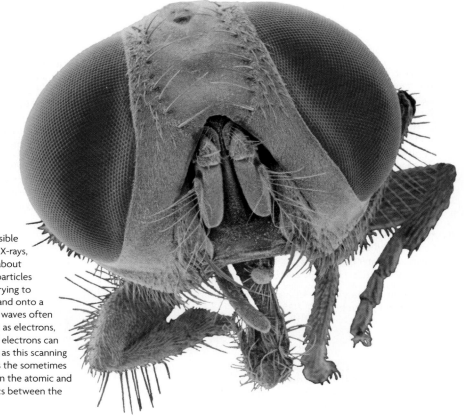

Most of the images that we encounter are made with visible light. Even those that are not, such as thermograms and X-rays, employ other kinds of electromagnetic radiation. Until about 80 years ago, the idea of making a picture by means of particles rather than radiation would have seemed absurd—like trying to create a portrait by bouncing paintballs off the subject and onto a canvas. Yet by the 1920s, physicists discovered that light waves often behave like particles and, conversely, that particles, such as electrons, often behave like waves. Indeed, the wave properties of electrons can be exploited to create remarkably detailed images, such as this scanning electron micrograph of a housefly. This chapter explores the sometimes odd-seeming laws that describe the behavior of nature in the atomic and subatomic realms, and the series of revolutions in physics between the 1890s and the 1930s that uncovered them.

To understand the behavior of nature at the atomic level, it is necessary to introduce a number of new concepts to physics and to modify many others. In this chapter we consider the basic ideas of quantum physics and show that they lead to a deeper understanding of microscopic systems—in much the same way that relativity extends physics into the realm of high speeds. Relativity and quantum physics, taken together, provide the basis for what we refer to today as modern physics.

We begin this chapter by introducing the concept of *quantization,* in which a physical quantity—such as energy—varies in discrete steps rather than continuously, as in classical physics. This concept leads to the idea of the *photon,* which can be thought of as a "particle" of light. Next, we find that just as light can behave like a particle, particles—such as electrons, protons, and neutrons—can behave like waves. Finally, the wave nature of matter introduces a fundamental uncertainty to our knowledge of physical quantities and allows for such classically "forbidden" behavior as *quantum tunneling.*

30-1 Blackbody Radiation and Planck's Hypothesis of Quantized Energy

If you have ever looked through a small opening into a hot furnace, you have seen the glow of light associated with its high temperature. As unlikely as it may seem, this light played a central role in the revolution of physics that occurred in the early 1900s. It was through the study of such systems that the idea of *energy quantization*—energy taking on only discrete values—was first introduced to physics.

More precisely, physicists in the late 1800s were actively studying the electromagnetic radiation given off by a physical system known as a **blackbody.** An example of a blackbody is illustrated in **Figure 30–1**. Note that this blackbody has a cavity with a small opening to the outside world—much like a furnace. Light that enters the cavity through the opening is reflected multiple times from the interior walls until it is completely absorbed. It is for this reason that the system is referred to as "black," even though the material from which it is made need not be black at all.

> An ideal blackbody absorbs all the light that is incident on it.

Objects that absorb much of the incident light—though not all of it—are reasonable approximations to a blackbody; objects that are highly reflective and shiny are poor representations of a blackbody.

As we saw in Section 16–6, objects that are effective at absorbing radiation are also effective at giving off radiation. Thus an ideal blackbody is also an ideal radiator. In fact, the basic experiment performed with a blackbody is the following: Heat the blackbody to a fixed temperature, T, and measure the amount of electromagnetic radiation it gives off at a given frequency, f. Repeat this measurement for a number of different frequencies, then plot the intensity of radiation versus frequency. The results of a typical blackbody experiment are shown in **Figure 30–2** for a variety of different temperatures. Note that there is little radiation at low frequencies, a peak in the radiation at intermediate frequencies, and finally a fall-off to little radiation again at high frequencies.

Now, what is truly remarkable about the blackbody experiment is the following:

> The distribution of energy in blackbody radiation is *independent* of the material from which the blackbody is constructed—it depends only on the temperature, T.

Therefore, a blackbody of steel and one of wood give precisely the same results when held at the same temperature. When physicists observe a phenomenon that is independent of the details of the system, it is a clear signal that they are observing something of fundamental significance. This was certainly the case with blackbody radiation.

Two aspects of the blackbody curves in Figure 30–2 are of particular importance. First, note that as the temperature is increased, the area under the curve increases. Since the total area under the curve is a measure of the total energy emitted by the blackbody, it follows that an object radiates more energy as it becomes hotter.

Second, note that the peak in the blackbody curve moves to higher frequency as the absolute temperature T is increased. This movement, or displacement, of the peak with temperature is described by **Wien's displacement law:**

Wien's Displacement Law

$$f_{peak} = (5.88 \times 10^{10}\ \text{s}^{-1} \cdot \text{K}^{-1})T$$

30–1

SI unit: Hz = s^{-1}

Thus, there is a direct connection between the temperature of an object and the frequency of radiation it emits most strongly.

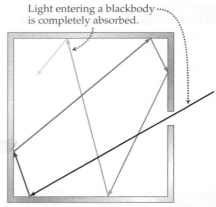

Light entering a blackbody is completely absorbed.

▲ **FIGURE 30–1 An ideal blackbody**
In an ideal blackbody, incident light is completely absorbed. In the case shown here, the absorption occurs as the result of multiple reflections within a cavity. The blackbody, and the electromagnetic radiation it contains, are in thermal equilibrium at a temperature T.

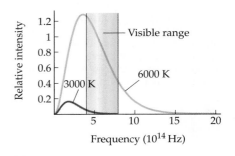

(a)

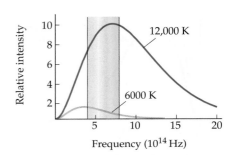

(b)

▲ **FIGURE 30–2 Blackbody radiation**
Blackbody radiation as a function of frequency for various temperatures: **(a)** 3000 K and 6000 K; **(b)** 6000 K and 12,000 K. Note that as the temperature is increased, the peak in the radiation shifts toward higher frequency.

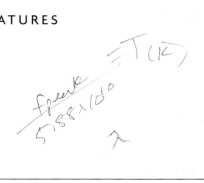

CONCEPTUAL CHECKPOINT 30–1   COMPARE TEMPERATURES

Betelgeuse is a red-giant star in the constellation Orion; Rigel is a bluish white star in the same constellation. Is the surface temperature of Betelgeuse **(a)** higher than, **(b)** lower than, or **(c)** the same as the surface temperature of Rigel?

REASONING AND DISCUSSION
Recall that red light has a lower frequency than blue light, as can be seen in Figure 25–8. It follows, from Wien's displacement law, that a red star has a lower temperature than a blue star. Therefore, Betelgeuse has the lower surface temperature.

ANSWER
(b) The surface temperature of Betelgeuse is lower than that of Rigel.

▲ All objects emit electromagnetic radiation over a range of frequencies. The frequency that is radiated most intensely depends on the object's temperature, as specified by Wien's law. The glowing bolt in this picture radiates primarily in the infrared part of the spectrum, but it is hot enough (a few thousand kelvin) so that a significant portion of its radiation falls within the red end of the visible region. The other bolts are too cool to radiate any detectable amount of visible light.

REAL-WORLD PHYSICS
Measuring the temperature of a star

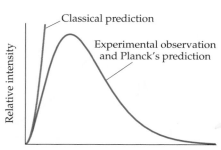

▲ **FIGURE 30–3** The ultraviolet catastrophe

Classical physics predicts a blackbody radiation curve that rises without limit as the frequency increases. This outcome is referred to as the ultraviolet catastrophe. By assuming energy quantization, Planck was able to derive a curve in agreement with experimental results.

To be more specific about the conclusion given in Conceptual Checkpoint 30–1, let's consider Figure 30–2 in greater detail. At the lowest temperature shown, 3000 K, the radiation is more intense at the red end of the visible spectrum than at the blue end. An object at this temperature—like the heating coil on a stove, for example—would appear "red hot" to the eye. Even so, most of the radiation at this temperature is in the infrared, and thus is not visible to the eye at all. A blackbody at 6000 K, like the surface of the Sun, gives out strong radiation throughout the visible spectrum, though there is still more radiation at the red end than at the blue end. As a result, the light of the Sun appears somewhat yellowish. Finally, at 12,000 K a blackbody appears bluish white, and most of its radiation is in the ultraviolet. The temperature of the star Rigel is determined from the location of its radiation peak in the following Exercise.

EXERCISE 30–1

Find the surface temperature of Rigel, given that its radiation peak occurs at a frequency of 1.17×10^{15} Hz.

SOLUTION
Solving Equation 30–1 for T, we find

$$T = \frac{f_{\text{peak}}}{5.88 \times 10^{10}\ \text{s}^{-1} \cdot \text{K}^{-1}} = \frac{1.17 \times 10^{15}\ \text{Hz}}{5.88 \times 10^{10}\ \text{s}^{-1} \cdot \text{K}^{-1}} = 19{,}900\ \text{K}$$

This is a little more than three times the surface temperature of the Sun. Thus blackbody radiation allows us to determine the temperature of a distant star that we may never visit.

Planck's Quantum Hypothesis

Although experimental understanding of blackbody radiation was quite extensive in the late 1800s, there was a problem. Attempts to explain the blackbody curves of Figure 30–2 theoretically, using classical physics, failed—and failed miserably. To see the problem, consider the curves shown in **Figure 30–3**. The green curve is the experimental result for a blackbody at a given temperature. In contrast, the blue curve shows the prediction of classical physics. Clearly, the classical result cannot be valid, since its curve diverges to infinity at high frequency, which in turn implies that the blackbody radiates an infinite amount of energy. This unphysical divergence at high frequencies is referred to as the *ultraviolet catastrophe*.

The German physicist Max Planck (1858–1947) worked long and hard on this problem. Eventually, he was able to construct a mathematical formula that agreed with experiment for all frequencies. His next problem was to "derive" the equation. The only way he could do this, it turned out, was to make the following bold and unprecedented assumption: The radiation energy in a blackbody at the frequency f must be an integral multiple of a constant (h) times the frequency; that is, energy is *quantized*:

Quantized Energy

$E_n = nhf \qquad n = 0, 1, 2, 3, \ldots$

30–2

The constant, h, in this expression is known as **Planck's constant,** and it has the following value:

Planck's Constant, h

$$h = 6.63 \times 10^{-34} \, \text{J} \cdot \text{s}$$

SI unit: $\text{J} \cdot \text{s}$

30–3

This constant is recognized today as one of the fundamental constants of nature, on an equal footing with other such constants as the speed of light in a vacuum and the rest mass of an electron.

The assumption of energy quantization is quite a departure from classical physics, in which energy can take on any value at all and is related to the amplitude of a wave rather than its frequency. In Planck's calculation, the energy can have only the discrete values hf, $2hf$, $3hf$, and so on. Because of this quantization, it follows that the energy can change only in *quantum jumps* of energy no smaller than hf as the system goes from one quantum state to another. The fundamental increment, or *quantum*, of energy, hf, is incredibly small, as can be seen from the small magnitude of Planck's constant. The next Example explores the size of the quantum and the value of the *quantum number, n*, for a typical macroscopic system.

EXAMPLE 30–1   QUANTUM NUMBERS

Suppose the maximum speed of a 1.2-kg mass attached to a spring with a force constant of 35 N/m is 0.95 m/s. **(a)** Find the frequency of oscillation and total energy of this mass–spring system. **(b)** Determine the size of one quantum of energy in this system. **(c)** Assuming the energy of this system satisfies $E_n = nhf$, find the quantum number, n.

PICTURE THE PROBLEM

Our sketch shows a 1.2-kg mass oscillating on a spring with a force constant of 35 N/m. The mass has its maximum speed of $v_{max} = 0.95$ m/s when it passes through the equilibrium position. At this moment, the total energy of the system is simply the kinetic energy of the mass.

STRATEGY

a. We can find the frequency of oscillation using $\omega = \sqrt{k/m}$ (Equation 13–10) and $\omega = 2\pi f$ (Equation 13–15). The total energy is simply the kinetic energy as the mass passes through equilibrium, $E = K_{max} = \frac{1}{2}mv_{max}^2$.

b. The energy of one quantum is hf, where f is the frequency found in part (a).

c. We determine the quantum number by solving $E_n = nhf$ for n.

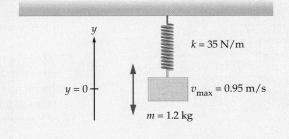

SOLUTION

Part (a)

1. Calculate the frequency of oscillation using $\omega = \sqrt{k/m} = 2\pi f$:

$$\omega = \sqrt{\frac{k}{m}} = 2\pi f$$

$$f = \frac{1}{2\pi}\sqrt{\frac{k}{m}} = \frac{1}{2\pi}\sqrt{\frac{35 \, \text{N/m}}{1.2 \, \text{kg}}} = 0.86 \, \text{Hz}$$

2. Calculate the maximum kinetic energy of the mass $\left(\frac{1}{2}mv_{max}^2\right)$ to find the total, E, of the system:

$$E = \frac{1}{2}mv_{max}^2 = \frac{1}{2}(1.2 \, \text{kg})(0.95 \, \text{m/s})^2 = 0.54 \, \text{J}$$

Part (b)

3. The energy of one quantum is hf, where $f = 0.86$ Hz:

$$hf = (6.63 \times 10^{-34} \, \text{J} \cdot \text{s})(0.86 \, \text{Hz}) = 5.7 \times 10^{-34} \, \text{J}$$

Part (c)

4. Set $E_n = nhf$ equal to the total energy of the system and solve for n:

$$E_n = nhf$$

$$n = \frac{E_n}{hf} = \frac{0.54 \, \text{J}}{5.7 \times 10^{-34} \, \text{J}} = 9.5 \times 10^{32}$$

CONTINUED ON NEXT PAGE

CONTINUED FROM PREVIOUS PAGE

INSIGHT

The numbers found in parts (b) and (c) are incredible for their size. For example, the quantum is on the order of 10^{-34} J, as compared with the energy required to break a bond in a DNA molecule, which is on the order of 10^{-20} J. Thus the quantum for a macroscopic system is about 10^{14} times smaller than the energy needed to affect a molecule. Similarly, the number of quanta in the system, roughly 10^{33}, is comparable to the number of atoms in four Olympic-size swimming pools.

PRACTICE PROBLEM

If the quantum of energy for a 1.5-kg mass on a spring is 0.80×10^{-33} J, what is the force constant of the spring?   [**Answer:** $k = 86$ N/m]

Some related homework problems: Problem 10, Problem 86

Clearly, then, the quantum numbers in typical macroscopic systems are incredibly large. As a result, a change of one in the quantum number is completely insignificant and undetectable. Similarly, the change in energy from one quantum state to the next is so small that it cannot be measured in a typical experiment; hence, for all practical purposes, the energy of a macroscopic system seems to change continuously, even though it actually changes by small increments. In contrast, in an atomic system, the energy jumps are of great importance, as we shall see in the next section.

Returning to the ultraviolet catastrophe for a moment, we can now see how Planck's hypothesis removes the unphysical divergence at high frequency predicted by classical physics. In Planck's theory, the higher the frequency f, the greater the quantum of energy, hf. Therefore, as the frequency is increased, the amount of energy required for even the smallest quantum jump increases as well. Since a blackbody has only a finite amount of energy, however, it simply cannot supply the large amount of energy required to produce an extremely high-frequency quantum jump. As a result, the amount of radiation at high frequency drops off toward zero.

Planck's theory of energy quantization leads to an adequate description of the experimental results for blackbody radiation. Still, the theory was troubling and somewhat unsatisfying to Planck and to many other physicists as well. Although the idea of energy quantization worked, at least in this case, it seemed ad hoc and more of a mathematical trick than a true representation of nature. With the work of Einstein, however, which we present in the next section, the well-founded misgivings about quantum theory began to fade away.

30–2 Photons and the Photoelectric Effect

From Max Planck's point of view, energy quantization in a blackbody was probably related to quantized vibrations of atoms in the walls of the blackbody. We are familiar, for example, with the fact that a string tied at both ends can produce standing waves at only certain discrete frequencies (Chapter 14), so perhaps atoms vibrating in a blackbody behave in a similar way, vibrating only with certain discrete energies. Certainly, Planck did not think the light in a blackbody had a quantized energy, since most physicists thought of light as being a wave, which can have any energy.

A brash young physicist named Albert Einstein, however, took the idea of quantized energy seriously and applied it to the radiation in the blackbody. Einstein proposed that light comes in bundles of energy, called **photons,** that obey Planck's hypothesis of energy quantization; that is, light of frequency f consists of photons with an energy given by the following relation:

Energy of a Photon of Frequency f

$$E = hf$$

SI unit: J

30–4

Thus the energy in a beam of light of frequency f can have only the values hf, $2hf$, $3hf$, and so on. Planck's initial reaction to Einstein's suggestion was that he had gone too far with the idea of quantization. As it turns out, nothing could have been further from the truth.

In Einstein's photon model, a beam of light can be thought of as a beam of particles, each carrying the energy hf, as indicated in **Figure 30–4**. If the beam of light is made more intense while keeping the frequency the same, the result is that the photons in the beam are more tightly packed, so that more photons pass a given point in a given time. In this way, more photons shine on a given surface in a given time, increasing the energy delivered to the surface per time. Even so, each photon in the more intense beam has exactly the same amount of energy as those in the less intense beam. The energy of a typical photon of visible light is calculated in the next Exercise.

EXERCISE 30–2

Calculate the energy of a photon of yellow light with a frequency of 5.25×10^{14} Hz. Give the energy in both joules and electron-volts.

SOLUTION

Applying Equation 30–4, we find

$$E = hf = (6.63 \times 10^{-34}\ \text{J} \cdot \text{s})(5.25 \times 10^{14}\ \text{s}^{-1}) = 3.48 \times 10^{-19}\ \text{J}$$

$$= 3.48 \times 10^{-19}\ \text{J}\left(\frac{1\ \text{eV}}{1.60 \times 10^{-19}\ \text{J}}\right) = 2.18\ \text{eV}$$

Note that the energy of a visible photon is on the order of an electron-volt (eV). This is also the typical energy scale for atomic and molecular systems, as we show in detail in the following Example.

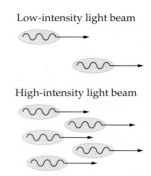

Low-intensity light beam

High-intensity light beam

▲ FIGURE 30–4 The photon model of light

In the photon model of light, a beam of light consists of many photons, each with an energy hf. The more intense the beam, the more tightly packed the photons.

EXAMPLE 30–2   WHEN OXYGENS SPLIT

Molecular oxygen (O_2) is a diatomic molecule. The energy required to dissociate 1 mol of O_2 to form 2 mol of atomic oxygen is 118 kcal. **(a)** Find the energy (in joules and electron-volts) required to dissociate one O_2 molecule. **(b)** Assuming the dissociation energy for one molecule is supplied by a single photon, find the frequency of the photon.

PICTURE THE PROBLEM

In our sketch we show a single photon dissociating an O_2 molecule to form two O atoms. The energy of the photon is $E = hf$.

STRATEGY

a. This part of the problem is simply a matter of converting from kcal per mole to joules per molecule. This can be accomplished by using the fact that 1 kcal = 4186 J and that Avogadro's number (Section 17–1) is 6.02×10^{23} molecules/mol. We can then convert to electron-volts using 1 eV = 1.60×10^{-19} J.

b. We can find the frequency of the photon by setting hf equal to the energy E found in part (a). Since Planck's constant is given in units of J·s, we must use the energy expressed in joules from part (a).

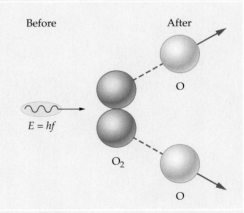

Before   After

O

$E = hf$

O_2

O

SOLUTION

Part (a)

1. Convert 118 kcal/mol to J/molecule:

$$\left(118\frac{\text{kcal}}{\text{mol}}\right)\left(\frac{4186\ \text{J}}{1\ \text{kcal}}\right)\left(\frac{1}{6.02 \times 10^{23}\ \text{molecules/mol}}\right)$$

$$= 8.21 \times 10^{-19}\ \text{J/molecule}$$

2. Convert the preceding result to eV/molecule:

$$8.21 \times 10^{-19}\frac{\text{J}}{\text{molecule}}\left(\frac{1\ \text{eV}}{1.60 \times 10^{-19}\ \text{J}}\right)$$

$$= 5.13\ \text{eV/molecule}$$

CONTINUED ON NEXT PAGE

CONTINUED FROM PREVIOUS PAGE

Part (b)

3. Use $E = 8.21 \times 10^{-19}$ J $= hf$ to solve for the frequency, f:

$$f = \frac{E}{h} = \frac{8.21 \times 10^{-19} \text{ J}}{6.63 \times 10^{-34} \text{ J} \cdot \text{s}} = 1.24 \times 10^{15} \text{ Hz}$$

INSIGHT

This frequency is in the ultraviolet. In fact, ultraviolet rays in Earth's upper atmosphere cause O_2 molecules to dissociate, freeing up atomic oxygen which can then combine with O_2 to form ozone, O_3.

PRACTICE PROBLEM

An infrared photon has a frequency of 1.00×10^{13} Hz. How much energy is carried by one mole of these photons?
[**Answer:** 3990 J = 0.953 kcal]

Some related homework problems: Problem 25, Problem 26

Since photons typically have rather small amounts of energy on a macroscopic scale, it follows that enormous numbers of photons must be involved in everyday situations, as demonstrated in the following Active Example.

REAL-WORLD PHYSICS: BIO
Dark-adapted vision

ACTIVE EXAMPLE 30–1 DARK VISION: FIND THE NUMBER OF PHOTONS

Dark-adapted (scotopic) vision is possible in humans with as little as 4.00×10^{-11} W/m^2 of 505-nm light entering the eye. If light of this intensity and wavelength enters the eye through a pupil that is 6.00 mm in diameter, how many photons enter the eye per second?

SOLUTION *(Test your understanding by performing the calculations indicated in each step.)*

1. Calculate the area of the pupil:	2.83×10^{-5} m^2
2. Multiply the intensity by the area of the pupil to find the energy entering the eye per second:	1.13×10^{-15} J/s
3. Calculate the energy of a photon:	3.94×10^{-19} J
4. Divide the energy of a photon into the energy per second to find the number of photons per second:	2870 photons/s

INSIGHT

Thus, even though a typical lightbulb gives off roughly 10^{18} photons per second, we need only about 10^3 photons per second to see. Our eyes are extraordinary instruments, sensitive to an incredibly wide range of intensities.

Finally, suppose an astronomer views a dim, distant galaxy with scotopic vision. The 2870 photons that enter the astronomer's eye each second are separated from one another by about 65 miles. It follows that only one photon at a time from the distant galaxy traverses the astronomer's telescope.

YOUR TURN

Suppose we consider light with a wavelength greater than 505 nm by a factor of 1.25. Will more or fewer photons be required per second with this new wavelength? By what factor will the required number of photons per second change?

*(Answers to **Your Turn** problems are given in the back of the book.)*

The Photoelectric Effect

Einstein applied his photon model of light to the **photoelectric effect,** in which a beam of light (photo-) hits the surface of a metal and ejects an electron (-electric). The effect can be measured using a device like that pictured in **Figure 30–5**. Note that incoming light ejects an electron—referred to as a photoelectron—from a metal plate called the emitter (E); the electron is then attracted to a collector plate (C), which is at a positive potential relative to the emitter. The result is an electric current that can be measured with an ammeter.

The minimum amount of energy necessary to eject an electron from a particular metal is referred to as the **work function,** W_0, for that metal. Work functions

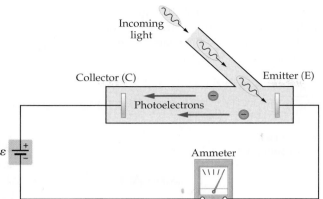

Photoelectrons

◀ **FIGURE 30-5 The photoelectric effect**
The photoelectric effect can be studied
with a device like that shown. Light
shines on a metal plate, ejecting electrons, which are then attracted to a
positively charged "collector" plate.
The result is an electric current that can
be measured with an ammeter.

vary from metal to metal but are typically on the order of a few electron-volts. If
an electron is given an energy E by the beam of light that is greater than W_0, the
excess energy goes into kinetic energy of the ejected electron. The maximum kinetic energy (K) a photoelectron can have, then, is

$$K_{max} = E - W_0 \qquad\qquad 30-5$$

Just as with blackbody radiation, the photoelectric effect exhibits behavior
that is at odds with classical physics. Two of the main areas of disagreement are
the following:

- Classical physics predicts that a beam of light of *any* color (frequency) can
  eject electrons, as long as the beam has sufficient intensity. That is, if a
  beam is intense enough, the energy it delivers to an electron will exceed
  the work function and cause it to be ejected.
- Classical physics also predicts that the maximum kinetic energy of an
  ejected electron should increase as the intensity of the light beam is increased. In particular, the more energy the beam delivers to the metal, the
  more energy that any given electron can have as it is ejected.

Although both of these predictions are reasonable—necessary, in fact, from the
classical physics point of view—they simply do not agree with experiments on
the photoelectric effect. In fact, experiments show the following behavior:

- To eject electrons, the incident light beam must have a frequency greater
  than a certain minimum value, referred to as the **cutoff frequency,** f_0. If
  the frequency of the light is less than f_0, it will not eject electrons, no matter how intense the beam.
- If the frequency of light is greater than the cutoff frequency, f_0, the effect of
  increasing the intensity is to increase the *number* of electrons that are emitted per second. The maximum kinetic energy of the electrons does not increase with the intensity of the light; the kinetic energy depends only on
  the frequency of the light.

As we shall see, these observations are explained quite naturally with the photon
model.

First, in Einstein's model each photon has an energy determined solely by its
frequency. Therefore, making a beam of a given frequency more intense simply
means increasing the number of photons hitting the metal in a given time—not increasing the energy carried by a photon. An electron, then, is ejected only if an incoming photon has an energy that is at least equal to the work function:
$E = hf_0 = W_0$. The *cutoff frequency* is thus defined as follows:

Cutoff Frequency, f_0

$$f_0 = \frac{W_0}{h} \qquad\qquad 30-6$$

SI unit: Hz = s^{-1}

If the frequency of the light is greater than f_0, the electron can leave the metal with a finite kinetic energy; if the frequency is less than f_0, no electrons are ejected, no matter how intense the beam. We determine a typical cutoff frequency in the next Exercise.

EXERCISE 30–3

The work function for a gold surface is 4.58 eV. Find the cutoff frequency, f_0, for a gold surface.

SOLUTION
Substitution in Equation 30–6 yields

$$f_0 = \frac{W_0}{h} = \frac{(4.58 \text{ eV})\left(\dfrac{1.60 \times 10^{-19} \text{ J}}{1 \text{ eV}}\right)}{6.63 \times 10^{-34} \text{ J} \cdot \text{s}} = 1.11 \times 10^{15} \text{ Hz}$$

This frequency is in the near ultraviolet.

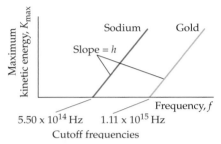

▲ **FIGURE 30–6 The kinetic energy of photoelectrons**
The maximum kinetic energy of photoelectrons as a function of the frequency of light. Note that sodium and gold have different cutoff frequencies, as one might expect for different materials. On the other hand, the slope of the two lines is the same, h, as predicted by Einstein's photon model of light.

Second, the fact that a more intense beam of monochromatic light delivers more photons per time to the metal just means that more electrons are ejected per time. Since each electron receives precisely the same amount of energy, however, the maximum kinetic energy is the same regardless of the intensity. In fact, if we return to Equation 30–5 and replace the energy, E, with the energy of a photon, hf, we find

$$K_{\text{max}} = hf - W_0 \qquad\qquad 30\text{–}7$$

Note that K_{max} depends linearly on the frequency but is independent of the intensity. A plot of K_{max} for sodium (Na) and gold (Au) is given in **Figure 30–6**. Clearly, both lines have the same slope, h, as expected from Equation 30–7, but have different cutoff frequencies. Therefore, with the result given in Equation 30–7, Einstein was able to show that Planck's constant, h, appears in a natural way in the photoelectric effect and is not limited in applicability to the blackbody.

EXAMPLE 30–3 WHITE LIGHT ON SODIUM

A beam of white light containing frequencies between 4.00×10^{14} Hz and 7.90×10^{14} Hz is incident on a sodium surface, which has a work function of 2.28 eV. **(a)** What is the range of frequencies in this beam of light for which electrons are ejected from the sodium surface? **(b)** Find the maximum kinetic energy of the "photoelectrons" that are ejected from this surface.

PICTURE THE PROBLEM
Our sketch shows a beam of white light, represented by photons with different frequencies, incident on a sodium surface. Photoelectrons are ejected from this surface with a kinetic energy that depends on the frequency of the photon that was absorbed.

STRATEGY

a. We can find the cutoff frequency, f_0, for sodium using $f_0 = W_0/h$, with $W_0 = 2.28$ eV. Frequencies between the cutoff frequency and the maximum frequency in the beam of light, 7.90×10^{14} Hz, will eject electrons.

b. We can obtain the maximum kinetic energy for a given frequency, f, from Equation 30–7: $K_{\text{max}} = hf - W_0$. Clearly, the higher the frequency, the greater the maximum kinetic energy. It follows, then, that the greatest possible maximum kinetic energy corresponds to the highest frequency in the beam, 7.90×10^{14} Hz.

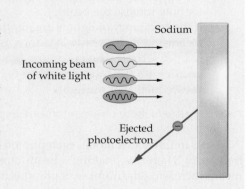

SOLUTION

Part (a)

1. Use $f_0 = W_0/h$ to calculate the cutoff frequency for sodium:

$$f_0 = \frac{W_0}{h} = \frac{(2.28 \text{ eV})\left(\dfrac{1.60 \times 10^{-19} \text{ J}}{1 \text{ eV}}\right)}{6.63 \times 10^{-34} \text{ J} \cdot \text{s}} = 5.50 \times 10^{14} \text{ Hz}$$

2. The frequencies that eject electrons are those between the cutoff frequency and the highest frequency in the beam:

frequencies in this beam that eject electrons:
5.50×10^{14} Hz to 7.90×10^{14} Hz

Part (b)

3. Using $K_{max} = hf - W_0$, calculate K_{max} for the maximum frequency in the beam, $f = 7.90 \times 10^{14}$ Hz:

$$K_{max} = hf - W_0$$
$$= (6.63 \times 10^{-34} \text{ J} \cdot \text{s})(7.90 \times 10^{14} \text{ Hz})$$
$$- (2.28 \text{ eV})\left(\frac{1.60 \times 10^{-19} \text{ J}}{1 \text{ eV}}\right) = 1.59 \times 10^{-19} \text{ J}$$

INSIGHT

Note that most of the photons in a beam of white light will eject electrons from sodium, and that the maximum kinetic energy of one of these photoelectrons is about 1 eV.

PRACTICE PROBLEM

What frequency of light would be necessary to give a maximum kinetic energy of 2.00 eV to the photoelectrons from this surface?   [**Answer:** 1.03×10^{15} Hz]

Some related homework problems: Problem 31, Problem 32, Problem 33

CONCEPTUAL CHECKPOINT 30-2   EJECTED ELECTRONS

Consider a photoelectric experiment such as the one illustrated in Figure 30–5. A beam of light with a frequency greater than the cutoff frequency shines on the emitter. If the frequency of this beam is increased while the intensity is held constant, does the number of electrons ejected per second from the metal surface **(a)** increase, **(b)** decrease, or **(c)** stay the same?

REASONING AND DISCUSSION

Increasing the frequency of the beam means that each photon carries more energy; however, we know that the intensity of the beam remains constant. It follows, then, that fewer photons hit the surface per time—otherwise the intensity would increase. Since fewer photons hit the surface per time, fewer electrons are ejected per time.

ANSWER

(b) The number of electrons ejected per second decreases.

Applications of the photoelectric effect are in common use all around us. For example, if you have ever dashed into an elevator as its doors were closing, you were probably saved from being crushed by the photoelectric effect. Many elevators and garage-door systems use a beam of light and a photoelectric device known as a *photocell* as a safety feature. As long as the beam of light strikes the photocell, the photoelectric effect generates enough ejected electrons to produce a detectable electric current. When the light beam is blocked—by a late arrival at the elevator, for example—the electric current produced by the photocell is interrupted and the doors are signaled to open. Similar photocells automatically turn on streetlights at dusk and measure the amount of light entering a camera.

REAL-WORLD PHYSICS
Photocells

◀ The photoelectric effect is the basic mechanism used by photovoltaic cells, which are now used to power both terrestrial devices such as pay phones (left) and the solar panels that supply electricity to the Hubble Space Telescope (right).

Photocells are also the basic unit in the *solar energy panels* that convert some of the energy in sunlight into electrical energy. A small version of a solar energy panel can be found on many pocket calculators. These panels are efficient enough to operate their calculators with nothing more than dim indoor lighting. Larger outdoor panels can operate billboards and safety lights in remote areas far from commercial power lines. Truly large solar panels, 240 ft in length, power the International Space Station and are so large that they make the station visible to the naked eye from Earth's surface. These applications of solar panels may only hint at the potential for solar energy in the future, however, especially when one considers that sunlight delivers about 200,000 times more energy to Earth each day than all the world's electrical energy production combined.

Finally, it is interesting to note that though Einstein is best known for his development of the theory of relativity, he was awarded the Nobel Prize in physics not for relativity but for the photoelectric effect.

30–3 The Mass and Momentum of a Photon

When we say that a photon is like a "particle" of light, we put the word *particle* in quotes because a photon is quite different from everyday particles in many important respects. For example, a typical particle can be held at rest in the hand. It can also be placed on a scale to have its mass determined. These operations are not possible with a photon.

First, photons travel with the speed of light, which means that all observers see them as having the same speed. It is not possible to stop a photon and hold it in your hand. In contrast, particles with a finite rest mass can never attain the speed of light. It follows, then, that photons must have *zero rest mass*. This condition can be seen mathematically by rewriting Equation 29–7 for the total energy, E, as follows:

$$E\sqrt{1 - \frac{v^2}{c^2}} = m_0 c^2 \qquad \text{30–8}$$

Since photons travel at the speed of light, $v = c$, the left side of the equation is zero. As a result, the right side of the equation must also be zero, which can happen only if the rest mass is zero:

Rest Mass of a Photon

$$m_0 = 0 \qquad \text{30–9}$$

Second, photons differ from everyday particles in that they have a finite momentum even though they have no mass. To see how this can be, note that the relativistic equation for momentum, Equation 29–5, can be rewritten as follows:

$$p\sqrt{1 - \frac{v^2}{c^2}} = m_0 v \qquad \text{30–10}$$

Dividing Equation 30–10 by Equation 30–8 we get

$$\frac{p}{E} = \frac{v}{c^2}$$

Once again setting $v = c$ for a photon, we find that the momentum of a photon is related to its total energy as follows:

$$p = \frac{E}{c}$$

Finally, recalling that $E = hf$, and that $f = c/\lambda$, we obtain the following result:

Momentum of a Photon

$$p = \frac{hf}{c} = \frac{h}{\lambda} \qquad \text{30–11}$$

Note that a photon's momentum increases with its frequency, and thus with its energy.

As one might expect, the momentum of a typical photon of visible light is quite small, as is illustrated in the next Exercise.

EXERCISE 30–4

Calculate the momentum of a photon of yellow light with a frequency of 5.25×10^{14} Hz.

SOLUTION

Substituting $f = 5.25 \times 10^{14}$ Hz in Equation 30–11, we obtain

$$p = \frac{hf}{c} = \frac{(6.63 \times 10^{-34}\,\text{J}\cdot\text{s})(5.25 \times 10^{14}\,\text{Hz})}{3.00 \times 10^{8}\,\text{m/s}} = 1.16 \times 10^{-27}\,\text{kg}\cdot\text{m/s}$$

As small as the momentum of a photon is, it can still have a significant impact if the number of photons is large. For example, NASA is studying the feasibility of constructing spaceships that would be powered by huge "light sails" that transfer the momentum of photons from the Sun to the ship. A sail would reflect photons, creating a change in momentum, and hence a reaction force on the sail due to the **radiation pressure** of the photons. With a large enough sail, it may one day be possible to cruise among the stars with space-faring sailboats. In fact, the designs under consideration by NASA would result in the largest and fastest spacecraft ever constructed—10 times faster than the space shuttle.

A more "down to Earth," though no less exotic, application of radiation pressure is the *optical tweezers*. Basically, an optical tweezers is a laser beam that is made more intense in the middle than near the edges. When a small, translucent object is placed in this beam, the recoil produced by photons passing through the object exerts a small force on it directed toward the center of the beam. The situation is similar to a ball suspended in the "beam" of air produced by a hair dryer. Although the force exerted by such tweezers is typically on the order of only 10^{-12} N, it is still large enough to manipulate cells, DNA, and other subcellular particles. In fact, optical tweezers can even capture, lift, and separate a single bacterium from a bacterial culture for further study.

30–4 Photon Scattering and the Compton Effect

Einstein's photon model of the photoelectric effect, published in 1905, focused on the energy of a photon, $E = hf$. The momentum of a photon, $p = hf/c = h/\lambda$, plays a key role in a different type of experiment, in which an X-ray photon undergoes a collision with an electron initially at rest. The result of this collision is that the photon is scattered, which changes its direction and energy. This type of process, referred to as the **Compton effect,** was explained in terms of the photon model of light by the American physicist Arthur Holly Compton (1892–1962) in 1923. The success of the photon model in explaining both the Compton effect and the photoelectric effect gained it widespread acceptance.

Figure 30–7 shows an X-ray photon striking an electron at rest and scattering at an angle θ with respect to the incident direction. To understand the behavior of a system like this, we treat the photon as a particle, with a certain energy and momentum, that collides with another particle (the electron) of mass m_e and initial speed zero. If we assume the electron is free to move, this collision conserves both energy and momentum.

First, to conserve energy the following relation must be satisfied:

energy of incident photon = energy of scattered photon +
final kinetic energy of electron

$$hf = hf' + K \qquad\qquad 30\text{–}12$$

Note that, in general, the frequency of the scattered photon, f', is less than the frequency of the incident photon, f, since part of the energy of the incident photon

▶ **FIGURE 30–7 The Compton effect**
An X-ray photon scattering from an electron at rest can be thought of as a collision between two particles. The result is a change of wavelength for the scattered photon. This is referred to as the Compton effect.

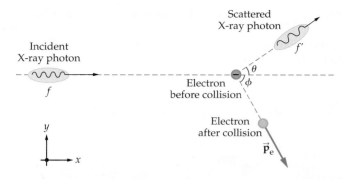

has gone into kinetic energy of the electron. Because the frequency of the scattered photon is reduced, its wavelength, $\lambda' = c/f'$, is increased.

Next, we conserve both the x and y components of momentum, using $p = h/\lambda$ as the magnitude of the incident photon's momentum, and $p' = h/\lambda'$ as the magnitude of the scattered photon's momentum. For the x component of momentum we have the following relation:

$$\frac{h}{\lambda} = \frac{h}{\lambda'} \cos \theta + p_e \cos \phi \qquad \text{30–13}$$

Note that we assume the electron has a momentum p_e at an angle ϕ to the incident direction. Conserving the y component of momentum, which initially is zero, we have

$$0 = \frac{h}{\lambda'} \sin \theta - p_e \sin \phi \qquad \text{30–14}$$

Given the initial wavelength and frequency, $\lambda = c/f$, and the scattering angle θ, we can use the three relations, Equations 30–12, 30–13, and 30–14, to solve for the three unknowns: λ', ϕ, and K.

Of particular interest is the change in wavelength produced by the scattering. We find the following result:

Compton Shift Formula

$$\Delta\lambda = \lambda' - \lambda = \frac{h}{m_e c}(1 - \cos \theta) \qquad \text{30–15}$$

SI unit: m

PROBLEM-SOLVING NOTE

Scattering Angle

The angle θ in the Compton shift formula is the angle between the incident direction of an X-ray and its direction of propagation after scattering.

In this expression, the quantity $h/m_e c = 2.43 \times 10^{-12}$ m is referred to as the *Compton wavelength of an electron*. The maximum change in the photon's wavelength occurs when it scatters in the reverse direction ($\theta = 180°$), in which case the change is twice the Compton wavelength, $\Delta\lambda = 2(h/m_e c)$. On the other hand, if the scattering angle is zero ($\theta = 0$)—in which case there really is no scattering at all—the change in wavelength is zero; $\Delta\lambda = 0$.

CONCEPTUAL CHECKPOINT 30–3   CHANGE IN WAVELENGTH

If X-rays are scattered from protons instead of from electrons, is the change in their wavelength for a given angle of scattering **(a)** increased, **(b)** decreased, or **(c)** unchanged?

REASONING AND DISCUSSION
As can be seen from Equation 30–15, the change in wavelength for a given angle is proportional to $h/m_e c$. If a proton is substituted for the electron, the change in wavelength will be proportional to $h/m_p c$ instead. Since protons have about 2000 times the mass of electrons, the change in wavelength will be reduced by a factor of about 2000.

ANSWER
(b) The change in wavelength is decreased when X-rays scatter from protons rather than from electrons.

The next Example gives a detailed analysis of a photon-electron collision.

EXAMPLE 30–4   SCATTERING X-RAYS

An X-ray photon with a wavelength of 0.650 nm scatters from a free electron at rest. After scattering, the photon moves at an angle of 152° relative to the incident direction. Find (a) the wavelength and (b) the energy of the scattered photon. (c) Determine the kinetic energy of the recoiling electron.

PICTURE THE PROBLEM

As shown in the sketch, the incoming photon is scattered at an angle of 152° relative to its initial direction; thus, it almost heads back the way it came. After the collision, the electron moves with a speed v at an angle ϕ relative to the forward direction.

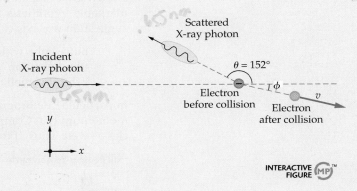

STRATEGY

a. We can find the wavelength after scattering, λ', by using the Compton shift formula, Equation 30–15. Note that we are given the initial wavelength, λ, as well as the scattering angle, θ.

b. The energy of the scattered photon is $E' = hf' = hc/\lambda'$.

c. Since energy is conserved, the kinetic energy of the electron is given by $hf = hf' + K$. We can solve this relation to yield the kinetic energy, K.

SOLUTION

Part (a)

1. Use Equation 30–15 to find the wavelength of the scattered photon:

$$\lambda' = \lambda + \frac{h}{m_e c}(1 - \cos \theta) = 0.650 \times 10^{-9}\,\text{m}$$

$$+ \left(\frac{6.63 \times 10^{-34}\,\text{J}\cdot\text{s}}{(9.11 \times 10^{-31}\,\text{kg})(3.00 \times 10^8\,\text{m/s})} \right)(1 - \cos 152°)$$

$$= 6.55 \times 10^{-10}\,\text{m} = 0.655\,\text{nm}$$

Part (b)

2. The energy of the scattered photon is given by $E' = hf' = hc/\lambda'$:

$$E' = hf' = \frac{hc}{\lambda'} = \frac{(6.63 \times 10^{-34}\,\text{J}\cdot\text{s})(3.00 \times 10^8\,\text{m/s})}{6.55 \times 10^{-10}\,\text{m}}$$

$$= 3.04 \times 10^{-16}\,\text{J}$$

Part (c)

3. Find the initial energy of the photon using $E = hf = hc/\lambda$:

$$E = hf = \frac{hc}{\lambda} = \frac{(6.63 \times 10^{-34}\,\text{J}\cdot\text{s})(3.00 \times 10^8\,\text{m/s})}{6.50 \times 10^{-10}\,\text{m}}$$

$$= 3.06 \times 10^{-16}\,\text{J}$$

4. Subtract the final energy of the photon from its initial energy to find the kinetic energy of the electron:

$$K = hf - hf'$$

$$= 3.06 \times 10^{-16}\,\text{J} - 3.04 \times 10^{-16}\,\text{J}$$

$$= 2 \times 10^{-18}\,\text{J} = 10\,\text{eV}$$

INSIGHT

Notice that the energy of the X-ray photon is roughly 1900 eV = 1.9 keV both before and after scattering. Only a very small fraction of its energy is delivered to the electron. Still, the electron acquires about 10 eV of energy—enough to ionize an atom.

PRACTICE PROBLEM

At what scattering angle will the photon have a wavelength of 0.652 nm?   [**Answer:** $\theta = 79.9°$]

Some related homework problems: Problem 54, Problem 55

As we have seen in both the photoelectric effect and the Compton effect, a photon can behave as a particle with a well-defined energy and momentum. We also know, however, that light exhibits wavelike behavior, as when it produces interference fringes in Young's two-slit experiment (Chapter 28). That light can have such seemingly opposite attributes is one of the deepest mysteries of quantum physics and, at the same time, one of its most significant insights. In the next section we expand on this insight and extend it to the behavior of matter.

30–5 The de Broglie Hypothesis and Wave–Particle Duality

As we have seen, the Compton effect was explained in terms of photons in 1923. Another significant advance in quantum physics occurred that same year, when a French graduate student, Louis de Broglie (1892–1987), put forward a most remarkable hypothesis that would later win him the Nobel Prize in physics. His suggestion was basically the following:

> Since light, which we usually think of as a wave, can exhibit particle-like behavior, perhaps a particle of matter, like an electron, can exhibit wavelike behavior.

In particular, de Broglie proposed that the *same* relation between wavelength and momentum that Compton applied to the photon, $p = h/\lambda$, should apply to particles as well. Thus, if the momentum of a particle is p, its de Broglie wavelength is

de Broglie Wavelength

$$\lambda = \frac{h}{p}$$

30–16

SI unit: m

Note that the greater a particle's momentum, the smaller its de Broglie wavelength.

How can the idea of a wavelength for matter make sense, however, when we know that objects like baseballs and cars behave like particles, not like waves? To see how this is possible, we calculate the de Broglie wavelength of a typical macroscopic object: a 0.13-kg apple moving with a speed of 5.0 m/s. Substituting these values into $p = mv$, and using $\lambda = h/p$, we find the following wavelength: $\lambda = 1.0 \times 10^{-33}$ m. Clearly, this wavelength, which is smaller than the diameter of an atom by a factor of 10^{23}, is much too small to be observed in any macroscopic experiment. Thus, an apple could have a wavelength as given by the de Broglie relation, and we would never notice it.

In contrast, consider an electron with a kinetic energy of 10.0 eV, a typical atomic energy. Using $K = \frac{1}{2}mv^2 = p^2/2m$ to solve for the momentum p, we find that the de Broglie wavelength in this case is $\lambda = 3.88 \times 10^{-10}$ m = 3.88 Å. Now this wavelength, which is comparable to the size of an atom or molecule, would clearly be significant in such systems. Therefore, the de Broglie wavelength may be unobservable in macroscopic systems but all-important in atomic systems.

ACTIVE EXAMPLE 30–2   THE SPEED AND WAVELENGTH OF AN ELECTRON

How fast is an electron moving if its de Broglie wavelength is 3.50×10^{-7} m?

SOLUTION *(Test your understanding by performing the calculations indicated in each step.)*

1. Write Equation 30–16 in terms of the electron's mass and speed: $\lambda = h/mv$

2. Rearrange to solve for the speed, v: $v = h/m\lambda$

3. Substitute numerical values: $v = 2080$ m/s

INSIGHT

The relatively small value obtained for the electron's speed justifies using the nonrelativistic expression for momentum, $p = mv$.

YOUR TURN

If the electron's speed is doubled, does its de Broglie wavelength increase or decrease? By what factor?

*(Answers to **Your Turn** problems are given in the back of the book.)*

In order for the de Broglie wavelength to be taken seriously, however, it must be observed experimentally. We next consider ways in which this can be done.

Diffraction of X-rays and Particles by Crystals

An especially powerful way to investigate wave properties is through the study of interference patterns. Consider, for example, directing a coherent beam of X-rays onto a crystalline substance composed of regularly spaced planes of atoms, as indicated in **Figure 30–8**. Notice that the reflected beam combines rays that have followed different paths, with different path lengths. If the difference in path lengths is half a wavelength, destructive interference occurs; if the difference in path lengths is one wavelength, constructive interference results, and so on. From Figure 30–8 it is clear that the difference in path lengths for rays reflecting from adjacent planes that have a spacing d is $2d \sin \theta$. Thus, constructive interference occurs when the following conditions are met:

Constructive Interference When Scattering from a Crystal

$$2d \sin \theta = m\lambda \quad m = 1, 2, 3, \ldots \qquad 30\text{–}17$$

This is very similar to the condition for constructive interference in a diffraction grating, as derived in Equation 28–16. The resulting pattern of interference maxima—referred to as a **diffraction pattern**—can be projected onto photographic film, as indicated in **Figure 30–9**, and used to determine the geometric properties of a crystal. Examples of diffraction patterns are shown in **Figure 30–10**.

Now, if particles—like electrons, for example—have a wavelength comparable to atomic distances, it should be possible to produce diffraction patterns with them in much the same way as with X-rays. This is indeed the case, as was shown by the American physicists C. J. Davisson and L. H. Germer, who, in 1928, produced diffraction patterns by scattering low-energy electrons (about 54 eV) from crystals of nickel. The spacing between spots in the electron diffraction pattern allowed the researchers to determine the electron's wavelength, which verified in

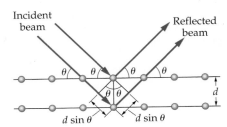

▲ **FIGURE 30–8 Scattering from a crystal**
Scattering of X-rays or particles from a crystal. Note that waves reflecting from the lower plane of atoms have a path length that is longer than the path of the upper waves by the amount $2d \sin \theta$.

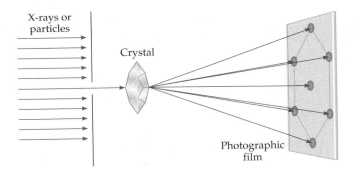

◀ **FIGURE 30–9 Diffraction patterns**
Diffraction patterns can be observed by passing a beam of X-rays or particles through a crystal. The beams emerging from the crystal are at specific angles, due to constructive interference, and can be recorded on photographic film.

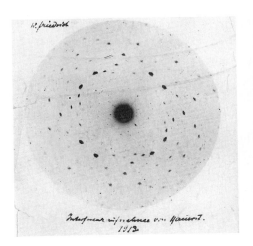
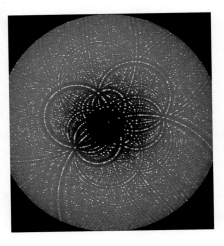

◀ **FIGURE 30–10 X-ray diffraction**
Because the spacing of atoms in a crystal is of the same order of magnitude as the wavelengths of X-rays, the planes of a crystal can serve as a diffraction grating for X-rays. When properly interpreted, the resulting diffraction patterns can provide remarkably detailed information about the structure of the crystal. The historic photograph at left, made by two of the pioneers of X-ray crystallography in 1912, records the diffraction pattern from a simple inorganic salt. The photograph at right, made nearly a century later, shows the much more intricate pattern produced by a large protein molecule.

The neutron diffraction pattern of the
protein lysozyme, a digestive enzyme.

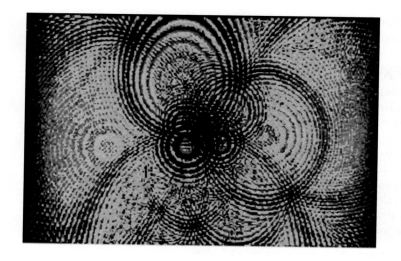

detail the de Broglie relation, $\lambda = h/p$. Similar diffraction patterns have since been
observed with neutrons, hydrogen atoms, and helium atoms. An example of a
neutron diffraction pattern is shown in **Figure 30–11**.

EXAMPLE 30–5 NEUTRON DIFFRACTION

A beam of neutrons moving with a speed of 1450 m/s is diffracted from a crystal of table salt (sodium chloride), which has an
interplanar spacing of $d = 0.282$ nm. **(a)** Find the de Broglie wavelength of the neutrons. **(b)** Find the angle of the first interference maximum.

PICTURE THE PROBLEM
Our sketch shows a beam of neutrons reflecting from two adjacent atomic planes of
table salt. The distance d between the planes is 0.282 nm.

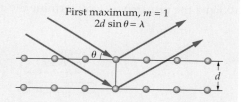

First maximum, $m = 1$
$2d \sin \theta = \lambda$

STRATEGY

a. We can find the de Broglie wavelength using Equation 30–16: $\lambda = h/p$. Since the
speed of the neutrons is much less than the speed of light, we can write the momentum as $p = m_n v$, where m_n is the mass of a neutron.

b. Referring to Equation 30–17, we see that the first interference maximum occurs
when $m = 1$ in the relation $2d \sin \theta = m\lambda$. Thus we set $2d \sin \theta$ equal to λ and
solve for θ.

SOLUTION

Part (a)

1. Calculate the de Broglie wavelength using $\lambda = h/p = h/m_n v$:

$$\lambda = \frac{h}{p}$$

$$= \frac{h}{m_n v} = \frac{6.63 \times 10^{-34} \text{ J} \cdot \text{s}}{(1.67 \times 10^{-27} \text{ kg})(1450 \text{ m/s})} = 0.274 \text{ nm}$$

Part (b)

2. Set $m = 1$ in $2d \sin \theta = m\lambda$:

$$2d \sin \theta = \lambda$$

3. Solve for the angle θ:

$$\theta = \sin^{-1}\left(\frac{\lambda}{2d}\right) = \sin^{-1}\left(\frac{0.274 \text{ nm}}{2(0.282 \text{ nm})}\right) = 29.1°$$

INSIGHT
Note that if we change the speed of the neutrons, we will change their wavelength. This, in turn, changes the angle of the interference maxima. This connection between the neutron speed and the interference maxima provides a detailed and precise way
of verifying the de Broglie relation.

PRACTICE PROBLEM
Find the angle of the first interference maximum if the neutron speed is increased to 2250 m/s. **[Answer: 18.2°]**

Some related homework problems: Problem 62, Problem 65

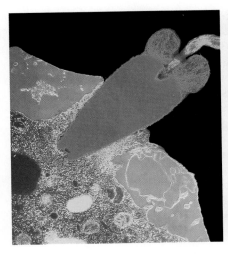
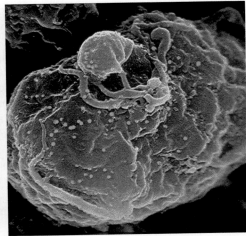

◀ **FIGURE 30–12 Electron micrography**
Electron micrographs can be produced in several different ways. In a transmission electron micrograph (TEM), such as the image at left, the beam of electrons passes through the specimen, as in an ordinary light microscope. This colorized photo shows a sea urchin sperm cell penetrating the membrane of an egg. In a scanning electron micrograph (SEM), like the image at right, the electrons are reflected from the surface of the specimen (which is usually coated with a thin layer of metal atoms first) and used to produce a startlingly detailed three-dimensional image. This photo, also colorized for greater clarity, shows particles of the AIDS virus HIV (purple) emerging from an infected human cell.

REAL-WORLD PHYSICS

Electron microscopes

In *electron microscopes*, a beam of electrons is used to form an image of an object in much the same way that a beam of light is used in a light microscope. One of the differences, however, is that the wavelength of the electrons can be much shorter than the wavelength of visible light. For example, the shortest visible wavelength is that of blue light, which is about 380 nm. In comparison, we have seen that the de Broglie wavelength of an electron with an energy of only 10.0 eV is 0.388 nm—about 1000 times smaller than the wavelength of blue light. Since the ability to resolve small objects depends on using a wavelength that is smaller than the object to be imaged, an electron microscope can see much finer detail than a light microscope, as **Figure 30–12** indicates.

Wave–Particle Duality

Notice that we have now come full circle in our study of waves and particles, from considering light as a wave, and then noting that it has particle-like properties, to investigating particles of matter like electrons, and finding that they have wave-like properties. This type of behavior is referred to as the **wave–particle duality.** Thus, as strange as it may seem, light is a wave, but it also comes in discrete units called photons. Electrons come in discrete units of well-defined mass and charge, but they also have wave properties.

To illustrate the wave–particle duality, we consider a two-slit experiment, as shown in **Figure 30–13**. If light is passed through these slits it forms interference patterns of dark and light fringes, as we saw in Chapter 28. If the intensity of the light is reduced to a very low level, it is possible to have only a single photon at a time passing through the apparatus. This photon lands on the distant screen. Eventually, as more and more photons land on the screen, the interference pattern emerges.

Similar behavior can be observed with electrons passing through a pair of slits. The results, as shown in **Figure 30–14**, are the direct analog of the results obtained with light. Notice that the dark portions of the interference pattern are

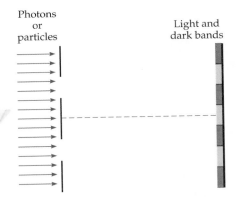

Photons or particles   Light and dark bands

▲ **FIGURE 30–13 The two-slit experiment**
When photons or particles are passed through a screen with two slits, the result is an interference pattern of light and dark bands.

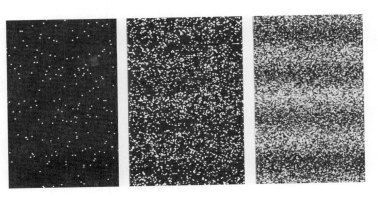

◀ **FIGURE 30–14 Creation of an interference pattern by electrons passing through two slits**
At first, the electrons seem to arrive at the screen in random locations. As the number of electrons increases, however, the interference pattern becomes more evident. (Compare Figures 28–3 and 28–6.)

those places where electron waves passing through the two slits have combined to produce destructive interference—that is, at these special points an electron is essentially able to "cancel itself out" to produce a dark fringe.

Clearly, then, subatomic particles and light are quite different from objects and waves that we observe on a macroscopic level. In fact, one of the most profound and unexpected insights of quantum physics is simply that even though baseballs, apples, and people are composed of electrons, protons, and other particles, the behavior of these subatomic particles is nothing like the behavior of baseballs, apples, and people. In short, an electron is not like a BB reduced in size. An electron has properties that are different from those of any object we experience on the macroscopic level. To try to force light and electrons into categories like waves and particles is to miss the essence of their existence—they are neither one nor the other, though they have characteristics of both. Again, one is reminded of the saying "The universe is not only stranger than we imagine, it is stranger than we can imagine."

30–6 The Heisenberg Uncertainty Principle

One of the more interesting aspects of Figure 30–14 is that the dots, corresponding to observations of individual electrons, appear on the screen in random order. To be specific, each time this experiment is run, the dots appear in different locations and in different sequence—all that remains the same is the final pattern that emerges after a large number of electrons are observed. The point is that as any given electron passes through the two-slit apparatus, *it is not possible to predict exactly where that one electron will land on the screen*. We can give the probability that it will land at different locations, but the fate of each individual electron is uncertain.

This kind of "uncertainty" is inherent in quantum physics and is due to the fact that matter has wavelike properties. As a simple example, consider a beam of electrons moving in the x direction and passing through a single slit, as in **Figure 30–15**. The result of this experiment is a diffraction pattern similar to that found for light in Chapter 28 (see Figure 28–18). In particular, if the beam passes through a slit of width W, it produces a large central maximum—where the probability of detecting an electron is high—with a dark fringe on either side at an angle θ given by Equation 28–12:

$$\sin \theta = \frac{\lambda}{W} \qquad \qquad 30\text{–}18$$

If the incoming electrons have a momentum p_x, the wavelength in Equation 30–18 is given by the de Broglie relation: $\lambda = h/p_x$.

We interpret this experiment as follows: By passing the beam of electrons through the slit, we have determined their location in the y direction to within an uncertainty of roughly $\Delta y \sim W$. After passing through the slit, however, the beam spreads out to form a diffraction pattern—with some electrons acquiring a y component of momentum. Thus there is now an uncertainty in the y component of momentum, Δp_y.

▶ **FIGURE 30–15 Diffraction pattern of electrons**

Central region of the diffraction pattern formed by electrons passing through a single slit in a screen. The curve to the right is a measure of the number of electrons detected at any given location. Note that no electrons are detected at the dark fringes. This diffraction pattern is identical in form with that produced by light passing through a single slit.

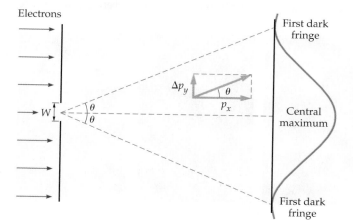

There is a reciprocal relationship between Δy and Δp_y, as we show in **Figure 30–16**. As the width W of the slit is made smaller, to decrease Δy, the angle in Equation 30–18 increases, and the diffraction pattern spreads out, increasing Δp_y. Conversely, if the width of the slit is increased, the diffraction pattern becomes narrower; that is, increasing Δy results in a decrease in Δp_y. To summarize;

> If we know the position of a particle with greater precision, its momentum is more uncertain; if we know the momentum of a particle with greater precision, its position is more uncertain.

We can give this conclusion in mathematical form by returning to Equation 30–18 and Figure 30–15. First, assuming small angles, θ, we can use the approximation that $\sin \theta \sim \theta$ to write Equation 30–18 as $\theta \sim \lambda / W$. Similarly, from Figure 30–15 we see that $\tan \theta = \Delta p_y / p_x$. Since $\tan \theta \sim \theta$ for small angles, it follows that $\theta \sim \Delta p_y / p_x$. Equating these two expressions for θ we get

$$\theta \sim \frac{\lambda}{W} \sim \frac{\Delta p_y}{p_x}$$

Setting $p_x = h/\lambda$ and $W \sim \Delta y$, we find

$$\frac{\lambda}{\Delta y} \sim \frac{\Delta p_y}{h/\lambda}$$

Finally, canceling λ and rearranging, we obtain

$$\Delta p_y \, \Delta y \sim h$$

Thus, the product of uncertainties in position and momentum cannot be less than a certain finite amount that is approximately equal to Planck's constant.

A thorough treatment of this system yields the more precise relation first given by the German physicist Werner Heisenberg (1901–1976) in 1927, known as the **Heisenberg uncertainty principle**:

The Heisenberg Uncertainty Principle: Momentum and Position

$$\Delta p_y \, \Delta y \geq \frac{h}{2\pi} \qquad\qquad 30\text{--}19$$

In fact, Heisenberg showed that this relation is a general principle and not restricted in any way to the single-slit system considered here. There is simply an unremovable, intrinsic uncertainty in nature that is the result of the wave behavior of matter. Since the wavelength of matter, $\lambda = h/p$, depends directly on the magnitude of h, so too does the uncertainty.

One way of stating the uncertainty principle is that it is impossible to know both the position and momentum of a particle with arbitrary precision at any given time. For example, if the position is known precisely, so that Δy approaches zero, it follows that Δp_y must approach infinity, as shown in Figure 30–16. This result implies that the y component of momentum is completely uncertain. Similarly, complete knowledge of p_y ($\Delta p_y \rightarrow 0$) implies that the position y is completely uncertain.

As one might expect, the uncertainty restrictions given in Equation 30–19 have negligible impact on macroscopic systems, but are all-important in atomic and nuclear systems. This is illustrated in the following Example.

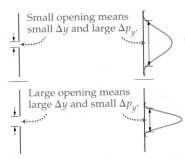

▲ **FIGURE 30–16   Uncertainty in position and momentum**
Reciprocal relationship between the uncertainty in position (Δy) and the uncertainty in momentum (Δp_y). As in Figure 30–15, the curves at the right indicate the number of electrons detected at any given location.

EXAMPLE 30–6   WHAT IS YOUR POSITION?

If the speed of an object is 35.0 m/s, with an uncertainty of 5.00%, what is the minimum uncertainty in the object's position if it is **(a)** an electron or **(b)** a volleyball ($m = 0.350$ kg)?

PICTURE THE PROBLEM
As shown in the sketch, the electron and the volleyball have the same speed. The fact that the electron has the smaller mass means that the values of its momentum, and of its uncertainty in momentum, are less than the corresponding values for the volleyball. As a result, the uncertainty in position will be larger for the electron.

CONTINUED ON NEXT PAGE

CONTINUED FROM PREVIOUS PAGE

STRATEGY

The 5.00% uncertainty in speed means that the magnitude of the momentum (mass times speed) of the electron and the volleyball are also uncertain by 5.00%. The minimum uncertainty in position, then, is given by the Heisenberg uncertainty principle: $\Delta y = h/(2\pi \, \Delta p_y)$.

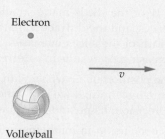

Electron

v

Volleyball

SOLUTION

Part (a)

1. Calculate the uncertainty in the electron's momentum:

$$\Delta p_y = 0.0500 m_e v = 0.0500(9.11 \times 10^{-31} \text{ kg})(35.0 \text{ m/s})$$
$$= 1.59 \times 10^{-30} \text{ kg} \cdot \text{m/s}$$

2. Use the uncertainty principle to find the minimum uncertainty in position:

$$\Delta y = \frac{h}{2\pi \, \Delta p_y}$$

$$= \frac{6.63 \times 10^{-34} \text{ J} \cdot \text{s}}{2\pi(1.59 \times 10^{-30} \text{ kg} \cdot \text{m/s})} = 6.64 \times 10^{-5} \text{ m}$$

Part (b)

3. Calculate the uncertainty in the volleyball's momentum:

$$\Delta p_y = 0.0500 m_{\text{volleyball}} v$$
$$= 0.0500(0.350 \text{ kg})(35.0 \text{ m/s}) = 0.613 \text{ kg} \cdot \text{m/s}$$

4. As before, use the uncertainty principle to find the minimum uncertainty in position:

$$\Delta y = \frac{h}{2\pi \, \Delta p_y}$$

$$= \frac{6.63 \times 10^{-34} \text{ J} \cdot \text{s}}{2\pi(0.613 \text{ kg} \cdot \text{m/s})} = 1.72 \times 10^{-34} \text{ m}$$

INSIGHT

To put these results in perspective, the minimum uncertainty in the position of the electron is roughly 100,000 times the size of an atom. Clearly, this is a significant uncertainty on the atomic level. On the other hand, the minimum uncertainty of the volleyball is smaller than an atom by a factor of about 10^{24}. It follows, then, that the uncertainty principle does not have measurable consequences on typical macroscopic objects.

PRACTICE PROBLEM

Repeat this problem, assuming the uncertainty in speed is reduced by a factor of 2 to 2.50%.   [**Answer:** The uncertainties in position increase by a factor of 2. Electron, $\Delta y = 1.33 \times 10^{-4}$ m; volleyball, $\Delta y = 3.44 \times 10^{-34}$ m.]

Some related homework problems: Problem 70, Problem 72

The Heisenberg uncertainty principle also sets the typical energy scales in atomic and nuclear systems. For example, if an electron is known to be confined to an atom, the uncertainty in its position Δy, will be roughly 1 Å. This implies a finite uncertainty for Δp_y, which in turn implies a finite kinetic energy—even though the average value of p_y is zero.

EXAMPLE 30-7 AN ELECTRON IN A BOX

Suppose an electron is confined to a box that is about 0.50 Å on a side. If this distance is taken as the uncertainty in position of the electron, **(a)** calculate the corresponding minimum uncertainty in momentum. **(b)** Because the electron is confined to a stationary box, its average momentum is zero. The magnitude of the electron's momentum is nonzero, however. Assuming the magnitude of the electron's momentum is the same as its uncertainty in momentum, calculate the corresponding kinetic energy.

PICTURE THE PROBLEM

In our sketch we show an electron bouncing around inside a box that is 0.50 Å on a side. The rapid motion of the electron is a quantum effect, due to the uncertainty principle.

STRATEGY

a. Letting $\Delta y = 0.50$ Å, we can find the minimum uncertainty in momentum by using the Heisenberg uncertainty principle: $\Delta p_y = h/(2\pi \, \Delta y)$.

b. We can calculate the kinetic energy of the electron from its momentum by using $K = p^2/2m$. For the momentum, p, we will use the value of Δp_y obtained in part (a).

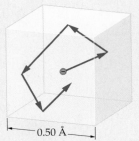

$\vdash$ —————0.50 Å—————$\dashv$

SOLUTION

Part (a)

1. Use $\Delta y = 0.50$ Å and the uncertainty principle to find Δp_y:

$$\Delta p_y = \frac{h}{2\pi \, \Delta y}$$

$$= \frac{6.63 \times 10^{-34} \, \text{J} \cdot \text{s}}{2\pi(0.50 \times 10^{-10} \, \text{m})} = 2.1 \times 10^{-24} \, \text{kg} \cdot \text{m/s}$$

Part (b)

2. Set $p = \Delta p_y$, and use $K = p^2/2m$ to find the kinetic energy:

$$K = \frac{p^2}{2m} = \frac{(2.1 \times 10^{-24} \, \text{kg} \cdot \text{m/s})^2}{2(9.11 \times 10^{-31} \, \text{kg})} = 2.4 \times 10^{-18} \, \text{J} = 15 \, \text{eV}$$

INSIGHT

The important point of this rough estimate of an electron's kinetic energy is that it is of the order of typical atomic energies. Thus, the fact that an electron is confined to an object the size of an atom means that its energy must be on the order of 10 eV. This is why the electron in our sketch is bouncing around inside the box rather than resting on its floor—as would be the case for a tennis ball inside a cardboard box. The uncertainty principle implies that the electron must have about 10 eV of kinetic energy when it is confined in this way; hence, it must be moving quite rapidly. We consider the case of a tennis ball in a cardboard box in the following Practice Problem.

PRACTICE PROBLEM

A tennis ball ($m = 0.06$ kg) is confined within a cardboard box 0.5 m on a side. Estimate the kinetic energy of this ball and its corresponding speed.   [**Answer:** $\Delta K \sim 4 \times 10^{-67}$ J, $\Delta v \sim 4 \times 10^{-33}$ m/s. For all practical purposes, the tennis ball will appear to be at rest in the box.]

Some related homework problems: Problem 71, Problem 77, Problem 78

Finally, the uncertainty relation $\Delta y \Delta p_y \geq h/2\pi$ is just one of many forms the uncertainty principle takes. There are a number of quantities like y and p_y that satisfy the same type of relation between their uncertainties. Perhaps the most important of these uncertainty principles is the one relating the uncertainty in energy to the uncertainty in time:

The Heisenberg Uncertainty Principle: Energy and Time

$$\Delta E \Delta t \geq \frac{h}{2\pi} \qquad\qquad 30\text{–}20$$

For example, the shorter the half-life of an unstable particle, the greater the uncertainty in its energy.

CONCEPTUAL CHECKPOINT 30–4   UNCERTAINTIES

If Planck's constant were magically reduced to zero, would the uncertainties in position/momentum and energy/time be **(a)** increased or **(b)** decreased?

REASONING AND DISCUSSION

If h were zero, the product of uncertainties—$\Delta y \Delta p_y$ and $\Delta E \Delta t$—would be reduced to zero as well. The result is that position and momentum, for example, could be determined simultaneously with arbitrary accuracy; that is, with zero uncertainty. In this limit, particles would behave as predicted in classical physics, with well-defined positions and momenta at all times.

ANSWER

(b) The uncertainties would decrease to zero.

Thus, if h were zero, the classical description of particles moving along well-defined trajectories, and having no wavelike properties, would be valid. Notice, however, that Planck's constant has an extremely small magnitude, on the order of 10^{-34} in SI units. It is this small difference from zero that is responsible for the quantum behavior seen on the atomic scale.

If h were relatively large, however, the wavelike properties of matter would be apparent even on the macroscopic level. For example, consider pitching a baseball toward the catcher's glove in a universe where h is rather large. If the ball is thrown with a well-defined momentum toward the glove, its position will be uncertain, and the ball may end up anywhere. Similarly, if the catcher gives the glove a relatively small uncertainty in position, its uncertainty in momentum is large, meaning that the glove is moving rapidly about its average position. So, even if the pitcher could aim the ball to go where desired—which is not possible—there is no way to know where to aim it. If h were significantly larger than it actually is, our experience of the natural world would be very different indeed.

30–7 Quantum Tunneling

Because particles have wavelike properties, any behavior seen in waves can be found in particles as well, under the right conditions. An example is the phenomenon known as **tunneling,** in which a wave, or a quantum particle, "tunnels" through a region of space that would be forbidden to it if it were simply a small piece of matter like a classical particle.

Figure 30–17 shows a case of tunneling by light. In Figure 30–17 (a) we see a beam of light propagating within glass and undergoing total internal reflection (Chapter 26) when it encounters the glass–air interface. If light were composed of classical particles this would be the end of the story; the particles of light would simply change direction at the interface and continue propagating within the glass.

What can actually happen in such a system is more interesting, however. In Figure 30–17 (b) we show a second piece of glass brought near the first piece, but with a small vacuum gap between them. If the gap is small, but still finite, a weak beam of light is observed in the second piece of glass, propagating in the same direction as the original beam. We say that the light waves have "tunneled" across the gap. The strength of the beam of light in the second piece of glass depends very sensitively on the size of the gap through which it tunnels, decreasing exponentially as the width of the gap is increased.

Just as with light, electrons and other particles can tunnel across gaps that would be forbidden to them if they were classical particles. An electron, for instance, is not simply a point of mass; instead, it has wave properties that extend outward from it like ripples on a pond. These waves, like those of light, can "feel" the surroundings of an electron, allowing it to "tunnel" through a barrier to a region on the other side where it can propagate.

An example of electron tunneling is shown in **Figure 30–18,** where we illustrate the operation of a scanning tunneling microscope, or STM. In the lower part of the figure we see the material, or specimen, to be investigated with the microscope. The upper portion of the figure shows the key element of the microscope—a small, pointed tip of metal that can be moved up and down with piezoelectric supports. This tip is brought very close to the specimen being observed, leaving a small vacuum gap between them. Classically, electrons in the specimen are not able to move across the gap to the tip, but in reality they can tunnel to the tip and create a small electric current. The number of electrons that tunnel, and thus the magnitude of the tunneling current, depends on the width of the gap.

In one version of the STM, the tunneling current between the specimen and the tip is held constant. Imagine, for example, that the tip in Figure 30–18 is scanned horizontally from left to right. Since the tunneling current is very sensitive to the size of the gap between the tip and the specimen, the tip must be moved up and down with the contours of the specimen in order to maintain a gap of constant width. The tip is moved by sending electrical voltage to the piezoelectric

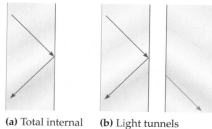

(a) Total internal reflection   **(b)** Light tunnels across the gap

▲ **FIGURE 30–17 Optical tunneling**

(a) An incident beam of light undergoes total internal reflection from the glass–air interface. **(b)** If a second piece of glass is brought near the first piece, a weak beam of light is observed continuing in the same direction as the incident beam. We say that the light has "tunneled" across the gap.

REAL-WORLD PHYSICS

Scanning tunneling microscopy

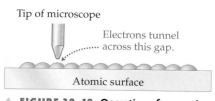

Tip of microscope

Electrons tunnel across this gap.

Atomic surface

▲ **FIGURE 30–18 Operation of a scanning tunneling microscope**

Schematic operation of a scanning tunneling microscope.

 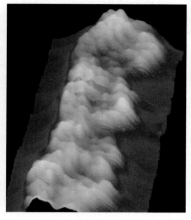

◀ **FIGURE 30–19 Scanning tunneling microscopy**
STM images are particularly good at recording the "terrain" of surfaces at the atomic level. The image at left shows clusters of antimony atoms on a silicon surface. At right, a DNA molecule sits on a substrate of graphite. Three turns of the double helix are visible, magnified about 2,000,000 times.

supports; thus, the voltage going to the supports is a measure of the height of the surface being scanned and can be converted to a visual image of the surface, as in **Figure 30–19**. The resolution of these microscopes is on the atomic level, showing the hills and valleys created by atoms on the surface of the material being examined.

THE BIG PICTURE   PUTTING PHYSICS IN CONTEXT

LOOKING BACK

The expression for total relativistic energy (Chapter 29) is used in Section 30–3 to find the momentum of a photon—even though a photon has zero rest mass. We then study collisions involving photons in Section 30–4 in the same way we studied collisions involving massive objects in Chapter 9.

Diffraction and two-slit interference (Chapter 28) play a key role in understanding neutron diffraction and electron interference in Section 30–5. In addition, diffraction forms the basis for our understanding of the uncertainty principle in Section 30–6.

LOOKING AHEAD

De Broglie waves, introduced in Section 30–5, appear again in Chapter 31 when we study the Bohr model of the atom. As we shall see, the allowed Bohr orbits correspond to standing waves formed from the de Broglie waves of electrons.

Photons are fascinating in that they have zero mass and yet have finite momentum and energy (Sections 30–3, 30–4 and 30–5). They also travel at the speed of light, which is not possible for any finite-mass object. We use the concept of photons again in Section 31–4 when we study atomic radiation.

CHAPTER SUMMARY

30–1 BLACKBODY RADIATION AND PLANCK'S HYPOTHESIS OF QUANTIZED ENERGY

An ideal blackbody is an object that absorbs all the light incident on it. The distribution of energy as a function of frequency within a blackbody is independent of the material from which the blackbody is made and depends only on the temperature, T.

Wien's Displacement Law

The frequency at which the radiation from a blackbody is maximum is given by the following relation:

$$f_{\text{peak}} = (5.88 \times 10^{10} \text{ s}^{-1} \cdot \text{K}^{-1})T \qquad \text{30–1}$$

Planck's Quantum Hypothesis

Planck hypothesized that the energy in a blackbody at a frequency f must be an integer multiple of the constant $h = 6.63 \times 10^{-34}$ J·s; that is,

$$E_n = nhf \quad n = 0, 1, 2, 3, \dots \qquad \text{30–2}$$

The constant h is known as Planck's constant.

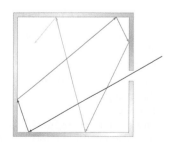

30–2 PHOTONS AND THE PHOTOELECTRIC EFFECT

Light is composed of particle-like photons, which carry energy in discrete amounts.

Energy of a Photon

The energy of a photon depends on its frequency. A photon with the frequency f has the energy

$$E = hf \qquad \text{30–4}$$

Noting the relation $\lambda f = c$, we can also express the energy of a photon in terms of its wavelength, $E = hc/\lambda$.

The Photoelectric Effect

The photoelectric effect occurs when photons of light eject electrons from the surface of a metal.

Work Function

The minimum energy required to eject an electron from a particular metal is the work function, W_0.

Cutoff Frequency

To eject an electron, a photon must have an energy at least as great as W_0, and thus the minimum, or cutoff, frequency to eject an electron is

$$f_0 = \frac{W_0}{h} \qquad \text{30–6}$$

If the frequency of the photon is greater than f_0, the ejected electron has a finite kinetic energy.

Maximum Kinetic Energy

The maximum kinetic energy of an electron ejected from a metal by a photon of frequency $f > f_0$ is

$$K_{\max} = hf - W_0 \qquad \text{30–7}$$

30–3 THE MASS AND MOMENTUM OF A PHOTON

A photon is like a "typical" particle in some respects but different in others. In particular, a photon has zero rest mass, yet it still has a nonzero momentum.

Rest Mass of a Photon

Photons, which travel through a vacuum at the speed of light, c, have zero rest mass

$$m_0 = 0 \qquad \text{30–9}$$

Only objects with zero rest mass can propagate at the speed of light.

Momentum of a Photon

The momentum, p, of a photon of frequency f and wavelength $\lambda = c/f$ is given by

$$p = \frac{hf}{c} = \frac{h}{\lambda} \qquad \text{30–11}$$

30–4 PHOTON SCATTERING AND THE COMPTON EFFECT

Photons can undergo collisions with particles, in much the same way that particles can collide with other particles. In order to conserve the total energy and momentum of a system during such a collision, the frequency and wavelength of a photon will change.

The Compton Effect

If a photon of wavelength λ undergoes a collision with an electron (mass $= m_e$) and scatters into a new direction at an angle θ from its incident direction, its new wavelength, λ', is given by the Compton shift formula:

$$\Delta\lambda = \lambda' - \lambda = \frac{h}{m_e c}(1 - \cos\theta) \qquad \text{30–15}$$

The Compton shift formula can be applied to scattering from particles other than the electron by changing the mass in Equation 30–15.

30–5 THE DE BROGLIE HYPOTHESIS AND WAVE–PARTICLE DUALITY

The de Broglie hypothesis is basically the following: Since light displays particle-like behavior, perhaps particles display wavelike behavior. In particular, de Broglie hypothesized that particles have wavelengths.

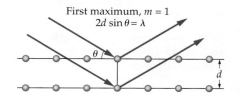

First maximum, $m = 1$
$2d \sin \theta = \lambda$

de Broglie Wavelength
According to de Broglie, the relationship between momentum and wavelength should be the same for both light and particles. Thus, the de Broglie wavelength of a particle of momentum p is

$$\lambda = \frac{h}{p} \qquad \text{30–16}$$

Diffraction of X-Rays and Particles by Crystals
X-rays and particles can show interference effects when reflected from the atomic layers of a crystal. If the wavelength of the X-rays or particles is λ, and the spacing between atomic layers is d, the angles at which constructive interference occurs are given by

$$2d \sin \theta = m\lambda \quad m = 1, 2, 3, \ldots \qquad \text{30–17}$$

The resulting patterns produced by the constructive interference are referred to as a diffraction pattern.

Wave–Particle Duality
Light and matter display both wavelike and particle-like properties.

30–6 THE HEISENBERG UNCERTAINTY PRINCIPLE

Because particles have wavelengths and can behave as waves, their position and momentum cannot be determined simultaneously with arbitrary precision.

The Heisenberg Uncertainty Principle: Momentum and Position
In terms of momentum and position, the Heisenberg uncertainty principle states the following:

$$\Delta p_y \Delta y \geq \frac{h}{2\pi} \qquad \text{30–19}$$

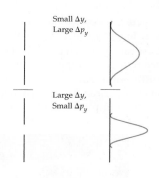

Small Δy,
Large Δp_y

Large Δy,
Small Δp_y

Thus, as momentum is determined more precisely, the position becomes more uncertain, and vice versa.

The Heisenberg Uncertainty Principle: Energy and Time
In terms of energy and time, the Heisenberg uncertainty principle states the following:

$$\Delta E \Delta t \geq \frac{h}{2\pi} \qquad \text{30–20}$$

30–7 QUANTUM TUNNELING

Particles, because of their wavelike behavior, can pass through regions of space that would be forbidden to a classical particle. This phenomenon is referred to as tunneling.

PROBLEM-SOLVING SUMMARY

Type of Problem	Relevant Physical Concepts	Related Examples
Relate the peak frequency of black-body radiation to the temperature of the blackbody.	A blackbody at the absolute temperature T has a peak in its radiation spectrum at the frequency $$f_{\text{peak}} = (5.88 \times 10^{10}\ \text{s}^{-1} \cdot \text{K}^{-1})T$$ This is true regardless of the blackbody's composition.	Exercise 30–1
Find the quantum number of a macroscopic system.	If a macroscopic system has the energy E and oscillates with frequency f, the corresponding quantum number is $n = E/hf$.	Example 30–1
Relate the energy and frequency of a photon.	A photon of frequency f has an energy hf, where h is Planck's constant.	Example 30–2 Active Example 30–1

Find the maximum kinetic energy of electrons given off in the photoelectric effect.	When light of frequency f illuminates a metal surface with a work function W_0, the maximum kinetic energy of ejected electrons is $K_{max} = hf - W_0$.	Example 30–3
Relate the change in wavelength of an X-ray to the angle through which it scatters from an electron.	The change in wavelength is related to the scattering angle θ by the following relation: $$\Delta\lambda = \lambda' - \lambda = \frac{h}{m_e c}(1 - \cos\theta)$$	Example 30–4
Determine the de Broglie wavelength of a particle.	The de Broglie wavelength of a particle with momentum p is $\lambda = h/p$.	Example 30–5 Active Example 30–2
Relate the uncertainty in position to the uncertainty in momentum.	The Heisenberg uncertainty principle states that the uncertainty in momentum, Δp_y, and the uncertainty in position, Δy, are related as follows: $\Delta p_y \Delta y \geq h/2\pi$. Similarly, the uncertainties in time and energy obey $\Delta E \, \Delta t \geq h/2\pi$.	Examples 30–6, 30–7

CONCEPTUAL QUESTIONS

(Answers to odd-numbered Conceptual Questions can be found in the back of the book.)

1. Give a brief description of the "ultraviolet catastrophe."
2. How does Planck's hypothesis of energy quantization resolve the "ultraviolet catastrophe"?
3. Is there a lowest temperature below which blackbody radiation is no longer given off by an object? Explain.
4. How can an understanding of blackbody radiation allow us to determine the temperature of distant stars?
5. **Differential Fading** Many vehicles in the United States have a small American flag decal in one of their windows. If the decal has been in place for a long time, the colors will show some

Differential fading. (Conceptual Question 5)

fading from exposure to the Sun. In fact, the red stripes are generally more faded than the blue background for the stars, as shown in the accompanying photo. Photographs and posters react in the same way, with red colors showing the most fading. Explain this effect in terms of the photon model of light.

6. A source of light is monochromatic. What can you say about the photons emitted by this source?
7. The relative intensity of radiation given off by a blackbody is shown in Figure 30–2. Notice that curves corresponding to different temperatures never cross one another. If two such curves did intersect, however, it would be possible to violate the second law of thermodynamics. Explain.
8. **(a)** Is it possible for a photon from a green source of light to have more energy than a photon from a blue source of light? Explain. **(b)** Is it possible for a photon from a green source of light to have more energy than a photon from a red source of light? Explain.
9. Light of a given wavelength ejects electrons from the surface of one metal but not from the surface of another metal. Give a possible explanation for this observation.
10. Why does the existence of a cutoff frequency in the photoelectric effect argue in favor of the photon model of light?
11. Why can an electron microscope resolve smaller objects than a light microscope?
12. A proton is about 2000 times more massive than an electron. Is it possible for an electron to have the same de Broglie wavelength as a proton? Explain.

PROBLEMS AND CONCEPTUAL EXERCISES

Note: Answers to odd-numbered Problems and Conceptual Exercises can be found in the back of the book. **IP** *denotes an integrated problem, with both conceptual and numerical parts;* **BIO** *identifies problems of biological or medical interest;* **CE** *indicates a conceptual exercise.* **Predict/Explain** *problems ask for two responses:* **(a)** *your prediction of a physical outcome, and* **(b)** *the best explanation among three provided. On all problems, red bullets (•, ••, •••) are used to indicate the level of difficulty.*

SECTION 30–1 BLACKBODY RADIATION AND PLANCK'S HYPOTHESIS OF QUANTIZED ENERGY

1. • **CE Predict/Explain** The blackbody spectrum of blackbody A peaks at a longer wavelength than that of blackbody B. **(a)** Is the temperature of blackbody A higher than or lower than the temperature of blackbody B? **(b)** Choose the *best explanation* from among the following:

   I. Blackbody A has the higher temperature because the higher the temperature the longer the wavelength.
   II. Blackbody B has the higher temperature because an increase in temperature means an increase in frequency, which corresponds to a decrease in wavelength.

2. • **The Surface Temperature of Betelgeuse** Betelgeuse, a red-giant star in the constellation Orion, has a peak in its radiation

at a frequency of 1.82×10^{14} Hz. What is the surface temperature of Betelgeuse?

3. • What is the frequency of the most intense radiation emitted by your body? Assume a skin temperature of 95 °F. What is the wavelength of this radiation?

4. • **The Cosmic Background Radiation** Outer space is filled with a sea of photons, created in the early moments of the universe. The frequency distribution of this "cosmic background radiation" matches that of a blackbody at a temperature near 2.7 K. **(a)** What is the peak frequency of this radiation? **(b)** What is the wavelength that corresponds to the peak frequency?

5. • The Sun has a surface temperature of about 5800 K. At what frequency does the Sun emit the most radiation?

6. •• **(a)** By what factor does the peak frequency change if the Kelvin temperature of an object is doubled from 20.0 K to 40.0 K? **(b)** By what factor does the peak frequency change if the Celsius temperature of an object is doubled from 20.0 °C to 40.0 °C?

7. •• **IP A Famous Double Star** Albireo in the constellation Cygnus, which appears as a single star to the naked eye, is actually a beautiful double-star system. The brighter of the two stars is referred to as A (or Beta-01 Cygni), with a surface temperature of $T_A = 4700$ K; its companion is B (or Beta-02 Cygni), with a surface temperature of $T_B = 13,000$ K. **(a)** When viewed through a telescope, one star is a brilliant blue color, and the other has a warm golden color, as shown in the accompanying photo. Is the blue star A or B? Explain. **(b)** What is the ratio of the peak frequencies emitted by the two stars, (f_A/f_B)?

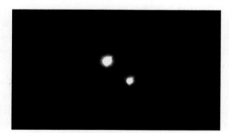

The double star Albireo in the constellation Cygnus. (Problem 7)

8. •• **IP Halogen Lightbulbs** Modern halogen lightbulbs allow their filaments to operate at a higher temperature than the filaments in standard incandescent bulbs. For comparison, the filament in a standard lightbulb operates at about 2900 K, whereas the filament in a halogen bulb may operate at 3400 K. **(a)** Which bulb has the higher peak frequency? **(b)** Calculate the ratio of peak frequencies (f_{hal}/f_{std}). **(c)** The human eye is most sensitive to a frequency around 5.5×10^{14} Hz. Which bulb produces a peak frequency closer to this value?

9. •• **IP** A typical lightbulb contains a tungsten filament that reaches a temperature of about 2850 K, roughly half the surface temperature of the Sun. **(a)** Treating the filament as a blackbody, determine the frequency for which its radiation is a maximum. **(b)** Do you expect the lightbulb to radiate more energy in the visible or in the infrared part of the spectrum? Explain.

10. •• **Exciting an Oxygen Molecule** An oxygen molecule (O_2) vibrates with an energy identical to that of a single particle of mass $m = 1.340 \times 10^{-26}$ kg attached to a spring with a force constant of $k = 1215$ N/m. The energy levels of the system are uniformly spaced, as indicated in **Figure 30–20**, with a separation given by hf. **(a)** What is the vibration frequency of this mol-

ecule? **(b)** How much energy must be added to the molecule to excite it from one energy level to the next higher level?

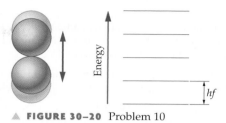

▲ **FIGURE 30–20** Problem 10

SECTION 30–2 PHOTONS AND THE PHOTOELECTRIC EFFECT

11. • **CE** A source of red light, a source of green light, and a source of blue light each produce beams of light with the same power. Rank these sources in order of increasing **(a)** wavelength of light, **(b)** frequency of light, and **(c)** number of photons emitted per second. Indicate ties where appropriate.

12. • **CE Predict/Explain** A source of red light has a higher wattage than a source of green light. **(a)** Is the energy of photons emitted by the red source greater than, less than, or equal to the energy of photons emitted by the green source? **(b)** Choose the *best explanation* from among the following:
   I. The photons emitted by the red source have the greater energy because that source has the greater wattage.
   II. The red-source photons have less energy than the green-source photons because they have a lower frequency. The wattage of the source doesn't matter.
   III. Photons from the red source have a lower frequency, but that source also has a greater wattage. The two effects cancel, so the photons have equal energy.

13. • **CE Predict/Explain** A source of yellow light has a higher wattage than a source of blue light. **(a)** Is the number of photons emitted per second by the yellow source greater than, less than, or equal to the number of photons emitted per second by the blue source? **(b)** Choose the *best explanation* from among the following:
   I. The yellow source emits more photons per second because (i) it emits more energy per second than the blue source, and (ii) its photons have less energy than those of the blue source.
   II. The yellow source has the higher wattage, which means its photons have higher energy than the blue-source photons. Therefore, the yellow source emits fewer photons per second.
   III. The two sources emit the same number of photons per second because the higher wattage of the yellow source compensates for the higher energy of the blue photons.

14. • **CE Predict/Explain** Light of a particular wavelength does not eject electrons from the surface of a given metal. **(a)** Should the wavelength of the light be increased or decreased in order to cause electrons to be ejected? **(b)** Choose the *best explanation* from among the following:
   I. The photons have too little energy to eject electrons. To increase their energy, their wavelength should be increased.
   II. The energy of a photon is proportional to its frequency; that is, inversely proportional to its wavelength. To increase the energy of the photons so they can eject electrons, one must decrease their wavelength.

15. • **CE** Light of a particular wavelength and intensity does not eject electrons from the surface of a given metal. Can electrons be ejected from the metal by increasing the intensity of the light? Explain.

16. • When a person visits the local tanning salon, they absorb photons of ultraviolet (UV) light to get the desired tan. What are the frequency and wavelength of a UV photon whose energy is 6.5×10^{-19} J?

17. • An AM radio station operating at a frequency of 880 kHz radiates 270 kW of power from its antenna. How many photons are emitted by the antenna every second?

18. • A photon with a wavelength of less than 50.4 nm can ionize a helium atom. What is the ionization potential of helium?

19. • A flashlight emits 2.5 W of light energy. Assuming a frequency of 5.2×10^{14} Hz for the light, determine the number of photons given off by the flashlight per second.

20. • Light of frequency 9.95×10^{14} Hz ejects electrons from the surface of silver. If the maximum kinetic energy of the ejected electrons is 0.180×10^{-19} J, what is the work function of silver?

21. • The work function of gold is 4.58 eV. What frequency of light must be used to eject electrons from a gold surface with a maximum kinetic energy of 6.48×10^{-19} J?

22. • (a) How many 350-nm (UV) photons are needed to provide a total energy of 2.5 J? (b) How many 750-nm (red) photons are needed to provide the same energy?

23. •• (a) How many photons per second are emitted by a monochromatic lightbulb ($\lambda = 650$ nm) that emits 45 W of power? (b) If you stand 15 m from this bulb, how many photons enter each of your eyes per second? Assume your pupil is 5.0 mm in diameter and that the bulb radiates uniformly in all directions.

24. •• IP Two 57.5-kW radio stations broadcast at different frequencies. Station A broadcasts at a frequency of 892 kHz, and station B broadcasts at a frequency of 1410 kHz. (a) Which station emits more photons per second? Explain. (b) Which station emits photons of higher energy?

25. •• **Dissociating the Hydrogen Molecule** The energy required to separate a hydrogen molecule into its individual atoms is 104.2 kcal per mole of H_2. (a) If the dissociation energy for a single H_2 molecule is provided by one photon, determine its frequency and wavelength. (b) In what region of the electromagnetic spectrum does the photon found in part (a) lie? (Refer to the spectrum shown in Figure 25–8.)

26. •• (a) How many photons are emitted per second by a He-Ne laser that emits 1.0 mW of power at a wavelength $\lambda = 632.8$ nm? (b) What is the frequency of the electromagnetic waves emitted by a He-Ne laser?

27. •• IP You have two lightbulbs of different power and color, as indicated in **Figure 30–21**. One is a 150-W red bulb, and the other is a 25-W blue bulb. (a) Which bulb emits more photons per second? (b) Which bulb emits photons of higher energy? (c) Calculate the number of photons emitted per second by each bulb. Take $\lambda_{red} = 650$ nm and $\lambda_{blue} = 460$ nm. (Most of the electromagnetic radiation given off by incandescent lightbulbs is in the infrared portion of the spectrum. For the purposes of this problem, however, assume that all of the radiated power is at the wavelengths indicated.)

28. •• The maximum wavelength an electromagnetic wave can have and still eject an electron from a copper surface is 264 nm. What is the work function of a copper surface?

29. •• IP Aluminum and calcium have photoelectric work functions of $W_{Al} = 4.28$ eV and $W_{Ca} = 2.87$ eV, respectively. (a) Which metal requires higher-frequency light to produce photoelectrons? Explain. (b) Calculate the minimum frequency that will produce photoelectrons from each surface.

30. •• IP Two beams of light with different wavelengths ($\lambda_A > \lambda_B$) are used to produce photoelectrons from a given metal surface. (a) Which beam produces photoelectrons with greater kinetic energy? Explain. (b) Find K_{max} for cesium ($W_0 = 1.9$ eV) if $\lambda_A = 620$ nm and $\lambda_B = 410$ nm.

31. •• IP Zinc and cadmium have photoelectric work functions given by $W_{Zn} = 4.33$ eV and $W_{Cd} = 4.22$ eV, respectively. (a) If both metals are illuminated by UV radiation of the same wavelength, which one gives off photoelectrons with the greater maximum kinetic energy? Explain. (b) Calculate the maximum kinetic energy of photoelectrons from each surface if $\lambda = 275$ nm.

32. •• White light, with frequencies ranging from 4.00×10^{14} Hz to 7.90×10^{14} Hz, is incident on a potassium surface. Given that the work function of potassium is 2.24 eV, find (a) the maximum kinetic energy of electrons ejected from this surface and (b) the range of frequencies for which no electrons are ejected.

33. •• Electromagnetic waves, with frequencies ranging from 4.00×10^{14} Hz to 9.00×10^{16} Hz, are incident on an aluminum surface. Given that the work function of aluminum is 4.28 eV, find (a) the maximum kinetic energy of electrons ejected from this surface and (b) the range of frequencies for which no electrons are ejected.

34. •• IP Platinum has a work function of 6.35 eV, and iron has a work function of 4.50 eV. Light of frequency 1.88×10^{15} Hz ejects electrons from both of these surfaces. (a) From which surface will the ejected electrons have a greater maximum kinetic energy? Explain. (b) Calculate the maximum kinetic energy of ejected electrons for each surface.

35. •• When light with a frequency $f_1 = 547.5$ THz illuminates a metal surface, the most energetic photoelectrons have 1.260×10^{-19} J of kinetic energy. When light with a frequency $f_2 = 738.8$ THz is used instead, the most energetic photoelectrons have 2.480×10^{-19} J of kinetic energy. Using these experimental results, determine the approximate value of Planck's constant.

36. •• BIO **Owl Vision** Owls have large, sensitive eyes for good night vision. Typically, the pupil of an owl's eye can have a diameter of 8.5 mm (as compared with a maximum diameter of about 7.0 mm for humans). In addition, an owl's eye is about 100

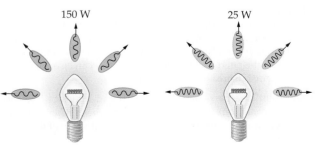

▲ **FIGURE 30–21** Problem 27

An apt pupil. (Problem 36)

times more sensitive to light of low intensity than a human eye, allowing owls to detect light with an intensity as small as 5.0×10^{-13} W/m². Find the minimum number of photons per second an owl can detect, assuming a frequency of 7.0×10^{14} Hz for the light.

SECTION 30–3   THE MASS AND MOMENTUM OF A PHOTON

37. • **CE** If the momentum of a particle with finite mass is doubled, its kinetic energy increases by a factor of 4. If the momentum of a photon is doubled, by what factor does its energy increase?

38. • The photons used in microwave ovens have a momentum of 5.1×10^{-33} kg·m/s. **(a)** What is their wavelength? **(b)** How does the wavelength of the microwaves compare with the size of the holes in the metal screen on the door of the oven?

39. • What speed must an electron have if its momentum is to be the same as that of an X-ray photon with a wavelength of 0.25 nm?

40. • What is the wavelength of a photon that has the same momentum as an electron moving with a speed of 1200 m/s?

41. • What is the frequency of a photon that has the same momentum as a neutron moving with a speed of 1500 m/s?

42. •• A hydrogen atom, initially at rest, emits an ultraviolet photon with a wavelength of $\lambda = 122$ nm. What is the recoil speed of the atom after emitting the photon?

43. •• A blue-green photon ($\lambda = 486$ nm) is absorbed by a free hydrogen atom, initially at rest. What is the recoil speed of the hydrogen atom after absorbing the photon?

44. •• **IP (a)** Which has the greater momentum, a photon of red light or a photon of blue light? Explain. **(b)** Calculate the momentum of a photon of red light ($f = 4.0 \times 10^{14}$ Hz) and a photon of blue light ($f = 7.9 \times 10^{14}$ Hz).

45. •• **IP** Photon A has twice the momentum of photon B. **(a)** Which photon has the greater wavelength? Explain. **(b)** If the wavelength of photon A is 333 nm, what is the wavelength of photon B?

46. ••• A laser produces a 5.00-mW beam of light, consisting of photons with a wavelength of 632.8 nm. **(a)** How many photons are emitted by the laser each second? **(b)** The laser beam strikes a black surface and is absorbed. What is the change in the momentum of each photon that is absorbed? **(c)** What force does the laser beam exert on the black surface?

47. ••• A laser produces a 7.50-mW beam of light, consisting of photons with a wavelength of 632.8 nm. **(a)** How many photons are emitted by the laser each second? **(b)** The laser beam strikes a mirror at normal incidence and is reflected. What is the change in momentum of each reflected photon? Give the magnitude only. **(c)** What force does the laser beam exert on the mirror?

SECTION 30–4   PHOTON SCATTERING AND THE COMPTON EFFECT

48. • **CE** In a Compton scattering experiment, the scattered electron is observed to move in the same direction as the incident X-ray photon. What is the scattering angle of the photon? Explain.

49. • An X-ray photon has 38.0 keV of energy before it scatters from a free electron, and 33.5 keV after it scatters. What is the kinetic energy of the recoiling electron?

50. • In the Compton effect, an X-ray photon scatters from a free electron. Find the *change* in the photon's wavelength if it scatters at an angle of **(a)** $\theta = 30.0°$, **(b)** $\theta = 90.0°$, and **(c)** $\theta = 180.0°$ relative to the incident direction.

51. • An X-ray scattering from a free electron is observed to change its wavelength by 3.13 pm. At what angle to the incident direction does the scattered X-ray move?

52. •• The maximum Compton shift in wavelength occurs when a photon is scattered through 180°. What scattering angle will produce a wavelength shift of one-fourth the maximum?

53. •• **IP** Consider two different photons that scatter through an angle of 180° from a free electron. One is a visible-light photon with $\lambda = 520$ nm, the other is an X-ray photon with $\lambda = 0.030$ nm. **(a)** Which (if either) photon experiences the greater change in wavelength as a result of the scattering? Explain. **(b)** Which photon experiences the greater percentage change in wavelength? Explain. **(c)** Calculate the percentage change in wavelength of each photon.

54. •• An X-ray photon with a wavelength of 0.240 nm scatters from a free electron at rest. The scattered photon moves at an angle of 105° relative to its incident direction. Find **(a)** the initial momentum and **(b)** the final momentum of the photon.

55. •• An X-ray photon scatters from a free electron at rest at an angle of 175° relative to the incident direction. **(a)** If the scattered photon has a wavelength of 0.320 nm, what is the wavelength of the incident photon? **(b)** Determine the energy of the incident and scattered photons. **(c)** Find the kinetic energy of the recoil electron.

56. •• **IP** An X-ray photon scatters through 180° from (i) an electron or (ii) a helium atom. **(a)** In which case is the change in wavelength of the X-ray greater? Explain. **(b)** Calculate the change in wavelength for each of these two cases.

57. ••• A photon has an energy E and wavelength λ before scattering from a free electron. After scattering through a 135° angle, the photon's wavelength has increased by 10.0%. Find the initial wavelength and energy of the photon.

58. ••• Find the direction of propagation of the scattered electron in Problem 51, given that the incident X-ray has a wavelength of 0.525 nm and propagates in the positive x direction.

SECTION 30–5   THE DE BROGLIE HYPOTHESIS AND WAVE–PARTICLE DUALITY

59. • **CE Predict/Explain (a)** As you accelerate your car away from a stoplight, does the de Broglie wavelength of the car increase, decrease, or stay the same? **(b)** Choose the *best explanation* from among the following:
  I. The de Broglie wavelength will increase because the momentum of the car has increased.
  II. The momentum of the car increases. It follows that the de Broglie wavelength will decrease, because it is inversely proportional to the wavelength.
  III. The de Broglie wavelength of the car depends only on its mass, which doesn't change by pulling away from the stoplight. Therefore, the de Broglie wavelength stays the same.

60. • **CE** By what factor does the de Broglie wavelength of a particle change if **(a)** its momentum is doubled or **(b)** its kinetic energy is doubled? Assume the particle is nonrelativistic.

61. • A particle with a mass of 6.69×10^{-27} kg has a de Broglie wavelength of 7.22 pm. What is the particle's speed?

62. • What speed must a neutron have if its de Broglie wavelength is to be equal to the interionic spacing of table salt (0.282 nm)?

63. • A 79-kg jogger runs with a speed of 4.2 m/s. If the jogger is considered to be a particle, what is her de Broglie wavelength?

64. • Find the kinetic energy of an electron whose de Broglie wavelength is 1.5 Å.

65. •• A beam of neutrons with a de Broglie wavelength of 0.250 nm diffracts from a crystal of table salt, which has an interionic spacing of 0.282 nm. **(a)** What is the speed of the neutrons? **(b)** What is the angle of the second interference maximum?

66. •• **IP** An electron and a proton have the same speed. **(a)** Which has the longer de Broglie wavelength? Explain. **(b)** Calculate the ratio (λ_e/λ_p).

67. •• **IP** An electron and a proton have the same de Broglie wavelength. **(a)** Which has the greater kinetic energy? Explain. **(b)** Calculate the ratio of the electron's kinetic energy to the kinetic energy of the proton.

68. •• Diffraction effects become significant when the width of an aperture is comparable to the wavelength of the waves being diffracted. **(a)** At what speed will the de Broglie wavelength of a 65-kg student be equal to the 0.76-m width of a doorway? **(b)** At this speed, how long will it take the student to travel a distance of 1.0 mm? (For comparison, the age of the universe is approximately 4×10^{17} s.)

69. ••• A particle has a mass m and an electric charge q. The particle is accelerated from rest through a potential difference V. What is the particle's de Broglie wavelength, expressed in terms of m, q, and V?

SECTION 30–6 THE HEISENBERG UNCERTAINTY PRINCIPLE

70. • A baseball (0.15 kg) and an electron both have a speed of 41 m/s. Find the uncertainty in position of each of these objects, given that the uncertainty in their speed is 5.0%.

71. • The uncertainty in position of a proton confined to the nucleus of an atom is roughly the diameter of the nucleus. If this diameter is 7.5×10^{-15} m, what is the uncertainty in the proton's momentum?

72. • The position of a 0.26-kg air-track cart is determined to within an uncertainty of 2.2 mm. What speed must the cart acquire as a result of the position measurement?

73. • The measurement of an electron's energy requires a time interval of 1.0×10^{-8} s. What is the smallest possible uncertainty in the electron's energy?

74. • A particle's energy is measured with an uncertainty of 0.0010 eV. What is the smallest possible uncertainty in our knowledge of when the particle had this energy?

75. • An excited state of a particular atom has a mean lifetime of 0.60×10^{-9} s, which we may take as the uncertainty Δt. What is the minimum uncertainty in any measurement of the energy of this state?

76. • The Σ^+ is an unstable particle, with a mean lifetime of 2.5×10^{-10} s. Its lifetime defines the uncertainty Δt for this particle. What is the minimum uncertainty in this particle's energy?

77. •• The uncertainty in an electron's position is 0.15 nm. **(a)** What is the minimum uncertainty Δp in its momentum? **(b)** What is the kinetic energy of an electron whose momentum is equal to this uncertainty $(\Delta p = p)$?

78. •• The uncertainty in a proton's position is 0.15 nm. **(a)** What is the minimum uncertainty Δp in its momentum? **(b)** What is the kinetic energy of a proton whose momentum is equal to this uncertainty $(\Delta p = p)$?

79. •• An electron has a momentum $p \approx 1.7 \times 10^{-25}$ kg·m/s. What is the minimum uncertainty in its position that will keep the relative uncertainty in its momentum $(\Delta p/p)$ below 1.0%?

GENERAL PROBLEMS

80. • **CE** Suppose you perform an experiment on the photoelectric effect using light with a frequency high enough to eject electrons. If the intensity of the light is increased while the frequency is held constant, describe whether the following quantities increase, decrease, or stay the same: **(a)** The maximum kinetic energy of an ejected electron; **(b)** the minimum de Broglie wavelength of an electron; **(c)** the number of electrons ejected per second; **(d)** the electric current in the phototube.

81. • **CE** Suppose you perform an experiment on the photoelectric effect using light with a frequency high enough to eject electrons. If the frequency of the light is increased while the intensity is held constant, describe whether the following quantities increase, decrease, or stay the same: **(a)** The maximum kinetic energy of an ejected electron; **(b)** the minimum de Broglie wavelength of an electron; **(c)** the number of electrons ejected per second; **(d)** the electric current in the phototube.

82. • **CE** An electron that is accelerated from rest through a potential difference V_0 has a de Broglie wavelength λ_0. What potential difference will double the electron's wavelength? (Express your answer in terms of V_0.)

83. • **CE** A beam of particles diffracts from a crystal, producing an interference maximum at the angle θ. **(a)** If the mass of the particles is increased, with everything else remaining the same, does the angle of the interference maximum increase, decrease, or stay the same? Explain. **(b)**. If the energy of the particles is increased, with everything else remaining the same, does the angle of the interference maximum increase, decrease, or stay the same? Explain.

84. • You want to construct a photocell that works with visible light. Three materials are readily available: aluminum ($W_0 = 4.28$ eV), lead ($W_0 = 4.25$ eV), and cesium ($W_0 = 2.14$ eV). Which material(s) would be suitable?

85. • **BIO Human Vision** Studies have shown that some people can detect 545-nm light with as few as 100 photons entering the eye per second. What is the power delivered by such a beam of light?

86. •• A pendulum consisting of a 0.15-kg mass attached to a 0.78-m string undergoes simple harmonic motion. **(a)** What is the frequency of oscillation for this pendulum? **(b)** Assuming the energy of this system satisfies $E_n = nhf$, find the maximum speed of the 0.15-kg mass when the quantum number is 1.0×10^{33}.

87. •• To listen to a radio station, a certain home receiver must pick up a signal of at least 1.0×10^{-10} W. **(a)** If the radio waves have a frequency of 96 MHz, how many photons must the receiver absorb per second to get the station? **(b)** How much force is exerted on the receiving antenna for the case considered in part (a)?

88. •• The latent heat for converting ice at 0 °C to water at 0 °C is 80.0 kcal/kg (Chapter 17). **(a)** How many photons of frequency 6.0×10^{14} Hz must be absorbed by a 1.0-kg block of ice at 0 °C to melt it to water at 0 °C? **(b)** How many molecules of H_2O can one photon convert from ice to water?

89. •• How many 550-nm photons would have to be absorbed to raise the temperature of 1.0 g of water by 1.0 C°?

90. •• A microwave oven can heat 205 mL of water from 20.0 °C to 90.0 °C in 2.00 min. If the wavelength of the microwaves is $\lambda = 12.2$ cm, how many photons were absorbed by the water? (Assume no loss of heat by the water.)

91. •• Light with a frequency of 2.11×10^{15} Hz ejects electrons from the surface of lead, which has a work function of 4.25 eV. What is the minimum de Broglie wavelength of the ejected electrons?

92. •• An electron moving with a speed of 2.7×10^6 m/s has the same momentum as a photon. Find **(a)** the de Broglie wavelength of the electron and **(b)** the wavelength of the photon.

93. •• **BIO The Cold Light of Fireflies** Fireflies are often said to give off "cold light." Given that the peak in a firefly's radiation occurs at about 5.4×10^{14} Hz, determine the temperature of a blackbody that would have the same peak frequency. From your result, would you say that firefly radiation is well approximated by blackbody radiation? Explain.

How cool is that? (Problem 93)

94. •• **IP** When light with a wavelength of 545 nm shines on a metal surface, electrons are ejected with speeds of 3.10×10^5 m/s or less. **(a)** Give a strategy that allows you to use the preceding information to calculate the work function and cutoff frequency for this surface. **(b)** Carry out your strategy and determine the work function and cutoff frequency.

95. •• **IP** A hydrogen atom absorbs a 486.2-nm photon. A short time later, the same atom emits a photon with a wavelength of 97.23 nm. **(a)** Has the net energy of the atom increased or decreased? Explain. **(b)** Calculate the change in energy of the hydrogen atom.

96. •• When a beam of atoms emerges from an oven at the absolute temperature T, the most probable de Broglie wavelength for a given atom is

$$\lambda_{mp} = \frac{h}{\sqrt{5mkT}}$$

In this expression, m is the mass of an atom, and k is Boltzmann's constant (Chapter 17). What is the most probable speed of a hydrogen atom emerging from an oven at 450 K?

97. •• **IP** **(a)** Does the de Broglie wavelength of a particle increase or decrease as its kinetic energy increases? Explain. **(b)** Show that the de Broglie wavelength of an electron in nanometers can be written as $\lambda = (1.23 \text{ nm})/\sqrt{K}$, where K is the kinetic energy of the electron in eV. Use classical expressions for momentum and kinetic energy.

98. ••• A jar is filled with monatomic helium gas at a temperature of 25 °C. The pressure inside the jar is one atmosphere; that is, 101 kPa. **(a)** Find the average de Broglie wavelength of the helium atoms. **(b)** Calculate the average separation between helium atoms in the jar. (*Note*: The fact that the spacing between atoms is much greater than the de Broglie wavelength means quantum effects are negligible, and the atoms can be treated as particles.)

99. ••• **The Compton Wavelength** The *Compton wavelength*, λ_C, of a particle of mass m is defined as follows: $\lambda_C = h/mc$. **(a)** Calculate the Compton wavelength of a proton. **(b)** Calculate the energy of a photon that has the same wavelength as found in part

(a). **(c)** Show, in general, that a photon with a wavelength equal to the Compton wavelength of a particle has an energy that is equal to the rest energy of the particle.

100. ••• **IP** Light of frequency 8.22×10^{14} Hz ejects electrons from surface A with a maximum kinetic energy that is 2.00×10^{-19} J greater than the maximum kinetic energy of electrons ejected from surface B. **(a)** If the frequency of the light is increased, does the difference in maximum kinetic energy observed from the two surfaces increase, decrease, or stay the same? Explain. **(b)** Calculate the difference in work function for these two surfaces.

PASSAGE PROBLEMS

Millikan and the Photoelectric Effect

Robert A. Millikan (1868–1953), best known for his "oil-drop experiment" that measured the charge of an electron, also performed pioneering research on the photoelectric effect. In fact, the 1923 Nobel Prize in physics was awarded to Millikan "for his work on the elementary charge of electricity and on the photoelectric effect." Initially convinced that Einstein's theory of the photoelectric effect was wrong—because of overwhelming evidence for the wave nature of light—Millikan undertook a decade-long experimental program to study the effect. In the end, his experiments confirmed Einstein's theory in every detail and ushered in the modern view of light as having a wave–particle duality.

Millikan carried out an exhaustive set of experiments on a variety of materials. In experiments on lithium, for example, Millikan observed a maximum kinetic energy of 0.550 eV when electrons were ejected with 433.9-nm light. When light of 253.5 nm was used, he observed a maximum kinetic energy of 2.57 eV. Using results like this, Millikan was able to measure the value of Planck's constant, and to show that the value obtained from the photoelectric effect is in complete agreement with the value obtained from blackbody radiation.

101. •• What is the work function, W_0, for lithium, as determined from Millikan's results?

   A. 0.0112 eV   **B.** 0.951 eV

   C. 1.63 eV   **D.** 2.29 eV

102. • What value does Millikan obtain for Planck's constant, based on the lithium measurements? (His value is close to, but not the same as, the currently accepted value.)

   A. 1.12×10^{-34} J·s   **B.** 3.84×10^{-34} J·s

   C. 6.14×10^{-34} J·s   **D.** 6.57×10^{-34} J·s

103. • What maximum kinetic energy do you predict Millikan found when he used light with a wavelength of 365.0 nm?

   A. 0.805 eV   **B.** 1.08 eV

   C. 2.29 eV   **D.** 2.82 eV

INTERACTIVE PROBLEMS

104. •• **IP Referring to Example 30–4** An X-ray photon with $\lambda = 0.6500$ nm scatters from an electron, giving the electron a kinetic energy of 7.750 eV. **(a)** Is the scattering angle of the photon greater than, less than, or equal to 152°? **(b)** Find the scattering angle.

105. •• **IP Referring to Example 30–4** An X-ray photon with $\lambda = 0.6500$ nm scatters from an electron. The wavelength of the scattered photon is 0.6510 nm. **(a)** Is the scattering angle in this case greater than, less than, or equal to 152°? **(b)** Find the scattering angle.

31 Atomic Physics

A number of animals, including many species of coral, sponges, and even scorpions, emit beautiful and sometimes eerie colors when illuminated by ultraviolet light. We do not ordinarily expect rabbits to fall into this category. However, scientists recently inserted a jellyfish gene that directs the production of green fluorescent protein (GFP) into the fertilized egg of a rabbit. The result was Alba, the "GFP bunny"—the world's first fluorescent rabbit. This chapter explores our modern understanding of the structure of the atom, which makes it possible to explain not only the phenomenon of fluorescence but many others as well, including the properties of the chemical elements and the generation of laser light.

In today's world it is taken for granted that we, along with everything else on Earth, are made of atoms. Although it may seem surprising at first, this belief in atoms has not always been universal. As recently as the first part of the twentieth century there was still serious debate about the microscopic nature of matter. With the advent of quantum physics, however, the debate quickly faded away as atomic structure came to be understood in ever greater detail.

In this chapter we begin by developing the quantum model of the simplest of all atoms—the hydrogen atom. We then show that the basic features of hydrogen apply to more complex atoms as well. As a result, we are able to understand—in detail—the arrangement of elements in the periodic table. That quantum physics can describe the structure of an atom, and show why the various elements have their characteristic properties, is one of the greatest successes of modern science.

31–1 Early Models of the Atom

Speculations about the microscopic structure of matter have intrigued humankind for thousands of years. Ancient Greek philosophers, including Leucippus and Democritus, considered the question of what would happen if you took a small object, like a block of copper, and cut it in half, then cut it in half again, and again, for many subsequent divisions. They reasoned that eventually you would reduce the block to a single speck of copper that could not be divided further. This smallest piece of an element was called the **atom** (a + tom), which means, literally, "without division."

It was not until the late nineteenth century, however, that the question of atoms began to yield to direct scientific investigation. We now consider some of the more important early developments in atomic models that helped lead to our current understanding.

The Thomson Model: Plum Pudding

In 1897 the English physicist J. J. Thomson (1856–1940) discovered a "particle" that is smaller in size and thousands of times less massive than even the smallest atom. The **electron,** as this particle was named, was also found to have a negative electric charge—in contrast with atoms, which are electrically neutral. Thomson proposed, therefore, that atoms have an internal structure that includes both electrons and a quantity of positively charged matter. The latter would account for most of the mass of an atom, and would have a charge equal in magnitude to the charge on the electrons.

The picture of an atom that Thomson settled on is one he referred to as the "plum-pudding model." In this model, electrons are embedded in a more or less uniform distribution of positive charge—like raisins spread throughout a pudding. This model is illustrated in **Figure 31–1.** Although the plum-pudding model was in agreement with everything Thomson knew about atoms at the time, new experiments were soon to rule out this model and replace it with one that was more like the solar system than a pudding.

The Rutherford Model: A Miniature Solar System

Inspired by the findings and speculations of Thomson, other physicists began to investigate atomic structure. In 1909, Ernest Rutherford (1871–1937) and his coworkers Hans Geiger (1882–1945) and Ernest Marsden (1889–1970) (at that time a twenty-year-old undergraduate) decided to test Thomson's model by directing a beam of positively charged particles, known as **alpha particles,** at a thin gold foil. Since alpha particles—which were later found to be the nuclei of helium atoms—carry a positive charge, they should be deflected as they pass through the positively charged "pudding" in the gold foil. The deflection should have the following properties: (i) it should be relatively small, since the alpha particles have a substantial mass and the positive charge in the atom is spread out; and (ii) all the alpha particles should be deflected in roughly the same way, since the positive pudding fills virtually all space.

When Geiger and Marsden performed the experiment, their results were not in agreement with these predictions. In fact, most of the alpha particles passed right through the foil as if it were not there—as if the atoms in the foil were mostly empty space. Because the results were rather surprising, Rutherford suggested that the experiment be modified to look not only for alpha particles with small angles of deflection—as originally expected—but for ones with large deflections as well.

This suggestion turned out to be an inspired hunch. Not only were large-angle deflections observed, but some of the alpha particles, in fact, were found to have practically reversed their direction of motion. Rutherford was stunned. In his own words, "It was almost as incredible as if you fired a fifteen-inch shell at a piece of tissue paper and it came back and hit you."

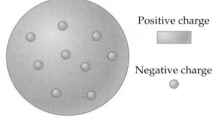

Positive charge

Negative charge

▲ **FIGURE 31–1 The plum-pudding model of an atom**

The model of an atom proposed by J. J. Thomson consists of a uniform positive charge, which accounts for most of the mass of an atom, with small negatively charged electrons scattered throughout, like raisins in a pudding.

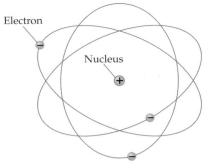

▲ **FIGURE 31–2 The solar system model of an atom**
Ernest Rutherford proposed that an atom is like a miniature solar system, with a massive positively charged nucleus orbited by lightweight negatively charged electrons.

▲ Emission nebulas, like the Lagoon Nebula in Sagittarius shown here, are masses of glowing interstellar gas. The gas is excited by high-energy radiation from nearby stars and emits light at wavelengths characteristic of the atoms present, chiefly hydrogen. Much of the visible light from such nebulas is contributed by the red Balmer line of hydrogen with a wavelength of 656.3 nm, known as H-alpha.

To account for the results of these experiments, Rutherford proposed that an atom has a structure similar to that of the solar system, as illustrated in **Figure 31–2**. In particular, he imagined that the lightweight, negatively charged electrons orbit a small, positively charged **nucleus** containing almost all the atom's mass. In this nuclear model of the atom, most of the atom is indeed empty space, allowing the majority of the alpha particles to pass right through. Furthermore, the positive charge of the atom is now highly concentrated in a small nucleus, rather than spread throughout the atom. This means that an alpha particle that happens to make a head-on collision with the nucleus can actually be turned around, as observed in the experiments.

To see just how small the nucleus must be in his model, Rutherford combined the experimental data with detailed theoretical calculations. His result was that the radius of a nucleus must be smaller than the diameter of the atom by a factor of about 10,000. To put this value into perspective, imagine an atom magnified in size until its nucleus is as large as the Sun. At what distance would an electron orbit in this "atomic" solar system? Using the factor given by Rutherford, we find that the orbit of the electron would have a radius essentially the same as the orbit of Pluto—inside this radius would be only empty space and the nucleus. Thus an atom must have an even larger fraction of empty space than the solar system!

Although Rutherford's nuclear model of the atom seems reasonable, it contains fatal flaws. First, an orbiting electron undergoes a centripetal acceleration toward the nucleus (Chapter 6). As we know from Section 25–1, however, any electric charge that accelerates gives off energy in the form of electromagnetic radiation. Thus, an electron continually radiating energy as it orbits is similar to a satellite losing energy to air resistance when it orbits too close to the Earth's atmosphere. Just as in the case of a satellite, an electron would spiral inward and eventually plunge into the nucleus. Since the entire process of collapse would occur in a fraction of a second (about 10^{-9} s in fact), the atoms in Rutherford's model would simply not be stable—in contrast with the observed stability of atoms in nature.

Even if we ignore the stability problem for a moment, there is another serious discrepancy between Rutherford's model and experiment. Maxwell's equations state that the frequency of radiation from an orbiting electron should be the same as the frequency of its orbit. In the case of an electron spiraling inward the frequency would increase continuously. Thus if we look at light coming from an atom, the Rutherford model indicates that we should see a continuous range of frequencies. This prediction is in striking contrast with experiments, which show that light coming from an atom has only certain discrete frequencies and wavelengths, as we discuss in the next section.

31–2 The Spectrum of Atomic Hydrogen

A red-hot piece of metal glows with a ruddy light that represents only a small fraction of its total radiation output. As we saw in Chapter 30, the metal gives off blackbody radiation that extends in a continuous distribution over all possible frequencies. This blackbody distribution, or spectrum, of radiation is characteristic of the entire collection of atoms that make up the metal—it is not characteristic of the spectrum of light that would be given off by a single, isolated metal atom.

To see the light produced by an isolated atom, we turn our attention from a solid—where the atoms are close together and strongly interacting—to a low-pressure gas—where the atoms are far apart and have little interaction with one another. Consider, then, an experiment in which we seal a low-pressure gas in a tube. If we apply a large voltage between the ends of the tube, the gas will emit electromagnetic radiation characteristic of the individual gas atoms. When this radiation is passed through a diffraction grating (Chapter 28), it is

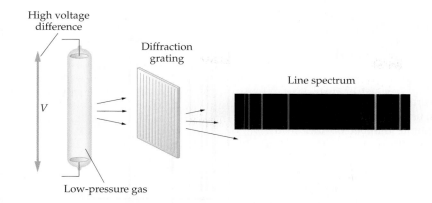

High voltage
difference

Diffraction
grating

Line spectrum

V

Low-pressure gas

◀ **FIGURE 31–3 The line spectrum
of an atom**
The light given off by individual atoms,
as in a low-pressure gas, consists of a series of discrete wavelengths corresponding to different colors.

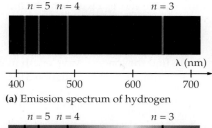

(a) Emission spectrum of hydrogen

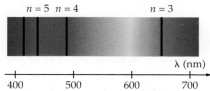

(b) Absorption spectrum of hydrogen

▲ **FIGURE 31–4 The line spectrum
of hydrogen**
The emission **(a)** and absorption **(b)** spectra of hydrogen. Note that the wavelengths absorbed by hydrogen (dark
lines) are the same as those emitted by
hydrogen (colored lines). The location of
these lines is predicted by the Balmer series (Equation 31–1) with the appropriate
values of n.

separated into its various wavelengths, as indicated in **Figure 31–3**. The result
of such an experiment is that a series of bright "lines" is observed, reminiscent of the bar codes used in supermarkets. The precise wavelength associated with each of these lines provides a sort of "fingerprint" identifying a particular type of atom, just as each product in a supermarket has its own unique
bar code.

This type of spectrum, with its bright lines in different colors, is referred to as
a **line spectrum.** As an example, we show the visible part of the line spectrum of
atomic hydrogen in **Figure 31–4 (a)**. Hydrogen produces additional lines in the infrared and ultraviolet parts of the electromagnetic spectrum.

The line spectrum shown in Figure 31–4 (a) is an *emission spectrum*, since it
shows light that is emitted by the hydrogen atoms. Similarly, if light of all colors
is passed through a tube of hydrogen gas, some wavelengths will be absorbed by
the atoms, giving rise to an *absorption spectrum*, which consists of dark lines
(where the atoms absorb the radiation) against an otherwise bright background.
The absorption lines occur at precisely the same wavelengths as the emission
lines. **Figure 31–4 (b)** shows the absorption spectrum of hydrogen.

The first step in developing a quantitative understanding of the hydrogen
spectrum occurred in 1885, when Johann Jakob Balmer (1825–1898), a Swiss
schoolteacher, used trial-and-error methods to discover the following simple formula that gives the wavelength of the visible lines in the spectrum:

$$\frac{1}{\lambda} = R\left(\frac{1}{2^2} - \frac{1}{n^2}\right) \qquad n = 3, 4, 5, \ldots \text{(Balmer series)} \qquad \text{31–1}$$

The constant, R, in this expression is known as the *Rydberg constant*. Its value is

$$R = 1.097 \times 10^7 \text{ m}^{-1}$$

Each integer value of n $(3, 4, 5, \ldots)$ in Balmer's formula corresponds to the wavelength, λ, of a different spectral line. For example, if we set $n = 5$ in Equation 31–1
we find

$$\frac{1}{\lambda} = (1.097 \times 10^7 \text{ m}^{-1})\left(\frac{1}{2^2} - \frac{1}{5^2}\right)$$

Solving for the wavelength, we have

$$\lambda = 4.341 \times 10^{-7} \text{ m} = 434.1 \text{ nm}$$

This is the bluish line, second from the left in Figure 31–4 (a).

The collection of all lines predicted by the Balmer formula is referred to as
the **Balmer series.** We consider the Balmer series in detail in the following
Example.

PROBLEM-SOLVING NOTE

**Calculating Wavelengths for
the Balmer Series**

Note that the formula for the Balmer series
gives the inverse of the wavelength, rather
than the wavelength itself.

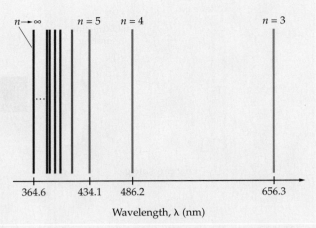

EXAMPLE 31–1 THE BALMER SERIES

Find **(a)** the longest and **(b)** the shortest wavelengths in the Balmer series of spectral lines.

PICTURE THE PROBLEM
In our sketch we indicate the first several lines in the Balmer series, along with their corresponding colors, using the results given in Figure 31–4 as a guide. There are an infinite number of lines in the Balmer series, as indicated by the ellipsis (three dots) to the right of the $n \to \infty$ line.

STRATEGY
By substituting the values $n = 3$, $n = 4$, and $n = 5$ in the Balmer series (Equation 31–1), we find that the wavelength decreases with increasing n. Hence, **(a)** the longest wavelength corresponds to $n = 3$, and **(b)** the shortest wavelength corresponds to $n \to \infty$.

SOLUTION

Part (a)

1. To find the longest wavelength in the Balmer series, substitute $n = 3$ in Equation 31–1:

$$\frac{1}{\lambda} = R\left(\frac{1}{2^2} - \frac{1}{3^2}\right) = (1.097 \times 10^7 \, \text{m}^{-1})\left(\frac{5}{36}\right)$$

2. Invert the result in Step 1 to obtain the corresponding wavelength, λ:

$$\lambda = \frac{36}{5(1.097 \times 10^7 \, \text{m}^{-1})} = 656.3 \, \text{nm}$$

Part (b)

3. The shortest wavelength is found in the limit $n \to \infty$ or, equivalently, $(1/n^2) \to 0$. Make this substitution in Equation 31–1:

$$\frac{1}{\lambda} = R\left(\frac{1}{2^2} - 0\right) = (1.097 \times 10^7 \, \text{m}^{-1})\left(\frac{1}{4}\right)$$

4. Invert the result in Step 3 to obtain the corresponding wavelength, λ:

$$\lambda = \frac{4}{(1.097 \times 10^7 \, \text{m}^{-1})} = 364.6 \, \text{nm}$$

INSIGHT
The longest wavelength corresponds to visible light with a reddish hue, whereas the shortest wavelength is well within the ultraviolet portion of the electromagnetic spectrum—it is invisible to our eyes.

PRACTICE PROBLEM
Which value of n corresponds to a wavelength of 377.1 nm in the Balmer series? **[Answer: $n = 11$]**

Some related homework problems: Problem 5, Problem 6

TABLE 31–1 Common Spectral Series of Hydrogen

n'	Series name
1	Lyman
2	Balmer
3	Paschen
4	Brackett
5	Pfund

PROBLEM-SOLVING NOTE

Correctly Applying Equation 31–2

Notice that n and n' are integers in Equation 31–2 and that the integer n must always be greater than n'.

Figure 31–5 shows that the Balmer series is not the only series of lines produced by atomic hydrogen. The series with the shortest wavelengths is the **Lyman series**—all its lines are in the ultraviolet. Similarly, the series with wavelengths just longer than those in the Balmer series is the **Paschen series.** The lines in this series are all in the infrared. The formula that gives the wavelength in all the series of hydrogen is

$$\frac{1}{\lambda} = R\left(\frac{1}{n'^2} - \frac{1}{n^2}\right) \quad n' = 1, 2, 3, \ldots$$

$$n = n' + 1, n' + 2, n' + 3, \ldots \qquad \text{31–2}$$

Referring to Equation 31–1, we see that the Balmer series corresponds to the choice $n' = 2$. Similarly, the Lyman series is given by Equation 31–2 with $n' = 1$, and the Paschen series corresponds to $n' = 3$. As we shall see later in this chapter, there is an infinite number of series in hydrogen, each corresponding to a different choice for the integer n'. The names of the most common spectral series of hydrogen are listed in Table 31–1.

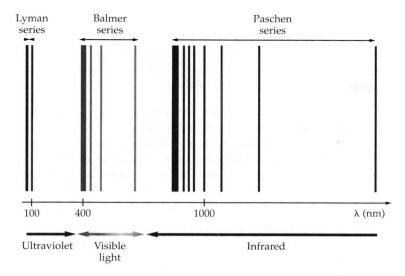

◀ **FIGURE 31–5 The Lyman, Balmer, and Paschen series of spectral lines**

The first three series of spectral lines in the spectrum of hydrogen. The shortest wavelengths appear in the Lyman series. There is no upper limit to the number of series in hydrogen or to the wavelengths that can be emitted.

EXERCISE 31–1

Find **(a)** the shortest wavelength in the Lyman series and **(b)** the longest wavelength in the Paschen series.

SOLUTION

a. Substitute $n' = 1$ and $n \to \infty$ in Equation 31–2:

$$\frac{1}{\lambda} = R\left(\frac{1}{1^2} - 0\right) = (1.097 \times 10^7 \, \text{m}^{-1})$$

$$\lambda = \frac{1}{(1.097 \times 10^7 \, \text{m}^{-1})} = 91.16 \, \text{nm}$$

b. Substitute $n' = 3$ and $n = 4$ in Equation 31–2:

$$\frac{1}{\lambda} = R\left(\frac{1}{3^2} - \frac{1}{4^2}\right) = (1.097 \times 10^7 \, \text{m}^{-1})\left(\frac{7}{144}\right)$$

$$\lambda = \frac{144}{7(1.097 \times 10^7 \, \text{m}^{-1})} = 1875 \, \text{nm}$$

As successful as Equation 31–2 is in giving the various wavelengths of radiation produced by hydrogen, it is still just an empirical formula. It gives no insight as to *why* these particular wavelengths, and no others, are produced. The goal of atomic physicists in the early part of the twentieth century was to *derive* Equation 31–2 from basic physical principles. The first significant step in that direction is the topic of the next section.

31–3 Bohr's Model of the Hydrogen Atom

Our scientific understanding of the hydrogen atom took a giant leap forward in 1913, when Niels Bohr (1885–1962), a Danish physicist who had just earned his doctorate in physics in 1911, introduced a model that allowed him to derive Equation 31–2. Bohr's model combined elements of classical physics with the ideas of quantum physics introduced by Planck and Einstein about ten years earlier. As such, his model is a hybrid that spanned the gap between the classical physics of Newton and Maxwell and the newly emerging quantum physics.

Assumptions of the Bohr Model

Bohr's model of the hydrogen atom is based on four assumptions. Two are specific to his model and do not apply to the full quantum mechanical picture of hydrogen that will be introduced in Section 31–5. The remaining two assumptions are quite general—they apply not only to hydrogen but to all atoms.

The two specific assumptions of the Bohr model are as follows:

- The electron in a hydrogen atom moves in a circular orbit about the nucleus.

▲ Niels Bohr, applying the principles of classical mechanics, with some members of his family.

- Only certain circular orbits are allowed. In these orbits the angular momentum of the electron is equal to an integer times Planck's constant divided by 2π. That is, the angular momentum of an electron in the nth allowed orbit is $L_n = nh/2\pi$, where $n = 1, 2, 3, \ldots$.

The next two assumptions are more general:

- Electrons do not give off electromagnetic radiation when they are in an allowed orbit. Thus, the orbits are stable.
- Electromagnetic radiation is given off or absorbed only when an electron changes from one allowed orbit to another. If the energy difference between two allowed orbits is ΔE, the frequency, f, of the photon that is emitted or absorbed is given by the relation $|\Delta E| = hf$.

Notice that Bohr's model retains the classical picture of an electron orbiting a nucleus, as in Rutherford's model. It adds the stipulations, however, that only certain orbits are allowed and that no radiation is given off from these orbits. Radiation is given off *only* when an electron shifts from one orbit to another, and then the radiation is in the form of a photon that obeys Einstein's quantum relation $E = hf$. Thus, as mentioned before, the Bohr model is a hybrid that includes ideas from both classical and quantum physics. We now use this model to determine the behavior of hydrogen.

Bohr Orbits

We begin by determining the radii of the allowed orbits in the Bohr model and the speed of the electrons in these orbits. There are two conditions that we must apply. First, for the electron to move in a circular orbit of radius r with a speed v, as depicted in **Figure 31–6**, the electrostatic force of attraction between the electron and the nucleus must be equal in magnitude to the mass of the electron times its centripetal acceleration, mv^2/r. Recalling Coulomb's law (Equation 19–5), we see that the electrostatic force between the electron (with charge $-e$) and the nucleus (with charge $+e$) has a magnitude given by ke^2/r^2. It follows that $mv^2/r = ke^2/r^2$, or, canceling one power of r,

$$mv^2 = k\frac{e^2}{r} \qquad \text{31–3}$$

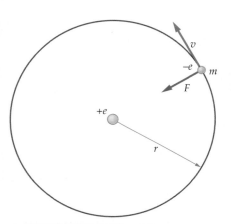

▲ **FIGURE 31–6 A Bohr orbit**
In the Bohr model of hydrogen, electrons orbit the nucleus in circular orbits. The centripetal acceleration of the electron, v^2/r, is produced by the Coulomb force of attraction between the electron and the nucleus.

Note that this relation is completely analogous to the one that was used to derive Kepler's third law (Chapter 12), except in that case the force of attraction was provided by gravity.

The second condition for an allowed orbit is that the angular momentum of the electron must be a nonzero integer n times $h/2\pi$. Since the electron moves with a speed v in a circular path of radius r, its angular momentum is $L = rmv$ (Equation 11–12). Hence, the condition for the nth allowed orbit is $L_n = r_n m v_n = nh/2\pi$, or, solving for v_n, we have

$$v_n = \frac{nh}{2\pi m r_n} \qquad n = 1, 2, 3, \ldots \qquad \text{31–4}$$

Combining these two conditions allows us to solve for the two unknowns, r_n and v_n.

For example, if we substitute v_n from Equation 31–4 into Equation 31–3, we can solve for r_n. Specifically, we find the following:

$$m\left(\frac{nh}{2\pi m r_n}\right)^2 = k\frac{e^2}{r_n}$$

Rearranging and solving for r_n, we get

$$r_n = \left(\frac{h^2}{4\pi^2 mke^2}\right)n^2 \qquad n = 1, 2, 3, \ldots \qquad \text{31–5}$$

The quantity in parentheses is the radius for the smallest ($n = 1$) orbit. Substitution of numerical values for h, m, e, and k gives the following value for r_1:

$$r_1 = \frac{h^2}{4\pi^2 m k e^2}$$

$$= \frac{(6.626 \times 10^{-34}\,\text{J} \cdot \text{s})^2}{4\pi^2 (9.109 \times 10^{-31}\,\text{kg})(8.988 \times 10^9\,\text{N} \cdot \text{m}^2/\text{C}^2)(1.602 \times 10^{-19}\,\text{C})^2}$$

$$= 5.29 \times 10^{-11}\,\text{m}$$

This radius, which is about half an angstrom and is referred to as the **Bohr radius,** is in agreement with the observed size of hydrogen atoms. Note the n^2 dependence in the radii of allowed orbits: $r_n = r_1 n^2 = (5.29 \times 10^{-11}\,\text{m})n^2$. This dependence is illustrated in **Figure 31–7.**

To complete the solution, we can substitute our result for r_n (Equation 31–5) into the expression for v_n (Equation 31–4). This yields

$$v_n = \frac{nh}{2\pi m}\left(\frac{4\pi^2 m k e^2}{n^2 h^2}\right) = \frac{2\pi k e^2}{nh} \qquad n = 1, 2, 3, \ldots \qquad \text{31–6}$$

Note that the speed of the electron is smaller in orbits farther from the nucleus.

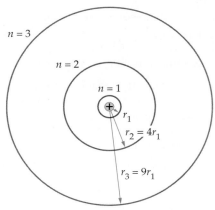

▲ **FIGURE 31–7 The first three Bohr orbits**
The first Bohr orbit has a radius $r_1 = 5.29 \times 10^{-11}$ m. The second and third Bohr orbits have radii $r_2 = 2^2 r_1 = 4r_1$, and $r_3 = 3^2 r_1 = 9r_1$, respectively. (*Note:* For clarity, the nucleus is drawn larger than its true scale relative to the size of the atom.)

EXAMPLE 31–2 FIRST AND SECOND BOHR ORBITS

Find the speed and kinetic energy of the electron in **(a)** the first Bohr orbit ($n = 1$) and **(b)** the second Bohr orbit ($n = 2$).

PICTURE THE PROBLEM
The first two orbits of the Bohr model are shown in the sketch. Note that the second orbit has a radius four times greater than the radius of the first orbit. In addition, the speed of the electron in the second orbit is half its value in the first orbit.

STRATEGY
The speed of the electron can be determined by direct substitution in $v_n = 2\pi k e^2/nh$ (Equation 31–6). Once v is determined, the kinetic energy is simply $K = \frac{1}{2}mv^2$.

Note that because the speed in the second orbit is half the speed in the first orbit, the kinetic energy is smaller by a factor of 4.

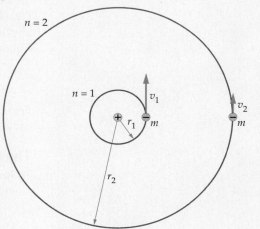

SOLUTION

Part (a)

1. Substitute $n = 1$ in Equation 31–6:

$$v_1 = \frac{2\pi k e^2}{h}$$

$$= \frac{2\pi (8.99 \times 10^9\,\text{N} \cdot \text{m}^2/\text{C}^2)(1.60 \times 10^{-19}\,\text{C})^2}{(6.63 \times 10^{-34}\,\text{J} \cdot \text{s})}$$

$$= 2.18 \times 10^6\,\text{m/s}$$

2. The corresponding kinetic energy is $K_1 = \frac{1}{2}mv_1^2$:

$$K_1 = \frac{1}{2}mv_1^2 = \frac{1}{2}(9.11 \times 10^{-31}\,\text{kg})(2.18 \times 10^6\,\text{m/s})^2$$

$$= 2.16 \times 10^{-18}\,\text{J}$$

Part (b)

3. Divide the speed found in part (a) by 2:

$$v_2 = \frac{1}{2}v_1 = \frac{1}{2}(2.18 \times 10^6\,\text{m/s}) = 1.09 \times 10^6\,\text{m/s}$$

4. Similarly, divide the kinetic energy found in part (a) by 4:

$$K_2 = \frac{1}{2}mv_2^2 = \frac{1}{2}m(v_1/2)^2$$

$$= \frac{1}{4}K_1 = \frac{1}{4}(2.16 \times 10^{-18}\,\text{J}) = 5.40 \times 10^{-19}\,\text{J}$$

CONTINUED ON NEXT PAGE

CONTINUED FROM PREVIOUS PAGE

INSIGHT
The speed of the electron in the first Bohr orbit is smaller than the speed of light by a factor of about 137. In higher orbits, the electron's speed is even less. It follows that relativistic effects are small for the hydrogen atom.

In addition, note that the kinetic energy of the electron in the first Bohr orbit is approximately 13.6 eV. We shall encounter this particular energy again later in the section.

PRACTICE PROBLEM
An electron in a Bohr orbit has a kinetic energy of 8.64×10^{-20} J. Find the radius of this orbit.   [**Answer:** $r = 1.32 \times 10^{-9}$ m, corresponding to $n = 5$.]

Some related homework problems: Problem 15, Problem 20, Problem 21

Bohr's model applies equally well to singly ionized helium, doubly ionized lithium, and other ions with only a single electron. In the case of singly ionized helium (one electron removed) the charge on the nucleus is $+2e$, for doubly ionized lithium (two electrons removed) it is $+3e$, and so on. In the general case, we may consider a nucleus that contains Z protons and has a charge of $+Ze$, where Z is the atomic number associated with that nucleus. Hydrogen, which has only a single proton in its nucleus, corresponds to $Z = 1$.

To be more explicit about the Z dependence, notice that the electrostatic force between an electron and a nucleus with Z protons has a magnitude of $k(e)(Ze)/r^2 = kZe^2/r^2$. Thus the results derived earlier in this section can be applied to the more general case if we simply replace e^2 with Ze^2. For example, Equation 31–3 becomes $mv^2 = kZe^2/r$. Similarly, the radius of an allowed orbit is

$$r_n = \left(\frac{h^2}{4\pi^2 mkZe^2}\right)n^2 \qquad n = 1, 2, 3, \ldots \qquad \text{31–7}$$

For example, the radius of the $n = 1$ orbit of singly ionized helium is half the radius of the $n = 1$ orbit of hydrogen.

The Energy of a Bohr Orbit

To find the energy of an electron in a Bohr orbit, we simply note that its total mechanical energy, E, is the sum of its kinetic and potential energies:

$$E = K + U = \tfrac{1}{2}mv^2 + U$$

Using the fact that $mv^2 = kZe^2/r$ for a hydrogen-like atom of atomic number Z (Equation 31–3), and the fact that the electrostatic potential energy of a charge $-e$ and a charge $+Ze$ a distance r apart is $U = -kZe^2/r$ (Equation 20–8), we find that the total mechanical energy is

$$E = \tfrac{1}{2}\left(\frac{kZe^2}{r}\right) - \frac{kZe^2}{r} = -\frac{kZe^2}{2r}$$

Finally, substituting the radius of a Bohr orbit, as given in Equation 31–7, we obtain the corresponding energy for the nth orbit:

$$E_n = -\frac{kZe^2}{2r_n} = -\left(\frac{kZe^2}{2}\right)\left(\frac{4\pi^2 mkZe^2}{h^2}\right)\frac{1}{n^2}$$

$$= -\left(\frac{2\pi^2 mk^2 e^4}{h^2}\right)\frac{Z^2}{n^2} \qquad n = 1, 2, 3, \ldots \qquad \text{31–8}$$

PROBLEM-SOLVING NOTE

The Z Dependence of Hydrogen-like Ions

Note that the energy of a hydrogen-like ion depends on Z^2, the square of the atomic number.

Using the numerical values for m, k, e, and h, as we did in calculating the Bohr radius, we have

$$E_n = -(13.6 \text{ eV})\frac{Z^2}{n^2} \qquad n = 1, 2, 3, \ldots \qquad \text{31–9}$$

Let's first consider the specific case of hydrogen. With $Z = 1$ we find that the energy of the orbits in hydrogen are given by the relation $E_n = -(13.6 \text{ eV})/n^2$. We plot these energies in **Figure 31–8** for various values of n. This type of plot is referred to as an **energy-level diagram.** Notice that the ground state ($n = 1$) corresponds to the lowest possible energy of the system. The higher energy levels are referred to as **excited states.** As the integer n tends to infinity, the energy of the excited states approaches zero—the energy the electron and proton would have if they were at rest and separated by an infinite distance. Thus to **ionize** hydrogen—that is, to remove the electron from the atom—requires a minimum energy of 13.6 eV. This value, which is a specific prediction of the Bohr model, is in complete agreement with experiment.

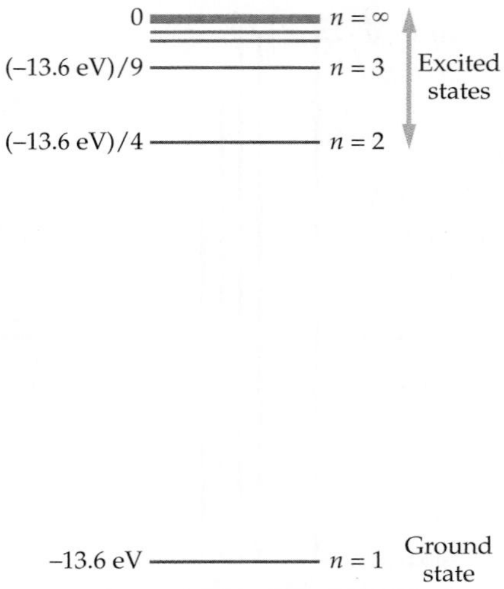

EXERCISE 31–2

In doubly ionized lithium, a single electron orbits a lithium nucleus. Calculate the minimum energy required to remove this electron.

SOLUTION

The nucleus of lithium has a charge of $+3e$. Substitution of $Z = 3$ and $n = 1$ in Equation 31–9 yields

$$E_1 = -(13.6 \text{ eV})\left(\frac{3^2}{1^2}\right) = -122 \text{ eV}$$

Therefore, 122 eV must be added to remove the electron.

The electron in doubly ionized lithium experiences a stronger attractive force than the single electron in hydrogen. In fact, the force is greater by a factor of 3. In addition, the $n = 1$ orbit of doubly ionized lithium has one-third the radius of the $n = 1$ orbit in hydrogen. As a result, nine times as much energy is required to remove the electron from the lithium ion.

At room temperature, most hydrogen atoms are in the ground state. This is because the typical thermal energy of such atoms is too small to cause even the lowest-energy excitation from the ground state. Specifically, a typical thermal energy, $k_B T$, corresponding to room temperature ($T \sim 300$ K) is only about $\frac{1}{40}$ eV. In comparison, the energy required to excite an electron from the ground state of hydrogen to the first excited state is roughly 10 eV; that is, $\Delta E = E_2 - E_1 = (-3.40 \text{ eV}) - (-13.6 \text{ eV}) = 10.2 \text{ eV}$. Excitations to higher excited states require even more energy. As a result, typical intermolecular collisions are simply not energetic enough to produce an excited state in hydrogen.

The Spectrum of Hydrogen

To find a formula describing the spectrum of hydrogen, we use Bohr's assumption that the frequency of emitted radiation for a change in energy equal to ΔE is given by $|\Delta E| = hf$. Since $\lambda f = c$ for electromagnetic radiation (Equation 25–4), we can rewrite this relation in terms of the wavelength as $|\Delta E| = hc/\lambda$. To find $|\Delta E|$, we recall that the energy for hydrogen is given by Equation 31–9 with $Z = 1$. Therefore, the *change* in energy as the electron moves from an excited outer orbit with $n = n_i$ to a lower orbit with $n = n_f < n_i$ has the following magnitude:

$$|\Delta E| = \left(\frac{2\pi^2 mk^2 e^4}{h^2}\right)\left(\frac{1}{n_f^2} - \frac{1}{n_i^2}\right)$$

Using $|\Delta E| = hc/\lambda$ we can now solve for $1/\lambda$:

$$\frac{1}{\lambda} = \left(\frac{2\pi^2 mk^2 e^4}{h^3 c}\right)\left(\frac{1}{n_f^2} - \frac{1}{n_i^2}\right) \tag{31–10}$$

Comparing Equation 31–10 with Equation 31–2, we see that the expressions have precisely the same form, provided we identify n_f with n', and n_i with n. In addition, we see that the Rydberg constant in Equation 31–2, $R = 1.097 \times 10^7 \text{ m}^{-1}$, has

▲ **FIGURE 31–8 Energy-level diagram for the Bohr model of hydrogen**

The energy of the ground state of hydrogen is -13.6 eV. Excited states of hydrogen approach zero energy. Note that the difference in energy from the ground state to the first excited state is $\frac{3}{4}(13.6 \text{ eV})$, and the energy difference from the first excited state to the zero level is only $\frac{1}{4}(13.6 \text{ eV})$.

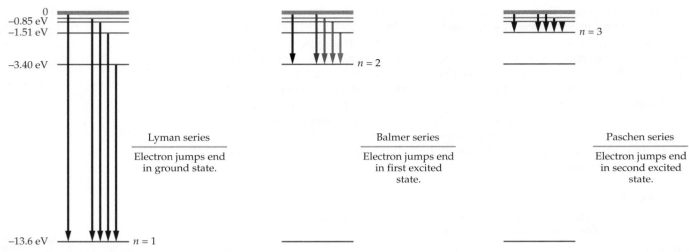

▲ **FIGURE 31–9 The origin of spectral series in hydrogen**
Each series of spectral lines in hydrogen is the result of electrons jumping from an excited state to a particular lower level. For the Lyman series the lower level is the ground state. The lower level for the Balmer series is the first excited state ($n = 2$), and the lower level for the Paschen series is the second excited state ($n = 3$).

been replaced by the rather unusual constant $2\pi^2 m k^2 e^4/h^3 c$ in Equation 31–10. It is remarkable that when the known values of the fundamental constants m, k, e, h, and c are substituted into $2\pi^2 m k^2 e^4/h^3 c$, the resulting value is precisely $1.097 \times 10^7 \text{ m}^{-1}$. This completes the derivation of Equation 31–2, one of the most significant accomplishments of the Bohr model.

The origin of the line spectrum of hydrogen can be visualized in **Figure 31–9**. Notice that transitions involving an electron jumping from an excited state ($n_i = n > 1$) to the ground state ($n_f = n' = 1$) result in the Lyman series of lines in the ultraviolet. Jumps ending in the $n = 2$ level give rise to the Balmer series, and jumps ending in the $n = 3$ level give the Paschen series. The largest energy jump in each series occurs when an electron falls from $n = \infty$ to the final level. Thus each series of spectral lines has a well-defined shortest wavelength.

CONCEPTUAL CHECKPOINT 31–1 COMPARE WAVELENGTHS

The wavelength of the photon emitted when an electron in hydrogen jumps from the $n_i = 100$ state to the $n_f = 2$ state is **(a)** greater than, **(b)** less than, or **(c)** equal to the wavelength of the photon when the electron jumps from the $n_i = 2$ state to the $n_f = 1$ state.

REASONING AND DISCUSSION
We begin by noting that wavelength is inversely proportional to the energy difference between levels ($|\Delta E| = hc/\lambda$), thus the smaller the energy difference, the greater the wavelength.

Next, we note that the full range of energies in the hydrogen atom extends from -13.6 eV to 0, and that the difference in energy between $n_i = 2$ and $n_f = 1$ is three-quarters of this range. It follows that the energy difference between $n_i = 2$ and $n_f = 1$ is greater than the energy difference between *any* state with $n_i > 2$ and $n_f = 2$. In fact, the maximum energy difference ending in the state 2 is $\frac{1}{4}(-13.6 \text{ eV})$, corresponding to $n_i = \infty$ and $n_f = 2$.

Since the energy difference for $n_i = 100$ and $n_f = 2$ is less than the energy difference for $n_i = 2$ and $n_f = 1$, the wavelength for $n_i = 100$ and $n_f = 2$ is the greater of the two.

ANSWER
(a) The wavelength for $n_i = 100$, $n_f = 2$ is greater than the wavelength for $n_i = 2$, $n_f = 1$.

We calculate a variety of specific wavelengths in the following Active Example.

ACTIVE EXAMPLE 31-1   FIND THE WAVELENGTHS

An electron in a hydrogen atom is in the initial state $n_i = 4$. Calculate the wavelength of the photon emitted by this electron if it jumps to the final state **(a)** $n_f = 3$, **(b)** $n_f = 2$, or **(c)** $n_f = 1$. (*Note*: To simplify the calculations, replace the constant $2\pi^2 mk^2 e^4/h^3 c$ in Equation 31–10 with its numerical equivalent, the Rydberg constant, $R = 1.097 \times 10^7 \text{ m}^{-1}$.)

SOLUTION *(Test your understanding by performing the calculations indicated in each step.)*

Part (a)

1. Substitute $n_i = 4$ and $n_f = 3$ in Equation 31–10: $\lambda = 1875 \text{ nm}$

Part (b)

2. Repeat with $n_f = 2$: $\lambda = 486.2 \text{ nm}$

Part (c)

3. Finally, use $n_f = 1$: $\lambda = 97.23 \text{ nm}$

INSIGHT

Other wavelengths are possible when an electron in the $n_i = 4$ state jumps to a lower state. For example, after dropping from the $n = 4$ state to the $n = 3$ state, the electron might then jump from the $n = 3$ state to the $n = 2$ state, and finally from the $n = 2$ state to the $n = 1$ state. Alternatively, the electron might first jump from the $n = 4$ state to the $n = 2$ state, and then from the $n = 2$ state to the $n = 1$ state. Thus an electron in an excited state may result in the emission of a variety of different wavelengths.

YOUR TURN

A hydrogen atom with its electron in the initial state $n_i = 5$ emits a photon with a wavelength of 434 nm. To which state did the electron jump?

(Answers to **Your Turn** *problems are given in the back of the book.)*

Just as an electron can *emit* a photon when it jumps to a lower level, it can also *absorb* a photon and jump to a higher level. This process occurs, however, only if the photon has the proper energy. In particular, the photon must have an energy that precisely matches the energy difference between the lower level of the electron and the higher level to which it is raised. This situation is explored in more detail in the next Active Example.

ACTIVE EXAMPLE 31-2   ABSORBING A PHOTON: WHAT IS THE FREQUENCY?

Find the frequency a photon must have if it is to raise an electron in a hydrogen atom from the $n = 3$ state to the $n = 5$ state.

SOLUTION *(Test your understanding by performing the calculations indicated in each step.)*

1. Calculate the energy of the $n = 5$ state in joules: $-8.70 \times 10^{-20} \text{ J}$
2. Calculate the energy of the $n = 3$ state: $-2.42 \times 10^{-19} \text{ J}$
3. Calculate the difference in energy between these states: $1.55 \times 10^{-19} \text{ J}$
4. Set the energy of a photon equal to this energy difference: $hf = 1.55 \times 10^{-19} \text{ J}$
5. Solve for the frequency of the photon: $f = 2.34 \times 10^{14} \text{ Hz}$

INSIGHT

Of course, this frequency corresponds to one of the lines in the absorption spectrum of hydrogen. In this case, the line is in the infrared.

YOUR TURN

Suppose an electron jumps from the $n = 3$ state to the $n = 5$ state in singly ionized helium. Does the required frequency of the absorbed photon increase or decrease? By what factor?

(Answers to **Your Turn** *problems are given in the back of the book.)*

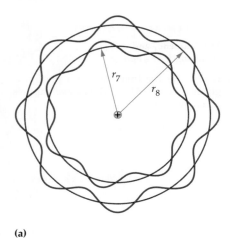

(a)

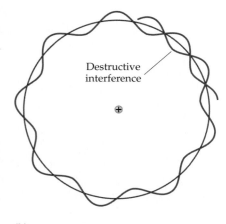

Destructive interference

(b)

▲ **FIGURE 31–10 de Broglie wavelengths and Bohr orbits**
Bohr's condition that the angular momentum of an allowed orbit must be an integer n times $h/2\pi$ is equivalent to the condition that n de Broglie wavelengths must fit into the circumference of an orbit. **(a)** The de Broglie waves for the $n = 7$ and $n = 8$ orbits. **(b)** If an integral number of wavelengths do not fit the circumference of an orbit, the result is destructive interference.

31–4 de Broglie Waves and the Bohr Model

The fact that hydrogen emits radiation only at certain well-defined wavelengths is reminiscent of the harmonics of standing waves on a string. Recall that a vibrating string tied down at both ends produces a standing wave only if an integral number of half-wavelengths fit within its length (Section 14–8). Perhaps the behavior of hydrogen can be understood in similar terms.

In 1923 de Broglie used his idea of matter waves (Section 30–5) to show that one of Bohr's assumptions could indeed be thought of as a condition for standing waves. As we saw earlier in this chapter, Bohr assumed that the angular momentum of an electron in an allowed orbit must be a nonzero integer times $h/2\pi$. Specifically,

$$rmv = n\frac{h}{2\pi} \qquad n = 1, 2, 3, \ldots$$

In Bohr's model there is no particular reason for this condition other than that it produces results in agreement with experiment.

Now, de Broglie imagined his matter waves as analogous to a wave on a string—except that in this case the "string" is not tied down at both ends. Instead, it is formed into a circle of radius r representing an electron's orbit about the nucleus, as illustrated in **Figure 31–10**. The condition for a standing wave in this case is that an integral number of wavelengths fit into the circumference of the orbit. Stated mathematically, the condition is $n\lambda = 2\pi r$, as shown in Figure 31–10 (a) for the cases $n = 7$ and $n = 8$. Other wavelengths would result in destructive interference, as Figure 31–10 (b) shows.

Finally, de Broglie combined the standing wave condition with his matter–wave relationship $p = h/\lambda$ (Equation 30–16). The result is as follows:

$$p = mv = \frac{h}{\lambda} = \frac{h}{(2\pi r/n)} = \frac{nh}{2\pi r} \qquad n = 1, 2, 3, \ldots$$

Multiplying both sides of this equation by r to obtain the angular momentum, we find

$$L = rmv = \frac{nh}{2\pi} \qquad n = 1, 2, 3, \ldots$$

This is precisely the Bohr orbital condition, now understood as a reflection of the wave nature of matter.

EXAMPLE 31–3 THE WAVELENGTH OF AN ELECTRON

Find the wavelength associated with an electron in the $n = 4$ state of hydrogen.

PICTURE THE PROBLEM
Our sketch shows that four wavelengths fit around the circumference of the $n = 4$ orbit. Recall that the radius of this orbit is $4^2 = 16$ times the radius of the ground-state orbit.

STRATEGY
To find the wavelength of this matter wave we simply calculate the circumference of the $n = 4$ orbit, then divide by 4.

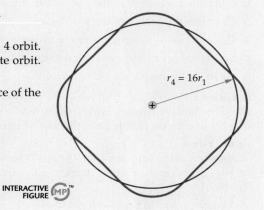

$r_4 = 16r_1$

INTERACTIVE FIGURE MP

SOLUTION

1. Calculate the radius of the $n = 4$ orbit: $r_4 = 4^2 r_1 = 16(5.29 \times 10^{-11}\ \text{m}) = 8.46 \times 10^{-10}\ \text{m}$

2. Use this result to find the circumference of the orbit: $2\pi r_4 = 2\pi(8.46 \times 10^{-10}\ \text{m}) = 5.32 \times 10^{-9}\ \text{m}$

3. Divide the circumference by 4 to find the wavelength: $\lambda_4 = \frac{1}{4}(2\pi r_4) = \frac{1}{4}(5.32 \times 10^{-9}\ \text{m}) = 1.33 \times 10^{-9}\ \text{m}$

INSIGHT

An equivalent way of determining the wavelength is to use the de Broglie relation, $p = mv = h/\lambda$. Solving for the wavelength yields $\lambda_4 = h/mv_4$, and substituting v_4 from Equation 31–6 yields the same result given in Step 3.

PRACTICE PROBLEM

In which state of hydrogen does the electron have a wavelength of $2.66 \times 10^{-9}\ \text{m}$? [**Answer:** $n = 8$]

Some related homework problems: Problem 35, Problem 36

The striking success of de Broglie's matter waves in deriving Bohr's angular momentum condition encouraged physicists to give the idea of matter waves serious consideration. If we accept matter waves as being real, however, a large number of new questions must be addressed. For example, if particles like electrons can be described by matter waves, how do the matter waves behave? What determines the value of a matter wave at a particular location? What is the physical significance of a matter wave having a large value at one location and a small value at another location?

These questions were answered by Erwin Schrödinger (1887–1961), Max Born (1882–1970), and others. In particular, Schrödinger introduced an equation—similar in many respects to the equation that describes sound waves—to describe the behavior of matter waves. Today, this equation, known as **Schrödinger's equation**, forms the basis for quantum mechanics, which is the quantum physics version of classical mechanics. In fact, Schrödinger's equation plays the same role in quantum mechanics that Newton's laws play in classical mechanics and Maxwell's equations play in electromagnetism.

After the introduction of the Schrödinger equation, Born developed an interpretation of the matter waves that was quite different from that for mechanical waves. For example, in the case of a wave on a string, the amplitude of a wave simply represents the displacement of the string from its equilibrium position. For a matter wave, on the other hand, the amplitude is related to the *probability* of finding a particle in a particular location. Thus, matter waves do not tell us precisely where a particle is located; rather, they give the probability of finding the particle at a given place, as we shall see in detail in the next section.

31–5 The Quantum Mechanical Hydrogen Atom

Although Schrödinger's equation and its solution for the hydrogen atom are beyond the scope of this text, we present here some of the main features obtained by this analysis. Other than relativistic effects, the Schrödinger equation presents our most complete understanding of the hydrogen atom and of behavior at the atomic level in general. As we shall see, many aspects of the Bohr model survive in this analysis, though there are also significant differences.

To begin, we note that whereas the Bohr model was characterized by a single quantum number, n, the quantum mechanical description of the hydrogen atom requires four quantum numbers. They are as follows:

- **The principal quantum number, n:** The quantum number $n = 1, 2, 3, \dots$ plays a similar role in the quantum mechanical hydrogen atom and in Bohr's model. In Bohr's model, n is the only quantum number, and it determines the radius of an orbit, its angular momentum, and its energy. In particular, the energy in the Bohr model is $E = (-13.6\ \text{eV})/n^2$. The energy

▲ Although we can't see the de Broglie waves associated with electrons, the standing waves they produce are of great importance because they correspond to allowed states in atoms and molecules. One way to visualize de Broglie waves, however, is to make an analogy with mechanical standing waves. In the photos shown above, a loop of wire is oscillated vertically about the support point at the bottom of the loop. The oscillations set up waves that travel on the circumference of the wire—like de Broglie waves on a Bohr orbit in hydrogen. If the wavelength of the mechanical wave is tuned to an appropriate value—by adjusting the frequency of the oscillator—the result is a variety of different standing wave patterns, analogous to different energy levels of de Broglie waves in the Bohr model.

given by Schrödinger's equation is precisely the same, if we neglect small relativistic effects and small magnetic interactions within the atom.

- **The orbital angular momentum quantum number, ℓ:** In the Bohr model an electron's orbital angular momentum is determined by the quantum number n. In particular, $L_n = nh/2\pi$, where $n = 1, 2, 3, \ldots$ In the quantum mechanical solution, there is a separate quantum number, ℓ, for the orbital angular momentum. This quantum number can take on the following values for any given value of the principal quantum number, n:

$$\ell = 0, 1, 2, \ldots, (n - 1) \qquad \text{31–11}$$

The magnitude of the angular momentum for any given value of ℓ is given by the following relation:

$$L = \sqrt{\ell(\ell + 1)}\,\frac{h}{2\pi} \qquad \text{31–12}$$

Note that the angular momentum of the electron can have a range of values for a given n, in contrast with the Bohr model, where the angular momentum has just a single value. In particular, the electron in a hydrogen atom can have zero angular momentum; an orbiting electron in the Bohr model always has a nonzero angular momentum.

Finally, although the energy of the hydrogen atom does not depend on ℓ, the energy of more complex atoms does have an ℓ dependence, as we shall see in the next section.

- **The magnetic quantum number, m_ℓ:** If a hydrogen atom is placed in an external magnetic field, its energy is found to depend not only on n but on an additional quantum number as well. This quantum number, m_ℓ, is referred to as the magnetic quantum number. The allowed values of m_ℓ are as follows:

$$m_\ell = -\ell, -\ell + 1, -\ell + 2, \ldots, -1, 0, 1, \ldots, \ell - 2, \ell - 1, \ell \qquad \text{31–13}$$

This quantum number gives the component of orbital angular momentum vector along a specified direction, usually chosen to be the z axis. With this choice, L_z has the following values:

$$L_z = m_\ell \frac{h}{2\pi}$$

Only a single component of the orbital angular momentum can be known precisely; it is not possible to know all three components of the angular momentum simultaneously, due to the Heisenberg uncertainty principle.

- **The electron spin quantum number, m_s:** The final quantum number needed to describe the hydrogen atom is related to the angular momentum of the electron itself. Just as Earth spins on its axis at the same time that it orbits the Sun, the electron can be thought of as having both an orbital and a "spin" angular momentum. The spin quantum number for an electron takes on just two values:

$$m_s = -\tfrac{1}{2}, \tfrac{1}{2}$$

These two values correspond to the electron's spin being "up" $\left(m_s = \tfrac{1}{2}\right)$ or "down" $\left(m_s = -\tfrac{1}{2}\right)$ with respect to the z axis.

Spin angular momentum is an *intrinsic* property of an electron, like its mass and its charge—*all* electrons have exactly the *same* mass, the *same* charge, and the *same* spin angular momentum. Thus, we do not imagine the electron to be a small spinning sphere, like a microscopic planet. You can't speed up or slow down the spin of an electron. Instead, spin is simply one of the properties that defines an electron.

The energy-level structure of hydrogen in zero magnetic field is shown in **Figure 31–11**, along with the corresponding quantum numbers. Since the energies are the same as in the Bohr model, it follows that the spectrum will be the same as the Bohr model, and experiment.

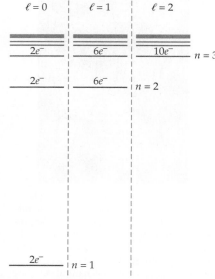

▲ **FIGURE 31–11 Energy-level structure of hydrogen**

The values of the quantum mechanical energy levels for hydrogen are in complete agreement with the Bohr model. In the quantum model, however, each energy level has associated with it a specific number of "quantum states" determined by the values of all four quantum numbers, as specified in Table 31–2. In multielectron atoms, these states lead to the formation of the periodic table of elements, as we show in Section 31–6.

TABLE 31–2 States of Hydrogen for $n = 1$ and $n = 2$

$n = 1, \ell = 0$			Two states
$n = 1$	$\ell = 0$	$m_\ell = 0$	$m_s = \frac{1}{2}$
$n = 1$	$\ell = 0$	$m_\ell = 0$	$m_s = -\frac{1}{2}$
$n = 2, \ell = 0$			Two states
$n = 2$	$\ell = 0$	$m_\ell = 0$	$m_s = \frac{1}{2}$
$n = 2$	$\ell = 0$	$m_\ell = 0$	$m_s = -\frac{1}{2}$
$n = 2, \ell = 1$			Six states
$n = 2$	$\ell = 1$	$m_\ell = 1$	$m_s = \frac{1}{2}$
$n = 2$	$\ell = 1$	$m_\ell = 1$	$m_s = -\frac{1}{2}$
$n = 2$	$\ell = 1$	$m_\ell = 0$	$m_s = \frac{1}{2}$
$n = 2$	$\ell = 1$	$m_\ell = 0$	$m_s = -\frac{1}{2}$
$n = 2$	$\ell = 1$	$m_\ell = -1$	$m_s = \frac{1}{2}$
$n = 2$	$\ell = 1$	$m_\ell = -1$	$m_s = -\frac{1}{2}$

We define a **state** of hydrogen to be a specific assignment of values for each of the four quantum numbers. For example, there are two states that correspond to the lowest possible energy level of hydrogen. These are $n = 1, \ell = 0, m_\ell = 0$, $m_s = \frac{1}{2}$; and $n = 1, \ell = 0, m_\ell = 0, m_s = -\frac{1}{2}$. Similarly, there are two states corresponding to the $n = 2, \ell = 0$ energy level, and six states corresponding to the $n = 2, \ell = 1$ level. These states are listed in Table 31–2, and the corresponding numbers are shown in **Figure 31–11**. When we consider multielectron atoms in the next section, we shall see that the number of states associated with a given energy level determines the number of electrons (e^-) it can accommodate. Once an energy level is "filled," additional electrons must occupy higher levels. This progressive filling of energy levels ultimately leads to the periodic table of elements.

Electron Probability Clouds: Three-Dimensional Standing Waves

As mentioned in the previous section, the solution to Schrödinger's equation gives a matter wave, or **wave function,** as it is known, corresponding to a particular physical system. The wave function for hydrogen gives the probability of finding the electron at a particular location. The best way to visualize this probability distribution is in terms of a "probability cloud," as shown in **Figure 31–12**. The probability of finding the electron is greatest where the cloud is densest. In the case shown in Figure 31–12, corresponding to the ground state, $n = 1, \ell = 0, m_\ell = 0$, the probability of finding the electron is distributed with spherical symmetry. Note that the probability decreases rapidly far from the nucleus, as one would expect.

Figure 31–13 gives a different way of looking at the probability distribution for the ground state. Here we plot the probability versus distance from the nucleus. It is interesting to note that the maximum probability occurs at a distance from the nucleus equal to the Bohr radius. Thus certain aspects of the Bohr model find their way into the final solution for hydrogen. The difference, however, is that in the Bohr model the electron is always at a particular distance from the nucleus as it moves in its circular orbit. In the quantum solution to hydrogen, the electron can be found at virtually any distance from the nucleus, not just one distance.

States of higher quantum number have increasingly complex probability distributions, as indicated in **Figure 31–14**. As the quantum number n is increased, for example, the basic shape of the distribution remains the same, but additional nodes appear. This is illustrated in Figure 31–14 (a) for the case $n = 2, \ell = 0$. Note that the distribution is spherically symmetric, as in the state $n = 1, \ell = 0$, but now there is a node—where the probability is zero—separating an inner and an outer portion of the distribution. As the quantum number ℓ is increased, the distributions become more complex in their shape. An example is given in Figure 31–14 (b) for the case $n = 2, \ell = 1$, and $m_\ell = 0$.

▲ **FIGURE 31–12 Probability cloud for the ground state of hydrogen**

In the quantum mechanical model of hydrogen, the electron can be found at any distance from the nucleus. The probability of finding the electron at a given location is proportional to the density of the "probability cloud."

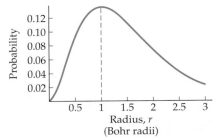

▲ **FIGURE 31–13 Probability as a function of distance**

This plot shows the probability of finding an electron in the ground state of hydrogen at a given distance from the nucleus. Note that the probability is greatest at a distance equal to the Bohr radius, r_1.

The probability cloud for the $n = 2$, $\ell = 1$ state of hydrogen is shown in Figure 31–14 (b). Notice that this cloud consists of two lobes of high probability separated by a plane of zero probability. Given that an electron in this state can never be halfway between the two lobes, how is it possible that the electron is as likely to be found in the upper lobe as in the lower lobe?

REASONING AND DISCUSSION
The probability lobes of an electron are the result of a standing wave pattern, analogous to the standing waves found on a string tied down at both ends. Both the node on a string and the region of zero probability of an electron are the result of destructive interference. For example, the displacement of a string may be of equal amplitude on either side of a node—where the displacement is always zero—just as the electron probability may be high on either side of a probability node and zero in the middle. In summary, an electron does not simply move from place to place in an atom, like a small ball of charge; instead, it forms a standing wave pattern with nodes in certain locations.

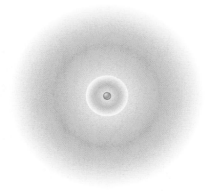

(a) $n = 2$, $\ell = 0$

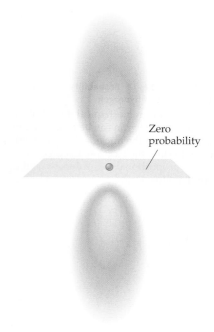

Zero probability

(b) $n = 2$, $\ell = 1$

▲ **FIGURE 31–14 Probability clouds for excited states of hydrogen**
The "probability cloud" for an electron in hydrogen increases in complexity as the quantum numbers increase.
(a) $n = 2$, $\ell = 0$; **(b)** $n = 2$, $\ell = 1$.

31–6 Multielectron Atoms and the Periodic Table

In this section we extend our considerations of atomic physics to atoms with more than one electron. As we shall see, certain regularities arise in the properties of multielectron atoms, and these regularities are intimately related to the quantum numbers described in the previous section.

Multielectron Atoms

One of the great simplicities of the hydrogen atom is that the only electrostatic force in the atom is the attractive force between the electron and the proton. In multielectron atoms the situation is more complex. Specifically, the electrons in such atoms experience repulsive electrostatic interactions with one another, in addition to their interaction with the nucleus. Thus the simple expression for the energy of a single-electron atom given in Equation 31–9 cannot be applied to atoms with multiple electrons, since it does not include energy contributions due to the forces between electrons.

Although no simple formula analogous to Equation 31–9 exists for multielectron atoms, the energy levels of these atoms can be understood in terms of the four quantum numbers (n, ℓ, m_ℓ, m_s) used to describe hydrogen. In fact, by applying Schrödinger's equation to such atoms, we have discovered that the energy levels of multielectron atoms depend on the principal quantum number, n, and on the orbital quantum number, ℓ. For example, increasing n for a fixed value of ℓ results in an increase in energy. This relationship is illustrated in **Figure 31–15**, where we see that the energy for $n = 2$ and $\ell = 0$ is greater than the energy for $n = 1$ and $\ell = 0$. Similarly, the energy for $n = 3$ and $\ell = 1$ is greater than the energy for $n = 2$ and $\ell = 1$. It is also found that the energy increases with increasing ℓ for fixed n. Thus the energy for $n = 2$ and $\ell = 1$ is greater than the energy for $n = 2$ and $\ell = 0$.

Figure 31–15 also shows that in some cases the energy levels corresponding to different values of n can cross. For example, the energy for $n = 3$ and $\ell = 2$ is greater than the energy for $n = 4$ and $\ell = 0$. A similar crossing occurs with the $n = 4$, $\ell = 2$ state and the $n = 5$, $\ell = 0$ state. Note in all cases, however, that the energy still increases with n for fixed ℓ, and increases with ℓ for fixed n. We shall see the effect of these energy-level crossings in terms of the periodic table later in this section.

Because the energy levels of multielectron atoms depend on n and ℓ, we give specific names and designations to the various values of these quantum numbers. For example, all electrons that have the same value of n are said to be in the same **shell**. Specifically, electrons with $n = 1$ are said to be in the K shell, those with $n = 2$ are in the L shell, those with $n = 3$ are in the M shell, and so on. These designations are summarized in Table 31–3 and displayed in Figure 31–15.

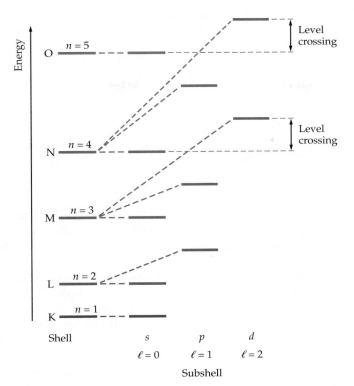

**◀ FIGURE 31–15 Energy levels in
multielectron atoms**

In multielectron atoms the energy increases with n for fixed ℓ and increases with ℓ for fixed n. Note the possibility of energy-level crossing. For example, the $n = 3$, $\ell = 2$ energy level is higher than the $n = 4$, $\ell = 0$ energy level.

Similarly, electrons in a given shell with the same value of ℓ are said to be in the same **subshell,** and different values of ℓ have different alphabetical designations. For example, electrons with $\ell = 0$ are said to be in the s subshell, those with $\ell = 1$ are in the p subshell, and those with $\ell = 2$ are in the d subshell. These names, though not particularly logical, are used for historical reasons. After the f subshell ($\ell = 3$), the names of subsequent subshells continue in alphabetical order, as indicated in Table 31–3.

The Pauli Exclusion Principle

As mentioned in Section 31–3, most hydrogen atoms are in their ground state at room temperature, since typical thermal energies are not great enough to excite the electron to higher energy levels. The same is true of multielectron atoms—they too are generally found in their ground state at room temperature. The question is this: What is the ground state of a multielectron atom?

The answer to this question involves an entirely new fundamental principle of physics put forward by the Austrian physicist Wolfgang Pauli (1900–1958) in 1925. Pauli's "exclusion principle" states that no two electrons in an atom can be in the same state at the same time. That is, once an electron occupies a given state, as defined by the values of its quantum numbers, other electrons are *excluded* from occupying the same state:

> **The Pauli Exclusion Principle**
>
> Only one electron at a time may have a particular set of quantum numbers, n, ℓ, m_ℓ, and m_s. Once a particular state is occupied, other electrons are excluded from that state.

Because of the exclusion principle, the ground state of a multielectron atom is *not* obtained by placing all the electrons in the lowest possible energy state, as one might at first suppose. Once the lowest-energy states are occupied, additional electrons in the atom must occupy levels of higher energy. As more electrons are added to an atom, they fill up one subshell after another until all the electrons are accommodated. The situation is analogous to placing marbles

TABLE 31–3 Shell and Subshell Designations

n	Shell
1	K
2	L
3	M
4	N
. . .	. . .

ℓ	Subshell
0	s
1	p
2	d
3	f
4	g
. . .	. . .

▶ **FIGURE 31–16 Filling a jar with marbles**
As marbles are added to a jar, they fill in one level, then another, and another. Once a given level is filled with marbles, additional marbles are excluded from that level, analogous to the filling of energy levels in multielectron atoms.

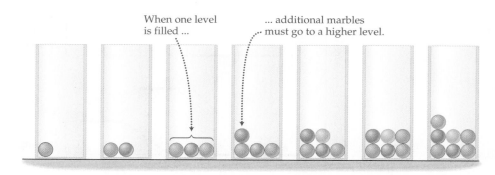

When one level is filled ...   ... additional marbles must go to a higher level.

into a jar, as illustrated in **Figure 31–16**. The first few marbles occupy the lowest level of the jar, but as more marbles are added they must occupy levels of higher gravitational potential energy, simply because the lower levels are already occupied.

In the case of atoms, the lowest energy level corresponds to $n = 1$ and $\ell = 0$; that is, to the s subshell of the K shell. To completely define a state, however, we must specify all four quantum numbers: n, ℓ, m_ℓ, and m_s. First, recall that $m_\ell = 0, \pm 1, \pm 2, \ldots, \pm \ell$ for general ℓ. It follows that in the $\ell = 0$ state the only possible value of m_ℓ is 0. The quantum number m_s, however, can always take on two values, $m_s = +\frac{1}{2}$ and $m_s = -\frac{1}{2}$. Therefore, two electrons can occupy the $n = 1$, $\ell = 0$ energy level, since two different states—$n = 1$, $\ell = 0$, $m_\ell = 0$, $m_s -\frac{1}{2}$; and $n = 1$, $\ell = 0$, $m_\ell = 0$, $m_s = +\frac{1}{2}$—correspond to that level, as indicated in **Figure 31–17**.

The next higher energy level corresponds to $n = 2$ and $\ell = 0$. Again, two states correspond to this level, allowing it to hold two electrons. Lest we conclude that all levels can hold two electrons, consider the next level up: $n = 2$ and $\ell = 1$.

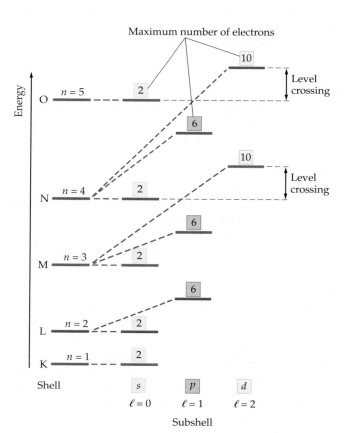

▶ **FIGURE 31–17 Maximum number of electrons in an energy level**
The maximum number of electrons that can occupy a given energy level in a multielectron atom is $2(2\ell + 1)$. Occupancy by more than this number of electrons would violate the Pauli exclusion principle.

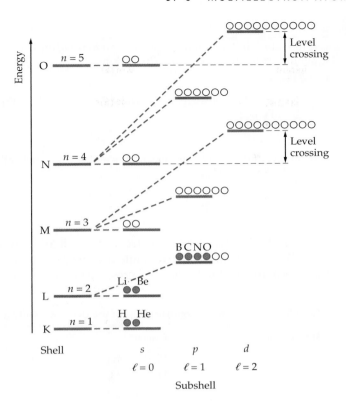

The elements hydrogen and helium fill the $n = 1$, $\ell = 0$ level; the elements lithium and beryllium fill the $n = 2$, $\ell = 0$ level; the elements boron, carbon, nitrogen, and oxygen fill four of the six available states in the $n = 2$, $\ell = 1$ level.

In this case, m_ℓ can take on three values: $m_\ell = 0, \pm 1$. For each of these three values, m_s can take on two values. Therefore, the $n = 2$, $\ell = 1$ energy level can accommodate 6 electrons, as shown in Figure 31–17. For general ℓ, the number of possible values of m_ℓ ($0, \pm 1, \pm 2, \ldots, \pm \ell$) is $2\ell + 1$. When we multiply by 2 (for the number of values of m_s), we find a total number of states equal to $2(2\ell + 1)$. For example, note in Figure 31–17 that the $n = 3$, $\ell = 2$ energy level can hold $2(2 \cdot 2 + 1) = 10$ electrons.

Figure 31–18 presents the ground-state electron arrangements for the following elements: hydrogen (1 electron); helium (2 electrons); lithium (3 electrons); beryllium (4 electrons); boron (5 electrons); carbon (6 electrons); nitrogen (7 electrons); and oxygen (8 electrons). Notice that the energy levels are filled from the bottom upward, like marbles in a jar, with each level having a predetermined maximum number of electrons.

Electronic Configurations

Indicating the arrangements of electrons as in Figure 31–18 is instructive, but also somewhat cumbersome. This is especially so when we consider elements with a large number of electrons. To streamline the process, we introduce a shorthand notation that can be applied to all elements.

As an example of this notation, consider the element lithium, which has two electrons in the $n = 1$, $\ell = 0$ state and one electron in the $n = 2$, $\ell = 0$ state. This arrangement of electrons—referred to as an **electronic configuration**—will be abbreviated as follows:

$$1s^2 2s^1$$

In this expression, the $1s^2$ part indicates $n = 1$ ($1s^2$), $\ell = 0$ ($1s^2$), and an occupancy of two electrons ($1s^2$). Similarly, the $2s^1$ part indicates one electron ($2s^1$) in the $n = 2$ ($2s^1$) and $\ell = 0$ ($2s^1$) state. The general labeling is indicated in **Figure 31–19**.

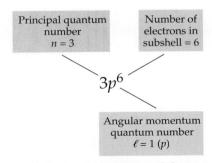

▲ **FIGURE 31–19 Designation of electronic configuration**

The example shown here indicates six electrons in the $n = 3$, $\ell = 1$ (p) energy level.

EXERCISE 31–3

Use the shorthand notation just introduced to write the electronic configuration for the ground state of **(a)** nitrogen and **(b)** sodium.

SOLUTION

a. The $1s$ and $2s$ subshells of nitrogen $(Z = 7)$ are filled; three electrons are in the $2p$ subshell:

$$1s^2\,2s^2\,2p^3$$

b. In sodium $(Z = 11)$, one electron is in the $3s$ subshell; all the inner subshells are fully occupied:

$$1s^2\,2s^2\,2p^6\,3s^1$$

Table 31–4 presents a list of electronic configurations for the elements hydrogen (H) through potassium (K). Note that the $3d$ levels in potassium have no electrons, in agreement with the level crossing shown in Figure 31–18.

TABLE 31–4 Electronic Configurations of the Elements Hydrogen through Potassium

Atomic number	Element	Electronic configuration
1	Hydrogen (H)	$1s^1$
2	Helium (He)	$1s^2$
3	Lithium (Li)	$1s^2 2s^1$
4	Beryllium (Be)	$1s^2 2s^2$
5	Boron (B)	$1s^2 2s^2 2p^1$
6	Carbon (C)	$1s^2 2s^2 2p^2$
7	Nitrogen (N)	$1s^2 2s^2 2p^3$
8	Oxygen (O)	$1s^2 2s^2 2p^4$
9	Fluorine (F)	$1s^2 2s^2 2p^5$
10	Neon (Ne)	$1s^2 2s^2 2p^6$
11	Sodium (Na)	$1s^2 2s^2 2p^6 3s^1$
12	Magnesium (Mg)	$1s^2 2s^2 2p^6 3s^2$
13	Aluminum (Al)	$1s^2 2s^2 2p^6 3s^2 3p^1$
14	Silicon (Si)	$1s^2 2s^2 2p^6 3s^2 3p^2$
15	Phosphorus (P)	$1s^2 2s^2 2p^6 3s^2 3p^3$
16	Sulfur (S)	$1s^2 2s^2 2p^6 3s^2 3p^4$
17	Chlorine (Cl)	$1s^2 2s^2 2p^6 3s^2 3p^5$
18	Argon (Ar)	$1s^2 2s^2 2p^6 3s^2 3p^6$
19	Potassium (K)	$1s^2 2s^2 2p^6 3s^2 3p^6 4s^1$

The Periodic Table

Referring to the elements listed in Table 31–4, we observe a number of interesting patterns. For example, notice that the elements hydrogen, lithium, sodium, and potassium all have the same type of electronic configuration for the final (and outermost) electron in the atom. In particular, hydrogen's outermost (and only) electron is the $1s^1$ electron. In the case of lithium, we see that the outermost electron is $2s^1$, the outermost electron of sodium is $3s^1$, and for potassium it is $4s^1$. Continuing through the list of elements, we note that rubidium has a $5s^1$ outer electron, cesium has a $6s^1$ outer electron, and francium has a $7s^1$ outer electron. In each case the outermost electron is a single electron in an otherwise empty s subshell.

What makes this similarity in *electronic configuration* of particular interest is that these elements have similar *chemical properties* as well. In particular, the metallic members of this group of elements (lithium, sodium, potassium, rubidium, cesium, and francium) are referred to as the *alkali metals*. In each of these metals, the outermost electron is easily removed from the atom, leading to a stable, positively charged ion, as in the familiar case of the sodium ion, Na$^+$. Thus the regular pattern in the filling of shells leads to a regular pattern in the properties of the elements.

Grouping various elements with similar chemical properties was the motivation behind the development of the **periodic table** of elements by the Russian chemist Dmitri Mendeleev (1834–1907). The periodic table is presented in Appendix E. Notice that the elements just mentioned form Group I in the leftmost column of the table. Although Mendeleev grouped these elements strictly on the basis of their chemical properties, we can now see that the grouping also corresponds to the filling of shells, in accordance with the Pauli exclusion principle.

Many groups of elements appear in the periodic table. As another example, Group VII consists of the *halogens:* fluorine, chlorine, bromine, and iodine. Note that these elements also have similar configurations of their outer electrons. In this case, the outer electrons are $2p^5$, $3p^5$, and so on. Thus we see that these elements are just one electron short of filling one of the p subshells. As a result, halogens are highly reactive—they can readily acquire a single electron from another element to form a stable negative ion, as in the case of chlorine, which forms the chloride ion, Cl$^-$.

Group VIII consists of the *noble gases*. These elements all have completely filled subshells. Thus they do not readily gain or lose an electron. It is for this reason that the noble gases are relatively inert.

Finally, the *transition elements* represent those cases where a crossing of energy levels occurs. For example, the $4s$ subshell is filled at the element calcium ($Z = 20$). The next electron, rather than going into the $4p$ subshell, goes into the $3d$ subshell. In fact, the 10 elements from scandium ($Z = 21$) to zinc ($Z = 30$) correspond to filling of the $3d$ subshell. After this subshell is filled, additional electrons go into the $4p$ subshell in the elements gallium ($Z = 31$) to krypton ($Z = 36$).

Figure 31–20 shows the meaning of the various symbols used in the periodic table. Note that each box in the table gives the symbol for the element, its atomic mass, and its atomic number. Each box also includes the configuration of the outermost electrons in the element. In the case of iron, shown in Figure 31–20, this configuration is $3d^6 4s^2$.

▲ Neon (atomic number 10), one of the "noble gases," is most widely known for its role in neon tubes. Electrons in neon atoms are excited by electrical discharge through the tube. When they return to the ground state, they emit electromagnetic radiation, much of it in the red part of the visible spectrum. The other colors familiarly found in "neon signs" or "neon lights" are produced by adding other elements of the same chemical family: argon, krypton, or xenon.

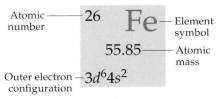

▲ **FIGURE 31–20 Designation of elements in the periodic table**
An explanation of the various entries to be found for each element in the periodic table of Appendix E. The example shown here is for the element iron (Fe).

CONCEPTUAL CHECKPOINT 31–3 COMPARE THE ENERGY

The energy required to remove the outermost electron from sodium is 5.1 eV. Is the energy required to remove the outermost electron from potassium **(a)** greater than, **(b)** less than, or **(c)** equal to 5.1 eV?

REASONING AND DISCUSSION
Referring to the periodic table, we see that potassium has one electron in its outermost shell, just like sodium. In this respect the two elements are alike. The difference between the elements is that the outermost electron in potassium is in a higher energy state; that is, its electron is in an $n = 4$ state, as opposed to the $n = 3$ state in sodium. Less energy is required to remove an electron from a state of higher energy; hence, the outermost electron of potassium can be removed with less than 5.1 eV. In fact, the required energy is only 4.3 eV.

ANSWER
(b) Less energy is required to remove the outermost electron from potassium.

31–7 Atomic Radiation

We conclude this chapter with a brief investigation of various types of radiation associated with multielectron atoms. Examples range from X-rays that are energetic enough to pass through a human body, to the soft white light of a fluorescent lightbulb.

X-rays

X-rays were discovered quite by accident by the German physicist Wilhelm Roentgen (1845–1923) on November 8, 1895. Within months of their discovery they were being used in medical applications, and they have played an important role in medicine ever since. Today, the X-rays used to give diagnostic images in hospitals and dentist offices are produced by an X-ray tube similar to the one shown in **Figure 31–21.** The basic operating principle of this device is that an energetic beam of electrons is generated and directed at a metal target—when the electrons collide with the target, X-rays are emitted. **Figure 31–22** shows a typical plot of X-ray intensity per wavelength versus the wavelength for such a device.

REAL-WORLD PHYSICS: BIO
Medical X-ray tubes

▶ **FIGURE 31–21 X-ray tube**
An X-ray tube accelerates electrons in a vacuum through a potential difference and then directs the electrons onto a metal target. X-rays are produced when the electrons decelerate in the target, and as a result of the excitations they cause when they collide with the metal atoms of the target.

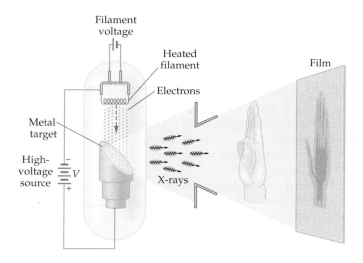

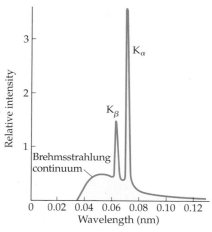

▲ **FIGURE 31–22 X-ray spectrum**
The spectrum produced by an X-ray tube in which electrons are accelerated from rest through a potential difference of 35,000 V and directed against a molybdenum target.

The radiation produced by an X-ray tube is created by two completely different physical mechanisms. The first mechanism is referred to as **bremsstrahlung,** which is German for "braking radiation." What is meant by this expression is that as the energetic electrons impact the target, they undergo a rapid deceleration. This is the "braking." As we know from Chapter 25, an accelerated charge gives off electromagnetic radiation; hence, as the electrons suddenly come to rest in the target they give off high-energy radiation in the form of X-rays. These X-rays cover a wide range of wavelengths, giving rise to the continuous part of the spectrum shown in Figure 31–22.

The sharp peaks in Figure 31–22 are produced by the second physical mechanism. To understand their origin, imagine what happens if one of the electrons in the incident beam is energetic enough to knock an electron out of a target atom. In addition, suppose the ejected electron comes from the lowest energy level of the atom, that is, from the K shell. This "vacancy" is filled almost immediately when an electron from an outer shell drops to the K shell, with the energy difference given off as a photon. In an atom of large atomic number [molybdenum ($Z = 42$) is often used for a target, as is tungsten ($Z = 74$)], the photon that emerges is an X-ray. If an electron drops from the $n = 2$ level to the K shell, the sharp peak of radiation that results is called the K_α line. Similarly, if an electron drops from the $n = 3$ level to the K shell, the resulting peak is called the K_β line. These lines are shown in Figure 31–22, and the corresponding electron jumps are indicated in **Figure 31–23.** Because the wavelengths of these lines vary from element to element—that is, they are characteristic of a certain element—they are referred to as **characteristic X-rays.** Similar characteristic X-rays can be emitted when electrons drop in energy to fill a vacancy in the L shell, the M shell, and so on.

The energy an incoming electron must have to dislodge a K-shell electron from an atom can be estimated using results from the Bohr model. The basic idea is that the energy of a K-shell electron in an atom of atomic number Z is given approximately by Equation 31–9 [$E_n = -(13.6\,\text{eV})Z^2/n^2$], with one minor modification: since there are two electrons in the K shell, each electron shields the other from the nucleus. That is, the negative charge on one electron, $-e$, partially cancels the positive charge of the nucleus, $+Ze$, giving an effective charge experienced by the second electron of $+(Z - 1)e$. Thus, replacing Z with $Z - 1$ in Equation 31–9, and setting $n = 1$, we obtain a reasonable estimate for the energy of a K-shell electron:

$$E_K = -(13.6 \text{ eV}) \frac{(Z-1)^2}{1^2} \qquad 31\text{–}14$$

We apply this result in the following Active Example.

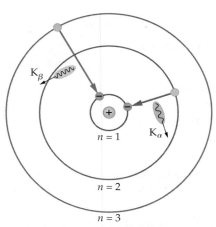

▲ **FIGURE 31–23 Production of characteristic X-rays**

When an electron strikes a metal atom in the target of an X-ray tube, it may knock one of the two K-shell ($n = 1$) electrons out of the atom. The resulting vacancy in the K shell will be filled by an electron dropping from a higher shell. If an electron drops from the $n = 2$ shell to the $n = 1$ shell, we say that the resulting photon is a K_α X-ray. Similarly, if an electron in the $n = 3$ shell drops to the $n = 1$ shell, the result is a K_β X-ray. Clearly, the K_β X-ray has the greater energy and the shorter wavelength, as we see in Figure 31–22.

ACTIVE EXAMPLE 31–3 **FIND THE VOLTAGE OF AN X-RAY TUBE**

Estimate the minimum energy an incoming electron must have to knock a K-shell electron out of a tungsten atom ($Z = 74$).

SOLUTION *(Test your understanding by performing the calculations indicated in each step.)*

1. Calculate $(Z - 1)^2$:   5329
2. Multiply -13.6 eV by $(Z - 1)^2$: $-72{,}500$ eV

INSIGHT

An electron at rest an infinite distance from a tungsten atom has an energy of zero. Therefore, a minimum energy of 72,500 eV must be supplied to the electron to remove it from the K shell of tungsten.

Recall that an electron gains an energy of 1 eV when it accelerates through a potential difference of 1 V. Thus a potential difference of at least 72,500 V = 72.5 kV is required to produce characteristic X-rays with a tungsten target. Typical X-ray tubes, like those used in dental offices, operate in the 100-kV range.

YOUR TURN

Suppose an X-ray tube has a voltage of only 35 kV. What is the largest value of Z for which this tube can knock out a K-shell electron?

(Answers to **Your Turn** *problems are given in the back of the book.)*

Equation 31–14 can also be used to obtain an estimate of the K_α wavelength for a given element. We show how this can be done for the case of molybdenum in the next Example.

EXAMPLE 31–4 K_α **FOR MOLYBDENUM**

Estimate the K_α wavelength for molybdenum ($Z = 42$).

PICTURE THE PROBLEM

Our sketch indicates the electron jump that is responsible for the K_α X-ray; that is, from $n = 2$ to $n = 1$. Note that the net charge from the K shell outward is $+(Z - 1)e$.

STRATEGY

To find the wavelength of a K_α X-ray, we start with the relationship between the change in energy of an electron and the wavelength of the corresponding photon, $|\Delta E| = hf = hc/\lambda$. Once we have calculated ΔE, we can find the wavelength using $\lambda = hc/|\Delta E|$.

To find ΔE, we first calculate the energy of an electron in the K shell of molybdenum, using $E_K = -(13.6 \text{ eV})(Z - 1)^2/1^2$ with $Z = 42$.

Next, we calculate the energy of an electron in the L shell ($n = 2$) of molybdenum, since it is an L-shell electron that fills the vacancy in the K shell and emits a K_α X-ray. Note that an L-shell electron sees a nucleus with an effective charge of $+(Z - 1)e$. This follows because there is one electron in the K shell, and this electron partially screens the nucleus. As a result, the energy of an L-shell electron is given by the following expression: $E_L = -(13.6 \text{ eV})(Z - 1)^2/2^2$.

Finally, with these two energies determined, the change in energy is simply $\Delta E = E_K - E_L$.

One electron in K shell
$-e$
$n = 1$
$+Ze$
K_α
$n = 2$
$n = 3$

CONTINUED ON NEXT PAGE

CONTINUED FROM PREVIOUS PAGE

SOLUTION

1. Calculate the energy of a K-shell electron, E_K:

$$E_K = -(13.6 \text{ eV})\frac{(Z-1)^2}{1^2} = -(13.6 \text{ eV})\frac{(42-1)^2}{1^2}$$
$$= -22,900 \text{ eV}$$

2. Calculate the energy of an L-shell electron, E_L:

$$E_L = -(13.6 \text{ eV})\frac{(Z-1)^2}{2^2} = -(13.6 \text{ eV})\frac{(42-1)^2}{2^2}$$
$$= -5720 \text{ eV}$$

3. Determine the change in energy, ΔE, of an electron that jumps from the L shell to the K shell:

$$\Delta E = E_K - E_L = -22,900 \text{ eV} - (-5720 \text{ eV}) = -17,200 \text{ eV}$$

4. Calculate the wavelength corresponding to ΔE:

$$\lambda = \frac{hc}{|\Delta E|} = \frac{(6.63 \times 10^{-34} \text{ J} \cdot \text{s})(3.00 \times 10^8 \text{ m/s})}{(17,200 \text{ eV})(1.60 \times 10^{-19} \text{ J/eV})}$$
$$= 7.23 \times 10^{-11} \text{ m} = 0.0723 \text{ nm}$$

INSIGHT

Comparing our result with Figure 31–22, which shows the X-ray spectrum for molybdenum, we see that our approximate wavelength of 0.0723 nm is in good agreement with experiment.

PRACTICE PROBLEM

Which element has a K_α peak at a wavelength of approximately 0.155 nm?   [**Answer:** Copper, $Z = 29$]

Some related homework problems: Problem 61, Problem 62, Problem 81

As mentioned earlier in this section, X-rays were put to medical use as soon as people found out about their properties and how to produce them. In fact, Roentgen himself, whose first article about X-rays was published in late December 1895, produced an X-ray image of his wife's left hand, clearly showing the bones in her fingers and her wedding ring. Less than two months later, in February 1896, American physicians were starting to test them on patients. One of the earliest patients was a young boy named Eddie McCarthy, who had his broken forearm X-rayed. A New Yorker by the name of Tolson Cunningham had a bullet removed from his leg after its position was determined by a 45-minute X-ray exposure.

Standard X-rays can be difficult to interpret, however, since they cast shadows of all the body materials they pass through onto a single sheet of film. It is somewhat like placing several transparencies on top of one another and trying to decipher their individual contents. With the advent of high-speed computers, a new type of X-ray image is now possible. In a **computerized axial tomography scan** (CAT scan), thin beams of X-rays are directed through the body from a variety of directions. The intensity of the transmitted beam is detected for each direction, and the results are sent to a computer for processing. The result is an image that shows the physician a "cross-sectional slice" through the body. In this way, each part of the body can be viewed individually and with clarity. If a series of such slices are stacked together in the computer, they can give a three-dimensional view of the body's interior.

REAL-WORLD PHYSICS: BIO

Computerized axial tomography

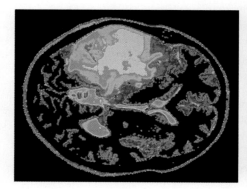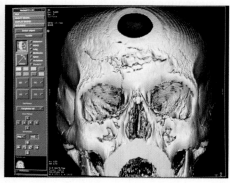

▶ The false-color CAT scan at left represents a horizontal section through the brain, revealing a large benign tumor (the white and orange area at top). A series of CAT scans can also be combined to create remarkable three-dimensional images such as the one at right.

Lasers

The production of light by humans advanced significantly when flames were replaced by the lightbulb. An even greater advancement occurred in 1960, however, when the first laser was developed. Lasers produce light that is intense, highly collimated (uni-directional), and pure in its color. Because of these properties, lasers are used in a multitude of technological applications, ranging from supermarket scanners to CDs, from laser pointers to eye surgery. In fact, lasers are now almost as common in everyday life as the lightbulb.

To understand just what a laser is and what makes it so special, we start with its name. The word **laser** is an acronym for **l**ight **a**mplification by the **s**timulated **e**mission of **r**adiation. As we shall see, the properties of stimulated emission lead directly to the amplification of light.

To begin, consider two energy levels in an atom, and suppose an electron occupies the higher of the two levels. If the electron is left alone, it will eventually drop to the lower level in a time that is typically about 10^{-8} s, giving off a photon in the process referred to as **spontaneous emission.** The photon given off in this process can propagate in any direction.

In contrast, suppose the electron in the excited state just described is not left alone. For example, a photon with an energy equal to the energy difference between the two energy levels might pass near the electron, which *enhances* the probability that the electron will drop to the lower level. That is, the incident photon can *stimulate* the emission of a second photon by the electron. The photon given off in this process of **stimulated emission** has the same energy as the incident photon, the same phase, and propagates in the same direction, as indicated in **Figure 31–24**. This accounts for the fact that laser light is highly focused and of a single color.

As for the amplification of light, notice that a single photon entering an excited atom can cause two identical photons to exit the atom. If each of these two photons encounters another excited atom and undergoes the same process, the number of photons increases to four. Continuing in this manner, the photons undergo a sort of "chain reaction" that doubles the number of photons with each generation. It is this property of stimulated emission that results in **light amplification.**

In order for the light amplification process to work, it is necessary that photons continue to encounter atoms with electrons in excited states. Under ordinary conditions this will not be the case, since most electrons are in the lowest possible energy levels. For laser action to occur, atoms must first be prepared in an excited state. Then, before the electrons have a chance to drop to a lower level by way of spontaneous emission, the process of stimulated emission can proceed. This requires what is known as a **population inversion,** in which more electrons are found in the excited state than in a lower state. In addition, the excited state must be one that lasts for a relatively long time so that photons will continue to encounter excited atoms. A long-lived excited state is referred to as a **metastable state.** So to produce a laser, one needs a metastable excited state with a population inversion.

A specific example of a laser is the **helium-neon laser,** shown schematically in **Figure 31–25 (a)**, in which the neon atoms produce the laser light. The appropriate energy-level diagrams for neon are shown in **Figure 31–25 (b)**. The excited state E_3 is metastable—electrons promoted to that level stay in the level for a relatively long time. Electrons are excited to this level by an electrical power supply connected to the tube containing the helium-neon mixture. The power supply causes electrons to move through the tube, colliding with neon atoms and exciting them. Similarly, excited helium atoms colliding with the neon atoms also cause excitations to the E_3 level. Since this level is metastable, the excitation processes can cause a population inversion, setting the stage for laser action. Stimulated emission then occurs, allowing the electrons to drop to the lower level, E_2. The electrons subsequently proceed through various intermediate steps to the ground state. It is the emission of light between the levels E_3 and E_2 that results in laser light. Since the difference in energy between these levels is 1.96 eV, the light coming out of the laser is red, with a wavelength of 633 nm.

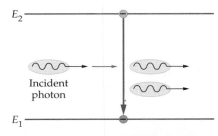

▲ **FIGURE 31–24** Stimulated emission
A photon with an energy equal to the difference $E_2 - E_1$ can enhance the probability that an electron in the state E_2 will drop to the state E_1 and emit a photon. When a photon stimulates the emission of a second photon in this way, the new photon has the same frequency, direction, and phase as the incident photon. If each of the two photons resulting from this process in turn produces two photons, the total number of photons can increase exponentially, resulting in an intense beam of light.

REAL-WORLD PHYSICS

Helium-neon laser

▶ **FIGURE 31–25 The helium-neon laser**
(a) A schematic representation of the basic features of a helium-neon laser. **(b)** The relevant energy levels in helium and neon that result in the laser action.

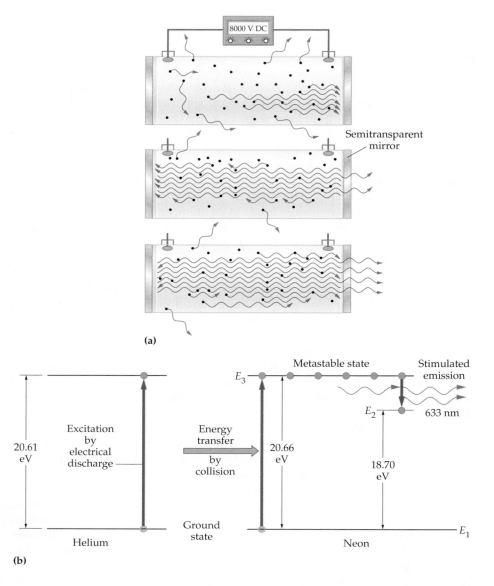

(a)

(b)

Helium

Ground state

Neon

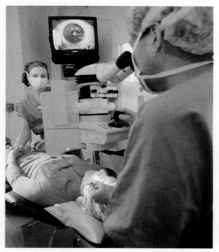

▲ A doctor uses a microscope and video monitor to ensure precise cutting of the corneal flap in the LASIK procedure. A UV laser will then remove some of the material under the flap, flattening the cornea and correcting the patient's nearsightedness.

REAL-WORLD PHYSICS: BIO

Laser eye surgery

To enhance the output of a laser, the light is reflected back and forth between mirrors. In fact, this reflection produces a resonant condition in the light, much like standing waves in an organ pipe. A schematic of a laser in operation is shown in **Figure 31–25 (a)**.

A recent medical application of lasers involves several types of *laser eye surgery*. In these techniques, a laser that emits high-energy photons in the UV range (typically at wavelengths of 193 nm) is used to reshape the cornea and correct nearsightedness. For example, in LASIK (**la**ser **i**n **si**tu **k**eratomileusis) eye surgery the procedure begins with a small mechanical shaver known as a microkeratome cutting a flap in the cornea, leaving a portion of the cornea uncut to serve as a hinge. After the mechanical cut is made, the corneal flap is folded back, exposing the middle portion of the cornea as shown in **Figure 31–26 (a)**. Next, an *excimer laser* sends pulses of UV light onto the cornea, each pulse vaporizing a small layer of corneal material (0.1 to 0.5 μm in thickness) with no heating. This process continues until the cornea is flattened just enough to correct the nearsightedness, after which the corneal flap is put back into place.

Photorefractive keratectomy (PRK) is similar to LASIK eye surgery, except that material is removed directly from the surface of the cornea, without the use of a corneal flap, as shown in **Figure 31–26 (b)**. To correct nearsightedness, the laser beam is directed onto the central portion of the cornea (left), resulting in a flattening of the cornea. To correct farsightedness, it is necessary to increase the

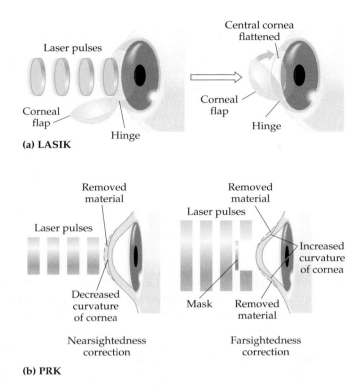

(a) LASIK

Laser pulses

Corneal
flap

Hinge

Central cornea
flattened

Corneal
flap

Hinge

Removed
material

Laser pulses

Decreased
curvature
of cornea

Nearsightedness
correction

Removed
material

Laser pulses

Mask   Removed
material

Increased
curvature
of cornea

Farsightedness
correction

(b) PRK

◀ **FIGURE 31–26 Laser vision correction**
(a) In LASIK eye surgery, a flap of the cornea is cut and folded back. Next, an ultraviolet excimer laser is used to vaporize some of the underlying corneal material. When the flap is replaced, the cornea is flatter than it was, correcting the patient's nearsightedness. **(b)** In the PRK procedure, the laser removes material directly from the corneal surface. If the cornea is too curved, producing nearsightedness, the laser beam is directed at its center and the cornea is flattened. If the cornea is too flat, producing farsightedness, the central region is masked and material is removed from the periphery, increasing the curvature.

curvature of the cornea. This is accomplished by masking the central portion of the cornea so that the laser removes only peripheral portions of the cornea (right). In both cases, it is necessary to keep the beam focused at the desired location on the eye. This is difficult, because the eye routinely moves by small amounts roughly every 15 ms. In the most sophisticated application of PRK, these eye movements can be tracked and the aiming of the laser beam corrected accordingly.

Another medical application of lasers is known as *photodynamic therapy,* or PDT. In this type of therapy, light-sensitive chemicals (such as porphyrins) are injected into the bloodstream and are taken up by cells throughout the body. These chemicals are found to remain in cancerous cells for greater periods of time than in normal cells. Thus, after an appropriate time interval, the light-sensitive chemicals are preferentially concentrated in cancerous cells. If a laser beam with the precise wavelength absorbed by the light-sensitive chemicals illuminates the cancer cells, the resulting chemical reactions kill the cancer cells without damaging the adjacent normal cells. The laser beam can be directed to the desired location using a flexible fiber-optic cable in conjunction with a bronchoscope to treat lung cancer or with an endoscope to treat esophageal cancer.

PDT is also used to treat certain types of age-related macular degeneration (AMD). For most people, macular degeneration results when abnormal blood vessels behind the retina leak fluid and blood into the central region of the retina, or macula, causing it to degenerate. In this case, light-sensitive chemicals are preferentially taken up by the abnormal blood vessels; hence, a laser beam of the proper wavelength can destroy these blood vessels without damaging the normal structures in the retina.

Finally, lasers can also be used to take three-dimensional photographs known as **holograms.** A typical setup for taking a hologram is shown in **Figure 31–27**. Notice that the hologram is produced with no focusing lenses, in contrast with normal photography. The basic procedure in holography begins with the splitting of a laser beam into two separate beams. One beam, the reference beam, is directed onto the photographic film. The second beam, called the object beam, is directed onto the object to be recorded in the hologram. The object beam reflects from various parts of the object and then combines with the reference beam on the film. Because the laser light is coherent, and because the object and reference beams travel

REAL-WORLD PHYSICS: BIO
Photodynamic therapy

▲ A hologram creates a three-dimensional image in empty space. The image can be viewed from different angles to reveal different parts of the original subject.

REAL-WORLD PHYSICS
Holography

▶ **FIGURE 31–27 Holography**

To create a hologram, laser light is split into two beams. One, the reference beam, is directed onto the photographic film. The other is reflected from the surface of an object onto the film, where it combines with the reference beam to create an interference pattern. When this pattern is illuminated with laser light of the same wavelength, a three-dimensional image of the original object is produced.

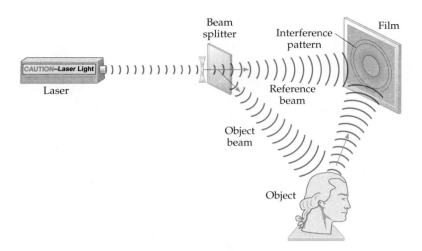

different distances, the combined light results in an interference pattern. In fact, if you look at a hologram in normal light it simply looks like a confusing mass of swirls and lines. When a laser beam illuminates the hologram, however, the interference pattern causes the laser light to propagate away from the hologram in exactly the same way as the light that originally produced the interference pattern. Thus, a person viewing the hologram sees precisely the same patterns of light that would have been observed when the hologram was recorded.

Holograms give a true three-dimensional image. When you view a hologram, you can move your vantage point to see different parts of the scene that is recorded. In particular, viewing the hologram from one angle may obscure an object in the background, but by moving your head, you can look around the foreground objects to get a clear view of the obscured object. In addition, you have to adjust your focus as you shift your gaze from foreground to background objects, just as in the real world. Finally, if you cut a hologram into pieces, each piece still shows the entire scene! This is analogous to your ability to see everything in your front yard through a small window just as you can through a large window—both show the entire scene.

The holograms on your credit cards are referred to as *rainbow holograms*, because they are designed to be viewed with white light containing all the colors of the rainbow. Although these holograms give an impression of three-dimensionality, they do not compare in quality to a hologram viewed with a laser.

Fluorescence and Phosphorescence

In the Insight to Active Example 31–1 it was noted that an electron in an excited state can emit photons of various energies as it falls to the ground state. This type of behavior is at the heart of fluorescence and phosphorescence.

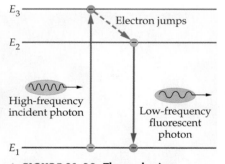

▲ **FIGURE 31–28   The mechanism of fluorescence**

In fluorescence, a high-frequency photon raises an electron to an excited state. When the electron drops back to the ground state, it may do so by way of various intermediate states. The jumps between intermediate states produce photons of lower frequency, which are observed as the phenomenon of fluorescence.

REAL-WORLD PHYSICS

Fluorescent lightbulbs

Consider the energy levels shown in **Figure 31–28**. If an atom with these energy levels absorbs a photon of energy $E_3 - E_1$, it can excite an electron from state E_1 to state E_3. In some atoms, the most likely way for the electron to return to the ground state is by first jumping to level E_2 and then jumping to level E_1. The photons emitted in these jumps have less energy than the photon that caused the excitation in the first place. Hence, in a system like this, an atom is illuminated with a photon of one frequency, and it subsequently emits photons of lower energy and lower frequency. The emission of light of lower frequency after illumination by a higher frequency is referred to as **fluorescence.** In essence, fluorescence can be thought of as a conversion process, in which photons of high frequency are converted to photons of lower frequency.

Perhaps the most common example of fluorescence is the *fluorescent lightbulb*. This device uses fluorescence to convert high-frequency ultraviolet light to lower-frequency visible light. In particular, the tube of a fluorescent light contains mercury vapor. When electricity is applied to one of these tubes, a filament is heated, producing electrons. These electrons are accelerated by an applied voltage. The

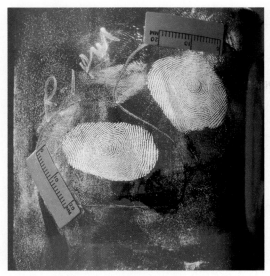

▲ The fingerprints on this mug become clearly visible when treated with fluorescent dye and illuminated by ultraviolet light.

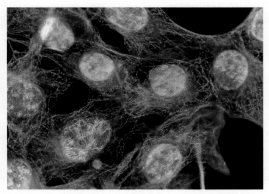

▲ In the technique known as immunofluorescence, an antibody molecule is linked to a fluorescent dye molecule, making it possible to visualize cellular structures and components that are otherwise largely invisible. This photograph utilized antibodies that bind to a protein found in intermediate filaments. Such filaments (seen here as lacy fibers resembling a spiderweb) are a component of the cytoskeleton, the cellular scaffolding that enables cells to maintain or change their shape and move materials about.

electrons strike mercury atoms in the tube, exciting them, and they give off ultraviolet light as they decay to their ground state. This process is not particularly useful in itself, since the ultraviolet light is invisible to us. However, the inside of the tube is coated with a phosphor that absorbs the ultraviolet light and then emits a lower-frequency light that is visible. Thus, several different physical processes must take place before a fluorescent lightbulb produces visible light.

Fluorescence finds many other less familiar applications as well. In forensics, the analysis of a crime scene is enhanced by the fact that human bones and teeth are fluorescent. Thus, illuminating a crime scene with ultraviolet light can make items of interest stand out for easy identification. In addition, the use of a fluorescent dye can make fingerprints visible with great clarity.

Many creatures produce fluorescence in their bodies, as well. For example, several types of coral glow brightly when illuminated with ultraviolet light. It is also well known that scorpions are strongly fluorescent, giving off a distinctive green light. In fact, it is often possible to discern a greenish cast when viewing a scorpion in sunlight. At night in the desert, scorpions stand out with a bright green glow when a person illuminates the area with a portable ultraviolet light. This aids researchers who would like to find certain scorpions for study, and campers who are just as interested in avoiding scorpions altogether.

The green fluorescence produced by the jellyfish *Aequorea victoria* finds many uses in biological experiments. The gene that produces the green fluorescent protein (GFP) can serve as a marker to identify whether an organism has incorporated a new segment of DNA into its genome. For example, bacterial colonies that incorporate the GFP gene can be screened by eye simply by viewing the colony

REAL-WORLD PHYSICS: BIO
Applications of fluorescence in forensics

REAL-WORLD PHYSICS: BIO
Detecting scorpions at night

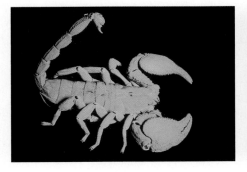

◀ A variety of creatures, including the scorpion at left, are naturally fluorescent when illuminated with ultraviolet light. So too are many minerals, such as the one at right.

REAL-WORLD PHYSICS: BIO
The GFP bunny

under an ultraviolet light. Recently, GFP has been inserted into the genome of a white rabbit, giving rise to the "GFP bunny." The bunny appears normal in white light, but when viewed under light with a wavelength of 392 nm, it glows with a bright green light at 509 nm (see p. 1078). The fluorescence spectrum of GFP is shown in **Figure 31–29**.

Phosphorescence is similar to fluorescence, except that phosphorescent materials continue to give off a secondary glow long after the initial illumination that excited the atoms. In fact, phosphorescence may persist for periods of time ranging from a few seconds to several hours, as on the hands of a watch that glow in the dark.

▶ **FIGURE 31–29 The fluorescence spectrum of GFP**
The green fluorescent protein (GFP) strongly absorbs light with a wavelength of about 400 nm (violet). It reemits green light with a wavelength of 509 nm.

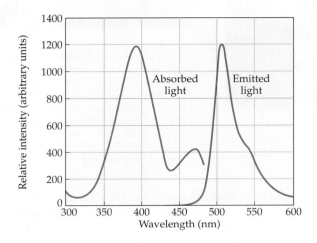

THE BIG PICTURE PUTTING PHYSICS IN CONTEXT

LOOKING BACK

We've come a long way in our study of physics, from its earliest beginnings with Galileo and Newton to the quantum revolution of the twentieth century with Einstein, Planck, Bohr, Schrödinger, Heisenberg, and others. As we look back on this journey, we see that even the most exotic predictions and discoveries of modern physics are grounded in the fundamentals that have been presented throughout this book. For example:

Bohr orbits (Section 31–3) for the hydrogen atom are calculated in the same way that gravitational orbits of planets and satellites were calculated in Chapter 12.

The energy of a Bohr orbit (Section 31–3) is determined using the electric potential energy introduced in Chapter 20.

Photons emitted from a hydrogen atom in the Bohr model obey the relation $E = hf$, which was used to understand blackbody radiation and the photoelectric effect in Chapter 30.

And finally, we showed in Section 31–4 that Bohr orbits can be understood in terms of standing waves, just like the standing waves on a string studied in Chapter 14, only this time using the de Broglie waves introduced in Chapter 30.

CHAPTER SUMMARY

31–1 EARLY MODELS OF THE ATOM

Atoms are the smallest unit of a given element. If an atom is broken down into smaller pieces, it loses the properties that characterized the element.

The Thomson Model: Plum Pudding
In Thomson's model, an atom is imagined to be like a positively charged pudding with negatively charged electrons scattered throughout.

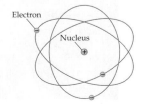

The Rutherford Model: A Miniature Solar System
Rutherford discovered that an atom is somewhat like an atomic-scale solar system: mostly empty space, with most of its mass concentrated in the nucleus. The electrons were thought to orbit the nucleus.

31–2 THE SPECTRUM OF ATOMIC HYDROGEN

Excited atoms of hydrogen in a low-pressure gas give off light of specific wavelengths. This is referred to as the spectrum of hydrogen.

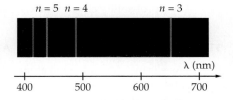

Line Spectra
The spectrum of hydrogen is a series of bright lines of well-defined wavelengths.

Series
Hydrogen's line spectrum is formed by a series of lines that are grouped together. The wavelengths of these series are given by the expression

$$\frac{1}{\lambda} = R\left(\frac{1}{n'^2} - \frac{1}{n^2}\right) \qquad n' = 1, 2, 3, \ldots$$
$$n = n' + 1, n' + 2, n' + 3, \ldots$$

31–2

Each line in a given series corresponds to a different value of n. The different series correspond to different values of n'. For example, $n' = 1$ is the Lyman series, $n' = 2$ is the Balmer series, and $n' = 3$ is the Paschen series.

31–3 BOHR'S MODEL OF THE HYDROGEN ATOM

Bohr's model of hydrogen is basically a solar-system model, with the electron orbiting the nucleus. In Bohr's model, however, only certain orbits are allowed.

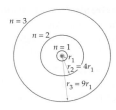

Assumptions of the Bohr Model
The Bohr model assumes the following: (i) Electrons move in circular orbits about the nucleus; (ii) allowed orbits must have an angular momentum equal to $L_n = nh/2\pi$, where $n = 1, 2, 3, \ldots$; (iii) electrons in allowed orbits do not give off electromagnetic radiation; and (iv) radiation is emitted only when electrons jump from one orbit to another.

Bohr Orbits
The radii of allowed orbits in the Bohr model are given by

$$r_n = \left(\frac{h^2}{4\pi^2 mkZe^2}\right)n^2 = (5.29 \times 10^{-11}\text{m})n^2 \quad n = 1, 2, 3 \ldots$$

31–7

The Energy of a Bohr Orbit
The energy of an allowed Bohr orbit is

$$E_n = -(13.6 \text{ eV})\frac{Z^2}{n^2} \quad n = 1, 2, 3, \ldots$$

31–9

These expressions correspond to hydrogen when $Z = 1$.

31–4 de BROGLIE WAVES AND THE BOHR MODEL

De Broglie was able to show that the allowed orbits of the Bohr model correspond to standing matter waves of the electrons. In particular, an allowed orbit in Bohr's model has a circumference equal to an integer times the wavelength of the electron in that orbit.

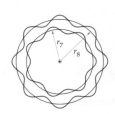

31–5 THE QUANTUM MECHANICAL HYDROGEN ATOM

The correct description of the hydrogen atom is derived from Schrödinger's equation. It agrees in many key ways with the Bohr model but has significant differences as well.

Quantum Numbers
The quantum mechanical hydrogen atom is described by four quantum numbers. They are as follows: (i) the principal quantum number, n, which is analogous to n in the Bohr model; (ii) the orbital angular momentum quantum number, ℓ, which takes on the values $\ell = 0, 1, 2, \ldots, (n - 1)$; the orbital angular momentum has a magnitude given by $L = \sqrt{\ell(\ell + 1)}(h/2\pi)$; (iii) the magnetic quantum number, m_ℓ, for which the allowed values are $m_\ell = -\ell, -\ell + 1, -\ell + 2, \ldots,$

$-1, 0, 1, \ldots, \ell - 2, \ell - 1, \ell$; the z component of the orbital angular momentum is $L_z = m_\ell(h/2\pi)$; and (iv) the electron spin quantum number, m_s, which can have the values $m_s = -\frac{1}{2}, \frac{1}{2}$.

Electron Probability Clouds: Three-Dimensional Standing Waves

In the quantum mechanical hydrogen atom, the electron does not orbit at a precise distance from the nucleus. Instead, the electron distribution is represented by a probability cloud, where the densest regions of the cloud correspond to regions of highest probability.

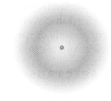

31–6 MULTIELECTRON ATOMS AND THE PERIODIC TABLE

As electrons are added to atoms, the properties of the atoms change in a regular and predictable way.

Multielectron Atoms

Energy levels in a multielectron atom depend on n and ℓ. The energy increases with increasing n for fixed ℓ, and with increasing ℓ for fixed n.

Shells and Subshells

Electrons with the same value of n are said to be in the same shell. Electrons in a given shell with the same value of ℓ are said to be in the same subshell.

The Pauli Exclusion Principle

The Pauli exclusion principle states that only a single electron may have a particular set of quantum numbers. This means that it is not possible for all the electrons in a multielectron atom to occupy the lowest energy level.

Electronic Configurations

The arrangement of electrons is indicated by the electronic configuration. For example, the configuration $1s^2$ indicates that 2 electrons $(1s^2)$ are in the $n = 1$ $(1s^2)$, $\ell = 0$ $(1s^2)$ state.

The Periodic Table

As electrons fill subshells of progressively higher energy, they produce the elements of the periodic table. Atoms with the same configuration of outermost electrons generally have similar chemical properties.

31–7 ATOMIC RADIATION

Atoms can give off radiation ranging from X-rays to visible light to infrared rays.

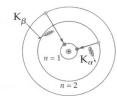

X-rays

X-rays characteristic of a particular element are given off when an electron in an inner shell is knocked out of the atom, and an electron from an outer shell drops down to take its place.

Lasers

A laser is a device that produces light amplification by the stimulated emission of radiation.

Fluorescence and Phosphorescence

When an electron in an atom is excited to a high energy level, it may return to the ground state through a series of lower-energy jumps. These jumps give off radiation of longer wavelength than the radiation that caused the original excitation.

PROBLEM-SOLVING SUMMARY

Type of Problem	Relevant Physical Concepts	Related Examples
Find the wavelength of a spectral line in hydrogen.	All the spectral lines of hydrogen are described by Equation 31–2. Note that n must be greater than n', and that $n' = 1$ gives the Lyman series, $n' = 2$ gives the Balmer series, $n' = 3$ gives the Paschen series, and so on.	Example 31–1 Exercise 31–1
Find the radius and energy of a Bohr orbit in hydrogen. Also determine the speed of the electron.	The radius of a Bohr orbit varies with n^2 as follows: $r_n = r_1 n^2 = (5.29 \times 10^{-11} \text{ m})n^2$. The speed of the electron varies as $1/n$, as shown in Equation 31–6. The energy associated with an orbit is the kinetic energy of the electron plus the potential energy of the system, $U = -ke^2/r$.	Example 31–2 Active Example 31–2

Determine the wavelength (or frequency) of a photon that is emitted or absorbed when an electron jumps from one Bohr orbit to another.

When an electron jumps from an initial state n_i to a final state n_f, the wavelength of the associated photon is given by

$$\frac{1}{\lambda} = \left(\frac{2\pi^2 m k^2 e^4}{h^3 c}\right)\left(\frac{1}{n_f^2} - \frac{1}{n_i^2}\right) = (1.097 \times 10^7 \text{ m}^{-1})\left(\frac{1}{n_f^2} - \frac{1}{n_i^2}\right)$$

If $n_i > n_f$, the photon is emitted; if $n_i < n_f$, the photon is absorbed. The photon's wavelength and frequency are related by the expression $\lambda f = c$.

Active Examples 31–1, 31–2

Calculate the wavelength of an electron in a Bohr orbit.

An integral number of wavelengths fit around the circumference of a Bohr orbit, starting with one wavelength for the ground state.

Example 31–3

Estimate the energy of a K-shell electron and the wavelength of a K_α X-ray for an element with atomic number Z.

The nuclear charge of $+Ze$ is partially screened by a K-shell electron, giving an effective nuclear charge of $+(Z-1)e$. Therefore, we use $Z - 1$ in Equation 31–9, which results in Equation 31–14.

Example 31–4
Active Example 31–3

CONCEPTUAL QUESTIONS

For instructor-assigned homework, go to www.masteringphysics.com

(Answers to odd-numbered Conceptual Questions can be found in the back of the book.)

1. Give a reason why the Thomson plum-pudding model does not agree with experimental observations.

2. Give a reason why the Rutherford solar-system model does not agree with experimental observations.

3. Cite one example of how the Bohr model disagrees with the quantum mechanical model of the hydrogen atom.

4. What observation led Rutherford to propose that atoms have a small nucleus containing most of the atom's mass?

5. Do you expect the light given off by (a) a neon sign or (b) an incandescent lightbulb to be continuous in distribution or in the form of a line spectrum? Explain.

6. In principle, how many spectral lines are there in any given series of hydrogen? Explain.

7. Is there an upper limit to the radius of an allowed Bohr orbit? Explain.

8. (a) Is there an upper limit to the wavelength of lines in the spectrum of hydrogen? Explain. (b) Is there a lower limit? Explain.

9. The principal quantum number, n, can increase without limit in the hydrogen atom. Does this mean that the energy of the hydrogen atom also can increase without limit? Explain.

10. For each of the following configurations of outermost electrons, state whether the configuration is allowed by the rules of quantum mechanics. If the configuration is not allowed, give the rule or rules that are violated. (a) $2d^1$, (b) $1p^7$, (c) $3p^5$, (d) $4g^6$.

11. (a) In the quantum mechanical model of the hydrogen atom, there is one value of n for which the angular momentum of the electron must be zero. What is this value of n? (b) Can the angular momentum of the electron be zero in states with other values of n? Explain.

12. Would you expect characteristic X-rays to be emitted by (a) helium atoms or (b) lithium atoms in their ground state? Explain.

13. The elements fluorine, chlorine, and bromine are found to exhibit similar chemical properties. Explain.

PROBLEMS AND CONCEPTUAL EXERCISES

Note: Answers to odd-numbered Problems and Conceptual Exercises can be found in the back of the book. **IP** *denotes an integrated problem, with both conceptual and numerical parts;* **BIO** *identifies problems of biological or medical interest;* **CE** *indicates a conceptual exercise.* **Predict/Explain** *problems ask for two responses:* **(a)** *your prediction of a physical outcome, and* **(b)** *the best explanation among three provided. On all problems, red bullets (•, ••, •••) are used to indicate the level of difficulty.*

SECTION 31–1 EARLY MODELS OF THE ATOM

1. • The electron in a hydrogen atom is typically found at a distance of about 5.3×10^{-11} m from the nucleus, which has a diameter of about 1.0×10^{-15} m. If you assume the hydrogen atom to be a sphere of radius 5.3×10^{-11} m, what fraction of its volume is occupied by the nucleus?

2. • Referring to Problem 1, suppose the nucleus of the hydrogen atom were enlarged to the size of a baseball (diameter = 7.3 cm). At what typical distance from the center of the baseball would you expect to find the electron?

3. •• Copper atoms have 29 protons in their nuclei. If the copper nucleus is a sphere with a diameter of 4.8×10^{-15} m, find the work required to bring an alpha particle (charge = $+2e$) from rest at infinity to the "surface" of the nucleus.

4. •• In Rutherford's scattering experiments, alpha particles (charge = $+2e$) were fired at a gold foil. Consider an alpha particle with an initial kinetic energy K heading directly for the nucleus of a gold atom (charge = $+79e$). The alpha particle will come to rest when all its initial kinetic energy has been converted to electrical potential energy. Find the distance of closest approach between the alpha particle and the gold nucleus for the case $K = 3.0$ MeV.

SECTION 31–2   THE SPECTRUM OF ATOMIC HYDROGEN

5. • Find the wavelength of the Balmer series spectral line corresponding to $n = 15$.

6. • What is the smallest value of n for which the wavelength of a Balmer series line is less than 400 nm?

7. • Find the wavelength of the three longest-wavelength lines of the Lyman series.

8. • Find the wavelength of the three longest-wavelength lines of the Paschen series.

9. • Find **(a)** the longest wavelength in the Lyman series and **(b)** the shortest wavelength in the Paschen series.

10. •• In Table 31–1 we see that the Paschen series corresponds to $n' = 3$ in Equation 31–2, and that the Brackett series corresponds to $n' = 4$. **(a)** Show that the ranges of wavelengths of these two series overlap. **(b)** Is there a similar overlap between the Balmer series and the Paschen series? Verify your answer.

SECTION 31–3   BOHR'S MODEL OF THE HYDROGEN ATOM

11. • **CE Predict/Explain** **(a)** If the mass of the electron were magically doubled, would the ionization energy of hydrogen increase, decrease, or stay the same? **(b)** Choose the *best explanation* from among the following:
    I. The ionization energy would increase because the increased mass would mean the electron would orbit closer to the nucleus and would require more energy to move to infinity.
    II. The ionization energy would decrease because a more massive electron is harder to hold in orbit, and therefore it is easier to remove the electron and leave the hydrogen ionized.
    III. The ionization energy would be unchanged because, just like in gravitational orbits, the orbit of the electron is independent of its mass. As a result, there is no change in the energy required to move it to infinity.

12. • **CE** Consider the Bohr model as applied to the following three atoms: (A) neutral hydrogen in the state $n = 2$; (B) singly ionized helium in the state $n = 1$; (C) doubly ionized lithium in the state $n = 3$. Rank these three atoms in order of increasing Bohr radius. Indicate ties where appropriate.

13. • **CE** Consider the Bohr model as applied to the following three atoms: (A) neutral hydrogen in the state $n = 3$; (B) singly ionized helium in the state $n = 2$; (C) doubly ionized lithium in the state $n = 1$. Rank these three atoms in order of increasing energy. Indicate ties where appropriate.

14. • **CE** An electron in the $n = 1$ Bohr orbit has the kinetic energy K_1. In terms of K_1, what is the kinetic energy of an electron in the $n = 2$ Bohr orbit?

15. • Find the ratio v/c for an electron in the first excited state $(n = 2)$ of hydrogen.

16. • Find the magnitude of the force exerted on an electron in the ground-state orbit of the Bohr model.

17. • How much energy is required to ionize hydrogen when it is in the $n = 4$ state?

18. • Find the energy of the photon required to excite a hydrogen atom from the $n = 2$ state to the $n = 5$ state.

19. •• **CE** In the Bohr model, the potential energy of a hydrogen atom in the nth orbit has a value we will call U_n. What is the potential energy of a hydrogen atom when the electron is in the $(n + 1)$th Bohr orbit? Give your answer in terms of U_n and n.

20. •• A hydrogen atom is in its second excited state, $n = 3$. Using the Bohr model of hydrogen, find **(a)** the linear momentum and **(b)** the angular momentum of the electron in this atom.

21. •• Referring to Problem 20, find **(a)** the kinetic energy of the electron, **(b)** the potential energy of the atom, and **(c)** the total energy of the atom. Give your results in eV.

22. •• Initially, an electron is in the $n = 3$ state of hydrogen. If this electron acquires an additional 1.23 eV of energy, what is the value of n in the final state of the electron?

23. •• Identify the initial and final states if an electron in hydrogen emits a photon with a wavelength of 656 nm.

24. •• **IP** An electron in hydrogen absorbs a photon and jumps to a higher orbit. **(a)** Find the energy the photon must have if the initial state is $n = 3$ and the final state is $n = 5$. **(b)** If the initial state was $n = 5$ and the final state $n = 7$, would the energy of the photon be greater than, less than, or the same as that found in part (a)? Explain. **(c)** Calculate the photon energy for part (b).

25. •• **IP** Consider the following four transitions in a hydrogen atom:

    (i) $n_i = 2, n_f = 6$ (ii) $n_i = 2, n_f = 8$
    (iii) $n_i = 7, n_f = 8$ (iv) $n_i = 6, n_f = 2$

    Find **(a)** the longest- and **(b)** the shortest-wavelength photon that can be emitted or absorbed by these transitions. Give the value of the wavelength in each case. **(c)** For which of these transitions does the atom lose energy? Explain.

26. •• **IP Muonium** Muonium is a hydrogen-like atom in which the electron is replaced with a muon, a fundamental particle with a charge of $-e$ and a mass equal to $207m_e$. (The muon is sometimes referred to loosely as a "heavy electron.") **(a)** What is the Bohr radius of muonium? **(b)** Will the wavelengths in the Balmer series of muonium be greater than, less than, or the same as the wavelengths in the Balmer series of hydrogen? Explain. **(c)** Calculate the longest wavelength of the Balmer series in muonium.

27. •• **IP** **(a)** Find the radius of the $n = 4$ Bohr orbit of a doubly ionized lithium atom (Li^{2+}, $Z = 3$). **(b)** Is the energy required to raise an electron from the $n = 4$ state to the $n = 5$ state in Li^{2+} greater than, less than, or equal to the energy required to raise an electron in hydrogen from the $n = 4$ state to the $n = 5$ state? Explain. **(c)** Verify your answer to part (b) by calculating the relevant energies.

28. •• Applying the Bohr model to a triply ionized beryllium atom (Be^{3+}, $Z = 4$), find **(a)** the shortest wavelength of the Lyman series for Be^{3+} and **(b)** the ionization energy required to remove the final electron in Be^{3+}.

29. •• **(a)** Calculate the time required for an electron in the $n = 2$ state of hydrogen to complete one orbit about the nucleus. **(b)** The typical "lifetime" of an electron in the $n = 2$ state is roughly 10^{-8} s—after this time the electron is likely to have dropped back to the $n = 1$ state. Estimate the number of orbits an electron completes in the $n = 2$ state before dropping to the ground state.

30. •• **IP** The kinetic energy of an electron in a particular Bohr orbit of hydrogen is 1.35×10^{-19} J. **(a)** Which Bohr orbit does the electron occupy? **(b)** Suppose the electron moves away from the nucleus to the next higher Bohr orbit. Does the kinetic energy of the electron increase, decrease, or stay the same? Explain. **(c)** Calculate the kinetic energy of the electron in the orbit referred to in part (b).

31. •• **IP** The potential energy of a hydrogen atom in a particular Bohr orbit is -1.20×10^{-19} J. **(a)** Which Bohr orbit does the electron occupy in this atom? **(b)** Suppose the electron moves away from the nucleus to the next higher Bohr orbit. Does the potential energy of the atom increase, decrease, or stay the

same? Explain. **(c)** Calculate the potential energy of the atom for the orbit referred to in part (b).

32. ••• Consider a head-on collision between two hydrogen atoms, both initially in their ground state and moving with the same speed. Find the minimum speed necessary to leave both atoms in their $n = 2$ state after the collision.

33. ••• A hydrogen atom is in the initial state $n_i = n$, where $n > 1$. **(a)** Find the frequency of the photon that is emitted when the electron jumps to state $n_f = n - 1$. **(b)** Find the frequency of the electron's orbital motion in the state n. **(c)** Compare your results for parts (a) and (b) in the limit of large n.

SECTION 31–4   DE BROGLIE WAVES AND THE BOHR MODEL

34. • **CE Predict/Explain (a)** Is the de Broglie wavelength of an electron in the $n = 2$ Bohr orbit of hydrogen greater than, less than, or equal to the de Broglie wavelength in the $n = 1$ Bohr orbit? **(b)** Choose the *best explanation* from among the following:
   I. The de Broglie wavelength in the nth state is $2\pi r/n$, where r is proportional to n^2. Therefore, the wavelength increases with increasing n, and is greater for $n = 2$ than for $n = 1$.
   II. The de Broglie wavelength of an electron in the nth state is such that n wavelengths fit around the circumference of the orbit. Therefore, $\lambda = 2\pi r/n$ and the wavelength for $n = 2$ is less than for $n = 1$.
   III. The de Broglie wavelength depends on the mass of the electron, and that is the same regardless of which state of the hydrogen atom the electron occupies.

35. • Find the de Broglie wavelength of an electron in the ground state of the hydrogen atom.

36. •• Find an expression for the de Broglie wavelength of an electron in the nth state of the hydrogen atom.

37. •• What is the radius of the hydrogen-atom Bohr orbit shown in **Figure 31–30**?

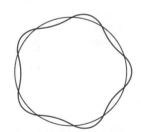

▲ **FIGURE 31–30** Problem 37

38. •• **(a)** Find the kinetic energy (in eV) of an electron whose de Broglie wavelength is equal to 0.5 Å, a typical atomic size. **(b)** Repeat part (a) for an electron with a wavelength equal to 10^{-15} m, a typical nuclear size.

SECTION 31–5   THE QUANTUM MECHANICAL HYDROGEN ATOM

39. • What are the allowed values of ℓ when the principal quantum number is $n = 5$?

40. • How many different values of m_ℓ are possible when the principal quantum number is $n = 4$?

41. • Give the value of the quantum number ℓ, if one exists, for a hydrogen atom whose orbital angular momentum has a magnitude of **(a)** $\sqrt{6}(h/2\pi)$, **(b)** $\sqrt{15}(h/2\pi)$, **(c)** $\sqrt{30}(h/2\pi)$, or **(d)** $\sqrt{36}(h/2\pi)$.

42. •• **IP** Hydrogen atom number 1 is known to be in the $4f$ state. **(a)** What is the energy of this atom? **(b)** What is the magnitude of this atom's orbital angular momentum? **(c)** Hydrogen atom number 2 is in the $5d$ state. Is this atom's energy greater than, less than, or the same as that of atom 1? Explain. **(d)** Is the magnitude of the orbital angular momentum of atom 1 greater than, less than, or the same as that of atom 2? Explain.

43. •• **IP** A hydrogen atom has an orbital angular momentum with a magnitude of $10\sqrt{57}(h/2\pi)$. **(a)** Determine the value of the quantum number ℓ for this atom. **(b)** What is the minimum possible value of this atom's principal quantum number, n? Explain. **(c)** If $10\sqrt{57}(h/2\pi)$ is the *maximum* orbital angular momentum this atom can have, what is its energy?

44. •• **IP** The electron in a hydrogen atom with an energy of -0.544 eV is in a subshell with 18 states. **(a)** What is the principal quantum number, n, for this atom? **(b)** What is the maximum possible orbital angular momentum this atom can have? **(c)** Is the number of states in the subshell with the next lowest value of ℓ equal to 16, 14, or 12? Explain.

45. •• **IP** Consider two different states of a hydrogen atom. In state I the maximum value of the magnetic quantum number is $m_\ell = 3$; in state II the corresponding maximum value is $m_\ell = 2$. Let L_I and L_{II} represent the magnitudes of the orbital angular momentum of an electron in states I and II, respectively. **(a)** Is L_I greater than, less than, or equal to L_{II}? Explain. **(b)** Calculate the ratio L_I/L_{II}.

SECTION 31–6   MULTIELECTRON ATOMS AND THE PERIODIC TABLE

46. • **CE** How many electrons can occupy **(a)** the $2p$ subshell and **(b)** the $3p$ subshell?

47. • **CE (a)** How many electrons can occupy the $3d$ subshell? **(b)** How many electrons can occupy the $n = 2$ shell?

48. • **CE** The electronic configuration of a given atom is $1s^2 2s^2 2p^6 3s^2 3p^1$. How many electrons are in this atom?

49. • Give the electronic configuration for the ground state of carbon.

50. • List the values of the four quantum numbers (n, ℓ, m_ℓ, m_s) for each of the electrons in the ground state of neon.

51. • Give the electronic configuration for the ground state of nitrogen.

52. • Give a list of all possible sets of the four quantum numbers (n, ℓ, m_ℓ, m_s) for electrons in the $3s$ subshell.

53. • Give a list of all possible sets of the four quantum numbers (n, ℓ, m_ℓ, m_s) for electrons in the $3p$ subshell.

54. •• List the values of the four quantum numbers (n, ℓ, m_ℓ, m_s) for each of the electrons in the ground state of magnesium.

55. •• The configuration of the outer electrons in Ni is $3d^8 4s^2$. Write out the complete electronic configuration for Ni.

56. •• Determine the number of different sets of quantum numbers possible for each of the following shells: **(a)** $n = 2$, **(b)** $n = 3$, **(c)** $n = 4$.

57. ••• Generalize the results of Problem 56 and show that the number of different sets of quantum numbers for the nth shell is $2n^2$.

58. •• Suppose that the $5d$ subshell is filled in a certain atom. Write out the 10 sets of four quantum numbers (n, ℓ, m_ℓ, m_s) for the electrons in this subshell.

SECTION 31–7 ATOMIC RADIATION

59. • **CE Predict/Explain** (a) In an X-ray tube, do you expect the wavelength of the characteristic X-rays to increase, decrease, or stay the same if the energy of the electrons striking the target is increased? (b) Choose the *best explanation* from among the following:
   I. Increasing the energy of the incoming electrons will increase the wavelength of the emitted X-rays.
   II. When the energy of the incoming electrons is increased, the energy of the X-rays is also increased; this, in turn, decreases the wavelength.
   III. The wavelength of characteristic X-rays depends only on the material used in the metal target, and does not change if the energy of incoming electrons is increased.

60. • **CE** Is the wavelength of the radiation that excites a fluorescent material greater than, less than, or equal to the wavelength of the radiation the material emits? Explain.

61. • Using the Bohr model, estimate the wavelength of the K_α X-ray in nickel ($Z = 28$).

62. • Using the Bohr model, estimate the energy of a K_α X-ray emitted by lead ($Z = 82$).

63. •• The K-shell ionization energy of iron is 8500 eV, and its L-shell ionization energy is 2125 eV. What is the wavelength of K_α X-rays emitted by iron?

64. •• An electron drops from the L shell to the K shell and gives off an X-ray with a wavelength of 0.0205 nm. What is the atomic number of this atom?

65. •• Consider an X-ray tube that uses platinum ($Z = 78$) as its target. (a) Use the Bohr model to estimate the minimum kinetic energy electrons must have in order for K_α X-rays to just appear in the X-ray spectrum of the tube. (b) Assuming the electrons are accelerated from rest through a voltage V, estimate the minimum voltage necessary to produce the K_α X-rays.

66. •• **BIO Photorefractive Keratectomy** A person's vision may be improved significantly by having the cornea reshaped with a laser beam, in a procedure known as photorefractive keratectomy. The excimer laser used in these treatments produces ultraviolet light with a wavelength of 193 nm. (a) What is the difference in energy between the two levels that participate in stimulated emission in the excimer laser? (b) How many photons from this laser are required to deliver an energy of 1.58×10^{-13} J to the cornea?

GENERAL PROBLEMS

67. • **CE** Consider the following three transitions in a hydrogen atom: (A) $n_i = 5, n_f = 2$; (B) $n_i = 7, n_f = 2$; (C) $n_i = 7, n_f = 6$. Rank the transitions in order of increasing (a) wavelength and (b) frequency of the emitted photon. Indicate ties where appropriate.

68. • **CE** Suppose an electron is in the ground state of hydrogen. (a) What is the highest-energy photon this system can absorb without dissociating the electron from the proton? Explain. (b) What is the lowest-energy photon this system can absorb? Explain.

69. • **CE** The electronic configuration of a particular carbon atom is $1s^2 2s^2 2p^1 3s^1$. Is this atom in its ground state or in an excited state? Explain.

70. • **CE** The electronic configuration of a particular potassium atom is $1s^2 2s^2 2p^6 3s^2 3p^6 3d^1$. Is this atom in its ground state or in an excited state? Explain.

71. • **CE** Do you expect the ionization energy of sodium (Na) to be greater than, less than, or equal to the ionization energy of lithium (Li)? Explain.

72. • Find the minimum frequency a photon must have if it is to ionize the ground state of the hydrogen atom.

73. •• It was pointed out in Section 31–3 that intermolecular collisions at room temperature do not have enough energy to cause an excitation in hydrogen from the $n = 1$ state to the $n = 2$ state. Given that the average kinetic energy of a hydrogen atom in a high-temperature gas is $\frac{3}{2}kT$ (where k is Boltzmann's constant), find the minimum temperature required for atoms to have enough thermal energy to excite electrons from the ground state to the $n = 2$ state.

74. •• The electron in a hydrogen atom makes a transition from the $n = 4$ state to the $n = 2$ state, as indicated in **Figure 31–31**. (a) Determine the linear momentum of the photon emitted as a result of this transition. (b) Using your result to part (a), find the recoil speed of the hydrogen atom, assuming it was at rest before the photon was emitted.

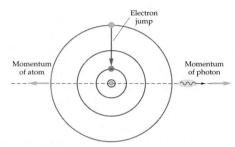

▲ **FIGURE 31–31** Problems 74 and 75

75. •• **IP** Referring to Problem 74, find (a) the energy of the emitted photon and (b) the kinetic energy of the hydrogen atom after the photon is emitted. (c) Do you expect the sum of the energies in parts (a) and (b) to be greater than, less than, or the same as the difference in energy between the $n = 4$ and $n = 2$ states of hydrogen? Explain.

76. •• **BIO Laser Eye Surgery** In laser eye surgery, the laser emits a 1.45-ns pulse focused on a spot that is 34.0 μm in diameter. (a) If the energy contained in the pulse is 2.75 mJ, what is the power per square meter (the irradiance) associated with this beam? (b) Suppose a molecule with a diameter of 0.650 nm is irradiated by the laser beam. How much energy does the molecule receive in one pulse from the laser? (The energy obtained in part (b) is more than enough to dissociate a molecule.)

77. •• Consider an electron in the ground-state orbit of the Bohr model of hydrogen. (a) Find the time required for the electron to complete one orbit about the nucleus. (b) Calculate the current (in amperes) corresponding to the electron's motion.

78. •• A particular Bohr orbit in a hydrogen atom has a total energy of −0.85 eV. What are (a) the kinetic energy of the electron in this orbit and (b) the electric potential energy of the system?

79. •• The element helium is named for the Sun because that is where it was first observed. (a) What is the shortest wavelength that one would expect to observe from a singly ionized helium atom in the atmosphere of the Sun? (b) Suppose light with a wavelength of 388.9 nm is observed from singly ionized helium. What are the initial and final values of the quantum number n corresponding to this wavelength?

80. •• An ionized atom has only a single electron. The $n = 6$ Bohr orbit of this electron has a radius of 2.72×10^{-10} m. Find (a) the atomic number Z of this atom and (b) the total energy E of its $n = 3$ Bohr orbit.

81. •• Find the approximate wavelength of K_β X-rays emitted by molybdenum ($Z = 42$), and compare your result with Figure

31–22. (*Hint*: An electron in the M shell is shielded from the nucleus by the single electron in the K shell, plus all the electrons in the L shell.)

82. •• Referring to the hint given in Problem 81, estimate the wavelength of L_α X-rays in molybdenum.

83. •• **IP The Pickering Series** In 1896, the American astronomer Edward C. Pickering (1846–1919) discovered an unusual series of spectral lines in light from the hot star Zeta Puppis. After some time, it was determined that these lines are produced by singly ionized helium. In fact, the "Pickering series" is produced when electrons drop from higher levels to the $n = 4$ level of He$^+$. Spectral lines in the Pickering series have wavelengths given by

$$\frac{1}{\lambda} = C\left(\frac{1}{16} - \frac{1}{n^2}\right)$$

In this expression, $n = 5, 6, 7, \ldots$. **(a)** Do you expect the constant C to be greater than, less than, or equal to the Rydberg constant R? Explain. **(b)** Find the numerical value of C. **(c)** Pickering lines with $n = 6, 8, 10, \ldots$ correspond to Balmer lines in hydrogen with $n = 3, 4, 5, \ldots$. Verify this assertion for the $n = 6$ Pickering line.

84. •• **IP Rydberg Atoms** There is no limit to the size a hydrogen atom can attain, provided it is free from disruptive outside influences. In fact, radio astronomers have detected radiation from large, so-called "Rydberg atoms" in the diffuse hydrogen gas of interstellar space. **(a)** Find the smallest value of n such that the Bohr radius of a single hydrogen atom is greater than 8.0 microns, the size of a typical single-celled organism. **(b)** Find the wavelength of radiation this atom emits when its electron drops from level n to level $n - 1$. **(c)** If the electron drops one more level, from $n - 1$ to $n - 2$, is the emitted wavelength greater than or less than the value found in part (b)? Explain.

85. ••• Consider a particle of mass m, charge q, and constant speed v moving perpendicular to a uniform magnetic field of magnitude B, as shown in **Figure 31–32**. The particle follows a circular path. Suppose the angular momentum of the particle about the center of its circular motion is quantized in the following way: $mvr = n\hbar$, where $n = 1, 2, 3, \ldots$, and $\hbar = h/2\pi$.

a. Show that the radii of its allowed orbits have the following values:

$$r_n = \sqrt{\frac{n\hbar}{qB}}$$

b. Find the speed of the particle in each allowed orbit.

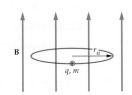

▲ **FIGURE 31–32** Problem 85

86. ••• Consider a particle of mass m confined in a one-dimensional box of length L. In addition, suppose the matter wave associated with this particle is analogous to a wave on a string of length L that is fixed at both ends. Using the de Broglie relationship, show that **(a)** the quantized values of the linear momentum of the particle are

$$p_n = \frac{nh}{2L} \qquad n = 1, 2, 3, \ldots$$

and **(b)** the allowed energies of the particle are

$$E_n = n^2\left(\frac{h^2}{8mL^2}\right) \qquad n = 1, 2, 3, \ldots$$

87. ••• Show that the time required for an electron in the nth Bohr orbit of hydrogen to circle the nucleus once is given by

$$T = T_1 n^3 \qquad n = 1, 2, 3, \ldots$$

where $T_1 = h^3/4\pi^2 mk^2 e^4$.

PASSAGE PROBLEMS

BIO Welding a Detached Retina

As a person ages, a normal part of the process is a shrinkage of the vitreous gel—the gelatinous substance that fills the interior of the eye. When this happens, the usual result is that the gel pulls away cleanly from the retina, with little or no adverse effect on the person's vision. This is referred to as posterior vitreous detachment. In some cases, however, the vitreous membrane that surrounds the vitreous gel pulls on the retina as the gel contracts, eventually creating a hole or a tear in the retina itself. At this point, fluid can seep through the hole in the retina and separate it from the underlying supporting cells—the retinal pigment epithelium. This process, known as rhegmatogenous retinal detachment, causes a blind spot in the person's vision. If not treated immediately, a retinal detachment can lead to permanent vision loss.

One way to treat a detached retina is to "weld" it back in place using a laser beam. This type of operation is performed with an argon laser because the blue-green light it produces passes through the vitreous gel with little absorption or damage, but is strongly absorbed by the red pigments in the retina and the retinal epithelium. An argon laser produces light consisting primarily of two wavelengths, 488.0 nm (blue-green) and 514.5 nm (green), and has a power output ranging between 1 W and 20 W.

88. •• Suppose an argon laser emits 1.49×10^{19} photons per second, half with a wavelength of 488.0 nm and half with a wavelength of 514.5 nm. What is the power output of this laser in watts?

   A. 1.49 W   **B.** 5.76 W

   C. 5.92 W   **D.** 6.07 W

89. •• A different type of laser also emits 1.49×10^{19} photons per second. If all of its photons have a wavelength of 414.0 nm, is its power output greater than, less than, or equal to the power output of the argon laser in Problem 88?

90. •• What is the power output of the laser in Problem 89?

   A. 1.23 W   **B.** 2.39 W

   C. 4.80 W   **D.** 7.16 W

91. •• What is the energy difference (in eV) between the states of an argon atom that are responsible for a photon with a wavelength of 514.5 nm?

   A. 2.13 eV   **B.** 2.42 eV

   C. 3.87 eV   **D.** 6.40 eV

INTERACTIVE PROBLEMS

92. •• **IP Referring to Example 31–3** Suppose the electron is in a state whose standing wave consisting of two wavelengths. **(a)** Is the wavelength of this standing wave greater than or less than 1.33×10^{-9} m? **(b)** Find the wavelength of this standing wave.

93. •• **Referring to Example 31–3 (a)** Which state has a de Broglie wavelength of 3.99×10^{-9} m? **(b)** What is the Bohr radius of this state?

32 Nuclear Physics and Nuclear Radiation

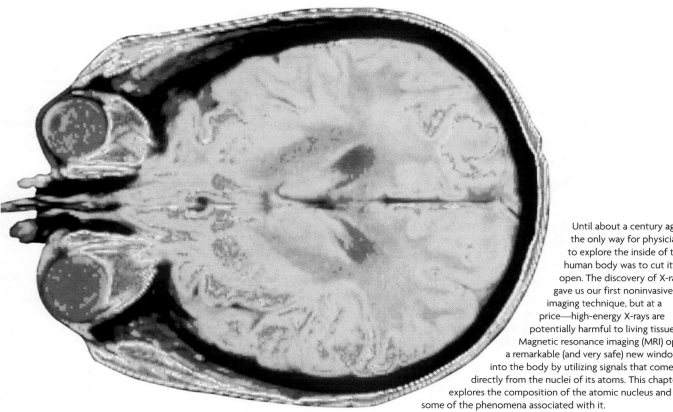

Until about a century ago, the only way for physicians to explore the inside of the human body was to cut it open. The discovery of X-rays gave us our first noninvasive imaging technique, but at a price—high-energy X-rays are potentially harmful to living tissues. Magnetic resonance imaging (MRI) opens a remarkable (and very safe) new window into the body by utilizing signals that come directly from the nuclei of its atoms. This chapter explores the composition of the atomic nucleus and some of the phenomena associated with it.

In the previous chapter our focus was almost entirely on the electrons in an atom. We studied their orbits and energies, their jumps from orbit to orbit, and the photons they emitted or absorbed. The nucleus played little role in these considerations. It was treated as a point object at the center of the atom, providing the electrostatic force necessary to hold the atom together.

The nucleus is much more than a point, however. Most nuclei contain a number of strongly interacting particles packed closely together in a more or less spherical assembly. The energies associated with changes inside a nucleus are orders of magnitude greater than those involved in chemical reactions, which involve only the electrons. It is for this reason that the Sun, which is powered by nuclear reactions, can burn for many billions of years—if the Sun were powered by chemical reactions it would have burned out after giving off light for only a few million years. This chapter considers the physics at play in the nucleus, discusses the nuclear reactions that occur in the Sun and in nuclear power plants, and describes a number of biomedical applications related to the nucleus.

32–1 The Constituents and Structure of Nuclei

The simplest nucleus is that of the hydrogen atom. This nucleus consists of a single **proton,** whose mass is about 1836 times greater than the mass of an electron and whose electric charge is $+e$. All other nuclei contain neutrons in addition to protons. The **neutron** is an electrically neutral particle (its electric charge is zero) with a mass just slightly greater than that of the proton. No other particles are found in nuclei. Collectively, protons and neutrons are referred to as **nucleons.**

Nuclei are characterized by the number and type of nucleons they contain. First, the **atomic number,** Z, is defined as the number of protons in a nucleus. In an electrically neutral atom, the number of electrons will also be equal to Z. Next, the number of neutrons in a nucleus is designated by the **neutron number,** N. Finally, the total number of nucleons in a nucleus is the **mass number,** A. These definitions are summarized in Table 32–1. Clearly, the mass number is the sum of the atomic number and the neutron number:

$$A = Z + N \qquad\qquad 32–1$$

A special notation is used to indicate the composition of a nucleus. Consider, for example, an unstable but very useful form of carbon known as carbon-14. The nucleus of carbon-14 is written as follows:

$$^{14}_{6}C$$

In this expression, C represents the chemical element carbon. The number 6 is the atomic number of carbon, $Z = 6$, and the number 14 is the mass number of this nucleus, $A = 14$. This means that carbon-14 has 14 nucleons in its nucleus. The neutron number can be found by solving Equation 32–1 for N: $N = A - Z = 14 - 6 = 8$. Thus the nucleus of carbon-14 consists of 6 protons and 8 neutrons. The most common form of carbon is carbon-12, whose nucleus is designated as follows: $^{12}_{6}C$. This nucleus has 6 protons and 6 neutrons.

In general, the nucleus of an arbitrary element, X, with atomic number Z and mass number A, is represented as

$$^{A}_{Z}X$$

Note that once a given element is specified, the value of Z is known. As a result, the subscript Z is sometimes omitted.

TABLE 32–1 Numbers That Characterize a Nucleus

Z	Atomic number = number of protons in nucleus
N	Neutron number = number of neutrons in nucleus
A	Mass number = number of nucleons in nucleus

EXERCISE 32–1

a. Give the symbol for a nucleus of aluminum that contains 14 neutrons.

b. Tritium is a type of "heavy hydrogen." The nucleus of tritium can be written as $^{3}_{1}H$. What is the number of protons and neutrons in a tritium nucleus?

SOLUTION

a. Looking up aluminum in the periodic table in Appendix E, we find that $Z = 13$. In addition, we are given that $N = 14$. Therefore, $A = Z + N = 27$, and hence the symbol for this nucleus is $^{27}_{13}Al$.

b. We obtain the number of protons from the subscript; therefore, $Z = 1$. The number of neutrons, from Equation 32–1, is $N = A - Z$, where A is the superscript. Therefore, the number of neutrons is $N = 3 - 1 = 2$.

All nuclei of a given element have the same number of protons, Z. They may have different numbers of neutrons, N, however. Nuclei with the same value of Z but different values of N are referred to as **isotopes.** For example, $^{12}_{6}C$ and $^{13}_{6}C$ are two isotopes of carbon, with $^{12}_{6}C$ being the most common one, constituting about 98.89% of naturally occurring carbon. About 1.11% of natural carbon is $^{13}_{6}C$. Values for the percentage abundance of various isotopes can be found in Appendix F.

Also given in Appendix F are the atomic masses of many common isotopes. These masses are given in terms of the **atomic mass unit,** u, defined so that the mass of one atom of $^{12}_{6}C$ is exactly 12 u. The value of u is as follows:

Definition of Atomic Mass Unit, u

$$1 \, u = 1.660540 \times 10^{-27} \, kg \qquad\qquad 32\text{-}2$$

SI unit: kg

Protons have a mass just slightly greater than 1 u, and the neutron is slightly more massive than the proton. The precise masses of the proton and neutron are given in Table 32–2, along with the mass of the electron.

TABLE 32–2 Mass and Charge of Particles in the Atom

Particle	Mass (kg)	Mass (MeV/c^2)	Mass (u)	Charge (C)
Proton	1.672623×10^{-27}	938.28	1.007276	$+1.6022 \times 10^{-19}$
Neutron	1.674929×10^{-27}	939.57	1.008665	0
Electron	9.109390×10^{-31}	0.511	0.0005485799	-1.6022×10^{-19}

When we consider nuclear reactions later in this chapter, an important consideration will be the energy equivalent of a given mass, as given by Einstein's famous relation, $E = mc^2$ (Equation 30–7). The energy equivalent of one atomic mass unit is

$$E = mc^2 = (1 \, u)c^2$$

$$= (1.660540 \times 10^{-27} \, kg)(2.998 \times 10^8 \, m/s)^2 \left(\frac{1 \, eV}{1.6022 \times 10^{-19} \, J} \right)$$

$$= 931.5 \, MeV$$

where $1 \, MeV = 10^6 \, eV$. When we consider that the ionization energy of hydrogen is only 13.6 eV, it is clear that the energy equivalent of nucleons is enormous compared with typical atomic energies. In general, energies involving the nucleus are on the order of MeV, and energies associated with the electrons in an atom are on the order of eV. Finally, because mass and energy can be converted from one form to the other, it is common to express the atomic mass unit in terms of energy as follows:

$$1 \, u = 931.5 \, MeV/c^2 \qquad\qquad 32\text{-}3$$

The masses of the proton, neutron, and electron are also given in units of MeV/c^2 in Table 32–2.

Nuclear Size and Density

To obtain an estimate for the size of a nucleus, Rutherford did a simple calculation using energy conservation. He considered the case of a particle of charge $+q$ and mass m approaching a nucleus of charge $+Ze$ with a speed v. He further assumed that the approach was head-on. At some distance, d, from the center of the nucleus, the incoming particle comes to rest instantaneously, before turning around. It follows that the radius of the nucleus is less than d. In Example 32–1 we obtain a symbolic expression for the distance d.

EXAMPLE 32–1 SETTING A LIMIT ON THE RADIUS OF A NUCLEUS

A particle of mass m, charge $+q$, and speed v heads directly toward a distant, stationary nucleus of charge $+Ze$. Find the distance of closest approach between the incoming particle and the center of the nucleus.

PICTURE THE PROBLEM

The incoming particle moves on a line that passes through the center of the nucleus. Far from the nucleus, the particle's speed is v. The particle turns around (comes to rest instantaneously) a distance d from the center of the nucleus.

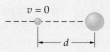

STRATEGY
We can find the distance d by applying energy conservation. In particular, the initial energy of the system is the kinetic energy of the particle, $\frac{1}{2}mv^2$, assuming the particle approaches the nucleus from infinity. The final energy is the electric potential energy, $U = kq_1q_2/r = k(Ze)q/d$. Setting these energies equal to each other allows us to solve for d.

SOLUTION

1. Write an expression for the initial energy of the system: $E_i = \frac{1}{2}mv^2$

2. Write an expression for the final energy of the system: $E_f = \dfrac{k(Ze)q}{d}$

3. Set the final energy equal to the initial energy and solve for d: $\frac{1}{2}mv^2 = \dfrac{k(Ze)q}{d}$

$$d = \dfrac{kZeq}{\left(\frac{1}{2}mv^2\right)}$$

INSIGHT
Notice that the distance of closest approach is inversely proportional to the initial kinetic energy of the incoming particle and directly proportional to the charge on the nucleus.

PRACTICE PROBLEM
Find the distance at which the speed of the incoming particle is equal to $\frac{1}{2}v$. $\left[\textbf{Answer: } \dfrac{kZeq}{\frac{3}{4}\left(\frac{1}{2}mv^2\right)}\right]$

Some related homework problems: Problem 6, Problem 7

   Using the result obtained in Example 32–1, Rutherford found that for an alpha particle approaching a gold nucleus in one of his experiments, the distance of closest approach was 3.2×10^{-14} m. A similar calculation for alpha particles fired at silver atoms gives a closest approach distance of 2.0×10^{-14} m. This suggests that the size of the nucleus varies from element to element; in particular, the nucleus of silver is smaller than that of gold. In fact, more careful measurements since Rutherford's time have established that the average radius of a nucleus of mass number A is given approximately by the following expression:

$$r = (1.2 \times 10^{-15}\ \text{m})A^{1/3} \qquad \text{32–4}$$

Notice that the length scale of the nucleus is on the order of 10^{-15} m, as opposed to the length scale of an atom, which is on the order of 10^{-10} m. Recall that 10^{-15} m is referred to as a *femtometer* (fm). To honor the pioneering work of Enrico Fermi (1901–1954) in the field of nuclear physics, the femtometer is often referred to as the **fermi:**

Definition of the Fermi, fm
1 fermi = 1 fm = 10^{-15} m

SI unit: m

We now use Equation 32–4 to find the radius of a particular nucleus.

EXERCISE 32–2

Find the radius of a $^{14}_{6}\text{C}$ nucleus.

SOLUTION
Substitute $A = 14$ in Equation 32–4:

$$r = (1.2 \times 10^{-15}\ \text{m})(14)^{1/3} = 2.9 \times 10^{-15}\ \text{m} = 2.9\ \text{fm}$$

   The fact that the radius of a nucleus depends on $A^{1/3}$ has interesting consequences for the density of the nucleus, and these are explored in the next Example.

EXAMPLE 32–2 NUCLEAR DENSITY

Using the expression $r = (1.2 \times 10^{-15} \text{ m})A^{1/3}$, calculate the density of a nucleus with mass number A.

PICTURE THE PROBLEM
Our sketch shows a collection of neutrons and protons in a densely packed nucleus of radius r.

STRATEGY
To find the density of a nucleus, we must divide its mass, M, by its volume, V. Ignoring the small difference in mass between a neutron and a proton, we can express the mass of a nucleus as $M = Am$, where $m = 1.67 \times 10^{-27}$ kg. The volume of a nucleus is simply the volume of a sphere of radius r: $V = 4\pi r^3/3$.

SOLUTION

1. Write an expression for the mass of a nucleus:

$$M = Am = A(1.67 \times 10^{-27} \text{ kg})$$

2. Write an expression for the volume of a nucleus:

$$V = \tfrac{4}{3}\pi r^3 = \tfrac{4}{3}\pi[(1.2 \times 10^{-15} \text{ m})A^{1/3}]^3$$

$$= \tfrac{4}{3}\pi(1.7 \times 10^{-45} \text{ m}^3)A$$

3. Divide the mass by the volume to find the density:

$$\rho = \frac{M}{V} = \frac{Am}{\tfrac{4}{3}\pi(1.7 \times 10^{-45} \text{ m}^3)A}$$

$$= \frac{(1.67 \times 10^{-27} \text{ kg})}{\tfrac{4}{3}\pi(1.7 \times 10^{-45} \text{ m}^3)} = 2.3 \times 10^{17} \text{ kg/m}^3$$

INSIGHT
Note that the density of a nucleus is found to be *independent* of the mass number, A. This means that a nucleus can be thought of as a collection of closely packed nucleons, much like a group of marbles in a bag. The neutrons in a nucleus serve to separate the protons, thereby reducing their mutual electrostatic repulsion.

The density of a nucleus is incredibly large. For example, a single teaspoon of nuclear matter would weigh about a trillion tons.

PRACTICE PROBLEM
Find the surface area of a nucleus in terms of the mass number, A, assuming it to be a sphere. [**Answer:** area = $(1.8 \times 10^{-29} \text{ m}^2)A^{2/3}$]

Some related homework problems: Problem 8, Problem 9

Nuclear Stability

We know that like charges repel one another, and that the force of repulsion increases rapidly with decreasing distance. It follows that protons in a nucleus, with a separation of only about a fermi, must exert relatively large forces on one another. Applying Coulomb's law (Equation 19–5), we find the following force for two protons (charge $+e$) separated by a distance of 10^{-15} m:

$$F = \frac{ke^2}{r^2} = 230 \text{ N}$$

The acceleration such a force would give to a proton is $a = F/m = (230 \text{ N})/(1.67 \times 10^{-27} \text{ kg}) = 1.4 \times 10^{29} \text{ m/s}^2$, which is about 10^{28} times greater than the acceleration of gravity! Thus, if protons in the nucleus experienced only the electrostatic force, the nucleus would fly apart in an instant. It follows that large attractive forces must also act within the nucleus.

The attractive force that holds a nucleus together is called the **strong nuclear force.** This force has the following properties:

- The strong force is short range, acting only to distances of a couple fermis.
- The strong force is attractive and acts with nearly equal strength between protons and protons, protons and neutrons, and neutrons and neutrons.

In addition, the strong nuclear force does not act on electrons. As a result, it has no effect on the chemical properties of an atom.

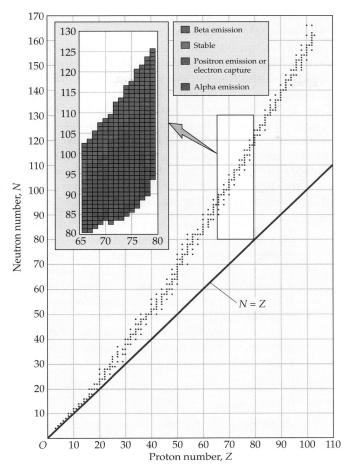

Stable nuclei with proton numbers less than 104 are indicated by small red dots. Notice that large nuclei have significantly more neutrons, *N*, than protons, *Z*. The inset shows unstable nuclei and their decay modes for proton numbers between 65 and 80.

It is the competition between the repulsive electrostatic forces and the attractive strong nuclear forces that determines whether a given nucleus is stable. **Figure 32–1** shows the neutron number, N, and atomic number (proton number), Z, for nuclei that are stable. Note that nuclei of relatively small atomic number are most stable when the numbers of protons and neutrons in the nucleus are approximately equal, $N = Z$. For example, $^{12}_{6}\text{C}$ and $^{13}_{6}\text{C}$ are both stable. As the atomic number increases, however, we see that the points corresponding to stable nuclei deviate from the line $N = Z$. In fact, we see that large stable nuclei tend to contain significantly more neutrons than protons, as in the case of $^{185}_{75}\text{Re}$. Since all nucleons experience the strong nuclear force, but only the protons experience the electrostatic force, the neutrons effectively "dilute" the nuclear charge density, reducing the effect of the repulsive forces that otherwise would make the nucleus disintegrate.

As the number of protons in a nucleus increases, however, a point is reached at which the strong nuclear forces are no longer able to compensate for the repulsive forces between protons. In fact, the largest number of protons in a stable nucleus is $Z = 83$, corresponding to the element bismuth. Nuclei with more than 83 protons are simply not stable, as can be seen by noting that all elements with $Z > 83$ in Appendix F decay in a finite time—that is, they have a finite half-life. (We shall discuss the half-life of unstable nuclei in detail in Section 32–3.) The nuclei of many well-known elements, such as radon and uranium, disintegrate—decay—in a finite time. We turn now to a discussion of the various ways in which an unstable nucleus decays.

32–2 Radioactivity

An unstable nucleus does not last forever—sooner or later it changes its composition by emitting a particle of one type or another. Alternatively, a nucleus in an excited state may rearrange its nucleons into a lower-energy state and emit a high-energy photon. We refer to such processes as the **decay** of a nucleus, and the various emissions that result are known collectively as **radioactivity.**

When a nucleus undergoes radioactive decay, the mass of the system decreases. That is, the mass of the initial nucleus before decay is greater than the mass of the resulting nucleus plus the mass of the emitted particle. The difference in mass, $\Delta m < 0$, appears as a release of energy, according to the relation $E = |\Delta m|c^2$. The mass difference for any given decay can be determined by referring to Appendix F. Note that the atomic masses listed in Appendix F are the masses of neutral atoms; that is, the values given in the table include the mass of the electrons in an atom. This factor must be considered whenever the mass difference of a reaction is calculated, as we shall see later in this section.

Three types of particles with mass are given off during the various processes of radioactive decay. They are as follows:

- Alpha (α) particles, which are the nuclei of $^4_2$He. Note that an alpha particle consists of two protons and two neutrons. When a nucleus decays by giving off alpha particles, we say that it emits α rays.
- Electrons, also referred to as beta (β) particles. The electrons given off by a nucleus are called β rays, or β^- rays to be more precise. (The minus sign is a reminder that the charge of an electron is $-e$.)
- **Positrons,** which have the same mass as an electron but a charge of $+e$. If a nucleus gives off positrons, we say that it emits β^+ rays. (A positron, which is short for "positive electron," is the **antiparticle** of the ordinary electron. Positrons will be considered in greater detail in Section 32–7.)

Finally, radioactivity may take the form of a photon rather than a particle with nonzero mass:

- A nucleus in an excited state may emit a high-energy photon, a gamma (γ) ray, and drop to a lower-energy state.

The following Conceptual Checkpoint examines the behavior of radioactivity in a magnetic field.

CONCEPTUAL CHECKPOINT 32–1   IDENTIFY THE RADIATION

A sample of radioactive material is placed at the bottom of a small hole drilled into a piece of lead. The sample emits α rays, β^- rays, and γ rays into a region of constant magnetic field. It is observed that the radiation follows three distinct paths, 1, 2, and 3, as shown in the sketch. Identify each path with the corresponding type of radiation: **(a)** path 1, α rays; path 2, β^- rays; path 3, γ rays; **(b)** path 1, β^- rays; path 2, γ rays; path 3, α rays; or **(c)** path 1, α rays; path 2, γ rays; path 3, β^- rays.

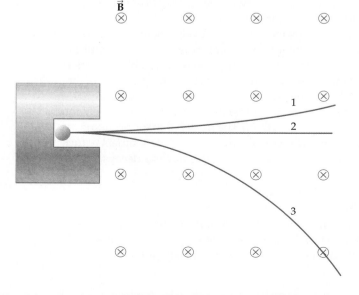

REASONING AND DISCUSSION

First, because γ rays are uncharged, they are not deflected by the magnetic field. It follows that path 2 corresponds to γ rays.

Next, the right-hand rule for the magnetic force (Section 22–2) indicates that positively charged particles will be deflected upward, and negatively charged particles will be deflected downward. As a result, path 1 corresponds to α rays, and path 3 corresponds to β^- rays.

ANSWER

(c) Path 1, α rays; path 2, γ rays; path 3, β^- rays

Radioactivity was discovered by the French physicist Antoine Henri Becquerel (1852–1908) in 1896 when he observed that uranium was able to expose photographic emulsion, even when the emulsion was covered. Thus radioactivity has the ability to penetrate various materials. In fact, the various types of radioactivity were initially named according to their ability to penetrate, starting with α rays, which are the least penetrating. Typical penetrating abilities for α, β, and γ rays are as follows:

- α rays can barely penetrate a sheet of paper.
- β rays (both β^- and β^+) can penetrate a few millimeters of aluminum.
- γ rays can penetrate several centimeters of lead.

We turn now to a detailed examination of each of these types of decay.

Alpha Decay

When a nucleus decays by giving off an α particle ($^4_2$He), it loses two protons and two neutrons. As a result, its atomic number, Z, decreases by 2, and its mass number decreases by 4. Symbolically, we can write this process as follows:

$$^A_Z X \longrightarrow \,^{A-4}_{Z-2} Y + \,^4_2 He$$

where X is referred to as the **parent nucleus,** and Y is the **daughter nucleus.** Notice that the sum of the atomic numbers on the right side of this process is equal to the atomic number on the left side; similar remarks apply to the mass numbers.

The next Example considers the alpha decay of uranium-238. We first use conservation of atomic number and mass number to determine the identity of the daughter nucleus. Next, we use the mass difference to calculate the amount of energy released by the decay.

PROBLEM-SOLVING NOTE

The Effects of Alpha Decay

In alpha decay, the total number of protons is the same before and after the reaction. The same is true of the total number of neutrons. On the other hand, the mass number of the daughter nucleus is 4 less than the mass number of the parent nucleus. Similarly, the atomic number of the daughter nucleus is 2 less than the atomic number of the parent nucleus.

PROBLEM-SOLVING NOTE

Atomic Masses Include Electrons

When calculating the mass difference in a nuclear reaction, be sure to note that the atomic masses in Appendix F include the electrons that would be present in a neutral atom. The only item in Appendix F to which this does *not* apply is the neutron.

EXAMPLE 32–3 URANIUM DECAY

Determine (a) the daughter nucleus and (b) the energy released when $^{238}_{92}$U undergoes alpha decay.

PICTURE THE PROBLEM

Our sketch shows the specified decay of $^{238}_{92}$U into a daughter nucleus plus an α particle. Note that the number of neutrons and protons is indicated for $^{238}_{92}$U. The α particle consists of two neutrons and two protons.

STRATEGY

a. We can identify the daughter nucleus by requiring that the total number of neutrons and protons be the same before and after the decay.

b. To find the energy, we first calculate the mass before and after the decay. The magnitude of the difference in mass, $|\Delta m|$, times the speed of light squared, c^2, gives the amount of energy released.

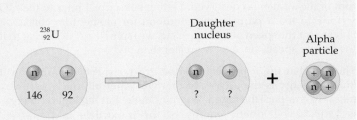

CONTINUED ON NEXT PAGE

CONTINUED FROM PREVIOUS PAGE

SOLUTION

Part (a)

1. Determine the number of neutrons and protons in the daughter nucleus. Add these numbers together to obtain the mass number of the daughter nucleus:

$N = 146 - 2 = 144$
$Z = 92 - 2 = 90$
$A = N + Z = 144 + 90 = 234$

2. Referring to Appendix F, we see that the daughter nucleus is thorium-234:

$^{234}_{90}\text{Th}$

Part (b)

3. Use Appendix F to find the initial mass of the system; that is, the mass of a $^{238}_{92}\text{U}$ atom:

$m_i = 238.050786 \text{ u}$

4. Use Appendix F to find the final mass of the system; that is, the mass of a $^{234}_{90}\text{Th}$ atom plus the mass of a ^4_2He atom:

$m_f = 234.043596 \text{ u} + 4.002603 \text{ u} = 238.046199 \text{ u}$

5. Calculate the mass difference and the corresponding energy release (recall that $1 \text{ u} = 931.5 \text{ MeV}/c^2$):

$\Delta m = m_f - m_i$
$\quad = 238.046199 \text{ u} - 238.050786 \text{ u} = -0.004587 \text{ u}$

$E = |\Delta m|c^2 = (0.004587 \text{ u})\left(\dfrac{931.5 \text{ MeV}/c^2}{1 \text{ u}}\right)c^2$

$\quad = 4.273 \text{ MeV}$

INSIGHT
Each of the masses used in this decay includes the electrons of the corresponding neutral atom. Since the number of electrons initially (92) is the same as the number of electrons in a thorium atom (90) plus the number of electrons in a helium atom (2), the electrons make no contribution to the total mass difference, Δm.

PRACTICE PROBLEM
Find the daughter nucleus and energy released when $^{226}_{88}\text{Ra}$ undergoes alpha decay.   [**Answer:** $^{222}_{86}\text{Rn}$, 4.871 MeV]

Some related homework problems: Problem 20, Problem 21

A considerable amount of energy is released in the alpha decay of uranium-238. As indicated in **Figure 32–2**, this energy appears as kinetic energy of the daughter nucleus and the α particle as they move off in opposite directions. The following Conceptual Checkpoint compares the kinetic energy of the daughter nucleus with that of the α particle.

CONCEPTUAL CHECKPOINT 32–2 COMPARE KINETIC ENERGIES

When a stationary $^{238}_{92}\text{U}$ nucleus decays into a $^{234}_{90}\text{Th}$ nucleus and an α particle, is the kinetic energy of the α particle **(a)** greater than, **(b)** less than, or **(c)** the same as the kinetic energy of the $^{234}_{90}\text{Th}$ nucleus?

REASONING AND DISCUSSION
Because no external forces are involved in the decay process, it follows that the momentum of the system is conserved. Letting subscript 1 refer to the $^{234}_{90}\text{Th}$ nucleus, and subscript 2 to the α particle, the condition for momentum conservation is $m_1 v_1 = m_2 v_2$. Solving for the speed of the α particle, we have $v_2 = (m_1/m_2)v_1$; that is, the α particle has the greater speed, since m_1 is greater than m_2.

The fact that the α particle has the greater speed does not, in itself, ensure that its kinetic energy is the greater of the two. After all, the α particle also has the smaller mass. To compare the kinetic energies, note that the kinetic energy of the $^{234}_{90}\text{Th}$ nucleus is $\frac{1}{2}m_1 v_1^2$. The kinetic energy of the α particle is $\frac{1}{2}m_2 v_2^2 = \frac{1}{2}m_2[(m_1/m_2)v_1]^2 = \frac{1}{2}m_1 v_1^2(m_1/m_2) > \frac{1}{2}m_1 v_1^2$. Therefore, the α particle carries away the majority of the kinetic energy released in the decay. In this particular case, the α particle has a kinetic energy that is $m_1/m_2 = 234/4 = 58.5$ times greater than the kinetic energy of the $^{234}_{90}\text{Th}$ nucleus.

ANSWER
(a) The α particle has the greater kinetic energy.

Although you may not be aware of it, many homes are protected from the hazards of fire by a small device—a smoke detector—that uses the alpha decay of a man-made radioactive isotope, $^{241}_{95}$Am. In this type of smoke detector, a minute quantity of $^{241}_{95}$Am is placed between two metal plates connected to a battery or other source of emf. The α particles emitted by the radioactive source ionize the air, allowing a measurable electric current to flow between the plates. As long as this current flows, the smoke detector remains silent. When smoke enters the detector, however, the ionized air molecules tend to stick to the smoke particles and become neutralized. This reduces the current and triggers the alarm. These "ionization" smoke detectors are more sensitive than the "photoelectric" detectors that rely on the thickness of smoke to dim a beam of light.

Beta Decay

The basic process that occurs in beta decay is the conversion of a neutron to a proton and an electron:

$$^1_0\text{n} \longrightarrow {}^1_1\text{p} + \text{e}^-$$

Thus when a nucleus decays by giving off an electron, its mass number is unchanged (since protons and neutrons count equally in determining A), but its atomic number increases by 1. This process can be represented as follows:

$$^A_Z X \longrightarrow {}^{A}_{Z+1} Y + \text{e}^-$$

Similarly, if a nucleus undergoes a different type of decay in which it gives off a positron, the process can be written

$$^A_Z X \longrightarrow {}^{A}_{Z-1} Y + \text{e}^+$$

In the next Example we determine the energy that is released as carbon-14 undergoes beta decay.

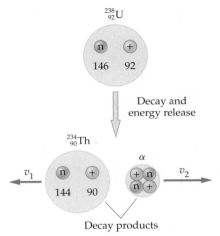

▲ **FIGURE 32–2 Alpha decay of uranium-238**
When $^{238}_{92}$U decays into $^{234}_{90}$Th and an alpha particle, the mass of the system decreases. The "lost" mass is actually converted into energy; it appears as the kinetic energy of the $^{234}_{90}$Th nucleus and the alpha particle.

EXAMPLE 32–4   BETA DECAY OF CARBON-14

Find **(a)** the daughter nucleus and **(b)** the energy released when $^{14}_6$C undergoes β^- decay.

PICTURE THE PROBLEM
In our sketch we show $^{14}_6$C giving off a β^- particle and converting into a daughter nucleus. The number of neutrons and protons in the $^{14}_6$C nucleus is indicated. Note that the β^- particle is not a nucleon.

STRATEGY

a. We can identify the daughter nucleus by requiring that the total number of nucleons be the same before and after the decay. The number of neutrons will be decreased by 1, and the number of protons will be increased by 1.

b. To find the energy, we begin by calculating the mass before and after the decay. The magnitude of the difference in mass, $|\Delta m|$, times the speed of light squared, c^2, gives the amount of energy released.

SOLUTION

Part (a)

1. Determine the number of neutrons and protons in the daughter nucleus. Add these numbers together to obtain the mass number of the daughter nucleus:

$$N = 8 - 1 = 7$$
$$Z = 6 + 1 = 7$$
$$A = N + Z = 7 + 7 = 14$$

2. Referring to Appendix F, we see that the daughter nucleus is nitrogen-14:

$$^{14}_7\text{N}$$

CONTINUED ON NEXT PAGE

CONTINUED FROM PREVIOUS PAGE

Part (b)

3. Use Appendix F to find the initial mass of the system; that is, the mass of a $^{14}_{6}$C atom:

$m_i = 14.003242$ u

4. Use Appendix F to find the final mass of the system, which is simply the mass of a $^{14}_{7}$N atom (the mass of the β^- particle is included in the mass of $^{14}_{7}$N, as we point out in the Insight):

$m_f = 14.003074$ u
$\Delta m = m_f - m_i$
$\quad = 14.003074$ u $- 14.003242$ u $= -0.000168$ u

5. Calculate the mass difference and the corresponding energy release (recall that 1 u $= 931.5$ MeV/c^2):

$E = |\Delta m| c^2 = (0.000168 \text{ u}) \left(\dfrac{931.5 \text{ MeV}/c^2}{1 \text{ u}} \right) c^2$

$\quad = 0.156$ MeV

INSIGHT

With regard to the masses used in this calculation, note that the mass of $^{14}_{6}$C includes the mass of its 6 electrons. Similarly, the mass of $^{14}_{7}$N includes the mass of 7 electrons in the neutral $^{14}_{7}$N atom. However, when the $^{14}_{6}$C nucleus converts to a $^{14}_{7}$N nucleus, the number of electrons orbiting the nucleus is still 6. In effect, the newly created $^{14}_{7}$N atom is missing one electron; that is, the mass of the $^{14}_{7}$N atom includes the mass of one too many electrons. Therefore, it is not necessary to add the mass of an electron (representing the β^- particle) to the final mass of the system, because this extra electron mass is already included in the mass of the $^{14}_{7}$N atom.

PRACTICE PROBLEM

Find the daughter nucleus and the energy released when $^{234}_{90}$Th undergoes β^- decay.   [**Answer:** $^{234}_{91}$Pa, 0.274 MeV]

Some related homework problems: Problem 22, Problem 24

PROBLEM-SOLVING NOTE

The Effects of Beta Decay

In β^- decay, the number of neutrons decreases by 1, and the number of protons increases by 1. The mass number is unchanged.

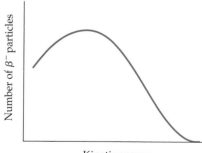

Number of β^- particles

Kinetic energy

▲ **FIGURE 32–3 Energy of electrons emitted in β^- decay**

When electrons are emitted during β^- decay, they come off with a range of energies. This indicates that another particle (the neutrino) must also be taking away some of the energy.

Referring to Conceptual Checkpoint 32–2, we would expect the kinetic energy of an electron emitted during beta decay to account for most of the energy released by the decay process. In fact, energy conservation allows us to predict the precise amount of kinetic energy the electron should have. It turns out, however, that when the kinetic energy of emitted electrons is measured, a range of values is obtained, as indicated in **Figure 32–3**. Specifically, we find that all electrons given off in beta decay have energies that are less than would be predicted by energy conservation. On closer examination it is found that beta decay seems to violate conservation of linear and angular momentum as well! For these reasons, beta decay was an interesting and intriguing puzzle for physicists.

The resolution of this puzzle was given by Pauli in 1930, when he proposed that the "missing" energy and momentum were actually carried off by a particle that was not observed in the experiments. For this particle to have been unobserved, it must have zero charge and little or no mass. Fermi dubbed Pauli's hypothetical particle the **neutrino,** meaning, literally, "little neutral one." We now know that neutrinos do in fact exist and that they account exactly for the missing energy and momentum. They interact so weakly with matter, however, that it wasn't until 1950 that they were observed experimentally. Recent experiments on neutrinos given off by the Sun provide the best evidence yet that the mass of a neutrino is in fact finite—though extremely small. In fact, the best estimate of the neutrino mass at this time is that it is less than about 7 eV/c^2. For comparison, the mass of the electron is 511,000 eV/c^2.

To give an indication of just how weakly neutrinos interact with matter, only one in every 200 million neutrinos that pass through the Earth interacts with it in any way. As far as the neutrinos are concerned, it is almost as if the Earth did not exist. Right now, in fact, billions of neutrinos are passing through your body every second without the slightest effect.

We can now write the correct expression for the decay of a neutron. Indicating the electron neutrino with the symbol v_e, we have the following:

$$^1_0\text{n} \longrightarrow {}^1_1\text{p} + \text{e}^- + \bar{v}_e \qquad \qquad 32\text{–}5$$

The bar over the neutrino symbol indicates that the neutrino given off in β^- decay is actually an **antineutrino,** the antiparticle counterpart of the neutrino (just as the positron is the antiparticle of the electron). The neutrino itself is given off in β^+ decay.

Gamma Decay

An atom in an excited state can emit a photon when one of its electrons drops to a lower-energy level. Similarly, a nucleus in an excited state can emit a photon as it decays to a state of lower energy. Since nuclear energies are so much greater than typical atomic energies, the photons given off by a nucleus are highly energetic. In fact, these photons have energies that place them well beyond X-rays in the electromagnetic spectrum. We refer to such high-energy photons as **gamma (γ) rays.**

As an example of a situation in which a γ ray can be given off, consider the following beta decay:

$$^{14}_{6}\text{C} \longrightarrow {}^{14}_{7}\text{N}^* + e^- + \bar{\nu}_e$$

The asterisk on the nitrogen symbol indicates that the nitrogen nucleus has been left in an excited state as a result of the beta decay. Subsequently, the nitrogen nucleus may decay to its ground state with the emission of a γ ray:

$$^{14}_{7}\text{N}^* \longrightarrow {}^{14}_{7}\text{N} + \gamma$$

Notice that neither the atomic number nor the mass number is changed by the emission of a γ ray.

▲ Smoke detectors like this one make use of a synthetic radioactive isotope, americium-241. The alpha particles emitted when this isotope decays ionize air molecules, making them able to conduct a small current. Smoke particles neutralize the ions, interrupting the current and setting off an alarm.

ACTIVE EXAMPLE 32–1   GAMMA-RAY EMISSION: FIND THE CHANGE IN MASS

A $^{226}_{88}\text{Ra}$ nucleus in an excited state emits a γ ray with a wavelength of 6.67×10^{-12} m. Find the decrease in mass of the $^{226}_{88}\text{Ra}$ nucleus as a result of this process.

SOLUTION *(Test your understanding by performing the calculations indicated in each step.)*

1. Find the frequency of the γ ray: $f = c/\lambda = 4.50 \times 10^{19}$ Hz

2. Calculate the energy of the γ ray photon: $E = hf = 0.186$ MeV

3. Determine the mass difference corresponding $|\Delta m| = E/c^2 = 0.000200$ u
   to the energy of the photon:

INSIGHT
As a result of emitting this γ ray, the mass of the $^{226}_{88}\text{Ra}$ nucleus decreases by an amount that is about one-third the mass of the electron.

YOUR TURN
If the wavelength of the emitted gamma ray is doubled, by what factor does the mass difference change?

*(Answers to **Your Turn** problems are given in the back of the book.)*

Radioactive Decay Series

Consider an unstable nucleus that decays and produces a daughter nucleus. If the daughter nucleus is also unstable, it will eventually decay and produce its own daughter nucleus, which may in turn be unstable. In such cases, an original parent nucleus can produce a series of related nuclei referred to as a **radioactive decay series.** An example of a radioactive decay series is shown in **Figure 32–4**. In this case, the parent nucleus is $^{235}_{92}\text{U}$, and the final nucleus of the series is $^{207}_{82}\text{Pb}$, which is stable.

Notice that several of the intermediate nuclei in this series can decay in two different ways—either by alpha decay or by beta decay. Thus there are various "paths" a $^{235}_{92}\text{U}$ nucleus can follow as it transforms into a $^{207}_{82}\text{Pb}$ nucleus. In addition, the intermediate nuclei in this series decay fairly rapidly, at least on a geological time scale. For example, any actinium-227 that was present when the Earth formed would have decayed away long ago. The fact that actinium-227 is still

▶ **FIGURE 32–4 Radioactive decay series of $^{235}_{92}$U**

When $^{235}_{92}$U decays, it passes through a number of intermediate nuclei before reaching the stable end of the series, $^{207}_{82}$Pb. Note that some intermediary nuclei can decay in only one way, whereas others have two decay possibilities.

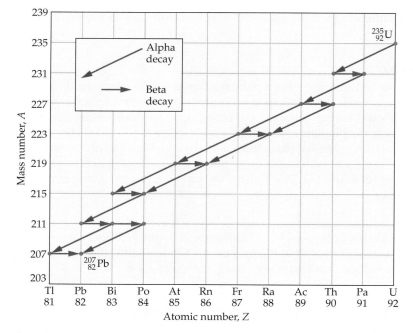

found on the Earth today in natural uranium deposits is due to its continual production in this and other decay series.

Activity

The rate at which nuclear decay occurs—that is, the number of decays per second—is referred to as the **activity.** A highly active material has many nuclear decays occurring every second. For example, a typical sample of radium (usually a fraction of a gram) might have 10^5 to 10^{10} decays per second.

The unit we use to measure activity is the curie, named in honor of Pierre (1859–1906) and Marie (1867–1934) Curie, pioneers in the study of radioactivity. The **curie (Ci)** is defined as follows:

$$1 \text{ curie} = 1 \text{ Ci} = 3.7 \times 10^{10} \text{ decays/s} \qquad 32\text{–}6$$

The reason for this choice is that 1 Ci is roughly the activity of 1 g of radium. In SI units, we measure activity in terms of the **becquerel** (Bq):

$$1 \text{ becquerel} = 1 \text{ Bq} = 1 \text{ decay/s} \qquad 32\text{–}7$$

The units of activity most often encountered in practical applications are the millicurie (1 mCi = 10^{-3} Ci) and the microcurie (1 μCi = 10^{-6} Ci).

EXERCISE 32–3

A sample of radium has an activity of 15 μCi. How many decays per second occur in this sample?

SOLUTION

Using the definition given in Equation 32–6, we find

$$15 \ \mu\text{Ci} = 15 \times 10^{-6} \text{ Ci}$$

$$= (15 \times 10^{-6} \text{ Ci})\left(\frac{3.7 \times 10^{10} \text{ decays/s}}{1 \text{ Ci}}\right) = 5.6 \times 10^5 \text{ decays/s}$$

32–3 Half-Life and Radioactive Dating

The phenomenon of radioactive decay, though fundamentally random in its behavior, has certain properties that make it useful as a type of "nuclear clock." In fact, it has been discovered that radioactive decay can be used to date numerous items of interest from the recent—and not so recent—past. In this section we

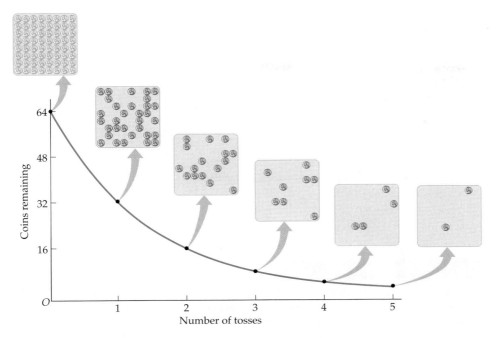

The points on this graph show the number of coins remaining (on average) if one starts with 64 coins and removes half with each round of tosses. The curve is a plot of the mathematical function $64e^{-(0.693)t}$, where $t = 1$ means one round of tosses, $t = 2$ means two rounds, and so on.

consider the behavior of radioactive decay as a function of time, introduce the concept of a half-life, and show explicitly how these concepts can be applied to dating.

To begin, consider an analogy in which coins represent nuclei, and the side that comes up when a given coin is tossed determines whether the corresponding nucleus decays. Suppose, for example, that we toss a group of 64 coins and remove any coin that comes up tails. We expect that—on average—32 coins will be removed after the first round of tosses. Which coins will be removed—and the precise number that will be removed—cannot be known, because the flip of a coin, like the decay of a nucleus, is a random process. When we toss the remaining 32 coins, we expect an average of 16 more to be removed, and so on, with each round of tosses decreasing the number of coins by a factor of 2. The results after the first few rounds are shown by the points in **Figure 32–5**.

Also shown in Figure 32–5 is a smooth curve representing the following mathematical function:

$$N = (64)e^{-(\ln 2)t} = (64)e^{-(0.693)t} \qquad \text{32–8}$$

where N represents the number of coins, and the time variable, t, represents the number of rounds of tosses. For example, if we set $t = 1$ in Equation 32–8, we find $N = 32$, and if we set $t = 2$, we find $N = 16$. This type of "exponential dependence" is a general feature whenever the number of some quantity increases or decreases by a constant factor with each constant interval of time. Examples include the balance in a bank account with compounding interest and the population of the Earth as a function of time.

When nuclei decay, their behavior is much like that of the coins in our analogy. Which nucleus will decay in a given interval of time, and the precise number that will decay, are controlled by a random process that causes the decay—on average—of a given fraction of the original number of nuclei. Thus the number of nuclei, N, remaining at time t is given by an expression analogous to Equation 32–8:

$$N = N_0 e^{-\lambda t} \qquad \text{32–9}$$

where N_0 is the number of nuclei present at time $t = 0$, and the constant λ is referred to as the **decay constant.** In the analogy of the coins, $N_0 = 64$ and $\lambda = 0.693 \text{ s}^{-1}$. Note that the larger the value of the decay constant, the more rapidly the number of nuclei decreases with time. **Figure 32–6** shows the dependence on λ graphically.

PROBLEM-SOLVING NOTE

Consistent Units and the Decay Constant

When calculating the number of nuclei present at a given time, be sure to use consistent units for the time, t, and the decay constant, λ. For example, if you express λ in units of y^{-1}, measure the time in units of y.

▶ **FIGURE 32–6 Dependence on the decay constant**
The larger the decay constant, λ, the more rapidly the population of a group of nuclei decreases. In this plot, the value of λ doubles as we move downward from one curve to the next.

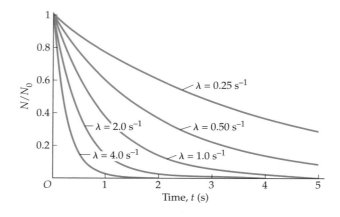

ACTIVE EXAMPLE 32–2   FIND THE RADON LEVEL

Radon can pose a health risk when high levels become trapped in the basement of a house. Suppose 4.75×10^7 radon atoms are in a basement at a time when it is sealed to prevent any additional radon from entering. Given that the decay constant of radon is 0.181 d^{-1}, how many radon atoms remain in the basement after **(a)** 7 d and **(b)** 14 d?

SOLUTION *(Test your understanding by performing the calculations indicated in each step.)*

Part (a)

1. Evaluate Equation 32–9 with $t = 7$ d: $N = 1.34 \times 10^7$

Part (b)

1. Evaluate Equation 32–9 with $t = 14$ d: $N = 3.77 \times 10^6$

INSIGHT

Notice that after 7 d the number of radon atoms has decreased by a factor of about 3.55. After 14 d, the number has decreased by a factor of about $3.55^2 = 12.6$. It follows that every 7 d the number of radon atoms decreases by another factor of 3.55.

YOUR TURN

What period of time is required for the number of radon atoms to decrease by a factor of 2.00?

*(Answers to **Your Turn** problems are given in the back of the book.)*

▲ Radon-222 ($^{222}_{86}$Rn) is an isotope produced in a radioactive decay series that includes uranium-238 and radium-226. Since uranium is naturally present in certain kinds of rocks and the soils derived from them, radon, which is a gas, can accumulate in basements and similar enclosed underground spaces that lack adequate ventilation. Radon-222 is itself radioactive, undergoing alpha decay with a half-life of about 4 days, and although its concentration is generally small, it may produce radiation levels great enough to be a health hazard if exposure is prolonged. Homeowners in many parts of the country use test kits to monitor the radiation levels produced by radon.

A useful way to characterize the rate at which a given type of nucleus decays is in terms of its half-life, which is defined as follows:

> The **half-life** of a given type of radioactive nucleus is the time required for the number of such nuclei to decrease by a factor of 2; that is, for the number to decrease from N_0 to $\frac{1}{2}N_0$, from $\frac{1}{2}N_0$ to $\frac{1}{4}N_0$, and so on.

We can solve for this time, call it $T_{1/2}$, by setting $N = \frac{1}{2}N_0$ in Equation 32–9:

$$\tfrac{1}{2}N_0 = N_0 e^{-\lambda T_{1/2}}$$

Canceling N_0 and taking the natural logarithm of both sides of the equation, we find

$$T_{1/2} = \frac{\ln 2}{\lambda} = \frac{0.693}{\lambda} \qquad \text{32–10}$$

Notice that a large decay constant corresponds to a short half-life, in agreement with the plots shown in Figure 32–6.

CONCEPTUAL CHECKPOINT 32-3   HOW MANY NUCLEI?

A system consists of N_0 radioactive nuclei at time $t = 0$. The number of nuclei remaining after *half* a half-life (that is, at time $t = \frac{1}{2}T_{1/2}$) is **(a)** $\frac{1}{4}N_0$, **(b)** $\frac{3}{4}N_0$, or **(c)** $\frac{1}{\sqrt{2}}N_0$?

REASONING AND DISCUSSION

Referring to the Insight following Active Example 32–2, we note that if the number of nuclei decreases by a factor f in the time $\frac{1}{2}T_{1/2}$, it will decrease by the factor f^2 in the time $2\left(\frac{1}{2}T_{1/2}\right) = T_{1/2}$. We know, however, that the number of nuclei remaining at the time $T_{1/2}$ is $\frac{1}{2}N_0$. It follows that $f^2 = \frac{1}{2}$, or that $f = \frac{1}{\sqrt{2}}$. Therefore, the number of nuclei remaining at the time $\frac{1}{2}T_{1/2}$ is $\frac{1}{\sqrt{2}}N_0$.

ANSWER

(c) At half a half-life, the number of nuclei remaining is $\frac{1}{\sqrt{2}}N_0$.

Now, the property that makes radioactivity so useful as a clock is that its **decay rate, R,** or **activity,** depends on time in a straightforward way. To see this, think back to the analogy of the coins. The number of coins that decay (are removed) on the first round of tosses is $32 = \frac{1}{2}(64)$, the number that decay on the second round is $16 = \frac{1}{2}(32)$, and so on. That is, the number that decay in any given interval of time is proportional to the number present at the beginning of the interval.

The same type of analysis applies to nuclei. Therefore, the number of nuclei that decay, ΔN, in a given time interval, Δt, is proportional to the number, N, that are present at time t:

$$R = \left|\frac{\Delta N}{\Delta t}\right| = \lambda N \qquad\qquad 32\text{–}11$$

There are two points to note regarding this equation. First, observe that the number of nuclei is decreasing, $\Delta N < 0$. It is for this reason that we take the absolute value of the quantity $\Delta N/\Delta t$. Second, notice that the proportionality constant is simply λ, the decay constant. That λ is the correct constant of proportionality can be shown using calculus.

Combining Equation 32–11 with Equation 32–9, we obtain the time dependence of the activity, R:

$$R = \lambda N_0 e^{-\lambda t} = R_0 e^{-\lambda t} \qquad\qquad 32\text{–}12$$

Note that $R_0 = \lambda N_0$ is the initial value of the activity. We apply this relation in the following Active Example.

ACTIVE EXAMPLE 32-3   FIND THE ACTIVITY OF RADON

Referring to Active Example 32–2, calculate how many radon atoms disintegrate per second **(a)** initially and **(b)** after 7 d.

SOLUTION *(Test your understanding by performing the calculations indicated in each step.)*

Part (a)

1. Calculate the activity for $N_0 = 4.75 \times 10^7$. Be sure to convert the decay constant to the unit s^{-1}:

   $R = \lambda N_0 = 99.5 \text{ decays/s}$

Part (b)

2. Repeat the calculation for $N = 1.34 \times 10^7$:

   $R = \lambda N = 28.1 \text{ decays/s}$

INSIGHT

We see that the initial activity (number of decays per time) is 99.5 Bq. In terms of the curie, the initial activity is 0.00269 μCi.

YOUR TURN

How long does it take after the basement is sealed for the activity of the radon to decrease to 10.0 Bq?

*(Answers to **Your Turn** problems are given in the back of the book.)*

Referring to Equation 32–12, we see that the basic idea of radioactive dating is simply this: If we know the initial activity of a sample, R_0, and we also know the sample's activity now, R, we can find the corresponding time, t, as follows:

$$\frac{R}{R_0} = e^{-\lambda t}$$

$$t = -\frac{1}{\lambda}\ln\frac{R}{R_0} = \frac{1}{\lambda}\ln\frac{R_0}{R}$$ 32–13

The current activity, R, can be measured in the lab, but how can we know the initial activity, R_0? We address the question next for the specific case of carbon-14.

Carbon-14 Dating

To determine the initial activity of carbon-14 requires a basic knowledge of the role it plays in Earth's biosphere. First, we note that carbon-14 is unstable, with a half-life of 5730 y. It follows that the carbon-14 initially present in any *closed* system will decay away to practically nothing in a time of several half-lives, yet the ratio of carbon-14 to carbon-12 in Earth's atmosphere remains approximately constant at the value 1.20×10^{-12}. Evidently, Earth's atmosphere is not a closed system, at least as far as carbon-14 is concerned.

This is indeed the case. Cosmic rays, which are high-energy particles from outer space, are continuously entering Earth's upper atmosphere and initiating nuclear reactions in nitrogen-14 (a stable isotope). These reactions result in a steady production of carbon-14. Thus, the steady level of carbon-14 in the atmosphere is a result of the balance between the *production rate* due to cosmic rays and the *decay rate* due to the properties of the carbon-14 nucleus.

We note that living organisms have the same ratio of carbon-14 to carbon-12 as the atmosphere, since they continuously exchange carbon with their surroundings. When an organism dies, however, the exchange of carbon ceases and the carbon-14 in the organism (wood, bone, shell, etc.) begins to decay. This process is illustrated in **Figure 32–7,** where we see that the carbon-14 activity of an organism is constant until it dies, at which point it decreases exponentially with a half-life of 5730 y.

All that remains to implement carbon-14 dating is to determine the initial activity, R_0. For convenience, we calculate R_0 for a 1-g sample of carbon. As we know from Chapter 31, 12 g of carbon-12 consists of Avogadro's number of atoms, 6.02×10^{23}. Therefore, 1 g consists of $(6.02 \times 10^{23})/12 = 5.02 \times 10^{22}$ carbon-12 atoms. Multiplying this number of atoms by the ratio of carbon-14 to carbon-12, 1.20×10^{-12}, shows that the number of carbon-14 atoms in the 1-g sample is

▲ The Iceman, found in the Italian Alps in 1991. His age was established by means of radiocarbon dating.

▶ **FIGURE 32–7 Activity of carbon-14**
While an organism is living and exchanging carbon with the atmosphere, its carbon-14 activity remains constant. When the organism dies, the carbon-14 activity decays exponentially with a half-life of 5730 years.

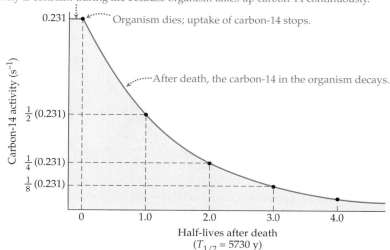

6.02×10^{10}. Next, we need the decay constant, which we obtain by rearranging Equation 32–10:

$$\lambda = \frac{\ln 2}{T_{1/2}} = \frac{0.693}{T_{1/2}}$$

Using $T_{1/2} = 5730 \text{ y} = 1.81 \times 10^{11} \text{ s}$, we find $\lambda = 3.83 \times 10^{-12} \text{ s}^{-1}$. Finally, the initial activity of a 1-g sample of carbon is

$$R_0 = \lambda N_0 = (3.83 \times 10^{-12} \text{ s}^{-1})(6.02 \times 10^{10}) = 0.231 \text{ Bq} \qquad \text{32–14}$$

This is the initial activity used in Figure 32–7. It follows that 5730 y after an organism dies, its carbon-14 activity per gram of carbon will have decreased to about $\frac{1}{2}(0.231 \text{ Bq}) = 0.116 \text{ Bq}$.

The next Example applies this basic idea to a real-world case of some interest— the Iceman of the Alps.

REAL-WORLD PHYSICS

Dating the Iceman

EXAMPLE 32–5 AGE OF THE ICEMAN: YOU DON'T LOOK A DAY OVER 5000

Early in the afternoon of September 19, 1991, a German couple hiking in the Italian Alps noticed something brown sticking out of the ice 8 to 10 m ahead of them. At first they took the object to be a doll or some rubbish. As they got closer, however, it became apparent that the object they had discovered was the body of a person trapped in the ice, with only the top part of the body exposed. Subsequent investigation revealed the remarkably well-preserved body to be that of a Stone Age man who had died in the mountains and become entombed in the ice. When the carbon-14 dating method was applied to the remains of the Iceman and some of the materials he had carried with him, it was found that the carbon-14 activity was about 0.121 Bq per gram of carbon. Using this information, date the remains of the Iceman.

PICTURE THE PROBLEM
Our sketch shows the decay of the carbon-14 activity of a gram of carbon as a function of time. The initial activity of carbon-14 in such a sample is 0.231 Bq.

STRATEGY
We can obtain the age of the remains directly from Equation 32–13: $t = (1/\lambda) \ln(R_0/R)$. In this case, $R_0 = 0.231 \text{ Bq}$, and $R = 0.121 \text{ Bq}$. Since an answer in years would be most useful, we express the decay constant as $\lambda = 0.693/T_{1/2}$, with $T_{1/2} = 5730 \text{ y}$.

Note that the observed activity of 0.121 Bq is slightly greater than $\frac{1}{2}(0.231 \text{ Bq}) = 0.116 \text{ Bq}$; hence, we expect the age of the remains to be slightly less than the half-life of 5730 y.

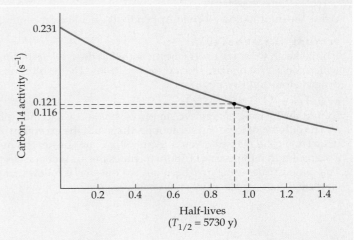

SOLUTION

1. Determine the value of the decay constant, λ, in units of y^{-1}:

$$\lambda = \frac{0.693}{5730 \text{ y}} = 1.21 \times 10^{-4} \text{ y}^{-1}$$

2. Substitute λ, R_0, and R into Equation 32–13:

$$t = \frac{1}{\lambda} \ln\left(\frac{R_0}{R}\right)$$

$$= \frac{1}{(1.21 \times 10^{-4} \text{ y}^{-1})} \ln\left(\frac{0.231 \text{ Bq}}{0.121 \text{ Bq}}\right) = 5340 \text{ y}$$

INSIGHT
We conclude that the Iceman, who has been dubbed Ötzi, died in the mountains during the Stone Age, some 5340 y ago. Detailed examination of Ötzi's body and possessions indicates he was probably an itinerant sheepherder and/or hunter. He met his end in a violent fashion, however. Recent CT scans confirm that Ötzi was killed by an arrow that entered through his shoulder blade and lodged less than an inch from his left lung.

PRACTICE PROBLEM
If the remains of another Iceman of the same age are found in the year 2991, what will be the carbon-14 activity of 1 g of carbon? [**Answer:** 0.107 Bq]

Some related homework problems: Problem 35, Problem 36

As useful as carbon-14 dating is, it is limited to time spans of only a few half-lives, say, 10,000 to 15,000 y. Beyond that range, the current activity will be so small that accurate measurements will be difficult. To measure dates on different time scales, different radioactive isotopes must be used. Other frequently used isotopes and their half-lives are $^{210}_{82}\text{Pb}$ (22.3 y), $^{40}_{19}\text{K}$ (1.28 × 10^9 y), and $^{238}_{92}\text{U}$ (4.468 × 10^9 y).

32–4 Nuclear Binding Energy

An α particle consists of two protons and two neutrons. Does it follow that the mass of an α particle is twice the mass of a proton plus twice the mass of a neutron? One would certainly think so, but in fact this is not the case. Alpha particles, and all other stable nuclei containing more than one nucleon, have a mass that is *less* than the mass of the individual nucleons added together.

As strange as this result may seem, it is just one more manifestation of Einstein's theory of relativity. In particular, the reduction in mass of a nucleus, compared with the mass of its constituents, corresponds to a reduction in its energy, according to the relation $E = mc^2$. This reduction in energy is referred to as the **binding energy** of the nucleus. To separate a nucleus into its individual nucleons requires an energy at least as great as the binding energy; therefore, the binding energy indicates how firmly a given nucleus is held together. We illustrate this concept in **Figure 32–8**.

The next Example uses $E = (\Delta m)c^2$ and the atomic masses given in Appendix F to calculate the binding energy of tritium.

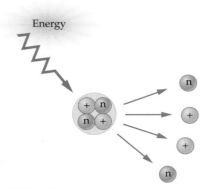

▲ **FIGURE 32–8 The concept of binding energy**

The minimum energy that must be supplied to a stable nucleus to break it into its constituent nucleons is referred to as the binding energy. Because the stable nucleus is the lower-energy state, it has less mass than the sum of the masses of its individual constituents.

EXAMPLE 32–6 THE BINDING ENERGY OF TRITIUM

Using the information given in Appendix F, calculate the binding energy of tritium, ^3_1H.

PICTURE THE PROBLEM
In our sketch we show energy being added to the nucleus of tritium, which consists of one proton and two neutrons. The result is one hydrogen atom and two neutrons.

STRATEGY
To find the binding energy, we simply calculate Δm and multiply by c^2. The only point to be careful about is that, with the exception of the neutron, all the masses given in Appendix F are for neutral atoms. This means that the mass of tritium includes the mass of one electron. The same is true of the hydrogen atom, however. It follows that the electron mass will cancel when we calculate Δm.

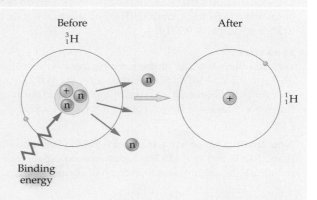

SOLUTION

1. Let the initial mass be the mass of tritium: $m_i = 3.016049 \text{ u}$

2. The final mass is the mass of hydrogen plus two neutrons: $m_f = 1.007825 \text{ u} + 2(1.008665 \text{ u}) = 3.025155 \text{ u}$

3. Calculate Δm:
$$\Delta m = m_f - m_i$$
$$= 3.025155 \text{ u} - 3.016049 \text{ u} = 0.009106 \text{ u}$$

4. Find the corresponding binding energy (recall that $1 \text{ u} = 931.5 \text{ MeV}/c^2$):
$$E = (\Delta m)c^2 = (0.009106 \text{ u})\left(\frac{931.5 \text{ MeV}/c^2}{1 \text{ u}}\right)c^2$$
$$= 8.482 \text{ MeV}$$

INSIGHT
We see that it takes about 8.5 MeV to separate the nucleons of tritium. Compare this with the fact that only 13.6 eV is required to remove the electron from tritium. As we have noted before, nuclear processes typically involve energies in the MeV range, whereas atomic processes require energies in the eV range.

PRACTICE PROBLEM
Calculate the binding energy of an α particle.   [**Answer:** 28.296 MeV]

Some related homework problems: Problem 41, Problem 42

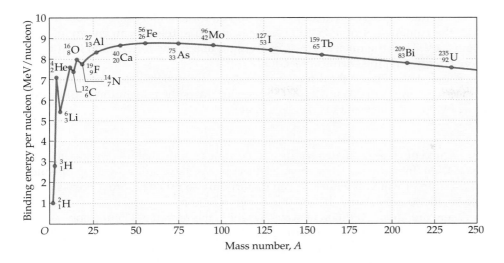

◄ **FIGURE 32–9** The curve of binding energy

This graph shows the binding energy per nucleon for a variety of nuclei. Notice that the binding energy per nucleon for tritium is approximately 2.8 MeV, in agreement with Example 32–6.

In addition to knowing the binding energy of a given nucleus, it is also of interest to know the binding energy per nucleon. In the case of tritium, there are three nucleons; hence, the binding energy per nucleon is $\frac{1}{3}(8.482 \text{ MeV}) = 2.827 \text{ MeV}$. **Figure 32–9** presents the binding energy per nucleon for various stable nuclei as a function of the mass number, A. Notice that the binding-energy curve rises rapidly to a maximum near $A = 60$ and then decreases slowly to a value near 7.4 MeV per nucleon for larger A.

It follows that nuclei with A in the range $A = 50$ to $A = 75$ are the most stable nuclei in nature. In addition, the fact that the binding energy per nucleon changes very little for large A means that the strong nuclear force "saturates," in the sense that adding more nucleons does not add to the binding energy per nucleon. This phenomenon can be understood as a consequence of the short range of the nuclear force. Because of this short range, each nucleon interacts with only a few nucleons that are close neighbors to it in the nucleus. Nucleons on the other side of the nucleus are too far away to interact. Thus, from the point of view of a given nucleon, the attractive energy it feels due to other nucleons in the nucleus is essentially the same whether the nucleus has 150 nucleons or 200 nucleons—only the nearby nucleons interact.

The fact that the binding-energy curve has a maximum has important implications, as will be shown in the next two sections. For example, energy can be released when large nuclei split into smaller nuclei (fission), or when small nuclei combine to form a larger nucleus (fusion).

PROBLEM-SOLVING NOTE

Interpreting the Mass Difference

If the mass difference due to a reaction is negative, $\Delta m = m_f - m_i < 0$, it follows that energy is *released* by the reaction. If Δm is positive, energy must be *supplied* to the system to make the reaction occur.

32–5 Nuclear Fission

A new type of physical phenomenon was discovered in 1939 when Otto Hahn and Fritz Strassman found that, under certain conditions, a uranium nucleus can split apart into two smaller nuclei. This process is called **nuclear fission.** As we shall see, nuclear fission releases an amount of energy that is many orders of magnitude greater than the energy released in chemical reactions. This fact has had a profound impact on the course of human events over the last several decades.

To obtain a rough estimate of just how much energy is released as a result of a fission reaction, consider the binding-energy curve in Figure 32–9. For a large nucleus like uranium-235, $^{235}_{92}\text{U}$, we see that the binding energy per nucleon is approximately 7.5 MeV. If this nucleus splits into two nuclei with roughly half the mass number ($A \sim 235/2 \sim 115$), we see that the binding energy per nucleon increases to roughly 8.3 MeV. If all 235 nucleons in the original uranium nucleus release (8.3 MeV $-$ 7.5 MeV) = 0.8 MeV of energy, the total energy release is

$$(235 \text{ nucleons})(0.8 \text{ MeV/nucleon}) = 200 \text{ MeV}$$

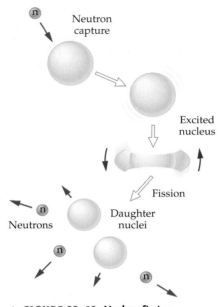

▲ **FIGURE 32–10 Nuclear fission**
When a large nucleus captures a neutron, it may become excited and ultimately split into two smaller nuclei. This is the process of fission.

Note that this is the energy released by a *single nucleus* undergoing fission. In contrast, the combustion of a *single molecule* of gasoline releases an energy of only about 2 eV. Thus the nuclear reaction gives off about one hundred million times more energy than the chemical reaction!

The first step in a typical fission reaction occurs when a slow neutron is absorbed by a uranium-235 nucleus. This step increases the mass number of the nucleus by one and leaves it in an excited state:

$$\,^1_0n + \,^{235}_{92}U \longrightarrow \,^{236}_{92}U^*$$

The excited nucleus oscillates wildly and becomes highly distorted, as depicted in **Figure 32–10**. In many respects, the nucleus behaves like a spinning drop of water. Like a drop of water, the nucleus can distort only so much before it breaks apart into smaller pieces; that is, before it undergoes fission.

There are about 90 different ways in which the uranium-235 nucleus can undergo fission. Typically, 2 or 3 neutrons (2.47 on average) are released during the fission process, in addition to the two smaller nuclei that are formed. The reason that neutrons are released can be seen by examining Figure 32–1, where we show N and Z for various nuclei. A large nucleus, like uranium-235, contains a higher percentage of neutrons than a smaller nucleus—that is, the larger nuclei deviate more from the $N = Z$ line. Thus, if $\,^{235}_{92}U$ were to simply break in two—keeping the same percentage of neutrons and protons in each piece—the smaller nuclei would have too many neutrons to be stable. As a result, neutrons are typically given off in a fission reaction.

One of the possible fission reactions for $\,^{235}_{92}U$ is considered in the following Example.

EXAMPLE 32–7 A FISSION REACTION OF URANIUM-235

When uranium-235 captures a neutron, it may undergo the following fission reaction: $\,^1_0n + \,^{235}_{92}U \longrightarrow \,^{236}_{92}U^* \longrightarrow \,^{141}_{56}Ba + ? + 3\,^1_0n$.
(a) Complete the reaction and **(b)** determine the energy it releases.

PICTURE THE PROBLEM
The indicated fission reaction is shown in our sketch, starting with the $\,^{236}_{92}U^*$ nucleus. The known numbers of protons and neutrons are indicated for both $\,^{236}_{92}U^*$ and $\,^{141}_{56}Ba$. The corresponding numbers for the unidentified nucleus are to be determined.

STRATEGY

a. The total number of protons must be the same before and after the reaction, as must the number of neutrons. By conserving protons and neutrons, we can determine the missing numbers on the unidentified nucleus.

b. As in other reactions, we calculate the difference in mass, $|\Delta m|$ and multiply by c^2. The same number of electrons appears on both sides of the reaction; hence, the electron mass will cancel when we calculate $|\Delta m|$.

SOLUTION

Part (a)

1. Determine the number of protons, Z, neutrons, N, and nucleons, A, in the unidentified nucleus. When calculating the number of neutrons, be sure to include the three individual neutrons given off in the reaction:

$Z = 92 - 56 = 36$
$N = 144 - 85 - 3 = 56$
$A = Z + N = 36 + 56 = 92$

2. Use Z and A to specify the unidentified nucleus:

$\,^{92}_{36}Kr$

Part (b)

3. Use Appendix F to find the initial mass of the system; that is, the mass of a $\,^{235}_{62}U$ atom plus the mass of a neutron:

$m_i = 235.043925 \text{ u} + 1.008665 \text{ u} = 236.052590 \text{ u}$

4. Use Appendix F to find the final mass of the system, including both nuclei and the three neutrons:

$$m_f = 140.914406 \text{ u} + 91.926111 \text{ u} + 3(1.008665 \text{ u})$$
$$= 235.866512 \text{ u}$$

5. Calculate the mass difference and the corresponding energy release (recall that 1 u = 931.5 MeV/c^2):

$$\Delta m = m_f - m_i$$
$$= 235.866512 \text{ u} - 236.052590 \text{ u} = -0.186078 \text{ u}$$

$$E = |\Delta m|c^2 = (0.186078 \text{ u})\left(\frac{931.5 \text{ MeV}/c^2}{1 \text{ u}}\right)c^2$$

$$= 173.3 \text{ MeV}$$

INSIGHT

Thus, as suggested by our crude calculation earlier in this section, the energy given off by a typical fission reaction is on the order of 200 MeV.

PRACTICE PROBLEM

Complete the following reaction, and determine the energy it releases: $_0^1\text{n} + _{92}^{235}\text{U} \longrightarrow _{92}^{236}\text{U}^* \longrightarrow _{54}^{140}\text{Xe} + ? + 2_0^1\text{n}$. [**Answer:** $_{38}^{94}\text{Sr}$, 184.7 MeV]

Some related homework problems: Problem 48, Problem 49

Chain Reactions

The fact that fission reactions of $_{92}^{235}\text{U}$ give off more than one neutron on average has significant implications. To see why, recall that the fission of $_{92}^{235}\text{U}$ is initiated by the absorption of a neutron in the first place. So the neutrons given off by one $_{92}^{235}\text{U}$ fission reaction may cause additional fission reactions in other $_{92}^{235}\text{U}$ nuclei. A reaction that proceeds from one nucleus to another in this fashion is referred to as a **chain reaction.**

Suppose, for the sake of discussion, that a fission reaction gives off two neutrons and that both neutrons induce additional fissions in other nuclei. These nuclei in turn give off two neutrons. Starting with one nucleus to begin the chain reaction, we have two nuclei in the second generation of the chain, four nuclei in the third generation, and so on, as indicated in **Figure 32–11.** After only 100 generations, the number of nuclei undergoing fission is 1.3×10^{30}. If each of these reactions gives off 200 MeV of energy, the total energy release after just 100 generations is 4.1×10^{19} J. To put this into everyday terms, this is enough energy to supply the needs of the entire United States for half a year. Clearly, a rapidly developing **runaway** chain reaction, like the one just described, would result in the explosive release of an enormous amount of energy.

A great deal of effort has gone into controlling the chain reaction of $_{92}^{235}\text{U}$. If the release of energy can be kept to a manageable and usable level—avoiding

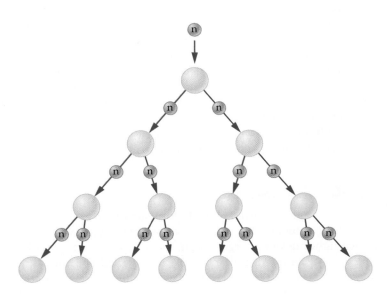

◀ **FIGURE 32–11 A chain reaction**
A chain reaction occurs when a neutron emitted by one fission reaction induces a fission reaction in another nucleus. In a runaway chain reaction, the number of neutrons given off by one reaction that cause an additional reaction is greater than one. In the case shown here, two neutrons from one reaction induce additional reactions. In a controlled chain reaction, like those used in nuclear power plants, only one neutron, on average, induces additional reactions.

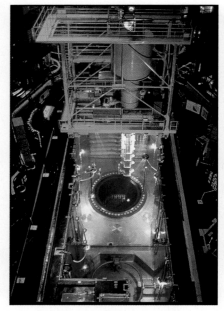

▲ (Left) The core of a nuclear reactor at a power plant. The core sits in a pool of water, which provides cooling while absorbing stray radiation. The blue glow suffusing the pool is radiation from electrons traveling through the water at relativistic velocities. Above the core is a crane used to replace the fuel rods when their radioactive material becomes depleted. (Center) A closer view of a reactor core. The structures projecting above the core house the mechanism that lowers and raises the control rods, thereby regulating the rate of fission in the reactor. The tubes that are nearly flush with the top of the core house the fuel rods. (Right) A technician loads pellets of fissionable material into fuel rods.

explosions or meltdowns—it follows that a powerful source of energy is at our disposal.

The first controlled nuclear chain reaction was achieved by Fermi in 1942, using a racquetball court at the University of Chicago for his improvised laboratory. His reactor consisted of blocks of uranium (the fuel) stacked together with blocks of graphite (the moderator) to form a large "pile." In such a **nuclear reactor,** the **moderator** slows the neutrons given off during fission, making it more likely that they will be captured by other uranium nuclei and cause additional fissions. The reaction rate is adjusted with **control rods** made of a material (such as cadmium) that is very efficient at absorbing neutrons. With the control rods fully inserted into the pile, any reaction that begins quickly dies out because neutrons are absorbed rather than allowed to cause additional fissions. As the control rods are pulled partway out of the pile, more neutrons become available to induce reactions. When, on average, one neutron given off by any fission reaction produces an additional reaction, the reactor is said to be **critical.** If the control rods are pulled out even farther from the pile, the number of neutrons causing fission reactions in the next generation becomes greater than one, and a runaway reaction begins to occur. Nuclear reactors that produce power for practical applications are operated near their critical condition by continuous adjustment of the placement of the control rods.

Fermi's original reactor, which was designed to test the basic scientific principles underlying fission reactions, produced a power of only about 0.5 W when operating near its critical condition. This is the power necessary to operate a flashlight bulb. In comparison, modern-day nuclear power reactors typically produce 1000 MW of power, enough to power an entire city.

REAL-WORLD PHYSICS
Nuclear reactors

32–6 Nuclear Fusion

When two light nuclei combine to form a more massive nucleus, the reaction is referred to as **nuclear fusion.** The binding-energy curve in Figure 32–9 shows that when light nuclei undergo fusion, the resulting nucleus will have a greater binding energy per nucleon than the original nuclei. This means that *the larger nucleus formed by fusion has less mass than the sum of the masses of the original light nuclei.* The mass difference appears as an energy ($E = mc^2$) given off by the reaction.

It is not easy to initiate a fusion reaction. The reason is that to combine two small nuclei, say two protons, into a single nucleus requires that the initial nuclei have a kinetic energy great enough to overcome the Coulomb repulsion the protons exert on each other. The temperature required to give protons the needed kinetic energy is about 10^7 K. When the temperature is high enough to initiate fusion, we say that the resulting process is a **thermonuclear fusion reaction.**

One place where such high temperatures may be encountered is at the core of a star. In fact, all stars generate energy by the process of thermonuclear reactions. Most stars (including the Sun) fuse hydrogen to produce helium. At this very moment, the Sun is converting roughly 600,000,000 tons of hydrogen into helium every second. Some stars also fuse helium or other heavier elements.

In its early stages a star begins as an enormous cloud of gas and dust. As the cloud begins to fall inward under the influence of its own gravity, it converts gravitational potential energy into kinetic energy. This means that the gas heats up. When the temperature becomes high enough to begin the fusion process, the resulting release of energy tends to stabilize the star, preventing its further collapse. A star like the Sun can burn its hydrogen for about 10 billion years, producing a remarkably stable output of energy so important to life on Earth. When its hydrogen fuel is depleted, the Sun will enter a new phase of its life, becoming a red giant and converting its fusion process to one that fuses helium. When the Sun enters its red giant phase, it will expand greatly in size, eventually engulfing the Earth and vaporizing it.

The Sun is powered by the **proton-proton cycle** of fusion reactions, as first described by Hans Bethe (1906–2005). This cycle consists of three steps. The first two steps are as follows:

REAL-WORLD PHYSICS

Powering the Sun: the proton-proton cycle

$$^1_1\text{H} + {}^1_1\text{H} \longrightarrow {}^2_1\text{H} + e^+ + v_e$$

$$^1_1\text{H} + {}^2_1\text{H} \longrightarrow {}^3_2\text{He} + \gamma$$

A number of reactions are possible for the third step, but the dominant one is

$$^3_2\text{He} + {}^3_2\text{He} \longrightarrow {}^4_2\text{He} + {}^1_1\text{H} + {}^1_1\text{H}$$

The overall energy production during this cycle is about 27 MeV.

In the next Active Example we show how the mass difference in a deuterium-deuterium reaction can be used to calculate the amount of energy released.

ACTIVE EXAMPLE 32–4   **THE DEUTERIUM-DEUTERIUM REACTION**

Find the energy released in the deuterium-deuterium reaction, $^2_1\text{H} + {}^2_1\text{H} \longrightarrow {}^3_1\text{H} + {}^1_1\text{H}$.

SOLUTION *(Test your understanding by performing the calculations indicated in each step.)*

1. Calculate the initial mass: $m_i = 4.028204$ u
2. Calculate the final mass: $m_f = 4.023874$ u
3. Find the difference in mass: $\Delta m = -0.004330$ u
4. Convert the mass difference to an energy release: $E = |\Delta m|c^2 = 4.033$ MeV

INSIGHT
The tritium produced in this reaction has a half-life of 12.33 y. Deuterium is stable.

YOUR TURN
Find the energy released in the reaction $^3_2\text{He} + {}^3_2\text{He} \longrightarrow {}^4_2\text{He} + {}^1_1\text{H} + {}^1_1\text{H}$.

(Answers to **Your Turn** *problems are given in the back of the book.)*

Just as nuclear fission reactions can be controlled, allowing for the generation of usable power, it would be desirable to control fusion as well. In fact, there are many potential advantages of fusion over fission. For example, fusion reactions yield more energy from a given mass of fuel than fission reactions. In addition, one type of fuel for a fission reactor, ^2_1H, is readily obtained from seawater. To date, however, more energy is required to produce sustained fusion reactions in the lab than is released by the reactions. Still, researchers are close to the break-even point, and many are confident that in time controlled fusion reactions that give off usable amounts of energy will be possible.

Most attempts at controlled nuclear fusion employ one of two basic methods. In both methods the basic idea is to overcome the repulsive electrical forces that keep nuclei apart by giving them sufficiently large kinetic energies, which can be accomplished by heating the fuel to temperatures on the order of 10^7 K.

At temperatures like these, all atoms are completely ionized, forming a gas of electrons and nuclei known as a *plasma*. To initiate fusion in the plasma, one must maintain the plasma long enough for collisions to occur between the appropriate nuclei. This can be accomplished using a technique known as **magnetic confinement,** in which powerful magnetic fields confine a plasma within a "magnetic bottle." The trapped plasma is kept away from the walls of the container—preventing their melting—and is heated until fusion begins.

Another approach to reaching the high temperatures and pressures required for fusion is by way of **inertial confinement.** In this technique, one begins by dropping a small solid pellet of fuel into a vacuum chamber. When the pellet reaches the center of the chamber it is bombarded from all sides by high-power laser beams. These beams heat and vaporize the surface of the pellet in an almost instantaneous event. The heating that causes fusion is still to come, however. As the vaporized exterior of the pellet expands rapidly outward, it exerts an inward "thrust" that causes the remainder of the pellet to implode. This implosion is so violent that the pellet's temperature and pressure rise to levels sufficient to ignite the desired fusion reactions.

REAL-WORLD PHYSICS
Man-made fusion

▲ Nuclear fusion, which powers the stars, may eventually turn out to be the clean, inexpensive, and renewable energy source that our society needs. But fusion requires enormous temperatures and pressures, and so far the problems involved in creating a practical fusion technology have not been overcome. Several different approaches have been explored in the effort to produce sustained nuclear fusion. One of them, embodied in the Z machine at Sandia National Laboratories in New Mexico, shown here, involves the compression of a plasma by an intense burst of X-rays. So far, temperatures of nearly 2 million degrees have been attained, but only for very brief periods of time.

32–7 Practical Applications of Nuclear Physics

Nuclear physics, though it involves objects none of us will ever touch or see, nevertheless has significant impact on our everyday life. In this section we consider a number of ways in which nuclear physics affects us all, either directly or indirectly.

Biological Effects of Radiation

Although nuclear radiation can be beneficial when used to image our bodies, or to treat a variety of diseases, it can also be harmful to living tissues. For example, the high-energy photons, electrons, and α particles given off by radioactive decay can ionize a neutral atom by literally "knocking" one or more of its electrons out of the atom. In fact, since typical nuclear decays give off energies in the MeV range, and typical ionization energies are in the 10 eV range, it follows that a single α, β, or γ particle can ionize thousands of atoms or molecules.

Such ionization can be harmful to a living cell by altering the structure of some of its molecules. The result can be a cell that no longer functions or behaves normally, or even a cell that will soon die. This effect is the basis for radiation treatments of cancer, which seek to concentrate high doses of radiation on a cancerous tumor in order to kill the malignant cells. But just what is a "high dose, " and how do we quantify radiation dosage?

The first radiation unit to be defined, called the **roentgen** (R), is directly related to the amount of ionization caused by X-rays or γ rays. Suppose such radiation is sent through a mass, m, of dry air at standard temperature and pressure (STP). If the radiation creates ions in the air with a total charge of q, we say that the dose of radiation delivered to the air is proportional to q/m. Specifically, we say that the dosage of X-rays or γ rays is 1 R if an ionization charge of 2.58×10^{-4} C is produced in 1 kg of dry air.

REAL-WORLD PHYSICS: BIO
Radiation and cells

Definition of the Roentgen, R

$$1\text{ R} = 2.58 \times 10^{-4}\text{ C/kg} \qquad \text{(X-rays or } \gamma \text{ rays in dry air at STP)} \qquad 32\text{–}15$$

SI unit: C/kg

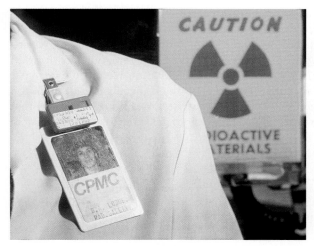 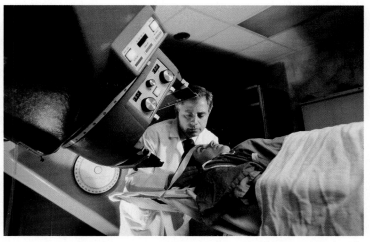

▲ Because radiation can kill or damage living cells, people who work with radioactive materials or other sources of ionizing radiation (such as X-ray tubes) must be careful to monitor their cumulative exposure. One way to do this is with a simple film badge (left). When the photographic film is developed, the degree of darkening records the amount of radiation it has received. Ironically, the fact that radiation can be lethal to cells also makes it a valuable therapeutic tool in the fight against cancer. At right, radiation from cobalt-60 is being used to attack a malignant tumor. Today, many hospitals have their own particle accelerators to manufacture short-lived radioactive isotopes for the treatment of different kinds of cancer or to use as tracers.

The roentgen is one key measure of radiation dosage, but there are others. The focus of the **rad,** an acronym for **r**adiation **a**bsorbed **d**ose, is the amount of energy that is absorbed by the irradiated material, regardless of the type of radiation that delivers the energy. For example, if a 1-kg sample of any material absorbs 0.01 J of energy, we say that it has received a dose of 1 rad:

Definition of the Rad

$$1 \text{ rad} = 0.01 \text{ J/kg} \qquad \text{(any type of radiation)} \qquad\qquad 32\text{–}16$$

SI unit: J/kg

Because the rad depends only on the amount of energy absorbed, and not on the type of radiation, more information is needed if we are to have an indication of the biological effect a certain dosage will produce. For example, a 1-rad dose of X-rays is far less likely to cause a cataract in the cornea of the eye than a 1-rad dose of neutrons. To take into account such differences, we introduce a quantity called the relative **b**iological **e**ffectiveness, or **RBE.** The standard used for comparison is the biological effect produced by a 1-rad dose of 200-keV X-rays. Thus, the RBE is defined as follows:

Definition of Relative Biological Effectiveness, RBE

$$\text{RBE} = \frac{\text{the dose of 200-keV X-rays necessary to produce a given biological effect}}{\text{the dose of a particular type of radiation necessary to produce the same biological effect}} \qquad 32\text{–}17$$

SI unit: dimensionless

Representative RBE values are given in Table 32–3. Note that the larger the RBE for a certain type of radiation, the greater the biological damage caused by a given dosage of that radiation.

Combining the dosage in rad and the RBE value for a given radiation yields a new unit of radiation referred to as the **biologically equivalent dose.** This unit is measured in **rem,** which stands for **r**oentgen **e**quivalent in **m**an. To be specific, dosage in rem is defined as follows:

Definition of Roentgen Equivalent in Man, rem

$$\text{dose in rem} = \text{dose in rad} \times \text{RBE} \qquad\qquad 32\text{–}18$$

SI unit: J/kg

TABLE 32–3 Relative Biological Effectiveness, RBE, for Different Types of Radiation

Type of radiation	RBE
Heavy ions	20
α rays	10–20
Protons	10
Fast neutrons	10
Slow neutrons	4–5
β rays	1.0–1.7
γ rays	1
200-kev X-Rays	1

TABLE 32–4 Typical Radiation Dosages

Source of radiation	Radiation dose (rem/y)
Inhaled radon	~0.200
Medical/dental examinations	0.040
Cosmic rays	0.028
Natural radioactivity in the Earth and atmosphere	0.028
Nuclear medicine	0.015

Defined in this way, 1 rem of radiation produces the same amount of biological damage, no matter which type of radiation is involved. In addition, the larger the dosage in rem, the greater the biological damage. Referring to Table 32–3, we see that 1 rad of 200-keV X-rays produces a radiation dose of 1 rem, whereas 1 rad of protons produces a dose of 10 rem. The dosage of radiation we receive per year from cosmic rays is about 28 mrem = 0.028 rem. Other typical values of radiation dosages are presented in Table 32–4. (For comparison, a dose of 50 to 100 rem damages blood-forming tissues, whereas a dose of 500 rem usually results in death.)

EXERCISE 32–4

A biological sample receives a dose of 456 rad from neutrons with an RBE of 6.20. **(a)** Find the dosage in rem. **(b)** If this same dosage in rem is to be delivered by α rays with an RBE of 13.0, what dosage in rad is required?

SOLUTION

    a. The dose in rem is the product of the dose in rad and the RBE:

$$(456 \text{ rad})(6.20) = 2830 \text{ ram}$$

    b. The dose in rad is the dose in rem divided by the RBE:

$$\frac{2830 \text{ rem}}{13.0} = 218 \text{ rad}$$

Radioactive Tracers

The phenomenon of radioactivity has found applications of many types in medicine and biology. One of these applications employs a radioactive isotope as a sort of "identification tag" or "tracer" to determine the location and quantity of a substance of interest.

REAL-WORLD PHYSICS: BIO

Radioactive tracers

For example, an artificially produced radioactive isotope of iodine, $^{131}_{53}\text{I}$, is used to determine the condition of a person's thyroid gland. A healthy thyroid plays an important role in the distribution of iodine—a necessary nutrient—throughout the body. To see if the thyroid is functioning as it should, a patient drinks a small quantity of radioactive sodium iodide, which incorporates the isotope $^{131}_{53}\text{I}$. A couple of hours later, the radiation given off by the patient's thyroid gland is measured, giving a direct indication of the amount of sodium iodide the gland has processed.

Because radioactive tracers differ from their nonradioactive counterparts only in the composition of their nuclei, they undergo the same chemical and metabolic reactions. This makes them useful in diagnosing and treating a variety of different conditions. For example, chromium-51 is used to label red blood cells and to quantify gastrointestinal protein loss, copper-64 is used to study genetic diseases affecting copper metabolism, and yttrium-90 is used for liver cancer therapy.

PET Scans

REAL-WORLD PHYSICS: BIO

Positron-emission tomography

Perhaps the most amazing medical use of radioactive decay is in the diagnostic technique known as **positron emission tomography** (PET). Like the images made using radioactive tracers, PET scans are produced with radiation that *emerges from within* the body, as opposed to radiation that is generated externally and is then passed through the body. Remarkably, the radiation in a PET scan is produced by the *annihilation of matter and antimatter* within the patient's body! It sounds like science fiction, but it is, in fact, a valuable medical tool.

To produce a PET scan, a patient is given a radiopharmaceutical, like fluorodeoxyglucose (FDG), which contains an atom that decays by giving off a positron (the antiparticle to the electron). In the case of FDG, a fluorine-18 atom is attached to a molecule of glucose, the basic energy fuel of cells. The fluorine-18 atom undergoes the following decay with a half-life of 110 minutes:

$$^{18}_{9}\text{F} \longrightarrow {}^{18}_{8}\text{O} + e^+ + v_e$$

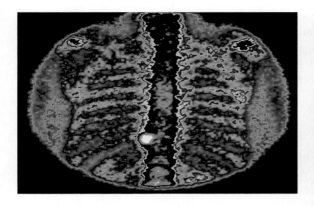

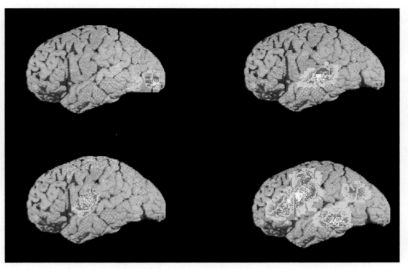

▲ Whereas X-ray images are created by passing a beam of radiation through the body, other imaging techniques make use of radiation produced within the body. One method involves the use of radioactive isotopes that tend to become concentrated in particular tissues or structures, such as the thyroid gland or the bones. Radiation from that component can then be measured, or even used to create an image, as in the bone scan at left, where areas of abnormally high radiation (white) indicate the presence of malignancy. A sophisticated variant of this approach is used to produce PET scans (see text), which are most commonly employed to visualize areas of the body where cellular energy production is most intense. The series of images at right records brain activity associated with seeing words (upper left), hearing words (upper right), speaking words (lower left), and thinking about words while uttering them (lower right). (The image of the brain, superimposed for reference, is not part of the PET scan.)

Almost immediately after the positron is emitted in this decay, it encounters an electron, and the two particles annihilate each other in a burst of energy. In fact, the annihilation process generates two powerful γ rays moving in opposite directions. As these γ rays emerge from the body, specialized detectors observe them and determine their point of origin. The resulting computerized image shows the areas in the body where glucose metabolism is most intense.

PET scans are particularly useful in examining the brain. For example, a PET scan using FDG can show which regions of the brain are most active when a person is performing a specific mental task, like counting, speaking, or translating a foreign language. The scans can also show abnormality in brain function, as when one side of a brain becomes more active than the other during an epileptic seizure. Finally, PET scans can be used to locate tumors in the brain and other parts of the body, and to monitor the progress of their treatment.

Magnetic Resonance Imaging

A diagnostic technique that gives images similar to those obtained with computerized tomography (Section 31–7) is **magnetic resonance imaging,** or MRI. The basic physical mechanism used in MRI is the interaction of a nucleus with an externally applied magnetic field.

Consider the simplest nucleus—a single proton. Protons have a magnetic moment, like a small bar magnet, and when they are placed in a constant magnetic field they precess about the field with a frequency proportional to the field strength. In addition, protons have a spin of one-half, like electrons; hence, their spin can be either "up" or "down" relative to the direction of the constant magnetic field. When an oscillating magnetic field is applied perpendicular to the constant magnetic field, it can cause protons to "flip" from one spin state to the other if the frequency of oscillation is in resonance with the proton's precessional frequency. These spin flips result in the absorption or release of energy in the radio portion of the electromagnetic spectrum, which can be detected electronically.

By varying the strength of the constant magnetic field as a function of position, spin flips can be detected in various parts of the body being examined. Using a computer to combine signals from various positions allows for the generation of detailed cross-sectional images, as in CAT scans.

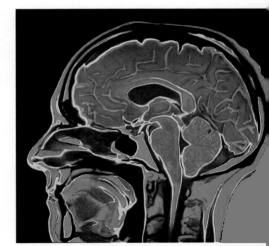

▲ Magnetic resonance imaging (MRI) is a safe, noninvasive technique for visualizing internal body structures. No ionizing radiation is involved, so the risk of tissue damage is minimal. MRI images generally show soft tissue with greater clarity than do ordinary X-rays.

REAL-WORLD PHYSICS: BIO

Magnetic resonance imaging (MRI)

One of the advantages of MRI, however, is that the photons associated with the magnetic fields used in the imaging process are very low in energy. In fact, typical energies are in the range of 10^{-7} eV, much less than typical ionization energies; hence, MRI causes very little cellular damage. In contrast, photons used in CAT scans have energies that can range from 10^4 to 10^6 eV, more than enough to produce cellular damage.

32–8 Elementary Particles

Scientists have long sought to identify the fundamental building blocks of all matter, the **elementary particles.** At one point, it was thought that atoms were elementary particles—one for each element. As we saw in the previous chapter, however, this idea was put to rest when atoms were discovered to be made up of electrons, protons, and neutrons. Of these three particles, only the electron is presently considered to be elementary—protons and neutrons are now known to be composed of still smaller particles. In addition, approximately 300 new particles were discovered in the last half of the twentieth century, most of which are unstable and have lifetimes of only 10^{-6} to 10^{-23} s.

Although a complete accounting of the current theories of elementary particles is beyond the scope of this text, we shall outline the basic insights that have been derived from these theories and from related experiments. We begin by describing the four fundamental forces of nature, since particles can be categorized according to which of these forces they experience.

The Fundamental Forces of Nature

Although nature presents us with a myriad of different physical phenomena—from tornadoes and volcanoes to sunspots and comets to galaxies and black holes—all are the result of just *four* fundamental forces. This is one example of the simplicity that physicists see in nature. These forces, in order of diminishing strength, are the strong nuclear force, the electromagnetic force, the weak nuclear force, and gravity. If we assign a strength of 1 to the strong nuclear force, for purposes of comparison, the strength of the electromagnetic force is 10^{-2}, the strength of the weak nuclear force is 10^{-6}, and the strength of the gravitational force is an incredibly tiny 10^{-43}. These results are summarized in Table 32–5.

TABLE 32–5 The Fundamental Forces

Force	Relative Strength	Range
Strong nuclear	1	≈ 1 fm
Electromagnetic	10^{-2}	Infinite $(\propto 1/r^2)$
Weak nuclear	10^{-6}	$\approx 10^{-3}$ fm
Gravitational	10^{-43}	Infinite $(\propto 1/r^2)$

All objects of finite mass experience gravitational forces. This is one reason why gravity is such an important force in the universe, even though it is spectacularly weak. Similarly, objects with a finite charge experience electromagnetic forces. As for the weak and strong nuclear forces, some particles experience only the weak force, whereas others experience both the weak and the strong force. We turn now to a discussion of particles that fall into these latter two categories.

Leptons

Particles that are acted on by the weak nuclear force but not by the strong nuclear force are referred to as **leptons.** There are only six leptons known to exist, all of which are listed in Table 32–6. The most familiar of these are the electron and its corresponding neutrino—both of which are stable. No internal structure has ever been detected in any of these particles. As a result, all six leptons have the status of true elementary particles.

TABLE 32–6 Leptons

Particle	Particle Symbol	Antiparticle Symbol	Rest Energy (MeV)	Lifetime (s)
Electron	e^- or β^-	e^+ or β^+	0.511	Stable
Muon	μ^-	μ^+	105.7	2.2×10^{-6}
Tau	τ^-	τ^+	1784	10^{-13}
Electron neutrino	ν_e	$\overline{\nu}_e$	≈ 0	Stable
Muon neutrino	ν_μ	$\overline{\nu}_\mu$	≈ 0	Stable
Tau neutrino	ν_τ	$\overline{\nu}_\tau$	≈ 0	Stable

The weak nuclear force is responsible for most radioactive decay processes, such as beta decay. It is also a force of extremely short range. In fact, the weak force can be felt only by particles that are separated by roughly one-thousandth the diameter of a nucleus. Beyond that range, the weak force has practically no effect at all.

Hadrons

Hadrons are particles that experience both the weak and the strong nuclear force. They are also acted on by gravity, since all hadrons have finite mass. The two most familiar hadrons, of course, are the proton and the neutron. A partial list of the hundreds of hadrons known to exist is given in Table 32–7. Notice that the proton is the only stable hadron (though some theories suggest that even it may decay with the incredibly long half-life of 10^{35} y).

TABLE 32–7 Hadrons

Particle	Particle Symbol	Antiparticle Symbol	Rest Energy (MeV)	Lifetime (s)
MESONS				
Pion	π^+	π	139.6	2.6×10^{-8}
	π^0	π^0	135.0	0.8×10^{-16}
Kaon	K^+	K^-	493.7	1.2×10^{-8}
	K_S^0	$\overline{K}_S^0$	497.7	0.9×10^{-10}
	K_L^0	$\overline{K}_L^0$	497.7	5.2×10^{-8}
Eta	η^0	η^0	548.8	$<10^{-18}$
BARYONS				
Proton	p	$\overline{p}$	938.3	Stable
Neutron	n	$\overline{n}$	939.6	900
Sigma	Σ^+	$\overline{\Sigma}^-$	1189	0.8×10^{-10}
	Σ^0	$\overline{\Sigma}^0$	1192	6×10^{-20}
	Σ^-	$\overline{\Sigma}^+$	1197	1.6×10^{-10}
Omega	Ω^-	Ω^+	1672	0.8×10^{-10}

The strong nuclear force is the only force powerful enough to hold a nucleus together. It is a short-range force, extending only to distances comparable to the diameter of a nucleus, but within that range it is strong enough to counteract the intense electromagnetic repulsion between positively charged protons. Outside the nucleus, however, the strong nuclear force is of negligible strength.

In striking contrast with leptons, none of the hadrons are elementary particles. In fact, all hadrons are composed of either two or three smaller particles called **quarks.** Hadrons formed from two quarks are referred to as **mesons;** those formed from three quarks are **baryons.** The properties of quarks will be considered next.

Quarks

To account for the internal structure observed in hadrons, Murray Gell-Mann (1929–) and George Zweig (1937–) independently proposed in 1963 that all

TABLE 32–8 Quarks and Antiquarks

Name	Rest Energy (MeV)	Quarks		Antiquarks	
		Symbol	Charge	Symbol	Charge
Up	360	u	$+\frac{2}{3}e$	$\bar{u}$	$-\frac{2}{3}e$
Down	360	d	$-\frac{1}{3}e$	$\bar{d}$	$+\frac{1}{3}e$
Charmed	1500	c	$+\frac{2}{3}e$	$\bar{c}$	$-\frac{2}{3}e$
Strange	540	s	$-\frac{1}{3}e$	$\bar{s}$	$+\frac{1}{3}e$
Top	173,000	t	$+\frac{2}{3}e$	$\bar{t}$	$-\frac{2}{3}e$
Bottom	5000	b	$-\frac{1}{3}e$	$\bar{b}$	$+\frac{1}{3}e$

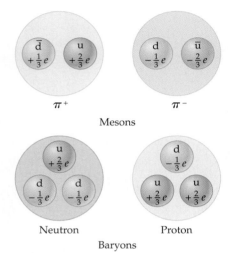

▲ **FIGURE 32–12 The quark composition of mesons and baryons**

Mesons and baryons are composed of various quark combinations—mesons always have a quark and an antiquark, baryons always have three quarks. Note that even though quarks have fractional charges (in units of the electron charge, e), the resulting mesons and baryons always have integer charges.

TABLE 32–9 Quark Composition of Some Hadrons

Particle	Quark Composition
MESONS	
π^+	$u\bar{d}$
π^-	$\bar{u}d$
K^+	$u\bar{s}$
K^-	$\bar{u}s$
K^0	$d\bar{s}$
BARYONS	
p	uud
n	udd
Σ^+	uus
Σ^0	uds
Σ^-	dds
Ξ^0	uss
Ξ^-	dss
Ω^-	sss

hadrons are composed of a number of truly elementary particles that Gell-Mann dubbed quarks. Originally, it was proposed that there are three types of quarks, arbitrarily named up (u), down (d), and strange (s). Discoveries of new and more massive hadrons, such as the J/ψ particle discovered in 1974, have necessitated the addition of three more quarks. The equally whimsical names for these new quarks are charmed (c), top or truth (t), and bottom or beauty (b). Table 32–8 lists the six quarks, along with some of their more important properties.

The antiparticles to the quarks are also given in Table 32–8. Notice that antiquarks are indicated with a bar over the symbol for the corresponding quark. For example, the symbol for the up quark is u; the symbol for the corresponding antiquark is $\bar{u}$.

Quarks are unique among the elementary particles in a number of ways. For example, they all have charges that are fractions of the charge of the electron. As can be seen in Table 32–8, some quarks have a charge of $+\left(\frac{2}{3}\right)e$ or $-\left(\frac{2}{3}\right)e$; others have a charge of $+\left(\frac{1}{3}\right)e$ or $-\left(\frac{1}{3}\right)e$. No other particles are known to have charges that differ from integer multiples of the electron's charge.

Now it might seem that the fractional charge of a quark would make it easy to identify experimentally. In fact, a number of experiments have searched for quarks in just that way, by looking for particles with fractional charge. No such particle has ever been observed, however. It is now believed that a free, independent quark cannot exist; quarks must always be bound with other quarks. This concept is referred to as **quark confinement.** The physical reason behind confinement is that the force between two quarks increases with separation—like two particles connected by a spring. Hence, an infinite amount of energy is required to increase the separation between two quarks to infinity.

The smallest system of bound quarks that can be observed as an independent particle is a pair of quarks. In fact, mesons consist of bound pairs of quarks and antiquarks, as illustrated schematically in **Figure 32–12**. For example, the π^+ meson is composed of a u$\bar{d}$ pair of quarks. Note that this combination of quarks gives the π^+ meson a net charge of $+e$. The π^- meson, the antiparticle to the π^+ meson, consists of a $\bar{u}$d pair with a charge of $-e$. Quarks are always bound in configurations that result in integer charges.

Baryons are bound systems consisting of three quarks, as shown in Figure 32–12. The proton, for example, has the composition uud, with a net charge of $+e$. The neutron, on the other hand, is formed from the combination udd, with a net charge of 0. A variety of hadrons and their corresponding quark compositions are given in Table 32–9.

Finally, not long after the quark model of elementary particles was introduced, it was found that some quark compositions implied a violation of the Pauli exclusion principle. To resolve these discrepancies, it was suggested that quarks must come in three different varieties, which were given the completely arbitrary but colorful names red, green, and blue. Though these quark "colors" have nothing to do with visible colors in the electromagnetic spectrum, they bring quarks into agreement with the exclusion principle and explain other experimental observations that were difficult to understand before the introduction of this new property. The

theory of how colored quarks interact with one another is called **quantum chromo-dynamics,** or QCD, in analogy with the theory describing interactions between charged particles, which is known as **quantum electrodynamics,** or QED.

32–9 Unified Forces and Cosmology

As discussed earlier, the universe as we see it today has four fundamental forces through which various particles interact with one another. This has not always been the case, however. Shortly after the Big Bang these four forces were combined into a single force sometimes referred to as the **unified force.** This situation lasted for only a brief interval of time. As the early universe expanded and cooled, it eventually underwent a type of "phase transition" in which the gravitational force took on a separate identity. This transition occurred at a time of approximately 10^{-43} s after the Big Bang, when the temperature of the universe was about 10^{32} K.

The phase transition just described was the first of three such transitions to occur in the early universe, as we see in **Figure 32–13**. At 10^{-35} s, when the temperature was 10^{28} K, the strong nuclear force became a separate force. Similarly, the weak nuclear force became a separate force at 10^{-10} s, when the temperature was 10^{15} K. From 10^{-10} s until the present, the situation has remained the same, even as the temperature of the universe has dipped to a chilly 2.7 K.

Let's look at these forces and the transitions between them more carefully. First, the electromagnetic force combines the forces associated with both electricity and magnetism. Although electricity and magnetism were originally thought to be separate forces, the work of Maxwell and others showed that these forces are simply different aspects of the same underlying force. For example, changing electric fields generate magnetic fields, and changing magnetic fields generate electric fields. In fact, the theory of electromagnetism can be thought of as the first *unified field theory,* in which seemingly different forces are combined into one all-encompassing theory.

At times earlier than 10^{-10} s the weak nuclear force was indistinguishable from the electromagnetic force. Thus, even though these forces seem very different today, we can recognize them as different aspects of the same underlying force—much like the two faces of a coin look very different but are part of the same physical object. The theory that encompasses the weak nuclear force and the electromagnetic force is called the **electroweak theory.** It was developed by Sheldon Glashow, Abdus Salam, and Steven Weinberg.

Going further back in time, the strong nuclear force was indistinguishable from the electroweak force before 10^{-35} s. Although no one has yet succeeded in producing a theory combining the electroweak force and the strong nuclear force, most physicists feel confident that such a theory exists. This hypothetical theory is referred to as the **grand unified theory,** or GUT.

Finally, a theory that encompasses gravity along with the other forces of nature is one of the ultimate goals of physics. Many physicists, including Einstein in his later years, have worked long and hard toward such an end, but so far with little success.

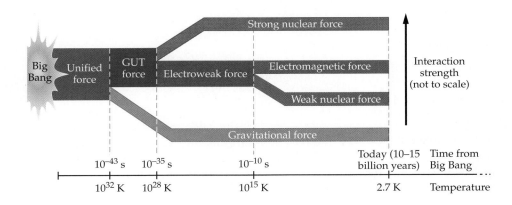

◀ **FIGURE 32–13 The evolution of the four fundamental forces**

The four forces we observe in today's universe began as a single unified force at the time of the Big Bang. As the universe cooled, the unified force evolved through a series of "transitions" in which the various forces took on different characteristics.

THE BIG PICTURE PUTTING PHYSICS IN CONTEXT

LOOKING BACK

In this, our final chapter, we've discussed some of the frontiers of modern physics, like fundamental forces, elementary particles, and the evolution of the universe. And yet, no matter how advanced or esoteric the topic, the fundamental principles of physics still form the basis for our understanding. A few examples from this chapter are as follows:

In considering the decay of a nucleus, we used the concept of electrostatic repulsion, momentum conservation, and energy conservation. For example, we compared the kinetic energies of nuclear decay products in Conceptual Checkpoint 32–2 with exactly the same methods used to study air carts in Chapter 9.

The most famous equation in physics, $E = mc^2$, plays a key role in nuclear reactions. In general, the mass of nuclear decay products does not add up to the mass of the initial nucleus—instead, the mass difference appears as energy given off during the decay. $E = mc^2$ also occurs in the study of matter/antimatter annihilation.

Finally, the evolution of the universe has been marked by a series of phase transitions as it cooled after the Big Bang. These phase transitions are similar to those studied in Chapter 17, and have resulted in a single force splitting into the four distinct types of forces we see in the universe today.

CHAPTER SUMMARY

32–1 THE CONSTITUENTS AND STRUCTURE OF NUCLEI

Nuclei are composed of just two types of particles: protons and neutrons. These particles are referred to collectively as nucleons.

Atomic Number, *Z*; Neutron Number, *N*; Mass Number, *A*
The atomic number, Z, is equal to the number of protons in a nucleus. The neutron number, N, is the number of neutrons in a nucleus. The mass number of a nucleus, A, is the total number of nucleons it contains. Thus, $A = N + Z$.

Designation of Nuclei
A nucleus with atomic number Z and mass number A is designated as follows:

$$^A_Z X$$

Isotopes
Isotopes are nuclei with the same atomic number but different neutron number.

Atomic Mass Unit, u
A convenient mass unit for nucleons and nuclei is the atomic mass unit, u, which is defined so that the mass of $^{12}_6C$ is exactly 12 u. The value of u is as follows:

$$1 \text{ u} = 1.660540 \times 10^{-27} \text{ kg} \qquad \text{32–2}$$

Neutrons and protons have masses slightly greater than 1 u, with the neutron slightly more massive than the proton.

Energy/Mass Equivalence
The atomic mass unit can be expressed in terms of energy (MeV) as follows:

$$1 \text{ u} = 931.5 \text{ MeV}/c^2 \qquad \text{32–3}$$

Nuclear Size and Density
The approximate radius of a nucleus of mass number A is given by

$$r = (1.2 \times 10^{-15} \text{ m})A^{1/3} \qquad \text{32–4}$$

All nuclei have roughly the same density, regardless of their mass number.

Nuclear Forces and Stability
Nuclei are held together by the strong nuclear force. This force is attractive between all nucleons and has a range of only a few fermis.

32–2 RADIOACTIVITY

Radioactivity refers to the emissions observed when an unstable nucleus changes its composition or when an excited nucleus decays to a lower-energy state.

Alpha Decay

An α particle (the nucleus of a helium atom) consists of two protons and two neutrons. A nucleus that emits an α particle decreases its mass number by 4 and its atomic number by 2:

$$^{A}_{Z}X \longrightarrow ^{A-4}_{Z-2}Y + ^{4}_{2}He$$

where X is the parent nucleus and Y is the daughter nucleus.

Beta Decay

Beta decay refers to the emission of an electron, as when a neutron decays into a proton, an electron, and an antineutrino:

$$^{1}_{0}n \longrightarrow ^{1}_{1}p + e^{-} + \bar{v}$$

This type of decay, which increases the atomic number by 1 but leaves the mass number unchanged, is referred to as β^{-} decay. If a positron and a neutrino are given off instead, we refer to the process as β^{+} decay.

Gamma Decay

Gamma decay occurs when an excited nucleus drops to a lower-energy state and emits a photon. In this case, neither the mass number nor the atomic number is changed.

Activity

The activity of a radioactive sample is equal to the number of decays per second. The units of activity are the curie (Ci) and the becquerel (Bq):

$$1 \text{ curie} = 1 \text{ Ci} = 3.7 \times 10^{10} \text{ decays/s} \qquad 32\text{–}6$$

$$1 \text{ becquerel} = 1 \text{ Bq} = 1 \text{ decay/s} \qquad 32\text{–}7$$

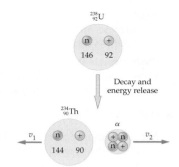

32–3 HALF-LIFE AND RADIOACTIVE DATING

Radioactive nuclei decay with time in a well-defined way. As a result, many radioactive nuclei can be used as a type of "nuclear clock."

Nuclei as a Function of Time and the Decay Constant, λ

If the number of radioactive nuclei in a sample at time $t = 0$ is N_0, the number, N, at a later time is

$$N = N_0 e^{-\lambda t} \qquad 32\text{–}9$$

The constant, λ, in this expression is referred to as the decay constant.

Half-life, $T_{1/2}$

The half-life of a radioactive material is the time required for half of its nuclei to decay. In terms of the decay constant, the half-life is

$$T_{1/2} = \frac{\ln 2}{\lambda} = \frac{0.693}{\lambda} \qquad 32\text{–}10$$

Decay Rate, or Activity, R

The rate at which radioactive nuclei decay is proportional to the number of nuclei present at any given time, and to the decay constant:

$$R = \left| \frac{\Delta N}{\Delta t} \right| = \lambda N = \lambda N_0 e^{-\lambda t} = R_0 e^{-\lambda t} \qquad 32\text{–}11, 32\text{–}12$$

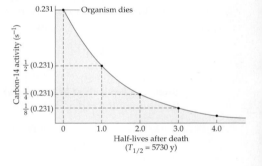

Carbon-14 Dating

Carbon-14 can be used to date organic materials with ages up to about 15,000 y. The age can be found using

$$t = -\frac{1}{\lambda} \ln \frac{R}{R_0} = \frac{1}{\lambda} \ln \frac{R_0}{R} \qquad 32\text{–}13$$

where $R_0 = 0.231$ is the initial activity, R is the present activity, and $\lambda = 1.21 \times 10^{-4} \text{ y}^{-1}$.

32–4 NUCLEAR BINDING ENERGY

The binding energy of a nucleus is the energy that must be supplied to separate it into its component nucleons.

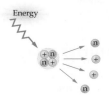

32–5 NUCLEAR FISSION

Nuclear fission is the process in which a large nucleus captures a neutron and then divides into two smaller "daughter" nuclei. Typical fission reactions emit two or three neutrons along with the daughter nuclei.

Chain Reactions

When neutrons given off by one fission reaction initiate additional fission reactions, we refer to the process as a chain reaction.

32–6 NUCLEAR FUSION

Nuclear fusion occurs when two small nuclei merge to form a larger nucleus. To initiate a fusion reaction, it is necessary to give the small nuclei enough energy to overcome their mutual Coulomb repulsion.

32–7 PRACTICAL APPLICATIONS OF NUCLEAR PHYSICS

Nuclear radiation can have both harmful and useful effects. An important way to characterize exposure to radiation is in terms of dosage, which can be defined in a number of ways.

Roentgen, R

The first unit of radiation, the roentgen, is related to the amount of ionization charge produced by 200-keV X-rays in 1 kg of dry air at STP:

$$1\,R = 2.58 \times 10^{-4}\,C/kg \quad \text{(X-rays or } \gamma \text{ rays in dry air at STP)} \qquad 32\text{–}15$$

Radiation Absorbed Dose (rad)

The rad is a measure of the amount of energy absorbed by an irradiated material, regardless of the type of radiation:

$$1\,rad = 0.01\,J/kg \quad \text{(any type of radiation)} \qquad 32\text{–}16$$

Relative Biological Effectiveness (RBE)

The RBE takes into account that different types of radiation produce different amounts of biological damage. It is defined as follows:

$$RBE = \frac{\text{the dose of 200-keV X-rays necessary to produce a given biological effect}}{\text{the dose of a particular type of radiation necessary to produce the same biological effect}} \qquad 32\text{–}17$$

Roentgen Equivalent in Man (rem)

Combining the rad and the RBE yields the rem:

$$\text{dose in rem} = \text{dose in rad} \times RBE \qquad 32\text{–}18$$

A dose of 1 rem of any type of radiation causes the same amount of biological damage.

32–8 ELEMENTARY PARTICLES

Elementary particles are the fundamental building blocks of all matter.

The Fundamental Forces of Nature

There are just four fundamental forces in nature. In order of decreasing strength, they are the strong nuclear force, the electromagnetic force, the weak nuclear force, and the gravitational force.

Leptons

Leptons are elementary particles that experience the weak nuclear force but not the strong nuclear force.

Hadrons

Hadrons are composite particles that experience both the weak and the strong nuclear force.

Quarks

Quarks are elementary particles that combine to form hadrons. Mesons are formed from quark-antiquark pairs; baryons are formed from combinations of three quarks.

32–9 UNIFIED FORCES AND COSMOLOGY

The four fundamental forces observed in the universe today began as a single force at the time of the Big Bang. As the universe expanded and cooled, the single force split into four different forces with different characteristics.

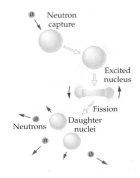

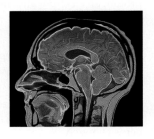

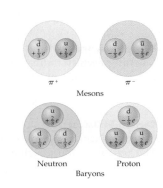

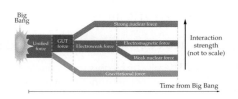

PROBLEM-SOLVING SUMMARY

Type of Problem	Relevant Physical Concepts	Related Examples
Identify a missing term in a nuclear reaction.	The number of protons and neutrons must be the same before and after an alpha decay. In a β^- decay, the number of neutrons after the decay is decreased by 1, and the number of protons is increased by 1.	Examples 32–3, 32–4, 32–7
Find the energy released in a nuclear reaction.	Calculate the difference in mass before and after the reaction. Multiply the difference in mass by the speed of light squared.	Examples 32–3, 32–4, 32–6, 32–7 Active Example 32–4
Determine the number of radioactive nuclei as a function of time.	The number of nuclei decreases exponentially with time according to the relation $N = N_0 e^{-\lambda t}$.	Active Example 32–2
Determine the activity of radioactive nuclei as a function of time.	Activity, which is the rate of decay, is the decay constant times the number of nuclei present at a given time	Active Example 32–3
Find the age of an organic sample using carbon-14 decay.	The age can be found using $t = \frac{1}{\lambda}\ln\frac{R_0}{R}$, where the initial activity of a 1-g sample of carbon-14 is $R_0 = 0.231$ Bq, and $\lambda = 1.21 \times 10^{-4}\ \text{y}^{-1}$.	Example 32–5

For instructor-assigned homework, go to www.masteringphysics.com

CONCEPTUAL QUESTIONS

(Answers to odd-numbered Conceptual Questions can be found in the back of the book.)

1. Nucleus A and nucleus B have different numbers of protons and different numbers of neutrons. Explain how it is still possible for these nuclei to have equal radii.

2. When α particles are emitted in a nuclear decay, they have well-defined energies. In contrast, β particles are found to be emitted with a range of energies. Explain this difference.

3. Is it possible for a form of heavy hydrogen to decay by emitting an α particle? Explain.

4. Which is more likely to expose film kept in a cardboard box, α particles or β particles? Explain.

5. It is not possible for a stable nucleus to contain more than one proton without also having at least one neutron. Explain why neutrons are necessary in a stable, multiparticle nucleus.

6. Different isotopes of a given element have different masses, but they have the same chemical properties. Explain why chemical properties are unaffected by a change of isotope.

7. **(a)** Give three examples of objects for which carbon-14 dating would give useful results. **(b)** Give three examples of objects for which carbon-14 dating would not be useful.

8. Explain why the large, stable nuclei in Figure 32–1 are found to lie above the $N = Z$ line, rather than below the line.

9. Suppose each of the following items is about 10,000 years old: a feather, a tooth, an obsidian arrowhead, a deer hide moccasin. Which of these items cannot be dated with carbon-14? Explain.

10. Can carbon-14 dating give the age of fossil dinosaur skeletons? Explain.

11. Two different samples contain the same radioactive isotope. Is it possible for these samples to have different activities? Explain.

12. Two samples contain different radioactive isotopes. Is it possible for these samples to have the same activity? Explain.

13. Two different types of radiation deliver the same amount of energy to a sample of tissue. Does it follow that each of these types of radiation has the same RBE? Explain.

PROBLEMS AND CONCEPTUAL EXERCISES

Note: Answers to odd-numbered Problems and Conceptual Exercises can be found in the back of the book. **IP** *denotes an integrated problem, with both conceptual and numerical parts;* **BIO** *identifies problems of biological or medical interest;* **CE** *indicates a conceptual exercise.* **Predict/Explain** *problems ask for two responses:* **(a)** *your prediction of a physical outcome, and* **(b)** *the best explanation among three provided. On all problems, red bullets (•, ••, •••) are used to indicate the level of difficulty.*

Refer to Appendix F for the masses and half-lives of relevant isotopes. Remember that the masses in Appendix F include the mass of the electrons associated with the neutral atoms.

SECTION 32–1   THE CONSTITUENTS AND STRUCTURE OF NUCLEI

1. • Identify Z, N, and A for the following isotopes: **(a)** $^{238}_{92}\text{U}$, **(b)** $^{239}_{94}\text{Pu}$, **(c)** $^{144}_{60}\text{Nd}$.

2. • Identify Z, N, and A for the following isotopes: **(a)** $^{202}_{80}\text{Hg}$, **(b)** $^{220}_{86}\text{Rn}$, **(c)** $^{93}_{41}\text{Nb}$.

3. • What are the nuclear radii of **(a)** $^{197}_{79}\text{Au}$ and **(b)** $^{60}_{27}\text{Co}$?

4. • A certain chlorine nucleus has a radius of approximately 4.0×10^{-15} m. How many neutrons are in this nucleus?

5. •• **IP** (a) What is the nuclear density of $^{228}_{90}\text{Th}$? **(b)** Do you expect the nuclear density of an alpha particle to be greater than,

less than, or the same as that of $^{228}_{90}$Th? Explain. **(c)** Calculate the nuclear density of an alpha particle.

6. •• **IP** **(a)** What initial kinetic energy must an alpha particle have if it is to approach a stationary gold nucleus to within a distance of 22.5 fm? **(b)** If the initial speed of the alpha particle is reduced by a factor of 2, by what factor is the distance of closest approach changed? Explain.

7. •• **IP** An α particle with a kinetic energy of 0.85 MeV approaches a stationary gold nucleus. **(a)** What is the speed of the α particle? **(b)** What is the distance of closest approach between the α particle and the gold nucleus? **(c)** If this same α particle were fired at a copper nucleus instead, would its distance of closest approach be greater than, less than, or the same as that found in part (b)? Explain. (To obtain the mass of an alpha particle, refer to Appendix F and subtract the mass of two electrons from the mass of $^4_2$He.)

8. •• Suppose a marble with a radius of 1.5 cm has the density of a nucleus, as given in Example 32–2. **(a)** What is the mass of this marble? **(b)** How many of these marbles would be required to have a mass equal to the mass of Earth?

9. •• **IP** **(a)** Find the nuclear radius of $^{30}_{15}$P. **(b)** What mass number would be required for a nucleus to have twice the radius found in part (a)? **(c)** Verify your answer to part (b) with an explicit calculation.

10. •• **IP** An alpha particle is the nucleus of a $^4_2$He atom. **(a)** How many nucleons are in a nucleus with twice the radius of an alpha particle? Explain. **(b)** Write the symbol for a phosphorus nucleus that has twice the radius of an alpha particle.

11. •• **IP** Suppose a uranium-236 nucleus undergoes fission by splitting into two smaller nuclei of equal size. **(a)** Is the radius of each of the smaller nuclei one-half, more than one-half, or less than one-half the radius of the uranium-236 nucleus? Explain. **(b)** Calculate the radius of the uranium-236 nucleus. **(c)** Calculate the radii of the two smaller nuclei.

12. •• A hypothetical nucleus weighs 1 lb. **(a)** How many nucleons are in this nucleus? **(b)** What is the radius of this nucleus?

SECTION 32–2 RADIOACTIVITY

13. • **CE Predict/Explain** Consider a nucleus that undergoes α decay. **(a)** Is the radius of the resulting daughter nucleus greater than, less than, or equal to the radius of the original nucleus? **(b)** Choose the *best explanation* from among the following:
  I. The decay adds an alpha particle to the nucleus, causing its radius to increase.
  II. When the nucleus undergoes decay it ejects two neutrons and two protons. This decreases the number of nucleons in the nucleus, and therefore its radius will decrease.
  III. An α decay leaves the number of nucleons unchanged. As a result, the radius of the nucleus stays the same.

14. • **CE Predict/Explain** Consider a nucleus that undergoes β decay. **(a)** Is the radius of the resulting daughter nucleus greater than, less than, or the same as that of the original nucleus? **(b)** Choose the *best explanation* from among the following:
  I. Capturing a β particle will cause the radius of a nucleus to increase. Therefore, the daughter nucleus has the greater radius.
  II. The original nucleus emits a β particle, and anytime a particle is emitted from a nucleus the result is a smaller radius. Therefore, the radius of the daughter nucleus is less than the radius of the original nucleus.
  III. When a nucleus emits a β particle a neutron is converted to a proton, but the number of nucleons is unchanged. As a

result, the radius of the daughter nucleus is the same as that of the original nucleus.

15. • **CE** Which of the three decay processes (α, β or γ) results in a new element? Explain.

16. • Complete the following nuclear reaction:

$$^7_3\text{Li} + ^1_1\text{H} \rightarrow ^4_2\text{He} + ?$$

17. • Complete the following nuclear reaction:

$$^{234}_{90}\text{Th} \rightarrow ^{230}_{88}\text{Ra} + ?$$

18. • Complete the following nuclear reaction:

$$? \rightarrow ^{14}_7\text{N} + \text{e}^- + \bar{v}$$

19. •• **CE** One possible decay series for $^{238}_{92}$U is $^{234}_{90}$Th, $^{234}_{91}$Pa, $^{234}_{92}$U, $^{230}_{90}$Th, $^{226}_{88}$Ra, $^{222}_{86}$Rn, $^{218}_{84}$Po, $^{218}_{85}$At, $^{218}_{86}$Rn, $^{214}_{84}$Po, $^{210}_{82}$Pb, $^{210}_{83}$Bi, $^{206}_{81}$Tl, and $^{206}_{82}$Pb. Identify, in the order given, each of the 14 decays that occur in this series.

20. •• Complete the following nuclear reaction and determine the amount of energy it releases:

$$^3_1\text{H} \rightarrow ^3_2\text{He} + ? + ?$$

Be sure to take into account the mass of the electrons associated with the neutral atoms.

21. •• The following nuclei are observed to decay by emitting an α particle: **(a)** $^{212}_{84}$Po and **(b)** $^{239}_{94}$Pu. Write out the decay process for each of these nuclei, and determine the energy released in each reaction. Be sure to take into account the mass of the electrons associated with the neutral atoms.

22. •• The following nuclei are observed to decay by emitting a β^- particle: **(a)** $^{35}_{16}$S and **(b)** $^{212}_{82}$Pb . Write out the decay process for each of these nuclei, and determine the energy released in each reaction. Be sure to take into account the mass of the electrons associated with the neutral atoms.

23. •• The following nuclei are observed to decay by emitting a β^+ particle: **(a)** $^{18}_9$F and **(b)** $^{22}_{11}$Na. Write out the decay process for each of these nuclei, and determine the energy released in each reaction. Be sure to take into account the mass of the electrons associated with the neutral atoms.

24. •• Find the energy released when $^{211}_{82}$Pb undergoes β^- decay to become $^{211}_{83}$Bi . Be sure to take into account the mass of the electrons associated with the neutral atoms.

25. ••• It is observed that $^{66}_{28}$Ni, with an atomic mass of 65.9291 u, decays by β^- emission. **(a)** Identify the nucleus that results from this decay. **(b)** If the nucleus found in part (a) has an atomic mass of 65.9289 u, what is the maximum kinetic energy of the emitted electron?

SECTION 32–3 HALF-LIFE AND RADIOACTIVE DATING

26. • **CE** The half-life of carbon-14 is 5730 y. **(a)** Is it possible for a particular nucleus in a sample of carbon-14 to decay after only 1 s has passed? Explain. **(b)** Is it possible for a particular nucleus to decay after 10,000 y? Explain.

27. • **CE** Suppose we were to discover that the ratio of carbon-14 to carbon-12 in the atmosphere was significantly smaller 10,000 years ago than it is today. How would this affect the ages we have assigned to objects on the basis of carbon-14 dating? In particular, would the true age of an object be greater than or less than the age we had previously assigned to it? Explain.

28. • **CE** A radioactive sample is placed in a closed container. Two days later only one-quarter of the sample is still radioactive. What is the half-life of this sample?

29. • Radon gas has a half-life of 3.82 d. What is the decay constant for radon?

30. • A radioactive substance has a decay constant equal to $8.9 \times 10^{-3}\,\text{s}^{-1}$. What is the half-life of this substance?

31. • The number of radioactive nuclei in a particular sample decreases over a period of 18 d to one-sixteenth the original number. What is the half-life of these nuclei?

32. • The half-life of $^{15}_{8}\text{O}$ is 122 s. How long does it take for the number of $^{15}_{8}\text{O}$ nuclei in a given sample to decrease by a factor of 10^{-4}?

33. •• **BIO A Radioactive Tag** A drug prepared for a patient is tagged with $^{99}_{43}\text{Tc}$, which has a half-life of 6.05 h. **(a)** What is the decay constant of this isotope? **(b)** How many $^{99}_{43}\text{Tc}$ nuclei are required to give an activity of 1.50 μCi ?

34. •• **BIO** Referring to Problem 33, suppose the drug containing $^{99}_{43}\text{Tc}$ with an activity of 1.50 μCi is injected into the patient 2.05 h after it is prepared. What is its activity at the time it is injected?

35. •• An archeologist on a dig finds a fragment of an ancient basket woven from grass. Later, it is determined that the carbon-14 content of the grass in the basket is 9.25% that of an equal carbon sample from present-day grass. What is the age of the basket?

36. •• The bones of a saber-toothed tiger are found to have an activity per gram of carbon that is 15.0% of what would be found in a similar live animal. How old are these bones?

37. •• Charcoal from an ancient fire pit is found to have a carbon-14 content that is only 17.5% that of an equivalent sample of carbon from a living tree. What is the age of the fire pit?

38. •• One of the many isotopes used in cancer treatment is $^{198}_{79}\text{Au}$, with a half-life of 2.70 d. Determine the mass of this isotope that is required to give an activity of 225 Ci.

39. •• **Smoke Detectors** The radioactive isotope $^{241}_{95}\text{Am}$, with a half-life of 432 y, is the active element in many smoke detectors. Suppose such a detector will no longer function if the activity of the $^{241}_{95}\text{Am}$ it contains drops below $\frac{1}{525}$ of its initial activity. How long will this smoke detector work?

40. •• **BIO Radioactivity in the Bones** Because of its chemical similarity to calcium, $^{90}_{38}\text{Sr}$ can collect in the bones and present a health risk. What percentage of $^{90}_{38}\text{Sr}$ present initially still exists after a period of **(a)** 50.0 y, **(b)** 60.0 y, and **(c)** 70.0 y?

SECTION 32–4   NUCLEAR BINDING ENERGY

41. • The atomic mass of gold-197 is 196.96654 u. How much energy is required to completely separate the nucleons in a gold-197 nucleus?

42. • The atomic mass of lithium-7 is 7.016003 u. How much energy is required to completely separate the nucleons in a lithium-7 nucleus?

43. •• Calculate the average binding energy per nucleon of **(a)** $^{56}_{26}\text{Fe}$ and **(b)** $^{238}_{92}\text{U}$.

44. •• Calculate the average binding energy per nucleon of **(a)** $^{4}_{2}\text{He}$ and **(b)** $^{64}_{30}\text{Zn}$.

45. •• Find the energy required to remove one neutron from $^{16}_{8}\text{O}$.

46. ••• **IP (a)** Consider the following nuclear process, in which a proton is removed from an oxygen nucleus:

$$^{16}_{8}\text{O} + \text{energy} \rightarrow ^{15}_{7}\text{N} + ^{1}_{1}\text{H}$$

Find the energy required for this process to occur. **(b)** Now consider a process in which a neutron is removed from an oxygen nucleus:

$$^{16}_{8}\text{O} + \text{energy} \rightarrow ^{15}_{8}\text{O} + ^{1}_{0}\text{n}$$

Find the energy required for this process to occur. **(c)** Which particle, the proton or the neutron, do you expect to be more tightly bound in the oxygen nucleus? Verify your answer.

SECTION 32–5   NUCLEAR FISSION

47. • Find the number of neutrons released by the following fission reaction:

$$^{1}_{0}\text{n} + ^{235}_{92}\text{U} \rightarrow ^{132}_{50}\text{Sn} + ^{101}_{42}\text{Mo} + (?)\,\text{neutrons}$$

48. •• Complete the following fission reaction and determine the amount of energy it releases:

$$^{1}_{0}\text{n} + ^{235}_{92}\text{U} \rightarrow ^{133}_{51}\text{Sb} + ? + 5^{1}_{0}\text{n}$$

49. •• Complete the following fission reaction and determine the amount of energy it releases:

$$^{1}_{0}\text{n} + ^{235}_{92}\text{U} \rightarrow ^{88}_{38}\text{Sr} + ^{136}_{54}\text{Xe} + (?)\,\text{neutrons}$$

50. •• A gallon of gasoline releases about 2.0×10^8 J of energy when it is burned. How many gallons of gas must be burned to release the same amount of energy as is released when 1.0 lb of $^{235}_{92}\text{U}$ undergoes fission. (Assume that each fission reaction in $^{235}_{92}\text{U}$ releases 173 MeV.)

51. •• Assuming a release of 173 MeV per fission reaction, determine the minimum mass of $^{235}_{92}\text{U}$ that must undergo fission to supply the annual energy needs of the United States. (The amount of energy consumed in the United States each year is 8.4×10^{19} J.)

52. •• Assuming a release of 173 MeV per fission reaction, calculate how many reactions must occur per second to produce a power output of 150 MW.

SECTION 32–6   NUCLEAR FUSION

53. • Consider a fusion reaction in which two deuterium nuclei fuse to form a tritium nucleus and a proton. How much energy is released in this reaction?

54. • Consider a fusion reaction in which a proton fuses with a neutron to form a deuterium nucleus. How much energy is released in this reaction?

55. • Find the energy released in the following fusion reaction:

$$^{1}_{1}\text{H} + ^{2}_{1}\text{H} \rightarrow ^{3}_{2}\text{He} + \gamma$$

56. •• **(a)** Complete the following fusion reaction and determine the energy it releases:

$$^{2}_{1}\text{H} + ^{3}_{1}\text{H} \rightarrow ? + ^{1}_{0}\text{n}$$

(b) How many of these reactions must occur per second to produce a power output of 25 MW?

57. ••• **The Evaporating Sun** The Sun radiates energy at the prodigious rate of 3.90×10^{26} W. **(a)** At what rate, in kilograms per second, does the Sun convert mass into energy? **(b)** Assuming that the Sun has radiated at this same rate for its entire lifetime of 4.50×10^9 y, and that the current mass of the Sun is 2.00×10^{30} kg, what percentage of its original mass has been converted to energy?

SECTION 32–7 PRACTICAL APPLICATIONS OF NUCLEAR PHYSICS

58. • **BIO Radiation Damage** A sample of tissue absorbs a 55-rad dose of α particles (RBE = 20). How many rad of protons (RBE = 10) cause the same amount of damage to the tissue?

59. • **BIO X-ray Damage** How many rad of 200-keV X-rays cause the same amount of biological damage as 50 rad of heavy ions?

60. •• **IP BIO** (a) Find the energy absorbed by a 78-kg person who is exposed to 52 mrem of α particles with an RBE of 15. (b) If the RBE of the α particles is increased, does the energy absorbed increase, decrease, or stay the same? Explain.

61. •• **BIO** A patient undergoing radiation therapy for cancer receives a 225-rad dose of radiation. (a) Assuming the cancerous growth has a mass of 0.17 kg, calculate how much energy it absorbs. (b) Assuming the growth to have the specific heat of water, determine its increase in temperature.

62. •• **BIO** Alpha particles with an RBE of 13 deliver a 32-mrad whole-body radiation dose to a 72-kg patient. (a) What dosage, in rem, does the patient receive? (b) How much energy is absorbed by the patient?

63. ••• **BIO A Radioactive Pharmaceutical** As part of a treatment program, a patient ingests a radioactive pharmaceutical containing $^{32}_{15}P$, which emits β rays with an RBE of 1.50. The half-life of $^{32}_{15}P$ is 14.28 d, and the initial activity of the medication is 1.34 MBq. (a) How many electrons are emitted over the period of 7.00 d? (b) If the β rays have an energy of 705 keV, what is the total amount of energy absorbed by the patient's body in 7.00 d? (c) Find the absorbed dosage in rem, assuming the radiation is absorbed by 125 g of tissue.

GENERAL PROBLEMS

64. • **CE** An α particle (charge $+2e$) and a β particle (charge $-e$) deflect in opposite directions when they pass through a magnetic field. Which particle deflects by a greater amount, given that both particles have the same speed? Explain.

65. • **CE** Radioactive samples A and B have equal half-lives. The initial activity of sample A is twice that of sample B. What is the ratio of the activity of sample A to that of sample B after two half-lives have elapsed?

66. • **CE** The initial activity of sample A is twice that of sample B. After two half-lives of sample A have elapsed, the two samples have the same activity. What is the ratio of the half-life of B to the half-life of A?

67. • **CE** To produce a given amount of electrical energy, is the amount of coal burned in a coal-burning power plant greater than, less than, or the same as the amount of $^{235}_{92}U$ consumed in a nuclear power plant? Explain.

68. • Determine the number of neutrons and protons in (a) $^{232}_{90}Th$, (b) $^{211}_{82}Pb$, and (c) $^{60}_{27}Co$.

69. • Identify the daughter nucleus that results when (a) $^{210}_{82}Pb$ undergoes α decay, (b) $^{239}_{92}U$ undergoes β^- decay, and (c) $^{11}_{6}C$ undergoes β^+ decay.

70. • Suppose it is desired to give a cancerous tumor a dose of 3800 rem. How many rads are needed if the tumor is exposed to alpha radiation?

71. • A patient is exposed to 260 rad of gamma rays. What is the dose the patient receives in rem?

72. • **CE** The two radioactive decay series that begin with $^{232}_{90}Th$ and end with $^{208}_{82}Pb$ are shown in **Figure 32–14**. Identify the ten intermediary nuclei that appear in these series.

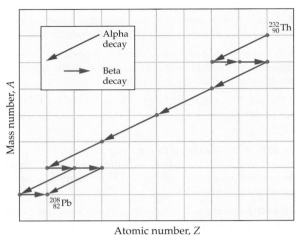

▲ **FIGURE 32–14** Problem 72

73. •• **Moon Rocks** In one of the rocks brought back from the Moon, it is found that 80.5% of the initial potassium-40 in the rock has decayed to argon-40. (a) If the half-life for this decay is 1.20×10^9 years, how old is the rock? (b) How much longer will it take before only 10.0% of the original potassium-40 is still present in the rock?

74. •• **IP Mantles in Gas Lanterns** Gas lanterns used on camping trips have mantles (small lacy bags that give off light) made from a rayon mesh impregnated with thorium and other materials. Thorium is used, even though it is radioactive, because it forms an oxide that can withstand being incandescent for long periods of time. Almost all natural thorium is $^{232}_{90}Th$, which has a half-life of 1.405×10^{10} y. (a) A typical mantle contains 325 mg of thorium. What is the activity of the mantle? (b) If the half-life of thorium had been double its actual value, by what factor would the activity of a mantle be changed? Explain.

The glowing mantle of a gas lantern. (Problem 74)

75. •• Identify the nucleus whose β^- decay produces the same nucleus as that produced by the α decay of $^{214}_{84}Po$.

76. •• An α particle fired head-on at a stationary nickel nucleus approaches to a radius of 15 fm before being turned around. **(a)** What is the maximum Coulomb force exerted on the α particle? **(b)** What is the electric potential energy of the α particle at its point of closest approach? **(c)** Find the initial kinetic energy of the α particle.

77. •• Calculate the number of disintegrations per second that one would expect from a 1.7-g sample of $^{226}_{88}$Ra. What is the activity of this sample in curies?

78. •• **IP** Initially, a sample of radioactive nuclei of type A contains four times as many nuclei as a sample of radioactive nuclei of type B. Two days later (2.00 d) the two samples contain the same number of nuclei. **(a)** Which type of nucleus has the longer half-life? Explain. **(b)** Determine the half-life of type B nuclei if the half-life of type A nuclei is known to be 0.500 d.

79. •• Stable nuclei have mass numbers that range from a minimum of 1 to a maximum of 209. **(a)** Find the corresponding range in nuclear radii. **(b)** Assuming all nuclei to be spherical, determine the ratio of the surface area of the largest stable nucleus to the surface area of the smallest nucleus. **(c)** Repeat part (b), only this time find the ratio of the volumes.

80. •• **Radius of a Neutron Star** Neutron stars are so named because they are composed of neutrons and have a density the same as that of a nucleus. Referring to Example 32–2 for the nuclear density, find the radius of a neutron star whose mass is 0.50 that of the Sun.

81. •• A specimen taken from the wrappings of a mummy contains 7.82 g of carbon and has an activity of 1.38 Bq. How old is the mummy? (Refer to pages 1132 and 1133 for relevant information regarding the isotopes of carbon.)

82. •• **(a)** How many fission reactions are required to light a 120-W lightbulb for 2.5 d? Assume an energy release of 212 MeV per fission reaction and a 32% conversion efficiency. **(b)** What mass of $^{235}_{92}$U corresponds to the number of fission reactions found in part (a)?

83. •• **IP** Energy is released when three α particles fuse to form carbon-12. **(a)** Is the mass of carbon-12 greater than, less than, or the same as the mass of three α particles? Explain. **(b)** Calculate the energy given off in this fusion reaction.

84. •• Find the dose of γ rays that must be absorbed by a block of ice at 0 °C to convert it to water at 0 °C. Give the dosage in rad.

85. •• **IP** **(a)** What dosage (in rad) must a 1.0-kg sample of water absorb to increase its temperature by 1.0 C°? **(b)** If the mass of the water sample is increased, does the dosage found in part (a) increase, decrease, or stay the same? Explain.

86. •• **BIO Chest X-rays** A typical chest X-ray uses X-rays with an RBE of 0.85. If the radiation dosage is 35 mrem, find the energy absorbed by a 72-kg patient, assuming one-quarter of the patient's body is exposed to the X-rays.

87. ••• A γ ray photon emitted by $^{226}_{88}$Ra has an energy of 0.186 MeV. Use conservation of linear momentum to calculate the recoil speed of a $^{226}_{88}$Ra nucleus after such a γ ray is emitted. Assume that the nucleus is at rest initially, and that relativistic effects can be ignored.

88. ••• The energy released by α decay in a 50.0-g sample of $^{239}_{94}$Pu is to be used to heat 4.75 kg of water. Assuming all the energy released by the radioactive decay goes into heating the water, find how much the temperature of the water increases in 1.00 h.

89. ••• Consider a solid sphere of $^{235}_{92}$U with a radius of 2.25 cm in a room with a temperature of 293 K. Assume that all the energy released by α decay goes into heating the sphere, and that the sphere radiates heat to its surroundings as a blackbody. What is the change in temperature of the sphere as a result of the α decay? (*Note*: The density of uranium is 18.95 g/cm^3.)

PASSAGE PROBLEMS

BIO Treating a Hyperactive Thyroid

Of the many endocrine glands in the body, the thyroid is one of the most important. Weighing only an ounce and situated just below the "Adam's apple, " the thyroid produces hormones that regulate the metabolic rate of every cell in the body. To produce these hormones the thyroid uses iodine from the food we eat—in fact, the thyroid specializes in absorbing iodine.

The central role played by the thyroid is evidenced by the symptoms produced when it ceases to function properly. For example, a person experiencing hyperthyroidism (an overactive thyroid) presents the internist with a wide range of indicators, including weight loss, ravenous appetite, anxiety, fatigue, hyperactivity, apathy, palpitations, arrhythmias, and nausea, just to mention a few.

The most common treatment for hyperthyroidism is to destroy the overactive thyroid tissues with radioactive iodine-131. This treatment takes advantage of the fact that only thyroid cells absorb and concentrate iodine. To begin the treatment, a patient swallows a single, small capsule containing iodine-131. The radioactive isotope quickly enters the bloodstream and is taken up by the overactive thyroid cells, which are destroyed as the iodine-131 decays with a half-life of 8.04 d. Other cells in the body experience very little radiation damage, which minimizes side effects. In one or two months the thyroid activity is reduced to an acceptable level. Sometimes too much—or even all—of the thyroid is killed, which can result in hypothyroidism, or underactive thyroid. This is easily treated, however, with dietary supplements to replace the missing thyroid hormones.

90. • What is the decay constant, λ, for iodine-131?

    A. 9.98×10^{-7} s^{-1} **B.** 1.44×10^{-6} s^{-1}

    C. 2.39×10^{-5} s^{-1} **D.** 5.99×10^{-5} s^{-1}

91. • If a sample of iodine-131 contains 4.5×10^{16} nuclei, what is the activity of the sample? Express your answer in curies.

    A. 0.27 Ci   **B.** 1.2 Ci

    C. 1.7 Ci   **D.** 4.5 Ci

92. • If the half-life of iodine-131 were only half of its actual value, would the activity of the sample in Problem 91 be increased or decreased?

Appendix A   Basic mathematical tools

This text is designed for students with a working knowledge of basic algebra and trigonometry. Even so, it is useful to review some of the mathematical tools that are of particular importance in the study of physics. In this Appendix we cover a number of topics related to mathematical notation, trigonometry, algebra, mathematical expansions, and vector multiplication.

MATHEMATICAL NOTATION

Common mathematical symbols

In Table A–1 we present some of the more common mathematical symbols, along with a translation into English. Though these symbols are probably completely familiar, it is worthwhile to be sure we all interpret them in the same way.

TABLE A–1 Mathematical Symbols

$=$	is equal to
$\neq$	is not equal to
$\approx$	is approximately equal to
$\propto$	is proportional to
$>$	is greater than
$\geq$	is greater than or equal to
$\gg$	is much greater than
$<$	is less than
$\leq$	is less than or equal to
$\ll$	is much less than
$\pm$	plus or minus
$\mp$	minus or plus
x_{av} or $\bar{x}$	average value of x
Δx	change in $x(x_f - x_i)$
$|x|$	absolute value of x
Σ	sum of
$\rightarrow 0$	approaches 0
∞	infinity

A couple of the symbols in Table A–1 warrant further discussion. First, Δx, which means "change in x," is used frequently, and in many different contexts. Pronounced "delta x," it is defined as the final value of x, x_f, minus the initial value of x, x_i:

$$\Delta x = x_f - x_i \qquad \text{A–1}$$

Thus, Δx is not Δ times x; it is a shorthand way of writing $x_f - x_i$. The same delta notation can be applied to any quantity—it does not have to be x. In general, we can say that

$$\Delta(anything) = (anything)_f - (anything)_i$$

For example, $\Delta t = t_f - t_i$ is the change in time, $\Delta \vec{v} = \vec{v}_f - \vec{v}_i$ is the change in velocity, and so on. Throughout this text, we use the delta notation whenever we want to indicate the change in a given quantity.

Second, the Greek letter Σ (capital sigma) is also encountered frequently. In general, Σ is shorthand for "sum." For example, suppose we have a system comprised of nine masses, m_1 through m_9. The total mass of the system, M, is simply

$$M = m_1 + m_2 + m_3 + m_4 + m_5 + m_6 + m_7 + m_8 + m_9$$

This is a rather tedious way to write M, however, and would be even more so if the number of masses were larger. To simplify our equation, we use the Σ notation:

$$M = \sum_{i=1}^{9} m_i \qquad \text{A–2}$$

With this notation we could sum over any number of masses, simply by changing the upper limit of the sum.

In addition, Σ is often used to designate a general summation, where the number of terms in the sum may not be known, or may vary from one system to another. In a case like this we would simply write Σ without specific upper and lower limits. Thus, a general way of writing the total mass of a system is as follows:

$$M = \sum m \qquad \text{A–3}$$

Vector notation

When we draw a vector to represent a physical quantity, we typically use an arrow whose length is proportional to the magnitude of the quantity, and whose direction is the direction of the quantity. (This and other aspects of vector notation are discussed in Chapter 3.) A slight problem arises, however, when a physical quantity points into or out of the page. In such a case, we use the conventions illustrated in **Figure A–1**.

Figure A–1 (a) shows a vector pointing out of the page. Note that we see only the tip. Below, we show the corresponding convention, which is a dot set off by a circle. The dot represents the point of the vector's arrow coming out of the page toward you.

A similar convention is employed in Figure A–1 (b) for a vector pointing into the page. In this case, the arrow moves directly away from you, giving a view of its "tail feathers." The feathers are placed in an X-shaped pattern, so we represent the vector as an X set off by a circle.

These conventions are used in Chapter 22 to represent the magnetic field vector, $\vec{B}$, and in other locations in the text as well.

(a)   (b)

▲ **FIGURE A–1 Vectors pointing out of and into the page**
(a) A vector pointing out of the page is represented by a dot in a circle. The dot indicates the tip of the vector's arrow. **(b)** A vector pointing into the page is represented by an X in a circle. The X indicates the "tail feathers" of the vector's arrow.

Scientific notation

In physics, the numerical value of a physical quantity can cover an enormous range, from the astronomically large to the microscopically small. For example, the mass of the Earth is roughly

$$M_E = 5970000000000000000000000 \text{ kg}$$

In contrast, the mass of a hydrogen atom is approximately

$$M_{hydrogen} = 0.00000000000000000000000000167 \text{ kg}$$

Clearly, representing such large and small numbers with a long string of zeros is clumsy and prone to error.

The preferred method for handling such numbers is to replace the zeros with the appropriate power of ten. For example, the mass of the Earth can be written as follows:

$$M_E = 5.97 \times 10^{24} \text{ kg}$$

The factor of 10^{24} simply means that the decimal point for the mass of the Earth is 24 places to the right of its location in 5.97. Similarly, the mass of a hydrogen atom is

$$M_{hydrogen} = 1.67 \times 10^{-27} \text{ kg}$$

In this case, the correct location of the decimal point is 27 places to the left of its location in 1.67. This type of representation, using powers of ten, is referred to as **scientific notation.**

Scientific notation also simplifies various mathematical operations, such as multiplication and division. For example, the product of the mass of the Earth and the mass of a hydrogen atom is

$$
\begin{aligned}
M_E M_{hydrogen} &= (5.97 \times 10^{24} \text{ kg})(1.67 \times 10^{-27} \text{ kg}) \\
&= (5.97 \times 1.67)(10^{24} \times 10^{-27}) \text{ kg}^2 \\
&= 9.99 \times 10^{24-27} \text{ kg}^2 \\
&= 9.99 \times 10^{-3} \text{ kg}^2
\end{aligned}
$$

Similarly, the mass of a hydrogen atom divided by the mass of the Earth is

$$
\begin{aligned}
\frac{M_{hydrogen}}{M_E} &= \frac{1.67 \times 10^{-27} \text{ kg}}{5.97 \times 10^{24} \text{ kg}} = \frac{1.67}{5.97} \times \frac{10^{-27}}{10^{24}} \\
&= 0.280 \times 10^{-27-24} \\
&= 0.280 \times 10^{-51} = 2.80 \times 10^{-52}
\end{aligned}
$$

Note the change in location of the decimal point in the last two expressions, and the corresponding change in the power of ten.

Exponents and their manipulation are discussed in greater detail later in this Appendix.

TRIGONOMETRY

Degrees and radians

We all know the definition of a degree; there are 360 degrees in a circle. The definition of a radian is somewhat less well known; there are 2π radians in a circle. An equivalent definition of the radian is the following:

A radian is the angle for which the corresponding arc length is equal to the radius.

To visualize this definition, consider a pie with a piece cut out, as shown in **Figure A–2 (a)**. Note that a piece of pie has three sides—two radial lines from the center, and an arc of crust. If a piece of pie is cut with an angle of one radian, all three sides are

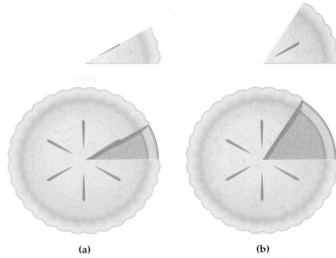

(a)   (b)

▲ **FIGURE A–2   The definition of a radian**
(a) This piece of pie is cut with an angle less than a radian. Thus, the two radial sides (coming out from the center) are longer than the arc of crust. **(b)** The angle for this piece of pie is equal to one radian (about 57.3°). Thus, all three sides of the piece are of equal length.

equal in length, as shown in **Figure A–2 (b)**. Since a radian is about 57.3°, this amounts to a fairly good-sized piece of pie. Thus, if you want a healthy helping of pie, just tell the server, "One radian, please."

Now, radians are particularly convenient when we are interested in the length of an arc. In **Figure A–3** we show a circular arc corresponding to the radius r and the angle θ. *If the angle θ is measured in radians*, the length of the arc, s, is given by

$$s = r\theta \qquad\qquad \text{A–4}$$

Note that this simple relation is *not valid* when θ is measured in degrees. For a full circle, in which case $\theta = 2\pi$, the length of the arc (which is the circumference of the circle) is $2\pi r$, as expected.

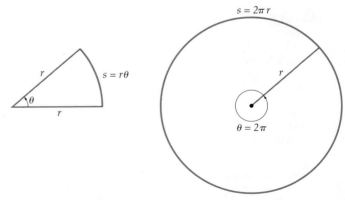

▲ **FIGURE A–3   The Length of an arc**
The arc created by a radius r rotated through an angle θ has a length $s = r\theta$. If the radius is rotated through a full circle, the angle is $\theta = 2\pi$, and the arc length is the circumference of a circle, $s = 2\pi r$.

Trigonometric functions and the Pythagorean theorem

Next, we consider some of the more important and frequently used results from trigonometry. We start with the right triangle, shown in **Figure A–4**, and the basic **trigonometric functions,** sin θ (sine theta), cos θ (cosine theta), and tan θ (tangent theta). The cosine of an angle θ is defined to be the side adjacent to the angle divided by the hypotenuse; cos $\theta = x/r$. Similarly, the sine is defined to be the opposite side divided by the hypotenuse, sin $\theta = y/r$, and the tangent is the opposite side divided by the adjacent side, tan $\theta = y/x$. These relations are summarized in the following equations:

$$\cos \theta = \frac{x}{r}$$

$$\sin \theta = \frac{y}{r} \qquad \text{A-5}$$

$$\tan \theta = \frac{y}{x} = \frac{\sin \theta}{\cos \theta}$$

Note that each of the trigonometric functions is the ratio of two lengths, and hence is dimensionless.

According to the **Pythagorean theorem,** the sides of the right triangle in Figure A–4 are related as follows:

$$x^2 + y^2 = r^2 \qquad \text{A-6}$$

Dividing by r^2 yields

$$\frac{x^2}{r^2} + \frac{y^2}{r^2} = 1$$

This can be re-written in terms of sine and cosine to give

$$\sin^2 \theta + \cos^2 \theta = 1$$

Figure A–4 also shows how sine and cosine are used in a typical calculation. In many cases, the hypotenuse of a triangle, r, and one of its angles, θ, are given. To find the short sides of the triangle we rearrange the relations given in Equation A–5. For example, in Figure A–4 we see that $x = r \cos \theta$ is the length of the short side adjacent to the angle, θ, and $y = r \sin \theta$ is the length of the short side opposite the angle. The following Example applies this type of calculation to the case of an inclined roadway.

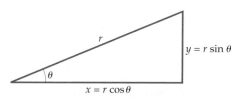

▲ **FIGURE A–4 Relating the short sides of a right triangle to its hypotenuse**

The trigonometric functions sin θ and cos θ and the Pythagorean theorem are useful in relating the lengths of the short sides of a right triangle to the length of its hypotenuse.

EXAMPLE A–1   HIGHWAY TO HEAVEN

You are driving on a long straight road that slopes uphill at an angle of 6.4° above the horizontal. At one point you notice a sign that reads, "Elevation 1500 feet." What is your elevation after you have driven another 1.0 mi?

PICTURE THE PROBLEM
From our sketch, we see that the car is moving along the hypotenuse of a right triangle. The length of the hypotenuse is one mile.

STRATEGY
The elevation gain is the vertical side of the triangle, y. We find y by multiplying the hypotenuse, r, by the sine of theta. That is, since sin $\theta = y/r$ it follows that $y = r \sin \theta$.

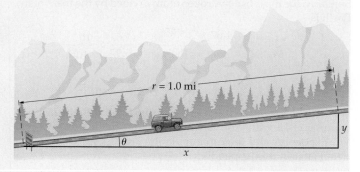

SOLUTION

1. Calculate the elevation gain, y:

$$y = r \sin \theta$$
$$= (1.0 \text{ mi}) \sin 6.4° = (1.0 \text{ mi})(0.11) = 0.11 \text{ mi}$$

2. Convert y from miles to feet:

$$y = (0.11 \text{ mi})\left(\frac{5280 \text{ ft}}{1 \text{ mi}}\right) = 580 \text{ ft}$$

3. Add the elevation gain to the original elevation to obtain the new elevation:

$$\text{elevation} = 1500 \text{ ft} + 580 \text{ ft} = 2100 \text{ ft}$$

INSIGHT
As surprising as it may seem, the horizontal distance covered by the car is $r \cos \theta = (5280 \text{ ft}) \cos 6.4° = 5200 \text{ ft}$, only about 80 ft less than the total distance driven by the car. At the same time, the car rises a distance of 580 ft.

PRACTICE PROBLEM
How far up the road from the first sign should the road crew put another sign reading "Elevation 3500 ft"?
[**Answer:** 18,000 ft = 3.4 mi]

In some problems, the sides of a triangle (x and y) are given and it is desired to find the corresponding hypotenuse, r, and angle, θ. For example, suppose that $x = 5.0$ m and $y = 2.0$ m. Using the Pythagorean theorem, we find $r = \sqrt{x^2 + y^2} = \sqrt{(5.0\text{ m})^2 + (2.0\text{ m})^2} = 5.4$ m. Similarly, to find the angle we use the definition of tangent: $\tan \theta = y/x$. The inverse of this relation is $\theta = \tan^{-1}(y/x) = \tan^{-1}(2.0\text{ m}/5.0\text{ m}) = \tan^{-1}(0.40)$. Note that the expression $\tan^{-1}$ is the *inverse tangent function*—it does not mean 1 divided by tangent, but rather "the angle whose tangent is——." Your calculator should have a button on it labeled $\tan^{-1}$. If you enter 0.40 and then press $\tan^{-1}$, you should get $22°$ (to two significant figures), which means that $\tan 22° = 0.40$. Inverse sine and cosine functions work in the same way.

Trigonometric identities

In addition to the basic definitions of sine, cosine, and tangent just given, there are a number of useful relationships involving these functions referred to as **trigonometric identities.** First, consider changing the sign of an angle. This corresponds to flipping the triangle in Figure A–4 upside-down, which changes the sign of y but leaves x unaffected. The result is that sine changes its sign, but cosine does not. Specifically, for a general angle A we find the following:

$$\sin(-A) = -\sin A$$
$$\cos(-A) = \cos A \qquad \text{A–7}$$

Next, we consider trigonometric identities relating to the sum or difference of two angles. For example, consider two general angles A and B. The sine and cosine of the sum of these angles, $A + B$, are given below:

$$\sin(A + B) = \sin A \cos B + \sin B \cos A$$
$$\cos(A + B) = \cos A \cos B - \sin A \sin B \qquad \text{A–8}$$

By changing the sign of B, and using the results given in Equation A–7, we obtain the corresponding results for the difference between two angles:

$$\sin(A - B) = \sin A \cos B - \sin B \cos A$$
$$\cos(A - B) = \cos A \cos B + \sin A \sin B \qquad \text{A–9}$$

Applications of these relations can be found in Chapters 4, 14, 23, and 24.

To see how one might use a relation like $\sin(A + B) = \sin A \cos B + \sin B \cos A$, consider the case where $A = B = \theta$. With this substitution we find

$$\sin(\theta + \theta) = \sin \theta \cos \theta + \sin \theta \cos \theta$$

Simplifying somewhat yields the commonly used double-angle formula

$$\sin 2\theta = 2 \sin \theta \cos \theta \qquad \text{A–10}$$

This expression is used in deriving Equation 4–16.

As a final example of using trigonometric identities, let $A = 90°$ and $B = \theta$. Making these substitutions in Equations A–9 yields

$$\sin(90° - \theta) = \sin 90° \cos \theta - \sin \theta \cos 90° = \cos \theta$$
$$\qquad \text{A–11}$$
$$\cos(90° - \theta) = \cos 90° \cos \theta + \sin 90° \sin \theta = \sin \theta$$

ALGEBRA

The quadratic equation

A well-known result that finds many uses in physics is the solution to the **quadratic equation**

$$ax^2 + bx + c = 0 \qquad \text{A–12}$$

In this equation, a, b, and c are constants and x is a variable. When we refer to the solution of the quadratic equation, we mean the values of x that satisfy Equation A–12. These values are given by the following expression:

Solutions to the Quadratic Equation

$$x = \frac{-b \pm \sqrt{b^2 - 4ac}}{2a} \qquad \text{A–13}$$

Note that there are two solutions to the quadratic equation, in general, corresponding to the plus and minus sign in front of the square root. In the special case that the quantity under the square root vanishes, there will be only a single solution. If the quantity under the square root is negative the result for x is not physical, which means a mistake has probably been made in the calculation.

To illustrate the use of the quadratic equation and its solution, we consider a standard one-dimensional kinematics problem, such as one might encounter in Chapter 2:

> A ball is thrown straight upward with an initial speed of 11 m/s. How long does it take for the ball to first reach a height of 4.5 m above its launch point?

The first step in solving this problem is to write the equation giving the height of the ball, y, as a function of time. Referring to Equation 2–11, we have

$$y = y_0 + v_0 t - \tfrac{1}{2} g t^2$$

To make this look more like a quadratic equation, we move all the terms onto the left-hand side, which yields

$$\tfrac{1}{2} g t^2 - v_0 t + y - y_0 = 0$$

This is the same as Equation A–12 if we make the following identifications: $x = t$; $a = \tfrac{1}{2} g$; $b = -v_0$; $c = y - y_0$. The desired solution, then, is given by making these substitutions in Equation A–13:

$$t = \frac{v_0 \pm \sqrt{v_0{}^2 - 2g(y - y_0)}}{g}$$

The final step is to use the appropriate numerical values; $g = 9.81$ m/s^2, $v_0 = 11$ m/s, $y - y_0 = 4.5$ m. Straightforward calculation gives $t = 0.54$ s and $t = 1.7$ s. Therefore, the time it takes to first reach a height of 4.5 m is 0.54 s; the second solution is the time when the ball is again at a height of 4.5 m, this time on its way down.

Two equations in two unknowns

In some problems, two unknown quantities are determined by two interlinked equations. In such cases it often seems at first that you have not been given enough information to obtain a solution. By patiently writing out what is known, however, you can generally use straightforward algebra to solve the problem.

As an example, consider the following problem: A father and daughter share the same birthday. On one birthday the father announces to his daughter, "Today I am four times older than you, but in 5 years I will be only three times older." How old are the father and daughter now?

You might be able to solve this problem by guessing, but here's how to approach it systematically. First, write what is given in the form of equations. Letting F be the father's age in years, and D the daughter's age in years, we know that on this birthday

$$F = 4D \qquad\qquad \text{A–14}$$

In 5 years, the father's age will be $F + 5$, the daughter's age will be $D + 5$, and the following will be true:

$$F + 5 = 3(D + 5)$$

Multiplying through the parenthesis gives

$$F + 5 = 3D + 15 \qquad\qquad \text{A–15}$$

Now if we subtract Equation A–15 from Equation A–14 we can eliminate one of the unknowns, F:

$$F = 4D$$
$$\underline{-F + 5 = 3D + 15}$$
$$-5 = \ D - 15$$

The solution to this new equation is clearly $D = 10$, and thus the father's age is $F = 4D = 40$.

The following Example investigates a similar problem. In this case, we use the fact that if you drive with a speed v for a time t the distance covered is $d = vt$.

EXAMPLE A–2 HIT THE ROAD

It takes 1.50 h to drive with a speed v from home to a nearby town, a distance d away. Later, on the way back, the traffic is lighter, and you are able to increase your speed by 15 mi/h. With this higher speed, you get home in just 1.00 h. Find your initial speed v, and the distance to the town, d.

PICTURE THE PROBLEM
Our sketch shows home and the town, separated by a distance d. Going to town the speed is v, returning home the speed is $v + 15$ mi/h.

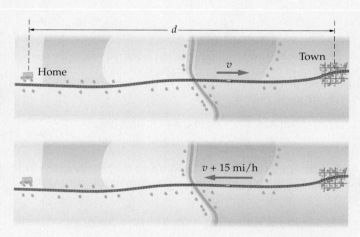

STRATEGY
To determine the two unknowns, v and d, we need two separate equations. One equation corresponds to what we know about the trip to the town, the second equation corresponds to what we know about the return trip.

SOLUTION

1. Write an equation for the trip to the town. Recall that this trip takes one and a half hours:

$$d = vt = v(1.50\ \text{h})$$

2. Write an equation for the trip home. This trip takes one hour, and covers the same distance d:

$$d = (v + 15\ \text{mi/h})t = (v + 15\ \text{mi/h})(1.00\ \text{h})$$

3. Subtract these two equations to eliminate d:

$$d = v(1.50\ \text{h})$$
$$\underline{-\quad d = (v + 15\ \text{mi/h})(1.00\ \text{h})}$$
$$0 = v(1.50\ \text{h}) - v(1.00\ \text{h}) - (15\ \text{mi/h})(1.00\ \text{h})$$

4. Solve this new equation for v:

$$0 = v(1.50\ \text{h}) - v(1.00\ \text{h}) - (15\ \text{mi/h})(1.00\ \text{h})$$
$$0 = v(0.50\ \text{h}) - (15\ \text{mi/h})(1.00\ \text{h})$$
$$v = \frac{(15\ \text{mi/h})(1.00\ \text{h})}{(0.50\ \text{h})} = 30\ \text{mi/h}$$

5. Use the first equation to solve for d:

$$d = vt = (30\ \text{mi/h})(1.50\ \text{h}) = 45\ \text{mi}$$

Exponents and logarithms

An **exponent** is the power to which a number is raised. For example, in the expression 10^3, we say that the exponent of 10 is 3. To evaluate 10^3 we simply multiply 10 by itself three times:

$$10^3 = 10 \times 10 \times 10 = 1000$$

Similarly, a negative exponent implies an inverse, as in the relation $10^{-1} = 1/10$. Thus, to evaluate a number like 10^{-4}, for example, we multiply $1/10$ by itself four times:

$$10^{-4} = \frac{1}{10} \times \frac{1}{10} \times \frac{1}{10} \times \frac{1}{10} = \frac{1}{10{,}000} = 0.0001$$

The relations just given apply not just to powers of 10, of course, but to any number at all. Thus, x^4 is

$$x^4 = x \times x \times x \times x$$

and x^{-3} is

$$x^{-3} = \frac{1}{x} \times \frac{1}{x} \times \frac{1}{x} = \frac{1}{x^3}$$

Using these basic rules, it follows that exponents add when two or more numbers are multiplied together:

$$x^2 x^3 = (x \times x)(x \times x \times x)$$
$$= x \times x \times x \times x \times x = x^5 = x^{2+3}$$

On the other hand, exponents multiply when a number is raised to a power:

$$(x^2)^3 = (x \times x) \times (x \times x) \times (x \times x)$$
$$= x \times x \times x \times x \times x \times x = x^6 = x^{2\times3}$$

In general, the rules obeyed by exponents can be summarized as follows:

$$x^n x^m = x^{n+m}$$

$$x^{-n} = \frac{1}{x^n}$$

$$\frac{x^n}{x^m} = x^{n-m} \qquad \text{A–16}$$

$$(xy)^n = x^n y^n$$

$$(x^n)^m = x^{nm}$$

Fractional exponents, such as $1/n$, indicate the nth root of a number. Specifically, the square root of x is written as

$$\sqrt{x} = x^{1/2}$$

For n greater than 2 we write the nth root in the following form:

$$\sqrt[n]{x} = x^{1/n} \qquad \text{A–17}$$

Thus, the nth root of a number, x, is the value that gives x when multiplied by itself n times: $(x^{1/n})^n = x^{n/n} = x^1 = x$.

A general method for calculating the exponent of a number is provided by the **logarithm.** For example, suppose x is equal to 10 raised to the power n:

$$x = 10^n$$

In this expression, 10 is referred to as the *base*. The exponent, n, is equal to the logarithm (log) of x:

$$n = \log x$$

The notation "log" is known as the *common logarithm*, and it refers specifically to base 10.

As an example, suppose that $x = 1000 = 10^n$. Clearly, we can write x as 10^3, which means that the exponent of x is 3:

$$\log x = \log 1000 = \log 10^3 = 3$$

When dealing with a number this simple, the exponent can be determined without a calculator. Suppose, however, that $x = 1205 = 10^n$. To find the exponent for this value of x we use the "log" button on a calculator. The result is

$$n = \log 1205 = 3.081$$

Thus, 10 raised to the 3.081 power gives 1205.

Another base that is frequently used for calculating exponents is $e = 2.718\ldots$. To represent $x = 1205$ in this base we write

$$x = 1205 = e^m$$

The logarithm to base e is known as the *natural logarithm*, and it is represented by the notation "ln." Using the "ln" button on a calculator, we find

$$m = \ln 1205 = 7.094$$

Thus, e raised to the 7.094 power gives 1205. The connection between the common and natural logarithms is as follows:

$$\ln x = 2.3026 \log x \qquad \text{A–18}$$

In the example just given, we have $\ln 1205 = 7.094 = 2.3026 \log 1205 = 2.3026(3.081)$.

The basic rules obeyed by logarithms follow directly from the rules given for exponents in Equation A–16. In particular,

$$\ln(xy) = \ln x + \ln y$$

$$\ln\!\left(\frac{x}{y}\right) = \ln x - \ln y \qquad \text{A–19}$$

$$\ln x^n = n \ln x$$

Though these rules are stated in terms of natural logarithms, they are satisfied by logarithms with any base.

MATHEMATICAL EXPANSIONS

We conclude with a brief consideration of small quantities in mathematics. Consider the following equation:

$$(1 + x)^3 = 1 + 3x + 3x^2 + x^3$$

This expression is valid for all values of x. However, if x is much smaller than one, $x \ll 1$, we can say to a good approximation that

$$(1 + x)^3 \approx 1 + 3x$$

Now, just how good is this approximation? After all, it ignores two terms that would need to be included to produce an equality. In the case $x = 0.001$, for example, the two terms that are neglected, $3x^2$ and x^3, have a combined contribution of only about 3 ten-thousandths of a percent! Clearly, then, little error is made in the approximation $(1 + 0.001)^3 \sim 1 + 3(0.001) = 1.003$. This can be seen visually in **Figure A–5 (a)**, where we plot $(1 + x)^3$ and $1 + 3x$ for x ranging from 0 to 1. Note that there is little difference in the two expressions for x less than about 0.1.

This is just one example of a general result in mathematics that can be derived from the **binomial expansion.** In general, we can say that the following approximation is valid for $x \ll 1$:

$$(1 + x)^n \approx 1 + nx \qquad \text{A–20}$$

This result holds for arbitrary n, not just for the case of $n = 3$. For example, if $n = -1$ we have

$$(1 + x)^{-1} = \frac{1}{1 + x} \approx 1 - x$$

We plot $(1 + x)^{-1}$ and $1 - x$ in **Figure A–5 (b)**, and again we see that the results are in good agreement for x less than about 0.1.

An example of an expansion that arises in the study of relativity concerns the following quotient:

$$\frac{1}{\sqrt{1 - \dfrac{v^2}{c^2}}}$$

In this expression v is the speed of an object and c is the speed of light. Since objects we encounter generally have speeds much less than the speed of light, the ratio v/c is much less than one, and v^2/c^2 is even smaller than v/c. Thus, if we let $x = v^2/c^2$ we have

$$\frac{1}{\sqrt{1 - x}}$$

We can apply the binomial expansion to this result if we replace n with $-1/2$ and x with $-x$ in Equation A–20. This yields

$$\frac{1}{\sqrt{1 - \dfrac{v^2}{c^2}}} \approx 1 + \tfrac{1}{2}\frac{v^2}{c^2}$$

The two sides of this approximate equality are plotted in **Figure A–5 (c)**, showing the accuracy of the approximation for small v/c.

Another type of mathematical expansion leads to the following useful results:

$$\sin \theta \approx \theta$$

$$\cos \theta \approx 1 - \tfrac{1}{2}\theta^2 \qquad \text{A–21}$$

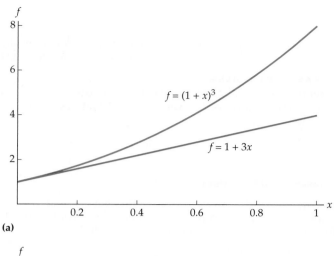

(a)

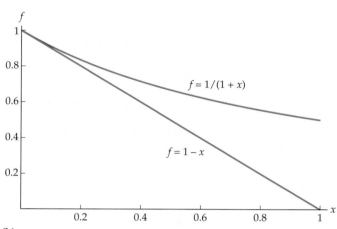

(b)

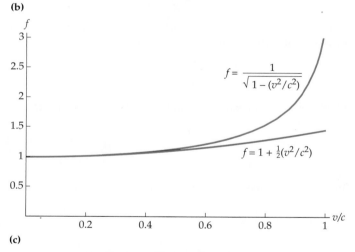

(c)

▲ **FIGURE A–5 Examples of mathematical expansions**
(a) A comparison between $(1 + x)^3$ and the result obtained from the binomial expansion, $1 + 3x$. **(b)** A comparison between $1/(1 + x)$ and the result obtained from the binomial expansion, $1 - x$. **(c)** A comparison between $1/\sqrt{1 - (v^2/c^2)}$ and the result obtained from the binomial expansion, $1 + \tfrac{1}{2}(v^2/c^2)$.

These expansions are valid for small angles θ measured in radians. Note that the result $\sin \theta \approx \theta$ is used to derive Equations 6–13 and 13–19. (See Table 6–2, p. 170, and Figure 13–15, p. 435, for more details on this expansion.)

VECTOR MULTIPLICATION

There are two distinct ways to multiply vectors, referred to as the **dot product** and the **cross product.** The difference between these two types of multiplication is that the dot product yields a scalar (a number) as its result, whereas the cross product results in a vector. Both types of product have important applications in physics. In what follows, we present the basic techniques associated with dot and cross products, and point out places in the text where they are used.

The dot product

Consider two vectors, $\vec{A}$ and $\vec{B}$, as shown in **Figure A–6 (a)**. The magnitudes of these vectors are A and B, respectively, and the angle between them is θ. We define the dot product of $\vec{A}$ and $\vec{B}$ as follows:

$$\vec{A} \cdot \vec{B} = AB \cos \theta \qquad \text{A–22}$$

In words, the dot product of two vectors is a scalar equal to the magnitude of one vector times the magnitude of the second vector times the cosine of the angle between them.

A geometric interpretation of the dot product is presented in **Figure A–6 (b)**. We begin by projecting the vector $\vec{A}$ onto the direction of vector $\vec{B}$. This is done by dropping a perpendicular from the tip of $\vec{A}$ onto the line that passes through $\vec{B}$, as shown in **Figure A–6 (b)**. Note that the projection of $\vec{A}$ on the direction of $\vec{B}$ has a length given by $A \cos \theta$. It follows that the dot product is simply the projection of $\vec{A}$ onto $\vec{B}$ times the magnitude of $\vec{B}$; that is, $(A \cos \theta)B = AB \cos \theta = \vec{A} \cdot \vec{B}$. Equivalently, the dot product can be thought of as the projection of $\vec{B}$ onto $\vec{A}$ times the magnitude of $\vec{A}$.

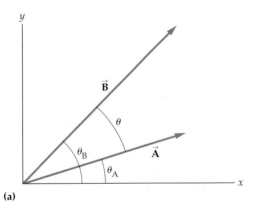

(a)

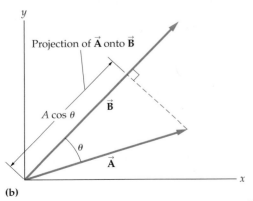

(b)

▲ **FIGURE A–6 The dot product between vectors $\vec{A}$ and $\vec{B}$.**

A few special cases will help to clarify the dot product. Suppose, for example, that $\vec{A}$ and $\vec{B}$ are parallel. In this case, $\theta = 0$ and $\vec{A} \cdot \vec{B} = AB$. Therefore, when vectors are parallel, the dot product is simply the product of their magnitudes. On the other hand, suppose $\vec{A}$ and $\vec{B}$ point in opposite directions. Now we have $\theta = 180$, and therefore $\vec{A} \cdot \vec{B} = -AB$. In general, the sign of $\vec{A} \cdot \vec{B}$ is positive if the angle between $\vec{A}$ and $\vec{B}$ is less than $90°$, and is negative if the angle between them is greater than $90°$. Finally, if $\vec{A}$ and $\vec{B}$ are perpendicular to one another—that is, if $\theta = 90°$—we see that $\vec{A} \cdot \vec{B} = AB \cos 90° = 0$. In this case, neither vector has a nonzero projection onto the other vector.

Dot products have a particularly simple form when applied to unit vectors. Recall, for example, that $\hat{x}$ and $\hat{y}$ have unit magnitude and are perpendicular to one another. It follows that

$$\hat{x} \cdot \hat{x} = 1, \quad \hat{x} \cdot \hat{y} = \hat{y} \cdot \hat{x} = 0, \quad \hat{y} \cdot \hat{y} = 1 \qquad \text{A–23}$$

These results can be applied to the general two-dimensional vectors $\vec{A} = A_x\hat{x} + A_y\hat{y}$ and $\vec{B} = B_x\hat{x} + B_y\hat{y}$ to give

$$\begin{aligned}
\vec{A} \cdot \vec{B} &= (A_x\hat{x} + A_y\hat{y}) \cdot (B_x\hat{x} + B_y\hat{y}) \\
&= A_xB_x\hat{x} \cdot \hat{x} + A_xB_y\hat{x} \cdot \hat{y} + A_yB_x\hat{y} \cdot \hat{x} + A_yB_y\hat{y} \cdot \hat{y} \qquad \text{A–24} \\
&= A_xB_x + A_yB_y
\end{aligned}$$

Thus, the dot product of two-dimensional vectors is simply the product of their x components plus the product of their y components.

At first glance the result $\vec{A} \cdot \vec{B} = AB \cos \theta$ looks quite different from the result $\vec{A} \cdot \vec{B} = A_xB_x + A_yB_y$. They are identical, however, as we now show. Suppose that $\vec{A}$ is at an angle θ_A to the positive x axis, and that $\vec{B}$ is at an angle $\theta_B > \theta_A$ to the x axis, from which it follows that the angle between $\vec{A}$ and $\vec{B}$ is $\theta = \theta_B - \theta_A$. Noting that $A_x = A \cos \theta_A$ and $A_y = A \sin \theta_A$, and similarly for B_x and B_y, we have

$$\vec{A} \cdot \vec{B} = AB(\cos \theta_A \cos \theta_B + \sin \theta_A \sin \theta_B)$$

The second trigonometric identity in Equation A–9 can be applied to the quantity in brackets, with the result that $\vec{A} \cdot \vec{B} = A_xB_x + A_yB_y = AB \cos(\theta_B - \theta_A) = AB \cos \theta$, as desired.

The most prominent application of dot products in this text is in Chapter 7, where in Equation 7–3 we define the work to be $W = Fd \cos \theta$, with θ the angle between $\vec{F}$ and $\vec{d}$. Clearly, this is simply a statement that work is the dot product of force and displacement:

$$W = \vec{F} \cdot \vec{d} = Fd \cos \theta$$

Later in the text, in Equation 19–11, we define the electric flux to be $\Phi = EA \cos \theta$. If we let $\vec{A}$ represent a vector that has a magnitude equal to the area, A, and points in the direction of the normal to the area, we can write the electric flux as a dot product:

$$\Phi = \vec{E} \cdot \vec{A} = EA \cos \theta$$

Similar remarks apply to the magnetic flux, defined in Equation 23–1.

The cross product

When two vectors are multiplied with the cross product, the result is a third vector that is perpendicular to both original vectors. An example is shown in **Figure A–7**, where we see a vector $\vec{A}$, a vector $\vec{B}$, and their cross product, $\vec{C}$:

$$\vec{C} = \vec{A} \times \vec{B} \qquad \text{A–25}$$

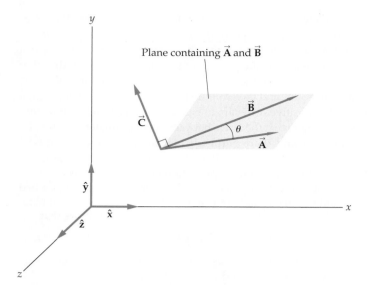

Plane containing $\vec{\mathbf{A}}$ and $\vec{\mathbf{B}}$

▲ **FIGURE A–7** The vector cross product of $\vec{\mathbf{A}}$ and $\vec{\mathbf{B}}$.

Notice that $\vec{\mathbf{C}}$ is perpendicular to the plane formed by the vectors $\vec{\mathbf{A}}$ and $\vec{\mathbf{B}}$. In addition, the direction of $\vec{\mathbf{C}}$ is given by the following right-hand rule:

To find the direction of $\vec{\mathbf{C}} = \vec{\mathbf{A}} \times \vec{\mathbf{B}}$, point the fingers of your right hand in the direction of $\vec{\mathbf{A}}$ and curl them toward $\vec{\mathbf{B}}$. Your thumb is now pointing in the direction of $\vec{\mathbf{C}}$.

It is clear from this rule that if $\vec{\mathbf{A}} \times \vec{\mathbf{B}} = \vec{\mathbf{C}}$, then $\vec{\mathbf{B}} \times \vec{\mathbf{A}} = -\vec{\mathbf{C}}$.

The magnitude of $\vec{\mathbf{C}} = \vec{\mathbf{A}} \times \vec{\mathbf{B}}$ depends on the magnitudes of the vectors $\vec{\mathbf{A}}$ and $\vec{\mathbf{B}}$, and on the angle θ between them. In particular,

$$C = AB \sin \theta \qquad \text{A–26}$$

Comparing with Equation A–22, we see that the cross product involves a sin θ, whereas the dot product depends on cos θ. As a result, it follows that the cross product has zero magnitude when $\vec{\mathbf{A}}$ and $\vec{\mathbf{B}}$ point in the same direction ($\theta = 0°$) or in opposite directions ($\theta = 180°$). On the other hand, the cross product has its greatest magnitude, $C = AB$, when $\vec{\mathbf{A}}$ and $\vec{\mathbf{B}}$ are perpendicular to one another ($\theta = 90°$).

When we apply these rules to unit vectors, which are at right angles to one another and of unit magnitude, the results are particularly simple. For example, consider the cross product $\hat{\mathbf{x}} \times \hat{\mathbf{y}}$. Referring to Figure A–7, we see that this cross product points in the positive z direction. In addition, the magnitude of $\hat{\mathbf{x}} \times \hat{\mathbf{y}}$ is $(1)(1) \sin 90° = 1$. It follows, therefore, that $\hat{\mathbf{x}} \times \hat{\mathbf{y}} = \hat{\mathbf{z}}$. On the

other hand, $\hat{\mathbf{x}} \times \hat{\mathbf{x}} = \hat{\mathbf{y}} \times \hat{\mathbf{y}} = \hat{\mathbf{z}} \times \hat{\mathbf{z}} = 0$ because $\theta = 0°$ in each of these cases. To summarize:

$$\begin{aligned} \hat{\mathbf{x}} \times \hat{\mathbf{y}} &= \hat{\mathbf{z}}, & \hat{\mathbf{y}} \times \hat{\mathbf{z}} &= \hat{\mathbf{x}}, & \hat{\mathbf{z}} \times \hat{\mathbf{x}} &= \hat{\mathbf{y}} \\ \hat{\mathbf{y}} \times \hat{\mathbf{x}} &= -\hat{\mathbf{z}}, & \hat{\mathbf{z}} \times \hat{\mathbf{y}} &= -\hat{\mathbf{x}}, & \hat{\mathbf{x}} \times \hat{\mathbf{z}} &= -\hat{\mathbf{y}} \\ \hat{\mathbf{x}} \times \hat{\mathbf{x}} &= 0, & \hat{\mathbf{y}} \times \hat{\mathbf{y}} &= 0, & \hat{\mathbf{z}} \times \hat{\mathbf{z}} &= 0 \end{aligned} \qquad \text{A–27}$$

As an example of how to use these unit-vector results, consider the cross product of the two-dimensional vectors shown in Figure A–6, $\vec{\mathbf{A}} = A_x\hat{\mathbf{x}} + A_y\hat{\mathbf{y}}$ and $\vec{\mathbf{B}} = B_x\hat{\mathbf{x}} + B_y\hat{\mathbf{y}}$. Straightforward application of Equation A–27 yields

$$\begin{aligned} \vec{\mathbf{C}} = \vec{\mathbf{A}} \times \vec{\mathbf{B}} &= (A_x\hat{\mathbf{x}} + A_y\hat{\mathbf{y}}) \times (B_x\hat{\mathbf{x}} + B_y\hat{\mathbf{y}}) \\ &= A_xB_x(\hat{\mathbf{x}} \times \hat{\mathbf{x}}) + A_xB_y(\hat{\mathbf{x}} \times \hat{\mathbf{y}}) \\ &\quad + A_yB_x(\hat{\mathbf{y}} \times \hat{\mathbf{x}}) + A_yB_y(\hat{\mathbf{y}} \times \hat{\mathbf{y}}) \\ &= (A_xB_y - A_yB_x)\hat{\mathbf{z}} \end{aligned} \qquad \text{A–28}$$

Notice that $\vec{\mathbf{C}}$ is perpendicular to both $\vec{\mathbf{A}}$ and $\vec{\mathbf{B}}$, as required for a cross product. In addition, the magnitude of $\vec{\mathbf{C}}$ is

$$A_xB_y - A_yB_x = AB(\cos \theta_A \sin \theta_B - \sin \theta_A \cos \theta_B)$$

If we now apply the first trigonometric identity in Equation A–9, we recover the result given in Equation A–26:

$$\begin{aligned} C &= AB(\cos \theta_A \sin \theta_B - \sin \theta_A \cos \theta_B) \\ &= AB \sin(\theta_B - \theta_A) = AB \sin \theta \end{aligned}$$

The first application of cross products in this text is torque, which is discussed in Chapter 11. In fact, Equation 11–2 defines the magnitude of the torque, τ, as follows: $\tau = r(F \sin \theta)$. As one might expect by referring to Equations A–25 and A–26, the torque vector, $\vec{\tau}$, can be written as the following cross product:

$$\vec{\tau} = \vec{\mathbf{r}} \times \vec{\mathbf{F}}$$

Similarly, the angular momentum vector, $\vec{\mathbf{L}}$, whose magnitude is given in Equation 11–13, is simply

$$\vec{\mathbf{L}} = \vec{\mathbf{r}} \times \vec{\mathbf{p}}$$

Finally, the cross product appears again in magnetism. In fact, the magnitude of the magnetic force on a charge q with a velocity $\vec{\mathbf{v}}$ in a magnetic field $\vec{\mathbf{B}}$ is $F = qvB \sin \theta$, as given in Equation 22–1. As one might guess, the vector form of this force is

$$\vec{\mathbf{F}} = q\vec{\mathbf{v}} \times \vec{\mathbf{B}}$$

The advantage of a cross product expression like this is that it contains both the direction and magnitude of a vector in one compact equation. In fact, we can now see the origin of the right-hand rule for magnetic forces given in Section 22–2.

Appendix B   Typical values

Mass

Sun	2.00×10^{30} kg
Earth	5.97×10^{24} kg
Moon	7.35×10^{22} kg
747 airliner (maximum takeoff weight)	3.5×10^{5} kg
blue whale	178,000 kg = 197 tons
elephant	5400 kg
mountain gorilla	180 kg
human	70 kg
bowling ball	7 kg
half gallon of milk	1.81 kg = 4 lbs
baseball	0.141–0.148 kg
golf ball	0.045 kg
female calliope hummingbird (smallest bird in North America)	3.5×10^{-3} kg = $\frac{1}{8}$ oz
raindrop	3×10^{-5} kg
antibody molecule (IgG)	2.5×10^{-22} kg
hydrogen atom	1.67×10^{-27} kg

Length

orbital radius of Earth (around Sun)	1.5×10^{8} km
orbital radius of Moon (around Earth)	3.8×10^{5} km
altitude of geosynchronous satellite	35,800 km = 22,300 mi
radius of Earth	6370 km
altitude of Earth's ozone layer	50 km
height of Mt. Everest	8848 m
height of Washington Monument	169 m = 555 ft
pitcher's mound to home plate	18.44 m
baseball bat	1.067 m
CD (diameter)	120 mm
aorta (diameter)	18 mm
period in sentence (diameter)	0.5 mm
red blood cell	7.8 μm = $\frac{1}{3300}$ in.
typical bacterium (*E. coli*)	2 μm
wavelength of green light	550 nm
virus	20–300 nm
large protein molecule	25 nm
diameter of DNA molecule	2.0 nm
radius of hydrogen atom	5.29×10^{-11} m

Time

estimated age of Earth	approx. 4.6 billion y $\approx 10^{17}$ s
estimated age of human species	approx. 150,000 y $\approx 5 \times 10^{12}$ s
half life of carbon-14	5730 y = 1.81×10^{11} s
period of Halley's comet	76 y = 2.40×10^{9} s
half life of technetium-99	6 h = 2.16×10^{4} s
time for driver of car to apply brakes	0.46 s
human reaction time	60–180 ms
air bag deployment time	10 ms

period of middle C sound wave	3.9 ms
collision time for batted ball	2 ms
decay of excited atomic state	10^{-8} s
period of green light wave	1.8×10^{-15} s

Speed

light	3×10^8 m/s
meteor	35–95 km/s
space shuttle (orbital velocity)	8.5 km/s = 19,000 mi/h
rifle bullet	700–750 m/s
sound in air (STP)	340 m/s
fastest human nerve impulses	140 m/s
747 at takeoff	80.5 m/s
kangaroo	18.1 m/s = 40.5 mi/h
200-m dash (Olympic record)	10.1 m/s
butterfly	1 m/s
blood speed in aorta	0.35 m/s
giant tortoise	0.076 m/s = 0.170 mi/h
Mer de Glace glacier (French Alps)	4×10^{-6} m/s

Acceleration

protons in particle accelerator	9×10^{13} m/s^2
ultracentrifuge	3×10^6 m/s^2
meteor impact	10^5 m/s^2
baseball struck by bat	3×10^4 m/s^2
loss of consciousness	$7.14g = 70$ m/s^2
acceleration of gravity on Earth (g)	9.81 m/s^2
braking auto	8 m/s^2
acceleration of gravity on the moon	1.62 m/s^2
rotation of Earth at equator	3.4×10^{-2} m/s^2

Appendix C Planetary data

Name	Equatorial Radius (km)	Mass (Relative to Earth's)*	Mean Density (kg/m³)	Surface Gravity (Relative to Earth's)	Orbital Semimajor Axis × 10⁶ km	A. U.	Escape Speed (km/s)	Orbital Period (Years)	Orbital Eccentricity
Mercury	2440	0.0553	5430	0.38	57.9	0.387	4.2	0.240	0.206
Venus	6052	0.816	5240	0.91	108.2	0.723	10.4	0.615	0.007
Earth	6370	1	5510	1	149.6	1	11.2	1.000	0.017
Mars	3394	0.108	3930	0.38	227.9	1.523	5.0	1.881	0.093
Jupiter	71,492	318	1360	2.53	778.4	5.203	60	11.86	0.048
Saturn	60,268	95.1	690	1.07	1427.0	9.539	36	29.42	0.054
Uranus	25,559	14.5	1270	0.91	2871.0	19.19	21	83.75	0.047
Neptune	24,776	17.1	1640	1.14	4497.1	30.06	24	163.7	0.009
Pluto	1137	0.0021	2060	0.07	5906	39.84	1.2	248.0	0.249

*Mass of Earth = 5.97×10^{24} kg

Appendix D   Elements of electrical circuits

Circuit Element	Symbol	Physical Characteristics
resistor		Resists the flow of electric current. Converts electrical energy to thermal energy.
capacitor		Stores electrical energy in the form of an electric field.
inductor		Stores electrical energy in the form of a magnetic field.
incandescent lightbulb		A device containing a resistor that gets hot enough to give off visible light.
battery		A device that produces a constant difference in electrical potential between its terminals.
ac generator		A device that produces a potential difference between its terminals that oscillates with time.
switches (open and closed)		Devices to control whether electric current is allowed to flow through a portion of a circuit.
ground		Sets the electric potential at a point in a circuit equal to a constant value usually taken to be $V = 0$.

Appendix E   Periodic table of the elements

PERIODS — s / d / p / f blocks · Transition elements

Legend:
- Atomic number: 26
- Symbol: Fe
- Atomic mass: 55.85
- Outer electron configuration: $3d^6 4s^2$

Period	GROUP I	GROUP II	3	4	5	6	7	8	9	10	11	12	GROUP III	GROUP IV	GROUP V	GROUP VI	GROUP VII	GROUP VIII
1	1 H 1.01 $1s^1$																	2 He 4.00 $1s^2$
2	3 Li 6.94 $2s^1$	4 Be 9.01 $2s^2$											5 B 10.81 $2p^1$	6 C 12.01 $2p^2$	7 N 14.01 $2p^3$	8 O 16.00 $2p^4$	9 F 19.00 $2p^5$	10 Ne 20.18 $2p^6$
3	11 Na 22.99 $3s^1$	12 Mg 24.31 $3s^2$											13 Al 26.98 $3p^1$	14 Si 28.09 $3p^2$	15 P 30.97 $3p^3$	16 S 32.07 $3p^4$	17 Cl 35.45 $3p^5$	18 Ar 39.95 $3p^6$
4	19 K 39.10 $4s^1$	20 Ca 40.08 $4s^2$	21 Sc 44.96 $3d^1 4s^2$	22 Ti 47.88 $3d^2 4s^2$	23 V 50.94 $3d^3 4s^2$	24 Cr 52.00 $3d^5 4s^1$	25 Mn 54.94 $3d^5 4s^2$	26 Fe 55.85 $3d^6 4s^2$	27 Co 58.93 $3d^7 4s^2$	28 Ni 58.69 $3d^8 4s^2$	29 Cu 63.55 $3d^{10} 4s^1$	30 Zn 65.39 $3d^{10} 4s^2$	31 Ga 69.72 $4p^1$	32 Ge 72.61 $4p^2$	33 As 74.92 $4p^3$	34 Se 78.96 $4p^4$	35 Br 79.90 $4p^5$	36 Kr 83.80 $4p^6$
5	37 Rb 85.47 $5s^1$	38 Sr 87.62 $5s^2$	39 Y 88.96 $4d^1 5s^2$	40 Zr 91.22 $4d^2 5s^2$	41 Nb 92.91 $4d^4 5s^1$	42 Mo 95.94 $4d^5 5s^1$	43 Tc (98) $4d^5 5s^2$	44 Ru 101.07 $4d^7 5s^1$	45 Rh 102.91 $4d^8 5s^1$	46 Pd 106.42 $4d^{10} 5s^6$	47 Ag 107.87 $4d^{10} 5s^1$	48 Cd 112.41 $4d^{10} 5s^2$	49 In 114.82 $5p^1$	50 Sn 118.71 $5p^2$	51 Sb 121.76 $5p^3$	52 Te 127.60 $5p^4$	53 I 126.90 $5p^5$	54 Xe 131.29 $5p^6$
6	55 Cs 132.91 $6s^1$	56 Ba 137.33 $6s^2$	57 La* 138.91 $5d^1 6s^2$	72 Hf 178.49 $5d^2 6s^2$	73 Ta 180.95 $5d^3 6s^2$	74 W 183.85 $5d^4 6s^2$	75 Re 186.21 $5d^5 6s^2$	76 Os 190.2 $5d^6 6s^2$	77 Ir 192.22 $5d^7 6s^2$	78 Pt 195.08 $5d^9 6s^1$	79 Au 196.97 $5d^{10} 6s^1$	80 Hg 200.59 $5d^{10} 6s^2$	81 Tl 204.36 $6p^1$	82 Pb 207.2 $6p^2$	83 Bi 208.98 $6p^3$	84 Po (209) $6p^4$	85 At (210) $6p^5$	86 Rn (222) $6p^6$
7	87 Fr (223) $7s^1$	88 Ra 226.03 $7s^2$	89 Ac† 227.03 $6d^1 7s^2$	104 Rf (261) $6d^2 7s^2$	105 Db (262) $6d^3 7s^2$	106 Sg (266) $6d^4 7s^2$	107 Bh (264) $6d^5 7s^2$	108 Hs (269) $6d^6 7s^2$	109 Mt (268) $6d^7 7s^2$	110 (271)	111 (272)	112 (277)		114 (289)		116 (289)		118 (293)

Lanthanides (*)

58 Ce 140.12 $5d^1 4f^1 6s^2$	59 Pr 140.91 $4f^3 6s^2$	60 Nd 144.24 $4f^4 6s^2$	61 Pm (145) $4f^5 6s^2$	62 Sm 150.36 $4f^6 6s^2$	63 Eu 151.96 $4f^7 6s^2$	64 Gd 157.25 $5d^1 4f^7 6s^2$	65 Tb 158.93 $4f^9 6s^2$	66 Dy 162.50 $4f^{10} 6s^2$	67 Ho 164.93 $4f^{11} 6s^2$	68 Er 167.26 $4f^{12} 6s^2$	69 Tm 168.93 $4f^{13} 6s^2$	70 Yb 173.04 $4f^{14} 6s^2$	71 Lu 174.97 $5d^1 4f^{14} 6s^2$

Actinides (†)

90 Th 232.04 $6d^2 7s^2$	91 Pa 231.04 $5f^2 6d^1 7s^2$	92 U 238.03 $5f^3 6d^1 7s^2$	93 Np 237.05 $5f^4 6d^1 7s^2$	94 Pu (244) $5f^6 6d^0 7s^2$	95 Am (243) $5f^7 6d^0 7s^2$	96 Cm (247) $5f^7 6d^1 7s^2$	97 Bk (247) $5f^8 6d^1 7s^2$	98 Cf (251) $5f^{10} 6d^0 7s^2$	99 Es (252) $5f^{11} 6d^0 7s^2$	100 Fm (257) $5f^{12} 6d^0 7s^2$	101 Md (258) $5f^{13} 6d^0 7s^2$	102 No (259) $5f^{14} 6d^0 7s^2$	103 Lr (262) $5f^{14} 6d^1 7s^2$

Appendix F   Properties of selected isotopes

Atomic Number (Z)	Element	Symbol	Mass Number (A)	Atomic Mass*	Abundance (%) or Decay Mode† (if radioactive)	Half-Life (if radioactive)
0	(Neutron)	n	1	1.008665	β^-	10.6 min
1	Hydrogen	H	1	1.007825	99.985	
	Deuterium	D	2	2.014102	0.015	
	Tritium	T	3	3.016049	β^-	12.33 y
2	Helium	He	3	3.016029	0.00014	
			4	4.002603	≈ 100	
3	Lithium	Li	6	6.015123	7.5	
			7	7.016003	92.5	
4	Beryllium	Be	7	7.016930	EC, γ	53.3 d
			8	8.005305	2α	6.7×10^{-17} s
			9	9.012183	100	
5	Boron	B	10	10.012938	19.9	
			11	11.009305	80.1	
			12	12.014353	β^-	20.2 ms
6	Carbon	C	11	11.011433	β^+, EC	20.3 min
			12	12.000000	98.89	
			13	13.003355	1.11	
			14	14.003242	β^-	5730 y
7	Nitrogen	N	13	13.005739	β^-	9.96 min
			14	14.003074	99.63	
			15	15.000109	0.37	
8	Oxygen	O	15	15.003065	β^+, EC	122 s
			16	15.994915	99.76	
			18	17.999159	0.204	
9	Fluorine	F	19	18.998403	100	
			18	18.000938	EC	109.77 min
10	Neon	Ne	20	19.992439	90.51	
			22	21.991384	9.22	
11	Sodium	Na	22	21.994435	β^+, EC, γ	2.602 y
			23	22.989770	100	
			24	23.990964	β^-, γ	15.0 h
12	Magnesium	Mg	24	23.985045	78.99	
13	Aluminum	Al	27	26.981541	100	
14	Silicon	Si	28	27.976928	92.23	
			31	30.975364	β^-, γ	2.62 h
15	Phosphorus	P	31	30.973763	100	
			32	31.973908	β^-	14.28 d
16	Sulfur	S	32	31.972072	95.0	
			35	34.969033	β^-	87.4 d
17	Chlorine	Cl	35	34.968853	75.77	
			37	36.965903	24.23	
18	Argon	Ar	40	39.962383	99.60	
19	Potassium	K	39	38.963708	93.26	
			40	39.964000	β^-, EC, γ, β^+	1.28×10^9 y

Atomic Number (Z)	Element	Symbol	Mass Number (A)	Atomic Mass*	Abundance (%) or Decay Mode[†] (if radioactive)	Half-Life (if radioactive)
20	Calcium	Ca	30	39.962591	96.94	
24	Chromium	Cr	52	51.940510	83.79	
25	Manganese	Mn	55	54.938046	100	
26	Iron	Fe	56	55.934939	91.8	
27	Cobalt	Co	59	58.933198	100	
			60	59.933820	β^-, γ	5.271 y
28	Nickel	Ni	58	57.935347	68.3	
			60	59.930789	26.1	
			64	63.927968	0.91	
29	Copper	Cu	63	62.929599	69.2	
			64	63.929766	β^-, β^+	12.7 h
			65	64.927792	30.8	
30	Zinc	Zn	64	63.929145	48.6	
			66	65.926035	27.9	
33	Arsenic	As	75	74.921596	100	
35	Bromine	Br	79	78.918336	50.69	
36	Krypton	Kr	84	83.911506	57.0	
			89	88.917563	β^-	3.2 min
			92	91.926153	β^-	1.84 s
38	Strontium	Sr	86	85.909273	9.8	
			88	87.905625	82.6	
			90	89.907746	β^-	28.8 y
39	Yttrium	Y	89	89.905856	100	
41	Niobium	Nb	98	97.910331	β^-	2.86 s
43	Technetium	Tc	98	97.907210	β^-, γ	4.2×10^6 y
47	Silver	Ag	107	106.905095	51.83	
			109	108.904754	48.17	
48	Cadmium	Cd	114	113.903361	28.7	
49	Indium	In	115	114.90388	95.7; β^-	5.1×10^{14} y
50	Tin	Sn	120	119.902199	32.4	
51	Antimony	Sb	133	132.915237	β^-	2.5 min
53	Iodine	I	127	126.904477	100	
			131	130.906118	β^-, γ	8.04 d
54	Xenon	Xe	132	131.90415	26.9	
			136	135.90722	8.9	
55	Cesium	Cs	133	132.90543	100	
56	Barium	Ba	137	136.90582	11.2	
			138	137.90524	71.7	
			141	140.914406	β^-	18.27 min
			144	143.92273	β^-	11.9 s
61	Promethium	Pm	145	144.91275	EC, α, γ	17.7 y
74	Tungsten (Wolfram)	W	184	183.95095	30.7	
76	Osmium	Os	191	190.96094	β^-, γ	15.4 d
			192	191.96149	41.0	
78	Platinum	Pt	195	194.96479	33.8	
79	Gold	Au	197	196.96656	100	

Atomic Number (Z)	Element	Symbol	Mass Number (A)	Atomic Mass*	Abundance (%) or Decay Mode† (if radioactive)	Half-Life (if radioactive)
81	Thallium	Tl	205	204.97441	70.5	
			210	209.990069	β^-	1.3 min
82	Lead	Pb	204	203.973044	β^-, 1.48	1.4×10^{17} y
			206	205.97446	24.1	
			207	206.97589	22.1	
			208	207.97664	52.3	
			210	209.98418	α, β^-, γ	22.3 y
			211	210.98874	β^-, γ	36.1 min
			212	211.99188	β^-, γ	10.64 h
			214	213.99980	β^-, γ	26.8 min
83	Bismuth	Bi	209	208.98039	100	
			211	210.98726	α, β^-, γ	2.15 min
			212	211.991272	α	60.55 min
84	Polonium	Po	210	209.98286	α, γ	138.38 d
			212	211.988852	α	0.299 μs
			214	213.99519	α, γ	164 μs
86	Radon	Rn	222	222.017574	α, β	3.8235 d
87	Francium	Fr	223	223.019734	α, β^-, γ	21.8 min
88	Radium	Ra	226	226.025406	α, γ	1.60×10^3 y
			228	228.031069	β^-	5.76 y
89	Actinium	Ac	227	227.027751	α, β^-, γ	21.773 y
90	Thorium	Th	228	228.02873	α, γ	1.9131 y
			231	231.036297	α, β^-	25.52 h
			232	232.038054	100; α, γ	1.41×10^{10} y
			234	234.043596	β^-	24.10 d
91	Protactium	Pa	234	234.043302	β^-	6.70 h
92	Uranium	U	232	232.03714	α, γ	72 y
			233	233.039629	α, γ	1.592×10^5 y
			235	235.043925	0.72; α, γ	7.038×10^8 y
			236	236.045563	α, γ	2.342×10^7 y
			238	238.050786	99.275; α, γ	4.468×10^9 y
			239	239.054291	β^-, γ	23.5 min
93	Neptunium	Np	239	239.052932	β^-, γ	2.35 d
94	Plutonium	Pu	239	239.052158	α, γ	2.41×10^4 y
95	Americium	Am	243	243.061374	α, γ	7.37×10^3 y
96	Curium	Cm	245	245.065487	α, γ	8.5×10^3 y
97	Berkelium	Bk	247	247.07003	α, γ	1.4×10^3 y
98	Californium	Cf	249	249.074849	α, γ	351 y
99	Einsteinium	Es	254	254.08802	α, γ, β^-	276 d
100	Fermium	Fm	253	253.08518	EC, α, γ	3.0 d
101	Mendelevium	Md	255	255.0911	EC, α	27 min
102	Nobelium	No	255	255.0933	EC, α	3.1 min
103	Lawrencium	Lr	257	257.0998	α	$\approx$35 s

*The masses given throughout this table are those for the neutral atom, including the Z electrons.
†EC stands for electron capture.

Answers to Your Turn Problems

CHAPTER 19

Active Example 19–1 The point of zero force remains in the same place. This can be seen most clearly in step 3, where we see that doubling each charge simply yields an additional factor of two on each side of the equation. Since the factor of two appears on both sides of the equation, it cancels.

Active Example 19–2 The new sphere exerts less force than the original sphere. Specifically, the new sphere has half the radius of the original sphere, and one quarter its surface area. Therefore, it has one quarter the total charge, and exerts one quarter as much force.

Active Example 19–3 The new Gaussian surface has zero electric flux—there is no flux through either end cap, nor through the curved sides of the cylindrical surface. This means that the net charge contained within the Gaussian surface is zero, which is evident when we note that the two plates of the capacitor have opposite charge densities of equal magnitude.

CHAPTER 20

Active Example 20–1 We are given that the electric potential at point A is *higher* than at point B; therefore, the electric potential energy of an electron (with its negative charge) is *less* at point A than at point B. As an electron moves from point A to point B its electric potential energy increases by 7.2×10^{-19} J.

Active Example 20–2 The initial speed must be considerably greater than 5.00 m/s. In fact, it must be 14.6 m/s.

Active Example 20–3 There is no change in the electric potential energy. The reason is that moving the charge as described does not change the *separation* between any pair of charges in the system. Therefore, the total electric potential energy is as given in step 4.

CHAPTER 21

Active Example 21–1 The required time is 230 s.

Active Example 21–2 In this case, we find $I_1 = 0.13$ A, $I_2 = 0.11$ A, and $I_3 = 0.020$ A. Notice that each current flows in the direction indicated in the sketch of the circuit.

Active Example 21–3 The 5.00-μF capacitor stores more energy than the 10.0-μF capacitor because more work is required to force a given amount of charge onto its plates. In fact, we find that the 5.00-μF capacitor stores 1.60×10^{-4} J of energy, twice as much (to three significant figures) as the 7.98×10^{-5} J stored in the 10.0-μF capacitor.

CHAPTER 22

Active Example 22–1 The reason is that the orbital speed is directly proportional to the radius, as we see in Equation 22–3. Therefore, reducing the radius (or circumference) by a given factor reduces the speed by the same factor. It follows that the time required to travel a distance equal to one circumference is independent of the radius.

Active Example 22–2 In this case, the fields produced by the two wires are in the same direction; namely, into the page. The net field is 9.1×10^{-6} T, into the page.

CHAPTER 23

Active Example 23–1 One way to calculate the current is to note that the light bulb consumes a power of 5.0 W and has a resistance of 12 Ω. We can solve $P = I^2R$ to find $I = 0.65$ A. A second way is to note from Example 23–3 that the external force acting on the rod has a magnitude of 1.6 N. The rod moves with constant speed, however, and hence the magnetic force acting on it, $F = ILB$, must have the same magnitude. If we equate these magnitudes, we find $I = 0.64$ A. The slight discrepancy between these answers is due to round-off error, as can be verified by repeating the calculations with more significant figures.

Active Example 23–2 The value of the inductance changes by a greater factor if we double the number of turns. This is because the inductance depends on the square of the number of turns, but depends only linearly on the cross-sectional area. Specifically, we find the following: (a) doubling N quadruples the inductance; (b) tripling A triples the inductance.

Active Example 23–3 From Equation 23–22 we see that the voltage in the secondary circuit is $V_s = V_p(N_s/N_p)$. Therefore, if $V_p \to 2V_p$ and $N_p \to 4N_p$, it is clear that we must double N_s to keep V_s the same.

CHAPTER 24

Active Example 24–1 The capacitive reactance is equal to 64 Ω when the frequency is reduced to 9.2 Hz. At this frequency, the current is 1.2 A.

Active Example 24–2 (a) 40 V. **(b)** 58 V.

Active Example 24–3 Setting the impedance of the circuit equal to 2.50 V/1.50 A = 1.67 Ω, we find $f = 68.0$ Hz and $f = 108$ Hz, to three significant figures.

CHAPTER 25

Active Example 25–1 If the beam spreads out to twice its initial diameter, its area quadruples. This means, in turn, that the intensity of the beam decreases by a factor of four. The intensity, however, depends on the fields squared. Therefore, it follows that both E_{max} and B_{max} decrease by a factor of two.

Active Example 25–2 With three equally rotated polarizers, we find a transmitted intensity of $0.689I_0$. This is considerably greater than the intensity found with two polarizers. In general, the more smoothly and continuously the plane of polarization is rotated, the greater the transmitted intensity.

CHAPTER 26

Active Example 26–1 The magnification of the tooth will decrease. After all, as the object moves closer to the mirror, the mirror behaves more and more like a plane mirror, in which case the magnification is 1. With $f = 1.38$ cm and $d_o = 1.00$ cm we find $d_i = -3.63$ cm and $m = 3.63$.

Active Example 26–2 Referring to Figure 26–34, it is clear that to obtain a larger magnification we must move the object closer to the lens. To obtain $m = 0.75$, we find that the object distance must be reduced from 12 cm to $d_o = 2.633$ cm, to four significant figures. The corresponding image distance is -1.975 cm.

CHAPTER 27

Active Example 27–1 The camera is now focussed at a distance of 1.72 m.

Active Example 27–2 The only change from the analysis given in the text for Figure 27–5 is that the concave mirror has a positive focal length ($+12.5$ cm) rather than a negative focal length. Therefore, the image distance for the mirror is 21.4 cm. This image is then an object for the lens, producing the final image of the system 15.4 cm to the left of the lens. The final magnification is -0.384, indicating an inverted image 38.4% of its original height.

Active Example 27–3 In this case, the far point is 202 cm from the person's eyes. Note that this far point is closer to the eye than the far point in Example 27–2; therefore, the required refractive power has a magnitude that is greater than 0.312.

Active Example 27–4 The second person's vision needs more correction, since the near point is farther from the eyes. Therefore, the refractive power of the second person's contacts must be greater, which, in turn, means that the focal length must be smaller. In fact, we find $f = 28.6$ cm.

CHAPTER 28

Active Example 28–1 The desired minimum thickness is one-quarter the wavelength in the material. Recall, however, that the wavelength in a material with an index of refraction n is $\lambda_n = \lambda/n$ (Equation 28–4). Therefore, if the index of refraction is increased, the minimum thickness will be decreased. In this case, we find a minimum thickness of 97.4 nm.

Active Example 28–2 The second dark fringe corresponds to $m = 2$ in Equation 28–12. With this substitution, we find that the linear distance is $y = 6.96$ cm.

Active Example 28–3 As the aperture of a telescope increases, the minimum angular separation that can be resolved decreases, as can be seen from Equation 28–15. If a telescope can resolve smaller angular separations, it follows that its maximum resolution distance is greater. For the case $D = 3.0$ m, we find $L = 1.1 \times 10^{10}$ m.

Active Example 28–4 First, convert the wavelength 486.2 nm to the frequency $f = 6.170 \times 10^{14}$ Hz and the wavelength 563.0 nm to the frequency $f' = 5.329 \times 10^{14}$ Hz. With these results, we can now apply Equation 25–3 to find $u = 4.089 \times 10^{7}$ m/s. Since this is only 13.6% of the speed of light, the approximations used to derive Equation 25–3 should be valid.

CHAPTER 29

Active Example 29–1 The speed in this case is $v = 0.953c$.

Active Example 29–2 In this case, the velocity of the probe relative to the planet is $v = 0.306c$. The corresponding length is $L = 9.52$ m.

CHAPTER 30

Active Example 30–1 Increasing the wavelength by a factor of 1.25 results in a reduction in the frequency by a factor of 1.25. Similarly, the energy of a photon (which is proportional to frequency) is reduced by a factor of 1.25. Since each photon carries less energy, it follows that more photons will be required per second at this new wavelength. In fact, the minimum number of photons per second will be increased by the factor 1.25.

Active Example 30–2 The de Broglie wavelength is inversely proportional to speed; therefore, doubling the speed results in a wavelength that is reduced by a factor of two.

CHAPTER 31

Active Example 31–1 In this case, the final state is $n_f = 2$.

Active Example 31–2 In singly ionized helium, the charge of the nucleus is $+Ze = +2e$; therefore, $Z = 2$. Referring to Equation 31–9, we see that the energy of any given energy level depends on Z^2. It follows that the energy—and frequency—of the absorbed photon increases by a factor of $2^2 = 4$.

Active Example 31–3 Direct substitution in Equation 31–14 shows that $Z = 51$ is the largest value of Z that requires an acceleration voltage less than 35 kV.

CHAPTER 32

Active Example 32–1 Doubling the wavelength reduces the frequency of the gamma ray (and its energy) by a factor of two. Since the mass difference is proportional to the energy of the gamma ray, it too will be reduced by a factor of two.

Active Example 32–2 Direct substitution in Equation 32–9 shows that the number of radon atoms has decreased by a factor of two when $t = 3.83$ d. Note that this result is in agreement with Equation 32–10.

Active Example 32–3 Using Equation 32–12, we see that the activity of the radon decreases to 10.0 Bq after 12.7 d.

Active Example 32–4 This reaction releases 12.9 MeV.

Answers to Odd-Numbered Conceptual Questions

CHAPTER 19

1. No. When an object becomes charged it is because of a transfer of charge between it and another object.

3. The charged comb causes the paper to become polarized, with the side nearest the comb acquiring a charge opposite to the charge of the comb. The result is an attractive interaction between the comb and the paper.

5. Yes. If the suspended object were neutral, it would be attracted to the charged rod by polarization effects. The fact that the suspended object is repelled indicates it has a charge of the same sign as that of the rod.

7. Both force laws depend on the product of specific properties of the objects involved; in the case of gravity it is the mass that is relevant, in the case of electrostatics it is the electric charge. In addition, both forces decrease with increasing distance as $1/r^2$. The extremely important difference between the forces, however, is that gravity is always attractive, whereas electrostatic forces can be attractive or repulsive.

9. No. Only for very special displacements will the electrostatic force act in a direction that points back toward the equilibrium point. For a general displacement the electrostatic force does not point toward the equilibrium point, and the fifth charge would move farther from equilibrium, making the system unstable.

11. One difference is that when an object is charged by induction, there is no physical contact between the object being charged and the object used to do the charging. In contrast, charging by contact—as the name implies—involves direct physical contact to transfer charge from one object to another. The main difference is that when an object is charged by induction, the sign of the charge the object acquires is opposite to that of the object used to do the charging. Charging by contact gives the object being charged the same sign of charge as the original charged object.

13. No. The direction of the forces might be different simply because the sign of the charges are different. The magnitude of the forces might be different simply because the magnitudes of the charges are different.

15. By definition, electric field lines point in the direction of the electric force on a positive charge at any given location in space. This force can point in only one direction at any one location, however. Therefore, electric field lines cannot cross, because if they did, it would imply two different directions for the electric force at the same location.

17. The electric field depends on both charges. The total electric field at any point is simply the superposition of the electric field produced by each charge separately.

19. Gauss's law is useful as a calculational tool only in cases of high symmetry, where one can produce a gaussian surface on which the electric field is either constant or has no perpendicular component. It is not possible to do this in any simple way for the case of a charged disk. Gauss's law still applies—it's just not particularly useful.

CHAPTER 20

1. The electric field is a measure of how much the electric potential changes from one position to another. Therefore, the electric field in each of these regions is zero.

3. Not necessarily. The electric field is related to the rate of change of electric potential, not to the value of the electric potential. Therefore, if the electric field is zero in some region of space, it follows that the electric potential is constant in that region. The constant value of the electric potential may be zero, but it may also be positive or negative.

5. Zero. The electric field is perpendicular to an equipotential, therefore the work done in moving along an equipotential is zero.

7. If the electric field is not perpendicular to an equipotential, the field would do work on a charge that moves along the equipotential. In this case, the potential energy of the charge would change, and the surface would not in fact be an equipotential.

9. When the capacitor is disconnected from the battery, the charge on the capacitor plates simply remains where it is—there is no way for it to go anywhere else. When the terminals are connected to one another the charges flow from plate to plate until both plates have zero charge.

11. The capacitance of a capacitor depends on **(b)** the separation of the plates and **(e)** the area of the plates. The capacitance does not depend on **(a)** the charge on the plates, **(c)** the voltage difference between the plates, or **(d)** the electric field between the plates.

13. No. As an example, note that the volume of a milk container is not zero just because the container happens to be empty of milk. The same can be said about the capacitance of a capacitor that happens to be uncharged.

CHAPTER 21

1. Electric current is in the opposite direction to the motion of negative charge, therefore the electric current of the falling electron is upward.

3. No. By rubbing the comb through your hair you have transferred charge from your hair to the comb, but the net charge of you and the comb together is still zero. Therefore, no current is produced when you walk.

5. No. An electron may have a fairly large velocity at any given time, but because its direction of motion keeps changing—due to its collisions with atoms in the wire—its average velocity is almost zero.

7. Connect the four resistors in a parallel arrangement with two branches, each branch containing two resistors connected in series. In this way, the equivalent resistance of each branch is $2R$, and the equivalent resistance of two resistors of $2R$ in parallel is simply R, as desired.

9. Resistors connected in series have the same current flowing through them.

11. Each electron in the wire affects its neighbors by exerting a force on them, causing them to move. Thus, when electrons begin to move out of a battery their motion sets up a propagating influence that moves through the wire at nearly the speed of light, causing electrons everywhere in the wire to begin moving.

13. A number of factors come into play here. First, the bottom of a bird's foot is tough, and definitely not a good conductor of electricity. Second, and more important, is the fact that a potential difference is required for there to be a flow of current. Just being in contact with a high-voltage wire isn't enough to cause a problem; somewhere else there must be contact with a lower voltage. But the bird is in contact with essentially the same high voltage in two different places (where its feet touch the wire), which doesn't lead to a potential difference. The only potential difference the bird experiences is due to the very small voltage drop along the segment of wire between the bird's two feet.

15. The junction rule is based on conservation of electric charge; the loop rule is based on the conservation of energy.

17. Capacitors in parallel have the same potential difference between their plates.

CHAPTER 22

1. No. The particles may have charge of the same sign but move in opposite directions along the same line. In this way, they would both move perpendicular to the field, but would deflect in opposite directions.

3. No. If the electron moves in the same direction as the magnetic field, or opposite to the direction of the field, the magnetic force exerted on it will be zero. As a result, its velocity will remain constant.

5. The radius of curvature is proportional to the speed of the particle. It follows that the particle moving in a circle of large radius (and large circumference) has a proportionally larger speed than the particle

moving in a circle of small radius (and small circumference). Therefore, the time required for an orbit ($t = d/v$) is the same for both particles.

CHAPTER 23

1. The magnetic field indicates the strength and direction of the magnetic force that a charged particle moving with a certain velocity would experience at a given point in space. The magnetic flux, on the other hand, can be thought of as a measure of the "amount" of magnetic field that passes through a given area.

3. As the magnet falls, it induces eddy currents in the copper tube. These induced currents produce a magnetic field opposite in direction to that of the magnet. This results in a magnetic repulsion that slows the fall of the magnet. In fact, the motion of the magnet is much the same as if it had been dropped into a tube filled with honey.

5. When the switch is closed, a magnetic field is produced in the wire coil and in the iron rod. This results in an increasing magnetic flux through the metal ring, and a corresponding induced emf. The current produced by the induced emf generates a magnetic field opposite in direction to the field in the iron rod. The resulting magnetic repulsion propels the ring into the air.

7. Initially, the rod accelerates to the left, due to the downward current it carries. As it speeds up, however, the motional emf it generates will begin to counteract the emf of the battery. Eventually the two emfs balance one another, and current stops flowing in the rod. From this point on, the rod continues to move with constant speed.

9. As the shuttle orbits, it moves through the Earth's magnetic field at high speed. A long conducting wire moving through the field can generate an induced emf. In fact, the emf is given by the product of the length of the wire, the speed of the shuttle, and the perpendicular component of the magnetic field. With such large values for the speed and length, the induced emf can be great enough to provide substantial electrical power.

11. The final current in an RL circuit is determined only by the resistor R and the emf of the battery. The reason is that when the current stops changing, the back emf in an inductor vanishes. Thus, the inductor behaves like an ideal wire of zero resistance when the current reaches its final value.

CHAPTER 24

1. The average voltage in an ac circuit is zero because it oscillates symmetrically between positive and negative values. To calculate the rms voltage, however, one first squares the voltage. This gives values that are always greater than or equal

to zero. Therefore, the rms voltage will be nonzero unless the voltage in the circuit is zero at all times.

3. An LC circuit consumes zero power because ideal inductors and capacitors have no resistance. From a different point of view, the phase angle in an LC circuit is either $\phi = +90°$ or $\phi = -90°$, depending on whether the frequency is less than or greater than the resonance frequency, respectively. In either case, the power factor, $\cos\phi$, is zero; hence the power consumed by the circuit is zero.

5. In the phasor diagram for an LC circuit, the impedance is always perpendicular to the current—the only question, therefore, is whether ϕ is +90° or −90°. At a frequency less than the resonance frequency the capacitive reactance ($X_C = 1/\omega C$) is greater than the inductive reactance ($X_L = \omega L$). Therefore, the situation is similar to that of a circuit with only a capacitor, in which case the phase angle is $\phi = -90°$.

7. As frequency is increased there is no change in resistance, R. On the other hand, the capacitive reactance ($X_C = 1/\omega C$) decreases and the inductive reactance ($X_L = \omega L$) increases.

9. Mass resists changes in its motion due to its inertia. Similarly, an inductor resists changes in the current flowing through it due to its inductance. Therefore, mass and inductance are analogous. As for the spring constant, a stiff spring (large spring constant) gives little stretch for a given force. Similarly, a capacitor with a small capacitance stores little charge for a given voltage. Since charge is what moves in a circuit, it follows that displacement and charge are analogous. Therefore, the spring constant and the inverse of the capacitance are analogous.

11. Yes. Recall that the impedance of an RLC circuit is $Z = \sqrt{R^2 + (X_L - X_C)^2}$. Because the quantity in parentheses is squared, we get the same impedance when X_L is greater than X_C by a certain amount as when X_C is greater than X_L by the same amount.

CHAPTER 25

1. Presumably, an "invisible man" would be invisible because light passes through his body unimpeded, just as if it were passing through thin air. If some light were deflected or absorbed, we would see this effect and the person would no longer be invisible. For a person to see, however, some light must be absorbed by the retina. This absorption would cause the invisible man to be visible.

3. As a grain of dust becomes smaller, its volume—and therefore its mass—decreases more rapidly than does its area. It follows that radiation pressure, which acts on the surface of the grain, becomes increasingly important as the size of the

grain is decreased. Gravity, which acts on the mass of the grain, becomes less important in this limit.

5. Your watch must have an LCD display. In such watches, the light coming from the display is linearly polarized. If the polarization direction of the display and the sunglasses align, you can read the time. If these directions are at 90° to one another no light will pass through the sunglasses, and the display will appear black.

7. Radio stations generate their electromagnetic waves with large vertical antennas. As a result, their waves are polarized in the vertical direction, as in Figure 25–1. On the other hand, the light we see from the sun and from light bulbs is unpolarized because the atoms emitting the light can have any orientation relative to one another. Hence, even if individual atoms emit polarized light, the net result from a group of atoms is light with no preferred direction of polarization.

9. Sound waves cannot be polarized. This is because sound is a longitudinal wave, and the molecules can move in only a single direction—back and forth along the direction of propagation. In contrast, the electric field in an electromagnetic wave, which is transverse, can point in any direction within the plane that is transverse to the direction of propagation. This means that the electromagnetic wave can have different polarizations.

11. The two projected images give a view of a scene from slightly different angles, just as your eyes view a three-dimensional object from different angles. Without the headsets, the screen is a confusing superposition of the two images; with the headsets, your right eye sees one view of the scene and your left eye sees the other view. As these views are combined in your brain, you experience a realistic three-dimensional effect.

CHAPTER 26

1. Three images are formed of object A. One extends from $(-2\,m, 2\,m)$ to $(-1\,m, 2\,m)$ to $(-1\,m, 3\,m)$. Another image forms an "L" from $(1\,m, -3\,m)$ to $(1\,m, -2\,m)$ to $(2\,m, -2\,m)$. Finally, the third image extends from $(-1\,m, -3\,m)$ to $(-1\,m, -2\,m)$ to $(-2\,m, -2\,m)$.

3. A plane mirror is flat, which means that its radius of curvature is infinite. This means that the focal length of the mirror is also infinite. Whether you consider the focal length to be positive infinity (the limit of a concave mirror) or negative infinity (the limit of a convex mirror) doesn't matter, because in either case the term $1/f$ in the mirror equation will be zero.

5. We can consider the Sun to be infinitely far from the mirror. As a result, its parallel rays will be focused at the focal point of the mirror. Therefore, the distance from the mirror to the paper should be $f = \frac{1}{2}R$.

7. The key to this system is the fact that the lifeguards can run much faster on the sand than they can swim in the water. Therefore, the path ACB—even though it is longer—has a shorter travel time because more time is spent on the sand and less time is spent in the water.

9. In a real image, light passes through the location of the image before reaching the eye. In a virtual image, light propagates *as if* it were coming from the image—though the reflected or refracted light never actually passes through the image location.

11. When the mug is filled with water, light coming upward from the bottom of the mug is bent toward the horizontal when it passes from water to air. If the light had not been bent, it would have passed over your head—placing the bottom of the mug out of sight. With the bending, however, the light can now propagate to your eyes—making the bottom of the mug visible.

13. No. In order to see, a person's eyes must first bring light to a focus, and then must absorb the light to convert it to nervous impulses that can travel to the brain. Both the bending of the light and its absorption would give away the presence of the invisible man. Therefore, if the man were truly invisible, he would be unable to see.

CHAPTER 27

1. When you focus on an object at infinity your ciliary muscles are relaxed. It takes muscular effort to focus your eyes on nearby objects.

3. If you are a distance D in front of a mirror your image is a distance D behind the mirror. Therefore, you can see your image clearly if the distance from you to your image, $2D$, is equal to N. In other words, the minimum distance to the mirror is $D = N/2$.

5. The person with the smaller near-point distance can examine an object at closer range than the person with the larger near-point distance. Therefore, the person with the larger near-point distance benefits more from the magnifier.

7. The image you see when looking through a telescope is virtual. First, the objective forms a real image of a distant object, as shown in Figure 27–16. Next, the eyepiece forms an upright and enlarged image of the objective's image. The situation with the eyepiece is essentially the same as that shown in Figure 26–35 (b). Therefore, it is clear that the final image is virtual in this case.

CHAPTER 28

1. At a point where destructive interference is complete, there is no energy. The total energy in the system is unchanged, however, because those regions with constructive interference have increased amounts of energy. One can simply think of the energy as being redistributed.

3. No. The net signal could be near zero if the waves from the two antennas interfere destructively.

5. In this case, the interference pattern would be reversed. That is, bright fringes in the ordinary pattern would now be the location of dark fringes, and dark fringes would be replaced with bright fringes. For example, in the usual case there is a bright fringe halfway between the slits because the paths lengths are equal. Changing the phase of the light from one slit by 180°, however, results in destructive interference (a dark fringe) at the center of the pattern.

7. A ray of light reflected from the lower surface of the curved piece of glass has no phase change. On the other hand, a ray of light reflected from the top surface of the flat piece of glass undergoes a phase change of half a wavelength. Near the center of the pattern the path difference for these two rays goes to zero. As a result, they are half a wavelength out of phase there, and undergo destructive interference.

9. In general, the larger the aperture in an optical instrument, the greater the resolution. This follows directly from Equation 28–14, where we see that a large aperture diameter D implies a small angle θ. The angular separation that can be resolved is decreased—by making θ smaller—and the resolution is increased.

11. The soap film in the photograph is thinnest near the top (as one might expect) because in that region the film appears black. Specifically, the light reflected from the front surface of the film has its phase changed by 180°; light that reflects from the back surface of the film has no change in phase. Therefore, light from the front and back surfaces of the film will undergo destructive interference as the path length between the surfaces goes to zero. This is why the top of the film, where the film is thinnest, appears black in the photo.

CHAPTER 29

1. The light received from these galaxies moves with the speed c, as is true for all light in a vacuum.

3. Velocities add by simple addition in the limit $c \to \infty$. This is evident from Equation 29–4 when one notices that the second term in the denominator vanishes in this limit, making the denominator equal to 1.

5. Yes. All light, regardless of its wavelength, has the same speed in a vacuum. The *frequency* of this "red shifted" light will be affected, however. Recalling that $v = \lambda f$, we see that a longer wavelength also implies a smaller frequency.

7. Since the total energy of an object of finite mass goes to infinity in the limit that $v \to c$, an infinite amount of energy is required to accelerate an object to the speed c.

CHAPTER 30

1. The "ultraviolet catastrophe" refers to the classical prediction that the intensity of light emitted by a blackbody increases without limit as the frequency is increased.

3. No, all objects with finite temperature give off blackbody radiation. Only an object at absolute zero—which is unattainable—gives off no blackbody radiation.

5. If you look at a painting, photo, decal, or similar object, you are viewing light that it reflects to you. Therefore, when an area appears red, it is because the pigments there absorb blue photons and reflect red photons. Similarly, a blue area absorbs red photons. Blue photons carry considerably more energy than red photons, however, and hence blue photons are more likely to cause damage to the pigment molecules, or to alter their structure. It follows that red pigments (which absorb blue photons) are more likely to become faded when exposed to intense light.

7. If two blackbody curves intersected, there would be a range of frequencies where the low-temperature blackbody gives off more energy than the high-temperature blackbody. In this frequency range, then, it would be possible for energy to be spontaneously transferred from the low-temperature body to the high-temperature body, in violation of the second law of thermodynamics.

9. The likely explanation is that the second metal has a greater work function than the first metal. In this case, a shorter-wavelength photon—that is, a photon with higher frequency and higher energy—would be required to supply the additional energy needed to eject an electron.

11. The resolution of a microscope is determined by the wavelength of the imaging radiation—the smaller the wavelength the greater the resolution. Since the typical wavelength of an electron in an electron microscope is much smaller than the wavelength of visible light, the electron microscope has the greater resolution.

CHAPTER 31

1. Rutherford's alpha particle scattering experiments indicate that the positive charge in an atom is concentrated in a small volume, rather than spread throughout the atom, as in Thomson's model.

3. In the Bohr model, the electron orbits at a well-defined radius; in the quantum mechanical model, the electron can be found at virtually any distance from the nucleus.

5. **(a)** The glass tube of a neon sign contains a low-pressure gas. Therefore, we expect the light from the sign to be in the form of a line spectrum. **(b)** The light from an incandescent lightbulb is basically black-

body radiation from a hot object; therefore, its radiation is distributed continuously as a function of frequency.

7. No, there is no upper limit to the radius of a Bohr orbit. In fact, the radius increases as n^2 for $n = 1, 2, 3, \ldots$.

9. No, the energy does not increase without limit. The energy of a given level in hydrogen ranges from a low of -13.6 eV to a maximum of 0.

11. **(a)** The angular momentum in the quantum mechanical model of the hydrogen atom is zero if the quantum number ℓ is zero. In the $n = 1$ state, the only allowed value for ℓ is 0, and hence the angular momentum must be zero for $n = 1$. **(b)** Yes. For $n > 1$, there are n allowed values for ℓ. One of these values is always zero, therefore the angular momentum can be zero for any value of n.

13. These elements all have similar configurations of their outermost electrons. In fact, the outermost electrons in fluorine, chlorine, and bromine are $2p^5$, $3p^5$, and $4p^5$, respectively. Therefore, each of these atoms is one electron shy of completing the p subshell. This accounts for their similar chemical behavior.

CHAPTER 32

1. The radius of a nucleus is given by the following expression:
$r = (1.2 \times 10^{-15} \text{ m})A^{1/3}$. Therefore, the radius depends only on the total number of nucleons in the nucleus, A, and not on the number of protons and neutrons separately. If the number of protons plus the number of neutrons is the same for two nuclei, their radii will be equal as well.

3. No. An alpha particle contains two protons, whereas any form of hydrogen contains only a single proton. Therefore, hydrogen cannot give off an alpha particle.

5. A nucleus that contained more than one proton and no neutrons would be unstable because the electrostatic repulsion between the protons would blow the nucleus apart. Neutrons tend to push the protons farther apart, reducing their mutual repulsion, and at the same time add more to the attractive strong nuclear force that holds nuclei together.

7. Carbon-14 dating is useful for objects that are of biological origin and—at most— are on the order of thousands of years old. **(a)** It is useful for dating such things as human or animal remains, plant tissue, clothing, and charcoal from a fire. **(b)** It is not useful for dating inorganic materials, like rocks and minerals, or biological materials that are millions of years old, like dinosaur fossils.

9. The obsidian arrowhead cannot be dated with carbon-14, because it is not of biological origin.

11. Yes. The two samples may contain different quantities of the radioactive isotope, and hence their activities may be different.

13. No. The RBE is related to the amount of biological effect produced by a given type of radiation, not to the amount of energy it delivers.

Answer to Odd-Numbered Problems and Conceptual Exercises

Note: In cases where an ambiguity might arise, numbers in this text are assumed to have the smallest possible number of significant figures. For example, a number like 150 is assumed to have just two significant figures—the zero simply indicates the location of the decimal point. To represent this number with three significant figures, we would use the form 1.50×10^2.

CHAPTER 19

1. (a) decrease
   (b) I
3. (a) negative
   (b) toward
5. -7.8×10^{-12} C
7. -1×10^6 C
9. 0.178 C
11. D < A < C < B
13. the center
15. C < A < B
17. 1.37 m
19. 4.7×10^{-10} m
21. 3.0 electrons
23. $(63 \text{ N})\hat{x}$
25. (a) $F_1 = F_3 < F_2$
   (b) 0°
   (c) 150°
   (d) 300°
27. 0.12 m
29. 5.5 km
31. 174.6°; 4.2 N
33. (a) 35 cm (b) No. The forces would reverse direction but would still balance.
35. (a) 248°; 58 N (b) The direction would not change but the magnitude would be cut to a fourth.
37. (a) greater than (b) 3.09×10^6 m/s
39. (a) 3.7×10^{-7} C (b) no
   (c) The tension will be zero.
41. $q_1 = 9q_2$
43. (a) 6.74×10^4 N/C (b) 1.69×10^4 N/C
45. (a) $(-3.0 \times 10^7 \text{ N/C})\hat{x}$
   (b) $(5.9 \times 10^7 \text{ N/C})\hat{x}$
47. (a) $(-3.3 \times 10^4 \text{ N/C})\hat{y}$
   (b) $(9.81 \text{ m/s}^2)\hat{y}$
49. 3.5 pC
51. (a) configuration (2)
   (b) case 1: $\vec{E}_{1,\text{net}} = 0$;
   case 2:
   $$\vec{E}_{2,\text{net}} = \left(-\frac{4\sqrt{2}\,kq}{a^2}\right)\hat{y}$$
53. (a) positive (b) 5.00 μC (c) 5.00 μC
55.

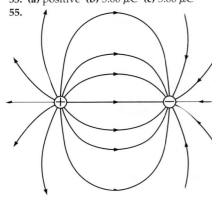

57.

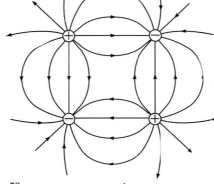

59.

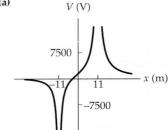

61. (a) the same as (b) II
63. D < C < B < A
65. 6.8×10^5 N·m²/C
67. 6.2×10^{-6} C/m²
69. (a) -1.14×10^5 N·m²/C
   (b) 1.46×10^5 N·m²/C
   (c) 3.28×10^5 N·m²/C
   (d) -2.90 μC
71. (a) greater than (b) II
73. the same as
75. negative
77. $+2q$ and $+q$
79. 8.4×10^8 electrons
81. (a) $-Q$ (b) $+Q$ (c) zero
83. 1.2×10^{25} N; 10^{22} times stronger
85. (a) (ii) to the left of $x = 0.30$ m
   (b) 0.297 m
87. (a) 9.39×10^{-7} C/m (b) 0.954 m
89. (a) 0.55 N; $-\hat{y}$ direction
   (b) greater than
   (c) $(-4.4 \text{ N})\hat{y}$
91. (a) 6.7×10^5 N/C (b) stay the same
93. (a) 5.71×10^{13} C (b) no change
95. 0.254 m
97. 8.85×10^{-6} C/m²
99. (a) 1.3×10^3 N/C (b) 5.3×10^{-2} N

101. (a) 4.13×10^3 N/C (b) 6.22×10^6 m/s
103. B. 5.81×10^8 electrons
105. C. 4.4 cm
107. (a) greater than (b) less than
   (c) 1.83×10^5 N/C (d) 11.8°
109. (a) greater than (b) -4.50 μC

CHAPTER 20

1. increasing
3. (a) 0 (b) -4.1×10^6 V
   (c) -4.1×10^6 V
5. 2.43×10^6 V/m
7. (a) 90 V (b) 18 V
9. 2×10^4 V
11. (a) 1.9 kV (b) increase (c) 3.8 kV
13. 110 m/s
15. (a) region 4, region 4
   (b) 1, 13 V/m; 2, 0; 3, -5.1 V/m;
   4, 68 V/m
17. (a) decrease (b) II
19. 9.4×10^7 m/s
21. (a) negative x direction (b) 3.8 cm/s
   (c) less than
23. (a) $-\sqrt{2}\ Q$ (b) negative
25. (a) $-Q$ (b) positive (c) negative
27. (a) -2.2×10^4 V (b) -2.2×10^4 V
   (c) -1.5×10^4 V
29. 1.34 cm
31. (a)

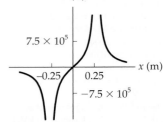

   (b) closer to the negative charge
   (c) -3.3 m
33. (a)

   V (V)

   7.5×10^5

   -0.25 0.25 x (m)

   -7.5×10^5

   (b) 0.21 m and 0.29 m
35. (a) 20.0 m/s (b) greater than (c) 28.3 m/s
37. (a) 0.86 J (b) less than (c) -0.54 J
39. (a) 76.7 kV (b) 14.1 m/s
41. $-(4 - \sqrt{2})\,(kQ^2/a)$
43. (a) greater than (b) III

45. (a)

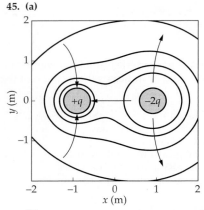

(b)

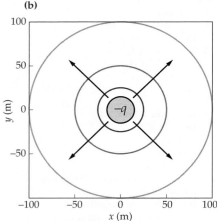

47. (a) to the left
(b) A, positive; B, positive; C, positive; D, negative; E, negative
(c) E < D < A < C < B
(d) less than
49. (a) 559 V/m; 243° **(b)** 8.95 mm
51. 1.8 V
53. 0.18 μF
55. (a) 19 kV **(b)** decrease **(c)** 9.7 kV
57. (a) 1.87 μm **(b)** 6.9 μm
59. 15 kV
61. (a) 4.0 nF **(b)** 6.6 C
63. 35 μJ
65. (a) 5.8×10^{-14} J **(b)** decrease
67. 10.6 J/m^3
69. (a) 0.29 C **(b)** 48 J
71. decreasing
73. (a) increase **(b)** increase
75. (a) decrease **(b)** decrease
(c) decrease **(d)** decrease
77. (a) remain the same **(b)** increase
(c) increase **(d)** increase
79. The capacitance is cut to a fourth.
81. (a) $x = -0.5$ m **(b)** region 3 **(c)** $x = -4.5$ m
83. 13.6 eV
85. (a) smallest at C, greatest at A
(b) A, 43 kV; B, 31 kV; C, 29 kV; D, 31 kV
87. (a) −6.96 J **(b)** 20.1 m/s
89. -5.9×10^{-14} J
91. (a) positive **(b)** -1.7×10^{-16} C
93. (a) from one end of its body to the other
(b) 5.3×10^{-9} C
95. (a) increase **(b)** 0.071 mm
97. (a) directed into the cell; 1.2×10^7 N/C
(b) 97 mV; outer wall

99. 3.58×10^{-14} m
101. (a)

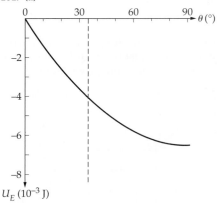

U_E (10^{-3} J)

(b)

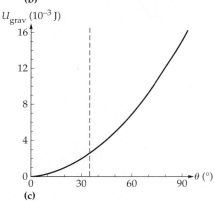

U_{grav} (10^{-3} J)

(c)

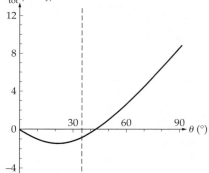

U_{tot} (10^{-3} J)

103. $(22\text{ kV})\left(1 - \dfrac{1}{\sqrt{5 + 4\cos\theta}}\right)$
105. 6.92 cm; 1.19 nC
107. B. 7.6×10^{-12} F
109. C. 1.6×10^{-6} J
111. (a) 1.50 m **(b)** −1.50 m

CHAPTER 21
1. 3600 C
3. 9.4×10^{19} electrons/s
5. 6.25×10^4 electrons/s
7. (a) 1.5 kC **(b)** 8.5 y
9. material A
11. 1/3
13. 51 Ω
15. 0.68 kΩ
17. (a) 2.5 mV **(b)** increase
19. (a) 7.2×10^{-13} A
(b) decrease by a factor of 2

21. (a) 0.11 Ω/m **(b)** decrease
(c) 0.074 Ω/m
23. $\left(\dfrac{C}{A}\right)^2 I_{AB}$
25. (a) bulb A **(b)** 4
27. 51 A
29. 0.58 kW
31. $0.072/kWh
33. 155-minute reserve capacity
35. (a) in series **(b)** III
37. (a) decrease **(b)** III
39. 6 resistors
41. (a) 3.1 W **(b)** 1.5 kW
43. (a) 71 mA
(b) $V_{42\,\Omega} = 3.0$ V; $V_{17\,\Omega} = 1.2$ V;
$V_{110\,\Omega} = 7.8$ V
45. (a) 29 V **(b)** $I_{65\,\Omega} = 0.45$ A;
$I_{25\,\Omega} = 1.2$ A; $I_{170\,\Omega} = 0.17$ A
47. 0.16 kV
49. 0.84 Ω
51. (a) $I_{7.1\,\Omega} = 0.29$ A, $I_{3.2\,\Omega} = 1.1$ A
(b) 1.4 A **(c)** 11.3 V
53. (a)

R (Ω)	1.5	2.5	4.8	3.3	8.1	6.3
I (A)	6.0	3.6	0.38	0.55	0.22	1.1

(b) less than
55. (a) 129 V **(b)** decrease
57. (a) stay the same **(b)** I_0
59. (a) decrease **(b)** 0.12 A ; clockwise
61. (a) $I_{11\,\Omega} = 0.92$ A, $I_{6.2\,\Omega} = I_{12\,\Omega} = 0.27$ A,
$I_{7.5\,\Omega} = 0.65$ A **(b)** same as (a)
63. (a) $I_{9.8\,\Omega} = I_{3.9\,\Omega} = 0.72$ A, $I_{1.2\,\Omega} = 1.8$ A,
$I_{6.7\,\Omega} = 1.0$ A
(b) greater than **(c)** 2.2 V
65. (a) C_1 **(b)** C_2
67. (a) increase **(b)** II
69. 1.1 V
71. (a) B < A < C **(b)** B < A = C
73. (a) 23 μF **(b)** the 15-μF capacitor
(c) $Q_{7.5\,\mu F} = 110$ μC, $Q_{15\,\mu F} = 230$ μC
75. 2.56 μF
77. 6.47 V
79. (a) 2.6×10^{-4} C **(b)** 28 mA
81. (a) 9.75 ms **(b)** 668 μC **(c)** 68.6 mA
83. 6.1 kΩ
85. (a) 6.7×10^{-4} s **(b)** 1.4 A **(c)** increased
87. charge
89. (a) increased **(b)** $\sqrt{2}$
91. (a) R_2 **(b)** R_1
93. (a) increase **(b)** III
95. (a) increase **(b)** stay the same
97. (a) increase **(b)** II
99. Connect the 146-Ω and 521-Ω resistors in series. Then connect this pair in parallel with the 413-Ω resistor.
101. 3.1×10^{-8} A
103. (a) A = B = C **(b)** A < C < B
105. 0.53 V
107. (a) 0.91 J **(b)** 4.0 min
109. (a) greater than **(b)** 0.82 A **(c)** 0.54 A
111. (a) $R_1 = 18$ Ω, $R_2 = 62$ Ω
113. $R/9$
115. (a) greater
(b) $I_{45\,\Omega} = 0.27$ A, $I_{35\,\Omega} = I_{82\,\Omega} = 0.103$ A
117. (a) 2.4 W/m **(b)** 2.0 W/m

119. (a) $P_{13\,\Omega} = 7.7$ W, $P_{6.5\,\Omega} = 3.8$ W,
     $P_{24\,\Omega} = 9.4$ W; all zero as $t \to \infty$
     (b) 0.38 mC (c) 7.0 mJ (d) It quadruples.
121. 44 V, 43 Ω
123. 7.50 Ω
125. A. $1.25 \times 10^7 \,\Omega$
127. increase
129. (a) 329 Ω (b) 794 Ω
131. (a) 273 μC (b) decrease (c) 58.7 ms

CHAPTER 22
  1. (a) less than (b) II
  3. positive z direction
  5. A, negative; B, negative; C, positive
  7. $9.9 \times 10^8 \,\text{m/s}^2$
  9. 0
 11. (a) 81° (b) 38° (c) 1.2°
 13. 4.4×10^{-16} N
 15. (a) particle 2 (b) 1/4
 17. (a) $(5.0 \times 10^6 \,\text{N/C})\hat{\mathbf{x}}$
     (b) $(-0.2\,\text{T})\hat{\mathbf{z}}$
 19. 5.4 μm
 21. 2.5 km/s
 23. (a) 1.1 m/s (b) bottom electrode; no
 25. (a) 9.83 m/s
     (b) 13.9 s
 27. 3.1 cm
 29. (a) 1.00 (b) 0.0233
 31. 3.3 N
 33. $(-0.34\,\text{T})\hat{\mathbf{z}}$
 35. 2.4 A
 37. $\tan^{-1}\dfrac{ILB}{mg}$
 39. 3.5 A
 41. 60°
 43. (a) less than (b) $\pi/4$
 45. 2.50×10^{-5} T
 49. 1.3 kA
 51. (a) point B
     (b) point A, 2.1 μT; point B, 13 μT
 53. (a) 2.57×10^{-5} N/m (b) the same as
 55. terminal A
 57. 35 mA
 59. 17.2 T
 61. east
 63. out of the page
 65. greater than $\theta = 45°$ and
     less than $\theta = 90°$
 67. toward wire 3
 69. 0
 71. 0.946
 73. 7.0 μm
 75. (a) $F_3 < F_2 = F_4 < F_1$ (b) $\vec{\mathbf{v}}_3$
 77. (a) $B_3 < B_1 < B_2$
     (b) B_1, out of the page; B_2, into the page;
     B_3, out of the page
 79. (a) stay the same (b) III
 81. (a) toward the wire (b) 3×10^{-5} N
 83. $(-2.0 \times 10^3 \,\text{N/C})\hat{\mathbf{x}} + (3.2 \times 10^3 \,\text{N/C})\hat{\mathbf{y}}$
 85. 2.3 mN; 65° measured from the positive z
     axis toward the negative y axis in the yz
     plane
 87. (a) 4.52×10^7 C/kg (b) less than
 89. $(-4\,\mu\text{T})\hat{\mathbf{z}}, (12\,\mu\text{T})\hat{\mathbf{z}}, (-4\,\mu\text{T})\hat{\mathbf{z}}$
 91. (a) less than (b) 1.2 A; to the left
 93. (a) 1.6×10^{11} N (b) The force from the
     magnetar is 280,000 times greater than
     the electron-proton force within a hydro-
     gen atom.

 95. (a) negative y axis (b) 43.0 cm
 97. (a) 4.3 mT (b) 230 A
 99. (a) 2.6 kT/s (b) increase by a factor of $\sqrt{2}$
101. $\frac{1}{2}\mu_0 \lambda \omega$
103. (a) clockwise (b) $\dfrac{2I}{3\pi}$
105. 64 mT
107. C. 2.5×10^{-10} A
109. A. 22 cm
111. (a) increase (b) 2 cm
113. 5.9 A to the right

CHAPTER 23
  1. 1.6×10^{-4} Wb
  3. 0.020 T
  5. 1.9 Wb
  7. (a) 11.7 A
     (b) The current would be cut to a fourth.
  9. 14 V
 11. (a) -0.1 kV (b) 0 (c) 0.04 kV
 13. 1, counterclockwise; 2, zero; 3, zero;
     4, clockwise
 15. (a) near $t = 0.5$ s (b) 0.1 s, 0.3 s, 0.5 s, . . .
     (c) -0.06 kV, 0, 0.06 kV
 17. 7.1 V
 19. 3.8×10^3
 21. (a) location 1, upward; location 2, zero;
     location 3, upward (b) III
 23. minimum
 25. counterclockwise
 27. (a) less than (b) less than
 29. (a) zero (b) clockwise
 31. (a) zero (b) zero (c) no
 33. bottom
 35. no
 37. 47 mV
 39. (a) 52.3 mN (b) 0.300 W (c) 0.300 W
 41. (a) 0.86 A (b) 4.1 m/s
 43. 33 mT
 45. (a) horizontal (b) 29.3 mV
 47. -1.40 V
 49. (a) $6.9 \times 10^{-3} \,\text{m}^2$ (b) 0.42 V
 51. -15.5 mV
 53. 3.6×10^{-4} s
 55. (a) 4.0×10^{-4} s (b) 57 mA (c) 65 mA
 57. (a) 3.8 H (b) 0.25 s (c) 1.6 A
 59. 8.2 mJ
 61. (a) $9.95 \times 10^8 \,\text{J/m}^3$ (b) 1.50×10^{10} V/m
 63. (a) 0.075 J (b) 0.14 J (c) decrease
 65. (a) 57 A (b) 0.15 T (c) $8.7 \,\text{kJ/m}^3$
 67. $6\,I_s$
 69. (a) less than (b) 9.4 turns
 71. 92
 73. 0.36 A; 0.16 kV
 75. less than
 77. either increasing and to the right or
     decreasing and to the left
 79. 4.7×10^{-10} Wb
 81. 4.0 mWb
 83. 0.29 V
 85. 0.15 kV
 87. 339 V
 89. 0.990 km
 91. (a) 66 μs (b) 43 μJ
 93. 1.1 μV
 95. (a) zero (b) $-vWB$ (c) zero
     (d) parts (a) and (c), zero; part (b),
     counterclockwise

 97. (a) $\dfrac{1}{\sqrt{\varepsilon_0 \mu_0}}$ (b) 3.00×10^8 m/s
 99. C. 2.1×10^{-4} V
101. B. 0.11 s
103. (a) zero (b) zero (c) counterclockwise
     (d) to the left
105. (a) to the right (b) clockwise (c) 2.2 T

CHAPTER 24
  1. 39 V
  3. 81 Ω
  5. (a) 2.99 W (b) 5.97 W
  7. $V_{\text{rms}} = V_{\text{avg}}$
  9. 86.9 Hz
 11. (a) 5.9 μA (b) 8.3 μA
 13. (a) 9.0 V (b) -6.7 V (c) 6.7 V
 15. (a) 1.16 kΩ (b) 0.137 μF
     (c) 0.860 mA (d) 4.3 mA
 17. $0 \le f < 60$ Hz
 19. 53.4 Ω
 21. 0.10 kΩ
 23. (a) 74.6 Hz (b) 4.61 W
 25. 0.967
 27.

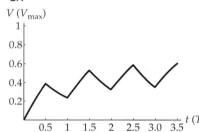

 29. 45.2 Ω
 31. 22 V
 33. (a) 79 Ω (b) 21 V (c) 14 V
 35. (a) 12.2 A (b) 0.500 (c) 6.09 A
 37. 29 mH
 39. (a)

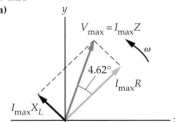

     (b) 136 W
 41. (a) 17 Ω (b) 14 A (c) 1.5 kW
 43. parallel
 45. (a) the same as (b) I
 47. (a) greater than (b) III
 49. 1.51 kΩ
 51. (a) 10 A (b) 5.0 A
 53. (a) 0.962 (b) increase (c) 0.998
 55. (a) 0.173 V (b) 7.84 V (c) 1.84 V
     (d) greater than
 57. 7.9 kΩ or 5.3 kΩ
 59. (a) $T/4$ (b) $T/2$
 61. decrease
 63. 43 Ω
 65. (a) decreased (b) decreased
 67. increased
 69. (a) 494 Hz (b) 180 Ω (c) 0.58
 71. (a) 9.7 pH (b) 10 Ω

73. 27 mJ
75. **(a)** the same as **(b)** I
77. **(a)** increase **(b)** II
79. 572 mA
81. 66.3 Ω
83. **(a)** greater than **(b)** 137 Hz **(c)** 175 Ω
85. 26 ms
87. 4.16 W
89. **(a)** 606 Ω **(b)** increase
91. **(a)** 29 Ω **(b)** 71 μF **(c)** decrease
93. **(a)** 535.6 Hz
    (b) decreased
95. **(a)** 0.10 kW **(b)** 0.55 H
97. **(a)** 1.3 pF **(b)** increase **(c)** 5.0 Ω **(d)** 15 Ω
99. $R = 76\ \Omega, C = 99$ nF
101. **(a)** 0.24 kHz **(b)** 93 V **(c)** 76 V
103. decrease
105. **C.** 8.06 A
107. **(a)** increase **(b)** 87.2 Hz **(c)** 0.634 A

CHAPTER 25
1. increasing
3. **(a)** x direction **(b)** z direction
   (c) positive y direction
5. **(a)** up **(b)** down **(c)** east **(d)** west
7. **(a)** zero
   (b) $(-9.57 \times 10^{-9}\ \text{T})\hat{\mathbf{x}} + (2.07 \times 10^{-8}\ \text{T})\hat{\mathbf{y}}$
9. case 1, positive x direction;
   case 2, positive z direction;
   case 3, negative x direction
11. 4.1×10^{16} m
13. 1.14
15. **(a)** away from **(b)** $0.12c$
17. 3.00×10^8 ms
19. the father
21. 4.945×10^{14} Hz
23. 2.82 kHz
25. **(a)** 8.238×10^{14} Hz
    (b) 8.203×10^{14} Hz
27. 1.0×10^{18} Hz
29. 1.7×10^3 waves
31. 1.71 m
33. **(a)** 9.38×10^{14} Hz to 1.07×10^{15} Hz
    (b) UV-A
35. **(a)** decrease **(b)** $\dfrac{3}{4}$
37. 97 s
39. 93 m
41. **(a)** 0.2 km **(b)** 2 m
43. **(a)** factor of 2 **(b)** factor of 4
45. 1.33×10^{-10} T
47. **(a)** 1.1 kV/m **(b)** 3.3 kW/m^2
    (c) 1.6 kW/m^2
49. 61.4 V/m
51. **(a)** 83 mW/m^2 **(b)** 0.83 mW/m^2
53. 6.6×10^{-4}
55. 1.7×10^{-11} J
57. 2.3 mN
59. **(a)** NIF **(b)** NIF **(c)** NIF
61. **(a)** 0.2 mJ **(b)** 3.8 kW/cm^2 **(c)** 3.8×10^5
63. **(a)** 0.28 kW/m^2 **(b)** 4.6 cm
65. **(a)** 34 mJ **(b)** 2.1 μPa
67. **(a)** 1.56×10^{12} W **(b)** 4.95×10^{17} W/m^2
    (c) 1.37×10^{10} V/m
69. **(a)** greater than **(b)** III
71. **(a)** green **(b)** red **(c)** blue
73. 0.134
75. $\cos^{-1}\dfrac{1}{\sqrt{5}}$

77. **(a)** C $<$ A $=$ B
    (b) case A, 9.25 W/m^2; case B, 9.25 W/m^2;
    case C, zero
79. **(a)** d-glutamic acid
    (b) l-leucine, 0.576 mW/m^2; d-glutamic
    acid, 0.732 mW/m^2
81. **(a)** $0.375 I_0 \cos^2(\theta - 30.0°)$
    (b) $I_{\max} = 0.375 I_0$
    (c) 30.0° or 210.0°
83. reflecting
85. 2.00×10^{-11} m
87. 0.40 km
89. **(a)** 6 MW/m^2 **(b)** less than **(c)** 1 MW/m^2
91. 1.2×10^7 m/s away from Earth
93. **(a)** Both arms show a red shift.
    (b) 8.149×10^{14} Hz
    (c) 8.114×10^{14} Hz
95. **(a)** 58.0° **(b)** 593 W/m^2
97. 6.6
99. **(a)** 1.0 mW/m^2 **(b)** 0.71 kN/C
101. **(a)** 1.08 W/m^2 **(b)** 543 W/m^2
103. **(a)** less than
    (b) $I_u = 9.4$ W/m^2; $I_p = 12.1$ W/m^2
105. 40 m^2
107. **(a)** 0.95 kW/m^2 **(b)** 1.9 kW/m^2
    (c) 3.2 μJ/m^3 **(d)** 1.0×10^{-11} N
    (e) The laser beam must be normal to the
    plane of the mirror.
109. **(a)** 50% **(b)** 0% **(c)** 63.4° **(d)** $0.138 I_0$
111. **C.** 6.45×10^{14} Hz
113. **A.** 3.25×10^{-5} N/m^2
115. 40°

CHAPTER 26
1. 38°
3. 2θ
5. 55 cm
7. **(a)** 36 cm below **(b)** 8.9°
   (c) You will still see the buckle.
9. counterclockwise
11. **(a)** 5.2 m/s **(b)** 4.1 m/s
13. 33 cm wide and 7.5 cm high
15. 23 cm
17. concave
19. -7.98 cm
21. **(a)** upright **(b)** reduced **(c)** virtual
23. $R/2$
25. toward the right
27. increases
29. 0.67 m; -0.33
31. -0.40 m; 0.20
33. **(a)** concave **(b)** 7.34 m **(c)** 16.4 m
35. **(a)** -1.9 **(b)** inverted **(c)** 1.2 m
37. **(a)** -84 cm **(b)** 42 cm
39. **(a)** -9.0 cm **(b)** -12 cm
41. 12 m
43. $d_o = 1.4$ m; $d_i = 0.48$ m
45. **(a)** point 1 **(b)** point 4
47. **(a)** equal to **(b)** greater than
49. medium A
51. 1.82
53. 1.5
55. **(a)** greater than **(b)** 55°
57. 14.4 ns
59. 21.6°
61. 4.7 ft
63. $1.71 \le n_{\text{prism}} < 2.02$
65. **(a)** 72° **(b)** No
67. 1.4

69. 30.5°
71. **(a)** $d_i \approx \frac{2}{3}|f|$ **(b)** upright **(c)** virtual
73. **(a)** $d_i \approx 2f$ **(b)** inverted **(c)** real
75. **(a)** The final image is located to the right
    of lens 2, just beyond F_2.
    (b) inverted **(c)** real
77. $d_i = -15$ cm, $m = 0.52$
79. **(a)** 0.98 m **(b)** 68 cm
81. **(a)** farther away **(b)** 2.8 cm
83. **(a)** 34 cm **(b)** -1.2
85. **(a)** farther away **(b)** 34 cm
87. 2.8 cm
89. **(a)** virtual **(b)** 69 cm **(c)** convex
91. 0.33°
93. 46.2 cm from the lens
95. $R/2$
97. **(a)** decrease **(b)** II
99. liquid B
101. **(a)** converge **(b)** diverge
103. 0.25 m
105. **(a)** 29 cm and 9.8 cm
    (b) 29 cm, real and inverted; 9.8 cm, vir-
    tual and upright
107. **(a)** 1.8 **(b)** No
109. **(a)** 40.6°
    (b) The answer to part (a) does not de-
    pend on the thickness of the oil film.
111. **(a)** 48.8° **(b)** No
113. **(a)** 1.69 m **(b)** 80.3 m
115. **(a)** 0.286 **(b)** 0.082
117. $\dfrac{f_1 f_2}{f_1 + f_2}$
119. $(t \sin\theta)\left(1 - \dfrac{\cos\theta}{\sqrt{n^2 - \sin^2\theta}}\right)$
121. **(a)** 506 ns **(b)** 510 ns
123. **(a)** 13.3° **(b)** No **(c)** 2.24
125. **B.** 105 cm
127. **B.** 134 cm
129. **(a)** 3.0 cm **(b)** -0.49 **(c)** increase
131. **(a)** 5.0 cm **(b)** -2.5 cm **(c)** increase

CHAPTER 27
1. **(a)** farther from **(b)** II
3. flower
5. **(a)** 2.58 cm **(b)** 2.38 cm
7. **(a)** 8.5 **(b)** 2.4
9. 3.0 m
11. 87 mm
13. **(a)** 1/25 s **(b)** 5.6
15. **(a)** 3.33 **(b)** 30.6
17. farsighted
19.

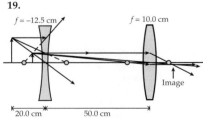

21. 0.80 m
23. 60.7 cm
25. **(a)** 5 mm in front of the lens closest to the
    object
    (b) 0.15
27. **(a)** increase **(b)** 3.5 mm
29. **(a)** 12 cm to the left of the converging lens
    (b) -0.12
31. -135 cm

33. 8.4 cm
35. 45 cm; 2.2 diopters
37. (a) nearsighted (b) diverging
    (c) −3.34 diopters
39. (a) 45.8 diopters (b) decrease
41. 24.2 cm
43. (a) diverging (b) converging
    (c) distant objects: −0.20 diopters; near
    objects: +2.3 diopters
45. far: 2.27 m; near: 42.0 cm
47. (a) 47 cm to the right of lens 2; 0.38
    (b) 3.0 cm to the right of lens 2; −0.61
    (c) 15.8 cm to the right of lens 2; 0.316
49. (a) 9.042×10^{-3} rad (b) 2.1 m
51. (a) f_1 (b) $M_1 = 6.0$; $M_2 = 2.9$
53. (a) 6.47 cm (b) 3.96
55. 3.72
57. lens 1
59. 3.3×10^{-3} rad
61. 0.324 mm
63. 9.1 mm
65. (a) 7.85 mm (b) 10 cm (c) 5.1 cm
67. (a) 1600 mm should be the objective
    (b) 50×
69. 2.8 m
71. 24 cm
73. 0.040
75. (a) −57 cm (b) 1.1 m
77. 1.8 cm
79. (a) decrease (b) II
81. microscope
83. (a) positive (b) I
85. (a) 41.7 diopters (b) increase
87. (a)

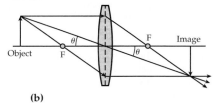

(b)

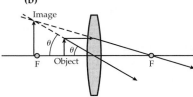

(c)

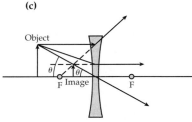

(d) A simple magnifier helps the eye view
an object that is closer than the near point,
thus making it appear larger.
89. (a) 2.42 cm (b) 2.42 cm
91. 4.6 cm
93. (a) 9.7 cm (b) increase
95. (a) 8.3 cm from the lens on the same side as
    the virtual object
    (b) real
97. 9.24 cm
99. 16 cm

101. (a) 12.2 cm to the right of the lens
     (b) virtual (c) 0.174 (d) upright
103. $\dfrac{f_1}{f_2}$
105. 19.8 cm; −63.0 cm
107. positive
109. A. 1.9 mm
111. (a) greater than (b) −351 cm
113. (a) greater than (b) 2.81 diopters

CHAPTER 28

1. destructively
3. 134 m
5. 0.18 kHz
7. 406 Hz; 812 Hz
9. 0.68 kHz; 2.0 kHz
11. (a) 600 m (b) maximum (c) 200 m
13. (a) increase (b) $\Delta\ell = m\lambda$
15. (a) increase (b) decrease
17. 623 nm
19. (a) 415 nm (b) increase (c) 587 nm
21. 154 μm
23. 658 nm
25. 86.9 cm
27. (a) 2.24 cm (b) increased
29. greater than
31. 488 nm and 627 nm
33. 493 nm and 658 nm
35. (a) destructive interference
    (b) 3.5 m
37. (a) destructive interference
    (b) constructive interference
    (c) constructive interference
39. (a) 102 nm (b) thinner
41. (a) 503.2 nm (b) 670.9 nm and 402.6 nm
43. (a) 2.00 (b) 0.500 (c) 2.00
45. 36°
47. 24.8 cm
49. (a) 12 μm (b) decrease
51. (a) 25.0 μm (b) less than
53. (a) dark cloudy day (b) I
55. (a) 0.78 m (b) 1.3 m
57. 5.5 cm
59. 17 cm
61. (a) 0.91 arc sec (b) 1.7 km
63. 2.95°, 5.92°, 8.90°
65. 3.4°
67. 371.3 cm^{-1}
69. one
71. 660 nm
73. (a) The $m = 3$ maximum of the 420-nm
    light overlaps the $m = 2$ maximum of the
    630-nm light.
    (b) 35°
75. 3.94 m
77. (a) outward (b) $W = \lambda$
79. decrease
81.

θ(°)	θ(rad)	$\sin\theta$	$\tan\theta$	$\sin\theta/\tan\theta$
0.0100°	0.000175	0.000175	0.000175	1.00
1.00°	0.0175	0.0175	0.0175	1.00
5.00°	0.0873	0.0872	0.0875	0.996
10.0°	0.175	0.174	0.176	0.985
20.0°	0.349	0.342	0.364	0.940
30.0°	0.524	0.500	0.577	0.866
40.0°	0.698	0.643	0.839	0.766

83. 24.4 km
85. (a) The bubble is thick enough for con-
    structive interference but not destructive
    interference at visible wavelengths.
    (b) 108 nm
87. 76 μm
89. 491 nm
91. (a) 496 nm (b) greater than
93. 401 nm and 515 nm
95. $\dfrac{2d}{\lambda} - \dfrac{1}{2} = m$ where $m = 0, 1, 2, \ldots$
97. 5.9 μm
99. C. 0.369 mm
101. C. 11.5 ft
103. (a) longer than (b) 0.63 μm
105. (a) increase (b) 15.1°

CHAPTER 29

1. (a) greater than (b) I
3. (a) proper time (b) proper time
   (c) proper time (d) dilated time
5. (a) according to the clock on Earth (b) I
7. 0.87c
9. 0.99986c
11. 748 m
13. 0.80c
15. 0.17 rad/s
17. (a) less than (b) 55 beats/min
19. 1.01 ms
21. 4:32 P.M.
23. less than
25. 1.13 m
27. 49 m
29. (a) 0.60c (b) 3.2 m
31. (a) less than (b) 0.583c (c) 2.84 cm
33. 72°
35. (a) (i) the *Picard* is longer (b) 2.3
37. 0.879c
39. 31 mi/h
41. 0.98c
43. 0.64c
45. (a) No (b) 0.98c
47. 2.86×10^8 m/s
49. 1.6 m/s
51. 9.4×10^7 kg
53. 0.999c
55. 0.19 nJ
57. 4.9×10^{-16} kg
59. 1.02 MeV
61. 0.88c
63. 0.35 kg
65. (a) 0.75% (b) 69%
67. (a) 43.5 J (b) greater than
    (c) 7.00×10^8 J
69. 2.4 kg
71. (a) 7.7×10^6 km
    (b) 5.8×10^6 m/s^2
    (c) decreases by a factor of 2
73. (a) the pilot (b) you (c) the pilot (d) you
75. greater than
77. greater than
79. 1.00c
    (b) $K_{\text{ant}} = 3.6 \times 10^9$ eV$\ll 1.0 \times 10^{20}$ eV
81. 1.20×10^{-18} kg·m/s
83. (a) 0.643c (b) 2.29×10^{-22} kg·m/s
    (c) -2.29×10^{-22} kg·m/s (d) 9.85 km/s
85. 0.33c
87. (a) 11 m (b) less than (c) 6.1 m
89. (a) 3.11×10^8 m/s (b) 2.28×10^8 m/s

91. (a) $\frac{1}{2}c$ (b) $\frac{1}{\sqrt{2}}c$
99. C. $0.312c$
101. A. $0.301c$
103. (a) $0.371c$ (b) $0.454c$ (c) $0.922c$

CHAPTER 30

1. (a) lower than (b) II
3. 1.81×10^{13} Hz; $16.6\ \mu\text{m}$
5. 3.4×10^{14} Hz
7. (a) star B (b) 0.36
9. (a) 1.68×10^{14} Hz (b) infrared
11. (a) blue < green < red
    (b) red < green < blue
    (c) blue < green < red
13. (a) greater than (b) I
15. No, because each photon has too little energy to eject an electron.
17. 4.6×10^{32} photons/s
19. 7.3×10^{18} photons/s
21. 2.08×10^{15} Hz
23. (a) 1.5×10^{20} photons/s
    (b) 1.0×10^{12} photons/s
25. (a) 1.09×10^{15} Hz; 275 nm
    (b) ultraviolet
27. (a) red (b) blue
    (c) red: 4.9×10^{20} photons/s;
    blue: 5.8×10^{19} photons/s
29. (a) aluminum
    (b) Al: 1.03×10^{15} Hz;
    Ca: 6.93×10^{14} Hz
31. (a) cadmium (b) Zn: 0.19 eV; Cd: 0.30 eV
33. (a) 369 eV
    (b) 4.00×10^{14} Hz $\le f < 1.03 \times 10^{15}$ Hz
35. 6.377×10^{-34} J·s
37. 2
39. 2.9×10^6 m/s
41. 1.1×10^{18} Hz
43. 0.815 m/s
45. (a) photon B (b) 666 nm
47. (a) 2.39×10^{16} photons/s
    (b) 2.10×10^{-27} kg·m/s
    (c) 5.00×10^{-11} N
49. 4.5 keV
51. 107°
53. (a) The change in wavelength is the same for both photons.
    (b) X-ray
    (c) visible: 9.3×10^{-4} %; X-ray: 16%
55. (a) 0.315 nm
    (b) incident: 3.94 keV; scattered: 3.88 keV
    (c) 60 eV
57. 41.4 pm; 4.80 fJ
59. (a) decrease (b) II
61. 13.7 km/s
63. 2.0×10^{-36} m
65. (a) 1.58 km/s (b) 62.4°
67. (a) electron (b) 1836
69. $\dfrac{h}{\sqrt{2mqV}}$
71. 1.4×10^{-20} kg·m/s
73. 1.1×10^{-26} J
75. 1.8×10^{-25} J
77. (a) 7.0×10^{-25} kg·m/s (b) 1.7 eV
79. 62 nm

81. (a) increase (b) decrease (c) decrease
    (d) decrease
83. (a) decrease (b) decrease
85. 3.65×10^{-17} W
87. (a) 1.6×10^{15} photons/s
    (b) 3.3×10^{-19} N
89. 1×10^{19} photons
91. 0.58 nm
93. 9200 K; No
95. (a) decreased (b) -10.2 eV
97. (a) decrease
99. (a) 1.32 fm (b) 1.51×10^{-10} J
101. D. 2.29 eV
103. B. 1.08 eV
105. (a) less than (b) 54°

CHAPTER 31

1. 8.4×10^{-16}
3. 5.6 pJ
5. 369.7 nm
7. 121.5 nm; 102.6 nm; 97.23 nm
9. (a) 121.5 nm (b) 820.4 nm
11. (a) increase (b) I
13. C < B < A
15. 2.42×10^{-3}
17. 0.544 eV
19. $\left(\dfrac{n}{n+1}\right)^2 U$
21. (a) 1.51 eV (b) -3.02 eV (c) -1.51 eV
23. $n_i = 3$ to $n_f = 2$
25. (a) (iii) $n_i = 7$ to $n_f = 8$; $19.06\ \mu\text{m}$
    (b) (ii) $n_i = 2$ to $n_f = 8$; 388.9 nm
    (c) only (iv), $n_i = 6$ to $n_f = 2$
27. (a) 2.83×10^{-10} m
    (b) greater than
    (c) Li^{2+}, 2.75 eV; H, 0.306 eV
29. (a) 1.22×10^{-15} s (b) 8×10^6 orbits
31. (a) 6 (b) increase (c) -0.555 eV
33. (a) $f_{\text{photon}} = \dfrac{2\pi^2 mk^2 e^4}{h^3}\left[\dfrac{1}{(n-1)^2} - \dfrac{1}{n^2}\right]$
    (b) $f_{\text{electron}} = \dfrac{4\pi^2 mk^2 e^4}{n^3 h^3}$
    (c) They are the same.
35. 0.332 nm
37. 1.32 nm
39. 0, 1, 2, 3, 4
41. (a) 2 (b) none (c) 5 (d) none
43. (a) 75 (b) 76 (c) -2.35 meV
45. (a) greater than (b) $\sqrt{2}$
47. (a) 10 (b) 8
49. $1s^2 2s^2 2p^2$
51. $1s^2 2s^2 2p^3$
53.

n	ℓ	m_ℓ	m_s
3	1	-1	$-\frac{1}{2}$
3	1	-1	$\frac{1}{2}$
3	1	0	$-\frac{1}{2}$
3	1	0	$\frac{1}{2}$
3	1	1	$-\frac{1}{2}$
3	1	1	$\frac{1}{2}$

55. $1s^2 2s^2 2p^6 3s^2 3p^6 3d^8 4s^2$
59. (a) stay the same (b) III
61. 0.167 nm
63. 0.195 nm
65. (a) 80.6 keV (b) 80.6 kV
67. (a) B < A < C (b) C < A < B
69. excited state
71. less than
73. 78,700 K
75. (a) 2.55 eV (b) 5.55×10^{-28} J
    (c) the same as
77. (a) 1.52×10^{-16} s (b) 0.00105 A
79. (a) 22.8 nm (b) $n_i = 16$ and $n_f = 4$
81. 0.0585 nm
83. (a) greater than
    (b) 4.38×10^7 m^{-1}
85. (b) $v_n = \dfrac{1}{m}\sqrt{nqB\hbar}$
89. greater than
91. B. 2.42 eV
93. (a) $n = 12$ (b) 7.62 nm

CHAPTER 32

1. (a) $Z = 92$; $N = 146$; $A = 238$
    (b) $Z = 94$; $N = 145$; $A = 239$
    (c) $Z = 60$; $N = 84$; $A = 144$
3. (a) 7.0 fm (b) 4.7 fm
5. (a) 2.3×10^{17} kg/m^3
    (b) the same as
    (c) 2.3×10^{17} kg/m^3
7. (a) 6.4×10^6 m/s (b) 0.27 pm
    (c) less than
9. (a) 3.7 fm (b) 240
11. (a) more than (b) 7.4 fm (c) 5.9 fm
13. (a) the same as (b) III
15. alpha (α) and beta (β)
17. ^4_2He
19. α decay; β decay; β decay; α decay;
    α decay; α decay; α decay; β decay;
    β decay; α decay; α decay; β decay;
    α decay; β decay
21. (a) $^{212}_{84}\text{Po} \longrightarrow {}^{208}_{82}\text{Pb} + {}^4_2\text{He}$; 8.95 MeV
    (b) $^{239}_{94}\text{Pu} \longrightarrow {}^{235}_{92}\text{U} + {}^4_2\text{He}$; 5.244 MeV
23. (a) $^{18}_{9}\text{F} \longrightarrow {}^{18}_{8}\text{O} + e^+ + v$; 0.634 MeV
    (b) $^{22}_{11}\text{Na} \longrightarrow {}^{22}_{10}\text{Ne} + e^+ + v$; 1.819 MeV
25. (a) $^{66}_{29}\text{Cu}$ (b) 0.2 MeV
27. less than
29. 0.181 d^{-1}
31. 4.5 d
33. (a) 0.115 h^{-1} (b) 1.7×10^9 nuclei
35. 1.97×10^4 y
37. 1.44×10^4 y
39. 3.90×10^3 y
41. 1559 MeV
43. (a) 8.790 MeV/nucleon
    (b) 7.570 MeV/nucleon
45. 15.66 MeV
47. 3 neutrons
49. $^1_0\text{n} + {}^{235}_{92}\text{U} \longrightarrow {}^{88}_{38}\text{Sr} + {}^{136}_{54}\text{Xe} + 12\,{}^1_0\text{n}$;
    126.5 MeV
51. 1.2×10^6 kg
53. 4.033 MeV
55. 5.494 MeV

ANSWERS TO ODD-NUMBERED PROBLEMS AND CONCEPTUAL EXERCISES **A-29**

57. **(a)** 4.33×10^9 kg/s **(b)** 0.0307%
59. 1.0×10^3 rad
61. **(a)** 0.38 J **(b)** 0.54 mK
63. **(a)** 6.9×10^{11} electrons **(b)** 0.078 J
 (c) 93 rem
65. 2
67. greater than

69. **(a)** $^{206}_{80}$Hg **(b)** $^{239}_{93}$Np **(c)** $^{11}_{5}$B
71. 260 rem
73. **(a)** 2.83×10^9 y **(b)** 1.16×10^9 y
75. $^{210}_{81}$T1
77. 1.7 Ci
79. **(a)** 1.2×10^{-15} m $\leq r \leq 7.1 \times 10^{-15}$ m
 (b) 35.2 **(c)** 209

81. 2220 y
83. **(a)** less than **(b)** 7.274 MeV
85. **(a)** 4.2×10^5 rad **(b)** stay the same
87. 264 m/s
89. 1.9 mK
91. **B.** 1.2 Ci

Photo Credits

FM—p. ii Donald Sutherland/Getty Images, Inc.

Chapter 1—**p. 1** Getty Images Inc. **p. 2 (top)** Oliver Meckes/Ottawa/Photo Researchers, Inc. **p. 2 (bottom)**NASA **p. 3** Courtesy of BIPM (Bureau International des Poids et Mesures/International Bureau of Weights and Measures, www.bipm.org) **p. 4 (top)** U.S. Department of Commerce **p. 4 (bottom)** © 2000 by Sidney Harris **p. 6** Jeff Greenberg/PhotoEdit **p. 7** AP Wide World Photos **p. 8** David Frazier Photolibrary/Photo Researchers, Inc. **p. 9** CNRI/Science Photo Library/Photo Researchers, Inc. **p. 10** Amaldi Archives, Dipartimento di Fisica, Universita "La Sapienza," Rome, courtesy AIP Emilio Segre Visual Archives **p. 16 (left)** Michael Mancuso/Omni-Photo Communications, Inc. **p. 16 (right)** Matt Moran/USDA Forest Service **p. 17** Young-Jin Son, Drexel University College of Medicine

Chapter 2—**p. 18** AP Wide World Photos **p. 24** Swerve/Alamy Images **p. 26** NASA Headquarters **p. 29** AP Wide World Photos **p. 39 (top)** James Sugar/Stockphoto.com/Black Star **p. 39 (bottom)** Alaska Stock **p. 42** CORBIS-NY **p. 51 (left)** Worlds of Fun Photo by Dan Feicht **p. 51 (right)** Stephen Dalton/Photo Researchers, Inc. **p. 52** Sidney Harris **p. 53** Frank Labua/Pearson Education/PH College

Chapter 3—**p. 57** National Maritime Museum, Greenwich, London **p. 58** Jeff Greenberg/Photo Researchers, Inc. **p. 63** EyeWire Collection/Getty Images-Photodisc **p. 68** George Whiteley/Photo Researchers, Inc. **p. 70** AP Wide World Photos **p. 77** © National Maritime Museum Picture Library, London, England **p. 79** US Department of Defense

Chapter 4—**p. 82** Bjorn Larsen/Alamy Images **p. 87 (left)** Loren M. Winters, North Carolina School of Science and Mathematics **p. 87 (right)** NASA Headquarters **p. 87** Richard Megna/Fundamental Photographs **p. 90 (top)** Krafft/Photo Researchers, Inc. **p. 90 (bottom)** Jerry Driendl/Getty Images Inc.-Taxi **p. 93 (left)** Michael Quinton/MINDEN PICTURES **p. 93 (middle)** John Terence Turner/Alamy Images **p. 93 (right)** Galen Rowell/Mountain Light/Explorer/Photo Researchers, Inc. **p. 97** Courtesy of the Archives, California Institute of Technology **p. 99** Stephen Dalton/Photo Researchers, Inc. **p. 107 (right)** Gregory G. Dimijian/Photo Researchers, Inc. **p. 107 (left)** NASA Headquarters

Chapter 5—**p. 111** © Wally McNamee/CORBIS **p. 113** James S. Walker **p. 115 (middle)** Joe Drivas/Getty Images **p. 115 (bottom)** Bruce Bennett Studios, Inc./Getty Images **p. 120** William J. Hughes Technical Center, Federal Aviation Administration **p. 129** © Pete Saloutos/CORBIS **P. 132** NASA/Johnson Space Center **p. 139** © Norbert Wu/DRK PHOTO **p. 145 (left)** AP Wide World Photos **p. 145 (right)** Mark Moffett/Minden Pictures

Chapter 6—**p. 147** Corbis RF **p. 149 (top)** Ian Aitken © Rough Guides **p. 149 (bottom)** Ernest Braun/Getty Images Inc.-Stone Allstock **p. 153** Jack Dykinga/Getty Images Inc.-Stone Allstock **p. 155 (left)** Thomas Florian/Index Stock Imagery, Inc. **p. 155 (right)** iStockphoto.com **p. 156 (top, bottom)** Richard Megna/Fundamental Photographs **p. 160 (top)** oote boe/Alamy Images **p. 160 (bottom)** Science Museum/The Image Works **p. 165** © Galen Rowell/CORBIS **p. 170** © Doug Wilson/CORBIS **p. 172 (left)** AP Wide World Photos **p. 172 (middle)** Official U.S. Navy photo by Mark Meyer **p. 172 (right)** Bombardier Transportation Sweden **p. 174** Chris Priest/Science Photo Library/Photo Researchers, Inc. **p. 177** Photofest **p. 183 (left)** Color Box/Getty Images Inc.-Taxi **p. 183 (right)** The British Library/Heritage Image Partnership/The Image Works **p. 187** Andy Lyons/Getty Images

Physics in Perspective: Force, Acceleration, and Motion—**p. 188 (top left)** Kinetic Imagery **p. 188 (top middle)** Christa Renee/Getty Images **p. 188 (top right)** Digital Vision/Getty Images **p. 188 (middle)** Corbis SuperRF/Alamy **p. 189 (top left)** Jupiter Images **p. 189 (top middle)** Shutterstock **p. 189 (top right)** David Madison/Getty Images

Chapter 7—**p. 190** Rich Pilling/MLB Photos/Getty Images **p. 193 (right)** AFP/Getty Images **p. 193 (left)** © Mimmo Jodice/Corbis **p. 203** Dr. T. J. McMaster/Wellcome Trust Medical Photographic Library **p. 206** Philippe Achache/Getty Images Inc.-Liaison **p. 214** John E. Bortle

Chapter 8—**p. 216** Photolibrary.Com **p. 221 (left)** Grant Heilman/Grant Heilman Photography, Inc. **p. 221 (right)** Jay Brousseau **p. 222** Philip H. Coblentz/World Travel Images, Inc. **p. 224** Barry Bland/Alamy Images

p. 227 (top) © Bill Nation/Corbis Sygma **p. 227 (bottom)** Salt River Project, Phoenix, Arizona **p. 232** Jeff Greenberg/PhotoEdit **p. 239** Federal Motor Carrier Safety Administration, Federal Highway Administration, Colorado Division **p. 244 (left)** Stephen Derr/Getty Images Inc.-Image Bank **p. 244 (right)** Lester Lefkowitz/Getty Images Inc.-Stone Allstock **p. 245** DESMETTE FREDE/AGE Fotostock

Physics in Perspective: Energy: A Breakthrough in Physics—**p. 252 (top left)** Alan Schein Photography **p. 352 (top middle)** John Giustina/Getty Images **p. 352 (middle)** Terry Oakley/Alamy **p. 352 (bottom)** ImageDJ/Alamy **p. 353** ImageState/Alamy

Chapter 9—**p. 254** INSADCO Photography/Alamy Images **p. 260 (top)** © Harold & Esther Edgerton Foundation, 2003, courtesy of Palm Press, Inc. **p. 260 (bottom)** © Wally McNamee/CORBIS **p. 261** J. Scott Altenbach, University of New Mexico **p. 264** NASA Headquarters **p. 252** NASA Headquarters **p. 267 (left)** J. McIsaac/Bruce Bennett Studios, Inc./Getty Images **p. 267 (right)** © Franck Seguin/CORBIS **p. 275 (top)** Richard Megna/Fundamental Photographs **p. 275 (bottom)** Richard Megna/Fundamental Photographs **p. 278** Alexander Calder (1898–1976). Myxomatose. 1953. Sheet metal, rod, wire, and paint. 256.5 ? 408.9 cm (101 × 161 in) © Copyright ARS, NY. Copyright Art Resource, NY. Private Collection, New York, NY, USA **p. 268** Richard Megna/Fundamental Photographs **p. 285 (top)** NASA Headquarters **p. 285 (bottom)** Fred Bavendam/Peter Arnold, Inc. **p. 289** © Jose Azel/Aurora/PNI **p. 293** Michael Rinaldi/U.S. Navy News Photo **p. 296** U.S. Geological Survey/U.S. Department of the Interior

Chapter 10—**p. 297** James Havey/Creative Eye/MIRA.com **p. 298 (left)** NASA Headquarters **p. 298 (middle)** Novastock/PhotoEdit **p. 298 (right)** Barry Dowsett/Photo Researchers, Inc. **p. 300** Fred Espenak/Photo Researchers, Inc. **p. 307 (top left)** Loren M. Winters **p. 307 (top right)** Karl Weatherly/Getty Images Inc.-Photodisc **p. 307 (bottom left)** Novosti/Sovfoto/Eastfoto **p. 307 (bottom right)** Tek Image/Photo Researchers **p. 311** Richard Megna/Fundamental Photographs **p. 315** Mehau Kulyk/Science Photo Library/Photo Researchers, Inc. **p. 324 (left)** Jeff Hester/Arizona State University/NASA Headquarters **p. 324 (right)** Dorling Kindersley **p. 325** Arthur Tilley/Getty Images Inc.-Taxi **p. 328** © Loren M. Winters, North Carolina School of Science and Mathematics **p. 330** NASA/Ames Research Center/Ligo Project **p. 329** Yomega

Chapter 11—**p. 332** Michael Svoboda/Shutterstock **p. 333** Kevin M. Law/Alamy Images **p. 335** Benelux Press B.V./Index Stock Imagery, Inc. **p. 339** Richard Megna/Fundamental Photographs **p. 345** Getty Images Inc.–Photodisc **p. 349 (top)** Murray Close/Paramount Pictures/Photofest **p. 349 (bottom left)** Dana White/PhotoEdit **p. 349 (bottom right)** Altrendo/Getty Images **p. 355** Glyn Kirk/Getty Images Inc.–Stone Allstock **p. 357 (top left)** Hasler, Pierce, Palaniappan, Manyin/NASA Goddard Laboratory for Atmospheres **p. 357 (top right)** Photo by Arthur Harvey, courtesy of the Miami Herald **p. 357 (bottom)** Chandra X-Ray Center/Smithsonian Astrophysical Observatory/NASA/Marshall Space Flight Center **p. 361 (left)** Pearson Education/PH College **p. 361 (right)** Photograph by Erik Gustafson, courtesy AIP Emilio Segre Visual Archives, Margrethe Bohr Collection **p. 362** Gravity Probe B at Stanford University

Physics in Perspective: Momentum: A Conserved Quantity—**p. 376** Bob Thomas **p. 277** Gene Lower/Getty Images

Chapter 12—**p. 378** NASA and The Hubble Heritage Team (STScI/AURA). R. G. French (Wellesley College), J. Cuzzi (NASA/Ames), L. Dones (SwRI), and J. Lissauer (NASA/Ames) **p. 379** The Burndy Library, Dibner Institute for the History of Science and Technology, Cambridge, Massachusetts, and the Grace K. Babson Collection of the Works of Sir Isaac Newton **p. 383** NASA/JPL/University of Texas' Center for Space Research/GeoForschungsZentrum (GFZ) Potsdam **p. 384 (left)** NASA Headquarters **p. 384 (right)** NASA **p. 390 (left and right)** NASA Headquarters **p. 392 (left)** NASA Headquarters **p. 392 (right)** NASA/Johnson Space Center **p. 398 (left)** E. R. Degginger/Color-Pic, Inc. **p. 398 (right)** © Mark Pilkington/Geological Survey of Canada/Science Photo Library/Photo Researchers, Inc. **p. 401** David Nunuk/Science Photo Library/Photo Researchers, Inc. **p. 405** David R. Frazier Photolibrary Inc./Alamy **p. 411 (left)** NASA/Johnson Space Center **p. 411 (right)** NASA Headquarters **p. 414 (left and right)** NASA/Jet Propulsion Laboratory

CORBIS **p. 965 (bottom)** Jeff J. Daly/Fundamental Photographs **Fig. 27–24b** © Aviation-images.com **p. 974 (middle)** Custom Medical Stock Photo, Inc. **p. 974 (bottom)** Reprinted with Permission of Bausch & Lomb Surgical

Chapter 28—p. 976 Corbis Digital Stock **p. 978** Richard Megna/Fundamental Photographs **p. 981** DESY in Hamburg **Fig. 28–11b** Fundamental Photos **p. 987** Michael Ventura/PhotoEdit **p. 989 (top)** Alamy Images **p. 989 (middle)** Jeremy Burgess/Science Photo Library/Photo Researchers, Inc. **p. 991** Richard Megna/Fundamental Photographs **p. 993** Ken Kay/Fundamental Photographs **Fig. 28–21** Reproduced by permission from Michel Cagnet, Maurice Franzon, and Jean Claude Thierr, *Atlas of Optical Phenomena.* New York: Springer-Verlag, 1962. © 1962 by Springer-Verlag GmbH & Co. **Fig. 28–22** Reproduced by permission from Michel Cagnet, Maurice Franzon, and Jean Claude Thierr, *Atlas of Optical Phenomena.* New York: Springer-Verlag, 1962. © 1962 by Springer-Verlag GmbH & Co. **p. 995** The Image Finders 2003 **Fig. 28–23** Museum of Fine Arts, Springfield, MA., Robert J. Freedman Collection. © 2005 Artists Rights Society (ARS), New York/ADAGP, Paris. **Fig. 28–28** Photo by M. H. F. Wilkins. Courtesy of Biophysics Department, King's College, London, England. **p. 1000 (left)** Phil Degginger/Color-Pic, Inc. **p. 1000 (right)** Charles Krebs/Corbis/Stock Market **p. 963** Richard Megna/Fundamental Photographs **p. 1005** E. R. Degginger/Color-Pic, Inc.

Physics in Perspective: Waves and Particles: A Theme of Modern Physics—p. 1010 (top left) Richard Hamilton Smith/Corbis **p. 1010 (top middle)** Brooks Martner/NOAA/Forecast Systems Laboratory **p. 1010 (top right)** NASA Jet Propulsion Laboratory **p. 1010 (middle right)** NASA/ESA/STSCI/SPL **p. 1010 (bottom left)** Matthias Kulka/Zefa/Corbis **p. 1010 (bottom right)** Chandra/GSFC/D. Berry **p. 1011 (top left)** Sabian Zigman **p. 1011 (top right)** Andrew Lambert Photography/Photo Researchers, Inc. **p. 1011 (middle top)** Wolfgang Rueckner **p. 1011 (middle bottom)** Hitachi, Ltd., Advanced Research Laboratory **p. 1011 (bottom left)** Pearson Science **p. 1011 (bottom right)** Dr. Gopal Murti/Visuals Unlimited

Chapter 29—p. 1012 Eric Reiman, University of Arizona/Science Photo Library/Photo Researchers, Inc. **p. 1013** Lotte Jacobi Collection, University of New Hampshire **Fig. 29-11** A. C. Searle and C. M. Savage, The Australian National University. Animations at http://www.anu.edu.au/Physics/Searle**p. 1029** SPL/Photo Researchers **Fig. 29–16** Michael Ventura/Alamy **Fig. 29–25a** Space Telescope Science Institute **Fig. 29–25b** European Space Agency/NASA Headquarters **p. 1036 (left)** Julian Baum/Photo Researchers, Inc. **p. 1036 (right)** NASA Headquarters

Chapter 30—p. 1046 Natural History Museum Picture Library, London/Getty Images Inc.-Stone Allstock **p. 1048** Getty Images Inc.-Photodisc **p. 1055 (left)** Eunice Harris/Photo Researchers, Inc. **p. 1055 (right)** Stock Trek/Getty Images Inc.-Photodisc **Fig. 30–10a** The Burndy Library, Cambridge, Massachusetts **Fig. 30–10b** DESY/HASYLAB, Hamburg, Germany **Fig. 30–11** John R. Helliwell **Fig. 30–12a** Dr. Everett Anderson/Science Photo Library/Photo Researchers, Inc. **Fig. 30–12b** Scott Camazine/Photo Researchers, Inc. **Fig. 30-14** Hitachi, Ltd., Advanced Research Laboratory **Fig. 30–19a** Courtesy of International Business Machines Corporation. Unauthorized use not permitted. **Fig. 30–19b** R. J. Driscoll, M. G. Youngquist, & J. D. Baldeschwieler, California Institute of Technology/Science Photo Library/Photo Researchers, Inc. **p. 1072** James S. Walker **p. 1073** John Chumack/Photo Researchers, Inc. **p. 1074** James Walker **p. 1077** Paul Zahl/Photo Researchers, Inc.

Chapter 31—p. 1078 Eduardo Kac, GFP Bunny, 2000. Alba, the fluorescent bunny. Photo: Chrystelle Fontaine. Courtesy Julia Friedman Gallery. **p. 1080** Kitt Peak National Observatory **p. 1083** AIP Emilio Segre Visual Archives, Uhlenbeck Collection **p. 1091** Richard Megna/Fundamental Photos **p. 1099** China Newphoto/Landov LLC **p. 1102 (left)** GCA/CNRI/Phototake NYC **p. 1102 (right)** Richard Shock/Getty Images Inc.-Stone Allstock **p. 1104** Yoav Levy/Phototake NYC **p. 1105** Philippe Plailly/SPL/Photo Researchers, Inc. **p. 1107 (top left)** Getty Images, Inc.–Photodisc **p. 1107 (top right)** Dr. Peter Dawson/Science Photo Library/Photo Researchers, Inc. **p. 1107 (bottom left)** Tom McHugh/Photo Researchers, Inc. **p. 1107 (bottom right)** Bjorn Bolstad/Photo Researchers, Inc.

Chapter 32—p. 1116 National Cancer Institute/Getty Images Inc.-Photodisc **p. 1127** Michael Dalton/Fundamental Photographs **p. 1138** Will & Deni McIntyre/Photo Researchers, Inc. **p. 1132** Corbis/Sygma **p. 1138 (left)** Catherine Pouedras/Science Photo Library/Photo Researchers, Inc. **p. 1138 (middle)** Martin Bond/Science Photo Library/Photo Researchers, Inc. **p. 1138 (right)** U.S. Department of Energy/Science Photo Library/Photo Researchers, Inc. **p. 1140** Randy Montoya/Sandia National Laboratories **p. 1141 (both)** Yoav Levy/Phototake NYC **p. 1143 (top left)** CNRI/Science Photo Library/Photo Researchers, Inc. **p. 1143 (right)** Wellcome Dept. of Cognitive Neurology/Science Photo Library/Photo Researchers, Inc. **p. 1143 (middle)** UHB Trust/Getty Images Inc.-Stone Allstock **p. 1154** Photo courtesy of The Coleman Company, Inc.

Index

For users of the two-volume edition, pages 1–651 are in Volume 1 and pages 652–1155 are in Volume 2.

MULTIPLES AND PREFIXES FOR METRIC UNITS

Multiple	Prefix (Abbreviation)	Pronunciation
10^{24}	yotta- (Y)	yot'ta (*a* as in *a*bout)
10^{21}	zetta- (Z)	zet'ta (*a* as in *a*bout)
10^{18}	exa- (E)	ex'a (*a* as in *a*bout)
10^{15}	peta- (P)	pet'a (as in *peta*l)
10^{12}	tera- (T)	ter'a (as in *terra*ce)
10^{9}	giga- (G)	ji'ga (*ji* as in *ji*ggle, *a* as in *a*bout)
10^{6}	mega- (M)	meg'a (as in *mega*phone)
10^{3}	kilo- (k)	kil'o (as in *kilo*watt)
10^{2}	hecto- (h)	hek'to (*heck-toe*)
10	deka- (da)	dek'a (*deck* plus *a* as in *a*bout)
10^{-1}	deci- (d)	des'i (as in *deci*mal)
10^{-2}	centi- (c)	sen'ti (as in *senti*mental)
10^{-3}	milli- (m)	mil'li (as in *milli*tary)
10^{-6}	micro- (μ)	mi'kro (as in *micro*phone)
10^{-9}	nano- (n)	nan'oh (*an* as in *an*nual)
10^{-12}	pico- (p)	pe'ko (*peek-oh*)
10^{-15}	femto- (f)	fem'toe (*fem* as in *fem*inine)
10^{-18}	atto- (a)	at'toe (as in an*ato*my)
10^{-21}	zepto- (z)	zep'toe (as in *zep*pelin)
10^{-24}	yocto- (y)	yock'toe (as in *sock*)

THE GREEK ALPHABET

Alpha	A	α
Beta	B	β
Gamma	Γ	γ
Delta	Δ	δ
Epsilon	E	ε
Zeta	Z	ζ
Eta	H	η
Theta	Θ	θ
Iota	I	ι
Kappa	K	κ
Lambda	Λ	λ
Mu	M	μ
Nu	N	ν
Xi	Ξ	ξ
Omicron	O	o
Pi	Π	π
Rho	P	ρ
Sigma	Σ	σ
Tau	T	τ
Upsilon	Υ	υ
Phi	Φ	ϕ
Chi	X	χ
Psi	Ψ	ψ
Omega	Ω	ω

SI BASE UNITS

Physical Quantity	Name of Unit	Symbol
Length	meter	m
Mass	kilogram	kg
Time	second	s
Electric current	ampere	A
Temperature	kelvin	K
Amount of substance	mole	mol
Luminous intensity	candela	cd

SOME SI DERIVED UNITS

Physical Quantity	Name of Unit	Symbol	SI Unit
Frequency	hertz	Hz	s^{-1}
Energy	joule	J	$kg \cdot m^2/s^2$
Force	newton	N	$kg \cdot m/s^2$
Pressure	pascal	Pa	$kg/(m \cdot s^2)$
Power	watt	W	$kg \cdot m^2/s^3$
Electric charge	coulomb	C	$A \cdot s$
Electric potential	volt	V	$kg \cdot m^2/(A \cdot s^3)$
Electric resistance	ohm	Ω	$kg \cdot m^2/(A^2 \cdot s^3)$
Capacitance	farad	F	$A^2 \cdot s^4/(kg \cdot m^2)$
Inductance	henry	H	$kg \cdot m^2/(A^2 \cdot s^2)$
Magnetic field	tesla	T	$kg/(A \cdot s^2)$
Magnetic flux	weber	Wb	$kg \cdot m^2/(A \cdot s^2)$

SI UNITS OF SOME OTHER PHYSICAL QUANTITIES

Physical Quantity	SI Unit
Density (ρ)	kg/m^3
Speed (v)	m/s
Acceleration (a)	m/s^2
Momentum, impulse (p)	$kg \cdot m/s$
Angular speed (ω)	rad/s
Angular acceleration (α)	rad/s^2
Torque (τ)	$kg \cdot m^2/s^2$ *or* $N \cdot m$
Specific heat (c)	$J/(kg \cdot K)$
Thermal conductivity (k)	$W/(m \cdot K)$ *or* $J/(s \cdot m \cdot K)$
Entropy (S)	J/K *or* $kg \cdot m^2/(K \cdot s^2)$ *or* $N \cdot m/K$
Electric field (E)	N/C *or* V/m

APPENDICES IN THE TEXT